Harnessing
AutoCAD®
2013 *and beyond*

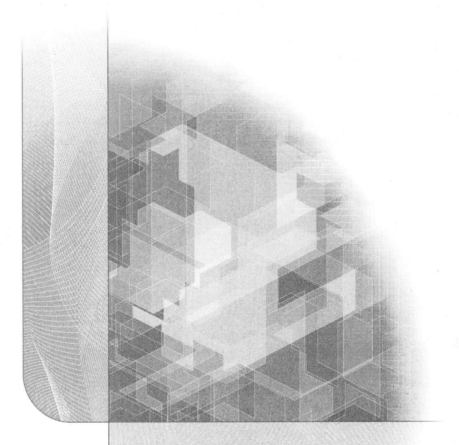

Harnessing
AutoCAD®
2013 *and beyond*

G. V. KRISHNAN • THOMAS A. STELLMAN

DELMAR
CENGAGE Learning·

Australia • Brazil • Japan • Korea • Mexico • Singapore • Spain • United Kingdom • United States

Harnessing AutoCAD® 2013 and beyond
G. V. Krishnan and Thomas A. Stellman

Vice President, Editorial: Dave Garza

Director of Learning Solutions: Sandy Clark

Acquisitions Editor: Stacy Masucci

Managing Editor: Larry Main

Senior Product Manager: John Fisher

Editorial Assistant: Kaitlin Murphy

Vice President, Marketing: Jennifer Baker

Marketing Director: Deborah Yarnell

Marketing Manager: Erin Brennan

Associate Marketing Manager: Jillian Borden

Production Director: Wendy Troeger

Content Project Management and
Design Management: PreMediaGlobal

Art Director: David Arsenault

Cover Image: Chris Harvey/Shutterstock.com

For product information and technology assistance, contact us at
**Professional Group Cengage Learning Customer &
Sales Support, 1-800-354-9706**

For permission to use material from this text or product,
submit all requests online at **www.cengage.com/permissions**
Further permissions questions can be e-mailed to
permissionrequest@cengage.com

Library of Congress Control Number: 2012941902
ISBN-13: 978-1-133-94659-5
ISBN-10: 1-133-94659-3

Delmar
5 Maxwell Drive
Clifton Park, NY 12065-2919
USA

Cengage Learning is a leading provider of customized learning solutions with office locations around the globe, including Singapore, the United Kingdom, Australia, Mexico, Brazil, and Japan. Locate your local office at:
international.cengage.com/region

Cengage Learning products are represented in Canada by
Nelson Education, Ltd.

To learn more about Delmar, visit **www.cengage.com/delmar**
Purchase any of our products at your local college store or at our preferred online store **www.cengagebrain.com**

Printed in the United States of America
2 3 4 5 6 7 16 15 14 13

CONTENTS

Note: Chapter 19 and the appendices are not printed in the text. They are located on the CAD Connect Website. See page xii of the Introduction for instructions on how to access the website.

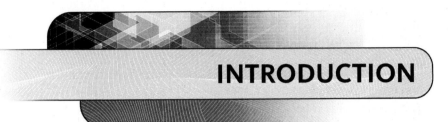

INTRODUCTION

HARNESSING THE POWER OF AUTOCAD 2013

The key phrase in AutoCAD 2013 is "Hidden Treasures." Not all of the changes in this new version are obvious (like new commands or features). AutoCAD 2013 has been "souped up." An already fast and powerful program is now faster in how it handles graphics and manipulates data. Experienced users should detect the improvement in performance. In addition to this, AutoCAD 2013 has tools for design aggregation, letting you streamline and connect your design and documentation workflow. Your drawings and folders can be synchronized with Autodesk Cloud where they can be accessed from almost anywhere.

Harnessing AutoCAD 2013 brings you comprehensive descriptions, explanations, and examples of the basics of AutoCAD 2013 and its new innovations. It has been written to be used both in the classroom as a textbook and in industry by the professional CADD designer/drafter as a reference and learning tool. Whether you're new to AutoCAD or a seasoned user upgrading your skills, *Harnessing AutoCAD 2013* will show you how to rein in the power of AutoCAD to improve your professional skills and increase your productivity. *Harnessing AutoCAD 2013* includes major updates to the chapter on 3D modeling.

Highlights and Features of AutoCAD 2013

- Import and aggregate models from almost any format
- Create layouts and views that automatically update
- Enhanced Command Line display
- Connectivity and synchronization to Autodesk Cloud
- Create new and open existing drawings from Welcome Screen
- Clickable options in Command window
- Dynamic In-canvas Property Preview
- Enhancements to the Array command
- Enhanced Point Cloud
- Enhanced Presspull command
- Surface curve extraction with new Extract Isolines tool
- Specify which 3D objects are represented with Base View tool

- Access to Section View Creation tools
- Detail Views can be created from existing drawing views
- Associative annotations
- Enhanced Content Explorer
- Import Inventor files

How to Use this Book
Overview

The first chapter of this text provides an overview of the AutoCAD program, its interface, the commands, special features and warnings, and AutoCAD 2013 enhancements. Specific commands are described in detail throughout the book, along with lessons on how to use them.

Fundamentals

Harnessing AutoCAD 2013 contains five chapters devoted to teaching the fundamentals of AutoCAD 2013. Fundamentals I introduces some of the basic commands and concepts, and Fundamentals II through V continue to build logically on that foundation until the student has a reasonable competency in the most basic functions of AutoCAD.

Intermediate

After mastering the fundamentals, you move on to the intermediate topics, which include dimensioning, plotting and printing, hatching and boundaries, design with parametric constraints, blocks and attributes, external references, and drawing environments. Other chapters teach you to make the most of AutoCAD using utility commands, scripts and slides, 3D commands, rendering, and the digitizing tablet.

Advanced

For the advanced AutoCAD user, this book offers a chapter on customizing AutoCAD 2013 (including tabs and panels customization) and Visual LISP. These two chapters teach you to make AutoCAD 2013 more individualized and powerful as you tailor it to your special needs. Chapter 19 Visual LISP is available as a PDF file on the CAD Connect Website.

Appendices

There are eight appendices posted on the CAD Connect Website. Appendix A is an introduction to hardware and software requirements of AutoCAD. Appendix B is a quick reference of AutoCAD commands with a brief description of their basic functions. Appendix C lists system variables, including default setting, type, whether or not it is read-only, and an explanation of the system variables. To see hatch and fill patterns, fonts, and linetypes provided with AutoCAD, refer to Appendices D, E, and F respectively. Appendix G lists AutoCAD command aliases. Appendix H addresses Express Tools. Use the Access Code attached to the back of this book to access the CAD Connect Website.

Style Conventions

In order to make this text easier for you to use, we have adopted certain typographic conventions that are used throughout the book:

Convention	Example
Command names are in small caps	The MOVE command
Shortcut menu names and Option names	Choose *Close* from the shortcut menu are in italics
Menu names appear with the first letter capitalized	Draw menu
Panel and Tab names appear with the first letter capitalized	Draw panel
Toolbar buttons and icons appear in boldface	**ORTHO**
Command sequences are indented. User inputs are indicated by boldface. Instructions are indicated by italics and are enclosed in parentheses	Command: **move** Enter variable name or [?]: **snapmode** Enter group name: *(enter group name)*

How to Invoke Commands

Like most Windows based programs, AutoCAD offers more than one method of invoking a command or accomplishing a particular task. As you progress through the different concepts and skill levels of using AutoCAD and as you use the program in your job, you will want to determine which method best suits your applications.

In early DOS-based versions, interfacing with AutoCAD usually involved typing the command names at the "Command line" in the "Command Window" or using the cursor to navigate through the nested levels of the "Side Screen Menu." The Side Screen Menu became obsolete with the migration from DOS to Windows as did its Menus and Toolbars for selecting commands. Now, with the new "Cursor-focused" interface, tabs and panels, the Command line has taken a back seat as the prime method of interface and command entry.

Almost all commands offer options to the default sequence of prompts and responses. For example, the CIRCLE command default method of drawing a circle is to select the center and then specify the radius. You can override the default with the center-diameter option or the tangent-tangent option. These options have always been "on display" in the Command Window. They have also been available in a Shortcut menu by right-clicking in the middle of the command. This means you can even turn off the display of the Command Window. But you can still access the list of options by right-clicking during a command.

For purposes of expediency, explanations and examples in *Harnessing AutoCAD 2013* assume that the user is entering commands at the On-Screen cursor, rather than entering them at the Command line.

Coordinate Input — What's the Point?

AutoCAD, like all bona fide Computer-Aided-Drafting programs, uses points in a coordinate system when drawing objects. Even if you select a point on the screen with your pointing device like you do in a paint program, that point contains highly accurate coordinate information. You are also able to type in the coordinates of the point you wish to specify when prompted. In AutoCAD, the default method of doing this is to type in the coordinates and see the values reported in text boxes following the new on-screen prompt referred to as Dynamic Input.

Caution! When you enter coordinates under certain conditions, AutoCAD 2013 automatically prefixes them with a symbol that forces them to be relative to the last point entered. If you wish to have the point coordinates you enter to be absolute (relative to the coordinate system origin), you must either prefix them with the proper symbol or reset the appropriate system variable. See the explanations in chapter 2 in the section on METHODS TO SPECIFY POINTS and in chapter 3 in the section on DYNAMIC INPUT, ON-SCREEN PROMPTS, AND GEOMETRIC VALUES DISPLAY.

Project Exercises and Drawing Exercises

Harnessing AutoCAD 2013, like its predecessors, still offers comprehensive learning exercises associated with the lessons in each chapter. These exercises are representative of the types of discipline-related drawing problems found in the design industry today. They can be found on the CAD Connect Website described below, and accessed using the Access Code attached to the back of this book. This site contains PDF files of each chapter in the *Harnessing AutoCAD 2013 Exercise Manual*. These chapters correspond to the text in this book and contain a project exercise and specific exercises for the following disciplines: mechanical, architectural, civil, electrical, and piping. Exercise icons and tabs identify the discipline sections (refer to the following table of exercise icons).

Step-by-step project exercises are identified by the special icon shown in the following table. Exercises that give you practice with types of drawings that are often found in a particular discipline are identified by the icons shown in the following table.

CAD Connect Website

We also include a CAD Connect website with this text. This website contains PDF files of exercises for various chapters, a PDF file of Chapter 19, and Appendices A through H. It will also contain future content addressing features of subsequent releases of the software. The access code included with this text enables you to access this web site. Instructions for redeeming your access code and accessing the website are below:

Redeeming an Access Code:

1. Go to http://www.CengageBrain.com
2. Enter the Access code in the Prepaid Code or Access Key field, **Register**.
3. Register as a new user or log in as an existing user if you already have an account with Cengage Learning or CengageBrain.com.
4. Open the product from the My Account page

Accessing CAD Connect from CengageBrain- My Account:

1. Sign in to your account at: http://www.CengageBrain.com
2. Go to My Account to view purchases
3. Locate the desired product
4. Click on the **Open** button next to the CAD Connect site entry, or Premium Website for [core title].

INSTRUCTOR SITE

An Instructor Companion Website containing supplementary material is available. This site contains an Instructor Guide, testbank, image gallery of text figures, and chapter presentations done in PowerPoint. Contact Delmar Cengage Learning or your local sales representative to obtain an instructor account.

Accessing an Instructor Companion Site from SSO Front Door

1. Go to: http://login.cengage.com and login using the Instructor email address and password
2. ENTER author, title or ISBN in **the Add a title to your bookshelf** search box, CLICK on **Search** button
3. CLICK **Add to My Bookshelf** to add Instructor Resources
4. At the Product page click on the **Instructor Companion site** link

New Users

If you're new to Cengage.com and do not have a password, contact your sales representative.

About the Authors

G.V. Krishnan is Director of the Applied Business and Technology Center, University of Houston-Downtown, a Premier Autodesk Training Center. He has used AutoCAD since the introduction of version 1.4 and writes about AutoCAD from the standpoint of a user, instructor, and general CADD consultant to area industries. Since 1985 he has taught courses ranging from basic to advanced levels of AutoCAD, including customizing, 3D AutoCAD, solid modeling, and AutoLISP programming.

Thomas A. Stellman received a B.A. degree in Architecture from Rice University and has over 20 years of experience in the architecture, engineering, and construction industry. He has taught at the college level for over ten years and has been teaching courses in AutoCAD since the introduction of version 2.0 in 1984. He conducts seminars covering both introductory and advanced AutoLISP. In addition, he develops and markets third-party software for AutoCAD.

Acknowledgments

We would like to thank and acknowledge the many professionals who reviewed the manuscript to help us publish this *Harnessing AutoCAD 2013* text. The authors would like to acknowledge and thank the following staff members of Delmar Cengage Learning:

- Acquisitions Editor: Stacy Masucci
- Senior Product Manager: John Fisher
- Senior Content Project Manager: Jim Zayicek
- Editorial Assistant: Kaitlin Murphy

The authors also would like to acknowledge and thank the following:

- Composition: PreMediaGlobal

Getting Started

INTRODUCTION

This chapter covers starting AutoCAD, entering commands, and finding your way around the AutoCAD screen.

OBJECTIVES

After completing this chapter, you will be able to do the following:

- Start AutoCAD
- Identify the various parts on the screen
- Use various methods of command and data input
- Obtain help about the AutoCAD commands and features
- Start a new drawing
- Open an existing drawing
- Set up the drawing environment

STARTING AUTOCAD

Designing and drafting is what AutoCAD (and this book) is all about. So how do you get into AutoCAD? Choose the **Start** button (Microsoft® Windows® 7 Enterprise, Ultimate, Professional, or Home Premium, Microsoft Windows Vista Home Premium, Ultimate, Business, and Enterprise, Windows XP Home and Professional SP2, Windows 2000 SP4 operating systems), which displays a menu that lets you easily access the most useful items on your computer. Clicking **All Programs** opens a list of programs currently installed on your computer. Navigate to the Autodesk program group, and then select the AutoCAD 2013 program. On some systems you can double-click on the AutoCAD 2013 shortcut icon on the Windows desktop.

The Startup Application Window

By default, when AutoCAD is started, it may display a blank drawing window surrounded by user interface objects such as a ribbon, command window, and status bar as shown in Figure 1–1 or various arrangements of menus and toolbars.

FIGURE 1–1 *AutoCAD 2013 Out of The Box (OTB) application window*

The window shown in Figure 1–1 is one of the possible windows that might appear when AutoCAD is opened; this one appears when the Initial Setup is set for General Design and Documentation drawing environment. The layout of the graphics area conforms to a particular set of drawing parameters. Other arrangements of the startup window and layout area are possible, depending on how AutoCAD has been configured.

Within the AutoCAD 2013 program window, you can create drawings for viewing, printing (referred to as *plotting in the trade*), solving geometry and engineering problems, accumulating data, creating 3D views of objects, and various other design, graphics, and engineering applications. Whatever your objective is, you will likely have to make changes in the layout and drawing parameters, or you can configure the startup arrangement to suit your needs.

The Workspace

The display of the ribbon, tabs, panels, menus, toolbars, palettes, and status bar control buttons that are shown and arranged depends on the current workspace. By default, AutoCAD provides four task-based workspaces (Drafting & Annotation, AutoCAD Classic, 3D Basics, and 3D Modeling), and you can easily switch between them from the Workspaces toolbar which is located in the top left of the AutoCAD application window. The default user options settings, drafting settings, paths, and other values depend on the current Profile. Profiles are updated each time you make a change to an option, setting, or other value.

If you double-click on an AutoCAD drawing file in the Windows Explorer window, AutoCAD will open the drawing, and the layout and settings affecting the drawing area will conform to that drawing's environment within the current workspace arrangement. Chapter 14 describes how to customize the initial setup and use workspaces. Chapter 18 describes how to customize workspaces.

The Drawing Layout

In this chapter, the significance of the width and height of the graphics area on the Startup window and how they correlate to a final plotted sheet will be discussed. The three key elements of drafting are location, direction, and distance. When you draw objects on a paper sheet, you choose a starting point on the drawing sheet for the location of a point on the object to be drawn and an orientation for the object. You must also consider the measurements of the object. In AutoCAD, the graphics area of the screen operates like a zoom lens on a camera through which you are viewing an imaginary drawing sheet. The imaginary drawing area itself is limitless, although you can limit the area in which points can be specified. The AutoCAD graphics area dimensions are relative to dimensions on the imaginary drawing sheet, depending on the "zoom" factor in effect.

One of the great advantages of computer-aided drafting is that objects can be drawn with real-world dimensions. You can zoom in to view a circuit on a 1-millimeter chip so that it fills the screen, or you can zoom out to view a map of the United States so that it fills the screen. In either case, AutoCAD allows you to draw the objects at their real size and then use the viewing commands and features to display all or part of the objects. Once completed, the true-size drawing can be scaled up or down to fit the final plotted sheet.

The initial AutoCAD startup graphics area depends on the video setup of your computer and monitor. It can range from a full view of a 12-unit-wide by 9-unit-high drawing sheet to a 60-unit-wide by 30-unit-high area or even greater. The startup graphics area relative to the startup coordinate system is explained later in this chapter. Features and commands in AutoCAD permit you to move your view around the drawing area and zoom in for a closer look or zoom out to see a broader area.

Startup with an Existing Drawing

You can start the AutoCAD program by choosing a drawing file, which has the extension of *.dwg*, from within Windows Explorer and double-clicking on its icon or filename. The AutoCAD program will start. This is similar to the way other Windows-based programs are started: double-click on one of the types of files that is created and edited by that particular program. When AutoCAD is started in this manner (double-clicking on a *.dwg* file), the initial screen will normally display the drawing that was double-clicked using the view in which it was last saved.

AUTOCAD APPLICATION WINDOW

The application window is the total window that AutoCAD occupies on your screen. It includes the title bar (at the top with the Application Menu icon, Quick Access Toolbar, and InfoCenter) and the current workspace with its interface features and the graphics window. The default workspace includes the ribbon with its tabs and panels, the graphics window, status bar, and status bar tray (see Figure 1–2). Other elements that appear from time to time are tool palettes, dialog boxes, on-screen prompts, the DesignCenter window, the Command window, and graphic geometric values. Some commands, like Help, may open their own application window that can be resized, relocated, and minimized/maximized independent from the main Auto-CAD application window. These can often be running in the background while you perform other tasks in AutoCAD.

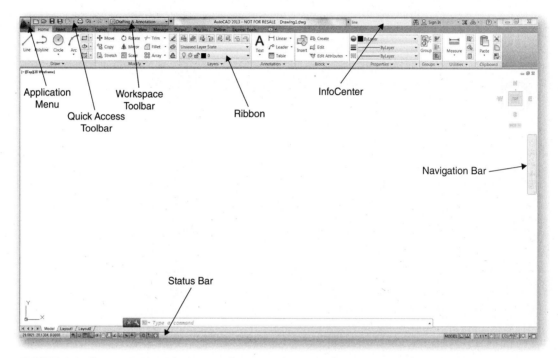

FIGURE 1–2 *The AutoCAD Application Window using the 2D Drafting and Annotation Workspace*

The AutoCAD application window can include a variety of user interface elements depending on the configuration of the current workspace. However, all four precon-figured workspaces that come with the program contain a Title Bar with an Application Menu icon, a Quick Access Toolbar, and an InfoCenter. Also, they each come with a Command Line and a Status Bar described later in this chapter.

Title Bar

The title bar displays the Application Menu, a Workspaces toolbar, Quick Access Toolbar, name of the active drawing, and the InfoCenter as shown in Figure 1–3.

FIGURE 1–3 *The Title Bar*

Application Menu

The Application menu is accessible by clicking the red "A" (Application Menu icon) at the left of the Title Bar. The Application Menu (see Figure 1–4) provides easy access for creating, opening, saving, printing, and publishing AutoCAD files, sending the current drawing as an email attachment, and producing electronic transmittal sets. In addition, you can perform drawing maintenance, such as audit and purge, and close drawings.

A search tool at the top of the Application menu enables you to query the Quick Access Toolbar, Application menu, and the currently loaded ribbon to locate commands, ribbon panel names, and other ribbon controls. For example, as you begin typing L-I-N-E in the search field, AutoCAD dynamically filters the search options to display all entries that include the word line (Linetype, Command Line, Line, Multiline, and so forth). You can then double-click a listed item to launch the associated command.

FIGURE 1–4 *The Menu Browser*

Image(s) © Cengage Learning 2013

In addition to command access, the Application menu enables you to view and access recently used or open documents. You can display the document names with icons or with small, medium, or large preview images, making them easy to identify. When viewing recent documents in the Application Menu, you can display them in an ordered list or group them by date or file type. Hovering the cursor over a document name automatically displays a preview image and other document information.

Workspaces Toolbar

The Workspaces Toolbar allows you to select from the available workspaces including **Drafting & Annotation, 3D Basics, 3D Modeling**, and **AutoCAD Classic**, and options including **Save Current As**, **Workspace Settings**, and **Customize**. A Workspace Switching button is also accessible on the right end of the Status Bar at the bottom of the drawing area. Workspaces and related commands and features are described in Chapter 14. Customizing and creating new workspaces is described in Chapter 18.

Quick Access Toolbar

By default, the Quick Access Toolbar has icons for file handling commands **New, Open, Save, Plot**, and commands **Undo** and **Redo**. In addition, the Quick Access Toolbar includes a flyout menu (see Figure 1–5), which displays a list of common tools that you can select to include in the Quick Access Toolbar, an option to display the menu bar, and an option to access the Command List pane to add additional commands to Quick Access Toolbar. You can further customize the Quick Access Toolbar using the Quick Access Toolbars node in the CUI (Customize User Interface) Editor (refer to Chapter 18 for detailed explanation). You can also create multiple versions of the Quick Access Toolbar and then add them to the appropriate workspaces.

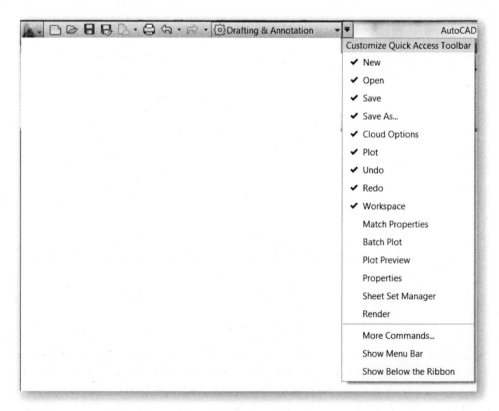

FIGURE 1–5 *The Quick Access Toolbar with flyout menu*

InfoCenter

The InfoCenter (see Figure 1–3) allows you to search for information through key words (or by entering a phrase). You can expand the search field or collapse it to save space on the title bar. And when searching the AutoCAD help system for a topic, you can specify which document to search, reducing the amount of time you spend trying to locate a topic.

You can also access the Communication Center panel or Favorites panel, which are explained in Chapter 15.

Viewport Controls and ViewCube Tool

Viewport controls are displayed at the top-left corner of each viewport to provide a convenient way of changing views, visual styles, and other settings. The labels display the current viewport settings. For example, the labels might read [-] [Top] [2D Wireframe].

You can click within each of the three bracketed areas to change the settings. Clicking on [-] displays options for maximizing the viewport, changing the viewport configuration, or controlling the display of navigation tools. Clicking on [Top] displays a menu to choose between several standard and custom views. And clicking on [2D Wireframe] displays a menu to choose one of several visual styles, which are mostly used for 3D visualization.

The ViewCube displayed at the top-right corner of the each viewport to control the orientation of 3D views.

Graphics Window

The graphics window is the area on the screen where you can view the objects that you create and modify. In this work area, AutoCAD displays the cursor, indicating your current working point. As you move your pointing device—usually a mouse or puck—around on a digitizing tablet, mouse pad, or other suitable surface, the cursor mimics your movements on the screen. On-screen prompts and graphic geometric values are displayed near the cursor.

When AutoCAD prompts you to select a point, the cursor shape is that of crosshairs. When you are required to select an object on the screen, the cursor changes to a small pick box. AutoCAD uses combinations of crosshairs, boxes, dashed rectangles, and arrows in various situations so that you can quickly see what type of selection or pick mode is in effect. After objects have been created and are visible on the screen, they become highlighted when the cursor passes over them to indicate that they can be selected for modifying or duplicating.

> It is possible to enter coordinates outside the viewing area for AutoCAD to use for creating objects. As you become more adept at AutoCAD, you may need to do this. Until then, working within the viewing area is recommended.

NOTE

Status Bar

The status bar displays the cursor's coordinate values along with several buttons for turning on and off drawing tools and displays several tools for scaling annotations. Different tools are displayed for model space and paper space (see Figure 1–6).

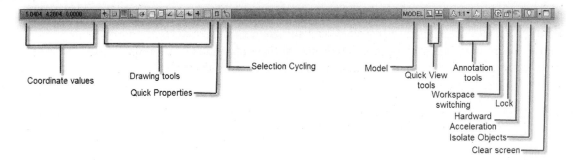

FIGURE 1–6 *Status Bar*

You can toggle ON/OFF for any of the available buttons on the status bar. The button changes from gray to color when it is toggled ON.

Setting DRAWING COORDINATE VALUES to ON causes the coordinate values to be displayed at the cursor location on the left end of the status bar.

The Quick View Layouts button on the status bar displays a horizontal row of Quick View Layout images (see Figure 1–7). This functionality provides a faster and more visual alternative to using the traditional Model and Layout tabs to switch between layouts within the current drawing. You can identify and select layouts not only by name but also by appearance. Move the mouse past the edge of the image strip to view layouts that extend beyond the screen. You can increase or decrease the size of the layout preview images by holding down the CTRL key while rolling the mouse wheel up or down.

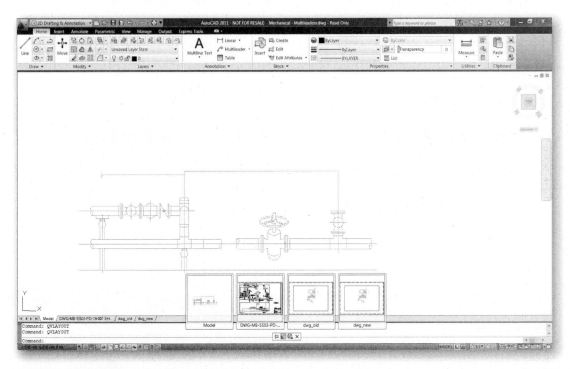

FIGURE 1–7 *Quick View Layouts*

Passing the cursor over a layout preview displays Plot and Publish buttons in the upper corners of the image. Additional tools are available in the Quick View control panel (see Figure 1–8), which is automatically displayed below the Quick View images. Using these tools you can pin Quick View open so that it remains visible while you work in the drawing editor. Then use the Close tool to turn off Quick View when you no longer want it displayed. The New Layout button creates a new layout in the current drawing and displays the preview at the end of the row of images. You can easily publish all the layouts in the drawing using the Publish tool.

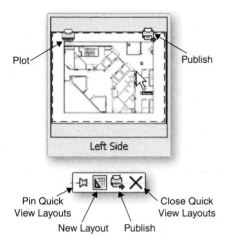

FIGURE 1–8 *Quick View Layouts Control Panel*

You can right-click a layout preview to display the Layout menu options, including access to the Page Setup Manager and exporting layout geometry to model space in a new drawing.

The Quick View Drawings button displays all currently open drawings in a row of Quick View drawing images (see Figure 1–9). This functionality is a faster and more visual alternative to using the ALT + TAB key combination or the Window menu for switching between open drawings. As with Quick View Layouts, you can move the mouse past the edge of the image strip to view images that extend beyond the screen. You can increase or decrease the size of the drawing preview images by holding down the CTRL key while rolling the mouse wheel up or down.

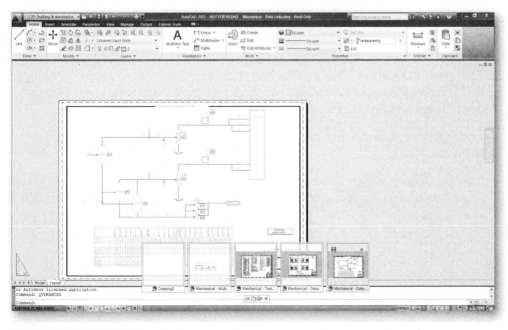

FIGURE 1–9 *Quick View Drawings*

Passing the cursor over a drawing preview displays Save and Close buttons in the upper corners of the image, enabling you to quickly save or close any open drawing, not just the current one. Additional tools are available in the Quick View Drawings control panel (see Figure 1–10), which is automatically displayed below the Quick View images. Using these tools you can pin Quick View Drawings open so that it remains visible while you work in the drawing editor. You can use the Close tool to turn off Quick View Drawings when you no longer want it displayed. The New button creates a new drawing and displays the preview at the end of the Quick View bar. You can easily open existing drawings using the Open tool.

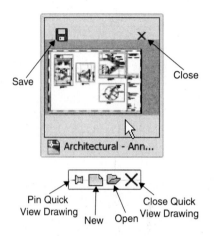

FIGURE 1–10 *Quick View Drawings Control Panel*

The navigation tools available in the status bar include icons for Pan, Zoom, Steering Wheel, and Show Motion. These are explained in Chapter 3.

Annotation tools include Annotation Scale, Annotation Visibility, and a tool to automatically add scales to annotative objects when the annotation scale changes. Annotative scaling is explained in Chapter 8.

The Workspace Switching button displays a shortcut menu where you can switch between workspaces and access commands for saving and customizing workspaces. Workspaces are discussed in Chapters 14 and 18.

The Toolbar/Window Position lock button locks the current positions of the toolbars/panels and windows. You can temporarily override the locking status by holding down CTRL while moving a toolbar that has been locked.

The Application Status Bar menu allows you to add or remove a button from the status bar.

Selecting **Clean Screen** clears the screen of toolbars and dockable windows. Clean screen can be also toggled on and off with CTRL + 0 (ZERO) or selected from the View Menu.

The Model and Layout tabs located just above the status bar allow you to change the drawing environment between model space for drawing objects and paper space for layouts for plotting. You generally create your designs in model space and then create layouts to plot your drawing in paper space. Refer to Chapter 8 for a detailed explanation of working with and plotting from layouts.

Ribbon

The ribbon is a palette that groups related task-based buttons and controls for easy access. The ribbon is included with the Drafting & Annotation Workspace, 3D Basic, and the 3D Modeling Workspace. By default, the ribbon displays across the top of the application window and has tabs that are titled **Home, Insert, Annotate, Layout, Parametric, View, Manage, Output, Plug-ins, Online**, and **Express Tools**. Each tab has panels that contain buttons for entering commands. For example, the LINE command and CIRCLE command are in the **Draw** panel located on the **Home** tab along with other commands used to create objects (see Figure 1–11). The panel can be expanded by clicking the arrow in the bottom middle to show additional draw commands (see Figure 1–12).

FIGURE 1–11 *Ribbon – Home tab selected*

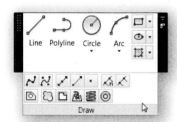

FIGURE 1–12 *Draw panel – expanded*

Selecting the push-pin icon in the lower-left corner of the extended panel causes the panel to remain extended. The icon changes to an embedded push-pin icon that, when selected, causes the panel to contract when the cursor in not hovering over it.

Ribbon Display Options

The large down-arrow icon on the title bar of the ribbon to the right of the tab titles allows you to toggle between hiding the ribbon and one of the display options or you can cycle through three display options: **Minimize to Tabs, Minimize to Panel Titles**, and **Minimize to Panel Buttons**.

Toggling to **Minimize to Tabs** causes the ribbon to be displayed with all of the panels for the current tab selection (see Figure 1–11). Toggling to **Minimize to Panel Titles** causes the ribbon to display only the titles of the panels for the current tab selection. Toggling to **Minimize to Panel Buttons** causes the ribbon panels to be displayed with large icons. When you hover the pointing device over one of the panels it will be expanded. Figure 1–13 shows the Draw panel expanded.

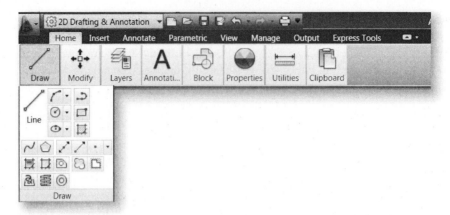

FIGURE 1–13 *Ribbon with Minimize to Panel Buttons selected and pointing device hovering over the Draw Panel*

Selecting the small down-arrow icon on the Title Bar displays a pull-down menu where you can choose whether the display will cycle through all options are just one of the display options when you choose the large down-arrow icon (Figure 1–14).

FIGURE 1–14 *Pull-down menu for Ribbon Display Control*

Right-clicking on the ribbon title bar (or in an area where there are no panels) causes the shortcut menu to be displayed as shown in Figure 1–15.

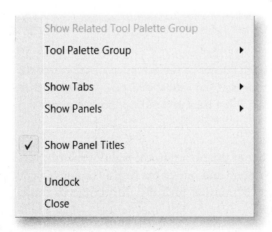

FIGURE 1–15 *Shortcut menu*

Choosing **Show Related Tool Palette Group** displays the Tool Palettes window with the selected Tool Palette Group. Refer to the Tool Palette window section for a detailed explanation of working with the tool palette window.

Choosing **Show Tabs** causes the flyout menu to be displayed with the names of the available tabs (see Figure 1–16). The default ribbon for the 2D Drafting & Annotation Workspace includes tabs for **Home, Insert, Annotate, Parametric, View, Manage, Output,** and **Express Tools**. Clicking on a tab name causes the tab to toggle between being displayed and not being displayed.

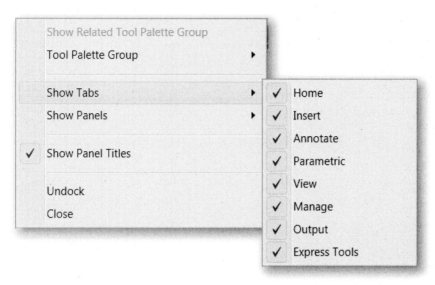

FIGURE 1–16 *Tabs flyout menu*

Choosing **Show Panels** causes the flyout menu to be displayed with the names of the available panels for the selected tab. The default ribbon Home tab includes panels for **Draw, Modify, Layers, Annotation, Block, Properties, Groups, Utilities,** and **Clipboard** (see Figure 1–17). Clicking on a panel name causes the panel to toggle between being displayed and not being displayed. Each tab has a different list of panels designed for its purpose.

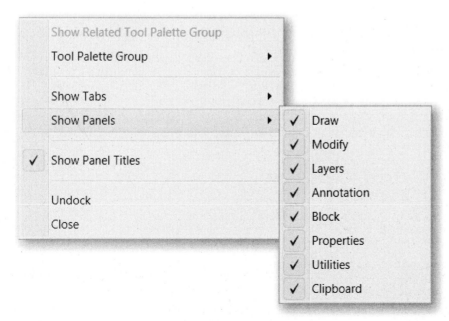

FIGURE 1–17 *Panels flyout menu when Home tab is current*

Choosing **Show Panel Titles** causes the panel titles on the tab to toggle between being displayed and not being displayed.

Choosing **Undock** causes the ribbon to be positioned on the screen where it was last undocked. It may even be all or partially off the application window. When undocked, it will assume a vertical position displaying **Panel Buttons**. When you hover the pointing device over one of the panels it will be expanded (see Figure 1–18).

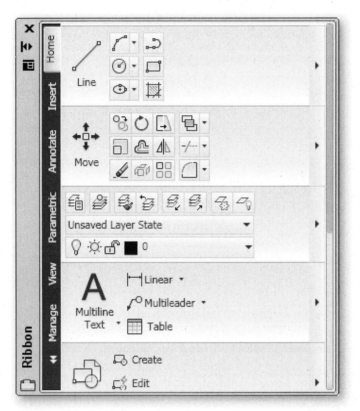

FIGURE 1–18 *Undocked Ribbon*

When undocked, you can change the width and height of the ribbon by dragging the edges inward or outward. If a panel is made too narrow to display all of its command icons, arrows will appear that, when clicked, will cause the hidden icons to appear.

One vertical edge of the undocked ribbon has a bar with icons for controlling the display of the ribbon. To move the undocked ribbon, click, hold, and drag it from the side vertical bar. Icons include **Close, Auto-hide,** and **Properties** (see Figure 1–19).

FIGURE 1–19 *Display controlling icons on ribbon side bar*

Choosing **Close** causes the ribbon to close.

> Entering **Ribbonclose** at the command prompt also causes the ribbon to close. Entering **Ribbon** causes the ribbon to be displayed again in the docked or undocked position it was in prior to closing. Choosing **Auto-hide** causes the ribbon to be hidden (except the side bar) until the cursor is placed over the bar and hovers over the ribbon.

Choosing **Properties** causes the Properties flyout menu to be displayed (see Figure 1–20).

FIGURE 1–20 *Properties flyout menu*

Choosing **Move** or **Size** causes the cursor to change to a move/size icon for moving or sizing the ribbon.

Choosing **Close** causes the ribbon to close. As noted above, entering **Ribbon** at the command prompt causes the ribbon to be displayed again.

Choosing **Allow Docking** enables you to dock the ribbon when dragged to the top, bottom, or one of the sides of the graphics screen.

Choosing **Anchor Left <** causes the ribbon to become docked on the left side of the graphics screen.

Choosing **Anchor Right >** causes the ribbon to become docked on the right side of the graphics screen.

Choosing **Auto-hide** causes the ribbon to be hidden (except the side bar) until the cursor is placed over the bar and hovers over the ribbon.

Choosing **Show Panel Titles** causes the panel titles on the tab to toggle between being displayed and not being displayed.

Choosing **Customize** causes the CUI dialog box to be displayed (see Chapter 18 for the explanation of the CUI).

Choosing **Help** opens the Help window.

Panel Display Options

The current tab on the ribbon may have some of the panels minimized due to limited space (see Figure 1–21).

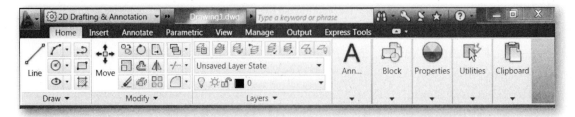

FIGURE 1–21 *Ribbon with some panels minimized*

Clicking the down-arrow in a minimized panel causes the panel to be expanded (see Figure 1–22).

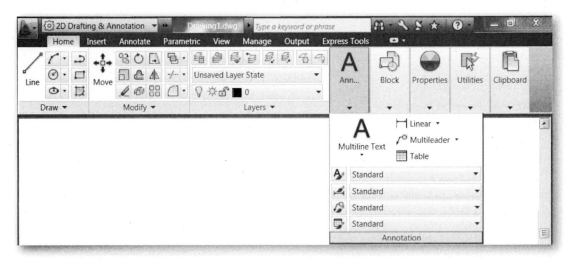

FIGURE 1–22 *Annotation panel expanded*

You can click and hold on the panel title bar and drag a panel onto the graphics screen (see Figure 1–23).

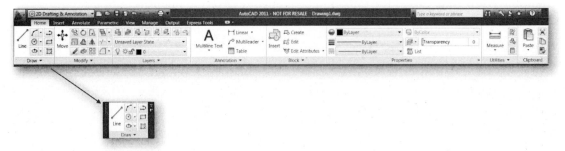

FIGURE 1–23 *Modify panel moved from ribbon to graphics screen*

When a panel is moved off of the ribbon, icons appear on the side bar for manipulating the panel (see Figure 1–24).

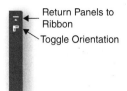

Return Panels to Ribbon

Toggle Orientation

FIGURE 1–24 *Icons on side bar of undocked panel*

Choosing **Return Panels to Ribbon** causes the panel to return to the ribbon.

Choosing **Toggle Orientation** causes the panel to toggle between horizontal and vertical orientations. The horizontal orientation has the title bar at the side of the panel and the vertical orientation has the title bar at the bottom. Clicking the expanding arrow on either orientation causes the panel to expand downward.

Navigation Bar

The navigation bar provides access to navigation tools which may be unified or product-specific. Unified navigation tools (such as Autodesk®, ViewCube®, Show-Motion®, 3Dconnexion®, and SteeringWheels®) are found in many Autodesk products. Product-specific navigation tools are unique to a specific product. The navigation bar can be found along one side of the drawing window. Figure 1–25 shows the default navigation bar.

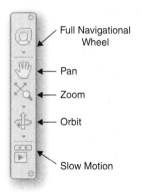

Full Navigational Wheel

Pan

Zoom

Orbit

Slow Motion

FIGURE 1–25 *Navigation bar*

In 2D Drafting & Annotation Workspace the **Full Navigational Wheel, Pan, Zoom, Orbit**, and **Slow Motion** options are available. In a 3D workspace options include **ViewCube, 3Dconnexion**, and **Orbit tools**.

To invoke a navigation tool click one of the buttons on the navigation bar. Or you can select a tool from a shortcut menu that is displayed when you click the down-arrow on one of the buttons. Right-clicking on one of the buttons displays a shortcut menu with the options to **Remove from navigation bar** or **Close navigation bar**.

You can hide the navigation bar by selecting the **X** at the top. If the navigation bar is not displayed, you can display it by checking the Navigation Bar option on the User Interface pull-down menu of the Windows panel of the View Tab while in the

Drafting & Annotation or 3D Modeling workspace as shown in Figure 1–26. You can also toggle the navigation bar on and off by entering NAVBAR and selecting On or Off.

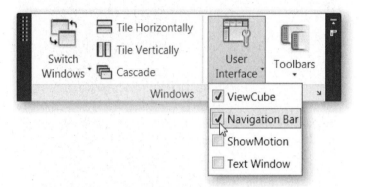

FIGURE 1–26 *Displaying the Navigation Bar*

Selecting the icon at the bottom of the navigation bar displays a shortcut menu where you can check the options that you wish to be displayed on the bar. Selecting **Docking positions** displays a shortcut menu that allows you to choose where the navigation bar is displayed and whether it is linked to the ViewCube as shown in Figure 1–27.

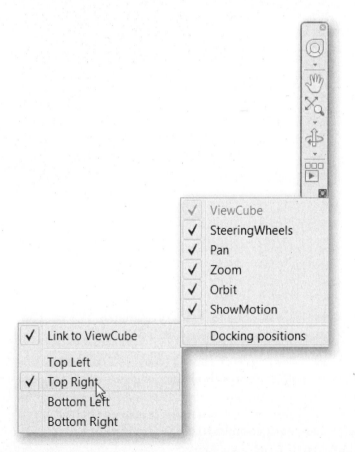

FIGURE 1–27 *Shortcut Menu for controlling the display of options and location of Navigation Bar*

The options available on the navigation bar are explained in other chapters of the book where applicable.

Toolbars

Toolbars are displayed with the AutoCAD Classic Workspace. Toolbars contain tools, represented by icons, from which you can invoke commands. Click a toolbar button to invoke a command, and then select options from a dialog box or respond to the on-screen prompts or the prompts on the command line in the Command window. If you position your pointer over a toolbar button and wait a moment, the name of the tool is displayed (see Figure 1–28). This is called the tooltip. In addition to the tooltip, AutoCAD displays a brief explanation of the command's function on the status bar.

FIGURE 1–28 *Toolbar with a tooltip displayed*

Some of the toolbar buttons have a small triangular symbol in the lower-right corner of the button indicating that there are *flyout* menus underneath that contain sub-commands. Figure 1–29 shows the ZOOM command flyout located on the Standard toolbar. The last option used from a flyout remains on top to become the default option.

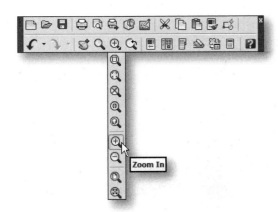

FIGURE 1–29 *Display of the Zoom flyout located on the Standard toolbar*

You can display multiple toolbars, resize them, and dock or float them. A *docked* tool-bar attaches to any edge of the graphics window. A *floating* toolbar can lie anywhere on the screen and can be resized. If the AutoCAD window does not fill your monitor screen, you can even locate a floating toolbar outside the window.

Docking and Undocking a Toolbar

To dock a floating toolbar, position the cursor on the caption; then press and hold the pick button on the pointing device. Drag the toolbar to a dock location at the top, bottom, or either side of the graphics window. When the outline of the toolbar appears in the docking area, release the pick button. To undock a toolbar, position the cursor on the left end, for horizontal toolbars, or the top end, for vertical toolbars, of the toolbar. Drag and drop it outside the docking regions. To place a toolbar in a docking region without docking it, press CTRL as you drag. By default, the Standard toolbar and the Properties toolbar are docked at the top of the graphics window when the AutoCAD Classic workspace has been selected.

Resizing a Floating Toolbar

If necessary, you can resize a floating toolbar. To do this, position the cursor anywhere on the border of the toolbar, and drag it in the direction you want to resize.

Closing a Floating Toolbar

To close a floating toolbar, position the cursor on the x located in the upper-right corner of the toolbar, and press the pick button on your pointing device (see Figure 1–30). The toolbar will disappear from the graphics window.

FIGURE 1–30 *Positioning the cursor to close a toolbar*

Opening a Toolbar or Closing a Docked Toolbar

AutoCAD 2013 comes with 44 regular toolbars and 3 "Express Tool" toolbars. To open any of the closed regular toolbars, place the cursor anywhere on any docked or floating toolbar (including the Quick Access Toolbar) that is already open, and press the right button on your pointing device. A shortcut menu appears, listing all the available toolbars with a check beside each open toolbar. Select the closed (unchecked) toolbar that you wish to open. You can also close an open toolbar by selecting it, thereby removing the checkmark next to its name. To open an "Express Tool" toolbar, instead of right-clicking on a toolbar, right-click on the blank area behind where toolbars can be placed just off the screen, not on the blank area to the right of the menus. Choose *Express* from the shortcut menu, and then choose ET:BLOCKS, ET:LAYERS, ET:STANDARD, or ET:TEXT.

If you select the customize option at the bottom of the shortcut menu that appears when you right-click on a toolbar, this causes the CUI dialog box to be displayed. The CUI dialog box allows you to create or modify toolbars. Refer to Chapter 18, "Customizing AutoCAD," for an explanation of this function.

Locking and Unlocking Toolbars and Windows

When you right-click on a toolbar, a shortcut menu is displayed. Selecting *Lock Location* at the bottom of the shortcut menu causes a flyout menu to appear with options to lock and unlock windows. You can lock and unlock floating and docked panels, toolbars, and windows. When one of the options is in locked mode, a check will be displayed beside that option. Panels, toolbars, and windows that are locked cannot be repositioned or resized, which prevents inadvertently changing a toolbar or window position or size. The All option allows you to lock or unlock all floating and docked panels, toolbars, and windows in one step.

Tool Palettes Window

Tool palettes are displayed with the AutoCAD Classic Workspace. Tool palettes are tabbed areas within the Tool Palettes window that provide an efficient method for organizing, sharing, and placing commands, blocks, and hatches. To open the Tool Palettes window, select View tab and choose Tool Palettes from the Palettes panel. Tool palettes can also contain custom tools provided by third-party developers. Blocks and hatch patterns are managed with tool palettes; see Chapters 9 and 11 for detailed explanations. You can also create a tool on a tool palette that executes a single AutoCAD command or a string of commands. The Tool Palettes window allows blocks and hatch patterns of similar usage and type to be grouped in their own tool palette. For example, one tool palette is named **Electrical** and contains, of course, blocks representing electrical symbols, as shown in Figure 1–31.

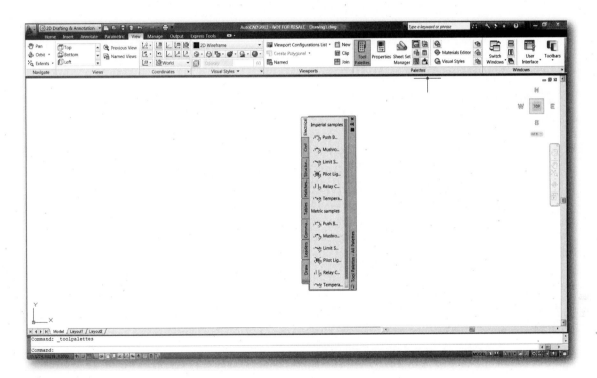

FIGURE 1–31 *The Tool Palettes window in the floating position*

Other tool palettes attached to the Tool Palettes window are **Modeling, Annotation, Architectural, Mechanical, Electrical, Civil, Structural, Hatches, Tables**, and **Command Tools**, all of which contain icons representing blocks, hatch patterns, or commands. These tabs appear when the Tool Palette is opened for the first time. Additional 22 tabs are available by clicking on the bottom of the tab row and selecting from the flyout menu that appears. The TOOLPALETTES command opens the Tool Palettes window.

The default position for the Tool Palettes window is floating on the right side of the screen. When the Tool Palettes window is undocked, it can be docked by double-clicking in the title bar, which may be on the left or right side of the window. Or, you can place the cursor over the title bar and drag the window all the way to the side where you wish to dock it. Its position can be changed by placing the cursor over the double-line bar at the top of the window and either double-clicking or dragging the window into the screen area or across to a docking position on the right side of the screen. Double-clicking causes the Tool Palettes window to alternate between docked and floating in the drawing area.

To insert a block from a tool palette, simply place the cursor on the block symbol in the tool palette, press the pick button, and drag the symbol into the drawing area. The block will be inserted at the point where the cursor is located when the pick button is released. This procedure is best implemented by using the appropriate Object Snap mode; to learn about Object Snap modes, see Chapter 3.

To use a tool created from a hatch, click a hatch tool and drag it to an object in the drawing.

Once you add a command to a tool palette, you can click the tool to execute the command. For example, clicking a New tool on a tool palette creates a new drawing just as the New button on the Standard toolbar does. You can also create a tool that executes a string of commands or customized commands such as an AutoLISP routine, a Visual Basic Applications macro or application, or a script.

Menu Bar

The Menu bar is displayed with the AutoCAD Classic Workspace. Drop-down menus are available from the menu bar at the top of the screen. To select any of the available commands, place the cursor into the menu bar area, and press the pick button on your pointing device, which pops that menu bar onto the screen (see Figure 1–32). To select a command or feature from the list, move the cursor until the desired item is highlighted, and press the pick button on the pointing device. If a menu item shows an arrow to the right, it has a cascading submenu. To display the submenu, place the pointer over the item, and press the pick button. Menu items that include ellipses (...) display dialog boxes. To select one of these, pick that menu item.

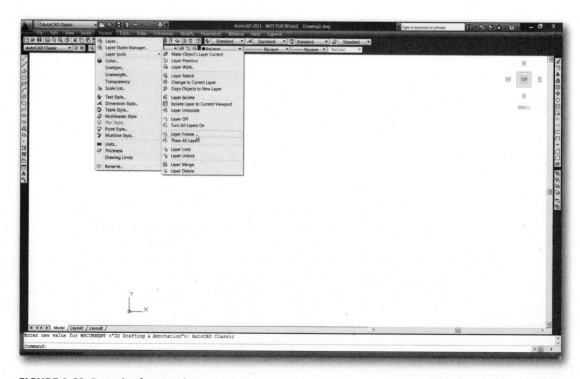

FIGURE 1–32 *Example of a menu bar*

Focusing on the Design—Dynamic Input

Dynamic input allows you to "keep your eyes on the road" when working by providing a prompt and input interface where you work, on the screen, at the cursor. Customizing dynamic input is explained in Chapter 3; see the section on "Dynamic Input, On-Screen Prompts, and Geometric Values Display." Earlier versions of AutoCAD emphasized the need to keep a constant vigil on the Command line in the Command window, as is discussed next. This meant having to look back and forth between the Command window and the point where you were working. Now, with an on-screen interface near the cursor, almost everything you need to know about what is going on with the program and your current work is right there, where you are working.

Dynamic Input and Feedback On-Screen

Figure 1–33 shows four stages of the information displayed on the screen during the process of drawing a circle. The first view is how the cursor appears when AutoCAD is ready for you to enter a command. The second view shows the text box that appears if you type in a command name from the keyboard (this step is skipped if you choose to initiate the command by another method such as selecting it from a toolbar). The third view shows the prompt that appears, asking you to specify a center point for the circle. The "or" and the down-arrow "⬇" indicate that options other than just entering the center point are available and that this is the time to choose one of them by right-clicking and selecting it from the shortcut menu. The fourth view shows the graphics feedback that appears while you are being prompted to input the circle radius. This type of feedback varies with the command in effect. The prompt for the second point of a line being drawn might include both the distance and the angle.

Make sure the Dynamic Input button located on the status bar, is toggled ON (changes from gray color) to enable the dynamic input feature.

NOTE

| CURSOR | COMMAND INPUT | COMMAND PROMPT | GRAPHIC FEEDBACK |

FIGURE 1–33 *On-screen input, prompts, and graphic feedback*

Locking in Values during Input

Responding to a prompt that specifies a point or vector normally requires you to enter two values: an X coordinate value and a Y coordinate value when using rectangular coordinate input or a distance and an angle when using polar coordinate input. If you wish to type in the first value, lock it in and then use the cursor to specify the second value. You can press the TAB key after typing the value, and AutoCAD displays a lock icon alongside the value. You can then move the cursor to specify the second required value.

Command Window

In the Command window, you can enter commands and AutoCAD displays prompts and messages. The Command window can be a floating window with a caption and frame. You can move the floating Command window anywhere on the screen and resize its width and height by dragging a side, bottom, or corner of the window. With the introduction of the on-screen input/prompt/graphics feedback feature in AutoCAD 2006, the Command window has taken a "back seat" and is no longer the primary place to interface with the program. It is now possible to do most of the design/drafting work with the Command window closed. However, if you close the window, you will need to press CTRL + 9 or invoke the COMMANDLINE command to reopen the Command window.

This dockable and resizable command window (see Figure 1–34) accepts commands and system variables and displays prompts that help you complete a command sequence (including commands that were initiated at another location such as the ribbon).

FIGURE 1–34 *Command window*

After you enter a command, you may see a series of prompts displayed at the command line. Each command has its own series of prompts. The prompts that appear when a particular command is used in one situation may differ from the prompts or sequence of prompts when invoked in another situation. You will become familiar with the prompts as you learn to use each command.

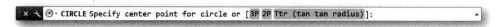

FIGURE 1–35 *Circle command prompt*

For example, if you enter CIRCLE, the following prompt is displayed (see Figure 1–35):

```
Specify center point for circle or [3P 2P Ttr (tan tan
radius)]:
```

In this case, the default is to specify the center point. You can either enter X,Y coordinate values or click a location in the drawing area.

To choose a different option, click the option. If you prefer to use the keyboard, specify the option by entering the capitalized, colored letter. You can enter uppercase or lowercase letters. For example, to choose the Ttr option, enter t.

Sometimes the default option (including the current value) is displayed after the angle-bracketed options:

```
POLYGON Enter number of sides <4>:
```

In this case, you can press Enter to retain the current setting (4). If you want to change the setting, type another number and press Enter.

By default, the name of a command or system variable is automatically completed as you type it as shown in Figure 1–36. Additionally, a list of valid choices is displayed from which you can choose. Use the AUTOCOMPLETE command to control which automatic features you want to use. If the automatic completion feature is turned off, you can type a letter on the command line and press the Tab key to cycle through all the commands and system variables that begin with that letter. Press Enter or SPACEBAR to start the command or system variable.

FIGURE 1–36 *Command window with list of available commands for the selected letter*

Some commands have abbreviated names, or command aliases, that you can enter at the command line. For example, instead of entering circle to start the CIRCLE command, you can enter c. The command Suggestion List (if displayed) indicates the alias in front of the command name:

C (CIRCLE)

Command aliases are defined in the .pgp file.

To set the number of lines of prompt history to display in the command window, click the Customize Command Line Settings button as shown in Figure 1–37 and select Lines of Prompt History. At the Command prompt, enter the number of rows to be displayed (0 to 10).

FIGURE 1–37

Similarly you can also control the transparency of the command window by clicking the Customize Command Line Settings button. In the Transparency dialog box, move the slider to the left to make the command window less transparent and to the right to make it more transparent. The range is from opaque to transparent.

To view a list of the recent commands used, click the arrow in the command window as shown in Figure 1–38.

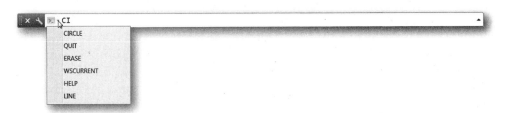

FIGURE 1–38 *List of recently used commands*

When you press F2, the command history text window is displayed and by pressing F2 again the window will be hidden. When the text window is displayed, you can scroll through the command history.

When you see "Type a command" displayed in the Command window, it signals that AutoCAD is ready to accept a command. After you enter a command name and press Enter or select a command from one of the menus or toolbars, the prompt area continues to inform you of the type of response(s) that you must furnish until the command is either completed or terminated. For example, when you pick the LINE command, the prompt displays "Specify first point." After selecting a starting point by the appropriate means, you will see "Specify next point or [Undo]," prompting for the endpoint of the line. This duplicates what is displayed in the on-screen prompt.

When you enter the command name or give any other response by typing from the keyboard, make sure to press Enter or SPACEBAR. Pressing Enter sends the input to the program for processing. For example, after you enter **line**, you must press Enter or SPACEBAR in order for AutoCAD to start the part of the program that lets you draw lines. If you type **lin** and press Enter or SPACEBAR, you will get an error message, unless someone has customized the program and created a command alias or command named "lin." Likewise, typing **lines** and pressing Enter or SPACEBAR is not a standard AutoCAD command.

Pressing SPACEBAR has the same function as Enter except when entering strings of words, letters, or numbers in response to the TEXT and MTEXT commands.

To repeat the previous command, you can press Enter or SPACEBAR at the "Command:" prompt. A few commands skip some of their normal prompts and assume default settings when repeated in this manner.

Terminating a Command

There are three ways to terminate a command:

- Complete the command sequence and return to the "Type a command" prompt.
- Press ESC to terminate the command before it is completed.
- Invoke another command from one of the menus or toolbars, which automatically cancels any command in progress.

NOTE

> All the examples of prompts and responses to commands in this book are based on what is displayed on-screen with dynamic input active.

AUTOCAD COMMANDS AND INPUT METHODS

This section introduces the methods available to initiate or invoke AutoCAD commands.

As much as possible, AutoCAD divides commands into related categories. For example, **Draw** is not a command but a category of commands used for creating primary objects such as lines, circles, arcs, text (lettering), and other useful objects that are visible on the screen. Categories include **Modify, View,** and **Tools,** and they include various commands and tools that help in managing AutoCAD drawings. The commands under **Format** menu are referred to as *drawing aids and utility commands* throughout this book. Learning the program can progress more quickly if the concepts and commands are mentally grouped into their proper categories. This step not only helps you to find them when you need them but it also helps you to grasp CAD fundamentals more quickly.

Input Methods

There are several ways to input an AutoCAD command: the keyboard, the ribbon/ tabs/panels, toolbars, menu bars, the side-screen menu, dialog boxes, the shortcut menu, or a digitizing tablet.

Keyboard

To invoke a command from the keyboard, type the command name at the on-screen prompt (the "Command:" prompt if you are using the Command window), and then press `Enter` or `SPACEBAR` —these are interchangeable except when entering a space in a text string. As you type in the command prompt, AutoCAD automatically completes the entry with an AutoCAD command or command alias. If you pause, it displays a list of all the commands whose prefix matches what you have typed enabling you to scroll and select from the list.

To repeat the last command used, press `Enter`, `SPACEBAR`, or right-click and choose *Repeat* <*LAST COMMAND*> from the shortcut menu. If you are using the Command window, you can also repeat a command by using the `UP-ARROW` and `DOWN-ARROW` keys to display the commands that you previously entered from the keyboard. Use the `UP-ARROW` key to display the previous line in the command history; use the `DOWN-ARROW` key to display the next line in the command history. Depending on the buffer size, AutoCAD stores all the information that you entered from the keyboard in the current session.

AutoCAD also allows you to use certain commands transparently, which means they can be entered on the command line while you are using another command. Transparent commands are usually commands that change drawing settings or drawing tools such as GRID, SNAP, and ZOOM. To invoke a command transparently, enter an apostrophe (') before the command name while you are using another command. After the transparent command is completed, the original command resumes.

Ribbon

The ribbon contains tabs that group related commands in panels. Select a tab and choose an icon to invoke the command from one of the available panels. Follow the prompts on-screen or on the command line in the Command window or select options from a dialog box.

Toolbars

The toolbars contain tools that represent commands. Click a toolbar button to invoke the command, and then select options from a dialog box or follow the prompts on-screen or on the command line in the Command window.

Menus

The menus are available from the menu bar at the top of the screen. You can display the menu bar by selecting Show Menu Bar from the drop down menu that appears when you click the down arrow at the right end of the Quick Access Toolbar. You can invoke almost all of the available commands from the menu bar. You can choose menu options in one of the following ways:

- First select the menu name to display a list of available commands, and then select the appropriate command.
- Press and hold down **ALT** and then enter the underlined letter in the menu name. For example, to invoke the LINE command, first hold down **ALT** and press D (**ALT** + D) to open the Draw menu, and then press L.

The default menu file, or the customized user interface, is *acad.cuix*. You can load a different menu file by invoking the MENU command.

Dialog Boxes

Many commands, when invoked, cause a dialog box to appear unless you prefix the command with a hyphen. For example, entering **insert** causes the dialog box to be displayed (see Figure 1–39), and entering **-insert** causes responses to be displayed in the on-screen prompt area. Dialog boxes display the lists and descriptions of options, long rectangles for receiving your input data, and, in general, are more convenient and user-friendly methods of communicating with the AutoCAD program for that particular command.

The commands listed in the menu bar that include ellipses (...), such as **Plot**... and **Hatch**..., display dialog boxes when selected. AutoCAD dialog boxes have features that are similar to Windows file-management dialog boxes.

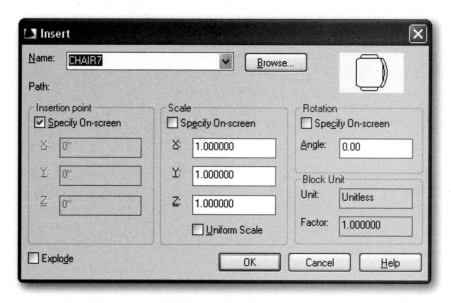

FIGURE 1–39 *Dialog box invoked from the insert command*

Cursor Menu

The AutoCAD cursor menu appears at the location of the cursor when you press the middle button on a mouse with three or more buttons (see Figure 1–40). On a two-button mouse, you can invoke this feature by pressing **SHIFT** and right-clicking. Also on a two-button mouse, the right button usually causes the shortcut menu to appear. The cursor menu, which is different from the shortcut menu, includes the handy Object Snap mode options along with the X, Y, Z filters. The reason that the Object Snap modes and Tracking are in such ready access will become evident when you learn the significance of these functions.

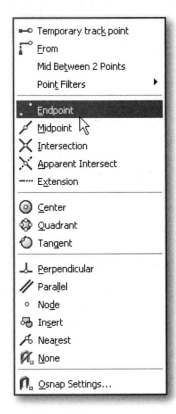

FIGURE 1–40 *Cursor menu*

Shortcut Menu

The AutoCAD shortcut menu appears at the location of the cursor when you right-click on the pointing device. The contents of the shortcut menu depend on the situation at hand.

If you right-click in the drawing window when no commands are in effect, the shortcut menu will include options to repeat the last command, a section for editing objects such as Cut and Copy, a section with Undo, Pan, and Zoom, and a section with Quick Select, Find, and Options (see Figure 1–41). Selections that are disabled under the current situation will appear in lighter text than those that are enabled.

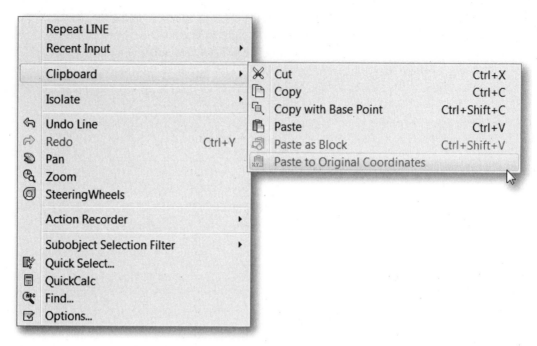

FIGURE 1–41 *Shortcut menu when no command is in effect*

If you select one or more objects under the default setup conditions and no commands are in effect, right-click, and the shortcut menu will include some of the editing commands (see Figure 1–42).

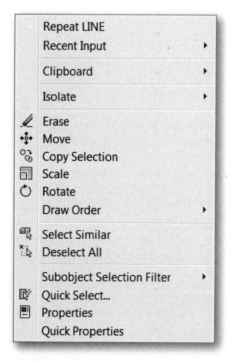

FIGURE 1–42 *Shortcut menu with one or more objects selected when no command is in effect*

After you have initiated a command and wish to use an option other than the default option, you can invoke the shortcut menu and select the desired option with the mouse. For example, if instead of the default center-radius method of drawing a circle, you wished to use the three-point (3P), two-point (2P), or tangent-tangent-radius (Ttr) option, you could select one of them from the shortcut menu, as shown in Figure 1–43. You could also transparently select the Pan and Zoom commands or cancel the command. If pressing 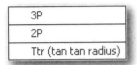 is required, that option is also available. If you are using dynamic input, you can also access the available options by pressing the DOWN-ARROW key.

3P
2P
Ttr (tan tan radius)

FIGURE 1–43 *Shortcut menu when the circle command is in effect*

If the Command window is open, right-click anywhere in it and the shortcut menu provides access to the six most recently used commands (see Figure 1–44).

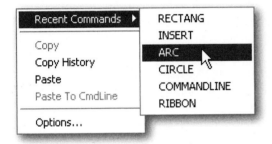

FIGURE 1–44 *Shortcut menu while the cursor is in the Command window*

Right-click on any of the buttons in the status bar, and the shortcut menu provides toggle options for drawing tools and ways to modify their settings. The options on the shortcut menu vary depending on the icon selected.

Right-click on any of the open AutoCAD dialog boxes and windows, and the shortcut menu provides context-specific options. Figure 1–45 shows an example for the Layer Properties Manager dialog box with a shortcut menu open.

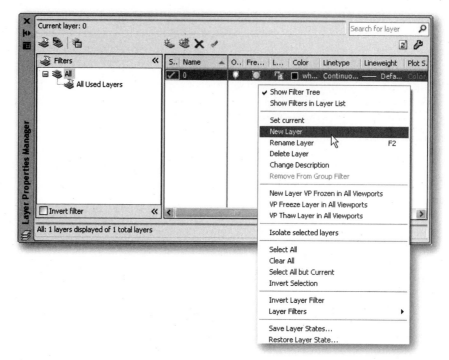

FIGURE 1–45 *Shortcut menu in the Layer Properties Manager dialog box*

Digitizing Tablet

In addition to the mouse, the digitizing tablet allows you to input information. It combines the screen cursor control of a mouse with its own printed menu areas for selecting items. However, with the new heads-up features and customizability in AutoCAD since release 2000, the tablet overlay with command entries is practically obsolete. One powerful feature of the tablet is its transfer function; it allows you to lay a map or other picture on the tablet and trace over it with the specific pointing device for a digitizing tablet (a "puck"), thereby transferring the image to the AutoCAD drawing.

GETTING HELP

Help is available through a traditional Windows-type help interface, by using the InfoCenter on the menu bar or title bar, by pressing F1 from within a command, or by clicking the question mark button in many dialog boxes.

Autodesk Help (see Figure 1–46) provides a web-based source of information directly within the product. The help window provides a wide variety of content, including announcements, expert tips, videos, and links to blogs. When enabled for online access, AutoCAD also includes access to the Knowledge Base, Communication Center, and Subscription Center.

FIGURE 1–46 *Autodesk Help Window*

The Help window can be opened transparently while you are in the middle of a command. For example, to use Help transparently, enter **'help** or **'?** in response to any prompt that does not require a text string. AutoCAD displays help for the current command. The help is often general in nature, but sometimes it is specific to the command's current prompt.

InfoCenter

On the Title bar, the InfoCenter allows you to search for information through key words (Figure 1–47). When you enter key words or type a question for help and then press **Enter** or click the search button, you search multiple Help resources in addition to any files that have been specified in the InfoCenter Settings dialog box. The results are displayed in the AutoCAD Help window.

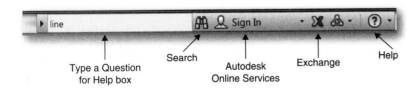

FIGURE 1–47 *InfoCenter located on the menu bar*

Help Pull-Down Menu

Use the Search box and button to quickly look up information in the online Help.

Sign In allows access to the Subscription services available to Autodesk subscription members. Services include access to: Latest releases of Autodesk software, Incremental product enhancements, Personalized web support from Autodesk technical experts, Self-paced e-Learning, and Autodesk 360 (cloud services).

Autodesk Exchange allows access to the Autodesk Exchange Apps page, where you can find various apps for use with your Autodesk applications.

The Help menu provides additional resources for using AutoCAD. Clicking the down-arrow next to the **?** icon causes the Help pull-down menu to be displayed (see Figure 1–48).

FIGURE 1–48 *Help pull-down menu*

From the Help pull-down menu you can choose **Help, Send Feedback, Customer Involvement Program**, and **About**.

Choosing **Help** causes the AutoCAD window to be displayed.

Choosing **Send Feedback** lets you contact Autodesk to tell them what you think about their products.

Choosing **Customer Involvement Program** lets you interact with the Autodesk team by providing system and usage data in exchange for information that will help optimize your software use.

Choosing **About** displays information about the AutoCAD software being used.

BEGINNING A NEW DRAWING

When the first new drawing is started in an AutoCAD drawing session, it is given the temporary name *drawing1.dwg*. It will not be saved with a name of your choice until you use a form of the SAVE command. The second drawing in a session is given the temporary name *drawing2.dwg* and so on.

To create a new drawing, invoke the NEW command from the Quick Access Toolbar by selecting **QNew** (see Figure 1–49). AutoCAD displays the Select template dialog box (see Figure 1–50).

FIGURE 1–49 *Invoking the New command from the Quick Access toolbar*

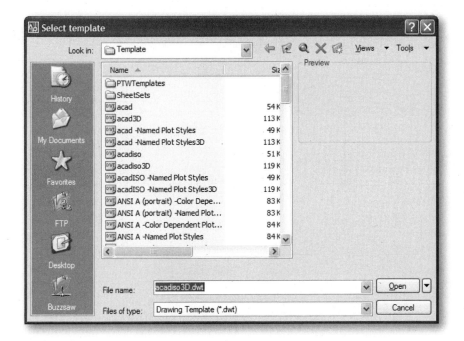

FIGURE 1–50 *Select template dialog box*

The Select template dialog box operates like a Windows file-management dialog box. It contains a **Preview** window that will show a thumbnail sketch, if it is available, of the template file selected. Select one of the available templates to create a new drawing.

A drawing template file is a drawing file with selected parameters preset to meet certain requirements so that you do not have to go through the process of setting them up each time you wish to begin drawing with those parameters. A template drawing might have the imaginary drawing sheet dimensions preset or the units of measurement preset, or it could contain objects already drawn. In many cases a blank title block has already been created on a standard sheet size. The type of coordinate system that is needed to make the drawing could already be set up with the origin (X, Y, and Z coordinates of 0, 0, and 0) located where needed relative to the edges of the envisioned drawing sheet.

The AutoCAD program files include over 60 templates for drawings of various standard sizes containing a predrawn title block conforming to standards such as ANSI, DIN, Gb, ISO, and JIS. You can create templates by making a drawing with the desired preset parameters and predrawn objects and then saving the drawing as a template file with the extension of *dwt*.

OPENING AN EXISTING DRAWING

The OPEN command allows you to open an existing drawing. Invoke the OPEN command from the Quick Access Toolbar by selecting **Open** (see Figure 1–51), and AutoCAD displays the Select File dialog box (Figure 1–52).

FIGURE 1–51 *Invoking the Open command from the Quick Access Toolbar*

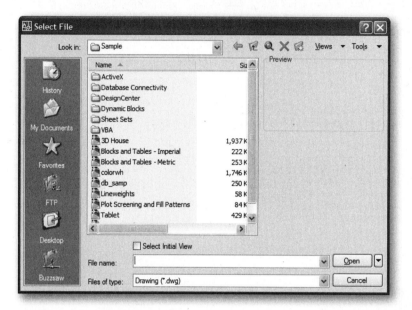

FIGURE 1–52 *Select File dialog box*

The Select File dialog box is similar to the standard file-selection dialog box, except that it includes options for selecting an initial view and for setting **Open Read-Only, Partial Open**, and **Partial Open Read-Only** modes. In addition, when you click on the filename, AutoCAD displays a bitmap image in the **Preview** section. A window on the left side of the dialog box displays Quick Access icons to folders on your computer: **Desktop** and **My Documents**; icons for **History** of recently opened drawings and **Favorites**; and locations on the Internet: **Buzzsaw** and **FTP**. Select the drawing from the appropriate folder and choose **Open** to open the drawing. You can also select multiple drawing files to open at the same time.

The **Select Initial View** checkbox permits you to specify a view name in the named drawing to be the startup view. If there are named views in the drawing, an M or P beside the name will tell you if the view is model or paper space, respectively, as shown in the Initial View dialog.

Buttons to the right of the **Look in** list box are **Back to <the last folder>, Up one level, Search the Web, Delete, Create New Folder, Views**, and **Tools**. The **Back to <the last folder>, Up one level, Search the Web, Delete**, and **Create New Folder** options are similar to most file-handling dialog boxes for reaching the location of the drawing you wish to open.

Choosing **Views** causes a menu to be displayed with the options of **List, Details**, or **Thumbnails** to determine how folders and files are displayed and a **Preview** option that opens a **Preview** area to show a thumbnail sketch of drawing selected.

Choosing **Tools** causes a menu to be displayed with the options of **Find, Locate, Add/ Modify FTP Locations, Add Current Folder to Places**, and **Add to Favorites**.

Choosing **Find** causes the **Find**: dialog box to be displayed (see Figure 1–53). Various drives and folders are searched using search criteria. The **Find**: dialog box combines the usual Windows file/path search of files by name and location in the **Name & Location** tab and by date ranges in the **Date Modified** tab.

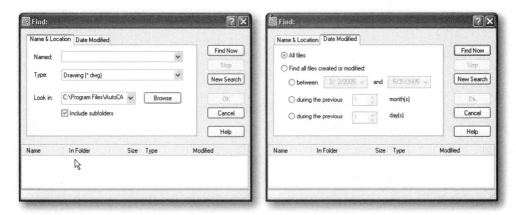

FIGURE 1–53 *Name and Location tab of the Find dialog box and Date Modified tab of the Find dialog box*

The **Date Modified** tab lets you specify dates and/or date ranges as criteria to search for drawings (see Figure 1–54).

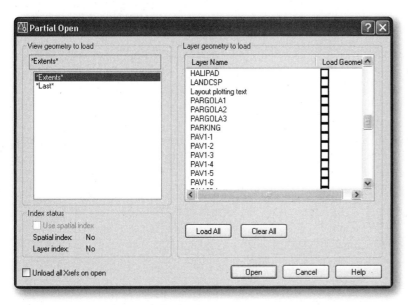

FIGURE 1–54 *Partial Open dialog box*

> **NOTE**
>
> You can open and edit an AutoCAD release 12, 13, or 14 drawing and, if necessary, save the drawing in other formats by using the SAVEAS command. Possible formats include 2010, 2007, 2004, 2000/LT2000, LT98 and LT97 Drawings [*.dwg], Drawing Standards [*.dws], Template [*.dwt], 2010, 2007, 2004, 2000/LT2000 DXF [*.dxf], and R12/LT 2 DXF [*.dxf]. Certain limitations apply when doing this, which are explained in the section about the SAVEAS command later in this chapter.

A drop-down menu is displayed when you select the down-arrow to the right of the **Open** button. From this menu you may open a drawing in the **Open Read-Only** option, which permits you to view the drawing but not save it with its current name. You can open a drawing with the **Partial Open** option, which when selected displays the Partial Open dialog box (see Figure 1–52). It displays the drawing views and layers available for specifying what geometry to load into the selected drawing. When working with large drawing files, you can select the minimal amount of geometry you need to load when opening a drawing. You can also select the **Partial Open Read-Only** option and save the file under a different name by invoking the SAVEAS command.

> **NOTE** Even though a drawing is partially open, all objects are still loaded. All layers are available, but only the specified layers will have their geometry appear in the drawing when it is opened.

SETTING UP THE DRAWING ENVIRONMENT

This section covers the commands and features used to communicate the physical appearance of objects. A rectangle might represent a small computer chip on a printed circuit, a building, or a map of the state of Colorado. Whichever object is depicted, the appropriate type of units (metric, architectural, surveyor's) should be used. The types of units include both linear and angular measurements. What shape and how much of the drawing area is to be set aside needs to be determined. The UNITS and LIMITS commands are used to accomplish these tasks.

Setting Drawing Units

The UNITS command lets you change the linear and angular units by means of the Drawing Units dialog box. In addition, it lets you set the display format measurement and precision of your drawing units. You can change any or all of the following:

Unit display format	Angle display precision
Unit display precision	Angle base
Angle display format	Angle direction

Invoke the UNITS command from the on-screen prompt by typing **units** and pressing [Enter], and AutoCAD displays the Drawing Units dialog box (see Figure 1–55).

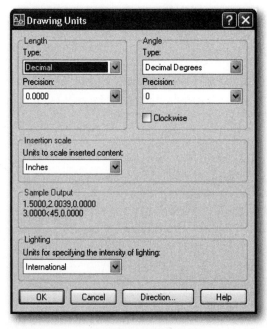

FIGURE 1–55 *The Drawing Units dialog box*

Setting Linear Units

The **Length** section of the Drawing Units dialog box allows you to change the units of linear measurement. From the **Type** list box, select one of the five types of report formats you prefer. For the selected report format, choose precision from the **Precision** list box.

The engineering and architectural units display as feet-and-inches displays. These units assume that each drawing unit represents 1 inch. The other units (scientific, decimal, and fractional) make no such assumptions, and they can represent whatever real-world units you like.

Drawing a 150-foot-long object might, however, differ depending on the units chosen. For example, if you use decimal units and decide that 1 unit = 1 foot, then the 150-foot-long object will be 150 units long. If you decide that 1 unit = 1 inch, then the 150-foot-long object will be drawn as $150 \times 12 = 1{,}800$ units long. In architectural and engineering units modes, the unit automatically equals 1 inch. You may then give the length of the 150-foot-long object as 150' or 1,800" or simply as 1,800.

Setting Angular Units

The **Angle** section of the Drawing Units dialog box allows you to set the drawing's angle measurement. From the **Type** list box, select one of the five types of report formats you prefer. The available formats include Decimal Degrees, Deg/Min/Sec, Grads, Radians, and Surveyor's Units. For the selected format, choose the precision from the **Precision** list.

Select the direction in which the angles are measured, clockwise or counterclockwise. If the **Clockwise** checkbox is checked, the angles will increase in value in the clockwise direction. If the box is not checked, the angles will increase in value in the counterclockwise direction (Figure 1–56).

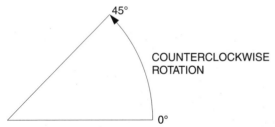

FIGURE 1–56 *The default counterclockwise direction of angle measurement*

The default of 0 degrees at east and angle values increasing in the counterclockwise direction is used for the angular prompts and responses throughout this book, unless otherwise noted.

NOTE

Setting Insertion Scale

The units setting that you select from the **Insertion Scale** list box determines the unit of measure used for block insertions from AutoCAD DesignCenter, Tool Palettes, or i-drop. If a block is created in units different from the units specified in the list box, the block will be inserted and scaled in the specified units. If you select Unitless, the block will be inserted as is, and the scale will not be adjusted to match the specified units.

Setting Lighting Units

The **Lighting** unit setting controls the unit of measurement for the intensity of photometric lights in the current drawing.

Setting the Base Angle for Angle Measurement

To set the base angle for angle measurement, choose **Direction**, and AutoCAD displays the Direction Control dialog box (see Figure 1–57).

FIGURE 1–57 *Direction Control dialog box*

By default, AutoCAD specifies that 0 degrees is to the right (east or 3 o'clock) and that angles increase in the counterclockwise direction (see Figure 1–58).

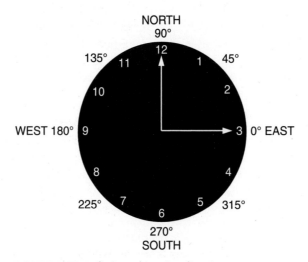

FIGURE 1–58 *Default angle-setting direction*

You can change angle measurement to start at any compass point by selecting one of the five available options. You can also determine the direction you want for angle 0 by specifying two points. This can be done by selecting **Other** and choosing **Angle**. AutoCAD prompts you for two points and sets the direction for angle 0. Choose **OK** to close the Direction Control dialog box.

Once you are satisfied with all of the settings in the Drawing Units dialog box, choose **OK** to accept the settings for the current working drawing and close the dialog box.

When AutoCAD prompts for a distance, displacement, spacing, or coordinates, you can enter numbers in integer, decimal, scientific, or fractional format. If the engineering or architectural format is in effect, you can also input feet, inches, or a combination of feet and inches. However, feet-and-inches input format differs slightly from the displayed format because it cannot contain a blank space. For example, a distance of 75.5 inches can be entered in the feet/inches/fractions format as 6'3-1/2". Note the hyphen in the unconventional location between the inches and the fraction and the absence of spaces. Normally, it will be displayed in the status area as 6'-3 1/2.

If you wish, you can use the SETVAR command to set the UNITMODE system variable to 1 to display feet-and-inches output in the accepted format; the default UNITMODE setting is 0. For example, if you set UNITMODE to 1, AutoCAD displays the fractional value of 45 1/4 as you enter it: 45-1/4. The feet input should be followed by an apostrophe ('). The trailing double quote (") after the inches input is optional.

When the engineering or architectural format is in effect, the drawing unit equals 1 inch, so you can omit the trailing double quote ("). When you enter feet-and-inches values combined, the inch values should immediately follow the apostrophe without an intervening space. Distance input does not permit spaces because, except when entering text, pressing the SPACEBAR functions the same as pressing Enter.

Setting Drawing Limits

The LIMITS command allows you to place an imaginary rectangular drawing sheet in the CAD drawing space. However, unlike the limitations of the drawing sheet of the board drafter, you can move or enlarge the CAD electronic sheet, or the limits, after you have started your drawing. The LIMITS command does not affect the current display on the screen. The defined area determined by the limits governs the portion of the drawing indicated by the visible grid if GRID is turned ON. For more on the GRID command, see Chapter 3. The rectangle specified by the limits also determines how much of the drawing is displayed by the ZOOM ALL command. For more on the ZOOM ALL command, see Chapter 3.

The limits are expressed as a pair of 2D points in the World Coordinate System: a lower-left limit and an upper-right limit. See Chapter 2 for further discussion. For example, to set limits for an A-size sheet, set lower-left as 0,0 and upper-right as 11,8.5 or 12,9; for a B-sized sheet, set lower-left as 0,0 and upper-right as 17,11 or 18,12. Many architectural floor plans are drawn at a scale of 1/4 = 1'-0. To set limits to plot on a C-sized (22 × 17) paper at 1/4 = 1'-0, the limits are set lower-left as 0,0 and upper-right as 88',68' (4 × 22, 4 × 17).

Invoke the LIMITS command from the on-screen prompt by typing **limits** and pressing Enter, and AutoCAD invokes the following prompts:

```
Reset Model space limits:
Specify lower left corner or ⬇ <current>: (press Enter to
accept the current setting, specify the lower-left corner,
or right-click and select one of the options)
Specify upper right corner <current>: (press Enter to accept
the current setting, or specify the upper-right corner)
```

The response you give for the upper-right corner gives the location of the upper-right corner of the imaginary rectangular drawing sheet.

Two additional options are available for the LIMITS command. When AutoCAD prompts for the lower-left corner, you may respond with the ON or OFF option. The

ON and OFF options determine whether or not you can specify a point outside the limits when prompted to do so.

When you select the ON option, limits checking is turned on, and you cannot start or end an object outside the limits, nor can you specify displacement points required by the MOVE or COPY command outside the limits. You can, however, specify two points (center and point on circle) that draw a circle, part of which might be outside the limits. The limits check is an aid to help you avoid drawing off the imaginary rectangular drawing sheet. Leaving the limits checking ON is a sort of safety net to keep you from inadvertently specifying a point outside the limits. On the other hand, limits checking is a hindrance if you need to specify such a point.

When you select the default OFF option, AutoCAD disables limits checking, allowing you to draw the objects and specify points outside the limits.

Whenever you change the limits, you will not see any change on the screen unless you use the ALL option of the ZOOM command. This option lets you see the newly set limits so they fill the screen. For example, if your current limits are 12 × 9 (lower-left corner 0,0 and upper-right corner 12,9) and you change the limits to 42 × 36 (lower-left corner 0,0 and upper-right corner 42,36), you still see the 12 × 9 area. You can draw the objects anywhere in the limits 42 × 36 area, but you will see on the screen the objects that are drawn only in the 12 × 9 area. To see the entire limits, invoke the ZOOM command using the ALL option.

When you invoke the ALL option of the ZOOM command, you see the entire limits or current extents, whichever is greater, on the screen. If objects are drawn outside the limits, all objects are displayed. For a detailed explanation of the ZOOM command, see Chapter 3.

Whenever you change the limits, you should always invoke the ALL option of the ZOOM command to see the entire limits or current extents on the screen.

For example, the following command sequence shows the steps to change the limits for an existing drawing (see Figures 1–59 and 1–60).

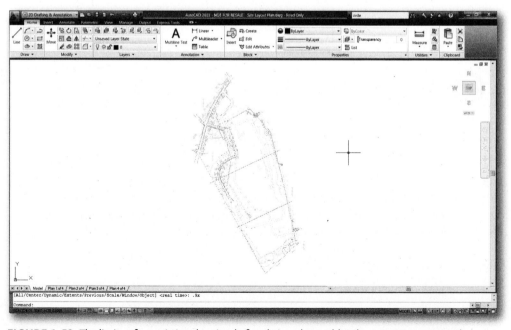

FIGURE 1–59 *The limits of an existing drawing before being changed by the* LIMITS *command*

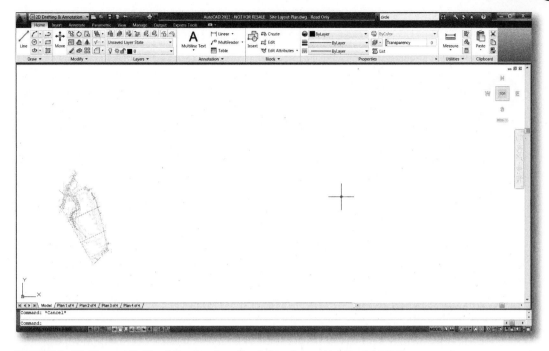

FIGURE 1–60 *The new limits of the drawing after being changed by the* LIMITS *command*

```
limits  Enter
Reset Model space limits:
Specify lower left corner or ☑ <current>: 0,0  Enter
Specify upper right corner <current>: 42,36
zoom  Enter
Specify corner of window, enter a scale factor (nX or nXP), or
☑ all  Enter
```

WORKING WITH MULTIPLE DRAWINGS

AutoCAD allows you to work on more than one drawing in a single AutoCAD session (Multiple Document). When multiple drawings are open, you can switch between them from the Quick View Drawings button located in the Status bar that displays all currently open drawings in a row of Quick View drawing images as explained earlier. This functionality is a faster and more visual alternative to using the CTRL + TAB key combination or the View menu for switching between open drawings.

You can copy and paste objects between drawings, and you can use the Properties palette or DesignCenter to transfer properties from objects in one drawing to objects in another. You can also use AutoCAD object snaps, the copy with base point command, and the paste to original coordinates command for accurate placement, especially when copying objects from one drawing to another.

SAVING A DRAWING

While working in AutoCAD, you should save your drawing once every 10 to 15 minutes without exiting AutoCAD. By saving your work periodically, you protect your work from possible power failures, editing errors, and other disasters. This step can be done automatically by setting the SAVETIME system variable to a specific interval (in minutes). In addition, you can manually save by using the SAVE, SAVEAS, and QSAVE commands.

The SAVE command saves an unnamed drawing with a filename that you specify in AutoCAD 2011 format. If the drawing is already named, the feature works like the SAVEAS command when the SAVE command is invoked from the on-screen prompt.

The SAVEAS command saves an unnamed drawing to a new filename or saves the current drawing to a specified new name, making it the new current drawing while retaining a copy of the drawing from which it is being created in the stage of its last save. If you specify a filename that already exists in the current folder, AutoCAD displays a message warning you that you are about to overwrite another drawing file. If you do not want to overwrite it, specify a different filename. The SAVEAS command also allows you to save in various formats. Possible formats include 2010, 2007, 2004/LT2004, 2000/LT2000, and R14/L98/L97 Drawings [*.*dwg*], Drawing Standards [*.*dws*], Template [*.*dwt*], 2010, 2007, 2004 and 2000/ LT2000 DXF [*.*dxf*], R12/LT 2 DXF [*.*dxf*]. AutoCAD 2004 is the drawing file format used by the AutoCAD 2004, AutoCAD 2005, and AutoCAD 2006 releases.

The QSAVE command saves an unnamed drawing with a filename that you specify. If the drawing is named, AutoCAD saves the drawing without requesting a filename.

Invoke the QSAVE command from the Quick Access Toolbar (see Figure 1–61), and AutoCAD displays the Save Drawing As dialog box.

FIGURE 1–61 *Invoke the SAVE command from the Quick Access Toolbar*

Select the appropriate folder in which to save the file, and type the name of the file in the **File name** text field. The window on the left side of the dialog box displays Quick Access icons to folders on your computer: **Desktop** and **My Documents**; icons for **History** of recently opened drawings and **Favorites**; and locations on the Internet: **Buzzsaw** and **FTP**. These steps let you quickly specify where to save the drawing. The filename can contain up to 255 characters, including embedded spaces and punctuation. Filenames cannot include any of the following characters: forward slash (/), backslash (\), greater than sign (>), less than sign (<), asterisk (*), question mark (?), quotation mark ("), pipe symbol (|), colon (:), or semicolon (;). The following are examples of valid filenames:

    this-is-my-first-drawing

    first_house

    machine part 123

AutoCAD automatically appends *.dwg* as a file extension. If you save the file as a template, AutoCAD appends *.dwt* as a file extension.

If you would like to save the drawing to a different drawing name, invoke the SAVEAS command by selecting **Save As** from the **Applications** menu, and Auto-CAD displays the Save Drawing As dialog box. Name and path restrictions are the same as in the SAVE command described earlier in this section.

If you want to save the current drawing to the given filename, invoke the SAVE command without changing the filename. AutoCAD saves the drawing without requesting a filename.

If you are working in multiple drawings, the CLOSE command closes the active drawing. Invoke the CLOSE command from the **Applications** menu, and AutoCAD closes the active drawing. If you have not saved the drawing since the last change, AutoCAD displays the AutoCAD alert: Save changes to *Filename.dwg*. If you select **No**, AutoCAD closes the drawing. If you select **Yes**, AutoCAD displays the Save Drawing As dialog box. Select the appropriate folder in which to save the file, and type the name of the file in the **File name** text field.

As mentioned earlier, the Quick View Drawings button located in the status bar displays all currently open drawings in a row of Quick View drawing images. Passing the cursor over a drawing preview displays Save and Close buttons in the upper corners of the image, enabling you to quickly save or close any open drawing, aside from the current one.

EXITING AUTOCAD

The EXIT or QUIT command allows you to exit AutoCAD. These commands exit the current drawing if there have been no changes since the drawing was last saved. If the drawing has been modified, AutoCAD displays the Drawing Modification dialog box to prompt you to save or discard the changes before quitting.

Invoke the EXIT command from the **Applications** menu by selecting **Exit**. If the drawing has been modified, AutoCAD displays the Drawing Modification dialog box to prompt you to save or discard the changes before exiting.

Invoke the QUIT command by typing **quit** at the On-Screen prompt or Command window. If the drawing has been modified, AutoCAD displays the Drawing Modification dialog box to prompt you to save or discard the changes before quitting.

REVIEW QUESTIONS

1. If you executed the following commands in order—LINE, CIRCLE, ARC, and ERASE—what would you need to do to reexecute the CIRCLE command?
   a. Press PGUP, PGUP, PGUP, Enter
   b. Press DOWN-ARROW, DOWN-ARROW, DOWN-ARROW, Enter
   c. Press UP-ARROW, UP-ARROW, UP-ARROW, Enter
   d. Press PGDN, PGDN, PGDN, Enter
   e. Press LEFT-ARROW, LEFT-ARROW, LEFT-ARROW, Enter

2. In a dialog box, when there is a set of mutually exclusive options, i.e., a list of several from which you must select only one, these are called:
   a. Text boxes
   b. Checkboxes
   c. Radio buttons
   d. Scroll bars
   e. List boxes .

3. What is the extension used by AutoCAD for template drawing files used by the setup wizard?

   a. *.dwg*

   b. *.dwt*

   c. *.dwk*

   d. *.tem*

   e. *.wiz*

4. To open and edit existing drawings what would you type in the Command Line?

   a. Open

   b. O

   c. Op

   d. Opn

   e. None of the above

5. If you select the ALL option of the ZOOM command and your drawing shrinks to a small portion of the screen, the problem might be caused by which of the following:

   a. Out of computer memory

   b. Misplaced drawing object

   c. Grid and snap are set incorrectly

   d. This should never happen in AutoCAD

   e. Limits are set much larger than current drawing objects

6. How do you reveal the shortcut menu?

   a. Right-click with your mouse

   b. Left-click with your mouse

   c. Right-click with your mouse and select shortcut menu

   d. Left-click with your mouse and select shortcut menu

7. What is an external tablet used to input absolute coordinate addresses to AutoCAD by means of a puck or stylus called?

   a. Digitizer

   b. Input pad

   c. Coordinate tablet

   d. Touch screen

8. The menu that can be made to appear at the location of the crosshairs is called a:

   a. Mouse menu

   b. Crosshair menu

   c. Cursor menu

   d. Shortcut menu

9. In order to save basic setup parameters, such as snap or grid for future drawings, you should:

    a. Create an AutoCAD Macro

    b. Create a prototype drawing template file

    c. Create a new configuration file

    d. Modify the ACAD.INI file

10. What happens when you press `Enter` when no command is active?

    a. It invokes the last AutoCAD command

    b. Nothing happens

    c. It undoes the last command

    d. It opens up the shortcut menu options

    e. None of the above

11. To cancel an AutoCAD command, press

    a. `CTRL` + A

    b. `CTRL` + X

    c. `ALT` + A

    d. `ESC`

    e. `CTRL` + `Enter`

12. In the Command Line, you can:

    a. Minimize your computer screen

    b. Search for help documents

    c. Type keyboard commands

    d. Display a series of tabs and panels

    e. None of the above

13. The SAVE command:

    a. Saves your work

    b. Does not exit AutoCAD

    c. Is a valuable feature for periodically storing information to disk

    d. All of the above

14. You can only open one drawing file at a time with the Open command.

    a. True

    b. False

15. From which direction does AutoCAD start measuring angles?

    a. 12 o'clock

    b. 3 o'clock

    c. 6 o'clock

    d. 9 o'clock

16. Is 300 degrees the same as −60 degrees in a drawing?

    a. Yes

    b. No

    c. Not always

    d. Never

17. Which one is NOT a vaild type of unit?

    a. Architectural

    b. Decimal

    c. Fractional

    d. Metric

2

Fundamentals I

INTRODUCTION

This chapter introduces some of the basic AutoCAD commands and concepts that can be used to complete a simple drawing. The exercises used in this chapter may appear relatively uncomplicated, but for the newcomer they present ample challenges. This chapter provides useful lessons in drawing setup as well as in how to create and edit objects. Once you learn how to access and use the basic commands, how to find your way around the screen, and how AutoCAD makes use of coordinate geometry, you can apply these skills to the chapters containing more advanced drawings and projects.

OBJECTIVES

After completing this chapter, you will be able to do the following:

- Construct geometric figures with `LINE`, `RECTANGLE`, `CIRCLE`, and `ARC` commands
- Use coordinate systems
- Use various object selection methods
- Use the `ERASE` command

CONSTRUCTING GEOMETRIC FIGURES

AutoCAD gives you a great variety of drawing elements called objects. It also provides you with many ways to generate each object in your drawing. You will learn about the properties of these objects as you progress through this text. It is important to keep in mind that the example instructions given to generate the lines, circles, arcs, and other objects are not always the only methods available. You are invited, even challenged, to find other more expedient methods to perform the tasks demonstrated in the lessons. You will progress at a faster rate if you make an effort to learn as much as possible, as soon as possible, about the descriptive properties of the individual objects. When you become familiar with how AutoCAD creates, manipulates, and stores the data that describes the objects, you will be able to create drawings more effectively.

Drawing Lines

A line is a primary drawing object; it may be defined as a path connecting two end-points. In AutoCAD, a series of line segments may be drawn by invoking the LINE command and then specifying a sequence of endpoints. AutoCAD connects the points with a line. The LINE command is one of the few AutoCAD commands that automatically repeats. It uses the ending point of the previous line as the starting point for the next segment and continues to prompt you for subsequent endpoints. To terminate this continuation behavior, press Enter or right-click and choose Enter from the shortcut menu. Even though a sequence of lines may be drawn using a single LINE command, each segment is a separate line object.

You can specify the endpoints using two-dimensional (2D) coordinates (X,Y), three-dimensional (3D) coordinates (X,Y,Z), or a combination of the two. If you enter 2D coordinates, AutoCAD uses the current elevation as the Z element of the point; zero is the default.

Invoke the LINE command from the Draw panel (Figure 2–1) located in the Home tab, and AutoCAD displays the on-screen prompt, "Specify first point:" (Figure 2–2).

FIGURE 2–1 *Invoking the LINE command from the Draw panel*

FIGURE 2–2 *On-screen prompt displayed near cursor*

Where to Start

You can designate the starting point of a line by entering absolute coordinates. See the following section on coordinate systems for a detailed explanation. Or, you can specify a location on the screen using your pointing device.

Where to from Here

After you specify the first point, AutoCAD displays the prompt "Specify next point or ⬇," as shown in Figure 2–3. You can specify the endpoint of a line by entering absolute coordinates, relative coordinates, or by using your pointing device.

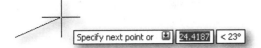

FIGURE 2-3 *On-screen prompt asking you to specify next point and displaying current geometric values*

As previously mentioned, to save time, the LINE command remains active and prompts "Specify next point or ⬇" after each point you specify. When you have finished, press Enter to terminate the LINE command.

If you are placing points with a pointing device instead of specifying coordinates, a rubber band preview line is displayed between the starting point and the crosshairs. This helps you to visualize where the line will be drawn. In Figure 2–4, the dotted lines represent previous cursor positions.

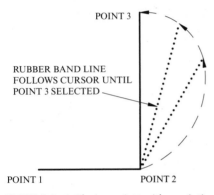

FIGURE 2-4 *Placing points with a pointing device rather than with keyboard coordinates input*

Continuing from a Previously Drawn Line or Arc

When you invoke the LINE command and respond to the "Specify first point:" prompt by pressing Enter or SPACEBAR, AutoCAD will automatically set the start of the new line sequence at the end of the most recently drawn line or arc. This action provides a simple method of constructing a tangentially connected line in an arc-line continuation.

The prompt sequence that follows will depend on whether a line or an arc was more recently drawn. If a line was drawn more recently, the starting point of the new line will be set at the endpoint of the previous line, the continuation prompt appears as usual. For instance, continuing with the example shown in Figure 2–4, three additional line segments can be drawn using the following sequence (Figure 2–5):

Specify first point: *(to continue the next line from Point 3, press* Enter *or* SPACEBAR *)*
Specify next point or ⬇ *(specify Point 4)*
Specify next point or ⬇ *(specify Point 5)*
Specify next point or ⬇ *(specify Point 6)*
Specify next point or ⬇ Enter

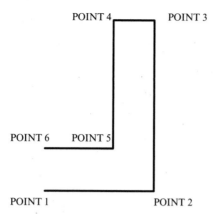

FIGURE 2–5 *Using the LINE command's Continue feature*

If an arc was drawn more recently, its endpoint and ending direction define the starting point and direction of the new line. In this case, AutoCAD displays the following prompt:

```
Length of line:
```

After specifying the length of the line to be drawn, AutoCAD continues with the normal "Specify next point or ⬇" prompt.

Line Command Options

Most of the AutoCAD commands have a variety of options. For the LINE command, two options are available: *Close* and *Undo*. You can access the available options from the shortcut menu that appears when you right-click your pointing device after you invoke the LINE command. You can also access the options from the on-screen prompt by pressing the DOWN-ARROW (⬇) key. If you are using the Command window, you can access the options by typing the key letter(s) of the option name that are shown in capital letters.

When drawing a sequence of lines to form a polygon, choose *Close* to automatically join the last and first points. AutoCAD performs two steps when you choose the *Close* option. The first step closes the polygon, and the second step terminates the LINE command.

The following command sequence shows an example of using the *Close* option (Figure 2–6).

```
line Enter
Specify first point: (specify Point 1)
Specify next point or ⬇ (specify Point 2)
Specify next point or ⬇ (specify Point 3)
Specify next point or ⬇ (specify Point 4)
Specify next point or ⬇ (specify Point 5)
Specify next point or ⬇ (specify Point 6)
Specify next point or ⬇ (choose Close from the shortcut menu)
```

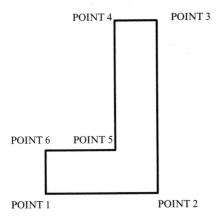

FIGURE 2–6 *Using the LINE command's Close option*

Using the *Close* option causes the LINE command to automatically draw the last line of a sequence from the endpoint of the most recent line to the starting point of the first line of the sequence. If the series of lines in Figure 2–6 had been drawn in two sequences, one from Point 1 through Point 3 and then continued by invoking the LINE command to draw the second sequence from Point 3 through Point 6, then the *Close* option would cause a line to be drawn from Point 6 to Point 3, not to Point 1.

Choosing *Undo* erases the most recent line without terminating the LINE command. For instance, while drawing a sequence of connected lines, you may wish to erase the most recently drawn line segment and continue from the endpoint of the previous line segment. You do so by selecting the *Undo* option from the shortcut menu. If necessary, you can select UNDO multiple times; this will erase previously drawn line segments one at a time. Once you exit the LINE command, its *Undo* option to erase the most recent line segments one at a time is no longer available.

The following command sequence shows an example of using the *Undo* option (Figure 2–7).

```
line Enter
Specify first point or ⊡ (specify Point 1)
Specify next point or ⊡ (specify Point 2)
Specify next point or ⊡ (specify Point 3)
Specify next point or ⊡ (specify Point 4)
Specify next point or ⊡ (specify Point 5)
Specify next point or ⊡ (choose undo from the shortcut menu)
Specify next point or ⊡ (choose undo from the shortcut menu)
Specify next point or ⊡ (specify New Point 4)
Specify next point or ⊡ (specify New Point 5)
Specify next point or ⊡ Enter
```

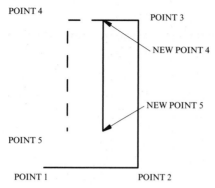

POINT 4

POINT 3

NEW POINT 4

NEW POINT 5

POINT 5

POINT 1   POINT 2

FIGURE 2–7 *Using the* LINE *command's Undo option*

Drawing Rectangles

The RECTANGLE command creates a closed polyline in a rectangular shape. Polylines are discussed in detail in Chapter 4. You can specify length, width, area, and rotation parameters. You can also control the type of corners on the rectangle, such as fillet, chamfer, or square.

Invoke the RECTANGLE command from the Draw panel (Figure 2–8) located in the Home tab, and AutoCAD displays the on-screen prompt shown in Figure 2–9.

FIGURE 2–8 *Invoking the* RECTANGLE *command from the Draw panel*

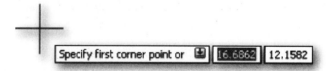

FIGURE 2–9 *On-screen prompt displayed near cursor*

AutoCAD displays the following prompts:

Specify first corner point or ⬇ (*specify first corner point to define the start of the rectangle, or right-click for the shortcut menu and choose one of the available options*)

Specify other corner point or ⬇ (*specify a point to define the opposite corner of the rectangle, or right-click for the shortcut menu and choose one of the available options*)

Rectangle Command Options

Choosing *Chamfer* sets the chamfer distance for the rectangle to be drawn. Refer to Chapter 4 for a detailed explanation of how to use the CHAMFER command and its available settings.

Choosing *Elevation* specifies the elevation of the rectangle to be drawn. Refer to Chapter 16 for a detailed explanation of how to use the ELEVATION setting.

Choosing *Fillet* sets the fillet radius for the rectangle to be drawn. Refer to Chapter 4 for a detailed explanation of how to use the FILLET command and its available settings.

Choosing *Thickness* sets the thickness of the rectangle to be drawn. Refer to Chapter 16 for a detailed explanation of how to use the THICKNESS setting.

Choosing *Width* allows you to set the line width for the rectangle to be drawn. The default width is set to 0.0.

Choosing *Area* allows you to create a rectangle of a specified area. You also specify either the length or the width value. When you choose the *Area* option after specifying the first corner point, AutoCAD displays the following prompts:

> Enter area of rectangle in current units <default>: *(specify area of the rectangle)*
>
> Calculate rectangle dimensions based on *(specify either length or width)*

If you have specified length, AutoCAD displays the following prompt:

> Enter rectangle length <default>: *(specify length of the rectangle)*

If you have specified width, AutoCAD displays the following prompt:

> Enter rectangle width <default>: *(specify width of the rectangle)*

AutoCAD draws a rectangle using the specified first corner, area, and length or width.

Choosing *Dimensions* allows you to create a rectangle by specifying length and width values. AutoCAD displays the following prompts:

> Specify length for rectangles <default>: *(specify length of the rectangle)*
>
> Specify width for rectangles <default>: *(specify width of the rectangle)*
>
> Specify other corner point or ⊡ *(specify a point by moving the cursor to one of the four possible locations for the diagonally opposite corner of the rectangle)*

Choosing *Rotation* allows you to create a nonorthogonal rectangle. AutoCAD displays the following prompts:

> Specify rotation angle or ⊡ *(specify angle of the base of the rectangle)*

> Specify other corner point or ⊞ *(specify a point by moving the crosshairs to one of the four possible locations for the diagonally opposite corner of the rectangle)*

AutoCAD draws a rectangle using the specified angle for the base and second corner.

Coordinate System

A coordinate system allows you to specify the locations of points in space or on a plane. Anytime you draw in AutoCAD, only one drawing plane is active. A drawing plane is an infinite 2D plane on which 2D drawing objects, such as lines and arcs, are drawn. All dimensions and points on the current drawing plane are expressed in numeric terms as pairs or triplets of distances (coordinate values) from an origin point (0,0), as shown in Figure 2–10. AutoCAD's Coordinate System allows for an infinite number of drawing planes that can be rotated or tilted as needed.

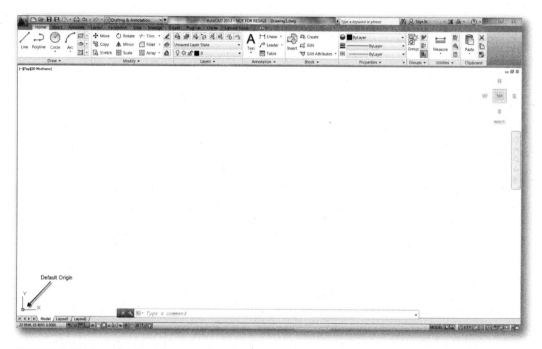

FIGURE 2–10 *AutoCAD Drawing Area showing the Coordinate System icon at the origin point (0,0)*

What's the Point?

A child's picture puzzle book guides you to create a "connect-the-dots" picture by drawing lines between points that are numbered 1, 2, 3, and so on. These numbers have nothing to do with the distances from one dot to another or their relative location in the picture. They specify only the sequence of starting and ending a series of lines. However, if the picture had been generated in AutoCAD, the points would have other numbers associated with them: their coordinates. Because an AutoCAD drawing plane has a built-in coordinate system, whether you elect to use it or not, all points in your drawing have pairs of numbers associated with them—triplets, when you advance to 3D (Figure 2–11). These pairs of numbers describe where the point is located as determined by its distances from two axes that intersect at the origin (0,0) in a specific coordinate system.

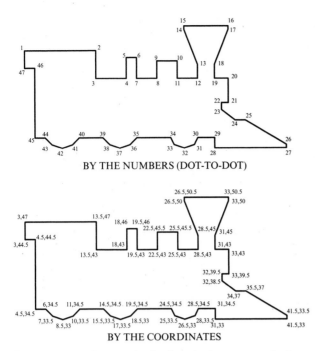

BY THE NUMBERS (DOT-TO-DOT)

BY THE COORDINATES

FIGURE 2–11 *"Connect-the-dots" method versus coordinates*

A coordinate system is not used for most board drafters' drawings, especially architectural and mechanical drawings, electrical schematics, and piping flow diagrams that are not concerned with dimensions and spatial relationships. However, surveyors' plats and layouts of manufacturing plants often tie the location of objects to some basic coordinate system. For instance, when a large petrochemical plant is situated on several hundred acres, those who design the original layout will establish two major axes, usually one east–west and the other north–south. These imaginary lines are perpendicular to each other (Figure 2–13). The point where they cross is called the "origin." If the origin is in the middle of the plant, then depending on which quadrant in which a point is situated, one or both coordinates might be negative. In Figure 2–12, the positive X and Y values are given as E (East) and N (North), respectively, while the negative X and Y values are given as W (West) and S (South), respectively.

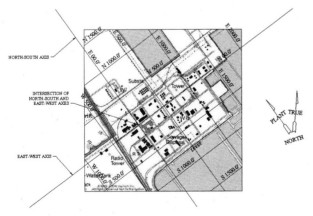

FIGURE 2–12 *Plant layout with axes*

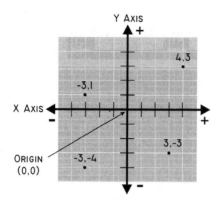

FIGURE 2–13 *A 2D drawing plane showing the X and Y axes, origin, and points in each quadrant*

AutoCAD's coordinate system is always present, and eventually you will need to interact with it. For instance, if you wish to plot a drawing, you must understand how the coordinate system functions and affects the arrangement and location of objects on the plotted sheet. The fact that every object in AutoCAD has a set of coordinates associated with it offers tremendous advantages. For example, if a grouping of objects is drawn correctly but not located as it should be, it can be moved en masse to the proper location with relative ease, thanks to the coordinate system.

Three Types of Coordinate Systems

In 3D space, there are three commonly used coordinate systems, and each system uses a set of three numbers to describe the location of a point.

The Spherical Coordinate System

The spherical coordinate system is used for specifying points on a sphere. It is the basis for latitude and longitude navigation on the surface of the earth. The first number of this coordinate system is the radius of the sphere (**r** in Figure 2–14). The second number (Θ in Figure 2–14) is the angle between a line through a zero point on the equator and a line that goes through where the point is projected onto the equatorial plane of the sphere. This number is given as the longitude in navigational terms. The third number (φ in Figure 2–14) is the angle between a line through the point and the equatorial plane. The number is given as the latitude in navigational terms. AutoCAD spherical coordinates should be entered as **r** (distance from the origin), Θ (angle from X axis), and Φ (angle from XY plane). In Figure 2–14, the illustration on the right shows how AutoCAD calculates the spherical coordinates 10 < 75 < 60.

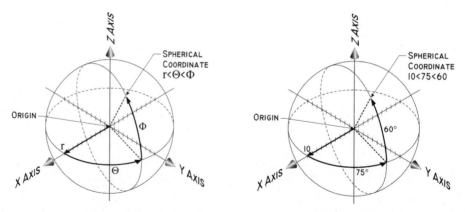

FIGURE 2–14 *Spherical coordinate system: mathematical model (left); AutoCAD's designation of point 10 < 75 < 60 (right)*

The Cylindrical Coordinate System

The cylindrical coordinate system is used for specifying a point on a cylinder. It has as its base a horizontal plane that is perpendicular to the centerline of the cylinder. On the base plane, a zero baseline exists from the center of the cylinder that extends to the surface of the cylinder. The first number (**r** in Figure 2–15) is the radius of the cylinder. The second number (Θ in Figure 2–15) is the angle of rotation from the zero baseline. The third point (**z** in Figure 2–15) is the distance along the Z axis from the base plane. In Figure 2–15, the illustration on the right shows how AutoCAD calculates the cylindrical coordinate 5 < 75,9.

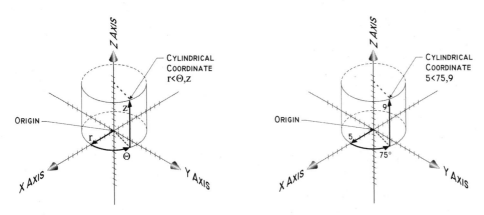

FIGURE 2–15 *Cylindrical coordinate system: mathematical model (left); AutoCAD's designation of point 5 < 75,9 (right)*

The Cartesian Coordinate System—AutoCAD's Default System

AutoCAD comes with a rectangular coordinate system that does not need curved surfaces, circles, arcs, or angles to describe points in 2D and 3D space. It consists of three mutually perpendicular planes. One plane is considered horizontal, which means that the other two are vertical. The three lines created by the intersections of the three pairs of planes are called the axes (Figure 2–16). The point where the three axes intersect is known as the origin, with the coordinates 0,0,0.

AutoCAD adheres to the conventions of the Cartesian coordinate system. From the origin (0,0,0), distances along the X axis to the right increase in value. On the Y axis, points above the origin have positive values. On the Z axis, points closer to the viewer than the origin have a positive value; this provides a sense of depth. These axes define the World Coordinate System (WCS).

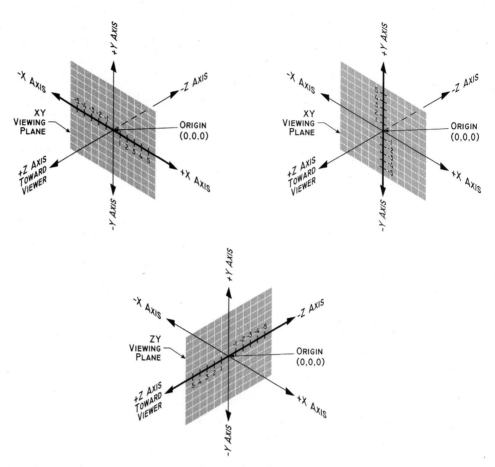

FIGURE 2–16 *Cartesian coordinate system showing the origin, XY, and ZY viewing planes*

World and User Coordinate Systems

The significance of the WCS is that it is always present in your drawing; it cannot be altered. However, an infinite number of other coordinate systems can be established relative to it. These User Coordinate Systems (UCS) are created with the UCS command. Even though the WCS is fixed, you can view it from any angle, side, or rotation without changing to another coordinate system. Even though you may be viewing the model from some other viewpoint than the plan view, the lines, circles, arcs, and other objects you draw will still be drawn relative to the plane of the current coordinate system.

AutoCAD provides a coordinate system icon to help you keep your bearings while working in different coordinate systems. The icon shows the orientation of your current UCS by indicating the positive directions of the X and Y axes. Figure 2–17 shows three examples of coordinate system icons.

FIGURE 2–17 *Examples of the UCS icon*

Scaling, Plotting, and Planning

The board drafter represents real-life objects with lines, circles, arcs, and so on that are scaled to fit a sheet of a certain size. The AutoCAD drafter does the opposite by drawing objects at their true sizes and then creating a plot configuration so that everything is reduced to fit on the actual-sized sheet. He or she then makes the border, title block, and other nonobject-associated features fit around the object. The completed combination is scaled to fit the sheet size required for plotting.

A more complicated situation arises when you need to draw objects at different scales on the same drawing. This can be handled easily by using the more advanced features and commands provided by AutoCAD.

Drawing a schematic that is not to scale is a simple task in AutoCAD. Even though the symbols and the distances between them have no relationship to real-life dimensions, you must still consider factors such as the sheet size, text size, line widths, and other visible characteristics of the drawing in order to make your schematic easy to read. Ultimately, some planning, including scaling and plotting, must be applied to all drawings.

Methods to Specify Points

When AutoCAD prompts you for the location of a point, you can use one of several point entry methods, including spherical, cylindrical, and Cartesian coordinates. Cartesian coordinates can be entered in the following formats: absolute rectangular, relative rectangular, absolute polar, and relative polar.

Absolute Rectangular Coordinates

Entering absolute rectangular coordinates specifies the location of a point by providing its distances from two axes (2D) or from three intersecting axes (3D). Each point's distance is measured along the X, Y, and Z axes relative to the intersection of the axes. The intersection of the axes is called the origin (2D = 0,0; 3D = 0,0,0) (Figure 2–13). In AutoCAD, by default the origin is located at the lower-left corner of the Grid display, as shown in Figure 2–18.

FIGURE 2–18 *Default location of the AutoCAD origin*

Image(s) © Cengage Learning 2013

As mentioned earlier, moving along the X axis, toward the right and away from the origin, increases the positive value of X. Movement along the Y axis, above and away from the origin, increases the value of Y. You may specify a point by entering its X and Y coordinates in decimal, architectural, fractional, or scientific notation, separated by commas. AutoCAD automatically assigns the current elevation as the Z coordinate. Unless it has been changed, the default value of Z is zero (0). Three-dimensional drafting involves specifying X, Y, and Z coordinates.

> **NOTE**
>
> If you use Dynamic Input, you should specify absolute coordinates with the # prefix (default setting) at the on-screen prompt. If you enter coordinates on the command line instead of in the on-screen prompt, the # prefix is not used.

Relative Rectangular Coordinates

Entering relative rectangular coordinates specifies the location of a point relative to the last specified point rather than the origin. In AutoCAD, to specify relative coordinates, @, or the "at" symbol, must precede your entry. This symbol is entered by holding down the SHIFT key and simultaneously pressing the number 2 key at the top of the keyboard. The following table shows examples of relative rectangular coordinate keyboard input when specifying a point. The first and second columns show the absolute coordinates of the last point specified, from which the newly specified point is offset. The fourth column shows the absolute coordinates resulting from the relative coordinate keyboard input shown in the third column.

ABSOLUTE COORDINATES OF LAST SPECIFIED POINT			
Specified at on-screen prompt	Specified at command line	Relative rectangular coordinates keyboard input	Resulting absolute coordinates of point specified by keyboard input
#3,4	3,4	@2,2	5,6
#5,5	5,5	@−7,0	−2,5
#3.25,8.0	3.25,8.0	@0,12.5	3.25,20.5

If you are working in a UCS and would like to enter points with reference to the WCS, prefix the coordinates with an asterisk (*). For example, to specify a point with an X coordinate of 3.5 and a Y coordinate of 2.57 as referenced to the WCS, regardless of the current UCS, enter the following:

```
*3.5,2.57
```

In the case of relative coordinates, the asterisk will be preceded by the @ symbol. For example:

```
@*4,5
```

This represents an offset of 4,5 from the previous point as referenced to the WCS.

Relative Polar Coordinates

Entering polar coordinates specifies the location of a point based on the distance from a fixed point at a given angle. In AutoCAD, a relative polar coordinate point is determined by the distance from the previous point and an angle measured from the X axis. By default, the angle is measured in the counterclockwise direction. It is important to remember that for points located using relative polar coordinates, they are positioned relative to the previous point and not the origin (0,0). You specify a point by entering its distance from the previous point and its angle from the X axis, separated by < rather than a comma. This symbol is selected by holding the SHIFT key and simultaneously pressing the COMMA (,) key at the bottom of the keyboard. Failure to use the @ symbol will cause the point to be located relative to the origin (0,0). The following table shows examples of relative polar coordinates keyboard input. The first and second columns show absolute coordinates of the last point specified, from which the newly specified point is offset. The fourth column shows the absolute coordinates resulting from the relative polar coordinates keyboard input shown in the third column.

ABSOLUTE COORDINATES OF LAST SPECIFIED POINT

Specified at on-screen prompt	Specified at command line	Relative polar coordinates keyboard input	Resulting absolute coordinates of point specified by keyboard input
#3,4	3,4	@2<0	5,4
#5,5	5,5	@4<180	1,5
#2.00,2.00	2.00,2.00	@1.4142135623<45	3.00,3.00

> **NOTE**
>
> The prompts and responses shown in examples throughout the text and the drawing exercises in this book assume that you are using Dynamic Input and typing the coordinates at the on-screen prompt. Absolute coordinates are indicated with the # prefix, and relative coordinates are indicated with the @ prefix. If you prefer to use the Command window, do not prefix with # for absolute coordinates.

Coordinates Display

The coordinates display located in the Status bar reports the cursor coordinates (Figure 2–19).

> 4.2935, -0.1125, 0.0000

FIGURE 2–19 *Coordinates display in the Status bar*

Drawing Circles

The CIRCLE command creates a circle and offers five different methods for drawing circles: Center-Radius (default), Center-Diameter, 3-Point, 2-Point, and Tangent, Tangent, Radius (TTR).

Center-Radius

The Center-Radius method allows you to draw a circle by specifying a center point and a radius. Invoke the CIRCLE command Center-Radius method from the Draw panel (Figure 2–20) located in the Home tab. AutoCAD displays the following prompts:

Specify center point for circle or ⊡ *(specify center point to define the center of the circle)*

Specify radius of circle or ⊡ *(specify the radius of the circle)*

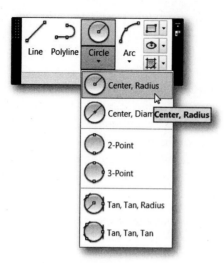

FIGURE 2-20 *Invoking the* CIRCLE *command from the Draw panel*

The following command sequence shows an example (Figure 2–21):

> circle [Enter]
> Specify center point for circle or ⊡ 2,2 [Enter]
> Specify radius of circle or ⊡ 1 [Enter]

The same circle can be drawn as follows (Figure 2–22):

> circle [Enter]
> Specify center point for circle or ⊡ 2,2 [Enter]
> Specify radius of circle or ⊡ #3,2 [Enter]

In the last example, AutoCAD used the distance between the center point and the second point given as the value for the radius of the circle.

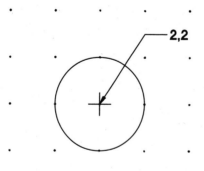

FIGURE 2-21 *A circle drawn with the* CIRCLE *command's default method: Center-Radius*

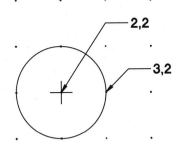

FIGURE 2–22 *A circle drawn with the Center-Radius method by specifying the coordinates*

Center-Diameter

The Center-Diameter method allows you to draw a circle by specifying a center point and a diameter. Invoke the `CIRCLE` command Center-Diameter method from the Draw panel (Figure 2–20). AutoCAD displays the following prompts:

> Specify center point for circle or ⬇ *(specify center point to locate the center of the circle)*
>
> Specify diameter of circle or ⬇ *(specify the diameter of the circle)*

The following command sequence creates a circle with a diameter of 2 units:

> circle `Enter`
> Specify center point for circle or ⬇ 2,2 `Enter`
> Specify radius of circle or ⬇ *(choose Diameter from the shortcut menu)*
> Specify diameter of circle: 2 `Enter`

The same circle can be generated as follows (Figure 2–23):

> circle `Enter`
> Specify center point for circle or ⬇ 2,2 `Enter`
> Specify radius of circle or ⬇ *(choose Diameter from the shortcut menu)*
> Specify diameter of circle: #4,2 `Enter`

When you specify a point coordinate as the diameter of a circle, AutoCAD calculates the distance from the center of the circle to the specified point and uses the resulting value for the diameter of the circle.

NOTE

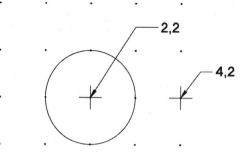

FIGURE 2–23 *A circle drawn using the Center-Diameter method by specifying coordinates*

Three-Point Circle

The Three-Point Circle method allows you to draw a circle by specifying three points on the circumference.

Invoke the CIRCLE command Three-Point method from the Draw panel (Figure 2–20). AutoCAD displays the following prompts:

> Specify first point on circle: *(specify a point or coordinates)*
>
> Specify second point on circle: *(specify a point or coordinates)*
>
> Specify third point on circle: *(specify a point or coordinates)*

The following command sequence shows an example (Figure 2–24):

> circle
>
> Specify center point for circle or ⊡ *(choose 3p from the shortcut menu)*
>
> Specify first point on circle: ⊡ 2,1 [Enter]
>
> Specify second point on circle: ⊡ #3,2 [Enter]
>
> Specify third point on circle: ⊡ #2,3 [Enter]

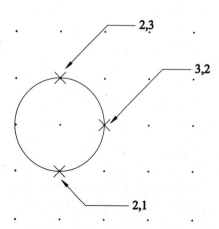

FIGURE 2–24 *A circle drawn with the 3-Point method*

Two-Point Circle

The Two-Point Circle method allows you to draw a circle by specifying two end-points of a diameter.

Invoke the `CIRCLE` command Two-Point method from the Draw panel (Figure 2–20). AutoCAD displays the following prompts:

> Specify first endpoint of circle's diameter: *(specify a point or coordinates)*
>
> Specify second endpoint of circle's diameter: *(specify a point or coordinates)*

The following command sequence shows an example (Figure 2–25):

> circle [Enter]
>
> Specify center point for circle or ⊡ *(choose 2p from the shortcut menu)*
>
> Specify first endpoint of circle's diameter: ⊡ 1,2 [Enter]
>
> Specify second endpoint of circle's diameter: ⊡ #3,2 [Enter]

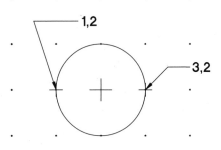

FIGURE 2–25 *A circle drawn with the 2-Point method*

Tangent, Tangent, Radius (TTR)

The Tangent, Tangent, Radius method allows you to draw a circle tangent to two objects—lines, arcs, or circles—with a specified radius.

Invoke the `CIRCLE` command Tangent, Tangent, Radius method from the Draw panel (Figure 2–20). AutoCAD displays the following prompts:

> Specify point on object for first tangent of circle: *(specify a point on an object for first tangent of circle)*
>
> Specify point on object for second tangent of circle: *(specify a point on an object for second tangent of circle)*
>
> Specify radius of circle: *(specify radius)*

When specifying the "tangent-to" objects, it normally does not matter where on the objects you make your selection. However, if more than one circle can be drawn to the specifications given, AutoCAD will draw the one whose tangent point is nearest to the selection made.

TIP	Until it is changed, the radius/diameter you specify in any one of the options becomes the default setting for subsequent circles.

Drawing Arcs

The ARC command offers four different combinations by which you can draw arcs: a combination of three points, a combination of two points and an included angle or starting direction, a combination of two points and a length of chord or radius, and a continuation from line or arc.

Three Points (Start, Point-On-Circumference or Center, and End)

There are three ways to use the 3-Points method of drawing arcs:

- Three-Point (3-Point)
- Start, Center, End (S,C,E)
- Center, Start, End (C,S,E)

Three-Point. The default 3-Point method draws an arc using three specified points on the arc's circumference. The first point specifies the start point, the second point specifies a point on the circumference of the arc, and the third point is the arc endpoint.

Invoke the ARC command 3-Point method from the Draw panel (Figure 2–26) located in the Home tab, and specify a 3-Point arc either clockwise or counterclockwise, as shown in the following command sequence:

```
Specify start point of arc or ⬇1,2  Enter
Specify second point of arc or ⬇#2,1  Enter
Specify endpoint of arc: ⬇#3,2  Enter
```

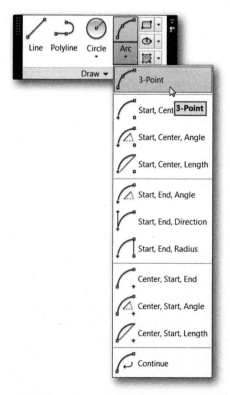

FIGURE 2–26 *Invoking the* ARC *command from the Draw panel*

AutoCAD draws an arc based on a start point at 1,2 with a circumference point at 2,1 and an endpoint at 3,2, as shown in Figure 2–27.

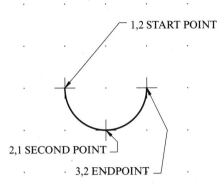

1,2 START POINT

2,1 SECOND POINT

3,2 ENDPOINT

FIGURE 2–27 *An arc drawn with the ARC command's default option, Three Points*

Start, Center, End (S,C,E). The Start, Center, End (S,C,E) method also draws an arc using three specified points. The first point specifies the start point, the second point is the center point, and the third point is the arc endpoint.

Invoke the ARC command S,C,E method from the Draw panel (Figure 2–26), and specify three points as shown in the following command sequence:

Specify start point of arc or ⊡ 1,2 ⌷Enter⌷
Specify center point of arc: #2,2 ⌷Enter⌷
Specify endpoint of arc or ⊡ #2,3 ⌷Enter⌷

AutoCAD draws an arc based on a starting point at 1,2 with the center point at 2,2 and an endpoint at 2,3 as shown in Figure 2–28.

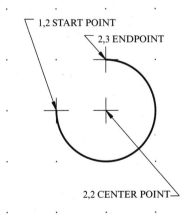

1,2 START POINT

2,3 ENDPOINT

2,2 CENTER POINT

FIGURE 2–28 *An arc drawn with the S,C,E method*

Arcs drawn by this method are always drawn counterclockwise from the starting point. The distance between the center point and the starting point determines the radius. Therefore, the point specified in response to the "endpoint" prompt needs only to be on the same radial line as the desired endpoint.

NOTE

Center, Start, End (C,S,E). The Center, Start, End (C,S,E) method is similar to the S,C,E method, except that the first point selected is the center point of the arc rather than the start point.

Two-Points and an Included Angle or Starting Direction

There are four ways to use a 2-Points and an Included Angle or Starting Direction method of drawing arcs:

- Start, Center, Angle (S,C,A)
- Center, Start, Angle (C,S,A)
- Start, End, Angle (S,E,A)
- Start, End, Direction (S,E,D)

Start, Center, Angle (S,C,A). The Start, Center, Angle (S,C,A) method draws an arc similar to the S,C,E method, but it places the endpoint on a radial line at the specified angle from the line between the center point and the start point. If you specify a positive number as the value of the included angle, the arc is drawn counterclockwise. When you specify a negative number for the value of the angle, the arc is drawn clockwise.

Invoke the ARC command S,C,A method from the Draw panel (Figure 2–26). Specify the first point as the start point, the second point as the center point of the arc to be drawn, and then specify the included angle, as shown in the following example:

```
Specify start point of arc or ⬇1,2 [Enter]
Specify center point of arc: #2,2 [Enter]
Specify included angle: 270 [Enter]
```

AutoCAD draws an arc based on a starting point at 1,2 with the center point at 2,2 and an included angle of 270 degrees, as shown in Figure 2–29.

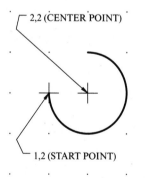

2,2 (CENTER POINT)

1,2 (START POINT)

FIGURE 2–29 *An arc drawn with the S,C,A method*

 NOTE In the previous example, if a point directly below the specified center were selected in response to the "Included angle:" prompt, AutoCAD would read the angle of the line (270 degrees from zero) as the included angle for the arc.

Center, Start, Angle (C,S,A). The Center, Start, Angle (C,S,A) method is similar to the S,C,A method, except that the first point selected is the center point of the arc rather than the start point.

Start, End, Angle (S,E,A). The Start, End, Angle (S,E,A) method draws an arc similar to the S,C,A method, but it places the endpoint on a radial line at the specified angle from the line between the center point and the start point. If you specify a positive angle as the included angle, the arc is drawn counterclockwise. Specify a negative included angle, and the arc is drawn clockwise.

Invoke the ARC command S,E,A method from the Draw panel (Figure 2–26). Specify the first point as the start point, the second point as the endpoint of the arc, and then specify the included angle, as shown in the following example:

```
Specify start point of arc or ⊡3,2 Enter
Specify endpoint of arc: #2,3 Enter
Specify included angle: 90 Enter
```

AutoCAD draws an arc based on a starting point at 3,2 with an endpoint at 2,3 and an included angle of 90 degrees, as shown in Figure 2–30.

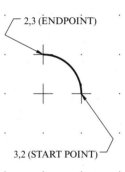

FIGURE 2–30 *An arc drawn counterclockwise with the S,E,A method*

The arc shown in Figure 2–31 is drawn with a negative angle using the following sequence:

```
arc Enter
Specify start point of arc or ⊡3,2 Enter
Specify endpoint of arc: #2,3 Enter
Specify included angle: −270 Enter
```

2,3 (ENDPOINT)

3,2 (START POINT)

FIGURE 2–31 *An arc drawn clockwise with the S,E,A method*

Start, End, Direction (S,E,D). The Start, End, Direction (S,E,D) method allows you to draw an arc between selected points by specifying the direction in which the arc will be drawn from the selected start point. Either the direction can be keyed in, or you can select a point on the screen with your pointing device. If you select a point on the screen, AutoCAD uses the angle from the start point to the endpoint as the starting direction.

Invoke the ARC command S,E,D method from the Draw panel (Figure 2–26). Specify the first point as the start point, the second point as the endpoint of the arc, and then specify the starting direction, as shown in the following example:

Specify start point of arc or ⬇3,2 Enter
Specify endpoint of arc: ⌗2,3 Enter
Specify tangent direction for the start point of arc: 90 Enter

AutoCAD draws an arc based on a starting point at 3,2 with an endpoint at 2,3 and direction set to 90 degrees, as shown in Figure 2–32.

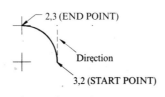

FIGURE 2–32 *An arc drawn with the S,E,D method*

Two Points and a Length of Chord or Radius

There are three ways to use a 2-Points and a Length of Chord or Radius method of drawing arcs:

- Start, Center, Length of Chord (S,C,L)
- Center, Start, Length of Chord (C,S,L)
- Start, End, Radius (S,E,R)

Start, Center, Length of Chord (S,C,L). The Start, Center, Length of Chord (S,C,L) method uses the specified chord length as the straight-line distance from the start point to the endpoint. With any chord length equal to or less than the diameter length, four possible arcs can be drawn: a major arc in either direction and a minor arc in either direction. Therefore, all arcs drawn by this method are counterclockwise from the start point. A positive value as the length of chord will cause AutoCAD to draw the minor arc; a negative value will result in the major arc being drawn. Invoke the ARC command S,C,L method from the Draw panel (Figure 2–26). Specify the first point as the start point, the second point as the center point of the arc to be drawn, and then specify the chord length, as shown in the following example:

Specify start point of arc or ⬇1,2 Enter
Specify center point of arc: ⌗2,2 Enter
Specify length of chord: 1.4142 Enter

AutoCAD draws the arc based on a starting point at 1,2 with the center point at 2,2 and a length of chord set to 1.4142, as shown in Figure 2–33.

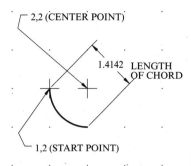

FIGURE 2–33 *A minor arc drawn with the S,C,L method*

The following example shows drawing a major arc:

```
arc Enter
Specify start point of arc or ⬇1,2 Enter
Specify center point of arc: ⬇⌗2,2 Enter
Specify length of chord: −1.414 Enter
```

AutoCAD draws an arc based on a starting point at 1,2 with the center point at 2,2 and a length of chord set to −1.414, as shown in Figure 2–34.

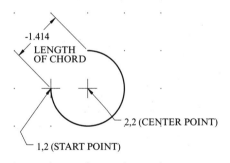

FIGURE 2–34 *A major arc drawn with the S,C,L method*

Center, Start, Length of Chord (C,S,L). The Center, Start, Length of Chord (C,S,L) method is similar to the S,C,L method, except that the first point selected is the center point of the arc rather than the start point.

Start, End, Radius (S,E,R). The Start, End, Radius (S,E,R) method allows you to specify a radius after selecting the start and endpoints of the arc. As with the Length of Chord method, four possible arcs can be drawn: a major arc in either direction and a minor arc in either direction. Therefore, all arcs drawn by this method are counterclockwise from the start point. A positive value for the radius causes AutoCAD to draw the minor arc; a negative value results in the major arc.

Invoke the ARC command S,E,R method from the Draw panel (Figure 2–26). Specify the first point as the start point, the second point as the endpoint, and then specify the radius, as shown in the following example:

```
Specify start point of arc or ⬇1,2 Enter
Specify endpoint of arc: #2,3 Enter
Specify radius of arc: -1 Enter
```

AutoCAD draws the arc based on a starting point at 1,2 with an endpoint at 2,3 and a radius of −1, as shown in Figure 2–35.

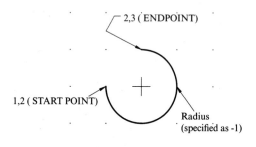

FIGURE 2–35 *A major arc drawn with the S,E,R method*

The following example shows how to draw a minor arc.

```
arc Enter
Specify start point of arc or ⬇2,3 Enter
Specify endpoint of arc: ⬇#1,2 Enter
Specify radius of arc: 1 Enter
```

AutoCAD draws an arc based on a starting point at 2,3 with an endpoint at 1,2 and a radius of 1 unit, as shown in Figure 2–36.

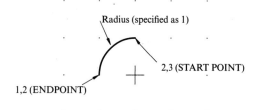

FIGURE 2–36 *A minor arc drawn with the S,E,R method*

Line-Arc and Arc-Arc Continuation

You can use an automatic Start Point, End Point, Starting Direction method to draw an arc by pressing Enter in response to the first prompt of the ARC command. After you press Enter, the only other input needed is to select or specify the endpoint of the arc. AutoCAD uses the endpoint of the previous line or arc, whichever was drawn last, as the start point of the new arc. To demonstrate these continuation methods, we must create the first arc. The next sequence creates an arc with the start point at 2,1 with the endpoint at 3,2 and a radius of 1. This action makes the ending direction of the existing arc 90 degrees, as shown in Figure 2–37.

Invoke the ARC command Start, End, Center method from the Draw panel (Figure 2–26), and AutoCAD displays the following prompts:

```
Specify start point of arc or ⊕2,1 Enter
Specify endpoint of arc: #3,2 Enter
Specify radius of arc: 1 Enter
```

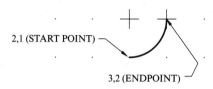

2,1 (START POINT)

3,2 (ENDPOINT)

FIGURE 2–37 *The initial arc drawn with a start point (2,1), an endpoint (3,2), and a radius of 1*

The next sequence demonstrates the Arc-Arc-Continuation method of drawing an arc from the last-drawn arc (Figure 2–37). The resulting arc is shown in Figure 2–38.

Invoke the ARC command Continue method from the Draw panel (Figure 2–26), and AutoCAD displays the following prompts:

```
Specify start point of arc or ⊕ Enter
Specify endpoint of arc: #2,3 Enter
```

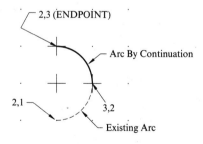

2,3 (ENDPOINT)

Arc By Continuation

2,1

3,2

Existing Arc

FIGURE 2–38 *An arc drawn by means of the Arc-Arc Continuation method*

The next example demonstrates the Arc-Arc Continuation method in a clockwise direction. The following sequence draws an initial arc clockwise from the start point at 3,2 to the endpoint at 2,1 (Figure 2–39).

Invoke the ARC command S,E,D method from the Draw panel (Figure 2–26), AutoCAD displays the following prompts:

```
Specify start point of arc or ⊕3,2 Enter
Specify endpoint of arc: #2,1 Enter
Specify tangent direction for the start point of arc: -90
Enter
```

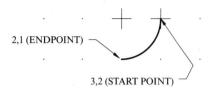

FIGURE 2–39 *An arc drawn clockwise with start point (3,2) and endpoint (2,1)*

The next sequence uses the Arc-Arc Continuation method to continue drawing a clockwise arc from the endpoint of the last-drawn arc (Figure 2–39). The resulting arc is shown in Figure 2–40.

Invoke the ARC command Continue method from the Draw panel (Figure 2–26), and AutoCAD displays the following prompts:

```
Specify start point of arc or ⬇ [Enter]
Specify endpoint of arc: #2,3 [Enter]
```

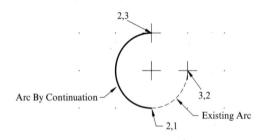

FIGURE 2–40 *A clockwise arc drawn by means of the Arc-Arc Continuation method*

In the last example, the direction used was 180 degrees. The same arc would have been drawn if the last object drawn was a line starting at 4,1 and ending at 2,1.

NOTE | This method uses the last drawn arc or a line. If you draw an arc, or a line, and then use this continuation method, AutoCAD will use the line as the basis for the start point and direction. This is because the line was the last of the "line-or-arc" objects drawn.

OBJECT SELECTION

Many of AutoCAD's modify and construct commands prompt you to select one or more objects for manipulation. When you select one or more objects, AutoCAD highlights them by displaying them as dashed lines. The group of objects selected for manipulation is called the selection set. There are several different ways to select objects. These include Window, Crossing, Previous, Last, Box, All, Fence, Wpolygon (WP), Cpolygon (CP), Group, Add, Remove, Multiple, Undo, Auto, and Single.

When a modify and construct command requires a selection set, AutoCAD displays the following prompt:

Select objects:

AutoCAD replaces the screen crosshairs with a small box called the object selection target. Using your pointing device or the keyboard's cursor keys, position the target box so that it touches only the desired object or a visible portion of it. The object selection target helps you to select the object without having to be very precise. For commands that allow multiple object selection, each time you select an object, the "Select objects:" prompt reappears. To indicate your acceptance of the selection set, press **Enter** at the "Select objects:" prompt.

Sometimes it is difficult to select objects that are close together or that lie directly on top of one another. You can cycle through objects for selection at the "Select Objects:" prompt by holding down the **CTRL** key and selecting a point as near as possible to the object. Press the pick button on your pointing device repeatedly until the object you want is highlighted, and then press **Enter** to select the object.

Selection by Window

The Window method allows you to select all objects that are contained completely within the rectangular area. You define this area by specifying two corner points. At the "Select objects:" prompt, use your pointing device to specify the first corner point, which must be above or below and to the left of the objects you want to select. AutoCAD then displays the following prompt:

Specify opposite corner: *(specify opposite corner)*

Move your cursor up or down and to the right to create the rectangular area. When the rectangle encompasses the target objects, press the pick button on your pointing device to specify the second corner point (Figure 2–41).

You can also invoke the window selection option by typing **w** in response to the "Select objects:" prompt, and AutoCAD displays the following prompts:

Specify first corner: *(specify first corner coordinate)*
Specify opposite corner: *(specify opposite corner coordinate)*

If there is an object that is partially included in the rectangular area, that object will not be included in the selection set. You can select only objects that are currently visible on the screen. You can also define a selection window by entering coordinates. The selection shown in Figure 2–41 will only include the lines, not the circles, because a portion of each of the circles is outside the rectangular area.

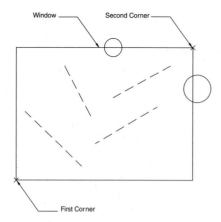

FIGURE 2–41 *Selecting objects by means of the Window method*

Selection by Crossing

The Crossing option allows you to select all objects that are contained completely or partially within the rectangular area. You define this area by specifying two corner points. At the "Select objects:" prompt, use your pointing device to specify the first corner point, which must be to the right of the objects you want to select. AutoCAD displays the following prompt:

```
Specify opposite corner: (specify opposite corner)
```

Move your cursor to the left to create the rectangular area. When the rectangle crosses a portion of the target objects, press the pick button on your pointing device to specify the second corner point (Figure 2–42).

You can also define a crossing window for selection of objects by entering **c** in response to the "Select objects:" prompt. AutoCAD displays the following prompts:

```
Specify first corner: (specify first corner coordinate)
Specify opposite corner: (specify opposite corner
coordinate)
```

The selection shown in Figure 2–42 includes all of the lines and circles, though parts of the circles and one line are outside the rectangle.

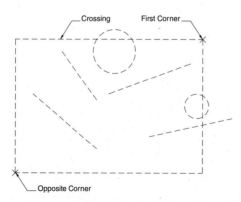

FIGURE 2–42 *Selecting objects by means of the Crossing option*

Previous Selection

The Previous option enables you to perform several operations on the same object or group of objects. AutoCAD remembers the most recent selection set and allows you to reselect it with the Previous option. For example, if you moved several objects and then wish to copy them elsewhere, you can invoke the COPY command and respond to the "Select objects:" prompt by entering **p** to select the same objects again. There is a command called SELECT that does nothing but create a selection set; you can then use the Previous option to select this set in subsequent commands.

Last Selection

The Last option provides an easy way to select the most recently created object currently visible. Only one object is selected, no matter how often you use the Last option. The Last option is invoked by entering **l** at the "Select objects:" prompt.

> You can remove objects from the current selection set by holding down SHIFT key and selecting them again.

NOTE

The Wpolygon (WP), Cpolygon (CP), Fence, All, Group, Box, Auto, Undo, Single, Multiple, Add, and Remove methods are explained in Chapter 5.

MODIFYING OBJECTS

AutoCAD not only allows you to draw objects easily but it also allows you to modify the objects you have drawn. Of the many modification commands available, the ERASE command will probably be the one you use most often. Everyone makes mistakes, and in AutoCAD it is easy to erase them. If you have created an object to aid in constructing other objects, and you are now finished with it, you can erase it.

Erasing Objects

To erase objects from a drawing, invoke the ERASE command from the Modify panel (Figure 2–43) located in the Home tab and AutoCAD displays the following prompt:

    Select objects: *(select objects to be erased, and then press*
    SPACEBAR *or* **Enter** *)*

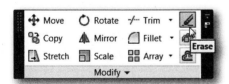

FIGURE 2–43 *Invoking the ERASE command from the Modify panel*

You can use one or more available object selection methods in response to the next "Select objects:" prompt. After selecting the object(s), press **Enter** to complete the ERASE command. All the objects that were selected will disappear.

The following command sequence shows an example of erasing individual objects (Figure 2–44):

    erase
    Select objects: *(select Line 2, the line is highlighted)*
    Select objects: *(select Line 4, the line is highlighted)*
    Select objects: **Enter**

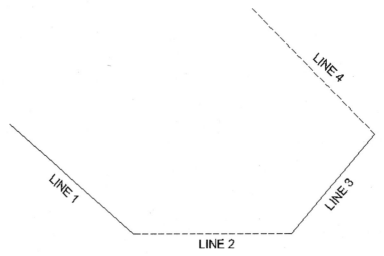

FIGURE 2–44 *Selection of individual objects to erase*

The following command sequence shows an example of erasing objects using the Window selection method (Figure 2–45):

```
erase
Select objects: (specify a point)
Specify opposite corner: (specify a point for the diagonally
opposite corner)
Select objects: Enter
```

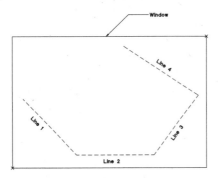

FIGURE 2–45 *Erasing a group of objects defined by the Window selection method*

Getting It Back

The OOPS command restores objects that have been unintentionally erased. Whenever the ERASE command is used, the last group of objects erased is stored in memory. The OOPS command will restore the objects. While it can be used at any time, it restores only the objects erased by the most recent ERASE command. Review the UNDO command in Chapter 3 if you need to step back further than one ERASE command.

To restore objects erased by the last ERASE command, invoke the OOPS command by typing **oops** at the on-screen prompt. AutoCAD restores the objects.

The following example shows the sequence for using the OOPS command in conjunction with the ERASE command:

```
erase Enter
Select objects: (specify a point)
Other corner: (specify a point for the diagonally opposite
corner)
Select objects: Enter
oops Enter (the last erased objects are restored)
```

Open the Exercise Manual PDF for Chapter 2 for discipline-specific exercises. Related files are downloaded from the student companion site mentioned in the Introduction (refer to page number xii for instructions).

REVIEW QUESTIONS

1. What Cartesian coordinates does AutoCAD use to indicate locations in a drawing?
   a. X,P
   b. X,Y
   c. X,Z
   d. @,X

2. The RECTANGLE command requests what information?
   a. An initial corner, the width, the height and the angle
   b. The coordinates of the four corners of the rectangle
   c. The coordinates of diagonally opposite corners of the rectangle
   d. The coordinates of three adjacent corners of the rectangle

3. When drawing a trace line, after you select the second point:
   a. Nothing appears on the screen
   b. You are prompted for the trace width
   c. The segment is drawn and the command terminates
   d. The first segment is drawn, and you are prompted for the next point

4. To draw multiple connected line segments, you must invoke the LINE command multiple times.
   a. True
   b. False

5. The file extension .BAK stands for:
   a. Backup drawing file
   b. Binary file
   c. Binary attribute file
   d. Drawing file
   e. Both b and c

6. The HELP command cannot be used:
   a. While in the LINE command
   b. While in the CIRCLE command option TTR
   c. To list commands
   d. For a text string

7. Points are located by relative rectangular coordinates in relation to:
   a. The last specified point or position
   b. The global origin
   c. The lower-left corner of the screen
   d. All of the above

8. Polar coordinates are based on a distance from:
   a. The global origin
   b. The last specified position at a given angle
   c. The center of the display
   d. All of the above

9. To enter a command from the keyboard, enter the command name at the on-screen prompt:
   a. In lowercase letters
   b. In uppercase letters
   c. Lowercase, uppercase, or mixed case
   d. Commands cannot be entered via the keyboard

10. To draw a line, what would you type in the Command Line?
    a. Li or L
    b. F2 or L
    c. Line or L
    d. Line or Lo

11. When you invoke the LINE command and respond to the "Specify first point:" prompt by pressing [Enter] or [SPACEBAR], AutoCAD will:
    a. Continue the line from the last line or arc that was drawn
    b. Close the previous set of line segments
    c. Display an error message

12. To draw a rectangle, what would you type in the Command Line?
    a. Rectangle or rec
    b. Rect or rec
    c. Rec or F4
    d. None of the above

13. A rectangle generated by the RECTANGLE command will always have horizontal and vertical sides.
    a. True
    b. False

14. Can you erase one side of a rectangle?

    a. Yes

    b. No

15. Which of the following coordinates will define a point at the screen default origin?

    a. 000

    b. 00

    c. 0.0, 0.0

    d. 112

    e. @0,0

16. Switching between graphics and text screen can be accomplished by:

    a. Pressing [Enter] twice

    b. Pressing [CTRL] and [Enter] at the same time

    c. Pressing the [ESC] key

    d. Pressing the F2 function key

    e. Both b and c

17. To draw a line a length of **8 feet, 4-5/8** inches in the **12** o'clock direction from the last point selected, enter:

    a. @8'4-5/8<90

    b. 8'-4-5/8<90

    c. @8-45/8<90

    d. #8'-45/8<90

    e. None of the above

18. By default, what direction does a positive number indicate when specifying angles in degrees?

    a. Clockwise

    b. Counterclockwise

    c. Has no impact when specifying angles in degrees

    d. None of the above

19. To erase an object, what would you type in the Command Line?

    a. Er or E

    b. Erase or E

    c. Delete or D

    d. Del or D

20. When erasing objects, if you select a point that is not on any object, AutoCAD will:

    a. Terminate the ERASE command

    b. Delete the selected objects and continue with the ERASE command

    c. Allow you to drag a window to select multiple objects within the area

    d. Ignore the selection and continue with the ERASE command

21. Which of the following methods of drawing circles using the CIRCLE command are correct? You can select more than one option.

    a. Center, Radius

    b. Two Points

    c. Three Points

    d. Tan, Tan, Tan

    e. Center, point on circumference

22. Regarding the arc options, what does S,C,E mean?

    a. Start, Center, End

    b. Second, Continue, Extents

    c. Second, Center, End

    d. Start, Continue, End

23. The number of different methods by which a circle can be drawn is:

    a. 1

    b. 3

    c. 4

    d. 7

    e. None of the above

24. When using the ERASE command, AutoCAD deletes each object from the drawing as you select it.

    a. True

    b. False

25. Once an object is erased from a drawing, which of the following commands could restore it to the drawing?

    a. OOPS

    b. RESTORE

    c. REPLACE

    d. CANCEL

26. When drawing a circle with the 2-Point option, the distance between the two points is equal to:

    a. The circumference

    b. The perimeter

    c. The shortest chord

    d. The radius

    e. The diameter

27. A circle may be created by any of the following options except:

    a. 2P

    b. 3P

    c. 4P

    d. Cen,Rad

    e. TTR

CHAPTER 3

Fundamentals II

INTRODUCTION

AutoCAD provides various drawing tools to make your drafting and design work easier. The tools described in this chapter will assist you in creating drawings rapidly while ensuring the highest degree of precision.

OBJECTIVES

After completing this chapter, you will be able to do the following:

- Use and control drafting settings such as GRID, SNAP, ORTHO, POLAR TRACKING, and OBJECT SNAP

- Use Tracking and Direct Distance

- Use display control commands such as ZOOM, PAN, REDRAW, and REGEN

- Create tiled viewports

- Use Dynamic Input and On-Screen Prompts

- Use layering techniques

- Use the UNDO and REDO commands

DRAFTING SETTINGS

The SNAP, GRID, ORTHO, POLAR/OBJECT SNAP TRACKING, OBJECT SNAP, DYNAMIC TRACKING, and LINEWEIGHT commands do not create objects. However, they make it possible to create and modify objects more easily and accurately. Each of these Drafting Settings commands can be readily toggled ON when needed and OFF when not. When ON, they function according to settings that can be changed easily. When used appropriately, they provide the power, speed, and accuracy associated with computer-aided design and drafting.

Snap Command

The SNAP command provides an invisible reference grid in the drawing area. When set to ON, the Snap feature forces the cursor to lock onto the nearest point on the specified Snap Grid. With the Snap value set appropriately, you can specify points quickly, which allows AutoCAD to ensure that they are placed precisely. You can always override the snap spacing by entering absolute or relative coordinate points from the keyboard or by simply turning the Snap mode OFF. When the Snap mode is OFF, it has no effect on the cursor. When it is ON, you cannot place the cursor on a point that is not on one of the specified Snap Grid locations.

Setting Snap On and Off

AutoCAD provides five methods for setting the Snap function to ON or OFF. Of the five explained in this section, the first method, selecting the **SNAP** icon on the status bar, is the easiest and is most commonly used with the pointing device. The second method, pressing the function key F9, is most commonly used from the keyboard. The other methods, while not quite as convenient as the first two, might be convenient in certain situations.

1. Snap can be toggled ON and OFF by choosing the **SNAP** icon on the status bar.

2. Snap can be toggled ON and OFF by pressing the function key F9.

3. Right-click the **SNAP** icon on the status bar, and choose GRID SNAP ON from the shortcut menu to set the Snap ON, or choose OFF to set the Snap OFF.

4. Right-click the **SNAP** icon on the status bar and choose SETTINGS from the shortcut menu. AutoCAD displays the Drafting Settings dialog box, Snap can be toggled ON and OFF by selecting **Snap On (F9)** (Figure 3–1).

5. At the on-screen prompt, type **snap,** and AutoCAD displays the following prompt:

> Specify snap spacing or ⊡ *(choose* ON *or* OFF *from the shortcut menu)*

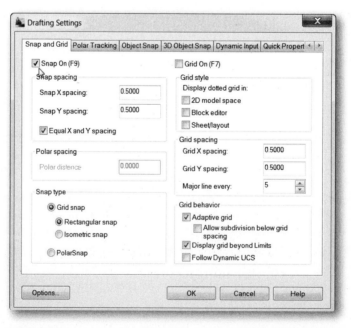

FIGURE 3–1 *Choosing the Snap On (F9) checkbox from the Snap and Grid tab of the Drafting Settings dialog box*

Changing Snap Spacing

AutoCAD has two methods for changing the Snap spacing:

1. Right-click the **SNAP** icon on the status bar and from the shortcut menu, choose SETTINGS. AutoCAD displays the Drafting Settings dialog box with the **Snap and Grid** tab selected. From the **Snap** section, you can change the settings of the *X* and *Y* spacing by entering the desired values in the **Snap X spacing** and **Snap Y spacing** boxes, respectively (Figure 3–2). To set the same value for X and Y, set the **Equal X and Y spacing** toggle to ON.

2. At the on-screen prompt, type **snap,** and AutoCAD displays the following prompt:

   Specify snap spacing or ⏎ (*specify distance to be used for both X and Y Snap spacings*)

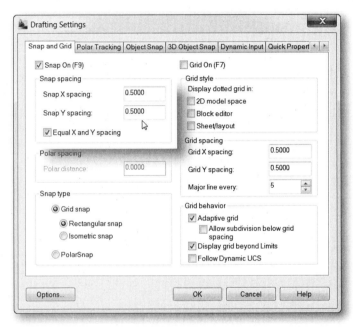

FIGURE 3–2 *Changing the Snap settings from the Snap and Grid section of the Drafting Settings dialog box*

Snap Options

Several snap options are available at the on-screen prompt. In addition to ON or OFF, the following options are available: *Aspect, Rotate* (SNAPANG system variable), *Style,* and *Type*.

Choose *Aspect* to set the Y Snap spacing different from the X Snap spacing. AutoCAD prompts for the X and Y values independently. This is handy if the X and Y modular dimensions of your design are of unequal multiples. At the on-screen prompt, type **snap,** and AutoCAD displays the following prompts:

   Specify snap spacing or ⏎ (*choose Aspect from the shortcut menu*)
   Specify horizontal spacing <current>: (*specify X Snap spacing*)
   Specify vertical spacing <current>: (*specify Y Snap spacing*)

For example, the following sequence describes how you can set a horizontal snap spacing of 0.5 and vertical snap spacing of 0.25.

snap [Enter]

Specify snap spacing or ⊡ *(choose Aspect from the shortcut menu)*

Specify horizontal spacing <Current>: **0.5** [Enter]

Specify vertical spacing <Current>: **0.25** [Enter]

Previous versions of AutoCAD included the Rotate option to specify an angle to rotate both the visible Grid and the invisible Snap grid. This is no longer included in the options as a response to the SNAP command. However, if you type **snapang** at the command prompt, you can change the angle of the snap and grid alignment. This method is similar to using the more complex user coordinate system (UCS). It permits you to set a Snap angle of rotation specified with respect to the default Zero-East system of direction. In conjunction with the X and Y spacing of the Snap grid, changing the SNAPANG value can make it easier to draw certain shapes.

The plot plan in Figure 3–3 is an example of where the SNAPANG system variable value setting can be applied. The property lines are drawn using surveyor's units of angular display. In this example, the architectural units and the surveyor's angular units are selected; both of these can be set in the Drawing Units dialog box (accessible by entering UNITS). The limits are set up with the lower-left corner at −20',−10' and the upper-right corner at 124',86' (initiated by entering LIMITS). After making sure the ORTHO mode is set to ON, the sequence for drawing the property lines is as follows:

snapang [Enter]

Enter new value for SNAPANG <0>: **4d45'08"** [Enter]

line [Enter] *(invoking the LINE command)*

Specify first point: **0,0** [Enter] *(selecting the first point)*

Specify next point or ⊡ **85'** [Enter] *(point selected with the cursor north of the first point)*

Specify next point or ⊡ **120'** [Enter] *(point selected with the cursor east of the previous point)*

Specify next point or ⊡ **85'** [Enter] *(point selected with the cursor south of the previous point)*

Specify next point or ⊡ **c** [Enter] *(choose CLOSE from the shortcut menu to complete the property boundary by closing the rectangle and exiting the LINE command)*

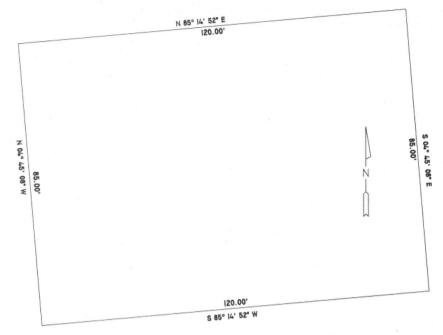

FIGURE 3–3 *Example drawing where the SNAPANG system variable can be used*

Figure 3–4 shows the property boundary drawn.

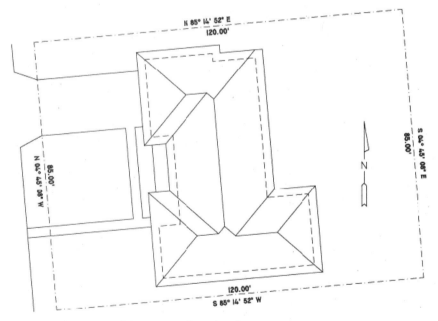

FIGURE 3–4 *Layout of the property boundary lines*

Choose *Style* to select one of two available formats: standard or isometric. *Standard* refers to the normal rectangular grid, the default, and *isometric* refers to a Grid and Snap designed for isometric drafting purposes (Figure 3–5).

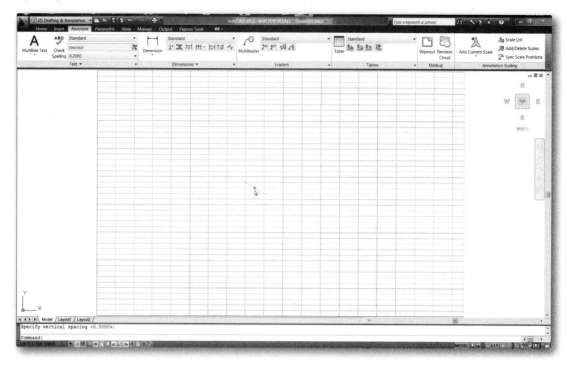

FIGURE 3–5 *Setting the snap for isometric drafting*

You can switch the isoplanes between left, or 90- and 150-degree angles, top, or 30- and 150-degree angles, and right, or 30- and 90-degree angles, by pressing CTRL + E—the combination keystrokes of holding down CTRL and then pressing E—or by simply pressing the function key F5.

Choose TYPE to select one of two available Snap types: Polar or Grid. Choosing the Polar type sets the snap to Polar Tracking angles. See the explanations on Polar Tracking later in this chapter. Choosing the Grid type sets the snap spacing equal to the grid spacing.

Grid Command

The GRID command is used to display a visible array of horizontal and vertical lines. AutoCAD creates a grid that is similar to a sheet of graph paper. You can set the grid display ON or OFF and can change the line spacing. The grid is a drawing tool and is not part of the drawing; it is for visual reference and is never plotted. In the World Coordinate System (WCS), the grid fills the area defined by the limits of the drawing area.

The grid has several uses within AutoCAD. First, it shows the extent of the drawing limits. For example, if you set the limits to 42 units by 36 units and the grid spacing is set to 0.5 units, there will be 85 columns and 73 rows of grid lines. This layout will give you a better sense of the drawing's size relative to the limits than if it were on a blank background.

Second, using the GRID command with the SNAP command is helpful when you create a design in terms of equally spaced units. For example, if your design uses multiples of 0.5 units, you can set the grid spacing as 0.5 to facilitate point entry. You could check your drawing visually by comparing the locations of the grid lines and the crosshairs. Figure 3–6 shows a drawing with a grid spacing of 0.5 units with limits set to 0,0 and 17,11.

> While the GRID, SNAP, ORTHO, POLAR TRACKING, and OBJECT SNAP commands do not create objects, they make it possible to create them more easily and accurately. Each of these drafting setting commands, when toggled ON, operates according to the value(s) to which you have set it and, when toggled OFF, has no effect. You are advised to identify and master the two skills involved in using these utilities: learn how to change the settings of the utility commands and learn how and when it is best to set them ON and OFF. Changing the settings of these utilities is normally done more easily by means of their associated dialog box(es). You can also set them ON and OFF from a dialog box. Because these features are normally switched ON and OFF frequently during a drawing session, special buttons are provided on the status bar at the bottom of the screen for this purpose.

NOTE

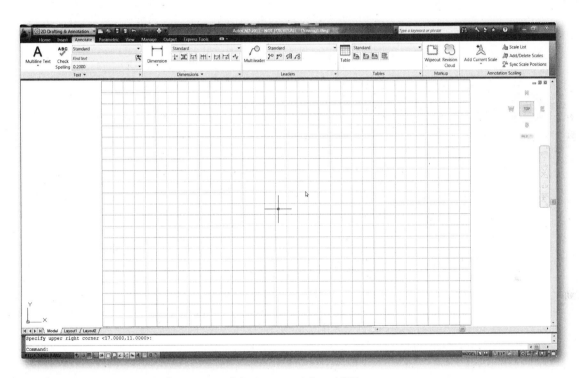

FIGURE 3–6 *A grid spacing of 0.5 units with limits set to 0,0 and 17,11*

Setting the Grid On and Off

AutoCAD provides five methods for setting the Grid ON and OFF. Of the five explained in this section, the first method, selecting the **GRID** icon on the status bar, is the easiest and is most commonly used with the pointing device. The second method, pressing the function key F7, is most commonly used from the keyboard. The other methods, while not quite as convenient as the first two, might be convenient in certain situations.

1. Grid can be toggled ON and OFF by choosing the **GRID** icon on the status bar.
2. Grid can be toggled ON and OFF by pressing the function key F7.
3. Right-click the **GRID** icon on the status bar and toggle the **Enabled** selection.
4. Right-click the **GRID** icon on the status bar and choose SETTINGS from the shortcut menu. AutoCAD displays the Drafting Settings dialog box. Grid can be toggled ON and OFF by selecting **Grid On (F7)** (Figure 3–7).

5. At the on-screen prompt, type **grid,** and AutoCAD displays the following prompt:

Specify grid spacing (X) or ⬇ *(choose* ON *or* OFF *from the shortcut menu)*

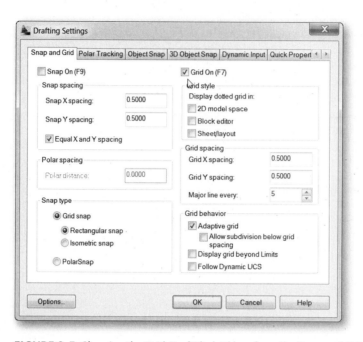

FIGURE 3–7 *Choosing the Grid On (F7) checkbox from the Snap and Grid tab of the Drafting Settings dialog box*

Changing Grid Styles

You can have the grid displayed in dots instead of lines in specified drawing venues. In the Grid Style section of the Drafting Settings dialog box check **2D model space, Block editor,** or **Sheet/layout** for changing from a grid of lines to a grid of dots in the specified venue. The 2D model space selection sets the grid style to dotted grid for 2D model space. The Block editor selection sets the grid style to dotted grid for the Block editor. The Sheet/layout selection sets the grid style to dotted grid for sheet and layout.

Changing Grid Spacing

AutoCAD offers two methods for changing the Grid spacing:

1. Right-click the **GRID** icon on the status bar at the bottom of the screen, and from the shortcut menu, choose SETTINGS. AutoCAD displays the Drafting Settings dialog box with the **Snap and Grid** tab selected. From the **Grid** section, you can change the settings of the X and Y spacings by entering the desired values in the **Grid X spacing** and **Grid Y spacing** boxes, respectively (Figure 3–8). To set the same value for X and Y, set the **Equal X and Y spacing** toggle to ON. A major grid line will appear at the spacing specified in the **Major line every** text box.

2. At the on-screen prompt, type **grid,** and AutoCAD displays the following prompt:

Specify grid spacing (X) or ⬇ *(specify distance to be used for both X and Y grid spacings)*

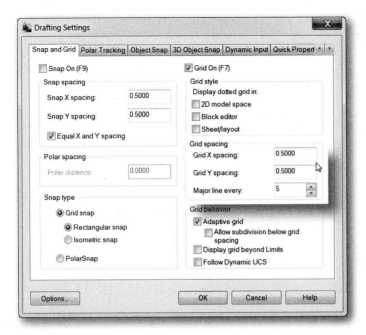

FIGURE 3–8 *Changing the Grid settings from the Snap and Grid section of the Drafting Settings dialog box*

Grid Options

Several grid options are available at the on-screen prompt.

Choose *Aspect* to set the Y Grid spacing different from the X Grid spacing. AutoCAD then prompts for the X and Y values independently. This is handy if the X and Y modular dimensions of your design are of unequal multiples. At the on-screen prompt, type **grid** and AutoCAD displays the following prompts:

> Specify grid spacing (X) or ⬇ *(choose* ASPECT *from the shortcut menu)*
>
> Specify the horizontal spacing (X) <current>: *(specify X Grid spacing)*
>
> Specify the vertical spacing (Y) <current>: *(specify Y Grid spacing)*

For example, the following sequence of commands shows how you can set a horizontal grid spacing of 0.5 and vertical grid spacing of 0.25.

> grid [Enter]
>
> Specify grid spacing (X) or ⬇ *(choose* ASPECT *from the shortcut menu)*
>
> Specify the horizontal spacing (X) <current>: 0.5 [Enter]
>
> Specify the vertical spacing (Y) <current>: 0.25 [Enter]

Applying the Aspect option in this example provides the grid dot spacing shown in Figure 3–9.

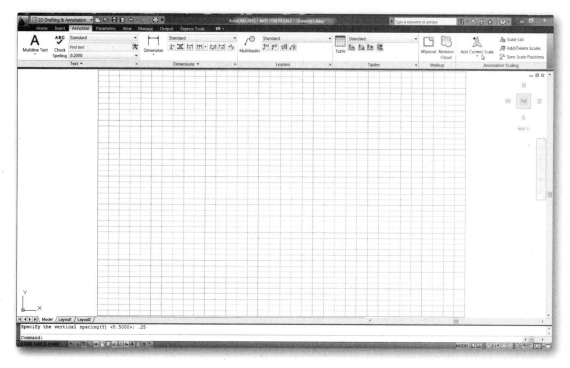

FIGURE 3–9 *Display after setting the grid aspect to 0.5 for horizontal and 0.25 for vertical spacing*

Choose SNAP from the shortcut menu to set the grid spacing equal to the snap resolution or make it a multiple of the snap spacing. To specify the grid spacing as a multiple of the snap value, enter **x** after the value. For example, to set up the grid value as three times the current snap value (snap = 0.5 units), enter **3x** for the prompt, as shown in the following example, which is the same as setting it to 1.5 units:

> grid [Enter]
> Specify grid spacing (X) or ⊡ 3x [Enter]

If the spacing of the visible grid is set too small, AutoCAD will not display the grid. To display the grid, invoke the GRID command again and specify a larger spacing.

NOTE

The relationship between the Grid setting and the Snap setting, when established as described in the previous section, is based on the current Snap setting. If the Snap setting is subsequently changed, the Grid setting does not change accordingly. For example, if the Snap setting is 1.00 and you enter **s** in response to the Specify grid spacing (X) or <current>: prompt, the Grid setting becomes 1.00 and remains 1.00 even if the Snap setting is later set to something else. Likewise, if you set the Grid setting to **3x**, it becomes 3.00 and will not change with a subsequent change in the Snap setting.

If the current Visual Style is set to something other than 2D Wireframe, the grid is displayed as lines rather than dots, and darker lines called major grid lines are displayed at preset intervals. Choose Major from the shortcut menu to set the value of the interval at which the major lines are spaced. Grid lines rather than grid dots are displayed in any visual style except 2D Wireframe.

Choose *Adaptive* from the shortcut menu to control the density of grid lines when zoomed in or out. For example, if you zoom far out, the density of displayed grid lines reduces automatically. Conversely, if you zoom close in, additional grid lines display in the same proportion as the major grid lines. You can also choose to turn on the **Allow subdivision below grid spacing** option, which generates additional, more closely spaced grid lines or dots when zoomed in. The frequency of these grid lines is determined by the frequency of the major grid lines.

By default, the Limits is set to OFF and the grid is displayed beyond the area specified by the LIMITS command. To have the grid restricted to the Limits area, uncheck the **Display grid beyond Limits** box in the Grid behavior section of the Snap and Grid tab of the Drafting Settings dialog box.

Choose *Follow* from the shortcut menu to change the grid plane to follow the XY plane of the dynamic UCS.

Ortho Command

The ORTHO command lets you draw lines and specify point displacements that are parallel to either the X or Y axis. Lines drawn with the Ortho mode set to ON are therefore either parallel or perpendicular to each other. This mode is helpful when you need to draw lines that are exactly horizontal or vertical. Additionally, when the Snap Style is set to Isometric, it forces lines to be parallel to one of the three isometric axes.

Setting Ortho ON and OFF

Four methods for setting the ORTHO command ON and OFF are available. Of the four explained in this section, the first method, selecting the **ORTHO** icon on the status bar, is the easiest and is most commonly used with the pointing device. The second method, pressing the function key F8, is most commonly used from the keyboard. The other methods, while not quite as convenient as the first two, might be convenient in certain situations.

1. Ortho can be toggled ON and OFF by choosing the **ORTHO** icon on the status bar.
2. Ortho can be toggled ON and OFF by pressing the function key F8.
3. Right-click the **ORTHO** icon on the status bar and toggle the **Enabled** from the shortcut menu.
4. At the on-screen prompt, type: **ortho**, and AutoCAD displays the following prompt:

   Enter mode *(choose* ON *or* OFF *from the shortcut menu)*

The Ortho and Polar Tracking modes, explained later in this chapter, cannot both be set to ON at the same time. They can both be set to OFF, or either one can be set to ON.	**NOTE**

When the Ortho mode is active, you can draw lines and specify displacements only in the horizontal or vertical directions, regardless of the cursor's on-screen position. The direction in which you draw is determined by the change in the X value of the cursor movement compared to the change in the cursor's distance to the Y axis. AutoCAD allows you to draw horizontally if the distance in the X direction is greater than the distance in the Y direction; conversely, if the change in the Y direction is greater than the change in the X direction, then it forces you to draw vertically. The Ortho mode does not affect keyboard entry of points.

Object Snap

The Object Snap (or OSNAP, for short) feature lets you specify points on existing objects in the drawing. For example, if you need to draw a line from the endpoint of an existing line, you can apply the Object Snap mode called ENDpoint. In response to the "Specify first point:" prompt, enter **end,** and place the cursor so that it touches the line nearer to the desired endpoint. AutoCAD will lock onto the endpoint of the existing line when you press the pick button of your pointing device. This endpoint becomes the starting point of the new line. This feature is similar to the basic SNAP command, which locks to invisible reference grid points.

You can invoke an Object Snap mode whenever AutoCAD prompts for a point. Object snap modes can be invoked while executing an AutoCAD command that prompts for a point such as the LINE, CIRCLE, MOVE, and COPY commands.

Applying Object Snap Modes

Object Snap modes can be applied in either of two ways:

1. Hold down the SHIFT key and press the right mouse button to display the Object Snap shortcut menu, then choose **Osnap Settings** (Figure 3–10), or right-click the **OSNAP** icon on the status bar and choose SETTINGS from the shortcut menu. The Drafting Settings dialog box will be displayed with the **Object Snap** tab selected, as shown in Figure 3–11. An Object Snap mode has been chosen if there is a check in its associated checkbox. Choose the desired Object Snap mode(s) by placing the cursor over its checkbox and pressing the pointing device pick button. This action makes it possible to use the checked mode(s) any time you are prompted to specify a point and is referred to as the "Running Osnap" method. Running Osnap can be set to ON and OFF by choosing the **OSNAP** icon on the status bar. You can also toggle the Running Osnap by pressing the function key F3.

FIGURE 3–10 *Choosing Object Snap Settings from the Object Snap shortcut menu*

2. When prompted to specify a point, enter the first three letters of the name of the desired Object Snap mode, or hold down the SHIFT key and press the right mouse button to display the Object Snap shortcut menu and select one of the available options. This is a *one-time-only* Osnap method, which will override any other Running Osnap mode for this one-point selection.

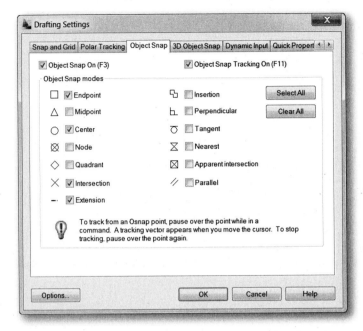

FIGURE 3–11 *Drafting Settings dialog box displaying the Object Snap tab*

To draw a line from an endpoint of an existing line, invoke the LINE command, and AutoCAD displays the following prompts:

```
line Enter
Specify first point: end (enter the first three letters of
the Osnap mode endpoint)
Specify first point: (move the cursor near the desired end of
a line, and press the pick button)
```

The first point of the new line is drawn from the endpoint of the selected object.

Osnap Markers and Tooltips

Whenever one or more Object Snap modes are activated and you move the cursor target box over a snap point, AutoCAD displays a geometric shape, or Marker, and a Tooltip. By displaying a Marker on the snap points with a Tooltip, you can see the point that will be selected and the Object Snap mode in effect. AutoCAD displays the Marker depending on the Object Snap mode selected. In the **Object Snap** tab of the Drafting Settings dialog box, each Marker is displayed next to the name of its associated Object Snap mode.

From the **Object Snap** tab of the Drafting Settings dialog box, choose **Options**, and AutoCAD displays the Options dialog box with the **Drafting** tab selected. This is where the settings for displaying the Markers and Tooltips can be changed by checking or clearing the appropriate checkboxes in the **AutoSnap Settings** section (Figure 3–12).

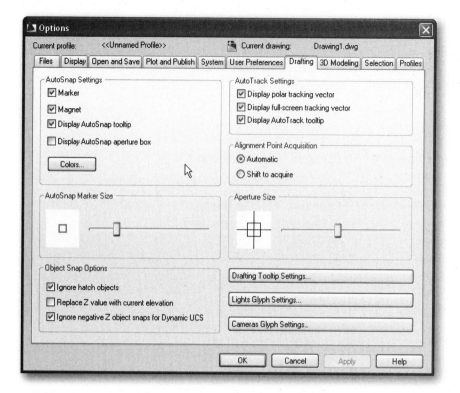

FIGURE 3–12 *Options dialog box with the Drafting tab selected*

The **AutoSnap Settings** section controls the size, appearance, action of the cursor, and whether or not a Tooltip is displayed while Object Snap is in effect.

Choosing **Marker** causes the geometric shape, unique for each OSNAP mode, to be displayed when the cursor moves over a snap point.

Choosing **Magnet** causes the cursor to lock onto the snap point when the cursor is near.

Choosing **Display AutoSnap Tooltip** causes a Tooltip to display the name of the Object Snap mode when AutoCAD is prompting for a point and an OSNAP mode is in effect.

Choosing **Display AutoSnap Aperture** box causes the square target box to be displayed when AutoCAD is prompting to specify a point.

Choosing **Colors** displays the Drawing Window Colors dialog box where you can change the color of the Marker. When initiated in this manner, the dialog box is displayed with the Autosnap marker option selected under the Interface element list box.

The **AutoSnap Marker Size** section controls the size of the AutoSnap Marker. To change the size of the AutoSnap Marker, press and hold the pick button of the pointing device while the cursor is over the slide bar. Move it to the right to make the Marker larger and to the left to make it smaller. The Image tile shows the current size of the Marker.

After making the necessary changes to the AutoSnap settings, choose **OK** to close the Options dialog box, and choose **OK** again to close the Drafting Settings dialog box.

Object Snap Modes

Sixteen Object Snap modes (see Figure 3–13) are available: Endpoint, Midpoint, Intersection, Apparent Intersection, Extension, Center, Quadrant, Tangent, Perpendicular, Parallel, Insert, Node, Nearest, None, From, and MTP, which allows snapping between specified points.

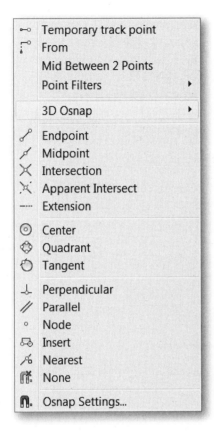

FIGURE 3–13 *The Object Snap shortcut menu*

> **NOTE**
>
> This action takes practice! You must be alert whenever you specify a point near objects when more than one Running Osnap mode is in effect. For example, if you are trying to specify an endpoint on a line that is near the circumference of a circle while the Center Osnap mode is active and the cursor pick box touches the circle, AutoCAD may select the center of the circle as the specified point. This kind of error is usually obvious, and you will have the opportunity to correct it. If a point is placed slightly off its intended location, due to a Running Osnap mode taking over unintentionally, it could introduce an undetected error into your drawing.

Endpoint, Intersection, Midpoint, and Perpendicular

The *Endpoint* mode allows you to snap to the closest endpoint of a line, arc, elliptical arc, multiline, polyline segment, spline, region, or ray or to the closest corner of a trace, solid, or 3D face. As shown in Figure 3–14, LINE B is drawn from the indicated starting point to the end of LINE A by using the Endpoint mode.

The *Intersection* mode allows you to snap to the intersection of two objects, which can include arcs, circles, ellipses, elliptical arcs, lines, multilines, polylines, rays, splines, or xlines. As shown in Figure 3–14, LINE C is drawn from the indicated starting point to the intersection of the CIRCLE C and LINE A by using the Intersection mode.

The *Midpoint* mode allows you to snap to the midpoint of a line, arc, elliptical arc, broken ellipse, multiline, polyline segment, xline, solid, or spline. As shown in Figure 3–14, LINE D is drawn from the indicated starting point to the midpoint of LINE A by using the Midpoint mode.

The *Perpendicular* mode allows you to snap to a point perpendicular to a line, arc, circle, elliptical arc, multiline, polyline, ray, solid, spline, or xline. Deferred Perpendicular snap mode is automatically turned on when more than one perpendicular snap is required by the object being drawn. As shown in Figure 3–14, LINE E is drawn from the indicated starting point to a point on and perpendicular to LINE A by using the Perpendicular mode. The location of the cursor may have to be adjusted to ensure that LINE A is selected; otherwise, if CIRCLE C is selected, the result will be LINE F, which is drawn perpendicular to the circle.

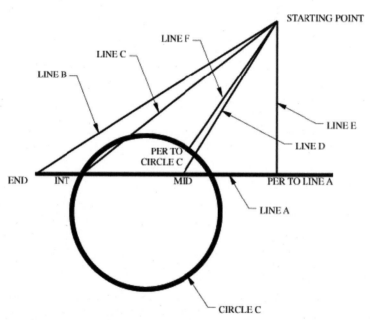

FIGURE 3–14 *Lines drawn to designated points of another line using the Endpoint, Intersection, Midpoint, and Perpendicular Object Snap modes*

Quadrant, Tangent, and Center

The *Quadrant* mode allows you to snap to one of the quadrant points of a circle, arc, ellipse, or elliptical arc. The quadrant points are located at 0°, 90°, 180°, and 270° from the center of the circle or arc. The quadrant points are determined by the zero-degree direction of the current coordinate system. As shown in Figure 3–15, LINE B is drawn from the indicated starting point to one of the quadrants of CIRCLE C by using Quadrant mode.

The *Tangent* mode allows you to snap to the tangent of an arc, circle, ellipse, or elliptical arc. Deferred Tangent snap mode is automatically turned on when more than one tangent is required by the object being drawn. As shown in Figure 3–15, LINE C is drawn from the indicated starting point to a point on and tangent to CIRCLE C by using Tangent mode.

The *Center* mode allows you to snap to the center of an arc, circle, ellipse, elliptical arc, or line. As shown in Figure 3–15, LINE D is drawn from the indicated starting point to the center of CIRCLE C by using Center mode.

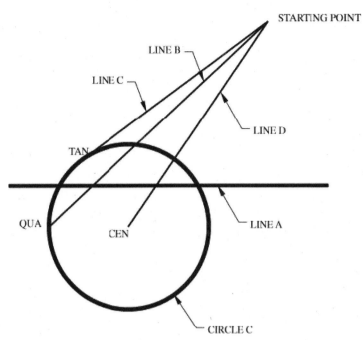

FIGURE 3–15 *Lines drawn to designated points of a circle using the Quadrant, Tangent, and Center Object Snap modes*

> The Quadrant point selected is one of four possible points on the circle, which include points at 0°, 90°, 180°, and 270°.

NOTE

Apparent Intersection, Extension, and Parallel

The *Apparent Intersection* mode allows you to snap to the apparent intersection of two objects, which can include an arc, circle, ellipse, elliptical arc, line, multiline, polyline, ray, spline, or xline. These objects may or may not actually intersect but would intersect if either or both objects were extended. In the following example, a rectangle box is drawn from the apparent intersection of two rectangles, as shown in Figure 3–16.

```
rectangle Enter
Specify first corner point or ⊞ app (invoke the Apparent
    Intersection object snap, and select LINE 1 and LINE 4)
Specify other corner point or ⊞ app (invoke the Apparent
    Intersection Object Snap mode, and select LINE 2 and LINE 3)
```

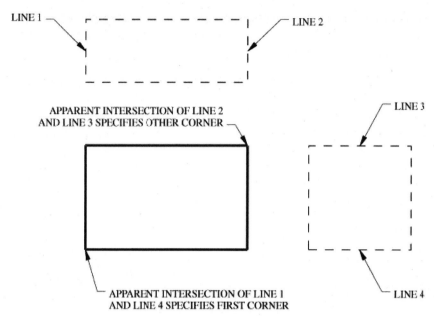

FIGURE 3–16 *Identifying the lines to draw a rectangle*

The *Extension* mode causes a temporary extension line to be displayed when you pass the cursor over the endpoint of objects, so you can draw objects to and from points on the extension line. As shown in Figure 3–17, LINE A was drawn using extensions of LINE B and ARC C.

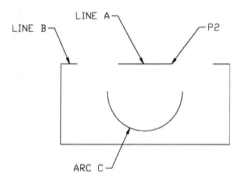

FIGURE 3–17 *LINE A is drawn using extensions of LINE B and ARC C*

The *Parallel* mode allows you to draw a line that is parallel to another object. Once the first point of the line has been specified and the Parallel Object Snap mode is selected, move the cursor over the object to which you wish to make the new line parallel. Move the cursor near a line from the first point that is parallel to the object selected, and a construction line will appear. While the construction line is visible, specify a point, and the new line will be parallel to the selected object.

Node, Insertion, Nearest, From, and None

The *Node* mode allows you to snap to a point on an object.

The *Insertion* mode allows you to snap to the insertion point of a block, text string, attribute, or shape.

The *Nearest* mode lets you select any object except text in response to a prompt for a point, and AutoCAD snaps to the point on that object nearest to the cursor.

The *From* mode locates a point offset from a reference point within a command. At an AutoCAD prompt for locating a point, select From Object snap mode, and then specify a temporary reference or base point from which you can specify an offset to locate the next point. Enter the offset location from this base point as a relative coordinate, or use direct distance entry.

The *None* mode temporarily overrides any running object snaps that may be in effect.

Snapping Between Specified Points

AutoCAD allows you to snap to a point that is midway between two specified points. You can enter **mtp** or **m2p** when prompted to specify a point. For example, if you want to start a line at the midpoint between the centers of two circles, you can invoke the Midway Between Two Points Object Snap mode as follows:

> line `Enter`
> Specify first point: mtp `Enter`
> First point of mid: cen
> First point of mid: *(specify the center of the first circle)*
> Second point of mid: cen
> Second point of mid: *(specify the center of the second circle)*
> Specify next point or ⊡ *(specify the endpoint of the line)*
> Specify next point or ⊡ `Enter`

The Midway Between Two Points Object Snap mode is not included on the **Object Snap** tab of the Drafting Settings dialog box nor on the Object Snap toolbar.

NOTE

3D Osnap

AutoCAD allows you to snap to points while working in 3D. The 3D mode can be turned on by checking the **3D Object Snap On** checkbox on the 3D Object Snap tab of the Drafting Settings dialog box (see Figure 3–18).

FIGURE 3–18 *Drafting Settings Dialog Box with 3D Object Snap tab displayed*

You can also access the 3D Osnap modes by holding down the SHIFT key and right-clicking to display the Object Snap shortcut menu, then choose **3D Osnap** to display the shortcut menu (see Figure 3–19).

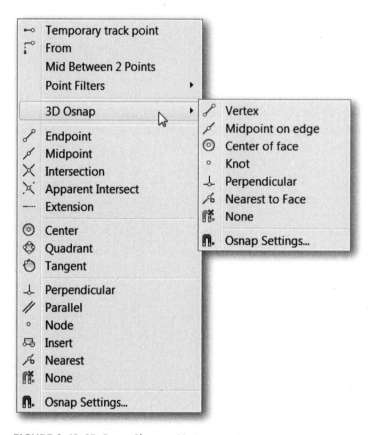

FIGURE 3–19 *3D Osnap Shortcut Menu*

Choosing *Vertex* allows you to snap to the closest vertex of a 3D object.

Choosing *Midpoint on Edge* allows you to snap to the midpoint of a face edge.

Choosing *Center of Face* allows you to snap to the center of a face.

Choosing *Knot* mode allows you to snap to a knot on a spline.

Choosing *Perpendicular* allows you to snap to a point perpendicular to a face.

Choosing *None* turns off all 3D object snap modes.

Drawing Objects Using Tracking

Tracking, or moving through nonselected point(s) to a selected point, could be called a command "enhancer." It can be used whenever a command prompts for a point. If the desired point can best be specified relative to some known point(s), you can "make tracks" to the desired point by invoking the Tracking option and then specifying one or more points relative to previous point(s) "on the way to" the actual point for which the command is prompting. These intermediate tracking points are not necessarily associated with the object being created or modified by the command. The primary significance of tracking points is that they are used to establish a path to the point you wish to specify as the response to the command prompt. At any prompt, to locate a tracking point, enter **tracking, track,** or **tk**. Some of the objects in the partial plan shown in Figure 3–20 can be drawn more easily by means of Tracking.

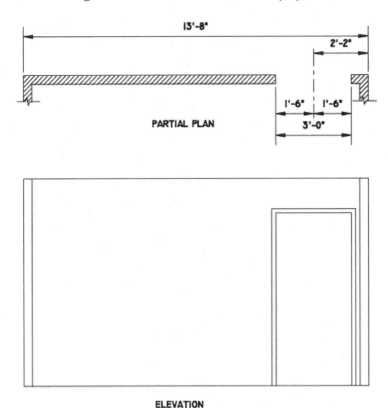

PARTIAL PLAN

ELEVATION

FIGURE 3–20 *Example of a partial plan to demonstrate the Tracking option*

The following example will use Tracking to draw lines A and B in the partial plan in Figure 3–21, leaving the 3'-0 door opening in the correct place. By means of Tracking, you can draw the lines with the given dimension information without having to calculate the missing information.

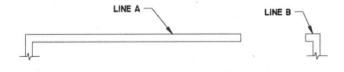

PARTIAL PLAN

FIGURE 3–21 *Lines A and B to be drawn using the Tracking option*

In Figure 3–22, line A from TK1/SP1 (tracking point 1 and starting point 1) to EP1 (ending point 1) and line B from SP2 (starting point 2) to TK2/EP2 (tracking point 2 and ending point 2) can be drawn by using the TRACKING command enhancer. To complete this example, change AutoCAD's drawing units to Architectural. Invoke the LINE command, and AutoCAD displays the following prompts:

> line [Enter]
> Specify first point: 0,12' [Enter] *(specify point TK1/SP1)*
> Specify next point or ⊡ track [Enter] *(invoking the Tracking feature)*

First tracking point: **0,12'** [Enter] *(specify point TK1/SP1 again as the first tracking point)*

Next point (press enter to end tracking): **@13'8,0** [Enter] *(locates the second tracking point, TK2)*

Next point (press enter to end tracking): **@2'2<180** [Enter] *(locates the third tracking point, TK3)*

Next point (press enter to end tracking): **@1'6<180** [Enter] *(locates the fourth tracking point, EP1)*

Next point (Press enter to end tracking): *(press* [Enter] *to exit Tracking; by this you are designating the point to which you have "made tracks" as the response to the prompt that was in effect when you entered Tracking)*

Specify next point or ⏎ *(press* [Enter] *to exit the LINE command)*

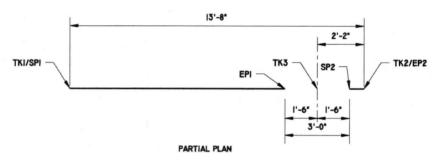

FIGURE 3–22 *Points to be drawn using the Tracking option*

Once the Tracking option is invoked, you establish a path to EP1 by specifying the initial tracking point, TK1, and then each subsequent point relative to the previous point—that is, TK2 relative to TK1, TK3 relative to TK2, and EP1 relative to TK3. The first track in this example is specified by the Relative Rectangular method, and the next two are Relative Polar. Because the tracking points are all on one horizontal line, you could use the direct distance feature, explained in the next section, by placing the cursor in the correct direction with ORTHO set to ON and entering the distance from the keyboard.

NOTE If you know the coordinates of one of the intermediate tracking points, it should probably be used as the initial tracking point. The idea behind Tracking is to establish a point by means of a path from and through other points. Thus, the shortest path is the best. If the coordinates of TK2 are known, or if you can specify it by some other method, it could become the initial tracking point. Keep this in mind as you learn to use Object Snap. You don't necessarily need to know the coordinates if you can use Object Snap to select a point from which a tracking path can be specified.

The line from SP2 to TK2/EP2 can be started and ended in a similar manner as how the line from TK1/SP1 to EP1 was started with some minor modifications. The sequence, which involves using Tracking twice, is as follows:

line [Enter]

Specify first point: **track** [Enter] *(invoking the Tracking option)*

First tracking point: **0,12'** [Enter] *(specify point TK1/SP1 as the first tracking point)*

Next point (press enter to end tracking): **@13'8,0** [Enter] *(locates the second tracking point, TK2)*

Next point (press enter to end tracking): **@2'2<180** [Enter] *(locates the third tracking point, TK3)*

Next point (press enter to end tracking): **@1'6<0** [Enter] *(locates the fourth tracking point, SP2)*

Next point (press enter to end tracking): *(press to exit Tracking; AutoCAD establishes point SP2)*

Specify next point or [Undo]: **track** [Enter] *(invoking the Tracking option again)*

First tracking point: **0,12'** [Enter] *(specify point TK1/SP1 as the first tracking point)*

Next point (press enter to end tracking): **@13'8,0** [Enter] *(locates the second tracking point, TK2/EP2)*

Next point (press enter to end tracking): *(press* [Enter] *to exit Tracking, AutoCAD establishes point EP2)*

Specify next point: *(press* [Enter] *to terminate the* LINE *command)*

This example shows an application of the Tracking option in which the points were established in reference to known points.

Drawing Objects Using Direct Distance

The Direct Distance feature for specifying a point relative to another point can be used with a command such as LINE as a variation of the Relative Coordinates mode. In the case of the Direct Distance option, the distance is keyed in, and the direction is determined by the current location of the cursor. This option is useful when you know the exact distance between two points, but specifying the exact angle is not as easy as placing the cursor on a point that forms the exact angle desired.

Figure 3–23 shows a shape that can be drawn more easily by using the Direct Distance option along with setting the Snap and Ortho modes to ON and OFF at the appropriate times. Points A, B, and C are on the Snap grid (X,Y coordinates 3,3 for A, 3,6 for B, and 9,6 for C). By setting the Snap mode to ON with the value set to 1 or perhaps 0.5, the cursor can be placed on the required points to draw lines A-B and B-C.

If you have exited the LINE command after drawing line B-C, you must invoke the LINE command and specify C as the first point before moving the cursor to A. The rubber-band line indicates that the next line is drawn from C to A, as shown in Figure 3–24. However, you wish to draw a line only 2 units long but in the same direction as a line from C to A. With the cursor placed on A, enter **2,** and press [Enter].

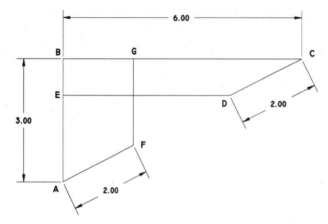

FIGURE 3–23 *Example of a Direct Distance application*

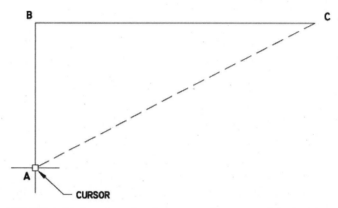

FIGURE 3–24 *Drawing a line from C to A by means of the Direct Distance option*

AutoCAD draws the line of 2 units, and Point D is established. To draw a line from D to E without exiting the LINE command, first set Ortho and Snap to ON, and then place the cursor on line A-B. The rubber-band line will indicate line D-E, as shown in Figure 3–25. In this case, you do not know the distance, but you do know that the line terminates on line A-B. Therefore, press the pick button on your pointing device, and line D-E is drawn.

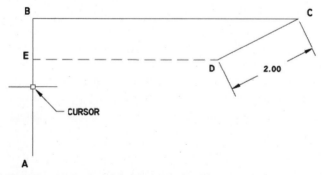

FIGURE 3–25 *Drawing a line from D to E*

Line A-F can be drawn in a similar manner as line C-D was drawn. Unlike line C-D, line A-F is not a continuation of another line. However, the starting point, A, can be selected with Snap set to ON. After starting point A has been specified, place the cursor on point C and enter **2**. From line A-F, line F-G can be drawn in a similar manner as line D-E was drawn. This example shows an application of the Direct Distance option in which the distance was entered and the direction was controlled by the cursor.

Polar Tracking and Object Snap Tracking

The Polar Tracking feature lets you draw lines and specify point displacements in directions that are multiples of a specified increment angle. You can specify whether the increment angles are measured from the zero point of the current coordinate system or from the angle of a previous object. You can also add up to 10 additional angles to which a line or displacement can be diverted from a base direction. The Object Snap Tracking feature lets you track your cursor from a strategic OSNAP point on an object in a specified direction, with either orthogonal or preset polar angles.

Turning Polar Tracking ON and OFF

Four primary methods are available to turn Polar Tracking ON and OFF to the specified increment angle:

1. Polar Tracking can be toggled ON and OFF by choosing the **POLAR** icon on the status bar.
2. Polar Tracking can be toggled ON and OFF by pressing the function key F10.
3. Right-click the **POLAR** icon on the status bar, and from the shortcut menu, choose SETTINGS. The Drafting Settings dialog box will be displayed with the **Polar Tracking** tab selected. You can toggle the setting by choosing the **Polar Tracking On (F10)** box.
4. Right-click the **POLAR** icon on the status bar and toggle the **Enabled** selection.

When the Polar Tracking mode is set to ON, it allows you to specify a displacement or draw in a direction that is a multiple of a specified increment angle. Even with Polar Tracking set to ON, you can still specify a "next point" of a line or displacement with the cursor that is not at one of the multiples of the specified Polar Tracking increment angle. This is different from the ORTHO and SNAP features where you are forced when specifying a point with the cursor to accept a point orthogonally or on the Snap grid when the respective mode is set to ON. You will notice, however, that when the Polar Tracking mode is set to ON and you move the cursor near one of the radials, a multiple of the increment angle, the cursor will snap to that radial; the radial will appear as a dotted construction line indicating the direction of the line that will be drawn if you specify that point by pressing the pick button on your pointing device.

Changing Polar Tracking Increment and Additional Angles

AutoCAD provides two primary methods to change the Polar Tracking increment angle and add additional angles:

1. From the status bar at the bottom of the screen, right-click on the **POLAR** icon and choose SETTINGS. AutoCAD displays the Drafting Settings dialog box with the **Polar Tracking** tab selected. From the **Polar Angle Settings** section, enter the desired increment angle in the **Increment angle** box. This is the base increment angle whose multiples are used by Polar Tracking. Up to 10 additional angles can be added in the text box under the **Additional angles** box. Select **New**, and then enter the desired additional angle(s). If no check appears in the **Additional angles** box, the

additional angles are not available for use when Polar Tracking is set to ON. To make the additional angle(s) available when the Polar Tracking mode is set to ON, set **Additional angles** box to ON (Figure 3–26).

2. At the on-screen prompt, type **dsettings,** and AutoCAD displays the Drafting Settings dialog box. Then select the **Polar Tracking** tab. See the explanation in Method 1 for changing settings.

NOTE The Polar Tracking and the Ortho modes cannot both be set to ON simultaneously. They can both be set to OFF, or either one can be set to ON.

FIGURE 3–26 *Selecting the Polar Tracking On (F10) checkbox and setting the increment angle to 18° and one additional angle to 9°*

With the 18° increment angle and the Polar Angle measurement set to Absolute, you can draw lines and specify displacements at 18°, 36°, 54°, 72°, 90°, 108°... ...324°, 342°, 360° around the compass by snapping to the construction lines when they appear. The accompanying AutoTrack Tooltip will appear when you have snapped to a multiple of the increment angle if the **Display AutoTrack Tooltip** box is set to ON in the **Drafting** tab of the Options dialog box. The Tooltip displays the cursor's distance and direction from the first specified point. With the cursor snapped to one of the Polar Tracking angles, you can use the Direct Distance option, enter a distance from the keyboard, and press ENTER. AutoCAD will draw the line or apply the displacement in accordance with the distance entered and the direction set by the cursor. For example, in Figure 3–27, the line from P1 to P2 was drawn 2 units long at 0°. This could have been done with Direct Distance and either ORTHO set to ON or with Polar Tracking. However, with P2 to P3 being at 18° and the measurement from P3 to P4 being at 36°, it is easier to use Polar Tracking with the settings shown in the example in Figure 3–26. Figure 3–28 shows the three line segments drawn using Polar Tracking with the increment angle set to 18° and the Polar Angle measurement set to **Absolute**. Figure 3–29 shows how you can use the Additional angles setting of 9° to draw a line with a length of 1 unit at 9° from absolute 0.

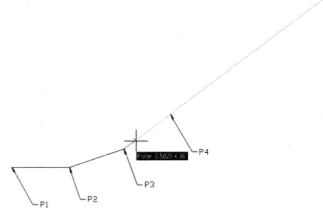

FIGURE 3–27 *Drawing line segments at multiples of the 18° increment angle*

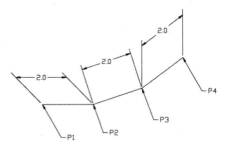

FIGURE 3–28 *Three line segments drawn using Polar Tracking with the increment angle set to 18° and the Polar Angle measurement set to absolute 0*

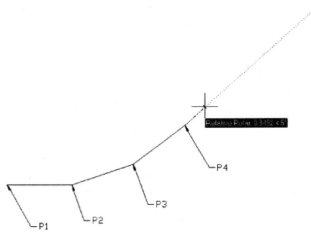

FIGURE 3–29 *Four line segments drawn using Polar Tracking with the increment angle set to 18° and at one additional angle at 9° from Polar Angle measurement set to absolute 0*

After completing the three segments drawn at multiples of 18°, instead of adding the segment at 9°, change the increment angle to 5° with the Polar Angle measurement set to **Relative to last segment** instead of **Absolute** to draw three additional line segments as shown in Figure 3–30. Continuing with three more segments, each 1 unit long, the first increment angle will be measured from the last segment, which is at 36°. The Tooltip displays "Relative Polar (distance) < 5°," but the resulting angle is drawn at 41°. The last two segments will be at 46° and 51°.

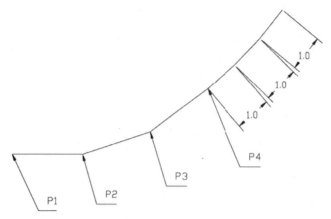

FIGURE 3–30 *Three line segments drawn using Polar Tracking with the increment angle set to 5° and the Polar Angle measurement set to Relative to last segment*

Polar Snap

The Polar Snap feature can be used to make drawing the 2-unit-long and 1-unit-long segments in the previous examples even easier. On the **Snap and Grid** tab of the Drafting Settings dialog box, choose the **Polar Snap** radio button in the **Snap type** section. In the **Polar distance** text box of the **Polar spacing** section, enter the desired distance. In the above-mentioned example, the 2-unit-long segments at 0°, 18°, and 36°, you can enter 2.0 for the distance. For the 41°, 46°, and 51° segment, with additions of 5° each in the **Relative to last segment** mode, you can use a distance of 1.0.

Object Snap Tracking

Object Snap Tracking (OTRACK) helps you to draw objects at specific angles or in specific relationships to other objects. When you set OTRACK to ON, temporary alignment paths help you create objects at precise positions and angles. OTRACK works in conjunction with object snaps and Polar Angle settings. You must set an object snap before you can track from an Object Snap Point. You can toggle OTRACK ON and OFF with OTRACK on the status bar or the function key F11.

OTRACK includes two tracking options: Object Snap Tracking and Polar Tracking. Use Object Snap Tracking to track along alignment paths that are based on Object Snap points. Acquired points display a small plus sign (+), and you can acquire up to seven tracking points at a time. After you acquire a point, horizontal, vertical, or polar alignment paths relative to the point are displayed as you move the cursor over their drawing paths. For example, you can select a point along a path based on an object endpoint or an intersection between objects.

The **Object Snap Tracking Settings** section in the **Polar Tracking** tab of the Drafting Settings dialog box allows you to select options for Object Snap Tracking. Choose **Track orthogonally only** to display orthogonal, or horizontal/vertical, Object Snap Tracking paths for acquired Object Snap points when Object Snap Tracking is set to ON. Choose **Track using all polar angle settings** to permit the cursor to track along any Polar Angle tracking path for an unacquired Object Snap Point when Object Snap Tracking is ON and while specifying points.

This next example shows how to apply Polar Tracking and Object Snap Tracking to draw the vertical line B and the horizontal line C from existing line A and arc D, as shown in Figure 3–31 and Figure 3–32.

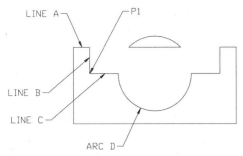

FIGURE 3–31 *Completed object including lines A, B, and C, and arc D*

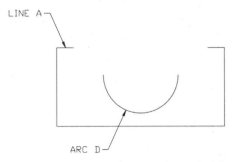

FIGURE 3–32 *Line A and arc D from which lines B and C will be drawn using Polar Tracking*

From the status bar at the bottom of the screen, right-click on the **POLAR** icon and choose SETTINGS. The Drafting Settings dialog box is displayed with the **Polar Tracking** tab selected. From the **Object Snap Tracking Settings** section, choose **Track orthogonal only**. Check to be sure that **OTRACK** is set to ON. Invoke the LINE command as follows:

> line Enter
>
> Specify first point: end *(enter the first three letters of the OSNAP mode Endpoint, press* Enter *, and then select the right end of LINE A to establish the start point of LINE B)*
>
> Specify next point or ⊡ *(move the aperture cursor to the right end of LINE A, keep it there until the Endpoint marker appears, and without pressing a button, move it downward from the line)*

As you move the cursor downward from line A, a dashed construction line will follow as long as the line between the end of line A and the cursor is vertical or horizontal, as shown in Figure 3–33.

Move the cursor to the left end of arc D, keep it there until the Endpoint marker appears, and then move it away from it toward the left, again without pressing the pick button. A horizontal construction line will follow until the cursor is near the vertical construction line through the end of line A, as shown in Figure 3–34. At this time, with both the vertical and horizontal construction lines being displayed, you can press the pick button on your pointing device and point P1, the endpoint of line B, will be established.

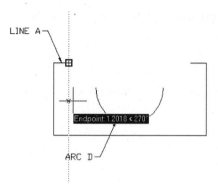

FIGURE 3–33 *A vertical construction line passes through the cursor and the end of LINE A*

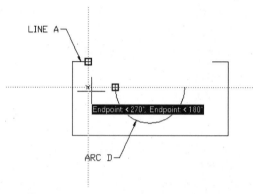

FIGURE 3–34 *A dashed arc follows the cursor indicating the extension of ARC D to the dashed extension of LINE B, at which point P1 is specified*

After pressing the pick button to specify point P1 and complete line B, move the cursor to the left end of arc D, invoke the Endpoint OSNAP mode, and select arc D. This action will complete line C, as shown in Figure 3–35.

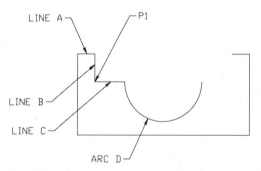

FIGURE 3–35 *Lines B and C are drawn with point P1 at their intersection*

Dynamic Input, On-Screen Prompts, and Geometric Values Display

Dynamic Input provides a command interface near the cursor to help you keep your focus in the drafting area. When Dynamic Input is set to ON, Tooltips display information near the cursor and are dynamically updated as the cursor moves. When a command is active, a Tooltip provides the field for user entry. The actions required to complete a command are the same as when you work from the command line. The difference is that your attention can stay near the cursor.

Dynamic Input can be toggled ON and OFF by choosing the **DYN** icon on the status bar. Dynamic Input has three components: pointer input, dimensional input, and dynamic prompts. To change the format and visibility of Pointer Input, Dimension Input, and Tooltip appearance, right-click the **DYN** icon on the status bar, and choose SETTINGS from the shortcut menu. AutoCAD will display the Drafting Settings dialog box with the **Dynamic Input** tab selected (Figure 3–36).

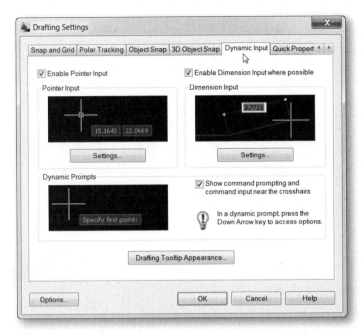

FIGURE 3–36 *Drafting Settings dialog box displaying the Dynamic Input tab*

Choosing **Enable Pointer Input** causes the cursor to be displayed with coordinate values in a Tooltip. When a command prompts you for a point, you can enter coordinate values in the Tooltip instead of on the command line. The Preview Area shows an example of pointer input.

Choosing **Enable Dimensional Input where possible** displays a dimension with Tooltips for distance value and angle value when a command prompts you for a second point or a distance. The values in the dimension Tooltips change as you move the cursor. When **Enable Pointer Input** and **Enable Dimensional Input where possible** are both checked, dimensional input supersedes pointer input when it is available.

The **Dynamic Prompts** section allows you to set how AutoCAD displays prompts in a Tooltip near the cursor when necessary in order to complete the command. Selecting **Show command prompting and command input near the crosshairs** causes prompts to be displayed in a Tooltip near the cursor. The Preview Area shows an example of dynamic prompts.

For examples of second-point input throughout the text, the ***Enable Dimensional Input where possible*** check box is not selected, providing a simpler display of coordinates.

NOTE

Pointer Input

The **Pointer Input** section controls how AutoCAD displays coordinate values in a Tooltip near the cursor.

Choose **Settings** in the **Pointer Input** section to display the Point Input Settings dialog box (Figure 3–37), which allows you to control the coordinate format in the Tooltips that are displayed and the visibility of the pointer when the pointer input is set to ON.

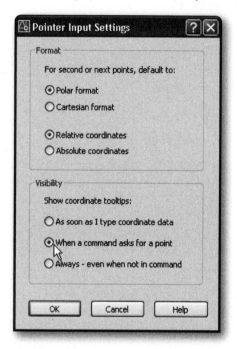

FIGURE 3–37 *Pointer Input Settings dialog box*

The **Format** section controls coordinate format in the Tooltips that are displayed when pointer input is turned on. Selecting **Polar format**, the default selection, causes the Tooltip to be displayed for the second or next point in polar coordinate format. Selecting **Cartesian format** causes the Tooltip to be displayed for the second or next point in Cartesian coordinate format. You can specify the coordinates in either Polar or Cartesian format.

Selecting **Relative coordinates**, the default selection, causes the Tooltip to be displayed for the second or next point in relative coordinate format. This selection allows you to specify the relative coordinates with or without the prefix "at" (@) sign, but to specify absolute coordinates, you have to prefix the coordinate with a pound (#) sign. Selecting **Absolute coordinates** causes the Tooltip to be displayed for the second or next point in absolute coordinate format. This selection allows you to specify the absolute coordinates with or without the prefix pound (#) sign, but to specify relative coordinates, you have to prefix the coordinate with an "at" (@) sign.

NOTE

> You can use the direct distance method when pointer input is set to absolute coordinates.

The **Visibility** section controls when pointer input is displayed.

Selecting **As soon as I type coordinate data** causes Tooltips to be displayed only when you start to enter coordinate data when pointer input is turned on. Selecting **When a command asks for a point** causes Tooltips to be displayed whenever a command prompts for a point. Selecting **Always–even when not in a command** causes Tooltips to always be displayed when pointer input is turned on.

After making the necessary changes, choose **OK** to exit the Pointer Input settings dialog box.

The prompts and responses shown in examples throughout the text and the drawing exercises in this book assume that you are using Dynamic Input and typing the coordinates at the on-screen prompt. Absolute coordinates are indicated with the # prefix, and relative coordinates are indicated with the @ prefix. If you prefer to use the Command window for coordinate input, do not prefix with # for absolute coordinates.

Dimension Input

The **Dimension Input** section allows you to set how AutoCAD displays a dimension with Tooltips for distance value and angle value when a command prompts for a second point or a distance. The values in the dimension Tooltips change as you move the cursor. You can enter values in the Tooltip instead of on the command line.

Choosing **Enable Dimension Input where possible** allows dimensional input whenever possible. Dimensional input is not available for some commands that prompt for a second point. Preview Area shows an example of dimensional input.

Choose **Settings** to display the Dimension Input Settings dialog box (Figure 3–38), which controls which Tooltips are displayed during grip stretching when dimensional input is turned ON.

FIGURE 3–38 *Dimension Input Settings dialog box*

Choose one of the three options in the **Visibility** section, which controls which Tooltips are displayed during grip stretching when dimensional input is turned ON.

Selecting **Show only 1 dimension input field at a time** causes only the Length Change dimensional input Tooltip to be displayed when you are using grip editing to stretch an object.

Selecting **Show 2 dimension input fields at a time** causes the Length Change and Resulting Dimension input Tooltips to be displayed when you are using grip editing to stretch an object.

Selecting **Show the following dimension input fields simultaneously** causes the selected dimensional input Tooltips to be displayed when you are using grip editing to stretch an object.

The input fields that can be displayed include the following: Resulting Dimension, Length Change, Absolute Angle, Angle Change, and Arc Radius.

Selecting **Resulting Dimension** causes a length dimension to be displayed and updated as you move the grip.

Selecting **Length Change** causes the change in length to be displayed as you move the grip.

Selecting **Absolute Angle** causes an angle dimension to be displayed and updated as you move the grip.

Selecting **Angle Change** causes the change in the angle to be displayed as you move the grip.

Selecting **Arc Radius** causes the radius of an arc to be displayed and updated as you move the grip.

After making the necessary changes, choose **OK** to exit the Dimension Input Settings dialog box.

Drafting Tooltip Appearance

Choose **Drafting Tooltip Appearance** to open the Tooltip Appearance dialog box (Figure 3–39), which can be used to control the appearance of Tooltips.

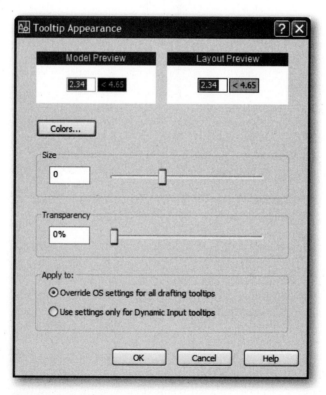

FIGURE 3–39 *Tooltip Appearance dialog box*

The **Model Preview** and **Layout Preview** show examples of the current Tooltip appearance settings in Model space and Layouts, respectively.

Choosing **Colors** causes the Drawing Window Colors dialog box to be displayed, where you can specify a color for Tooltips in model space and in layouts, respectively.

The **Size** slider bar and text box allow you to specify the size for Tooltips. The default size is zero. The slider bar is used to make Tooltips larger or smaller.

The **Transparency** slider bar and text box allow you to control the transparency of Tooltips. The lower the setting, the less transparent the Tooltip. A value of 0 sets the Tooltip to opaque. The slider bar is used to make Tooltips more or less transparent.

The **Apply to** section allows you to specify whether the settings apply to all drafting Tooltips or only to Dynamic Input Tooltips.

After making changes, choose **OK** to close the Tooltip Appearance dialog box. Choose **OK** again to close the Drafting Settings dialog box.

DISPLAY CONTROL

There are many ways to view a drawing in AutoCAD. These viewing options vary from on-screen viewing to hard-copy plots. The hard-copy options are discussed in Chapter 8. Using the display commands, you can select the portion of the drawing to be displayed, establish 3D perspective views, and much more. By enabling you to view the drawing in different ways, AutoCAD allows you to draw faster. The commands that are explained in this section are utility commands. They make your job easier and help you to draw more accurately.

Zoom Command

The ZOOM command is like a zoom lens on a camera. You can increase or decrease the viewing area while the actual size of objects remains constant. As you increase the visible size of objects, you view a smaller area of the drawing in greater detail as though you were closer to it. As you decrease the visible size of objects, you view a larger area as though you were farther away from it. This ability provides a close-up view for better accuracy and detail or a distant view to see the whole drawing.

AutoCAD provides smooth transition rather than instantaneous zooms and pans so that you can watch as the view changes from the current view to the selected view. This action allows you to better keep track of where you are going in relation to where you were. You can change to instantaneous transitions by changing the setting of the VTENABLE system variable. There are several methods by which the ZOOM command can be used.

Zoom Realtime

The *Zoom Realtime* selection, which is invoked from the shortcut menu (displayed when you are not in any command), as shown in Figure 3–40, lets you zoom interactively to a logical extent. The cursor changes to a magnifying glass with a "±" symbol. To zoom closer in, hold the pick button and move the cursor vertically toward the top of the window. To zoom farther out, hold the pick button and move the cursor vertically toward the bottom of the window. To discontinue zooming, release the pick button. To exit the *Zoom Realtime* option, press Enter, ESC, or from the shortcut menu select EXIT. In addition, you can perform other operations related to ZOOM and PAN by selecting appropriate commands from the shortcut menu.

FIGURE 3–40 *Invoking the zoom (Zoom Realtime) option from the shortcut menu*

The current drawing window is used to determine the zooming factor. If the cursor is moved by holding the pick button from the bottom of the window to the top of the window vertically, the zoom-in factor would be 200%. Conversely, when holding the pick button from the top of the window and moving vertically to the bottom of the window, the zoom-out factor would be 200%.

When you reach the zoom-out limit, the "−" symbol on the cursor disappears, indicating that you can no longer zoom out. Similarly, when you reach the zoom-in limit, the "+" symbol on the cursor disappears, indicating that you can no longer zoom in.

Zoom Window

The *Zoom Window* selection, invoked from the navigation bar (see Figure 3–41), lets you specify a smaller portion of the drawing and have that portion fill the drawing area. This step is done by specifying two diagonally opposite corners of a rectangle, similar to a selection window. The center of the area selected becomes the new display center, and the area inside the window is enlarged to fill the drawing area as completely as possible.

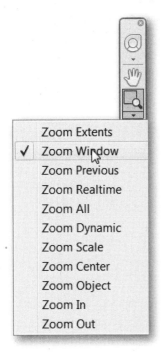

FIGURE 3–41 *Invoking the Zoom Window option from the navigation bar*

You can enter two opposite corner points to specify an area by means of entering their coordinates or using the pointing device to pick them on the screen. The following command sequence shows an example of using the Zoom Window selection (Figure 3–42).

AutoCAD displays the following prompts:

> Specify first corner: *(specify a point to define the first corner of the window, as shown in* Figure 3-42*)*
>
> Specify opposite corner: *(specify a point to define the diagonally opposite corner of the window)*

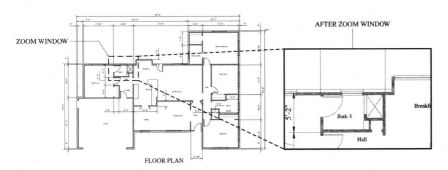

FIGURE 3–42 *Display of the drawing before and after Zoom Window selection*

Zoom All

The *Zoom All* selection invoked from the navigation bar, as shown in Figure 3–43, lets you see the entire drawing. In a plan view, it zooms to the drawing's limits or current extents, whichever is larger. If objects in the drawing extend outside the drawing limits, the display shows all objects in the drawing.

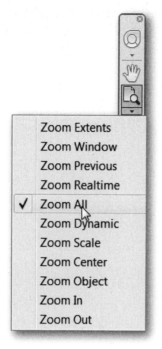

FIGURE 3–43 *Invoking the Zoom All option from the navigation bar*

Zoom Extents and Zoom Object

The *Zoom Extents* selection, invoked from the Zoom menu, as shown in Figure 3–44, lets you see the entire drawing on screen. Unlike the Zoom All method, the Extents method uses only the drawing extents of all objects, whether on visible layers or not, and not the drawing limits.

Figure 3–45 illustrates the difference between the *Zoom All* selection and the *Zoom Extents* selection.

FIGURE 3–44 *Invoking the Zoom Extents option from the Zoom menu of the Navigate panel*

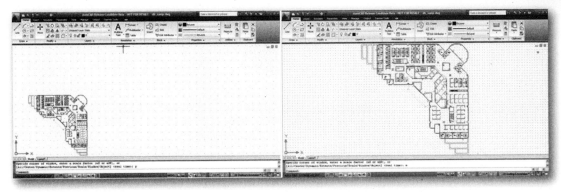

FIGURE 3–45 *Display of the drawing after Zoom All selection and Zoom Extents selection*

The *Zoom Object* selection from the Zoom menu (Figure 3–46) zooms to the largest possible display of only selected object(s) instead of all objects.

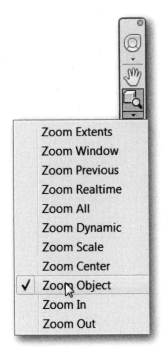

FIGURE 3–46 *Invoking the Zoom Object option from the Zoom menu of the Navigate panel*

Zoom Scale

The *Zoom Scale* selection, invoked from the Zoom menu, as shown in Figure 3–47, lets you enter a display scale or magnification factor. The scale factor, a numeric value that is not expressed in units of measure, is applied to the area covered by the drawing limits. For example, if you enter a scale factor of three, each object appears three times as large as in the *Zoom All* view. A scale factor of one displays the entire drawing, or the full view, which is defined by the established limits. If you enter a value less than one, AutoCAD decreases the magnification of the full view. For example, if you enter a scale factor of 0.5, each object appears half its size in the full view while the viewing area is twice the size in horizontal and vertical dimensions. When you use this option, the object in the center of the screen remains centered.

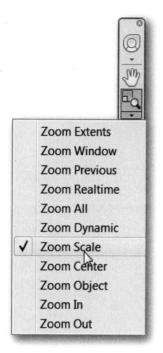

FIGURE 3–47 *Invoking the Zoom Scale option from the Zoom menu of the Navigate panel*

If you enter a number followed by X, the scale is determined relative to the current view. For instance, entering 2X causes each object to be displayed two times its current size on the screen. The scale factor XP option, related to the layout of the drawing, is explained in Chapter 7. Figure 3–48 shows the difference between a full view and a 0.5 zoom.

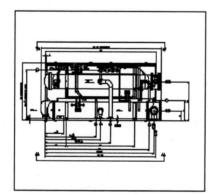
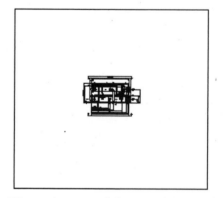

FIGURE 3–48 *Drawing display showing the difference between a full view and a 0.5 zoom display*

Zoom Center

The *Zoom Center* selection, invoked from the Zoom menu, as shown in Figure 3–49, lets you select a new view by specifying its center point and the magnification value or height of the view in current units. A smaller value for the height increases the magnification; a larger value decreases the magnification. Figure 3–50 shows a drawing before and after the *Zoom Center* selection.

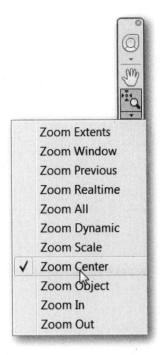

Zoom Extents
Zoom Window
Zoom Previous
Zoom Realtime
Zoom All
Zoom Dynamic
Zoom Scale
✓ Zoom Center
Zoom Object
Zoom In
Zoom Out

FIGURE 3–49 *Invoking the Zoom Center option from the Zoom menu of the Navigate panel*

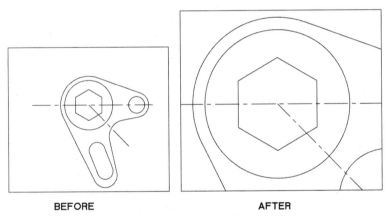

BEFORE   AFTER

FIGURE 3–50 *Display before and after Zoom Center selection*

In addition to providing coordinates for a center point, you can also specify the center point by specifying a point on the view window. The height can also be specified in terms of the current view height by specifying the magnification value followed by an X. A response of 3X will make the new view height three times as large as the current height. The fact that the coordinates of the specified center of the new view are at the center of the circle is coincidental. The coordinates you specify can be located anywhere on the drawing.

Zoom Dynamic

The AutoCAD *Zoom Dynamic* selection, invoked from the Zoom menu, as shown in Figure 3–51, provides a quick and easy method to move to another view of the drawing. With *Zoom Dynamic*, you can see the entire drawing and then simply select the location and size of the next view by means of cursor manipulations. Using *Zoom Dynamic* is one means by which you can visually select a new display area that is not entirely within the current display. The current viewport is then transformed into a selecting view that displays the drawing extents, as shown in the example in Figure 3–52.

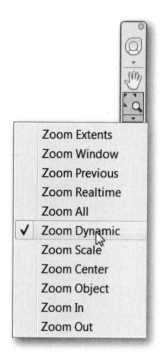

FIGURE 3–51 *Invoking the Zoom Dynamic option from the Zoom menu of the Navigate panel*

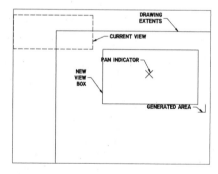

FIGURE 3–52 *Using the ZOOM DYNAMIC command to display the drawing extents*

When the selecting view is displayed, you see the drawing extents marked by a white or black box, and the current display is marked by a blue or magenta dotted box. A new view box, the same size as the current display, appears. Its location is controlled by the movement of the pointing device. Its size is controlled by a combination of the pick button and cursor movement. When the new view box has an X in the center, the box pans around the drawing in response to cursor movement. After you press the pick button on the pointing device, the X disappears, and an arrow appears at the right edge of the box. The new view box is now in Zoom mode. While the arrow is in the box, moving the cursor to the left decreases the box size; moving the cursor to the right increases the size.

When the desired size has been chosen, press the pick button again to pan, or press [Enter] to accept the view defined by the location and size of the new view box. Pressing ESC cancels the Zoom Dynamic and returns you to the current view.

Zoom Previous

The *Zoom Previous* selection, invoked from the Zoom menu, as shown in Figure 3–53, displays the last displayed view. While editing or creating a drawing, you may wish to zoom into a small area, zoom back out to view the larger area, and zoom into another small area. To do this, AutoCAD saves the coordinates of the current view whenever it is being changed by any of the zoom options or other view commands. You can return to the previous view by selecting the Previous option, which can restore the previous 10 views.

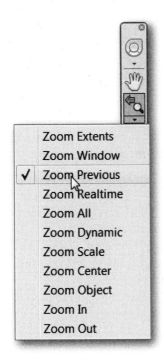

FIGURE 3–53 *Invoking the Zoom Previous option from the Zoom menu of the Navigate panel*

Pan Command

AutoCAD lets you view a different portion of the drawing in the current view without changing the magnification. You can move your viewing area to see details that are currently off the screen. Imagine that you are looking at your drawing through the display window and that you can slide the drawing left, right, up, and down without moving the window.

The *Pan Realtime* selection, invoked from the Navigate panel, located in the View tab, as shown in Figure 3–54, lets you pan interactively to the logical extent, the edge of the drawing space. Once you invoke the command, the cursor shape changes to a "hand" shape. To pan, hold the pick button on your pointing device to lock the cursor to its current location relative to your drawing, and move the cursor in any direction. Graphics within the window are moved in the same direction as the cursor. To discontinue the panning, release the pick button.

When you reach the logical extent of your drawing, a line-bar is displayed adjacent to the hand cursor. The line-bar is displayed at the top, bottom, left, or right side of the drawing, depending upon whether the logical extent is at the top, bottom, or side of the drawing.

FIGURE 3–54 *Invoking the* PAN REALTIME *command from the Navigate panel*

To exit *Pan Realtime*, press ESC or [Enter]. You can also exit by selecting *Exit* from the shortcut menu that is displayed when you right-click with your pointing device. In addition, you can perform other operations related to zoom and pan by selecting the appropriate commands from the shortcut menu. You can also access the PAN command from the Navigation Bar.

SteeringWheels

SteeringWheels are tracking menus that are divided into different sections called wedges. Each wedge represents one navigation tool. These allow you to zoom, pan, or manipulate the current view of a model in different ways. In most cases the command in effect is used by moving the cursor while the mouse button is depressed. For example, pressing and holding the mouse button over the Zoom wedge allows you to move the cursor up or down to increase or decrease the zoom factor respectively.

SteeringWheels combine the common navigation tools into a single interface. To invoke the NAVSWHEEL command, select SteeringWheels from the shortcut menu (displayed when you are not in any command), or choose **SteeringWheels** from the status bar. SteeringWheels can also be invoked from shortcut menus while in various drawing modes. The SteeringWheels navigator is displayed and when a section is clicked, the corresponding navigational feature is activated (see Figure 3–55).

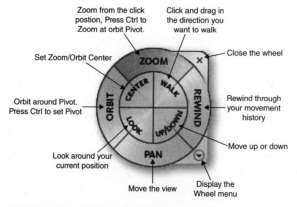

FIGURE 3–55 *Full SteeringWheels navigator*

You can activate one of the available navigation tools by either clicking over one of the wedges on the wheel or by clicking and holding down the button on the pointing device. Once the button is held down, dragging over the drawing window causes the current view to change. Releasing the button returns you to the wheel.

The Steering Wheel can be configured in several ways including **Full Navigation, View Object Wheel, Tour Building Wheel, 2D Navigation Wheel, Mini Full Navigation Wheel, Mini View Object Wheel,** and **Mini Tour Building Wheel** (see Figure 3–56).

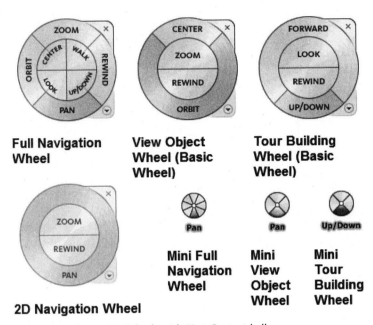

Full Navigation Wheel

View Object Wheel (Basic Wheel)

Tour Building Wheel (Basic Wheel)

2D Navigation Wheel

Mini Full Navigation Wheel

Mini View Object Wheel

Mini Tour Building Wheel

FIGURE 3–56 *SteeringWheels with First Contact balloon*

Tooltips are displayed for each wedge and button on a wheel as the cursor hovers over them. The tooltips appear below the wheel and identify what action will be performed if the wedge or button is clicked.

You can control the appearance of the wheels by changing the current mode, or by adjusting the size and opacity. Wheels are available in two different modes: big and mini. To change the current mode of a wheel, right-click the wheel and select a different mode.

In addition to changing the current mode, you can adjust the opacity and size for the wheels. The size of a wheel controls how large or small the wedges and labels appear on the wheel; the opacity level controls the visibility of the objects in the model behind the wheel. The settings used to control the appearance of the wheels are in the SteeringWheels Settings dialog box.

Various options are available for using the SteeringWheels and are accessible by right-clicking while the navigator is active (see Figure 3–57).

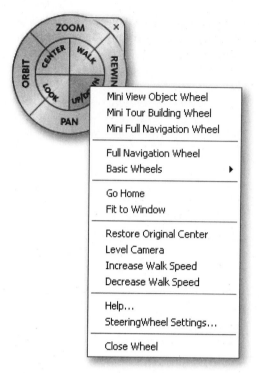

FIGURE 3–57 *SteeringWheels shortcut menu*

SteeringWheels Settings displays the dialog box where you can adjust the preferences for the wheels. Choosing Close Wheel from the shortcut menu or by pressing the ESC key closes the wheel.

Aerial View

The DSVIEWER command is used to activate the Aerial View window, which provides a quick method of visually panning and zooming. By default, AutoCAD displays the Aerial View window, with the entire drawing displayed in the window, as shown in Figure 3–58. You can select any portion of the drawing in the Aerial View window by visually panning and zooming. This behavior is similar to the Zoom Dynamic option; AutoCAD displays the selected portion in the view window, or current viewport.

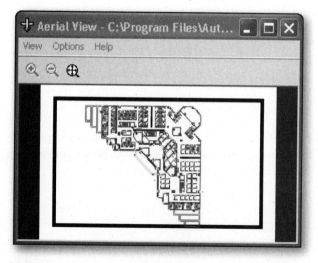

FIGURE 3–58 *The Aerial View window*

Invoke the DSVIEWER command by selecting Aerial View from the View menu located in the Menu Bar, and AutoCAD displays the Aerial View window. Two option menus are provided to pan and zoom visually.

The **View** menu in the Aerial View window has three options. The zoom-in option causes the view to appear closer, enlarging the details of objects but displaying a smaller area. The zoom-out option causes the view to appear farther away, decreasing the size of objects but displaying a larger area. The global option causes the entire drawing to be viewable in the Aerial View window. You can also select the three options from the toolbar provided in the Aerial View window.

The **Options** menu in the Aerial View window has three options. **Auto Viewport** causes the active viewport to be displayed in model space. **Dynamic Update** toggles whether or not the view is updated in response to editing. The **Realtime Zoom** controls whether or not the AutoCAD window updates in real time when you zoom using the Aerial View.

Controlling Display with Intellimouse

AutoCAD also allows you to control the display of the drawing with the small wheel provided with a two-button mouse called an IntelliMouse. You can use the wheel to zoom and pan in your drawing at any time without using any AutoCAD commands. By rotating the wheel forward, you can zoom in; by rotating it backward, you can zoom out. When double-clicking the wheel button, AutoCAD displays the drawing to the extent of the view window. To pan the display of the drawing, press the wheel button and drag the mouse. By default, each increment in the wheel rotation changes the zoom level by 10%. The ZOOMFACTOR system variable controls the incremental change, whether forward or backward. The higher the setting, the smaller the change.

AutoCAD also allows you to display the Object Snap shortcut menu when you click the wheel button. To do so, set the MBUTTONPAN system variable to 0. By default, the MBUTTONPAN is set to 1.

Redraw Command

The REDRAW command is used to refresh the on-screen image. You can use this command whenever you see an incomplete image of your drawing. You can use the REDRAW command to remove the blip marks—temporary markers created when in blipmode—from the screen. A redraw is considered a screen refresh as opposed to a database regeneration. You can invoke the REDRAW command from the View menu (accessed from the Menu Bar). AutoCAD provides no options for the REDRAW command.

Regen Command

The REGEN command is used to regenerate the drawing's on-screen data. In general, you should use the REGEN command if the image presented by REDRAW does not correctly reflect your drawing. REGEN goes through the drawing's entire database and projects the most up-to-date information on the screen; this command will give you the most accurate image possible. Because of the manner in which it functions, a REGEN takes significantly longer than a REDRAW. There are certain AutoCAD commands for which REGEN takes place automatically unless REGENAUTO is set to OFF.

You can invoke the REGEN command from the View menu (accessed from the Menu Bar). AutoCAD provides no options for the REGEN command.

SETTING MULTIPLE VIEWPORTS

The ability to divide the display into two or more separate viewports is one of the most useful features of AutoCAD. Multiple viewports divide your drawing screen into regions, permitting several different areas for drawing instead of just one. It is like having multiple zoom-lens cameras, with each camera being used to look at a different portion of the drawing.

Each viewport maintains a display of the current drawing independent of the other viewports. You can simultaneously display a viewport showing the entire drawing and another viewport showing a close-up of part of the drawing in greater detail. A view in one viewport can be from a different point of view than those in other viewports. You can begin drawing or modifying an object in one viewport and complete it in another viewport. For example, three viewports could be used in a 2D drawing, two of them to zoom in on two separate parts of the drawing, showing two widely separated features in great detail on the screen simultaneously, and the third to show the entire drawing, as shown in Figure 3–59. In a 3D drawing, four viewports could be used to display simultaneously four views of a wireframe model: top, front, right side, and isometric, as shown in Figure 3–60.

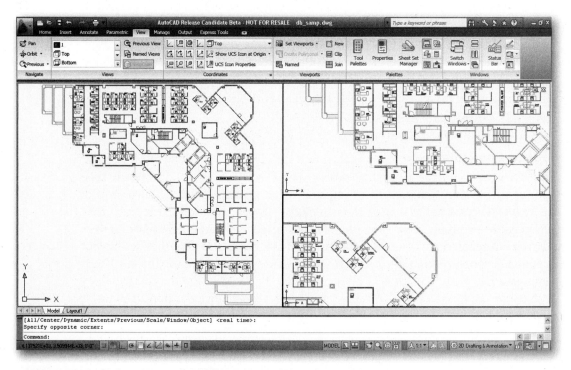

FIGURE 3–59 *Multiple viewports show different parts of the same 2D drawing*

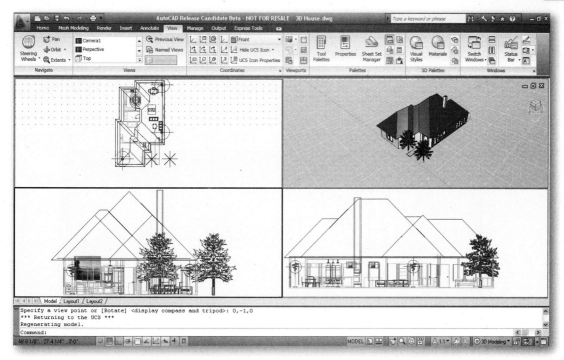

FIGURE 3–60 *Using viewports to show four views simultaneously for a 3D wireframe model*

AutoCAD allows you to divide the graphics area of your display screen into multiple, nonoverlapping, or tiled viewports, as shown in Figure 3–59 and Figure 3–60, when you are in model space and the system variable TILEMODE is set to 1. The maximum number of active tiled viewports that you can have is set by the system variable MAXACTVP, and the default is 64. In addition, you can also create multiple overlapping, or floating, viewports when the system variable TILEMODE is set to 0. For a detailed explanation on how to create floating viewports, refer to Chapter 8.

You can work in only one viewport at a time. It is considered the current viewport. A viewport is set current by moving the cursor into it with your pointing device and then pressing the pick button. You can even switch viewports during a command except during some of the display commands. For example, to draw a line using two viewports, you must start the line in the current viewport, make another viewport current by clicking in it, and then specify the endpoint of the line in the second viewport. When a viewport is current, its border will be thicker than the other viewport borders. The cursor is active for specifying points or selecting objects only in the current viewport; when you move your pointing device outside the current viewport, the cursor appears as an arrow pointer.

Display commands such as ZOOM and PAN and drawing tools such as GRID, SNAP, ORTHO, and UCS modes are set independently in each viewport. The most important thing to remember is that the images shown in multiple viewports are all of the same drawing. An object added to or modified in one viewport will affect its image in the other viewports. You are not making copies of your drawing but rather viewing its image in different viewports. When you are working in tiled viewports, the visibility of the layers is controlled globally in all the viewports. If you turn off a layer, AutoCAD turns it off in all viewports.

Creating Tiled Viewports

AutoCAD allows you to display tiled viewports in various configurations. Display of the viewports depends on the number and size of the views you need to see. By default, whenever you start a drawing, AutoCAD displays a single viewport that fills the entire drawing area. To create multiple viewports, at the on-screen prompt, type **vports**, and AutoCAD displays a Viewports dialog box similar to Figure 3–61.

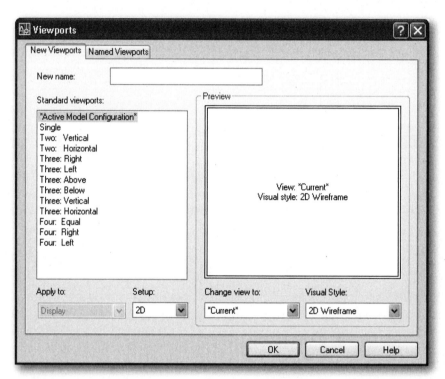

FIGURE 3–61 *Viewports dialog box*

Select the New Viewports tab and choose the name of the configuration you want to use from the **Standard viewports** list. AutoCAD displays the corresponding configuration in the **Preview** window. If necessary, you can save the selected configuration by providing a name in the **New name** box. Select the Display option from the **Apply to** menu, and select 2D from the **Setup** menu for 2D viewport setup or select 3D for 3D viewport setup. Choose **OK** to create the selected viewport configuration.

You can also create various viewport configurations from the Viewport Configurations list located in the Viewports panel on the View tab (Figure 3–62).

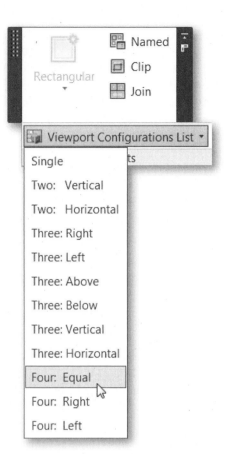

FIGURE 3–62 *Listing of the available viewport configurations from the Viewports Panel*

If you need additional viewports other than the standard configurations, you can subdivide a selected viewport. Open the Viewports dialog box, select one of the available configuration you want to use from the **Standard Viewports** list to divide the selected viewport. **Set Apply to**: to Current Viewport. The **Visual Style** drop-down list lets you choose from Current, 2D Wireframe, 3D Hidden, 3D Wireframe, Conceptual, Shaded, Shaded with edges, Shades of Gray, Sketch, X-Ray, and Realistic visual styles.

The **Named Viewports** tab lists all saved viewport configurations. To save the current viewport configuration, enter a name in the **New name** field, and select **OK**. You can restore one of the saved viewport configurations at any time.

CREATING AND MODIFYING LAYERS

AutoCAD offers a means of grouping objects in layers in a manner similar to the manual drafter's process of separating complex drawings into simpler ones on individual transparent sheets, superimposed in a single stack. In AutoCAD, you can draw only on the current layer. However, AutoCAD permits you to transfer selected objects from one layer to another, neither of which needs to be the current layer, with commands called CHANGE, CHPROP, PROPERTIES, and several others. Let's see the manual drafter try that!

A common application of the layer feature is to use one layer for construction, another one for drawing dimensions, text objects, and so forth. You can create geometric constructions with objects such as lines, circles, and arcs. These generate intersections, endpoints, centers, points of tangency, midpoints, and other useful data that might take the manual drafter considerable time to calculate with a calculator or to hand-measure on the board. From these, you can create other objects using intersections or other data generated from the layout. The layout layer can then be turned off, making it invisible, or set not to plot. The layer is not lost but can be recalled, or set to ON, for later viewing as required. The same drawing limits, coordinate system, and zoom factors apply to all layers in a drawing.

To draw an object on a particular layer, first make sure that the layer is set as the "current layer." There is one, and only one, current layer. Whatever you draw will be placed on the current layer. To draw an object on a particular layer, that layer must first have been created; if it is not the current layer, you must make it the current layer.

You can always move, copy, or rotate any object, whether or not it is on the current layer. When you copy an object that is not on the current layer, the copy is placed on the layer or the original object. This is also true with the mirror or array of an object or group of objects.

A layer can be visible (ON) or invisible (OFF). Only visible layers are displayed or plotted. If necessary, AutoCAD allows you to set a visible layer not to plot. The layer(s) that are visible and set not to plot will not be plotted. Invisible layers are still part of the drawing, though they are not displayed or plotted. You can turn layers ON and OFF at will, in any combination. It is possible to turn the current layer OFF. If this happens and you draw an object, it will not appear on the screen; it will be placed on the current layer and will appear on the screen when that layer is set to ON, provided that you are viewing the area in which the object was drawn. This is not a common occurrence, but it can cause concern to both the novice and the more experienced operator who has not faced the problem before. Do not turn the current layer OFF; the results can be very confusing. When the TILEMODE system variable is set to OFF, you can make specified layers visible only in certain viewports.

Each layer in a drawing has an associated name, color, lineweight, and linetype. The name of a layer may be up to 255 characters long. It may contain letters, digits, and these special characters: dollar ($), hyphen (−), underscore (_), and spaces. Always give the layer a descriptive name appropriate to your application such as "Floor Plan" or "Plumbing." The first several characters of the current layer's name are displayed in the layer list box located on the Layers panel located on the Home tab (Figure 3–63). You can change the name of a layer any time you wish, and you can delete unused layers except Layer 0.

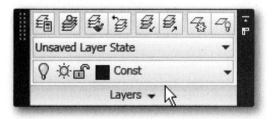

FIGURE 3–63 *The current layer name is displayed in the list box located on the Layers panel*

AutoCAD allows you to assign a color to a layer in a drawing. Later in this section, you will learn how to assign colors to layers from the range and types of colors available. If necessary, you can assign the same color to more than one layer.

Similar to assigning color, you can also assign a specific lineweight to a layer. Lineweights add width to your objects, both on the screen and on paper. Using lineweights, you can create heavy and thin lines to show varying object thicknesses in details. For example, by assigning varying lineweights to different layers, you can differentiate between new, existing, and demolition construction.

AutoCAD allows you to assign a lineweight to a specific layer or an object in either inches or millimeters, with millimeters being the default. If necessary, you can change the default setting by invoking the LINEWEIGHT command. A lineweight value of 0 is displayed as 1 pixel in model space and plots at the thinnest lineweight available on the specified plotting device. Appendix G lists all the available lineweights as well as associated industry standards.

Lineweights are displayed differently in model space than in a paper space layout. Refer to Chapter 8 for a detailed explanation on layouts. In model space, lineweights are displayed in relation to pixels. In a paper space layout, lineweights are displayed in the exact plotting width. You can recognize that an object has a thick or thin lineweight in model space, but the lineweight does not represent an object's real-world width. A lineweight of 0 will always be displayed on screen with the minimum display width of 1 pixel. All other lineweights are displayed using a pixel width in proportion to its real-world unit value. A lineweight displayed in model space does not change with the zoom factor. For example, a lineweight value that is represented by a width of 4 pixels is always displayed using 4 pixels regardless of how close you zoom in to your drawing.

If necessary, you can change the display scale of object lineweights in model space to appear thicker or thinner. Changing the display scale does not affect the lineweight plotting value. However, AutoCAD regeneration time increases with lineweights that are represented as more than 1 pixel wide. By default, all lineweight values that are less than or equal to 0.01 in. or 0.25 mm are displayed 1 pixel wide, and do not slow down performance in AutoCAD. If you want to optimize AutoCAD performance, when working in the Model Space, set the lineweight display scale to the minimum value. The LINEWEIGHT command allows you to change the display scale of lineweights.

In model space, AutoCAD allows you to turn ON and OFF the display of lineweights by toggling **LWT** on the status bar. With **LWT** set to OFF, AutoCAD displays all the objects with a lineweight of 0 and reduces regeneration time. When exporting drawings to other applications or cutting and copying objects to the clipboard, objects retain their lineweight information.

A linetype is a repeating pattern of dashes, dots, and blank spaces. AutoCAD adds the capability of creating custom linetypes. The assigned linetype is normally used to draw all objects on the layer unless you set the current linetype to another linetype instead of ByLayer.

The following linetypes are provided in AutoCAD in a library file called *acad.lin*:

Border	Dashdot	Dot
Center	Dashed	Hidden
Continuous	Divide	Phantom

See appendix G for examples of each of these linetypes. Linetypes provide another means of conveying visual information. You can assign the same linetype to any number of layers. In some drafting disciplines, conventions have been established giving specific meanings to particular dash-dot patterns. If a line is too short to hold even one dash-dot sequence, AutoCAD draws a continuous line between the endpoints. When you are working on large drawings, you may not see the gap between dash-dot patterns in a linetype unless the scaling for the linetype is set for a large value. This can be done by means of the LTSCALE command. This command is discussed in more detail later in this chapter.

Every drawing will have a Layer 0. By default, Layer 0 is set to ON and assigned the color white, the default lineweight, and the linetype of continuous. Layer 0 cannot be renamed or deleted.

If you need additional layers, you must create them. By default, each new layer is assigned the properties of the currently selected layer in the layer table. If necessary, you can always reassign the color, lineweight, and linetype of a newly created layer.

Creating and Managing Layers with the Layer Properties Manager

The Layer Properties Manager can be used to set up and control layers. Choose Layer Properties from the Layers panel on the Home tab (Figure 3–64) to open the Layer Properties Manager (Figure 3–65).

FIGURE 3–64 *Selecting the Layer Properties on the Layers panel*

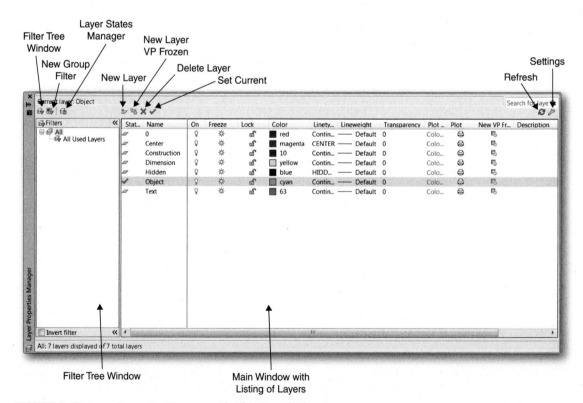

FIGURE 3–65 *Layer Properties Manager with Filter Tree displayed*

The main window lists each layer, along with the status of its associated properties. The Filter Tree window, as shown in Figure 3–63, lists the available filters. A layer filter limits the display of layers in the Layer Properties Manager and in the Layer control on the Layers toolbar. In a large drawing, you can use layer filters to display only the layers on which you need to work.

You can keep the Layer Properties Manager open while making multiple layer property changes. You can select the layers you want to be visible without reopening and closing the dialog box. All the changes made in the dialog box will be reflected in the drawing window immediately.

Creating New Layers

To create a new layer, choose **New Layer** from the Layer Properties Manager dialog box. AutoCAD then creates a new layer by assigning the name "Layer1," and the new layer inherits the properties of the currently selected layer in the layer list: color, ON or OFF state, and so on. When first listed in the layer list box, the name "Layer1" is highlighted and ready to be edited. Enter the desired name for the new layer. If you accept the name Layer1, you can still change it later. To rename a layer, click anywhere on the line where the layer is listed, and the whole line is highlighted. Click again on the layer name, and the entire name is highlighted—you can type a new name and press Enter. If you click again on the name, the highlighting is removed, and you can then change part of the name.

> The layer name cannot contain wildcard characters such as * and ?, and you cannot duplicate existing layer names.

NOTE

Making a Layer Current

To make a layer current, choose the layer name from the layer list box in the Layer Properties Manager dialog box. Then choose **Set Current (green checkmark)** located at the top of the dialog box, or double-click the icon corresponding to the layer name located under the **Status** column, in the first column from the left, or choose *Set Current* from the shortcut menu.

Visibility of Layers

When you turn a layer OFF, the objects on that layer are not displayed in the drawing area, and they are not plotted. The objects are still in the drawing, but they are not visible on the screen. They are still calculated during regeneration of the drawing, even though they are invisible.

To change the setting for the visibility of selected layer(s), select the icon corresponding to the layer name located under the **On** column (third column from the left) in the Layer Properties Manager dialog box. The icon is a toggle for setting layers ON or OFF. It is possible to turn the current layer OFF. If you draw an object while the current layer is turned OFF, it will not appear on the screen; it will be placed on the current layer and will appear on the screen when that layer is set to ON.

Freezing and Thawing Layers

In addition to turning the layers OFF, you can freeze layers. The layers that are frozen will not be visible in the view window, nor will they be plotted. In this respect, frozen layers are similar to layers that are OFF. However, layers that are simply turned OFF still go through screen regeneration each time the system regenerates your drawing, whereas the layers that are frozen are not affected during screen regeneration. If you want to see the frozen layer later, you thaw it, and automatic regeneration of the drawing area takes place.

To change the setting for the visibility of a layer by freezing or thawing, select the icon corresponding to the layer name located under the **Freeze** column (fourth column from the left) in the Layer Properties Manager dialog box. The icon is a toggle for the freezing or thawing of layers. You cannot freeze the current layer.

Locking and Unlocking Layers

Objects on locked layers are visible in the view window but cannot be modified by means of the modifying commands. However, it is still possible to draw on a locked layer by making it the current layer and using any of the inquiry commands or Object Snap modes on them. The objects on the locked layer(s) are dimmed to differentiate them from the objects on the unlocked layer.

To lock or unlock a layer, select the icon corresponding to the layer name located under the **Lock** column (fifth column from the left) in the Layer Properties Manager dialog box. The icon is a toggle for locking or unlocking layers.

Changing the Color of Layers

By default, AutoCAD assigns the color of the currently selected layer in the layer list to the newly created layer. To change the assigned color, choose the icon under the **Color** column (sixth column from the left) corresponding to the layer name in the Layer Properties Manager dialog box. AutoCAD displays the Select Color dialog box (Figure 3–66) with the **Index Color** tab displayed, which allows you to change the color of the selected layer(s). You can select one of 256 colors. Use the cursor to select the color you want, or enter its name or number in the **Color** box. Choose **OK** to accept the color selection. Two additional tabs are available in the Select Color dialog box: **True Color** (Figure 3–67) and **Color Books** (Figure 3–68). The **True Color** tab specifies color settings using true colors, or 24-bit colors, with the Hue, Saturation, and Luminance (HSL) color model or the Red, Green, and Blue (RGB) color model. Over 16 million colors are available when using true-color functionality. The **Color Books** tab specifies colors using third-party color books, such as PANTONE®, or user-defined color books. Once a color book is selected, the **Color Books** tab displays the name of the selected color book. **True Color** and **Color Books** make it easier to match colors in your drawing with colors of actual materials.

FIGURE 3–66 *Select Color dialog box with the Index Color tab displayed*

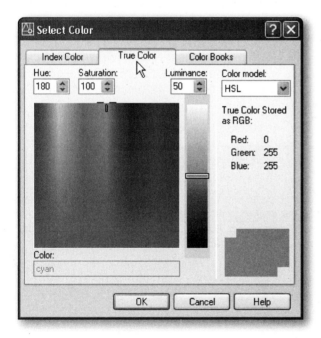

FIGURE 3–67 *Select Color dialog box with the True Color tab displayed*

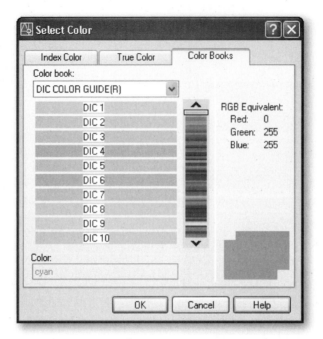

FIGURE 3–68 *Select Color dialog box with the Color Books tab displayed*

Changing the Linetype of Layers

By default, AutoCAD assigns the linetype of the currently selected layer in the layer list to the newly created layer. To change the assigned linetype, choose the linetype name corresponding to the layer name located under the **Linetype** column (seventh column from the left) in the Layer Properties Manager dialog box. AutoCAD displays the Select Linetype dialog box, as shown in Figure 3–69, which allows you to change the linetype of the selected layer(s). Select the appropriate linetype from the list box, and choose **OK** to accept the linetype selection.

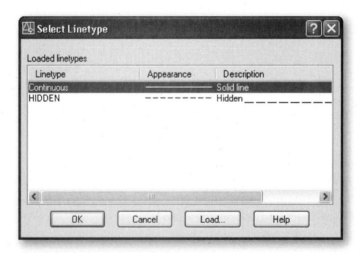

FIGURE 3–69 *Select Linetype dialog box*

Loading Linetypes. In the Select Linetype dialog box, AutoCAD lists only the linetypes that are loaded in the current drawing. To load additional linetypes in the current drawing, choose **Load**. AutoCAD displays the Load or Reload Linetype dialog box similar to Figure 3–70. AutoCAD lists the available linetypes from *acad.lin*, the default linetype file. Select all the linetypes that need to be loaded, and choose **OK** to load them into the current drawing. If necessary, you can change the default linetype file *acad.lin* to another file by choosing **File** and selecting the desired linetype file.

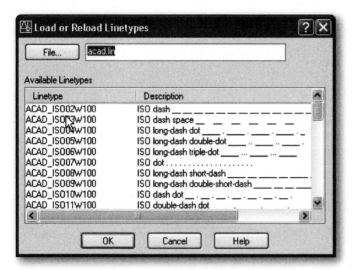

FIGURE 3–70 *Load or Reload Linetypes dialog box*

Setting Linetype Scale Factor. The linetype scale factor allows you to change the relative lengths of dashes and spaces between dashes and dots linetypes per drawing unit. The definition of the linetype instructs AutoCAD on how many units long to make dashes and the spaces between dashes and dots. As long as the linetype scale is set to 1.0, the displayed length of dashes and spaces coincides with the definition of the linetype.

The LTSCALE command allows you to set the linetype scale factor. Invoke the LTSCALE command by typing **ltscale** from the on-screen prompt, and AutoCAD displays the following prompt:

> Enter new linetype scale factor <current>: *(specify the scale factor)*

Changing the linetype scale affects all linetypes in the drawing. If you want dashes that have been defined as 0.5 units long in the dashed linetype to be displayed as 10 units long, set the linetype scale factor to 20. This action also makes the dashes that were defined as 1.25 units long in the center linetype display as 25 units long, and the short dashes defined as 0.25 units long display as 5 units long. Note that the 1.25-unit-long dash in the center linetype is 2.5 times longer than the 0.5-unit-long dash in the dashed linetype. This ratio will always remain the same, no matter what the setting of LTSCALE. If you wish to have some other ratio of dash and space lengths between different linetypes, you must change the definition of one of the linetypes in the *acad.lin* file.

Remember that linetypes are for visual effect. The actual lengths of dashes and spaces are bound more to how they should look on the final plotted sheet than to distances or sizes of any objects on the drawing. An object plotted full size can probably use an LTSCALE setting of 1.0. A 50'-long object plotted on an 18 × 24 sheet might be plotted at a 1/4 = 1'-0 scale factor. This would equate to 1 = 48. An LTSCALE setting of 48 would make dashes and spaces plot to the same lengths as the full-sized plot with a setting of 1.0. Changing the linetype scale factor causes the drawing to regenerate.

Changing the Lineweight of Layers

By default, AutoCAD assigns the lineweight of the currently selected layer in the layer list to the newly created layer. To change the assigned lineweight, choose the lineweight name corresponding to the layer name located under the **Lineweight** column (eighth column from the left) in the Layer Properties Manager dialog box. AutoCAD displays the Lineweight dialog box, similar to Figure 3–71, which allows you to change the lineweight of the selected layer(s). Select the appropriate lineweight from the list box, and choose **OK** to accept the lineweight selection.

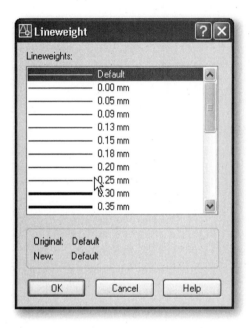

FIGURE 3–71 *Lineweight dialog box*

Setting Transparency Value

The transparency value controls the visibility of all objects on the selected layer. When transparency is applied to individual objects, the objects' transparency property overrides the transparency setting of the layer. To change the assigned transparency value, click the transparency value corresponding to the layer name located under the **Transparency** column (ninth column from the left) in the Layer Properties Manager dialog box. AutoCAD displays the Transparency dialog box, which allows you to change the transparency value of the selected layer(s). Select the appropriate transparency from the list box, and choose **OK** to accept the selection. By default the transparency value is set to 0. The value can be set between 0 and 90.

Assigning Plot Styles to Layers

Plot styles are a collection of property settings, such as color, linetype, and lineweight, that can be assigned to a layer or to individual objects. These property settings are contained in a named plot style table. For a detailed explanation on creating plot styles, refer to Chapter 8. When applied, the plot style can affect the appearance of the plotted drawing. By default, AutoCAD assigns the plot style Normal to a newly created layer if the default plot style is set to Named plot style. To change the assigned plot style, choose the plot style name corresponding to the layer name located under the **Plot Style** column (tenth column from the left) in the Layer Properties Manager dialog box. AutoCAD displays the Select Plot Style dialog box, which allows you to change the Plot Style of the selected layer(s). Select the appropriate plot style from the list box, and choose **OK** to accept the plot style selection. A layer that is assigned a Normal plot style assumes the properties that have already been assigned to that layer. You can create new plot style, name them, and assign them to individual layers. You cannot change the plot style if the default plot style is set to Color Dependent plot style.

Setting Plot Status for Layers

AutoCAD allows you to turn plotting ON or OFF for visible layers. For example, if a layer contains construction lines that need not be plotted, you can specify that the layer is not plotted. If you turn plotting for a layer OFF, the layer is displayed but is not plotted. You do not have to turn the layer OFF before you plot the drawing.

To change the setting for the plotting of layers, select the icon corresponding to the layer name located under the **Plot** column (tenth column from the left) in the Layer Properties Manager dialog box. The icon is a toggle for plotting or not plotting layers.

Freezing Layers in Viewports

AutoCAD allows you to freeze selected layers for new layout viewports. For example, freezing the DIMENSIONS layer in all new viewports restricts the display of dimensions on that layer in any newly created layout viewports, but it does not affect the DIMENSIONS layer in existing viewports. To change the setting, select the icon corresponding to the layer name located under the **New VP Freeze** column (eleventh column from the left) in the Layer Properties Manager dialog box. For details on the viewports and related settings, refer to Chapter 8.

Adding Description to Layers

In addition to assigning a layer name, AutoCAD allows you to add a description to individual layers. Click the appropriate layer name and add the description in the **Description** column (twelfth column from the left) in the Layer Properties Manager dialog box. You may edit a layer's description by selecting it and choosing CHANGE DESCRIPTION from the shortcut menu. The description is limited in length to 255 characters.

Managing the Main Window

The Main Window of the Layer Properties Manager dialog box can be customized to display at least 12 properties in columns; **Status, Name, On, Freeze, Lock, Color, Linetype, Lineweight, Transparency, Plot Style, Plot, New VP Freeze,** and **Description**. These properties were described earlier. Right-click in the heading of the Main Window and a shortcut menu is displayed (see Figure 3–72).

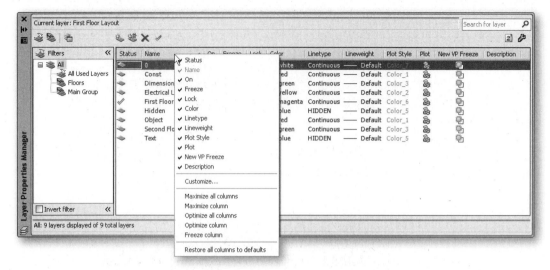

FIGURE 3–72 *Shortcut menu to manage Main Window of Layer Properties Manager dialog box*

Selecting one of the property names will toggle the corresponding column between being or not being displayed in the Main Window.

Choosing **Customize** causes the Customize Layer Columns dialog box to be displayed. From there you can toggle which properties will be displayed and move a property back and forth (Move Up/Move Down) in the list. There are options to Select All and Clear All property columns.

Choosing **Maximize all columns** causes all columns to be widened to accommodate the full title or data description.

Choosing **Maximize column** causes the selected column to be widened to accommodate the full title or data description of that column.

Choosing **Optimize all columns** causes all column widths to be changed to accommodate the data description.

Choosing **Optimize column** causes the selected column width to be changed to accommodate the data description.

Choosing **Freeze (or Unfreeze) column** freezes (or unfreezes) the selected column and any columns to the left of the selected column so that it will not scroll out of view.

Choosing **Restore All Columns to Defaults** causes all columns to be displayed in their default display and width settings.

Saving Layer Properties in Layer States

At any time during a drawing session, the collective status of all layer properties settings is known as the Layer State. This state can be saved and given a name by which it can be recalled later, thus causing every setting of the selected properties of every layer to revert to what they were when that particular Layer State was named and saved. Layer states are saved in files with the extension of *.las*.

To create a new Layer State, in the Layer Properties Manager dialog box choose **Layer States Manager.** This causes the Layer States Manager dialog box to be displayed, as shown in Figure 3–73. From this dialog box, choose **New**, which causes the New Layer State to Save dialog box to be displayed. In the **New Layer State name** and **Description** text boxes, enter the name and descriptions, respectively, and choose **OK**. This action will return you to the Layer States Manager dialog box, and the new Layer State will be saved based on settings of selected properties. Once the desired properties have been selected, choose **Save.** For example, you may wish to save and name a Layer State based only on the visibility of the layers. Select **On/Off**, and when the named Layer State is restored, all of the layers will revert to the visibility status they had when the Layer State was created. Other properties will not be changed.

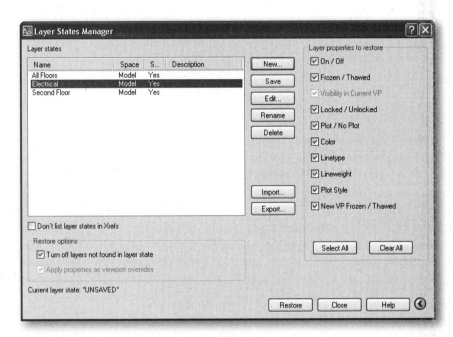

FIGURE 3–73 *The Layer States Manager dialog box*

Choose **Edit** to edit a saved Layer State, and AutoCAD displays the Edit Layer State dialog box. The procedure is the same as for creating a new Layer State.

Choose **Rename** to change the name of the selected Layer State.

To delete a named and saved Layer State, open the Layer States Manager dialog box. When the Layer State you wish to delete is highlighted in the **Layer states** list, choose **Delete**.

To import one or more Layer States into your drawing, in the Layer States Manager dialog box, choose **Import**. This action causes the Import Layer State dialog box to be displayed. This dialog box is similar to other Windows file search and management dialog boxes. From this dialog box, you can select a saved Layer State to import.

To export one or more Layer States from your drawing, in the Layer States Manager dialog box, choose **Export**. This action causes the Export Layer State dialog box to be displayed. This dialog box is similar to other Windows file search and management dialog boxes. From this dialog box, you can select a saved Layer State to export.

Layer states in xrefs, or external references, will not be listed if you choose **Don't list Layer States in Xrefs**.

If you choose **Turn Off Layers Not Found in Layer State**, when a Layer State is restored, AutoCAD turns off new layers for which settings were not saved so that the drawing looks as it did when the named Layer State was saved. Choose **Apply Properties as Viewport Overrides** so that when a Layer State is restored, it applies layer property overrides to the current viewport. This option is available when the Layer States Manager is accessed and when a layout viewport is active.

To restore a saved Layer State, select its name in the Layer States Manager dialog box, and then choose Restore.

Setting Filters for Listing Layers

AutoCAD allows you to create two kinds of filters: Layer Property and Layer Group. Layer Property filters include layers that have names or other properties in common. For example, you can define a filter that includes all layers that are blue and whose names include the letters "floor." Layer Group filters include the layers that are included into the filter when you define it, regardless of their names or properties.

The tree view in the Layer Properties Manager displays default layer filters and any named filters that you create and save in the current drawing. The icon next to a layer filter indicates the type of filter, as shown in Figure 3–74.

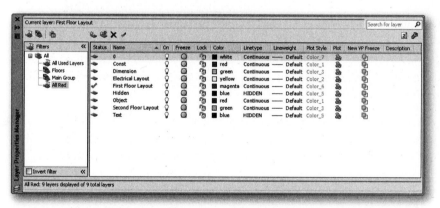

FIGURE 3–74 *Layer Properties Manager with listing of filters*

By default, AutoCAD creates three filters in a newly created drawing. The **All** filter selection displays all the layers in the current drawing. The **All Used Layers** filter selection displays all the layers on which objects in the current drawing are drawn. The **Xref** filter selection displays all the layers being referenced from other drawings if any xrefs are attached.

Once you have named and defined a layer filter, you can select it in the tree view to display the layers in the list view. You can also apply the filter to the Layers toolbar, so that the Layer control displays only the layers in the current filter. When you select a filter in the tree view and right-click, options on the shortcut menu can be used to *Delete*, *Rename*, or *Modify* filters.

Choosing **New Property Filter**, or choosing *New Properties Filter* from the shortcut menu, causes the Layer Filter Properties dialog box to be displayed, as shown in Figure 3–75.

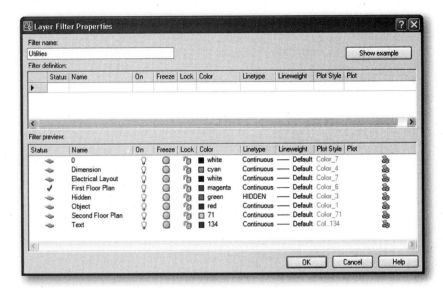

FIGURE 3–75 *Layer Filter Properties dialog box*

Specify the name of the filter in the **Filter name** box. In the **Filter definition** section, you can use one or more properties to define the filter. For example, you can define a filter that displays all layers that are either green or red and in use. To include more than one color, linetype, or lineweight, duplicate the filter on the next line and select a different setting. Choose **Show example** to display the examples from the help file of layer property filter definitions.

The **Status** column can be set to display the In Use icon or the Not In Use icon. In the **Name** column, you can use wildcard characters to filter layer names. For example, enter **floor*** to include all layers that start with "floor" in the name. Set any of the one or more corresponding properties to define the filter. AutoCAD displays the results of the filter as you define in the **Filter preview** section of the dialog box. The filter preview shows which layers will be displayed in the layer list in the Layer Properties Manager when you select this filter. To rename or delete a property filter, select the property filter in the tree view, and then from the shortcut menu, choose *Rename* or *Delete*, respectively. Choose **OK** to save the newly created or modified filter definition.

Choose **New Group Filter**, or choose *New Group Filter* from the shortcut menu, to create a new layer group filter. A new layer group filter named GROUP FILTER1 is created in the tree view. Rename the filter to an appropriate name. In the tree view, choose **All** or one of the other layer property filters to display layers in the list view. In the list view, select the layers you want to add to the filter, and drag them to the newly created Group filter name in the tree view. To rename or delete a group filter in the tree view, first select the group filter, and then from the shortcut menu, choose *Rename* or *Delete*, respectively.

Inverting a filter is another method for listing selected layers. When the **Invert filter** box on the Layer Properties Manager is OFF, or not checked, the layers that are shown in the main window are those that have all the matching characteristic(s) of the specified filter. For example, if you were to create and name a set of filters that specify the color yellow and the linetype dashed, the list will include only layers whose color is yellow and whose linetype is dashed. If you set the **Invert filter** box to ON, all layers will be listed *except* those with both the color yellow and the dashed linetype.

Choosing **Settings** causes the Layer Settings dialog box to be displayed (Figure 3–76), which allows you to control when notification occurs for new layers and if layer filters are applied to the Layers toolbar. You can also control the background color of viewport overrides in the Layer Properties Manager.

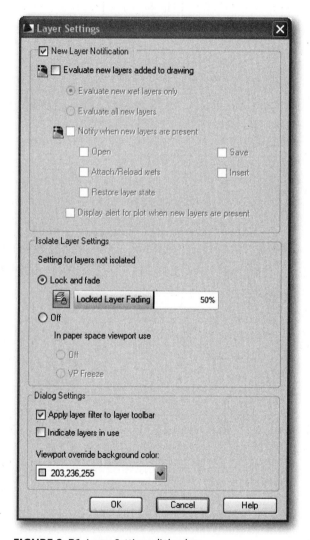

FIGURE 3–76 *Layer Settings dialog box*

The New Layer Notification Settings section controls the evaluation and notification of new layers based on LAYEREVAL setting in the *DWG* file. Choosing **Evaluate New Layers Added to Drawing** checks for new layers that have been added to the drawing. Choosing **Evaluate New Xref Layers Only** checks for new layers that have been added to attached xrefs. Choosing **Evaluate All New Layers** checks for new layers that have been added to the drawing, including new layers added to attached xrefs.

In the New Layer Notification Setting section, choosing **Notify when New Layers Are Present** turns on new layer notification. Choosing **Open** displays new layer notification when new layers are present. Choosing **Attach/Reload Xrefs** displays new layer notification when new layers are present when you are attaching or reloading xrefs. Choosing **Restore Layer State** displays new layer notification when you are restoring Layer States. Choosing **Save** displays new layer notification when new layers are present. Choosing **Insert** displays new layer notification when new layers are present.

Choosing **Display Alert for Plot When New Layers are Present** displays new layer notification when new layers are present when you use the PLOT command.

In the Isolate Layer Settings section, choosing **Lock and Fade** selects Lock and Fade as the isolation method. Choosing **Off** sets nonselected layers to Off. In paper space viewport, choosing **Off** sets nonselected layers to Off (in paper space) and choosing **VP Freeze** sets nonselected layers to Viewport Freeze (in paper space).

In the Dialog Settings section, choosing **Apply Layer Filter to Layer Toolbar** allows you to control the display of layers in the list of layers on the Layers toolbar and the Layers control panel on the ribbon by applying the current layer filter. Choosing **Indicate Layers in Use** displays icons in the list view to indicate whether layers are in use. You can improve performance in a drawing with many layers by clearing this option.

The **Viewport Override Background Color** drop-down list allows you to select a background color for viewport overrides from a list of colors.

Changing the Appearance of the Layer Properties Manager

If necessary, you can drag the widths of the column headings to see additional characters of the layer name, the full legend for each symbol and color name, or the number in the list box. You can sort the order in which layers are displayed in the list box by choosing the column headings, which causes AutoCAD to list the layers in descending order: from Z to A and then by numbers. Choosing the column heading again causes AutoCAD to list the layers in ascending order: first by numbers, then from A to Z. Choosing the **Status** column headers lists the layers by the property in the list.

Applying and Closing the Layer Properties Manager

Any changes made to Layer properties are applied instantly. You can keep the Layer Properties Manager open while making multiple layer property changes. You can even dock the dialog box (see Figure 3–77) and make necessary changes to the layer properties without reopening and closing the dialog box.

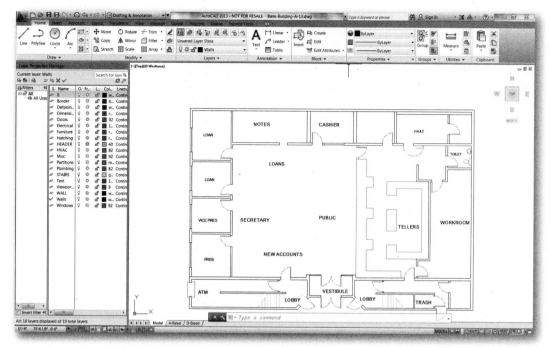

FIGURE 3–77 *Layer Properties Manager docked on the left side of the AutoCAD application window*

Changing Layer Status from the Layers Panel

You can toggle ON or OFF, freeze or thaw, lock or unlock, or plot or not plot, in addition to making a layer current in the Layer list box provided on the Layers toolbar. You can also change the color assigned to a specific layer directly from the Layer drop-down list by selecting the layer color swatch. Select the appropriate icon next to the layer name you wish to toggle, as shown in Figure 3–78.

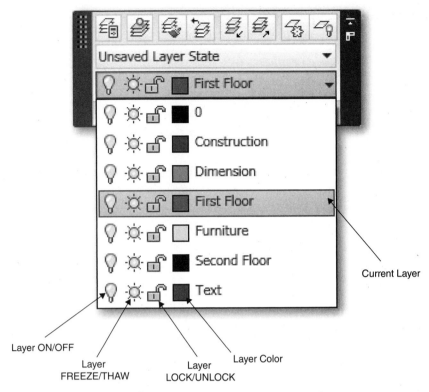

FIGURE 3–78 *Layer list box on the Layers panel on the Home tab*

Making an Object's Layer Current

AutoCAD allows you to select an object in the drawing and make its layer the current layer. To do so, choose Make Object's Layer Current (Figure 3–79) from the Layers panel. AutoCAD displays the following prompt:

> Select object whose layer will become current: *(select the object in the drawing to make its layer current)*

AutoCAD makes the selected object's layer current.

FIGURE 3–79 *Choosing the Make Object's Layer Current icon from the Layer panel*

Undoing Layer Settings

AutoCAD allows you to undo the last change or set of changes made to layer settings. To do so, choose Layer Previous (Figure 3–80) from the Layers panel.

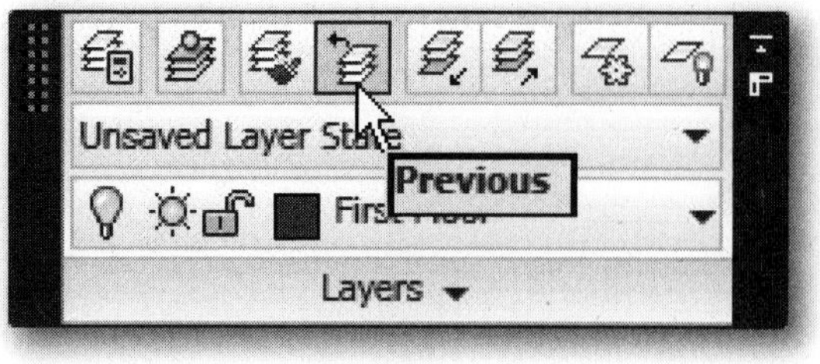

FIGURE 3–80 *Invoking the LAYERP command from the Layers panel*

This command undoes any changes you have made to layer settings such as color or linetype. The LAYERP command restores the original properties but not the original name if the layer is renamed, and it also does not restore a layer that you have deleted or purged.

Layer Tools

Following are the tools (accessed from the Layers panel) that are available to manipulate the display of the layers:

Isolate hides or locks all layers except those of the selected objects.

Unisolate restores all layers that were hidden or locked with the **Isolate** tool.

Freeze freezes the layer of selected objects.

Off turns off the layer of a selected object.

Match changes the layer of a selected object to match the destination layer.

Previous undoes the last change or set of changes made to layer settings.

Turn All Layers On turns on all layers in the drawing.

Thaw All Layers thaws all layers in the drawing.

Lock locks the layer of a selected object.

Unlock unlocks the layer of a selected object.

Change to Current Layer changes the layer of selected objects to the current layer.

Copy Objects to New Layer copies one or more objects to another layer.

Layer Walk displays objects on selected layers and hides objects on all other layers.

Merge merges selected layers into a target layer, removing the previous layers from the drawing.

Delete deletes all objects on a layer and purges the layer.

WILDCARDS AND NAMED OBJECTS

AutoCAD lets you use a variety of wildcards for specifying selected groups of named objects when responding to prompts during commands. By placing one or more of these wildcards in the string, that is, in your response, you can specify a group that includes or excludes all objects with certain combinations or patterns of characters.

The types of objects associated with a drawing that are referred to by name include blocks, layers, linetypes, text styles, dimension styles, named UCS, named views, shapes, plot style, and named viewport configurations.

The wildcard characters include the two most commonly used (* and ?) as well as eight more that come from the UNIX OS. The following table lists the wildcard characters and their uses.

Wildcard	Use
# (pound)	Matches any numeric digit
@ (at)	Matches any alphanumeric character
. (period)	Matches any character except alphabetic
* (asterisk)	Matches any string. It can be used anywhere in the search pattern: the beginning, middle, or end of the string.
? (question mark)	Matches any single character
~ (tilde)	Matches anything but the pattern
[...]	Matches any one of the characters enclosed
[~...]	Matches any character not enclosed
(hyphen)	Specifies single-character range
' (reverse quote)	Reads next character in string literally

The following table shows some examples of wildcard patterns.

Pattern	Will match or include...	But not...
ABC	Only ABC	
~ABC	Anything but ABC	
?BC	ABC through ZBC	AB, BC, ABCD, XXBC
A?C	AAC through AZC	AC, ABCD, AXXC, ABCX
AB?	ABA through ABZ	AB, ABCE, XAB
A*	Anything starting with A	XAAA
A*C	Anything starting with A and ending with C	XA, ABCDE
*AB	Anything ending with AB	ABCX, ABX
AB	AB anywhere in string	AXXXB
~*AB*	All strings without AB	AB, ABX, XAB, XABX
C	AC or BC	ABC, XAC
D	AD, BD, through KD	ABC, AKC, KD

U, UNDO, AND REDO COMMANDS

The UNDO command undoes the effects of the previous command or group of commands, depending on the option employed. The U command reverses the most recent operation, and the REDO command is a one-time reversal of the effects of the previous U and UNDO commands.

U Command

The U command undoes, or reverses, the effects of the previous command. Pressing Enter after using the U command undoes the next-previous command. It continues stepping back with each repetition until it reaches the state of the drawing at the beginning of the current editing session.

When an operation cannot be undone, AutoCAD performs no action. An operation external to the current drawing, such as plotting or writing to a file, cannot be undone. You can invoke the UNDO command by typing **u** at the on-screen prompt.

AutoCAD reverses the most recent operation. For example, if the previous command sequences drew a circle and then copied it, two U commands would undo the two previous commands in sequence, as follows:

> u [Enter]
> COPY
> [Enter]
> CIRCLE

Using the U command after commands that involve transparent commands or sub-commands causes the entire sequence to be undone. For example, when you set a dimension variable and then perform a dimension command, a subsequent U command nullifies the dimension drawn and the change in the setting of the dimension variable.

Undo Command

The UNDO command permits you to select a specified number or marked group of prior commands for undoing. You can invoke the UNDO command by typing **undo** at the on-screen prompt or from the Quick Access toolbar, as shown in Figure 3–81.

FIGURE 3–81 *Invoking the UNDO command from the Quick Access toolbar*

After invoking UNDO, AutoCAD displays the following prompt:

> Enter the number of operations to undo or ⊡ *(specify number of undo operations to undo or select one of the available options from the shortcut menu)*

The default number of undo operations is one. You can specify a higher number of undo operations. If you select a number higher than the available number of undo operations, AutoCAD will undo all of the operations available.

Choose *Control* from the shortcut menu to set the available undo options. By limiting the number of undo options, you can free up memory and disk space that is otherwise used to save undo operation information. AutoCAD prompts as follows when the *Control* option is selected:

> Enter an UNDO control option *(select one of the four available options)*

Selecting *All* (the default setting of the UNDO command) enables all Undo options.

Selecting *None* disables the U and UNDO commands but not the CONTROL option of the UNDO command that reenables the various options.

Selecting *One* sets the UNDO commands functionality to that of the U command.

Selecting *Combine* controls whether multiple, consecutive zoom and pan commands are combined as a single operation for UNDO and REDO operations, which are discussed later.

Choose *Mark* to set a mark in the undo information.

Choose *Back* to undo all the operations back to the last mark.

If you are at a point in the editing session where you would like to experiment but would still like the option of undoing the experiment, you can mark that point. An example of the use of the MARK and BACK options is as follows:

line [Enter] *(draw a line)*

circle [Enter] *(draw a circle)*

undo [Enter]

Enter number of operations to undo or ⬇ m [Enter]

text [Enter] *(enter text)*

arc [Enter] *(draw an arc)*

undo [Enter]

Enter number of operations to undo or ⬇ b [Enter]

The *Back* option returns you to the state of the drawing that has the line and the circle. Following this UNDO BACK with a U removes the circle. Another U command removes the line.

Using the *Back* option when no Mark has been established will prompt:

This will undo everything. OK? <Y>

Responding Y undoes everything done since the current editing session was begun or since the last SAVE command.

> The default is Y. Think twice before pressing [Enter] in response to the "This will undo everything" prompt.

NOTE

Choose *Begin* from the shortcut menu to group a sequence of actions into a set. All subsequent actions become part of this group until you choose the *End* option. Entering **undo begin** while a group is already active ends the current group and begins a new one. UNDO and U treat grouped actions as a single action. The *Begin* and *End* options are normally intended for use in strings of menu commands, where a menu pick involves several operations.

Choose *Auto* from the shortcut menu to group the actions of a single command, making them reversible by a single U command. When the *Auto* option is on, starting a command groups all actions until you exit that command.

The effects of the following commands cannot be undone: area, attext, dblist, delay, dist, dxfout, graphscr, help, hide, id, list, mslide, plot, quit, redraw, redrawall, regenall, resume, save, shade, shell, status, and textscr.

When the down-arrow to the right of the UNDO command icon on the Quick Access toolbar is selected, a drop-down menu is displayed with a list of commands that can be undone by using the UNDO command. They are listed from the most recent to the earliest. You can select any command on the list and AutoCAD will undo all of the commands back to and including the command selected.

Redo Command

The REDO command permits reversal of prior U or UNDO commands. All UNDO commands can be reversed with the REDO command. In order to function, the REDO command must be used following the U or UNDO command and prior to any other action. You can invoke the REDO command from the Quick Access toolbar (Figure 3–82).

FIGURE 3–82 *Invoking the REDO command from the Quick Access toolbar*

The REDO command has no options. When the down-arrow to the right of the REDO command icon on the Quick Access toolbar is selected, a drop-down menu is displayed with a list of commands that can be redone by using the REDO command. They are listed from the most recent to the earliest.

NOTE

The REDO command must immediately follow the U or UNDO command.

Open the Exercise Manual PDF for Chapter 3 for discipline-specific exercises. Related files are downloaded from the student companion site mentioned in the Introduction (refer to page number xii for instructions).

REVIEW QUESTIONS

1. The invisible grid, which the crosshairs lock onto, is called:
   a. SNAP
   b. GRID
   c. ORTHO
   d. Cursor lock

2. After you select the first point for the line and move the cursor what is displayed with the dynamic input switched on?
   a. The line's length and angle
   b. The line's color and length
   c. The line's color and angle
   d. The tooltips
   e. Coordinates

3. To globally change the sizes of the dashes for all dashed lines, you should adjust:
   a. Line scale
   b. ltscale
   c. scale
   d. Layer scale

4. To set the Snap or Grid spacing in a drawing you should:
   a. Select Properties from the Ribbon and select Snap or Grid
   b. Right-click on either button in the Status Bar and select Settings
   c. Press F6 and select Settings
   d. Open Layer Properties

5. To reverse the effect of the last 11 commands, you could:
   a. Use the UNDO command
   b. Use the U command multiple times
   c. Either A or B
   d. It is not possible because AutoCAD only retains the last 10 commands

6. The following are available AutoCAD tools except:
   a. GRID
   b. SNAP
   c. ORTHO
   d. TSNAP
   e. OSNAP

7. Drawing in ORTHO mode always forces the line to:
   a. 90 degree angles
   b. 45 degree angles
   c. 30 degree angles
   d. 15 degree angles

8. If you just used the U command, what command would restore the drawing to the state before the U command?
   a. RESTORE
   b. U
   c. REDO
   d. OOPS

9. Which of the following cannot be modified in the Drafting Settings dialog box?
   a. Snap
   b. Grid
   c. Ortho
   d. Limits
   e. All of the above

10. What does Extension Object Snap, snap to?
    a. A point on the continuation of an object
    b. Permanent reference points
    c. Quadrant point of a circle
    d. Centre point of a circle or arc

11. If the spacing of the visible grid is set too small, AutoCAD responds as follows:
    a. Does not accept the command
    b. Produces a "Grid too dense to display" message
    c. Produces a display that is distorted
    d. Automatically adjusts the size of the grid so it will display
    e. Displays the grid anyway

12. Why is Polar Tracking command helpful when drawing?
    a. It reduces the amount of typing
    b. Fewer commands can be used
    c. It saves the drawing automatically
    d. It predicts your next command

13. The smallest number that can be displayed in the denominator when setting units to architectural units is:
    a. 8
    b. 16
    c. 64
    d. 128
    e. None of the above

14. You can use Polar Tracking with or without Dynamic Input.
    a. True
    b. False

15. Which of the following is not a valid option of the Layer Properties Manager dialog box?
    a. Close
    b. Lock
    c. On
    d. Freeze
    e. Color

16. What type of objects does the Quadrant Object Snaps work with?
    a. Circles or Arcs
    b. Lines
    c. Rectangles
    d. Polylines

17. After having drawn a three-point circle, you want to begin a line at the exact center of the circle. What tool would you use?
    a. Snap
    b. Object Snap
    c. Entity Snap
    d. Geometric Calculator

18. Which function Key sets Object Snap Tracking?

    a. F11

    b. F2

    c. F5

    d. F6

19. How many previous zooms are available with the previous option of the ZOOM command?

    a. 4

    b. 6

    c. 8

    d. 10

20. Polar tracking enables you to:

    a. Draw lines horizontally and vertically only

    b. Draw lines at 90, 45, 30, 22.5, 18, 15, 10 and 5 degrees only

    c. Draw lines at any angle defined by the user

    d. Draw lines at any multiple of 30 degrees only

21. In general, a REDRAW is quicker than a REGEN.

    a. True

    b. False

22. What does object snaps enable you to do?

    a. Snap to exact points on objects while you are in a command

    b. Modify setting when you are in the middle of a command

    c. Open the dialog box as well as change the color of the Autosnap marker

    d. Draw horizontal and vertical lines, as well as lines at other angles

23. When a layer is ON and thawed:

    a. The objects on that layer are visible on the monitor

    b. The objects on that layer are not visible on the monitor

    c. The objects on that layer are ignored by a REGEN

    d. The drawing REDRAW time is reduced

    e. The objects on that layer cannot be selected

24. A Node is:

    a. A specific point in a drawing with an exact set of coordinates

    b. An Annotation

    c. Center of a circle

    d. Center of an arc

    e. All of the above

25. To ensure that the entire limits of the drawing are visible on the display, you should perform a ZOOM
    a. All
    b. Previous
    c. Extents
    d. Limits

26. A layer where objects may not be edited or deleted, but are still visible on the screen and may be OSNAPed to, is considered:
    a. Frozen
    b. Locked
    c. On
    d. Unset
    e. Fixed

27. Which OSNAP option allows you to select the closest endpoint of a line, arc, or polyline segment?
    a. Endpoint
    b. Midpoint
    c. Center
    d. Insertion point
    e. Perpendicular

28. Which OSNAP option allows you to select the point in the mid-point of a line?
    a. Endpoint
    b. Midpoint
    c. Center
    d. Insertion point
    e. Perpendicular

29. When drawing a line, which OSNAP option allows you to select the point on a line or polyline segment where the angle formed with the line is a 90° angle?
    a. Endpoint
    b. Midpoint
    c. Center
    d. Insertion point
    e. Perpendicular

30. Which OSNAP option allows you to select the location where two lines, arcs, or polyline segments cross each other?
    a. Node
    b. Quadrant
    c. Tangent
    d. Nearest
    e. Intersection

31. Which OSNAP option allows you to select the location where a "point" has been established?

    a. Node

    b. Quadrant

    c. Tangent

    d. Nearest

    e. Intersection

32. Which OSNAP option allows you to select a point on a circle that is 0°, 90°, 180°, or 270° from the circle's center?

    a. Node

    b. Quadrant

    c. Tangent

    d. Nearest

    e. Intersection

33. Which OSNAP option allows you to select a point on any object, except text, that is closest to the cursor's position?

    a. Node

    b. Quadrant

    c. Tangent

    d. Nearest

    e. Intersection

34. Which OSNAP option allows you to select the location where two lines or arcs may or may not cross each other in 3D space?

    a. Quadrant

    b. Apparent intersection

    c. Nearest

    d. Intersection

35. The _____ command returns your drawing to the state before the last undo. It is only available immediately after Undo.

CHAPTER
4

Fundamentals III

OBJECTIVES

After completing this chapter, you will be able to do the following:

- Draw construction lines with the XLINE and RAY commands

- Construct geometric figures with polygons, ellipses, and polylines

- Create single line text and multiline text using appropriate styles and sizes, to annotate drawings

- Use the construct commands: COPY, ARRAY, OFFSET, MIRROR, FILLET, and CHAMFER

- Use the modify commands: MOVE, TRIM, BREAK, and EXTEND

DRAWING CONSTRUCTION LINES

AutoCAD provides powerful tools for drawing lines—XLINE and RAY commands—that extend infinitely in one or both directions. These lines have no effect, however, on the ZOOM EXTENTS command. They can be moved, copied, and rotated like any other objects. If necessary, you can trim the lines, break them anywhere with the BREAK command, draw an arc between two nonparallel construction lines with the FILLET command, and draw a chamfer between two nonparallel construction lines with the CHAMFER command.

The construction lines can be used as reference lines for creating other objects. To keep the construction lines from being plotted, draw them on a separate layer, and set that layer to a nonprinting status.

XLINE Command

The XLINE command allows you to draw lines that extend infinitely in both directions from the point selected when being created.

Invoke the XLINE command by selecting Construction Line from the Draw panel (Figure 4–1) located on the Home tab, and AutoCAD displays the following prompt:

> Specify a point or ⊡ *(specify a point or right-click and choose one of the options)*

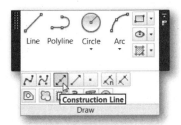

FIGURE 4–1 *Invoking the XLINE command by selecting Construction Line from the Draw panel*

When you specify a point, AutoCAD displays the following prompt:

> Specify through point: *(specify a point through which the construction line should pass)*

AutoCAD draws a line that passes through the two specified points and extends infinitely in both directions. AutoCAD continues to prompt for additional through points and uses these to draw construction lines through each of them and the first point specified. To terminate the command, press `Enter` or the `SPACEBAR`.

Xline Options

Choosing *Hor* (Horizontal) from the shortcut menu allows you to draw a construction line through a specified point that is parallel to the X axis of the current UCS.

Choosing *Ver* (Vertical) from the shortcut menu allows you to draw a construction line through a specified point that is parallel to the Y axis of the current UCS.

Choosing *Ang* (Angle) from the shortcut menu allows you to draw a construction line at a specified angle. AutoCAD displays the following prompts:

> Enter angle of xline (0) or ⊡ *(specify an angle at which to place the construction line or right-click and choose one of the options)*
>
> Specify through point: *(specify a point through which the construction line should pass)*

AutoCAD draws the construction line through the specified point, using the specified angle. You can specify the angle by typing it in or by selecting two points in the drawing.

Choosing *Reference* from the shortcut menu allows you to draw a construction line at a specific angle from a selected reference line. The angle is measured counterclockwise from the reference line.

Choosing *Bisect* from the shortcut menu allows you to draw a construction line through the first point bisecting the angle determined by the second and third points, with the first point being the vertex. AutoCAD displays the following prompts:

> Specify angle vertex point: *(specify a point for the vertex of an angle to be bisected and through which the construction line will be drawn)*
>
> Specify angle start point: *(specify a point to determine one boundary line of an angle)*
>
> Specify angle endpoint: *(specify a point to determine second boundary line of angle)*

The construction line lies in the plane determined by the three points.

Choosing *Offset* from the shortcut menu allows you to draw a construction line parallel to and at the specified distance from the line object selected and on the side selected. AutoCAD displays the following prompts:

> Specify offset distance or ⏎ *(specify an offset distance, or right-click and choose one of the available options)*
>
> Select a line object: *(select a line, pline, ray, or xline)*
>
> Specify side to offset: *(specify a point to draw a construction line parallel to the selected object)*
>
> Select a line object: *(press* **Enter** *to terminate this command sequence)*

Choosing *Through* allows you to specify a point through which an offset construction line is drawn.

RAY Command

The RAY command allows you to draw lines that extend infinitely in one direction from a specified point.

Invoke the RAY command from the Draw panel located on the Home tab, and AutoCAD displays the following prompts:

> Specify start point: *(specify the start point to draw the ray)*
>
> Specify through point: *(specify a point through which you want the ray to pass)*
>
> Specify through point: *(specify a point to draw additional rays, or press* **Enter** *to terminate the command sequence)*

A ray is drawn starting at the first point and extending infinitely in one direction through the second point. AutoCAD continues to prompt for through points until you terminate the command sequence.

DRAWING POLYGONS

The POLYGON command creates an equilateral (edges with equal length), closed polyline. It offers three different methods for drawing 2D polygons: Inscribed in Circle, Circumscribed about Circle, and Edge. The number of sides can vary from 3, which forms an equilateral triangle, to 1,024. Invoke the POLYGON command from the Draw panel (Figure 4–2) located on the Home tab, and AutoCAD displays the following prompts:

> Enter number of sides<current>: *(specify number of sides)*
>
> Specify center of polygon or ⊡ *(specify center of the polygon or right-click and choose one of the options)*

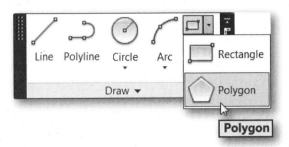

FIGURE 4–2 *Invoking the* POLYGON *command from the Draw panel*

After you specify the number of sides and the center of the polygon, AutoCAD prompts for one of two options, *Inscribed in circle* or *Circumscribed about circle*.

Polygon Options

Choosing *Inscribed in circle* draws the polygon of equal length for all sides inscribed inside an imaginary circle having the same diameter as the distance across opposite polygon corners, for an even number of sides, as shown in the following example:

> Enter number of sides<current>: 6 Enter
>
> Specify center of polygon or ⊡ 3,3 Enter
>
> Enter an option *(choose Inscribed in circle)*
>
> Specify radius of circle: 2 Enter

AutoCAD draws a polygon (Figure 4–3) with six sides, centered at 3,3, whose edge vertices are 2 units from the center of the polygon.

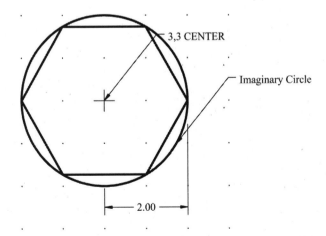

FIGURE 4–3 *Polygon drawn with six sides using the Inscribed in circle option*

Choosing *Circumscribed about circle* draws a polygon circumscribed around the outside of an imaginary circle having the same diameter as the distance across the opposite polygon sides, for an even number of sides, as shown in the following example:

```
Enter number of sides <current>: 8 Enter
Specify center of polygon or ⊡ 3,3 Enter
Enter an option (choose Circumscribed about circle)
Specify radius of circle: 2 Enter
```

AutoCAD draws a polygon with eight sides, centered at 3,3, whose edge midpoints are 2 units from the center of the polygon (Figure 4–4).

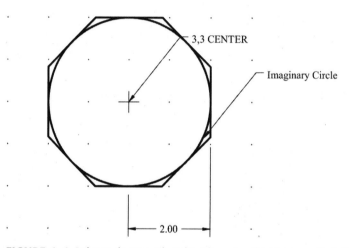

FIGURE 4–4 *Polygon drawn with eight sides using the Circumscribed about circle option*

Choosing *Edge* allows you to draw a polygon by specifying the endpoints of the first edge, as shown in the following example:

```
Enter number of sides <current>: 7 Enter
Specify center of polygon or ⊡ (choose Edge from the shortcut
menu)
Specify first endpoint of edge: 1,1 Enter
Specify second endpoint of edge: #3,1 Enter
```

AutoCAD draws a polygon with seven sides for the specified endpoints of one of the sides of the polygon (Figure 4–5).

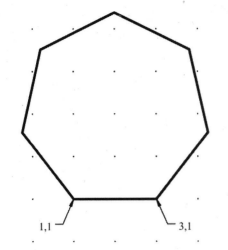

FIGURE 4–5 *Polygon drawn with seven sides using the* EDGE *option*

DRAWING ELLIPSES

AutoCAD allows you to draw an ellipse or an elliptical arc with the ELLIPSE command. Invoke the ELLIPSE command from the Draw panel (Figure 4–6) located on the Home tab, and AutoCAD displays the following prompt:

> Specify axis endpoint of ellipse or ⬇ *(specify axis end point of the ellipse to be drawn or right-click and choose one of the options)*

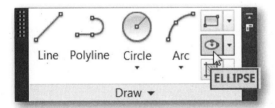

FIGURE 4–6 *Invoking the* ELLIPSE *command from the Draw panel on the Home tab*

Ellipse Options

The default option allows you to draw an ellipse by specifying the endpoints of the axes. AutoCAD prompts for two endpoints of the first axis. The first axis can define either the major or the minor axis of the ellipse. AutoCAD then prompts for an endpoint of the second axis as the distance from the midpoint of the first axis to the specified point.

For example, the following command sequence shows steps required to draw an ellipse by defining axis endpoints (Figure 4–7).

> Specify axis endpoint of ellipse or ⬇ 1,1 [Enter]
> Specify other endpoint of axis: #5,1 [Enter]
> Specify distance to other axis or ⬇ #3,2 [Enter]

AutoCAD draws an ellipse whose major axis is 4.0 units long in a horizontal direction and whose minor axis is 2.0 units long in a vertical direction (Figure 4–7).

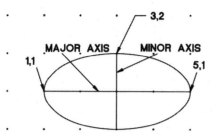

FIGURE 4–7 *An ellipse drawn by specifying the major and minor axes*

Choosing *Rotation* from the shortcut menu, after specifying two axis endpoints, allows you to draw an ellipse by specifying a rotation angle. The rotation angle defines the major-axis-to-minor-axis ratio of the ellipse by rotating a circle about the first axis. The greater the rotation angle value, the greater the ratio of major to minor axes. AutoCAD draws a circle if you set the rotation angle to 0°.

For example, the following command sequence shows the steps to draw an ellipse by specifying the rotation angle.

ellipse (Enter)

Specify axis endpoint of ellipse or ⊙ 3,-1 (Enter)

Specify other endpoint of axis: #3,3 (Enter)

Specify distance to other axis or ⊙ *(choose Rotation from the shortcut menu)*

Specify rotation around major axis: 30 (Enter)

See Figure 4–8 for examples of ellipses with various rotation angles.

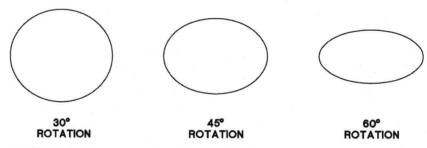

FIGURE 4–8 *Ellipses drawn with different rotation angles*

Choosing *Center* from the shortcut menu before specifying two axis endpoints allows you to draw an ellipse by defining the center point first and then the axis endpoints. First, AutoCAD prompts for the ellipse center point. AutoCAD then prompts for an endpoint of an axis as the distance from the center of the ellipse to the specified point. The first axis can define either the major or the minor axis of the ellipse. Next, AutoCAD prompts for an endpoint of the second axis as the distance from the center of the ellipse to the specified point.

For example, the following command sequence shows the steps required to draw an ellipse by defining the ellipse center point.

> `ellipse` [Enter]
> Specify axis endpoint of ellipse or ⬇ *(choose Center from the shortcut menu)*
> Specify center of ellipse: `3,1` [Enter]
> Specify endpoint of axis: `#1,1` [Enter]
> Specify distance to other axis or ⬇ `#3,2` [Enter]

AutoCAD draws an ellipse similar to the first example (Figure 4–7), with a major axis 4.0 units long in a horizontal direction and a minor axis 2.0 units long in a vertical direction.

Choosing *Arc* from the shortcut menu, before specifying two axis endpoints, allows you to draw an elliptical arc. After you specify the major and minor axis endpoints, AutoCAD prompts for the start and end angle points for the elliptical arc to be drawn. Instead of specifying the start angle or the end angle, you can toggle to the *Parameter* option, which prompts for the Start parameter and End parameter point locations. AutoCAD creates the elliptical arc using the following parametric vector equation:

$$p(u) = c + a*\cos(u) + b*\sin(u)$$

where *c* is the center of the ellipse, and *a* and *b* are its major and minor axes, respectively. Instead of specifying the end angle, you can specify the included angle of the elliptical arc to be drawn.

The `ELLIPSE` command, with the *Arc* option, is accessible directly from the Draw toolbar, or you can select the *Arc* option while in the `ELLIPSE` command.

For example, the following command sequence shows the steps to draw an elliptical arc:

> `ellipse` [Enter]
> Specify axis endpoint of ellipse or ⬇ *(choose Arc from the shortcut menu)*
> Specify axis endpoint of elliptical arc or ⬇ `1,1` [Enter]
> Specify other endpoint of axis: `#5,1` [Enter]
> Specify distance to other axis or ⬇ `#3,2` [Enter]
> Specify start angle or ⬇ `#4,1` [Enter]
> Specify end angle or ⬇ `#-2,1` [Enter]

AutoCAD draws an elliptical arc with the start angle at 3,2 and the ending angle at 1,1.

Isometric Circles (Isocircles)

By definition, Isometric Planes (*Iso* meaning "same" and *metric* meaning "measure") are all viewed at the same angle of rotation (Figure 4–9). The angle is approximately 54.7356°. AutoCAD uses this angle of rotation automatically when you wish to represent circles in one of the isoplanes by drawing ellipses with the *Isocircle* option.

Normally, a circle 1 unit in diameter viewed in one of the isoplanes will project a short axis dimension of 0.577350 units. Its diameter parallel to an isoaxis will project a dimension of 0.816497 units. A line drawn in isometric mode that is parallel to one of the three main axes will also project a dimension of 0.816497 units. Say that you would like these lines and circle diameters to project a dimension of exactly 1.0 unit. Therefore, you automatically increase the entire projection by a "fudge factor" of 1.22474, the reciprocal of 0.816497, in order to use true dimensioning parallel to one of the isometric axes.

This means that isocircles 1 unit in diameter will be measured along one of their isometric diameters rather than along their long axis. This action facilitates using true lengths as the lengths of distances projected from lines parallel to one of the isometric axes. Therefore, a 1-unit-diameter isocircle will project a long axis that is 1.224744871 units and a short axis that is 0.707107 (0.577350 × 1.22474) units. These "fudge factors" are built into AutoCAD isocircles.

To specify Isometric Snap mode, right-click on **SNAP**, located on the Status bar and then select Settings. In the **Snap type** section, select **Isometric snap**. The *Isometric Circle* option is available as one of the options of the ELLIPSE command when you are in Isometric Snap mode.

> ellipse [Enter]
>
> Specify axis endpoint of ellipse or ⊡ *(choose Isocircle from the shortcut menu)*
>
> Specify center of isocircle: *(select the center of the isometric circle)*
>
> Specify radius of isocircle or ⊡ *(specify the radius or right-click and choose one of the options)*

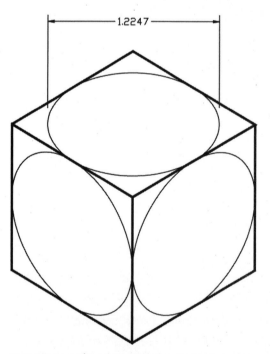

FIGURE 4–9 *Ellipses drawn using the Isocircle option of the* ELLIPSE *command*

Choosing *Diameter* causes AutoCAD to display the following prompt:

> Circle diameter: *(specify the desired diameter)*

NOTE

DRAWING POLYLINES

The "poly" in polyline refers to a single object with multiple connected straight-line and/or arc segments. The polyline is drawn by invoking the PLINE command and then selecting a series of points. In this respect, PLINE functions much like the LINE command. However, when completed, the segments act like a single object when operated on by modify commands. You specify the endpoints using only 2D (X,Y) coordinates. 3D Polylines are covered in Chapter 16.

The versatile PLINE command also draws lines and arcs of different widths, linetypes, and tapered lines. The area and perimeter of a closed 2D polyline can be calculated.

By default, polylines are drawn as optimized polylines. An optimized polyline provides most of the functionality of 2D polylines but with much improved performance and reduced drawing file size. The vertices are stored as an array of information on one object. When you use the PEDIT command to edit the polyline into a spline fitting or a curve fitting, the polyline loses its optimization feature, and vertices are stored as separate entities, but it still functions as a single object when operated on by modify commands.

Invoke the PLINE command from the Draw panel (Figure 4–10) located on the Home tab, and AutoCAD displays the following prompts:

> Specify start point: *(specify the start point of the polyline)*
>
> Specify next point or ⊡ *(specify next point or right-click and choose one of the options)*

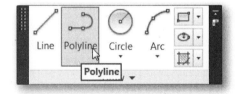

FIGURE 4–10 *Invoking the PLINE command from the Draw panel*

You can specify the end of the segment by means of absolute coordinates, relative coordinates, or by using your pointing device to specify the end of the segment on the screen. After completing this action, AutoCAD repeats the prompt:

> Specify next point or ⊡

Having drawn a connected series of segments, you can press Enter to terminate the PLINE command. The resulting figure is recognized by AutoCAD modify commands as a single object.

The following command sequence presents an example of connected segments drawn by means of the PLINE command (Figure 4–11).

```
pline Enter
Specify start point: 2,2 Enter
Specify next point or ⬇ #4,2 Enter
Specify next point or ⬇ #5,1 Enter
Specify next point or ⬇ #7,1 Enter
Specify next point or ⬇ #8,2 Enter
Specify next point or ⬇ #10,2 Enter
Specify next point or ⬇ #10,4 Enter
Specify next point or ⬇ #9,5 Enter
Specify next point or ⬇ #8,5 Enter
Specify next point or ⬇ #7,4 Enter
Specify next point or ⬇ #5,4 Enter
Specify next point or ⬇ #4,5 Enter
Specify next point or ⬇ #3,5 Enter
Specify next point or ⬇ #2,4 Enter
Specify next point or ⬇ (choose Close from the shortcut menu)
```

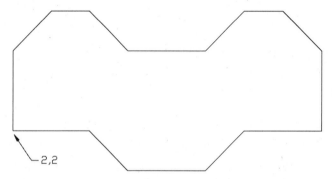

FIGURE 4–11 *Example of connected line segments drawn by means of the* PLINE *command*

Polyline Options

Choosing *Close* or *Undo* from the shortcut menu works similarly to the corresponding options in the LINE command.

Choosing *Width* from the shortcut menu allows you to specify a starting and an ending width for a wide segment. When you select this option, AutoCAD displays the following prompts:

```
Specify starting width <current>: (specify starting width)
Specify ending width <current>: (specify ending width)
```

You can specify a width by entering a value at the prompt or by selecting points on the screen. When you specify points on the screen, AutoCAD uses the distance from the starting point of the polyline to the point selected as the starting width. You can accept the default value for the starting width by pressing Enter or entering a new value. The starting width is the default for the ending width. If necessary, you can change the ending width to another value, which will result in a tapered segment or an arrow. The ending width, in turn, is the uniform width for all subsequent segments until you change the width again.

The following command sequence presents an example of connected lines with tapered width, drawn by means of the PLINE command (Figure 4–12).

```
Specify start point: 2,2 [Enter]
Specify next point or [↓] (choose Width from the shortcut menu)
Specify starting width <current>: 0 [Enter]
Specify ending width <current>: .25 [Enter]
Specify next point or [↓] #2,2.5 [Enter]
Specify next point or [↓] #2,3 [Enter]
Specify next point or [↓] (choose Width from the shortcut menu)
Specify starting width <0.2500>: [Enter]
Specify ending width <0.2500>: 0 [Enter]
Specify next point or [↓] #2,3.5 [Enter]
Specify next point or [↓] (choose Width from the shortcut menu)
Specify starting width <0.0000>: [Enter]
Specify ending width <0.0000>: .25 [Enter]
Specify next point or [↓] #2.5,3.5 [Enter]
Specify next point or [↓] #3,3.5 [Enter]
Specify next point or [↓] (choose Width from the shortcut menu)
Specify starting width <0.2500>: [Enter]
Specify ending width <0.2500>: 0 [Enter]
Specify next point or [↓] #3.5,3.5 [Enter]
Specify next point: (choose Width from the shortcut menu)
Specify starting width <0.0000>: [Enter]
Specify ending width <0.0000>: .25 [Enter]
Specify next point or [↓] #3.5,3 [Enter]
Specify next point or [↓] #3.5,2.5 [Enter]
Specify next point: (choose Width from the shortcut menu)
Specify starting width <0.2500>: [Enter]
Specify ending width <0.2500>: 0 [Enter]
Specify next point or [↓] #3.5,2 [Enter]
Specify next point or [↓] (choose Width from the shortcut menu)
Specify starting width <0.0000>: [Enter]
Specify ending width <0.0000>: .25 [Enter]
Specify next point or [↓] #3,2 [Enter]
Specify next point or [↓] #2.5,2 [Enter]
```

176 *Harnessing AutoCAD: 2013 and beyond*

Specify next point ⬇ *(choose Width from the shortcut menu)*

Specify starting width <0.2500>: `Enter`

Specify ending width <0.2500>: 0 `Enter`

Specify next point or ⬇ *(choose Close from the shortcut menu)*

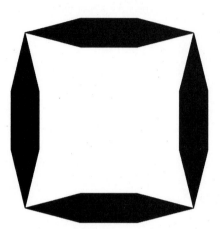

FIGURE 4–12 *Example of connected line segments with tapered width, drawn by means of the PLINE command*

Choosing *Halfwidth* from the shortcut menu operates similar to the *Width* option, including the prompts, except that it allows you to specify the width from the center of a wide polyline to one of its edges. In other words, you specify half of the total width. For example, it is easier to input 1.021756 as the half width than to figure out the total width by doubling. You can specify a half width by selecting points on the screen in the same manner used to specify the full width.

Choosing *Arc* allows you to draw a polyline arc. When you select the *Arc* option, AutoCAD displays the following prompt:

Specify endpoint of arc or ⬇ *(specify end point of arc or right-click and choose one of the available options)*

If you respond with a point, it is interpreted as the endpoint of the arc. The endpoint of the previous segment is the starting point of the arc, and the starting direction of the new arc will be the ending direction of the previous segment, whether the previous segment is a line or an arc. This action resembles the ARC command's *Start, End, Direction* (S, E, D) option but requires only the endpoints to be specified or selected on the screen.

Polyline Arc Options

The *Close, Width, Halfwidth,* and *Undo* options in the shortcut menu operate similar to the corresponding options for the straight-line segments described earlier.

Choosing *Angle* lets you specify the included angle. AutoCAD displays the following prompt:

Specify included angle: *(specify an angle)*

The arc is drawn counterclockwise if the value is positive, and clockwise if it is negative. After the angle is specified, AutoCAD prompts for the endpoint of the arc.

Image(s) © Cengage Learning 2013

Choosing *Center* lets you override with the location of the center of the arc. AutoCAD displays the following prompt:

> Specify center point: *(specify center point)*

When you provide the center point of the arc, AutoCAD prompts for additional information:

> Specify endpoint of arc or ⬇ *(specify end point of arc or right-click and choose one of the options)*

If you respond with a point, it is interpreted as the endpoint of the arc.

Choosing *Angle* or *Length* allows you to specify the arc's included angle or chord length.

Choosing *Direction* lets you override the direction of the last segment, and AutoCAD displays the following prompt:

> Specify the tangent direction for the start point of arc: *(specify the direction)*

If you respond with a point, AutoCAD interprets the starting point and the direction from this point and then prompts for the endpoint for the arc.

Choosing *Line* reverts to drawing straight-line segments.

Choosing *Radius* allows you to specify the radius by prompting:

> Specify radius of arc: *(specify the radius of the arc)*

After the radius is specified, you are prompted for the endpoint of the arc.

Choosing *Second pt* causes AutoCAD to use the 3-Point method of drawing an arc by prompting:

> Specify second point of arc: *(specify second point)*

If you respond with a point, it is interpreted as the second point, and you are then prompted for the endpoint of the arc. This resembles the ARC command's 3-Point option.

Choosing *Length* continues the polyline in the same direction as the last segment for a specified distance.

DRAWING TEXT

You have learned how to draw some geometric shapes that make up your design. Now it is time to learn how to annotate your design. When you draw on paper, adding descriptions of the design components and the necessary shop and fabrication notes is a tedious, time-consuming process. AutoCAD provides several text commands and tools, including a spell checker, that greatly reduces the tedium and time of text placement.

Text is used to label the various components of your drawing and to create the necessary shop or field notes needed for fabrication and construction of your design. AutoCAD includes a large number of text fonts. Text can be stretched, compressed, obliqued, mirrored, or drawn in a vertical column by applying a style. Each text string can be sized, rotated, and justified to meet your drawing needs. You should be aware that AutoCAD considers a text string, in which all the characters that comprise the line of characters use a single TEXT command, as one object.

The Text panel on the Annotate tab has text-related commands, which include the following: *Multiline Text, Single Line Text, Spell Check, Find and Replace, Text Style, Scale, Justify,* and setting *Text Height* (Figure 4–13).

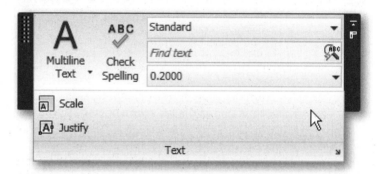

FIGURE 4–13 *Text panel*

Creating a Single Line of Text

The TEXT command allows you to create several lines of text in the current style. If necessary, you can change the current style. To modify a style or create a new style, refer to the "Creating and Modifying Text Styles" section later in this chapter.

Invoke the TEXT command for a single line of text from the Text panel (Figure 4–14), and AutoCAD displays the following prompt:

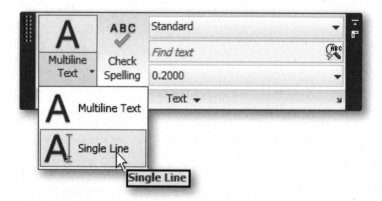

FIGURE 4–14 *Invoking the TEXT command from the Text panel*

> Specify start point of text or *(specify start point of text or right-click and choose one of the available options)*

The start point indicates the lower-left corner of the text. If necessary, you can change the location of the justification point. You can specify the starting point in absolute coordinates or by using your pointing device. After you specify the starting point, AutoCAD displays the following prompt:

> Specify height <current>: *(specify the text height)*

This step allows you to set the text height. You can accept the current text height by pressing `Enter`. You can set a new text height by using your pointing device or entering the appropriate text height. AutoCAD then displays the following prompt:

```
Specify rotation angle of text <current>: (specify the
rotation angle)
```

This step allows you to place the text at any angle in reference to 0° (default). A rotation angle of 0° causes the text to be placed horizontally at the specified start point.

A cursor appears on-screen at the starting point you have selected. After you enter the first line of text and press `Enter`, you will notice that the cursor drops down to the next line, anticipating that you wish to enter another line of text. If this is the case, enter the next line of text string. When you are through typing text strings, press `Enter` twice to terminate the command sequence.

If you are in the TEXT command and notice a mistake, or simply want to change a character or word, you can use the arrow keys on your keyboard to position the cursor and make the necessary changes.

The following command sequence shows placement of left-justified text by providing the starting point of the text (Figure 4–15).

```
text Enter
Specify start point of text or ⬇ (specify start point of text
as shown in Figure 4-15)
Specify height <current>: 0.25 Enter
Specify rotation angle of text <current>:0 Enter
(at the screen text cursor, type in the desired text)
Sample Text Left Justified (press Enter to accept first line
of text)
(press Enter again to accept single line text as entered)
```

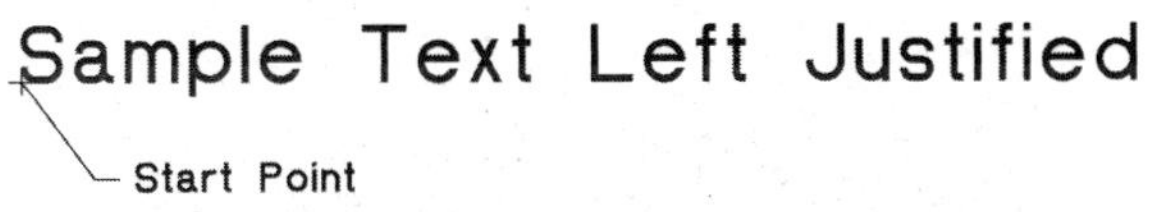

Sample Text Left Justified

└ Start Point

FIGURE 4–15 *Using the TEXT command to place left-justified text by specifying a start point*

Single-Line Text Options

When AutoCAD prompts for the start point, you can right-click for options to change justification and style. In addition to the standard justification options, you can choose to apply the *Align* and *Fit* options.

Choosing *Justify* allows you to place text in one of the 14 available justification points. When you select this option, AutoCAD displays the following prompt:

```
Enter an option (choose one of the options listed)
```

Options available from the on-screen prompt include *Align*, *Fit*, *Center*, *Middle*, *Right*, *TL*, *TC*, *TR*, *ML*, *MC*, *MR*, *BL*, *BC*, and *BR*.

Choosing *Center* allows you to select the center point for the baseline of the text. Baseline refers to the line along which the bases of the capital letters lie. Letters with descenders, such as g, q, and y, dip below the baseline. After providing the center point, enter the text height and rotation angle.

For example, the following command sequence shows placement of center-justified text, by providing the center point of the text (Figure 4–16).

```
text Enter
Specify start point of text or ⊡ (choose Justify from the
shortcut menu)
Enter an option (choose Center)
Specify center point of text: (specify center point)
Specify height <current>: .25 Enter
Specify rotation angle of text:<current>: 0 Enter
Enter text: Sample Text Center Justified Enter
Enter text: Enter
```

Sample Text Center Justified
──Center Point

Sample Text Middle Justified
── Middle Point

Sample Text Right Justified
End Point──

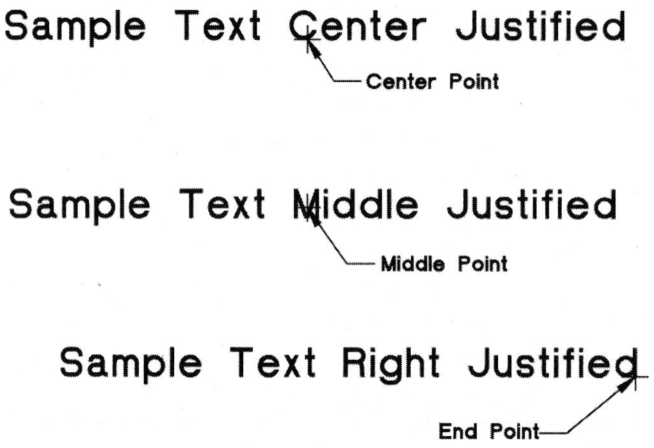

FIGURE 4–16 *Using the* TEXT *command to place text by specifying a center point (center justified), a middle point (middle justified), or an endpoint (right justified)*

Choosing *Middle* allows you to center the text both horizontally and vertically at a given point. After providing the middle point, enter the text height and rotation angle.

For example, the following command sequence shows placement of middle-justified text by providing the middle point of the text (Figure 4–16).

```
text Enter
Specify start point of text or ⊡ (choose Justify from the
shortcut menu)
Enter an option (choose Middle)
Specify middle point of text: (specify middle point)
Specify height <current>: .25 Enter
Specify rotation angle of text: 0 Enter
Enter text: Sample Text Middle Justified Enter
Enter text: Enter
```

Choosing *Right* allows you to place the text in reference to its lower-right corner (right justified). Here, you provide the point where the text will end. After providing the right point, enter the text height and rotation angle.

For example, the following command sequence shows placement of right-justified text (Figure 4–16).

```
text Enter
Specify start point of text or ⬇ (choose Justify from the
shortcut menu)
Enter an option (choose Right)
Specify right endpoint of text baseline: (specify right
endpoint)
Specify height <current>: .25 Enter
Specify rotation angle of text: 0 Enter
Enter text: Sample Text Right Justified Enter
Enter text: Enter
```

Other options are combinations of the previously mentioned options:

TL	Top left
TC	Top center
TR	Top right
ML	Middle left
MC	Middle center
MR	Middle right
BL	Bottom left
BC	Bottom center
BR	Bottom right

Choosing *Align* allows you to place the text by designating the endpoints of the baseline. AutoCAD computes the text height and orientation so that the text fits proportionately between two points. The overall character size adjusts in proportion to the height. The height and width of the character will be the same.

For example, the following command sequence shows how to place text using the *Align* option (Figure 4–17).

```
text Enter
Specify start point of text or ⬇ (choose Justify from the
shortcut menu)
Enter an option (choose Align)
Specify first endpoint of text baseline: (specify the first
point)
Specify second endpoint of text baseline: (specify the
second point)
Enter text: Sample Text Aligned Enter
Enter text: Enter
```

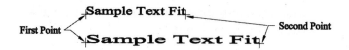

FIGURE 4–17 *Using the Align and Fit options of the TEXT command to place text*

Choosing *Fit* is similar to the *Align* option, but with the *Fit* option, AutoCAD uses the current text height and adjusts only the text's width, expanding or contracting it to fit between the points you specify.

For example, the following command sequence shows how to place text using the *Fit* option (Figure 4–17).

> text `Enter`
>
> Specify start point of text or ⬇ *(choose Justify from the shortcut menu)*
>
> Enter an option *(choose Fit)*
>
> Specify first endpoint of text baseline: *(specify the first point)*
>
> Specify second endpoint of text baseline: *(specify the second point)*
>
> Specify height <current>: 0.25 `Enter`
>
> Enter text: Sample Text Fit `Enter`
>
> Enter text: `Enter`

Choosing *Style* allows you to select one of the available styles in the current drawing. To modify a style or to create a new style, refer to the "Creating and Modifying Text Styles" section later in this chapter.

AutoCAD allows you to draw text with annotative scaling (display and plot at appropriate scale). For details refer to "Annotative Scaling" in Chapter 8.

Creating Multiline Text

The MTEXT command draws text by "processing" the words in paragraph form; the width of the paragraph is determined by the user-specified rectangular boundary. It is an easy way to have your text automatically formatted as a paragraph with left, right, or center justification as a group.

Each multiline text object is a single object, regardless of the number of lines it contains. If one character on one line is selected for editing, all characters on all lines are included in the selection set. The text boundary remains part of the object's framework, although it is not plotted or printed.

Multiline text features include indents and tabs, making it easier to correctly align your text for tables and numbered, lettered, or bulleted lists. At the top of the user-specified rectangle is a ruler similar to those in word processors (Figure 4–18).

Characters in the Multiline Text Editor can be selected individually or in a group for applying formatting styles such as bold, underline, and italic.

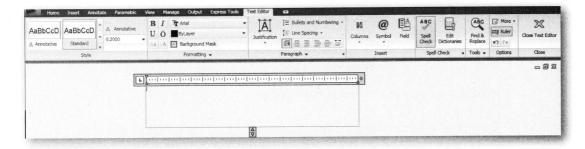

FIGURE 4–18 *Multiline Text tab displayed above the Multiline Text Editor*

Invoke the MTEXT command by selecting Multiline Text from the Text panel (Figure 4–19) on the Annotate tab, and AutoCAD displays the following prompts:

> Specify first corner: *(specify the first corner of the rectangular boundary)*
>
> Specify opposite corner or ⬇ *(specify the opposite corner of the rectangular boundary, or right-click and choose one of the options)*

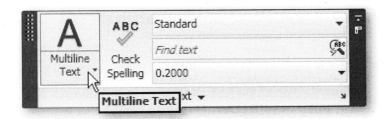

FIGURE 4–19 *Invoking the MTEXT command from the Text panel*

When you drag the cursor after specifying the first corner of the rectangular boundary, referred to as the bounding box, AutoCAD displays an arrow within the rectangle to indicate the direction of the paragraph's text flow. After you specify the opposite corner of the bounding box, AutoCAD displays the Multiline Text tab on the Ribbon and input (editing) area, as shown in Figure 4–18. Panels included on the Multiline Text tab include **Style, Formatting, Paragraph, Insert, Spell Check, Tools, Options,** and **Close**.

Style Panel

The *Style* panel (see Figure 4–20) allows you to select one of the available styles. The selected text style determines the text font, size, angle, orientation, and other text characteristics for next text or selected text. If necessary, you can override individual text characteristics for the selected text or new text. The default style is set to Standard style.

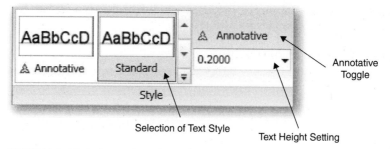

FIGURE 4-20 *Style panel on the Multiline Text tab*

Choosing *Annotative* toggles Annotative on or off for the current mtext object. The Annotative property allows you to automate the process of scaling annotations. Annotative objects are defined at a paper height and display in layout viewports and model space at the size determined by the annotation scale set for those spaces. Refer to Chapter 8 for a detailed explanation on Annotative scaling.

The *Height* text box lets you set the text height in drawing units for new text, or it changes the height of selected text. Multiline text objects can contain characters of various heights.

Formatting Panel

The *Formatting* panel (see Figure 4–21) allows you to select various formatting features to new text or selected text.

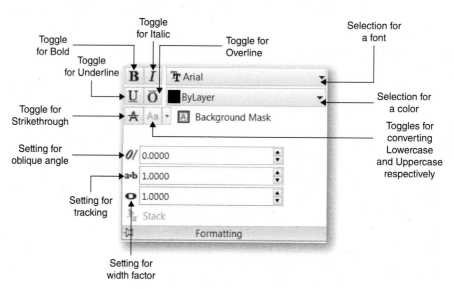

FIGURE 4-21 *Formatting panel*

Choosing *Bold* allows you to turn bold formatting ON and OFF for new text or selected text and is available only for the characters that belong to the TrueType font.

Choosing *Italic* allows you to turn italic formatting ON and OFF for new text or selected text and is available only for the characters that belong to the TrueType font.

Choosing *Underline* allows you to turn underlining ON and OFF for new text or selected text.

Choosing *Overline* causes new text or selected text to be drawn with an overscore line over the letters.

Choosing *Strikethrough* causes new text or selected text to be drawn with strikethrough line.

Choosing *Uppercase* causes selected text to uppercase.

Choosing *Lowercase* causes selected text to lowercase.

The *Font* selection menu allows you to specify a font for new text or change the font of selected text. All of the available TrueType fonts and SHX fonts are listed in the drop-down menu.

The *Color* selection menu allows you to specify the color for new texts or changes it for the selected text. You can assign the color as Bylayer, Byblock, or one of the available colors.

The setting for *Oblique Angle* determines the forward or backward slant of the text. The angle represents the offset from 90 degrees. Entering a value between −85 and 85 makes the text oblique. A positive obliquing angle slants text to the right. A negative obliquing angle slants text to the left.

The setting for *Tracking* decreases or increases the space between the selected characters. The 1.0 setting is normal spacing.

The setting for *Width Factor* widens or narrows the selected characters. The 1.0 setting represents the normal width of the letter in this font.

Background Mask allows you to set an opaque background behind the text.

Paragraph Panel

The *Paragraph* panel (see Figure 4–22) allows you to assign various settings related to a paragraph.

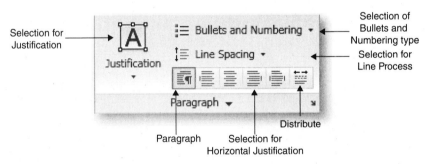

FIGURE 4–22 *Paragraph panel on the Multiline Text tab*

Choosing *Justification* allows you to place text in one of the nine available justification points, similar to the Single Line Text feature explained earlier.

Choosing *Line Spacing* sets the spacing between two lines of text. You can choose one of the predefined options that includes 1.0x, 1.5x, 2.0x, or 2.5x. Choosing More displays the Paragraph dialog box for additional settings. Choosing Clear Paragraph Spacing removes line spacing settings from selected or current paragraph.

Choosing *Left, Center,* or *Right* sets the horizontal justification for new text or selected text.

Choosing *Bullets and Numbering* causes a drop-down menu to appear, from which you can specify uppercase or lowercase, numbering, and bulleting. Choosing Uppercase letters causes AutoCAD to establish an alphabetized list using uppercase letters beginning with the letter "A" for new text or selected text. Choosing Lowercase letters causes AutoCAD to establish an alphabetized list using lowercase letters beginning with the letter "a" for new text or selected text.

Choosing the down arrow opens a Paragraph dialog box that allows you to set Tab stops, Indents, controls Paragraph Alignment, Paragraph Spacing, and Paragraph Line Spacing.

Choosing *Justify* sets the full justification for the selected paragraph.

Choosing *Distribute* includes spaces entered at the end of a line and sets the selected paragraph for full justification.

Insert Panel

The Insert panel (see Figure 4–23) allows you to set various symbols and fields.

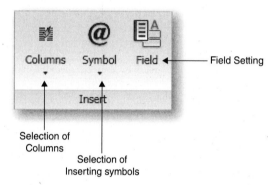

FIGURE 4–23 *Insert panel on the Multiline Text tab*

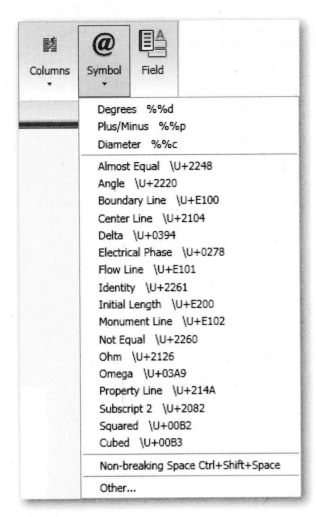

FIGURE 4–24 *Symbol shortcut menu*

Choosing *Symbol* causes a menu to be displayed (see Figure 4–24) from which symbols for Degrees, Plus/minus, and Diameter can be selected along with the Non-breaking Space. Also in the menu is the Other option, which causes the Character Map dialog box to be displayed. The Character Map dialog box operates in the same manner as the similar dialog box in other Windows-based word processors and text-handling programs. In addition to the options provided in the Multiline Text Editor dialog box for drawing special characters, you can also draw them by means of the control characters. The control characters for a symbol begin with a double percent sign (%%). The next character you enter represents the symbol. The control sequences defined by AutoCAD are presented in the following table.

CONTROL CHARACTER SEQUENCES FOR DRAWING SPECIAL CHARACTERS AND SYMBOLS

Special character or symbol	Control character sequence	Text string	Example of control character sequence
° (degree symbol)	%%d	104.5°F	104.5%%dF
± (plus/minus tolerance symbol)	%%p	34.5±3	34.5%%p3
Ø (diameter symbol)	%%c	56.06Ø	56.06%%c
% (single percent sign; necessary only when it must precede another control sequence)	%%%	34.67%±1.5	34.67%%%%% P1.5
Special coded symbols (where nnn stands for a three-digit code)	%%nnn	@	%%064

Choosing *Insert Field* causes the Field dialog box to be displayed (Figure 4–25), where you can select a field to insert in the text. When the dialog box closes, the current value of the field is displayed in the text.

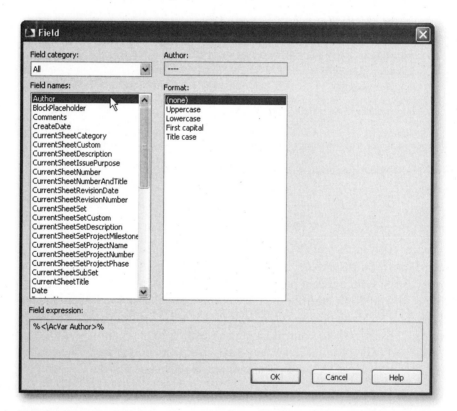

FIGURE 4–25 *The Field dialog box shown with Field category "All" and Field "Author" selected*

The *Field category* list box allows you to filter the fields that are displayed in the *Field names* list box into the following categories: **All, Date & Time, Document, Linked, Objects, Other, Plot,** and **SheetSet**. The *Field names* list box is visible for all categories, but the other textboxes vary with the category. The *Field names* list box allows you to select from the list of available fields according to the filter selected in the *Field category* list box.

Choose **All** from the *Field category* list box to list all the available fields in the *Field names* list box.

Choose **Date & Time** from the *Field category* list box to list fields related to the current drawing: CreateDate, Date, PlotDate, and SaveDate. The text boxes that appear when the **Date & Time** category is selected are *Date format* and *Examples*. The Date format text box displays the format of the example that is selected in the *Examples* text box. For example, if you choose 1/12/05 in the *Examples* text box, then m/d/yy is displayed in the *Date format* text box.

Choose **Document** from the *Field category* list box to list fields related to working in a document: Author, Comments, Filename, Filesize, HyperlinkBase, Keywords, Last-SavedBy, Subject, and Title. The top right text box that appears when the Document category is selected corresponds to the field selected in the Field names list box. The items listed in the *Format* list box are (none), Uppercase, Lowercase, First capital, and Title case. The Filesize item lists Bytes, Kilobytes, and Megabytes.

Choose **Linked** from the *Field category* list box to list fields related to text display for the hyperlinks. The top-right text box that appears when the Linked category is selected is *Text to display*, which allows to you enter a text string that will be inserted in the multiline text box and a link to the specified hyperlink when selected. Choosing *Hyperlink* causes the Insert Hyperlink dialog box to be displayed. Creation of hyperlinks is explained in Chapter 14.

Choose **Objects** from the *Field category* list box to list fields related to named objects: BlockPlaceholder, Formula, NamedObject, and Object.

Choose **Other** from the *Field category* list box to list fields related to SystemVariable, LispVariable, and DieselExpression.

Choose **Plot** from the *Field category* list box to list fields related to plot-related variables: DeviceName, Login, PageSetupName, PaperSize, PlotDate, PlotOrientation, PlotScale, and PlotStyleTable.

Choose **SheetSet** from the *Field category* list box to list fields related to sheet set variables: CurrentSheetCategory, CurrentSheetCustom, CurrentSheetDescription, CurrentSheet-IssuePurpose, CurrentSheetNumber, CurrentSheetNumberAnd Title, CurrentSheetRevisionDate, CurrentSheetRevisionNumber, Current Sheet-Set, CurrentSheetSetCustom, Current-SheetSetDescription, CurrentSheetSet ProjectMilestone, CurrentSheetSetProjectName, CurrentSheetSetProjectNumber, CurrentSheetSetProjectPhase, CurrentSheetSubSet, CurrentSheetTitle, SheetSet, SheetSetPlaceholder, SheetView.

Choosing *Columns* causes the Column flyout menu to be displayed (Figure 4–26), which provides three column options: No Columns, Dynamic Columns, and Static Columns.

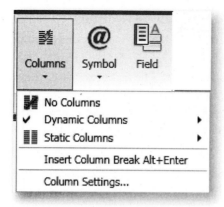

FIGURE 4–26 *Columns flyout menu*

Choosing *No Columns* specifies no columns for the current mtext object.

Choosing *Dynamic Columns* sets dynamic columns mode to the current mtext object which becomes text driven. Adjusting columns affects text flow and text flow causes columns to be added or removed. Options to dynamic columns are Auto height and Manual height.

Choosing *Static Columns* sets static columns mode to the current mtext object. This allows you to specify the total width and height of the mtext object and the number of columns. The columns share the same height and are aligned at both sides.

Choosing *Insert Column Break* causes a manual column break to be inserted. It is disabled when No Columns is selected.

Choosing *Column Settings* causes the Column Settings dialog box to be displayed from which you can select the column type, number, height, and width.

Spell Check

The *Spell Check* panel (see Figure 4–27) allows you to establish settings related to spell check.

FIGURE 4–27 *Spell Check panel on the Multiline Text tab*

Choosing *Spell Check* determines whether As-You-Type spell check is on or off.

Choosing the down arrow from the Spell Check panel opens the Check Spelling Settings dialog box (see Figure 4–28) which specifies the settings related to spell check.

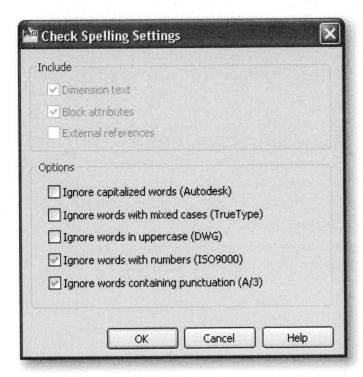

FIGURE 4–28 *Check Spelling Settings box*

The *Include* section specifies whether to include specific text options such as Dimension text, Block attributes, and External references in addition to the text objects that will be checked in your drawing.

The *Options* section specifies whether to include capitalized words, mixed cases, upper case, words with numbers, and words containing punctuation in the spell check.

Edit Dictionaries causes the Dictionaries dialog box to be displayed, which allows management of custom dictionaries.

Tools Panel

The *Tools* panel (see Figure 4–29) allows you to search for specified text strings and replace them with new text.

FIGURE 4–29 *Tools panel on the Multiline Text tab*

Choosing *Find and Replace* causes the Find and Replace dialog box to be displayed (Figure 4–30).

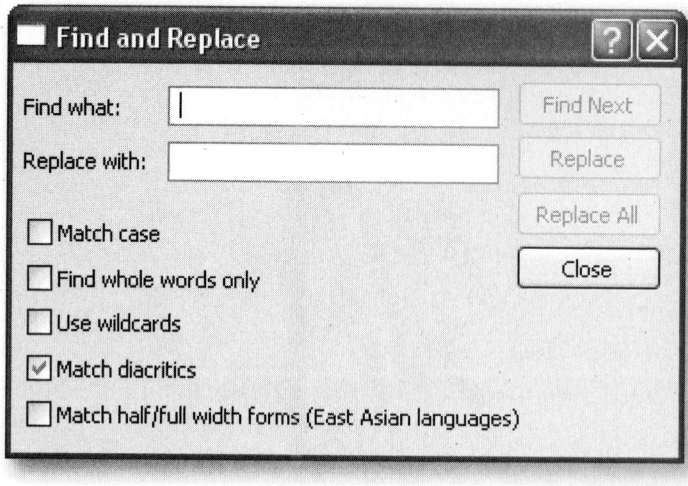

Find and Replace

Find what:		Find Next
Replace with:		Replace
		Replace All

☐ Match case

☐ Find whole words only

☐ Use wildcards

☑ Match diacritics

☐ Match half/full width forms (East Asian languages)

Close

FIGURE 4–30 *Find and Replace dialog box*

In the **Replace with** text box, type the text string that you want to replace the text string in the **Find what** text box. Then choose **Replace** to replace the highlighted text with the text in the **Replace with** text box. If you choose **Replace All**, all instances of the specified text will be replaced.

Selecting **Match case** causes AutoCAD to find text only if the case of all characters in the text object matches the text characters in the **Find what** text box. When it is not selected, AutoCAD finds a match for the specified text string regardless of the case of the characters.

Selecting **Match whole word only** causes AutoCAD to find text only if the text string is a single word. If the text is part of another text string, it is ignored. When it is not selected, AutoCAD finds a match for the specified text string whether it is a single word or part of another word.

Choosing **Import text** causes the Select File dialog box to be displayed, from which a file of ASCII or RTF format can be imported. The file imported is limited to 32 kilobytes.

Choosing **AutoCAPS** converts all new and imported text to uppercase without affecting existing text.

Options Panel

The *Options* panel (see Figure 4–31) allows you to assign various additional settings related to paragraph text.

FIGURE 4–31 *Options panel on the Multiline Text tab*

Choosing *More* causes a shortcut menu to be displayed (see Figure 4–32).

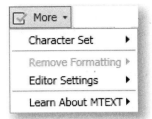

FIGURE 4–32 *More shortcut menu*

Choosing *Character Set* displays a menu of code pages from which a code page can be selected to apply to the selected text.

Choosing *Remove Formatting* causes character formatting to be removed for selected characters, paragraph formatting for selected paragraphs to be removed, or all formatting to be removed from a selected paragraph.

Choosing *Editor Settings* displays a list of options (see Figure 4–33) for the Text Formatting toolbar.

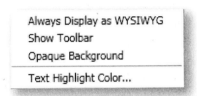

FIGURE 4–33 *Editor Settings shortcut menu*

Selecting *Always Display as WYSIWYG* **(What You See Is What You Get)** allows you to control the display of the In-Place Text Editor and the text within it. When unchecked, text that would otherwise be difficult to read (if it is very small, very large, or is rotated) is displayed at a legible size and is oriented horizontally so that you can easily read and edit it. This option is unchecked by default. When this option is checked, the MTEXTFIXED system variable will be set to 0. Otherwise, MTEXTFIXED will be set to 2.

Selecting *Show Toolbar* causes the Text Formatting toolbar to be displayed above the Text Editing box (see Figure 4–34). Tools and commands available on this traditional toolbar are now accessible in the Ribbon/Tab/Panel interface previously explained.

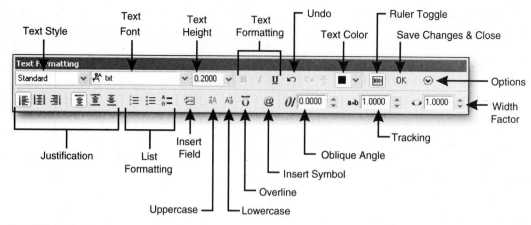

FIGURE 4–34 *Text Formatting Toolbar*

Selecting *Opaque Background* causes the background to become opaque.

Selecting *Text Highlight Color . . .* causes the Select Color dialog box to be displayed to choose a color for the text object.

Choosing *Ruler* causes a ruler to be displayed at the top of the text editor bounding box. The ruler allows you to set indents and tabs and to adjust the width of the bounding box. To set indents, use the arrows on the left of the ruler. To set the indent of the first line of a paragraph, drag the top arrow to the desired point. To set the indent of the rest of the lines in the paragraph, drag the bottom arrow to the desired point. To set a tab, specify a point on the ruler. To change the width of the bounding box, drag the double arrow on the right end of the ruler to the desired width. You can also right-click in the ruler and select *Indents and Tabs* or *Set MTEXT Width* from the shortcut menu. These menu items open dialog boxes in which numeric values can be entered for setting indents, tabs, and bounding-box width.

Choosing *Undo* undoes actions in the In-Place Text Editor, including changes to either text content or text formatting. You can also use CTRL + Z.

Choosing *Redo* causes the actions that were undone to be redone.

Close Panel

The *Close* panel (see Figure 4–35) allows you to close the Text Editor.

FIGURE 4–35 *Close panel on the Multiline Text tab*

Choosing **Close Text Editor** ends the editing session and closes the mtext ribbon.

EDITING TEXT

Multiline text can be edited with the MTEDIT command or by double-clicking the multiline text object.

AutoCAD displays the Multiline Text panel and the ruler above the selected text. Make the necessary changes in the text string in the same manner and with the same tools as when the selected text was originally created. Select **Close Text Editor** to accept the changes.

Single-line text can be edited with the DDEDIT command or by double-clicking the single line text object.

AutoCAD allows you to edit the selected text object. After making the necessary changes, press Enter and the prompt is repeated to select additional text objects to edit. When done with editing, provide a NULL response to terminate the command sequence.

The DDEDIT command can also be used to edit multi-line text.

Finding and Replacing Text

The FIND command is used to find a string of specified text and replace it with another string of specified text. Type the string of specified text in the Find edit field located in the Text panel (Figure 4–36) on the Annotation tab and press Enter. AutoCAD zooms to specified text and displays the Find and Replace dialog box, similar to Figure 4–37.

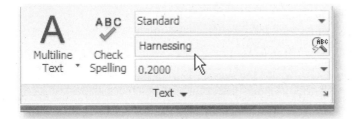

FIGURE 4–36 *Find edit field in the Text panel*

In the **Replace with** text box, enter the new text string. Select **Replace** to replace this instance with the string in the **Replace with** text box. To skip over this instance without replacing it, select **Find Next**. To replace all instances of the string entered in the **Find text string** text box, select **Replace All**.

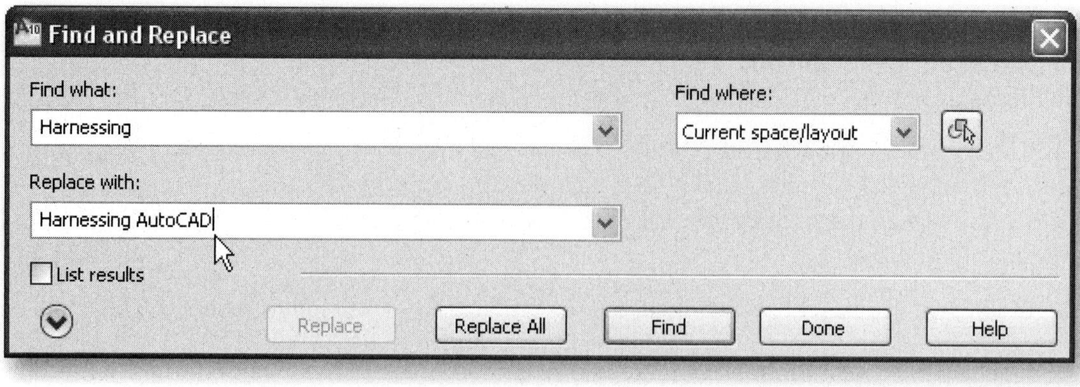

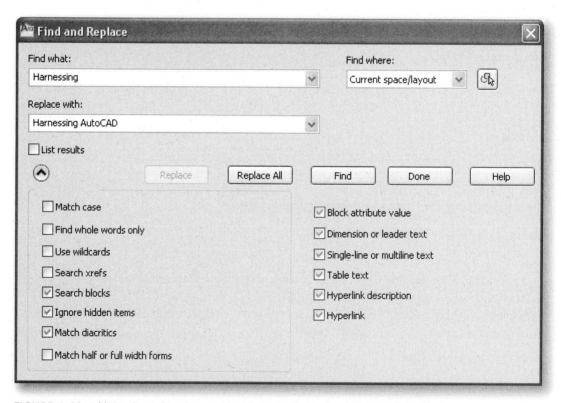

FIGURE 4–37 *Find and Replace dialog box*

In the **Find where** text box, you can direct AutoCAD to search either in the Entire drawing or the Current selection for the string to be replaced. The **Select objects** button returns you to the graphics screen to select text objects. AutoCAD searches for the string to be replaced. Selecting **More Options** button (located in the bottom left of the dialog box) expands the dialog box providing additional search options and text types (see Figure 4–38) to search and replace.

FIGURE 4–38 *Additional search options and text types in the Find and Replace dialog box*

Scaling Text Objects

The SCALETEXT command enlarges or reduces selected text objects without changing their locations. The SCALETEXT command is invoked by selecting *Scale* from the Text panel (Figure 4–39) on the Annotation tab.

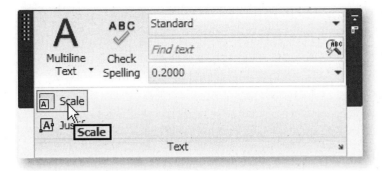

FIGURE 4–39 *Invoking the SCALETEXT command from the Text panel*

AutoCAD prompts:

> Select Objects: *(Use an object selection method, and press* Enter *when you finish)*
>
> Enter a base point option for scaling [Existing/Left/Center/ Middle/Right/TL/TC/TR/ML/MC/MR/BL/BC/BR]<Existing>: *(Specify a location to serve as a base point for scaling)*
> Specify new model height or [Paper height/Match object/Scale factor]: *(Specify a text height or select an option)*

With the base point prompt, you choose one of several locations to serve as base points for scaling, which is used individually for each selected text object. The base point for scaling is established on one of several insertion point locations for text options, but even though the options are the same as when you choose an insertion point, the justification of the text objects is not affected.

The *Paper Height* option selection scales the text height depending on the annotative property.

The *Match object* option selection scales the text objects that you originally selected to match the size of a selected text object.

The *Scale factor* option selection scales the selected text objects based on a reference length and a specified new length.

Justifying Text

The JUSTIFYTEXT command lets you change the justification point of a text string without having to change its location.

The JUSTIFYTEXT command is invoked by selecting *Justify* from the Text panel (Figure 4–40) on the Annotation tab, and AutoCAD displays the following prompts:

> Select objects: *(select a text object)*
>
> Select objects: Enter
>
> Enter a justification option *(select one of the available options to change the justification)*

AutoCAD changes the justification point of the selected text string without changing its location.

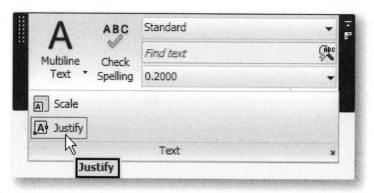

FIGURE 4–40 *Invoking the* JUSTIFYTEXT *command from the Text panel*

Spell Checking

The SPELL command is used to correct the spelling of text objects created with the TEXT or MTEXT command in addition to attribute values in Blocks. You can also check the spelling of words in dimension text and xrefs.

The SPELL command is invoked by selecting *Check Spelling* from the Text panel (Figure 4–41) on the Annotate tab.

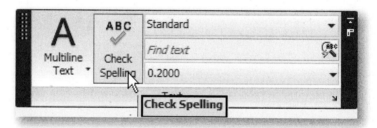

FIGURE 4–41 *Invoking the* SPELL *command from the Text panel*

AutoCAD displays the Check Spelling dialog box (Figure 4–42). Select the Start button to begin the spelling check.

If Check Spelling is set to Entire Drawing, spelling is checked on model space, then on available layouts in paper space. Regardless of what current drawing area is showing, once Check Spelling starts, the drawing window changes to display the content that is currently being checked. If a flagged word is identified, the drawing area highlights and zooms to that word.

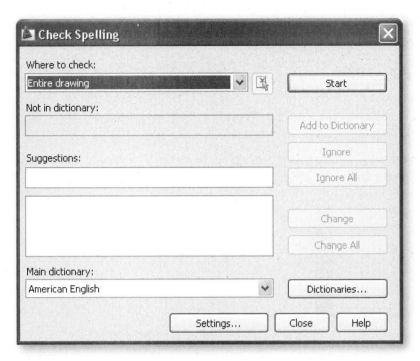

FIGURE 4–42 *Check Spelling dialog box*

The **Where to Check** text box displays the areas you want checked for spelling. Three options are available: Entire Drawing, Current Space/layout, and Selected Objects. When using **Selected objects,** click the button on the right of the list box to select the objects to spell check.

The **Not in Dictionary** text box displays the word identified as misspelled.

The **Suggestions** text box displays a list of suggested replacement words from the current dictionary. The first suggestion in the list box is highlighted in both Suggestions areas. You can select another replacement word from the list, or edit or enter a replacement word in the top Suggestions text area.

The **Main Dictionary** text box lists the default dictionary and if necessary you can switch to another dictionary.

Selecting **Ignore** skips the current word.

Selecting **Add to Dictionary** adds the current word to the current custom dictionary. The maximum word length is 63.

Selecting **Change** replaces the current word with the word in the Suggestions box.

Selecting **Change All** replaces the current word in all selected text objects in the spell check area.

Selecting **Dictionaries** causes the Dictionaries dialog box to be displayed, which allows you to manage custom dictionaries.

Choosing **Settings** causes the Check Spelling Settings dialog box to be displayed; from there, you can specify whether you wish to include Dimension text, Block attributes, and External references in the spell check. In the Options section, you can specify options to Ignore capitalized words, Ignore words with mixed cases, Ignore words in uppercase, Ignore words with numbers, or Ignore words containing punctuation.

After completion of the spell check, AutoCAD displays an AutoCAD message informing you that the spell check is complete.

Inserting a Field

The FIELD command creates a multiline text object with a field that is updated automatically as the field value changes. When the field is updated, the latest value of the field is displayed.

The FIELD command is invoked by selecting *Insert Field* from the Data panel (Figure 4–43) on the Insert tab.

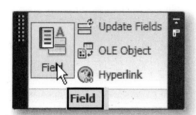

FIGURE 4–43 *Invoking the* FIELD *command from the Data panel on the Insert tab*

AutoCAD displays the **Field** dialog box. For detailed explanation, see the section on Creating Multiline Text.

Updating a Field

The UPDATEFIELD command allows you to manually update fields in selected objects in the drawing. By default, AutoCAD updates when you open the drawing, save the drawing, plot the drawing, regenerate the drawing, and eTransmit the drawing.

The UPDATEFIELD command is invoked by selecting *Update Fields* from the Data panel (Figure 4–44) on the Annotation tab.

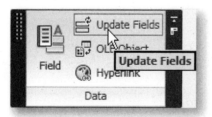

FIGURE 4–44 *Invoking the* UPDATEFIELD *command from the Text panel*

AutoCAD prompts:

Select Objects: *(Use an object selection method, and press* Enter *when you finish)*

Fields in selected objects are updated.

Controlling the Display of Text

The QTEXT command is a utility command for TEXT and MTEXT that is designed to reduce the redraw and regeneration times of a drawing. Regeneration time becomes a significant factor if the drawing contains a great amount of text and attribute information and/or if a fancy text font is used. Using QTEXT, the text is replaced with rectangular boxes of a height corresponding to the text height. These boxes are regenerated in a fraction of the time required for the actual text.

If a drawing contains many text and attribute items, it is advisable to set QTEXT to ON. However, before plotting the final drawing, or inspection of text details, the QTEXT command is set to OFF and is followed by the REGEN command.

Invoke the QTEXT command at the on-screen prompt, and AutoCAD displays the following prompt:

Enter mode *(Enter on or off)*

CREATING AND MODIFYING TEXT STYLES

The STYLE option of the TEXT and MTEXT commands, in conjunction with the STYLE command, lets you determine how text characters and symbols appear, other than the usual adjusting of height, slant, and angle of rotation. To specify a text style from the STYLE option of the TEXT and MTEXT commands, the style must have been defined by using the STYLE command. In other words, the STYLE command creates a new style or modifies an existing style. The STYLE option under the TEXT or MTEXT command allows you to choose a specific style from the styles available.

There are three things to consider in creating a new style with the STYLE command. First, you must name the newly defined style. Style names may contain up to 255 characters, numbers, and special characters, including $, –, and _. Names such as "title block," "notes," and "bill of materials" can remind you of the purpose for which the particular style was designed.

Second, you may apply a particular font to a style. The font that AutoCAD uses as a default is called TXT. It has blocky-looking characters that are economical to store in memory. The TXT.SHX font, made up entirely of straight-line (non-curved) segments, is not considered as attractive or readable. Other fonts offer many variations in characters, including those for foreign languages. All fonts are stored for use in files of their font name with an extension of .SHX. The most effective way to get a distinctive appearance in text strings is to use a specially designed font. You can also use TrueType and Type 2 PostScript fonts. See Appendix F for a list of fonts that come with AutoCAD. If necessary, you can buy additional fonts from third-party vendors, and AutoCAD can read hundreds of PostScript fonts available in the marketplace.

The third consideration of the STYLE command is in how AutoCAD treats the general physical properties of the characters, regardless of the font that is selected. These properties are the height, width-to-height ratio, obliquing angle, backwards, upside-down, and orientation (horizontal/vertical).

Invoke the STYLE command from the Text panel by clicking the down arrow (see Figure 4–45) on the Annotate tab, and AutoCAD displays the Text Style dialog box (Figure 4–46).

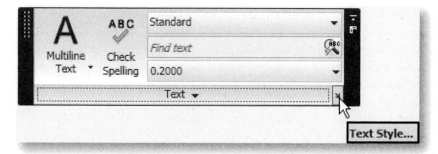

FIGURE 4–45 *Invoking the* STYLE *command from the Text panel*

Choose **New** to create a new style. AutoCAD displays the New Text Style dialog box (Figure 4–47). Enter an appropriate name for the text style, and choose **OK** to create the new style.

To rename an existing style, first select the style from the **Style Name** list box in the Text Style dialog box, and then choose **Rename**. AutoCAD displays the Rename Text Style dialog box. Make the necessary changes to the name of the style, and choose **OK** to rename the text style.

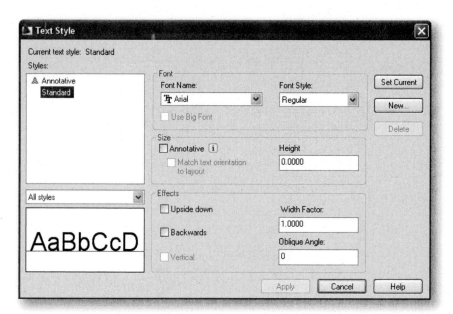

FIGURE 4–46 *Text Style dialog box*

FIGURE 4–47 *New Text Style dialog box*

To delete an existing style, select the style from the **Styles** list box in the Text Style dialog box, and then choose **Delete**. AutoCAD displays the AutoCAD Alert dialog box to confirm the deletion of the selected style. Click **Yes** to confirm the deletion or **No** to cancel the deletion of the selected style.

To assign a font to the selected text style, select the appropriate font from the **Font Name** list box. Similarly, select a font style from the **Font Style** list box. The font style specifies font character formatting such as italic, bold, or regular.

In the Size section, choosing **Annotative** causes the text created with the selected style to be annotative. Choosing **Match Text Orientation to Layout** causes the orientation of the text in paper space viewports to match the orientation of the layout. This option is available only if the Annotative option is chosen. The **Paper Text Height** text box sets the text height to the value you enter. If you set the height to zero, then when you use this style in the TEXT or MTEXT command, you are given an opportunity to change the text height with each occurrence of the command. If you set it to any other value, that value will be used for this style, and you will not be allowed to change the text height.

In the Effects section, selecting **Backwards** or **Upside Down** controls whether the text is drawn right to left, with the characters backward, or upside down, or left to right, respectively (Figure 4–48).

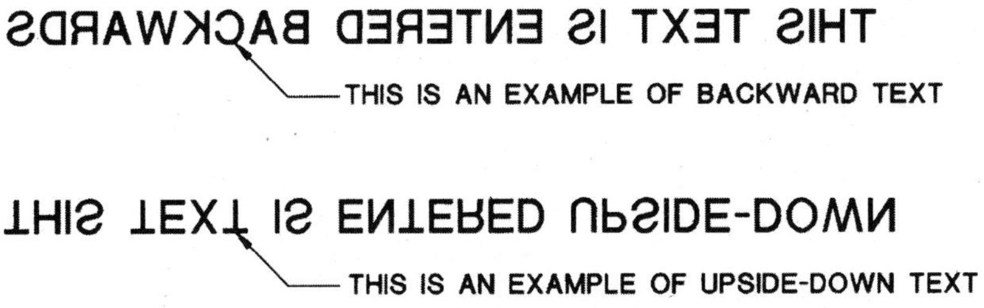

FIGURE 4–48 *Examples of Backward and Upside-down text*

Selecting **Vertical** controls the display of the characters aligned vertically. The Vertical option is available only if the selected font supports dual orientation. See Figure 4–49 for an example of vertically oriented text.

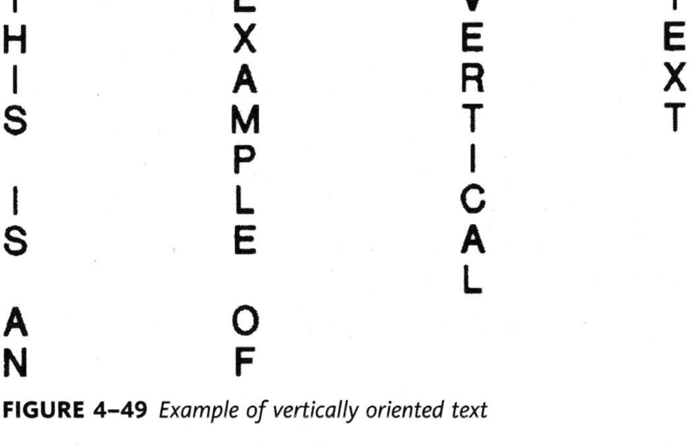

FIGURE 4–49 *Example of vertically oriented text*

The **Width Factor** text box sets the character width relative to text height. If it is set to more than 1.0, the text widens; if it is set to less than 1.0, the text narrows.

The **Oblique Angle** text box sets the obliquing angle of the text. If it is set to 0°, the text is drawn upright. Otherwise, it is set to 90°. A positive value slants the top of the characters toward the right or in clockwise direction. A negative value slants the characters in counterclockwise direction. See Figure 4–50 for examples of oblique angle settings applied to a text string.

OBLIQUING ANGLE SET TO 22°
— POSITIVE OBLIQUING ANGLE

OBLIQUING ANGLE SET TO -22°
— NEGATIVE OBLIQUING ANGLE

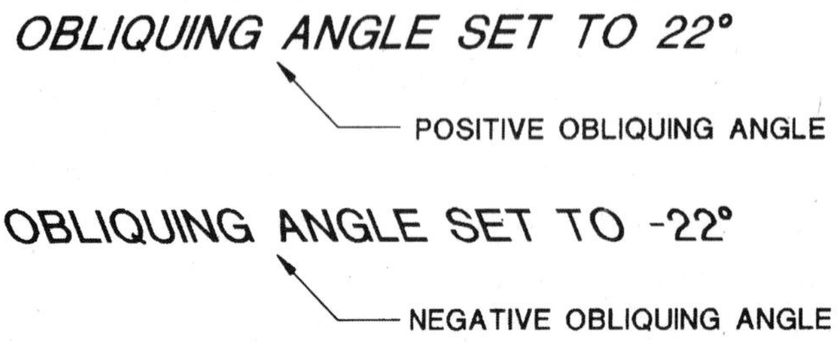

FIGURE 4–50 *Example of oblique angle settings applied to a text string*

The **Preview** section of the Text Style dialog box displays sample text that changes dynamically as you change fonts and modify the effects. To change the sample text, enter characters in the box below the larger preview image.

After making the necessary changes in the Text Style dialog box, choose **Apply** to apply the changes. Choose **Close** to close the Text Style dialog box.

CREATING AND MODIFYING TABLES

AutoCAD's TABLE command makes it easy to create tables that contain text in the row-and-column format customarily found on drawings for listing revisions, deadline schedules, specifications, and other structured textual information. A combination of table characteristics such as row and column sizes, border lineweights, text alignments and associated text styles, and colors can be saved in a Table Style with a specified name that will be recalled and applied to tables when required.

You can link table data to data in Microsoft Excel, including links to single cells, ranges of cells, or the entire spreadsheet.

Inserting Tables

Figures 4–51 and 4–52 show typical examples of tabular information included in drawings. Invoke the TABLE command from the Tables panel (see Figure 4–53) on the Annotate tab, and AutoCAD displays the Insert Table dialog box (Figure 4–54).

ROOM FINISH SCHEDULE				
ROOM NO.	ROOM DESCRIP	WALLS	FLOOR	CEILING
100	ENTRY	PLASTER	CARPET	SUSP. ACOUS.
101	HALL	GYPSUM	CARPET	GYPSUM
102	RECEPTION	PANELLING	VINYL TILE	GYPSUM
103	OFFICE	GYPSUM	CARPET	GYPSUM

FIGURE 4–51 *Example of Room Finish Schedule drawn as table from top down*

3	CLOSET DIM	T.A.S.	01/23/04
2	WDW DETAIL	G.V.K.	01/12/04
1	DOOR 101	A.B.C.	01/08/04
REVISION NO.	DESCRIPTION	BY	DATE
REVISIONS			

FIGURE 4–52 *Example of Revision History drawn as table from bottom up*

FIGURE 4–53 *Invoking the TABLE command from the Tables panel*

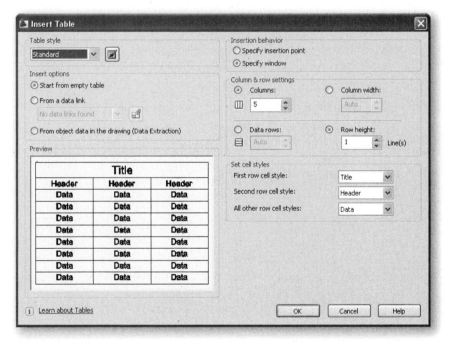

FIGURE 4–54 *The Insert Table dialog box*

The text box in the **Table Style** section of the Insert Table dialog box enables you to select the table style to be applied. Refer to the "Creating and Modifying Table Styles" section later in this chapter. You can create a new table style by clicking the button next to the drop-down list.

In the **Insert Options** section, choose **Start from empty table** to create an empty table that can be manually filled with data. Choose **From a data link** to create a table from data in an external spreadsheet. Choose **From object data in the drawing (Data Extraction),** and AutoCAD launches the Data Extraction wizard.

The Preview section displays an example of the current table style.

From the **Insertion behavior** section, choosing **Specify insertion point** creates a table with reference to the location of the upper-left corner of the table, the default table style, and based on the number of the columns, column width, number of data rows, and row height specified in the Column & Row Settings section of the dialog box. If the table style calls for the direction of the table to read from the bottom up, the insertion point is the lower-left corner of the table. Choosing **Specify window** creates a table based on the constraints set in the Column & Row Settings section of the table.

In the **Columns** text box you can specify the number of columns with the column width automatically set, or in the **Column Width** text box you can specify the column width with the number of columns automatically set. You can specify the number of **Data Rows** with the **Row Height** automatic, or you can specify the number of lines for the Row Height and the number of **Data Rows** automatic.

In the **Set cell styles** section, you can specify a cell style for rows in the new table for table styles that do not contain a starting table. In the **First row cell style** text box you can specify a cell style for the first row in the table with Title style as the default. In the **Second row cell style** text box you can specify a cell style for the second row in the table with as Header style the default. In the **All other row cell styles** text box you can specify a cell style for all other rows in the table with Data style as the default.

After setting up the Table Style and Column/Row configuration, choose **OK** to close the Insert Table dialog box. AutoCAD prompts you to "Specify first corner:" and shows a phantom image of the potential table attached to the cursor, following its movement.

Once you specify the insertion point, if you have chosen **Specify insertion point** in the Insertion Behavior section, AutoCAD draws the table, shows the text-entry cursor in the heading at the justification specified, and displays the Multiline Text tab. If you have chosen **Specify window** in the Insertion Behavior section, AutoCAD prompts you to "Specify second corner:" allowing you to drag the cursor to determine the number of columns and rows desired. Then AutoCAD displays the Multiline Text tab.

While the Multiline Text tab is displayed, text can be input into each cell by typing or pasting from the clipboard. You can move from a cell to the cell below it by pressing the Enter key, to the cell above by pressing SHIFT + Enter, to the cell on the right by pressing the TAB key, or to the cell on the left by pressing SHIFT+TAB.

When a table is active you can click in one of the cells and the Table Cell tab will be displayed (see Figure 4–55). This tab includes panels for **Rows, Columns, Merge, Cell Styles, Cell Form, Insert,** and **Data.** You can also right-click on a cell and use the options on the shortcut menu for similar functions.

FIGURE 4–55 *The Table Cell tab*

In the Rows and Columns panels you can insert rows above and below or columns left and right, and you can delete columns.

In the Merge panel you can merge and unmerge merged cells.

In the Cell Styles panel, choosing **Match Cell** applies the properties of a selected cell to other cells. Choosing **Cell Styles** lists all cell styles contained within the current table style. The cell styles Title, Header, and Data are always contained within any table style and cannot be deleted or renamed. Choosing **Cell Borders** lets you specify the properties of the borders of the selected table cells. Choosing **Alignment** lets you specify alignment for the content within cells. Content is middle-, top-, or bottom-aligned with respect to the top and bottom borders of the cell. Content is center-, left-, or right-aligned with respect to the left and right borders of the cell. Choosing **Background Fill** lets you specify the fill color. Select None or a background color, or click Select Color to display the Select Color dialog box.

In the Cell Format panel, choosing **Cell Locking** locks or unlocks cell content and/or formatting from editing. Choosing **Data Format** displays a list of data types including Angle, Date, Decimal Number, General, Percentage, Point, Text, and Whole Number.

In the Insert panel, choosing **Block** causes the Insert dialog box to be displayed. Choosing **Field** causes the Field dialog box to be displayed. Choosing **Formula** inserts a formula into the currently selected table cell. A formula must start with an equal sign (=). The formulas for sum, average, and count ignore empty cells and cells that do not resolve to a numeric value. Other formulas display an error (#) if any cell in the arithmetic expression is empty or contains nonnumeric data. Choosing **Manage Cell Contents** displays the content of the selected cell. You can change the order of cell content as well as change the direction in which cell content will appear.

In the Data panel, choosing **Link Cell** causes the New and Modify Excel Link dialog box to be displayed, where you can link data from a spreadsheet created in Microsoft Excel to a table within your drawing. Choosing **Download from Source** updates data in the table cell that is referenced by changed data in an established data link.

Table Modifying Options

Modifying options are available from the shortcut menu when a cell is selected. The options include the following:

Choosing *Cell Style* displays a shortcut menu. You can choose between *By Row/ Column*, *Data*, *Header*, or *Title*. Choosing *Save as New Cell Style* displays the Save as New Cell Style dialog box, which allows you to save the formatting of the cell with a specific style name.

Choosing *Background Fill* lets you specify the fill color.

Choosing *Alignment* allows you to change the alignment of the selected text in a cell.

Choosing *Borders* sets the properties of the borders of table cells.

Choosing *Locking* displays a shortcut menu. You can choose between *Unlocked*, *Content Locked*, *Format Locked*, or *Content and Format Locked*.

Choosing *Data Format* displays Data Cell Format dialog box where you can specify the data type for the cells. Types include Angle, Currency, Date, Decimal Number, General, Percentage, Point, Text, and Whole Number.

Choosing *Match Cell* applies the properties of a selected table cell to other table cells.

Choosing *Remove All Property Overrides* removes any property overrides applied to the selected table.

Choosing *Datalink* displays the Select a Data Link dialog box where you can create links to a spreadsheet.

Choosing *Insert* displays a shortcut menu. Choosing *Block* allows you to insert a block or drawing that is stored locally or in a network. Choosing *Field* allows you to insert a field. Choosing *Formula* displays a shortcut menu from which you can perform the functions *Sum*, *Average*, *Count*, *Cell*, or *Equation*.

Choosing *Edit Text* removes the existing text in the cell and allows you to retype it.

Choosing *Manage Content* displays the Manage Cell Content dialog box that allows you to list the cell content, move it up or down, or delete it. Other options include changing the layout mode with *Flow*, *Stacked horizontal*, and *Stacked vertical*. You can also specify the content spacing.

Choosing *Delete All Contents* deletes all of the contents of the selected column.

Choosing *Columns* displays a shortcut menu that allows you to insert a column right or left of the selected cell or delete the selected cell.

Choosing *Rows* displays a shortcut menu that allows you to insert a row above or below the selected cell or delete the selected cell.

Choosing Merge displays a shortcut menu that allows you to merge selected cells with the option of *All*, *By Row*, or *By Column*.

Choosing *Unmerge* causes selected merged cells to be become separate cells.

Similar to modifying the individual cells, AutoCAD allows you to modify the table. First select the table, and then select the available options from the shortcut menu. The options include the following:

Choosing *Properties* causes the Properties palette to display the properties of the selected cell.

Choosing *Quick Properties* causes the Quick Properties palette to display the properties of the selected cell.

Creating and Modifying Table Styles

The TABLESTYLE command is used to create a new table style or modify an existing one. The appearance of the table is controlled by its table style. The table style can specify a justification and appearance for the text and gridlines for the title, column heads, and data.

To create a new table style, invoke the TABLESTYLE command by selecting the down arrow from the Tables panel (Figure 4–56) on the Annotate tab. AutoCAD displays the Table Style dialog box, as shown in Figure 4–57.

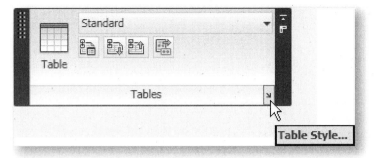

FIGURE 4–56 *Invoking the* TABLESTYLE *command from the Tables panel*

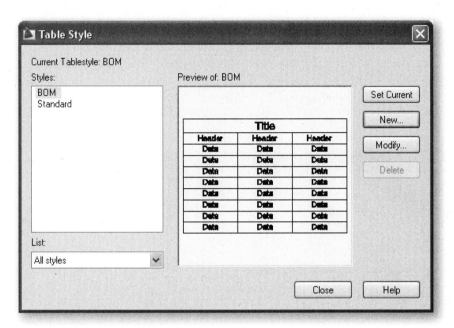

FIGURE 4–57 *Table Style dialog box*

The **Current Tablestyle** heading shows the name of the table style to which the next drawn table will conform.

The Styles section displays the name(s) of the current style(s) and any other style(s) available depending on the option chosen in the **List** selection box.

From the **List** selection box, you can choose to list all of the styles available in the drawing or only those in use.

The **Preview of** window shows how the table will appear using the current table style.

Choosing **Set Current** causes the table style chosen in the **Styles** section to become the current table style. To delete a table style, select the style in the **Styles** section and choose **Delete**. The Standard style cannot be deleted.

To create a new table style, choose **New**. AutoCAD displays the Create New Table Style dialog box, as shown in Figure 4–58. You can create a new table style based on an existing table style whose settings are the default for the new table style.

FIGURE 4–58 *Create New Table Style dialog box*

Specify the name of the new style in the **New Style Name** box. The **Start With** list box lets you choose the existing table style whose settings are the default for the new table style. Choose **Continue** to display the New Table Style dialog box, as shown in Figure 4–59, in which you define the new style properties. There are three tabs to choose from: **Data, Column Heads,** and **Title**. Options on each tab set the appearance of the data cells, the column heads, or the table title. The dialog box initially displays the properties of the table style that you selected as the **Start With** style.

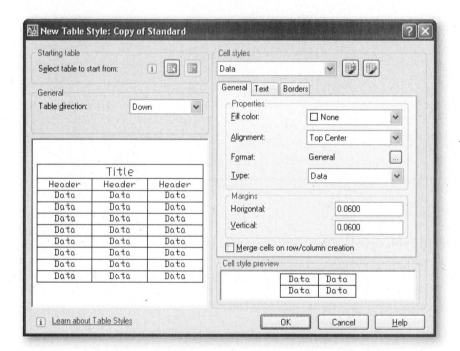

FIGURE 4–59 *New Table Style dialog box with the General tab selected*

The Starting Table section allows you to select a table in your drawing as a basis from which to start this table style. After selecting a table, you can specify the structure and contents you want copied from that table to the table style. Choose the **Remove Table** icon to remove a table from the current specified table style.

The **General** section changes the direction of the table. **Table Direction** allows you to set the direction of a table. Selecting **Down** creates a table that reads from top to bottom, and Selecting **Up** creates a table that reads from bottom to top.

The **Preview** section shows how the table will appear with the current table style settings.

The **Cell Styles** section allows you to set the appearance of the data cells, the cell text, and the cell borders, depending on which tab is active: General, Text, or Borders.

On the **General** tab of the Cell Styles section, you can specify cell properties such as Fill Color, Alignment, Format, and data Type in the Properties section. In the Margins section, you can specify the horizontal and vertical margins between the content and cell borders. This action applies to all cells. Choose **Merge cells on row/column creation** to merge new rows or columns created with the current cell style into one cell. This option can be used to create a title row at the top of your table.

On the **Text** tab of the Cell Styles section, you can specify the text properties for cell content such as Text Style, Text Height, Text Color, and Text Angle.

On the **Borders** tab of the Cell Styles section, you can specify the format of the cell borders such as Lineweight, Linetype, Color, Double line, and what the line spacing will be. Choose one of the Border buttons to specify where borders will be applied to the cell.

The **Cell style preview** section displays an example of the effect of the current table style settings.

CREATING OBJECTS FROM EXISTING OBJECTS

AutoCAD not only allows you to draw objects easily but also allows you to create additional objects from existing objects. This section discusses six important commands that will make drawing easier: COPY, ARRAY, OFFSET, MIRROR, FILLET, and CHAMFER.

Copying Objects

The COPY command places copies of the selected objects at the specified displacement, leaving the original objects intact. The copies are oriented and scaled the same as the original. If necessary, you can make multiple copies of selected objects and in a linear array. Each resulting copy is completely independent of the original and can be edited and manipulated like any other object.

Invoke the COPY command from the Modify panel (Figure 4–60) on the Home tab, and AutoCAD displays the following prompts:

> Select objects: *(select the objects and press* Enter *to complete the selection)*
>
> Specify base point or ⊡ *(specify base point or right-click and choose one of the options)*
>
> Specify second point or ⊡: *(specify a point for displacement, or right-click and choose one of the options)*

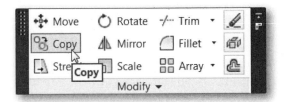

FIGURE 4–60 *Invoking the* COPY *command from the Modify panel*

If you specify two data points, AutoCAD computes the displacement and places a copy accordingly. If you press Enter in response to the second point of displacement, AutoCAD considers the point provided as the second point of a displacement vector with the origin (0,0,0) as the first point, indicating how far to copy the objects and in what direction.

The following command sequence shows an example of copying a group of objects by placing two data points (Figure 4–61).

> copy Enter
>
> Select objects: *(select objects with Crossing selection option)*
>
> Select objects: Enter
>
> Specify base point or ⊞ *(specify the base point)*
>
> Specify second point or ⊞: *(specify the second point)*
>
> Specify second point or ⊞ *(press Enter to complete the command sequence)*

AutoCAD copies the selected object(s), placing them at a new location displaced from the location of the original objects at a direction and distance determined by the base point/second displacement point vector. By default, AutoCAD allows you to make multiple copies. To terminate the command sequence, press ESC or choose *Exit* from the shortcut menu. To erase the most recent copy of the select objects, choose *Undo* from the shortcut menu.

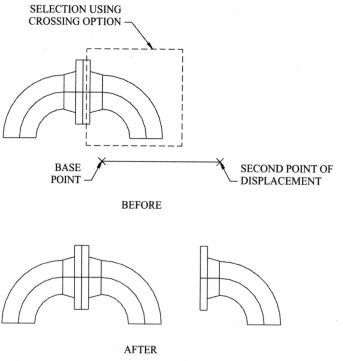

FIGURE 4–61 *Using the COPY command*

NOTE

The first and second points of displacement do not have to be on or near the object. For example, you can enter 1,1 and 3,4 as the first and second points, respectively, causing the objects to be moved or copied 2 units in the X direction and 3 units in the Y direction. To have the same result, you could have entered 0,0 at the first point and 2,3 at the second prompt. If you were moving or copying the object 24 to the right, you could select a point on the screen for the base point and then enter @24<0 for the displacement relative to the point specified. You can simplify specifying the move or copy displacement vector if you know how far in the X direction and how far in the Y direction you wish to move or copy the selected objects. To do this, enter the coordinates for the second point of displacement at the first prompt (specify base point of displacement). Then you press [Enter] in response to the second prompt. Auto-CAD uses the origin (0,0,0) as the first point. This is a stroke-saving procedure. For the example at the beginning of this note, you could have entered 2,3 at the first prompt and pressed [Enter] at the second prompt. It is as if you had entered 0,0,0 at the first prompt and 2,3 at the second prompt. If you wish to move the object in the direction opposite to that of the example, you could enter -2,-3 at the first prompt and press [Enter] at the second prompt.

The **Mode** option available as part of the first prompt (specify base point of displacement) controls whether the command repeats automatically or in single mode setting. By default, the COPY command is set to repeat automatically for the duration of the command.

The **Array** option arranges a specified number of copies in a linear array.

Creating a Pattern of Copies

The ARRAY command is used to make multiple copies of selected objects in either rectangular or polar arrays (patterns) or along a selected path. In the rectangular array, you can specify the number of rows, the number of columns, and the spacing between rows and columns. Row and column spacing may differ. The whole array can be skewed at an angle relative to the X-axis, the Y-axis or both. In the polar array, you can specify the angular intervals, the number of copies, the angle that the group covers, and whether the objects maintain their orientation or are rotated as they are arrayed. In the path array, AutoCAD creates multiple copies of the selected object(s) along a selected path such as a polyline, circle or rectangle. You can specify whether the objects maintain their orientation or are aligned with the path as they are copied.

Invoke the ARRAY command from the Modify panel (Figure 4–62). By default the rectangular option is selected.

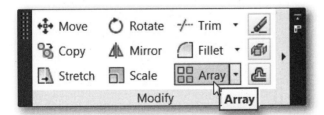

FIGURE 4–62 *Invoking the RECTANGULAR ARRAY command from the Modify panel*

You can change the type of Array command by selecting the down-arrow next to the current option (see Figure 4–63). From the drop-down menu you can select one of the other options, Path Array or Polar Array in this case. When a different option is selected it becomes the current option and its icon will be displayed in the Modify panel until a different option is selected.

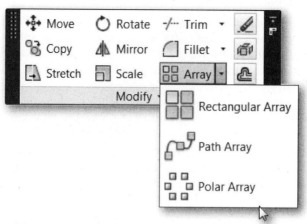

FIGURE 4–63 *Drop-down menu displayed when down-arrow is selected*

Rectangular Array

Invoke the RECTANGULAR ARRAY command from the Modify panel (Figure 4–62), and AutoCAD displays the following prompt:

> Select objects: *(select objects to array)*

After selecting one or more objects to be arrayed, AutoCAD prompts:

> Select grip to edit array or [ASsociative BasePoint COunt
> Spacing COLumns Rows Levels eXit].

By default, AutoCAD creates a rectangular pattern of 3 rows and columns as shown in Figure 4–64. Select one of the grips to make changes to the pattern or select one of the available options to make changes to the pattern.

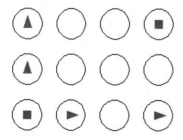

FIGURE 4–64 *Default rectangular array pattern*

By selecting the grip (blue box) of the Base Point (on the source object), AutoCAD allows you to reposition the array to a new location.

By selecting the grip of the last row and first column [see Figure 4–65(a)], AutoCAD allows you to add additional rows; and by selecting the grip of the first row and last column [see Figure 4–65 (b)], AutoCAD allows you to add additional columns.

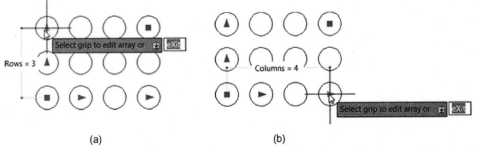

FIGURE 4–65 *(a) Selection of the grip of the last row and first column (b) Selection of the grip of the first row and last column*

To increase the spacing between the columns, select the grip of the first row and second column [see Figure 4–66(a)] and drag in the direction you want to increase the spacing.

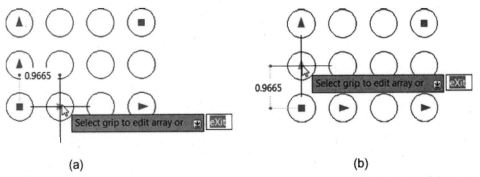

FIGURE 4–66 *(a) Selection of the grip of the first row and second column (b) Selection of the grip of the second row and first column*

Similarly, to increase the spacing between the rows, select the grip of the second row and first column [see Figure 4–66 (b)] and drag in the direction you want to increase the spacing.

To increase both number of rows and columns, select the grip of the last row and last column (Figure 4–67) and drag in the direction you want to increase the rows and columns.

Select the array and right-click and AutoCAD displays a drop down menu with options for editing the array.

Choosing ASsociative displays a menu for choosing Yes or No to determine if the array is created as associative or not. Whichever mode you choose (Yes or No) will remain the default until changed. If you choose Yes, the whole array of objects acts like a single object when operated on by commands such as MOVE, COPY, SCALE and ROTATE. The number of rows/columns and spaces, along with axis angles can be changed using control points (grips) placed at various points in the array.

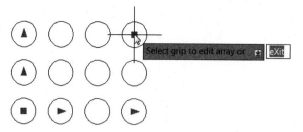

FIGURE 4–67 *Selection of the grip of the last row and last column*

Choosing *Base Point* lets you specify a base point for the array other than the default location at the source object(s). Right-clicking while being prompted for a Base Point displays a menu from which you can choose *Key Point*.

Choosing *Key Point* lets you specify a valid constraint (or *key point*) on the source objects to use as the base point. If you edit the source objects of the resulting array, the base point of the array remains coincident with the key point of the source objects.

NOTE

Choosing to specify a Key Point for a new location of the Base Point restricts you to available key points on the source object(s). If the source object is a line, available key points are the endpoints and the midpoint. For an arc, the endpoints and midpoint of the arc are available key points and if the arc angle is great enough, the center is available for selection. For a rectangle, the four corners and the midpoints of the four lines are available. If the source contains multiple objects, the key points of each of the objects are available. For a circle only the center is available. If you select a key point on a copied object in the array, the Base Point will be established at a corresponding point on the source object.

Choosing COunt lets you specify the number of rows and columns and provides a dynamic view of results as you move the cursor (a quicker alternative to the Rows and Columns options).

Choosing *Rows* lets you edit the number and spacing of rows in the array, and the incremental elevation between them.

Choosing *Columns* lets you edit the number and spacing of columns in the array.

Choosing *Levels* lets you specify the number and spacing of levels in the array, creating a 3D array.

Choosing *eXit* causes AutoCAD to draw the array in accordance with the numbers, distances and angles specified or their defaults.

When one of the objects in the array is selected AutoCAD shows an associative array of rectangles with the control points displayed (Figure 4–68).

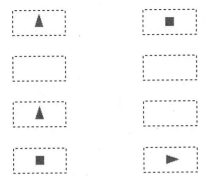

FIGURE 4–68 *Associative array of rectangles with control points displayed*

When you hover the cursor over the Base Point (on the source object) AutoCAD displays a shortcut menu (Figure 4–69).

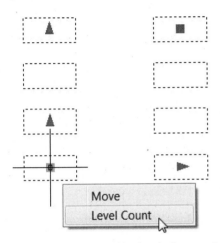

FIGURE 4–69 *Menu displayed when you hover the cursor over the Base Point*

Choosing *Move* lets you move the whole array in the same manner as the MOVE command.

Choosing *Level Count* lets you create a 3D array by specifying the number of levels in the array.

When you hover the cursor over the control point on the object in the last row/first column AutoCAD displays a shortcut menu (Figure 4–70).

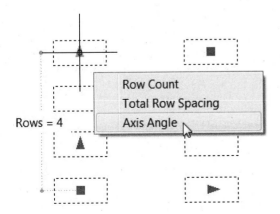

FIGURE 4–70 *Menu displayed when you hover the cursor over the first row/last column control point*

Choosing *Row Count* lets you specify the number of rows either with the cursor or entering the value from the keyboard.

Choosing *Total Row Spacing* lets you specify the distance between the first and last rows either with the cursor or entering the value from the keyboard.

Choosing *Axis Angle* lets you specify the angle that the columns deviate from the X-axis either with the cursor or entering the value from the keyboard.

When you hover the cursor over the control point on the object in the first row/last column AutoCAD displays a shortcut menu (Figure 4–71).

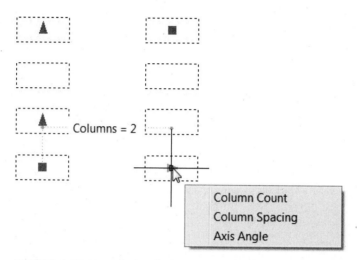

FIGURE 4–71 *Menu displayed when you hover the cursor over the last row/last column control point*

Choosing *Column Count* lets you specify the number of columns either with the cursor or entering the value from the keyboard.

Choosing *Total Column Spacing* lets you specify the distance between the first and last columns either with the cursor or entering the value from the keyboard.

Choosing *Axis Angle* lets you specify the angle that the rows deviate from the Y-axis either with the cursor or entering the value from the keyboard.

When you hover the cursor over the control point on the object in the last row/last column AutoCAD displays a shortcut menu (Figure 4–72).

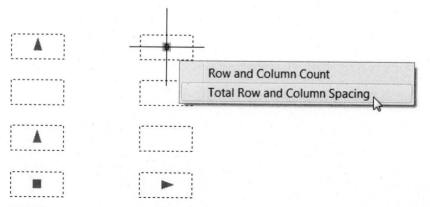

FIGURE 4–72 *Menu cursor when you hover the cursor over the last row/last column control point*

Choosing *Row and Column Count* lets you specify the number of rows and columns simultaneously with the cursor.

Choosing *Total Row and Column Spacing* lets you specify the distance between the first and last rows and first and last columns simultaneously with the cursor.

When you hover the cursor over the control point at the object on the second row/first column AutoCAD displays the distance between the rows (Figure 4–73).

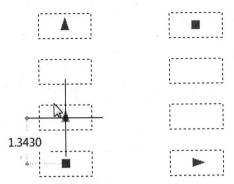

1.3430

FIGURE 4–73 *Distance between the rows is displayed when you hover the cursor over the second row/first column control point*

This option is *Row Spacing* and lets you change the distance between rows either with the cursor or entering the value from the keyboard.

NOTE

If there are more than 2 columns in the array, there will be a separate control point on the object at the second column/first row to adjust the column spaces in the same manner as the above option for rows.

Polar Array

Invoke the POLAR ARRAY command from the Modify panel (Figure 4–63), and AutoCAD displays the following prompt:

> Select objects: *(select objects to array)*

After selecting one or more objects to be arrayed, AutoCAD prompts:

> Select center point of array or *(select the center of the array)*

NOTE

If you right-click while being prompted for the center AutoCAD displays a menu containing the *Axis of rotation* option which allows you to specify an axis through the Base Point other than perpendicular to the XY plane. Doing this creates an array in 3D space, not in the XY plane.

After specifying the center of the array, AutoCAD prompts:

> Select grip to edit array or [ASsociative BasePoint Items
> Angle between Fill angle Rows Levels ROTate items eXit]:

By default, AutoCAD creates a polar pattern of 6 additional copies of the selected object as shown in Figure 4–74.

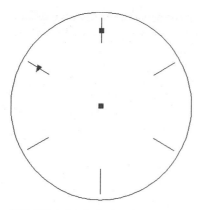

FIGURE 4–74 *Default Polar Pattern*

Select one of the grips to make changes to the pattern or select one of the available options to make changes to the pattern.

By selecting the grip (blue box) of the Base Point (on the source object) as shown in Figure in 4–75(a), AutoCAD allows you to change the array radius.

By selecting the grip located on one of the arrayed items as shown in Figure 4–75(b), AutoCAD allows you to change the angle between the items.

By selecting the grip of the center point of the array, AutoCAD allows you to reposition the array.

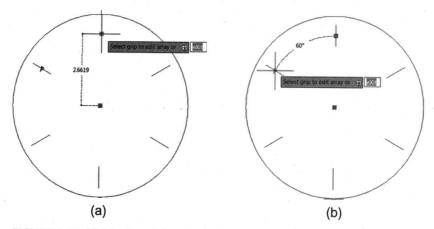

(a)   (b)

FIGURE 4–75 *(a) Selection of the girp on the Base Point (source object) (b) Selection of the grip on one of the arrayed items*

Select the array and right-click and AutoCAD displays a drop down menu with options for editing the array.

Choosing ASsociative displays a menu for choosing Yes or No to determine if the array is created as associative or not. Whichever mode you choose (Yes or No) will remain the default until changed. If you choose Yes, the whole array of objects acts like a single object when operated on by commands such as MOVE, COPY, SCALE and ROTATE. The radius of the array, location of the Base Point, number of items, angle between items, fill angle, rows and levels can be changed using control points placed at key points in the array.

Choosing *Base Point* lets you specify a base point for the array other than the default location at the source object(s). Right-clicking while being prompted for a Base Point displays a menu from which you can choose *Key Point*.

Choosing *Key Point* lets you specify a valid constraint (or *key point*) on the source objects to use as the base point. If you edit the source objects of the resulting array, the base point of the array remains coincident with the key point of the source objects.

NOTE

Choosing to specify a Key Point for a new location of the Base Point restricts you to available key points on the source object(s). If the source object is a line, available key points are the endpoints and the midpoint. For an arc, the endpoints and midpoint of the arc are available key points and if the arc angle is great enough, the center is available for selection. For a rectangle, the four corners and the midpoints of the four lines are available. If the source contains multiple objects, the key points of each of the objects are available. For a circle only the center is available. If you select a key point on a copied object in the array, the Base Point will be established at a corresponding point on the source object.

Choosing *Items* lets you edit the number of items in the array.

Choosing *Angle Between* lets you specify the angle between items in the array.

Choosing *Fill Angle* lets you specify the angle between the first and last item in the array.

Choosing *ROWs* lets you edit the number of rows of items in the array. AutoCAD prompts:

    Enter the number of rows or ⬇ (specify the number of rows)
    Enter the distance between rows or ⬇ (specify the distance between rows)
    Specify the incrementing elevation between rows or ⬇ (specify the incrementing elevation)
    Press Enter to accept or ⬇ (press Enter)

Figure 4–76 on the left shows a small circle to be selected as the object to be arrayed. The right view shows the circle arrayed with 2 rows.

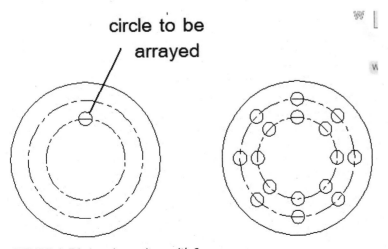

FIGURE 4–76 *Arraying an item with 2 rows*

Image(s) © Cengage Learning 2013

Choosing *Levels* lets you specify the number and spacing of levels in the array, creating a 3D array.

Choosing *ROTate items* displays a menu for choosing Yes or No to determine if the selected items are rotated as they are arrayed or if each item maintains the orientation of the source set. Whichever mode you choose (Yes or No) Yes will remain the default. You must change to No each time you wish to have the objects not rotated.

Figure 4–77 on the left shows a line and a circle to be selected as the objects to be arrayed. The middle view shows the array applying the *ROTate items* option (Yes). The right view shows the line and circle maintaining the original orientation while being arrayed. This is accomplished by choosing *ROTate items* and selecting No. Of course, in this example, rotating or not rotating the circle makes no difference.

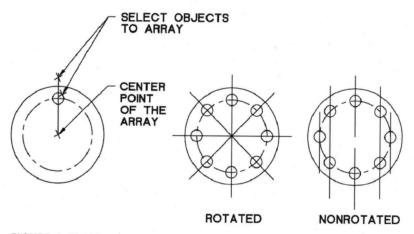

FIGURE 4–77 *Using the POLAR ARRAY command to create rotated and nonrotated polar arrays*

Choosing *eXit* causes AutoCAD to draw the array in accordance with the numbers, distances and angles specified or their defaults.

Figure 4–78 shows an associative array of items with the control points displayed when one of the objects in the array is selected.

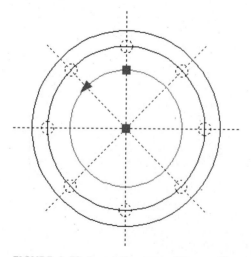

FIGURE 4–78 *Associative array of items with control points displayed*

Hovering the cursor over the Base Point does not display a menu as it does in the rectangular array. The Base Point can be used to move the array.

When you hover the cursor over the point on the source. AutoCAD displays a short-cut menu (Figure 4–79).

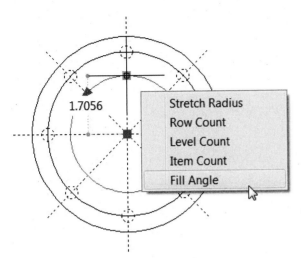

FIGURE 4–79 *Menu displayed when the Base Point (lower left control point) is selected*

NOTE

The five selections on the menu above are shown when the array angle fill is 360°. AutoCAD treats the 360° fill differently than fills that are less than 360°. For non-360° fills the fill angles are from the source item to the last item. When you specify 360° as the fill angle, AutoCAD does not place the last item on the source item but calculates the angle between items based on dividing 360° by the total number of items. There is a difference between how the control points are placed between specifying 360° and specifying the actual fill angle between the source item and last item that causes the array to be evenly spaced around 360°. For example, if you want 8 items evenly spaced around the circle, you can enter 360° or you can enter 315° as the fill angle. When you enter 360° there will be three control points with one at the source item as shown in the left view of Figure 4–80. When you enter 315° there will be four control points with separate points at the source item and the one at 315° as shown in the view on the right. When hovering the cursor over the source item in this case (having specified 315°) the menus will show options for Stretch Radius, Row Count and Level Count. And, the menu displayed when hovering the cursor over the last item will show options for Item Count and Fill Angle.

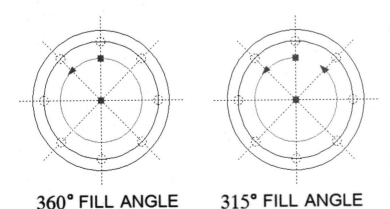

360° FILL ANGLE   315° FILL ANGLE

FIGURE 4–80 *The same array with different control point arrangements using 360° or 315°*

Choosing *Stretch Radius* lets you change the array radius.

Choosing *Row Count* lets you change the number of rows.

Choosing *Level Count* lets you change the number of levels.

Choosing *Item Count* lets you change the number of items.

Choosing *Fill Angle* lets you change the fill angle.

When you hover the cursor over the point on the first copied item AutoCAD displays the angle between items (Figure 4–81).

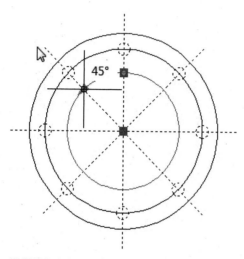

FIGURE 4–81 *Angle between items displayed when hovering the cursor over the control point at the first copied item*

Path Array

Invoke the PATH ARRAY command from the Modify panel (Figure 4–63), and AutoCAD displays the following prompt:

> Select objects: *(select objects to array)*

After selecting one or more objects to be arrayed, AutoCAD prompts:

> Select path curve: *(select an object to use as the path)*

After specifying the path of the array, AutoCAD prompts:

> Enter number of items along path or ⏎ *(select the number of items by moving the cursor along the path)*

Figure 4–82 shows a triangle being selected on the left in response to the PATH ARRAY command prompt. The polyline is then selected as the path. By default, AutoCAD creates a path array by making several copies of the selected object along the selected path.

AutoCAD prompts:

> Select grip to edit array or [ASsociative Method Base Point Tangent Direction Items Rows Levels Align Items z Direction eXit].

Select one of the grips to make changes to the pattern or select one of the available options to make changes to the pattern.

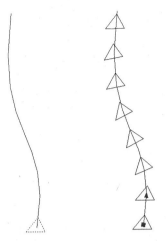

FIGURE 4–82 *Default Path Array.*

By selecting the grip (blue box) of the Base Point (on the source object) as shown in Figure in 4–83(a), AutoCAD allows you to create additional columns of arrays either to the left or to the right of the array.

By selecting the grip located on one of the arrayed items as shown in Figure 4–83(b), AutoCAD allows you to change the distance between the items.

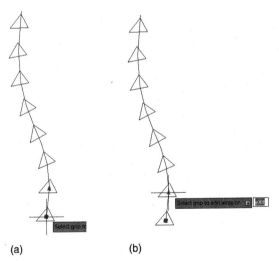

(a)   (b)

FIGURE 4–83 *(a) Selection of the grip on the Base Point (source object) (b) Selection of the grip located on one of the arrayed items*

After specifying the numbers of objects, AutoCAD prompts:

> `Specify the distance between items along path` ⏎ *(specify the distance by moving the cursor or select the end of the path)*

Choosing ASsociative displays a menu for choosing Yes or No to determine if the array is created as associative or not. Whichever mode you choose (Yes or No) will remain the default until changed. If you choose Yes, the whole array of objects acts like a single object when operated on by commands such as MOVE, COPY, SCALE and ROTATE. The location of the Base Point, number of items, distance between items, total item distance, rows and levels can be changed using control points placed at key points in the array.

Choosing *Base Point* lets you specify a base point for the array other than the default location at the source object(s). Right-clicking while being prompted for a Base Point displays a menu from which you can choose *Key Point*.

Choosing *Key Point* lets you specify a valid constraint (or *key point*) on the source objects to use as the base point. If you edit the source objects of the resulting array, the base point of the array remains coincident with the key point of the source objects.

> **NOTE**
>
> Choosing to specify a Key Point for a new location of the Base Point restricts you to available key points on the source object(s). If the source object is a line, available key points are the endpoints and the midpoint. For an arc, the endpoints and midpoint of the arc are available key points and if the arc angle is great enough, the center is available for selection. For a rectangle, the four corners and the midpoints of the four lines are available. If the source contains multiple objects, the key points of each of the objects are available. For a circle only the center is available. If you select a key point on a copied object in the array, the Base Point will be established at a corresponding point on the source object.

Choosing *Items* lets you edit the number of items in the array.

Choosing *Rows* lets you edit the number of rows of items in the array. AutoCAD prompts:

> Enter the number of rows or ⬇ *(specify the number of rows)*
> Enter the distance between rows or ⬇ *(specify the distance between rows)*
> Specify the incrementing elevation between rows or ⬇ *(specify the incrementing elevation)*
> Press **Enter** to accept or ⬇ *(press **Enter**)*

Figure 4–84 on the left shows a triangle to be selected as the object to be arrayed. The right view shows the triangle arrayed with 2 rows.

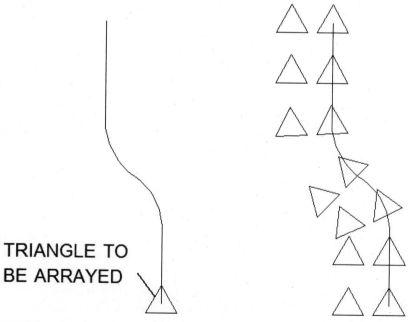

TRIANGLE TO BE ARRAYED

FIGURE 4–84 *Arraying an item with 2 rows*

Choosing *Levels* lets you specify the number and spacing of levels in the array, creating a 3D array.

Choosing *Align items* displays a menu for choosing Yes or No to determine if the selected items are aligned with the path as they are arrayed or if each set of objects maintains the orientation of the source set. Whichever mode you choose (Yes or No) will remain the default until changed.

Figure 4–85 on the left shows a triangle to be selected as the object to be arrayed. The middle view shows the array applying the *Align items* option (Yes). The right view shows the triangle maintaining the original orientation while being arrayed. This is accomplished by choosing *Align items* and selecting No.

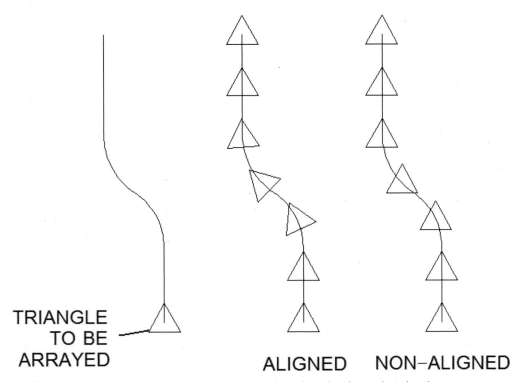

TRIANGLE TO BE ARRAYED

ALIGNED   **NON-ALIGNED**

FIGURE 4–85 *Using the* PATH ARRAY *command to place aligned and non-aligned path arrays*

Choosing *Z direction* controls whether to maintain the items' original Z direction or to naturally bank the items along a 3D path.

Choosing *eXit* causes AutoCAD to draw the array in accordance with the numbers, distances and angles specified or their defaults.

When you hover the cursor over the Base Point AutoCAD displays a shortcut menu (Figure 4–86).

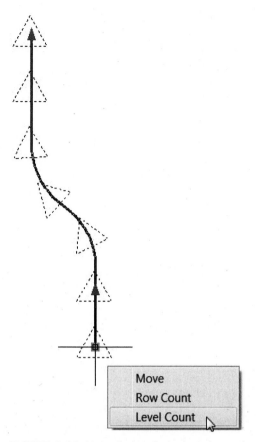

FIGURE 4–86 *Menu displayed when you hover the cursor over the Base Point*

Choosing *Move* lets you move the whole array in the same manner as the MOVE command.

Choosing *Row Count* lets you change the number of rows.

Choosing *Level Count* lets you change the number of levels.

Hovering the cursor over the point on the first copied item does not display a menu as it does in the polar array. Select this control point and move the cursor the desired distance from the Base Point (not the control point on the first item) in any direction. AutoCAD dynamically displays the array that will be created using that distance.

When you hover the cursor over the control point at the last item AutoCAD displays a shortcut menu (Figure 4–87).

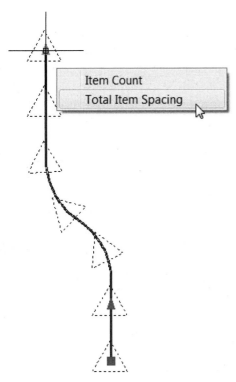

FIGURE 4–87 *Menu displayed when you hover the cursor over the control point at the last item*

Choosing *Item Count* lets you change the number of items.

Choosing *Total Item Spacing* lets you change the total item spacing.

Array Ribbons

Whenever you select an array object AutoCAD displays an editing ribbon with panels for editing the selected array. You can make the necessary changes to the pattern of the selected array. Figure 4–88 shows the ribbons for the rectangular, polar and path arrays respectively.

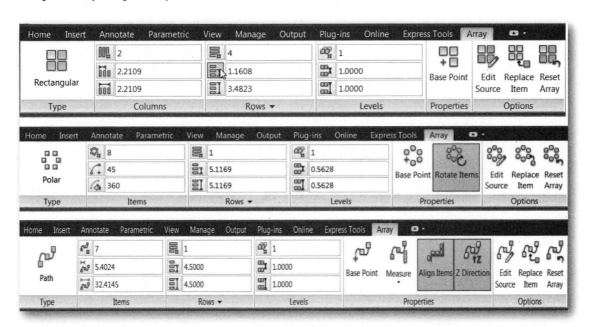

FIGURE 4–88 *Editing ribbons displayed when an array is selected*

If more than one object is selected (even two arrays of the same type) the applicable array editing ribbon will not be displayed.

Creating Parallel Lines, Parallel Curves, and Concentric Circles

The OFFSET command creates parallel lines, parallel curves, and concentric circles relative to existing objects (Figure 4–89). Special precautions must be taken when using the OFFSET command to prevent unpredictable results from occurring when using the command on arbitrary curve/line combinations in polylines.

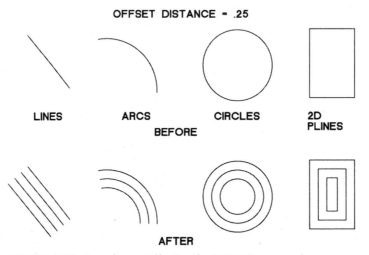

FIGURE 4–89 *Examples created using the* OFFSET *command*

Invoke the OFFSET command from the Modify panel (Figure 4–90), and AutoCAD displays the following prompt:

```
Specify offset distance or ⬇ (specify offset distance, or
right-click and choose one of the options)

Select object to offset: (select an object to offset)

Specify point on side to offset or ⬇ (specify a point to one
side of the object to offset)

Select object to offset or ⬇ (continue selecting additional
objects for offset, and specify the side of the object to
offset, or press Enter to terminate the command sequence)
```

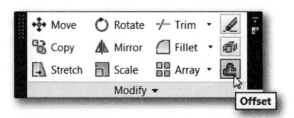

FIGURE 4–90 *Invoking the* OFFSET *command from the Modify panel*

Offset Options

Choosing *Through* from the shortcut menu causes AutoCAD to prompt for a through point, and an object is created passing through the specified point.

Choosing *Erase* from the shortcut menu causes AutoCAD to erase the source object after it is offset.

Choosing *Layer* from the shortcut menu lets you determine whether offset objects are created on the current layer or on the layer of the source object.

Valid Objects to Offset

Valid objects include the line, spline curve, arc, circle, and 2D polyline. The object selected for offsetting must be in a plane parallel to the current coordinate system.

Offsetting Miters and Tangencies

The OFFSET command affects single objects in a manner different from a polyline made up of the same objects. Polylines whose arcs join lines and other arcs in a tangent manner are affected differently than polylines with non-tangent connecting points. For example, in Figure 4–91, the seven lines are separate objects. When you specify a side to offset as shown, there are gaps and overlaps at the ends of the newly created lines.

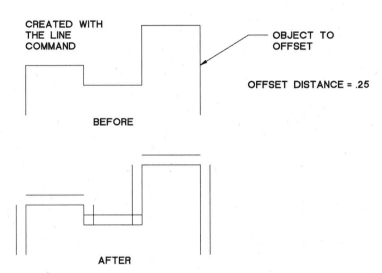

FIGURE 4–91 *Using the OFFSET command with single objects*

Figure 4–92 shows the lines joined together as a single polyline and how the OFFSET command affects the corners where the new polyline segments join. See the PEDIT feature in Chapter 5.

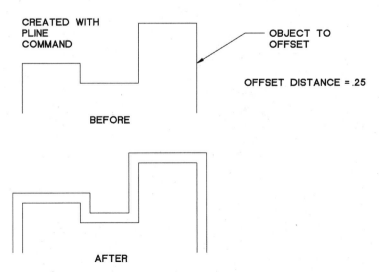

FIGURE 4–92 *Using the OFFSET command with polylines*

The results of offsetting polylines with arc segments that connect other arc segments and/or line segments in dissimilar (non-tangent) directions might be unpredictable. Examples of offsetting such polylines are shown in Figure 4–93.

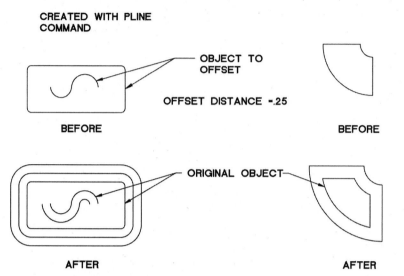

CREATED WITH PLINE
COMMAND

OBJECT TO
OFFSET

OFFSET DISTANCE = .25

BEFORE   BEFORE

ORIGINAL OBJECT

AFTER   AFTER

FIGURE 4–93 *Using the* OFFSET *command with non-tangent arc and/or line segments*

If you are not satisfied with the resulting polyline configuration, you can use the PEDIT command to edit it. You can also explode the polyline and edit the individual segments.

Creating a Mirror Copy of Objects

The MIRROR command creates a mirror image copy of selected objects. Invoke the MIRROR command from the Modify panel (Figure 4–94), and AutoCAD displays the following prompt:

Select objects: *(select the objects and then press* Enter *to complete the selection)*

Specify first point of mirror line: *(specify a point to define the first point of the mirror line)*

Specify second point of mirror line: *(specify a point to define the second point of the mirror line)*

Erase source objects? ⊡ *(enter* y, *for yes to delete the original objects, or* n, *not to delete the original objects, that is, to retain them)*

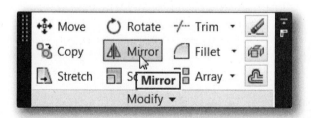

FIGURE 4–94 *Invoking the* MIRROR *command from the Modify panel*

The first and second points of the mirror line become the endpoints of an imaginary line about which the selected objects will be mirrored.

The following command sequence shows an example of mirroring a group of selected objects by means of the Window option, as shown in Figure 4–95.

> `mirror`
>
> `Select objects:` *(specify Point 1 to place one corner of a window)*
>
> `Specify opposite corner:` *(specify Point 2 to place the opposite corner of the window)*
>
> `Select objects:` [Enter]
>
> `Specify first point of mirror line:` *(specify Point 3)*
>
> `Specify second point of mirror line:` *(specify Point 4)*
>
> `Erase source objects?` ⊞ [Enter]

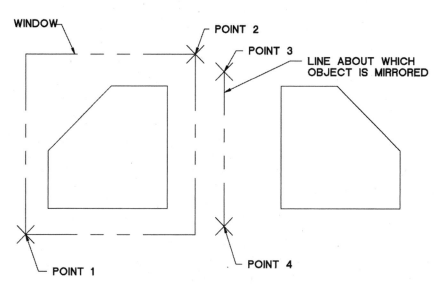

FIGURE 4–95 *Mirroring a group of objects selected by means of the Window option*

Text is mirrored relative to other objects within the selection group. Text will or will not retain its original orientation, depending on the setting of the system variable called `MIRRTEXT`. If the value of `MIRRTEXT` is set to 1, then text items in the selected group will have their orientations and location mirrored. That is, if their characters were normal and they read from left to right in the original group, in the mirrored copy they will read from right to left and the characters will be backwards. If `MIRRTEXT` is set to zero, the text strings in the group will have their locations mirrored, but the individual text strings will retain their normal, left-to-right character appearance. The `MIRRTEXT` system variable, like other system variables, can be changed by using the `SETVAR` command or by typing `MIRRTEXT` at the on-screen prompt, as follows:

> `mirrtext` [Enter]
>
> `Enter New Value for MIRRTEXT <1>: 0` [Enter]

This setting causes mirrored text to retain its readability. Figures 4–96 and 4–97 show the result of the mirror command when the `MIRRTEXT` variable is set to 1 and 0, respectively.

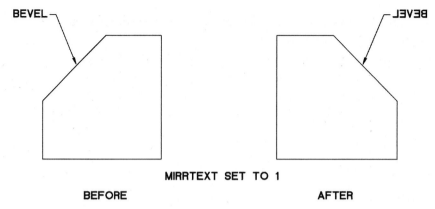

FIGURE 4–96 *The MIRROR command with the MIRRTEXT variable set to 1*

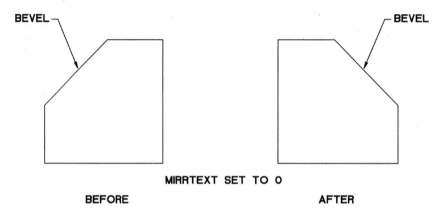

FIGURE 4–97 *The MIRROR command with the MIRRTEXT variable set to 0*

Creating a Fillet between Two Objects

The FILLET command fillets (rounds) the intersecting ends of two arcs, circles, lines, elliptical arcs, polylines, rays, xlines, or splines with an arc of a specified radius.

If the TRIMMODE system variable is set to 1 (default), the FILLET command trims the intersecting lines to the endpoints of the fillet arc. If TRIMMODE is set to 0, the FILLET command leaves the intersecting lines at the endpoints of the fillet arc.

Invoke the FILLET command from the Modify panel (Figure 4–98), and AutoCAD displays the following prompt:

> Select first object or ⏷ *(select one of the two objects to fillet, or right-click and choose one of the options)*

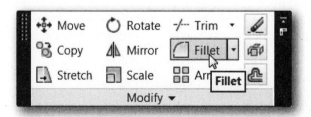

FIGURE 4–98 *Invoking the FILLET command from the Modify panel*

By default, AutoCAD prompts you to select an object. When you select an object to fillet, AutoCAD displays the following prompt:

> Select second object or shift-select to apply corner or ☑:
> *(select the second object to fillet, hold down SHIFT and*
> *select an object to create a sharp corner or choose Radius*
> *from the shortcut menu to specify radius for the fillet)*

AutoCAD joins the two objects with an arc of the specified radius. If the objects selected to be filleted are on the same layer, AutoCAD creates the fillet arc on the same layer. If not, AutoCAD creates the fillet arc on the current layer.

AutoCAD allows you to draw a fillet between parallel lines, xlines, and rays. The first selected object must be a line or ray, but the second object can be a line, xline, or ray.

Fillet Options

Selecting *Radius* allows you to change the current fillet radius. The following command sequence sets the fillet radius to 0.25 and draws the fillet between two lines (Figure 4–99).

> fillet [Enter]
>
> Select first object: *(choose Radius from the shortcut menu)*
>
> Specify fillet radius <default>: 0.25 [Enter]
>
> Select first object or ☑ *(select one of the lines)*
>
> Select second object or shift-select to apply corner or ☑:
> *(select the other line to fillet)*

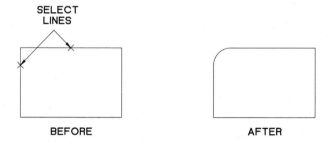

FIGURE 4–99 *Fillet drawn with a radius of 0.25*

If you select lines or arcs, AutoCAD extends these lines or arcs until they intersect or trims them at the intersection, keeping the selected segments if they cross. The following command sequence sets the fillet radius to 0 and draws the fillet between two lines (Figure 4–100).

> fillet [Enter]
>
> Select first object: *(choose Radius from the shortcut menu)*
>
> Specify fillet radius: 0 [Enter]
>
> Select first object or ☑ *(select the first object)*
>
> Select second object or shift-select to apply corner or ☑:
> *(select the second object to fillet)*

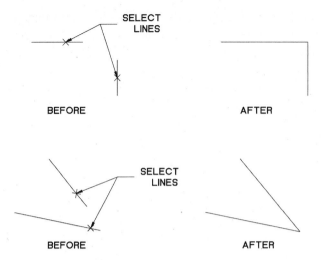

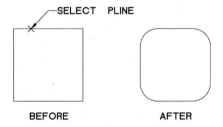

FIGURE 4–100 *Fillet drawn with a radius of 0*

Choosing *Polyline* causes AutoCAD to draw fillet arcs at each vertex of a selected 2D polyline where two line segments meet. The following command sequence sets the fillet radius to 0.5 and draws the fillet at each vertex of a 2D polyline (Figure 4–101).

> fillet `Enter`
>
> Select first object: *(choose Radius from the shortcut menu)*
>
> Specify fillet radius <default>: **0.5** `Enter`
>
> Select first object or ☒ *(choose Polyline from the shortcut menu)*
>
> Select 2D polyline: *(select the polyline)*

FIGURE 4–101 *Fillet with a radius of 0.5 drawn to a polyline*

Choosing *Trim* controls whether or not AutoCAD trims the selected edges to the fillet arc endpoints. This option is similar to setting the TRIMMODE system variable from 1 to 0 or 0 to 1, as explained earlier.

Choosing *Multiple* allows you to specify multiple pairs of objects to be filleted without exiting the FILLET command. To exit the command, select `Enter` from the shortcut menu.

Creating a Chamfer between Two Objects

The CHAMFER command allows you to draw an angled corner between two lines. The size of the chamfer is determined by the settings of the first and the second chamfer distances. If it is to be a 45° chamfer for perpendicular lines, the two distances are set to the same value.

If the TRIMMODE system variable is set to 1 (default), then the CHAMFER command trims the intersecting lines to the endpoints of the chamfer line. If TRIMMODE is set to 0, the CHAMFER command leaves the intersecting lines at the endpoints of the chamfer line.

Invoke the CHAMFER command from the Modify panel (Figure 4–102), and AutoCAD displays the following prompt:

```
Select first line or ⊥ (select one of the two lines to
chamfer, or right-click and choose one of the options)
```

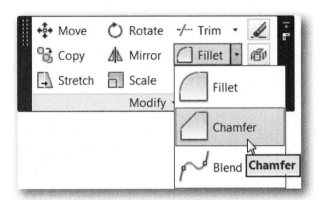

FIGURE 4–102 *Invoking the CHAMFER command from the Modify panel*

By default, AutoCAD prompts you to select the first line to chamfer. If you select a line to chamfer, AutoCAD displays the following prompt:

```
Select second line or shift-select to apply corner or ⊥:
(select the second line to chamfer, hold down SHIFT and
select an object to create a sharp corner or right-click and
choose one of the options)
```

AutoCAD draws a chamfer to the selected lines. If the selected lines to be chamfered are on the same layer, AutoCAD creates the chamfer on the same layer. If not, AutoCAD creates the chamfer on the current layer. You can also specify the distance or angle to chamfer after selecting the first line.

Chamfer Options

Choosing *Distance* allows you to set the first and second chamfer distances. The following command sequence sets the first chamfer and second chamfer distance to 0.5 and 1.0, respectively, and draws the chamfer between two lines (Figure 4–103).

```
chamfer Enter
Select first line or ⊥ (choose Distance from the shortcut
menu)
Specify first chamfer distance <default>: 0.5 Enter
Specify second chamfer distance <0.5000>: 1.0 Enter
Select first line or ⊥ (select the first line, as shown in
Figure 4-103)
Select second line or shift-select to apply corner or ⊥:
(select the second line, as shown in Figure 4-103)
```

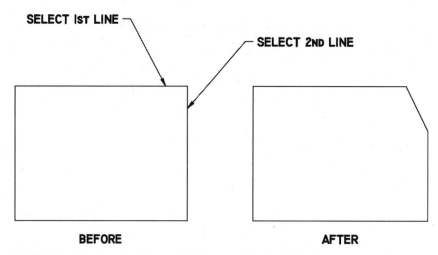

FIGURE 4–103 *Chamfer drawn with distances of 0.5 and 1.0*

Choosing *Polyline* causes AutoCAD to draw chamfers at each vertex of a selected 2D polyline where two line segments meet. The following command sequence sets the chamfer distances to 0.5 and draws the chamfer at each vertex of a 2D polyline (Figure 4–104).

```
chamfer Enter
Select first line or ⊡ (choose Distance from the shortcut
menu)
Specify first chamfer distance <default>: 0.5 Enter
Specify second chamfer distance <0.5000>: 0.5 Enter
Select first line or ⊡ (choose Polyline from the shortcut menu)
Select 2D polyline: (select the polyline, as shown in
Figure 4-104)
```

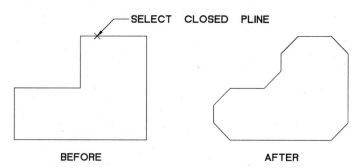

FIGURE 4–104 *Chamfer with distances of 0.5 drawn on a polyline*

Choosing *Angle* is similar to the *Distance* option, but instead of prompting for the first and second chamfer distances, AutoCAD prompts for the first chamfer distance and an angle from the first line.

Choosing *Method* controls whether AutoCAD uses two distances or a distance and an angle to create the chamfer line.

Choosing *Trim* controls whether or not AutoCAD trims the selected edges to the chamfer line endpoints. This option is similar to setting the TRIMMODE system variable from 1 to 0 or from 0 to 1, as explained earlier.

> The CHAMFER command set to 0 distance operates the same way the FILLET command operates when set to 0 radius.

Selecting *Multiple* allows you to specify multiple pairs of objects to be chamfered without exiting the CHAMFER command. To exit the command, select Enter from the shortcut menu.

MODIFYING OBJECTS

In this section, four additional MODIFY commands are explained: MOVE, TRIM, BREAK, and EXTEND. The ERASE command was explained in Chapter 2.

Moving Objects

The MOVE command lets you move one or more objects from their present location to a new one without changing orientation or size.

Invoke the MOVE command from the Modify panel (Figure 4–105), and AutoCAD displays the following prompt:

>   Select objects: *(select the objects, press* Enter *when done)*
>   Specify base point or ⬇ *(specify a point)*
>   Specify second point or <use first point as displacement>: *(specify a point for displacement, or press* Enter*)*

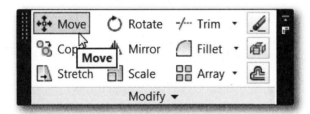

FIGURE 4–105 *Invoking the MOVE command from the Modify panel*

You can use one or more object selection methods to select the objects. If you specify two data points to move the selected objects, AutoCAD computes the displacement and moves the selected objects accordingly. If you specify the points on the screen, AutoCAD assists you in visualizing the displacement by drawing a rubber-band line from the first point to the second point, as you move the crosshairs. If you press Enter at the prompt for the second point of displacement, AutoCAD interprets the base point as relative X, Y, Z displacement.

The following command sequence shows an example of moving a group of objects, selected by means of the Window option, by relative displacement (Figure 4–106).

>   move Enter
>   Select objects: *(specify Point 1 to place one corner of a window)*
>   Specify opposite corner: *(specify Point 2 to place the opposite corner of the window)*
>   Select objects: Enter

Specify base point or ⬇ 2,3 [Enter]

Specify second point or <use first point as displacement>:
[Enter]

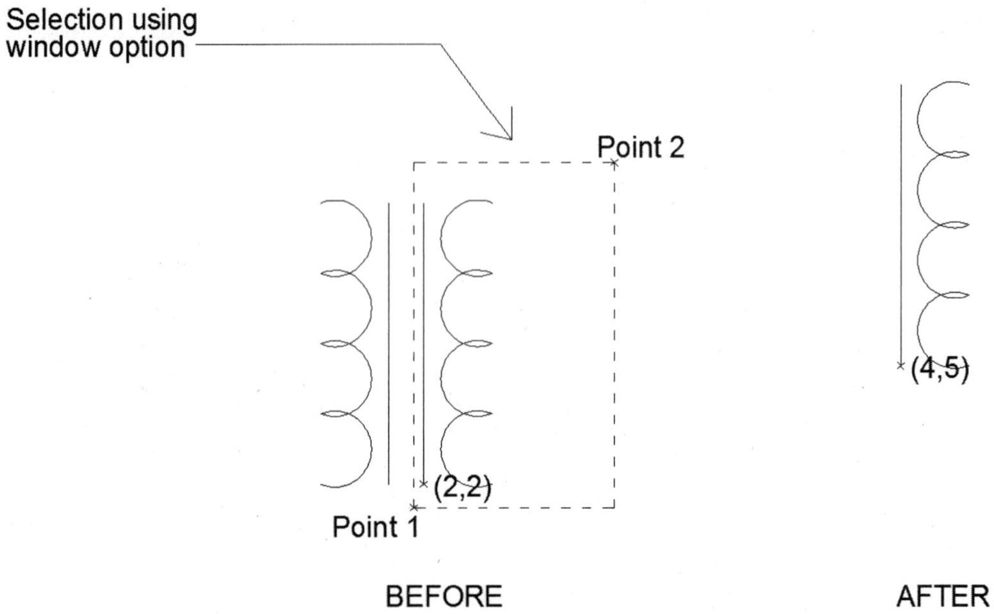

**Selection using
window option**

Point 2

×(4,5)

×(2,2)

Point 1

BEFORE   AFTER

FIGURE 4–106 *Using the Window option of the MOVE command to move a group of objects by means of
vector displacement*

The following command sequence shows an example of moving a group of objects
selected by the Window option and moving the objects by specifying two data points
(Figure 4–107).

move [Enter]

Select objects: *(pick Point 1 to place one corner for a
window)*

Specify opposite corner: *(pick Point 2 to place the opposite
corner of the window)*

Select objects: [Enter]

Specify base point or ⬇ *(specify Base Point)*

Specify second point or <use first point as displacement>:
(specify Second Point)

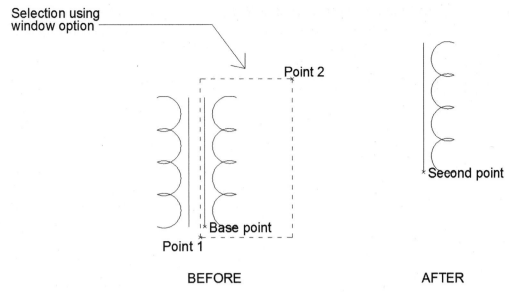

Selection using
window option

Point 2

Second point

Base point

Point 1

BEFORE

AFTER

FIGURE 4–107 *Using the Window option of the MOVE command to move a group of objects by specifying two data points*

Trimming Objects

The TRIM command is used to trim a portion of the selected object(s) that is drawn past a cutting edge or from an implied intersection defined by other objects. Objects that can be trimmed include lines, arcs, elliptical arcs, circles, 2D and 3D polylines, xlines, rays, and splines. Valid cutting-edge objects include lines, arcs, circles, ellipses, 2D and 3D polylines, floating viewports, xlines, rays, regions, splines, and text.

Invoke the TRIM command from the Modify panel (Figure 4–108), and AutoCAD displays the following prompts:

> Select objects or <select all>: *(select the objects, and press* Enter *to terminate the selection of the cutting edges)*
>
> Select object to trim or shift-select to extend or ⬇ *(select object(s) to trim, or right-click and choose one of the options)*

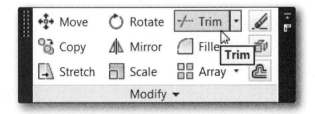

FIGURE 4–108 *Invoking the TRIM command from the Modify panel*

The TRIM command initially prompts you to "Select objects or <select all>:". After selecting one or more cutting edges to establish where to trim, press Enter. You are then prompted to "Select object to trim or shift-select to extend or ⬇." Select one or more objects to trim, and then press Enter to terminate the command. If you press Enter in response to the "Select objects or <select all>" prompt without selecting any objects, by default AutoCAD selects all the objects.

You can switch to the EXTEND command by holding the SHIFT key while selecting the objects, which will then be extended to the specified cutting edge instead of being trimmed. See the section on the EXTEND command in this chapter.

NOTE Don't forget to press Enter after selecting the cutting edge(s). Otherwise, the program will not respond as expected. In fact, trim continues to expect more cutting edges until you terminate the cutting edge selection mode.

Trim Options

Selecting *Fence* selects all objects that cross the selection fence. The selection fence is a series of temporary line segments that you specify with two or more fence points. The selection fence does not have to form a closed loop.

Selecting *Crossing* selects objects within and crossing a rectangular area defined by two points.

Selecting *Edge* determines whether or not objects that extend past a selected cutting edge or to an implied intersection are trimmed. AutoCAD prompts you as follows when the *Edge* option is selected:

```
Enter an implied edge extension mode: (select one of the two
options)
```

Selecting *Extend* from the *Edge* shortcut causes the object's cutting edge to extend along its natural path to intersect an object in 3D (implied intersection).

Selecting *No Extend* from the *Edge* shortcut specifies that the object is to be trimmed only at a cutting edge that intersects it in 3D space.

Selecting *Undo* reverses the most recent change made by the TRIM command.

Selecting *Erase* deletes selected objects. This option provides a convenient method to erase unneeded objects without leaving the TRIM command.

Selecting *Project* specifies the projection mode and coordinate system AutoCAD uses when trimming objects. By default, it is set to the current UCS.

Figure 4–109 shows examples of using the TRIM command.

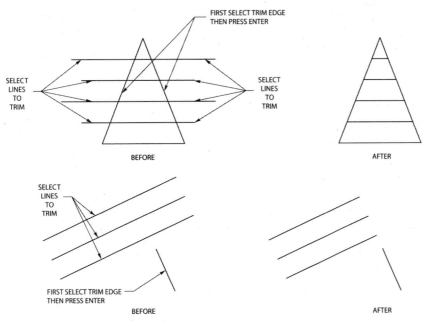

FIGURE 4–109 *Examples of using the* TRIM *command*

Erasing Parts of Objects

The BREAK command is used to remove parts of objects, to make a circle into an arc, or to split an object in two parts. It can be used on lines, xlines, rays, arcs, circles, ellipses, splines, donuts, traces, and 2D and 3D polylines.

Invoke the BREAK command from the Modify panel (Figure 4–110), and AutoCAD displays the following prompts:

> Select object: *(select an object)*
>
> Specify second break point or ☑ *(specify the second break point, or right-click and choose one of the options)*

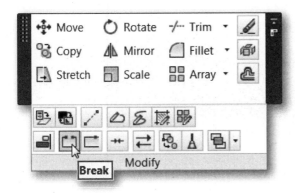

FIGURE 4–110 *Invoking the* BREAK *command from the Modify panel*

See Figure 4–111 for examples of applications of the BREAK command.

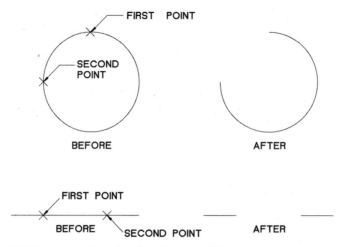

FIGURE 4–111 *Examples of applications of the BREAK command*

AutoCAD erases the portion of the object between the first point, the point where the object was selected, and second point. If the second point is not on the object, AutoCAD selects the nearest point on the object. If you need to erase an object to one end of a line, arc, or polyline, specify the second point beyond the end to be removed.

If instead of specifying the second point, you choose *First Point* from the shortcut menu, AutoCAD prompts for the first point and then for the second point.

An object can be split into two parts without removing any portion of the object by selecting the same point as the first and second points. You can do so by entering the "at" (@) symbol to specify the second point.

If you select a circle, then AutoCAD converts it to an arc by erasing a piece, moving counterclockwise from the first point to the second point. For a closed polyline, the part is removed between two selected points, moving in direction from the first to the last vertex. In the case of 2D polylines and traces with width, the BREAK command will produce square ends at the break points.

Extending Objects to Meet Another Object

The EXTEND command is used to change one or both endpoints of selected lines, arcs, elliptical arcs, open 2D and 3D polylines, and rays to extend to lines, arcs, elliptical arcs, circles, ellipses, 2D and 3D polylines, rays, xlines, regions, splines, text string, or floating viewports.

Invoke the EXTEND command from the Modify panel (Figure 4–112), and Auto-CAD displays the following prompts:

Select objects or <select all>: *(select the objects, and then press* Enter *to complete the selection)*

Select object to extend or shift-select to trim or ⬇ *(select the object(s) to extend, and press* Enter *to terminate the selection process, or right-click and choose one of the options)*

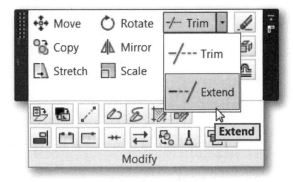

FIGURE 4–112 *Invoking the* EXTEND *command from the Modify panel*

The EXTEND command initially prompts you to "Select objects or <select all>:" After selecting one or more boundary edges, press **Enter** to terminate the selection process. Then AutoCAD prompts you to "Select object to extend or shift-select to trim or ⬇." Select one or more objects to extend to the selected boundary edges. After selecting the required objects to extend, press **Enter** to complete the selection process.

You can switch to the TRIM command by holding the SHIFT key while selecting the objects that will be trimmed to the specified cutting edge instead of being extended. See the section on the TRIM command in this chapter.

The EXTEND and TRIM commands are similar in this method of selecting. With EXTEND, you are prompted to select the boundary edge to extend to; with TRIM, you are prompted to select a cutting edge. If you press **Enter** in response to the "Select objects or <select all>:" prompt without selecting any objects, by default AutoCAD selects all the objects.

Extend Options

Selecting *Fence* selects all objects that cross the selection fence. The selection fence is a series of temporary line segments that you specify with two or more fence points. The selection fence does not have to form a closed loop.

Selecting *Crossing* selects objects within and crossing a rectangular area defined by two points.

Selecting *Edge* determines whether objects are extended past a selected boundary or to an implied edge. AutoCAD prompts as follows when the *Edge* option is selected:

        Enter an implied edge extension mode *(select one of the two options)*

Select *Extend* from the *Edge* shortcut to extend the boundary object along its natural path to intersect another object in 3D space (implied edge).

Select *No Extend* from the *Edge* shortcut to specify that the object is to extend only to a boundary object that actually intersects it in 3D space.

Select *Undo* to reverse the most recent change made by the EXTEND command.

Select *Project* to specify the projection and coordinate system that AutoCAD uses when trimming objects. By default, it is set to the current UCS.

Figure 4–113 shows examples of the use of the EXTEND command.

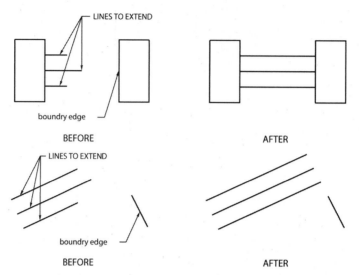

FIGURE 4–113 *Examples of applications of the* EXTEND *command*

Open the Exercise Manual PDF for Chapter 4 for discipline-specific exercises. Related files are downloaded from the student companion site mentioned in the Introduction (refer to page number xii for instructions).

REVIEW QUESTIONS

1. In order to draw two rays with different starting points, you must use the RAY command twice.
   a. True
   b. False

2. When you place xlines on a drawing they:
   a. May affect the limits of the drawing
   b. May affect the extents of the drawing
   c. Always appear as construction lines on Layer 0
   d. Can be constructed as offsets to an existing line
   e. None of the above

3. Filleting two nonparallel, nonintersecting line segments with a **0** radius will:
   a. Return an error message
   b. Have no effect
   c. Create a sharp corner
   d. Convert the lines to rays

4. Objects can be trimmed at the points where they intersect existing objects.
   a. True
   b. False

5. The default justification for text is:

    a. TL

    b. BL

    c. MC

    d. BR

    e. None of the above

6. Which command allows you to make a copy and keep the original to remain intact?

    a. CHANGE

    b. MOVE

    c. COPY

    d. MIRROR

7. The maximum number of sides accepted by the POLYGON command is:

    a. 8

    b. 32

    c. 128

    d. 1024

    e. infinite (limited by computer memory)

8. Portions of objects can be erased or removed by using the command:

    a. ERASE

    b. REMOVE

    c. BREAK

    d. EDIT

    e. PARERASE

9. Ellipses are drawn by specifying:

    a. The major and minor axes

    b. The major axis and a rotation angle

    c. Any three points on the ellipse

    d. Any of the above

    e. Both a and b

10. Which of the following commands can be used to place text on a drawing?

    a. TEXT

    b. CTEXT

    c. MTEXT

    d. Both a and c

    e. None of the above

11. Two lines are drawn. The first line is erased, and the second line is moved. Executing the OOPS command at the on-screen prompt will:

    a. Restore the erased line

    b. Execute the LINE command automatically

    c. Replace the second line at its original position

    d. Restore both lines to their original positions

    e. None of the above

12. To move multiple objects, which selection option would be more efficient?

    a. `OBJECTS`

    b. `LAST`

    c. `WINDOW`

    d. `ADD`

    e. `UNDO`

13. While you're using the `TEXT` command, AutoCAD will display the text you are typing:

    a. In the command prompt area

    b. In the drawing screen area

    c. Both a and b

    d. Neither a nor b

14. To insert the diameter symbol into a string of text, you should enter:

    a. `%%d`

    b. `%%c`

    c. `%%phi`

    d. `%%dia`

15. It is possible to force all text on a drawing to display as an open rectangle in order to speed up the redisplay of the drawing by using the:

    a. `TEXT` command

    b. `RTEXT` command

    c. `QTEXT` command

    d. `DTEXT` command

16. In order to select a different font for use in the `TEXT` command, a text style must be created.

    a. True

    b. False

17. Which of the following are not valid options when creating a text style?

    a. Width factor

    b. Upside down

    c. Vertical

    d. Backwards

    e. All answers are valid

18. To locate text such that it is centered exactly within a circle, a reasonable justification would be:

    a. Center

    b. Middle

    c. Full

    d. Both a and b

    e. None; simply zoom in and approximate it

19. To create a rectangular array of objects, you must specify:
    a. The number of items and the distance between them
    b. The number of rows, the number of items, and the unit cell size
    c. The number of rows, the number of columns, and the unit cell size
    d. None of the above

20. A polyline:
    a. Can have width
    b. Can be exploded
    c. Is one object
    d. All of the above

21. Polylines are:
    a. Made up of line and arc segments, each of which is treated as an individual object
    b. Are connected sequences of lines and arcs
    c. Both a and b
    d. None of the above

22. To create an arc that is concentric with an existing arc, what command could you use?
    a. ARRAY
    b. COPY
    c. OFFSET
    d. MIRROR

Directions: Answer the following questions based on Figure 4–114.

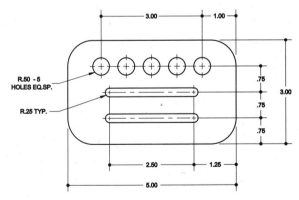

FIGURE 4–114 *Mechanical part*

23. The following are all options of the `PLINE` command except:
    a. UNDO
    b. HALFWIDTH
    c. ARC
    d. LTYPE
    e. WIDTH

24. If the RECTANGLE command were used to create the object outline with the first corner at **0,0**, what *"specify other corner point:"* coordinate location input would be required to complete the rectangle shape?

    a. @3.0, 3.0

    b. 5.0, 3.0

    c. 3.0, 5.0

    d. @3.0, 5.0

25. Which command would be the easiest and quickest to create rounded corners on the object?

    a. ARC

    b. ROUND

    c. FILLET

    d. CHAMFER

26. Creating the five holes, equally spaced, can be best achieved with which command?

    a. MIRROR

    b. OFFSET

    c. COPY

    d. ARRAY

27. Which array type would the five equally spaced holes be created with?

    a. POLAR

    b. RECTANGULAR

    c. COPY

    d. ARRAY

28. If the first corner of the object outline is at **0,0**, what coordinate location input is required to position the center of the circle on the left?

    a. @1.25, 2.5

    b. −2.25, 1.25

    c. @1.00, 2.25

    d. 1.25, 2.25

29. What is the distance between cells (holes) value required when using the ARRAY command to locate the five holes?

    a. 3.00

    b. 0.75

    c. 0.50

    d. 0.25

30. If the top horizontal line of the long slot is drawn with the LINE command, which command can easily establish the location of the remaining horizontal lines for the slots?

    a. XLINE

    b. MIRROR

    c. OFFSET

    d. MOVE

CHAPTER 5

Fundamentals IV

OBJECTIVES

After completing this chapter, you will be able to do the following:

- Construct geometric figures with the DONUT, SOLID, and POINT commands
- Create freehand line segments with the SKETCH command
- Use advanced object selection methods and modes to modify objects
- Use the modify commands: LENGTHEN, STRETCH, ROTATE, SCALE, PEDIT, JOIN, and MATCHPROP

CONSTRUCTING GEOMETRIC FIGURES

This section explains additional drawing commands (DONUT, SOLID, POINT, and SKETCH) that will help to construct geometric figures.

Drawing Solid-Filled Circles

The DONUT (or DOUGHNUT) command lets you draw solid-filled circles and rings by specifying the outer and inner diameters of the filled area. The fill display depends on the setting of the FILLMODE system variable.

Invoke the DONUT command from the Draw panel (Figure 5–1) located in the Home tab, and AutoCAD displays the following prompts:

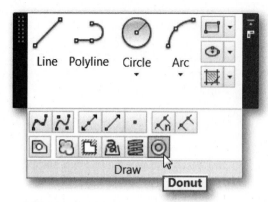

FIGURE 5–1 *Invoking the* DONUT *command from the Draw panel*

> Specify inside diameter of donut <current>: *(specify a distance, or press* Enter *to accept the current setting)*
>
> Specify outside diameter of donut <current>: *(specify a distance, or press* Enter *to accept the current setting)*
>
> Specify center of donut or <exit>: *(specify a point to draw the donut)*

You may specify the inside and outside diameters of the donut by using your pointing device to specify two points at the appropriate distance apart for either or both diameters, and AutoCAD will use the measured distance for the diameter(s). Or, you may enter the distances from the keyboard.

You can select the center point by entering its coordinates or by selecting it with your pointing device. After you specify the center point, AutoCAD prompts for the center of the next donut and continues prompting for subsequent center point. To terminate the command, press Enter.

NOTE

Be sure the FILLMODE system variable is set to ON, (a value of 1). If FILLMODE is set to OFF, (a value of 0), then the PLINE, TRACE, DONUT, and SOLID commands display only the outline of the shapes. With FILLMODE set to ON, the shapes you create with these commands appear solid. If FILLMODE is reset to ON after it has been set to OFF, you must use the REGEN command in order for the screen to display as filled any unfilled shapes created by these commands. Switching between ON and OFF affects only the appearance of shapes created with the PLINE, TRACE, DONUT, and SOLID commands. Solids can be selected or identified by specifying the outlines only. The interior of a solid area is not recognized as an object when specified with the pointing device.

For example, the following command sequence shows placement of a solid-filled circle by use of the DONUT command (Figure 5–2).

> donut Enter
>
> Specify inside diameter of donut <0.5000>: 0 Enter
>
> Specify outside diameter of donut <1.0000>:1 Enter
>
> Specify center of donut or <exit>: 3,2 Enter
>
> Specify center of donut or <exit>: Enter

INSIDE DIA.0   INSIDE DIA.5

FIGURE 5–2 *Using the* DONUT *command to place a solid-filled circle and a filled donut*

The following command sequence shows placement of a filled donut by using the
DONUT command (Figure 5–2).

> donut [Enter]
> Specify inside diameter of donut <0.0000>: 0.5 [Enter]
> Specify outside diameter of donut <1.0000>: 1 [Enter]
> Specify center of donut or <exit>: 6,2 [Enter]
> Specify center of donut or <exit>: [Enter]

Drawing Solid-Filled Polygons

The SOLID command creates a solid-filled, straight-sided polygon whose outline is
determined by the points you specify. Two important factors should be kept in mind
when using the SOLID command: (1) the points must be selected in a specified order
or else the four corners generate a "bow-tie" shape instead of a rectangle; and (2) the
polygon generated has straight sides. Closer study reveals that even filled donuts and
PLINE-generated curved areas are actually straight-sided, just as arcs and circles generate as straight-line segments, small enough in length to appear smooth.

Invoke the SOLID command from the on-screen prompt by typing **solid.** AutoCAD
displays the following prompts:

> Specify first point: *(specify a first point)*
> Specify second point: *(specify a second point)*
> Specify third point: *(specify a third point diagonally
> opposite the second point)*
> Specify fourth point or <exit>: *(specify a fourth point, or
> press* [Enter] *to exit)*

When you specify a fourth point, AutoCAD draws a quadrilateral area. If instead you
press [Enter], AutoCAD creates a filled triangle (see Figure 5–5).

For example, the following command sequence draws a quadrilateral shape
(Figure 5–3).

> solid [Enter]
> Specify first point: *(specify Point 1)*
> Specify second point: *(specify Point 2)*
> Specify third point: *(specify Point 3)*
> Specify fourth point or <exit>: *(specify Point 4)*
> Specify third point: *(press* [Enter] *to exit)*

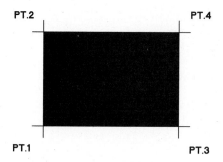

FIGURE 5-3 *The point-specification order for creating a quadrilateral area with the* SOLID *command*

In order for the point-specification order to create a polygonal shape with the SOLID command, the odd-numbered picks must be on one side and the even-numbered picks on the other side (Figure 5–6); if not, you get a bow-tie shape (Figure 5–4).

FIGURE 5-4 *Results of using the* SOLID *command when odd/even points are not specified correctly*

You can use the SOLID command to create an arrowhead or triangle (Figure 5–5).

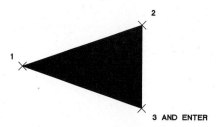

FIGURE 5-5 *Using the* SOLID *command to create a solid triangular shape*

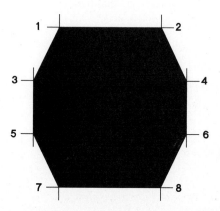

FIGURE 5-6 *Using the* SOLID *command to create a polygonal shape*

Drawing Point Objects

The POINT command creates a point object, and these points are drawn on the plotted drawing sheet with a single "pen down." The point object can be used as a reference point for object snapping. Points can be specified by 2D or 3D coordinates or with the pointing device.

Invoke the POINT command from the Draw panel (Figure 5–7) located in the Home tab, and AutoCAD displays the following prompt:

```
Point: (specify a point)
```

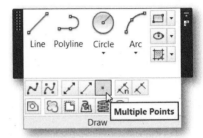

FIGURE 5–7 *Invoking the* POINT *command from the Draw panel*

AutoCAD continues to prompt for points until you terminate the command by pressing ESC.

When you draw the point, it appears on the display as a blip (+), if the BLIPMODE system variable is set to ON (default is set to OFF). You can make the point appear as a +, x, 0, or any of the available symbols by changing the PDMODE system variable. This step can be done by entering PDMODE at the on-screen prompt and entering the appropriate value. You can also change the PDMODE value by using the icon menu (Figure 5–8) invoked by typing DDPTYPE at the on-screen prompt. The default value of PDMODE is zero, which means that the point appears as a dot. If PDMODE is changed, all previous points drawn are replaced with the current setting.

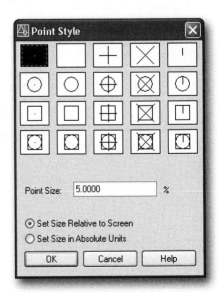

FIGURE 5–8 *The Point Style icon menu lets you select the shape and size of the point object*

The size that the point appears on screen depends on the value of the PDSIZE system variable. If necessary, you can change the size via the PDSIZE command. The default for PDSIZE is zero (one pixel in size). Any positive value larger than this will increase the size of the point accordingly.

Drawing Sketch Line Segments

The SKETCH command creates a series of freehand line segments, and it is useful for freehand drawings, contour mapping, and signatures. The sketched lines are not added to the drawing until they are recorded.

Invoke the SKETCH command from the on-screen prompt by typing **sketch.** Auto-CAD displays the following prompts:

Specify Sketch or [Type/Increment/tolerance]: *(specify a point to begin the sketching or select one of the available options)*

An imaginary pen follows the cursor movement. When the pen is down, AutoCAD sketches a connected segment whenever the cursor moves the specified increment distance from the previously sketched segment. When the pen is up, the pen follows the cursor movement without drawing.

The pen is raised (up) and lowered (down) either by pressing the pick button on the cursor/puck. When you invoke a *Pen Up* option, the current location of the pen will be the end point of the last segment drawn, which will be shorter than a standard increment length.

The *Pen Up* option selection does not take you out of the sketch mode, nor does it permanently record the lines drawn during the current sketch session.

While the pen is up, you cause AutoCAD to draw a straight line from the last segment to the current cursor location and return to the pen up status by entering a . (period) from the keyboard. This is convenient for long, straight lines that might occur in the middle of irregular shapes.

Lines being displayed while the cursor is moved, with the pen down, are temporary. They will appear green, or red if the current color for that layer or object is green, until they are permanently recorded. To record the temporarily drawn sketch lines, press either Enter or SPACEBAR. Pressing ESC exits sketch mode without recording any temporary lines. Once the sketch lines are recorded, the color of the sketch lines will change to the assigned color of the layer.

The *Type* option selection specifies the object type for the sketch line. The available selection includes Line, Polyline, and Spline.

The *Increment* option selection defines the length of each freehand line segment. You must move the pointing device a distance greater than the increment value to generate a line.

The *Tolerance* option selection specifies how closely the spline's curve fits to the freehand sketch.

OBJECT SELECTION

As mentioned earlier, all the modify commands initially prompt you to select objects. In most modify commands, the prompt allows you to select any number of objects. In some of the modify commands, however, AutoCAD limits your selection to only one object: the BREAK, DIVIDE, and MEASURE commands are examples. In the case of the FILLET and CHAMFER commands, AutoCAD requires you to select two objects. At the DIST and ID commands, AutoCAD requires you to select a point, but for the AREA command, AutoCAD permits you to select either a series of points or an object.

AutoCAD has various options to control the appearance of objects during selection preview. You can make the changes to the appearance of the selection preview from the Visual Effect Settings dialog box, which can be opened from the Selection tab of the Options dialog box (Figure 5–9).

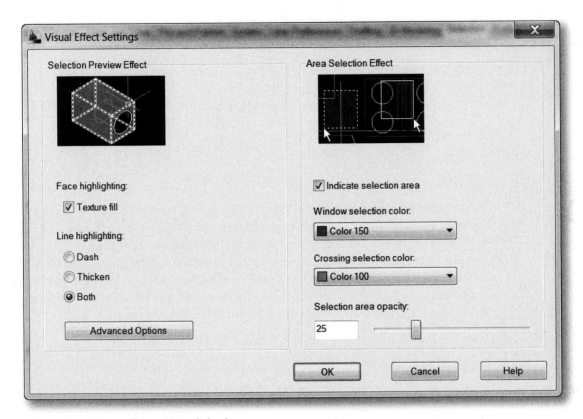

FIGURE 5–9 *Visual Effect Settings dialog box*

The preview area displays the effects of the current settings.

Selecting **Texture Fill** causes the selected face with texture fill effect.

Selecting **Dash** causes dashed lines to be displayed when the pick box cursor rolls over an object. This selection previewing indicates that the object would be selected if clicked.

Selecting **Thicken** causes thickened lines to be displayed when the pick box cursor rolls over an object. This selection preview indicates that the object would be selected if clicked.

Selecting **Both,** the default setting, causes thickened, dashed lines to be displayed when the pick box cursor rolls over an object. This selection previewing indicates that the object would be selected if clicked.

Choosing **Advanced Options** causes AutoCAD to open the Advanced Preview Options dialog box, from which you can select object types to exclude from selection previewing (Figure 5–10).

FIGURE 5–10 *Advanced Preview Options dialog box*

Selecting **Exclude Objects on Locked Layers** causes selection previewing to be excluded for objects on locked layers. The default is set to ON.

Select appropriate object types from the **Exclude** section to exclude selection previewing. The available object types include Xrefs, Tables, Groups, Multiline Text, and Hatches. By default, Xrefs and Tables are set to ON.

The **Area Selection Effect** section of the Visual Effect Settings dialog box controls the appearance of selection areas during selection preview. The preview area displays the effect of the current settings. Select **Indicate selection area** to indicate the selection area with the selected background color. Choose background colors for Window and Crossing selection area from the **Window selection color** and **Crossing selection color** drop-down lists, respectively. The slider bar for **Selection area opacity** controls the degree of transparency of background for window selection areas.

Choose **OK** to save and close the Visual Effect Settings dialog box.

Compared to the basic object selection options of *Pick, Window, Crossing, Previous,* and *Last,* the options covered in this section give you more flexibility and greater ease of use when you are prompted to select objects for use by the modify commands. The options explained in this section include *Wpolygon (WP), Cpolygon (CP), Fence, All, Multiple, Box, Auto, Undo, Add, Remove,* and *Single.*

Window Polygon (WP) Selection

The *Wpolygon* option is similar to the *Window* option, but it allows you to define a polygon-shaped window rather than a rectangular area. You define the selection area as you specify the points about the objects you want to select. The polygon can be of any shape but may not intersect itself. The polygon is formed as you select the points and includes rubber-band lines to the graphics cursor indicating the selection area. When the selected points define the desired polygon, press [Enter]. Only those objects that are totally inside the polygon shape are selected.

To select the *Wpolygon* option, type **WP**, and press [Enter] at the "Select objects:" prompt. The *Undo* option lets you undo the most recent polygon pick point.

Crossing Polygon (CP) Selection

The *Cpolygon* option is similar to the *Wpolygon* option, but it selects all objects within or crossing the polygon boundary. If an object is partially inside the polygon area, then the whole object is included in the selection set.

To select the *Cpolygon* option, type **CP**, and press (Enter) at the "Select objects:" prompt. The *Undo* option lets you undo the most recent polygon pick point.

Fence (F) Selection

The *Fence* option is similar to the *Cpolygon* option, except that you do not close the last vector of the polygon shape. The selection fence selects only those objects it crosses or intersects. Unlike *Wpolygon* and *Cpolygon*, the fence can cross over and intersect itself.

To select the *Fence* option, type **F**, and press (Enter) at the "Select objects:" prompt.

All Selection

The *All* option selects all the objects in the drawing excluding objects on frozen or locked layers. After selecting all the objects, you may use the *remove* (R) option to remove some of the objects from the selection set.

The *All* option must be spelled out in full as **All** and not applied as an abbreviation as you may do with the other options.

Multiple Selections

The *Multiple* option helps you overcome the limitations of the *Pick, Window*, and *Crossing* options. The *Pick* option is time consuming for use in selecting many objects. AutoCAD does a complete scan of the screen each time a point is picked. By using the *Multiple* modifier option, you can pick many points without delay, and when you press (Enter), AutoCAD applies all of the points during one scan.

Selecting one or more objects from a crowded group of objects is sometimes difficult with the *Point* option, and it is often impossible with the *Window* option. For example, if two objects are very close together and you wish to point to select them both, AutoCAD normally selects only one no matter how many times you select a point that touches them both. By using the *Multiple* option, AutoCAD excludes an object from being selected once it has been included in the selection set. As an alternative, use the *Crossing* option to cover both objects. If this is not feasible, then the *Multiple* modifier may be the best choice.

Box Selection

The *Box* option is usually employed in a menu macro to give the user an option of using both the *Window* and *Crossing* methods, depending on how and where the picks are made on the screen.

When *Box* is invoked as a response to a prompt to select an object, the options are applied as follows:

If the picks are made left to right—the first point is to the left of the second—then the two points become diagonally opposite corners of a rectangle that is used as a *window* option. All visible objects totally within the rectangle are part of that selection.

If the picks are right to left, the selection rectangle becomes the *crossing* option. All visible objects that are within or partially within the rectangle are part of that selection.

Auto Selection

The *Auto* option is actually a triple option. It combines the *Point* option with the two *Box* options. If the target box touches an object, that object is selected as you would in using the *Point* option. If the target box does not touch an object, the selection becomes either a *Window* or *Crossing* option, depending on where the second point is picked in relation to the first.

NOTE   The *Auto* option is the default when you are prompted to "Select objects:".

Undo Selection

The *Undo* option allows you to remove the last item(s) selected from the selection set without aborting the "Select objects:" prompt and then to continue adding to the selection set. It should be noted that if the last option to the selection process includes more than one object, the *Undo* option will remove all the objects from the selection set that were selected by that last option.

Add Selection

The *Add* option lets you exit the *Remove* mode in order to continue adding objects to the selection set by however many options you wish to use.

Remove Selection

The *Remove* option lets you remove objects from the selection set. The "Select objects:" prompt always starts in the ADD mode. The *Remove* mode is a switch from the *Add* mode, not a standard option. Once invoked, the objects selected by whatever and however many options you use will be removed from the selection set. It will remain in effect until reversed by the *Add* option.

Single Selection

The *Single* option causes the object selection to terminate and the command in progress to proceed after you use only one object selection option. It does not matter if one object is selected or a group is selected; only one selection opportunity is given.

OBJECT SELECTION MODES

AutoCAD provides six selection modes that will enhance object selection. You can toggle on/off one or more object selection modes from the Selection tab of the Options dialog box. By having the appropriate selection mode set to ON, you have access to several methods that give you more flexibility and greater ease in selecting objects.

To make changes to the selection modes settings, open the Options dialog box by selecting *Options* from the Shortcut menu when you are not in a command. Choose the Selection tab (Figure 5–11).

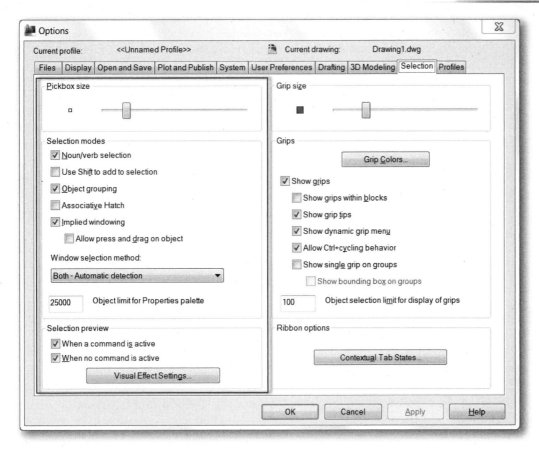

FIGURE 5–11 *Options dialog box with Selection tab selected*

You can toggle any one or more combinations of the settings provided under the Selection Modes. **Noun/verb selection**, **Implied windowing**, and **Object grouping** are the defaults.

In the **Pickbox Size** section, you can adjust the size of the pickbox using the Pickbox Size slider bar provided in the dialog box. In the **Selection Modes** section, you can customize how objects will be selected for modification or for use by AutoCAD commands.

Selecting **Noun/verb selection** allows the traditional verb-noun command syntax to be reversed for most modifying commands. When **Noun/verb selection** is checked, AutoCAD allows you to select objects when there is no command in progress and then invoke the modifying command that you want to use on the selection set. For example, instead of invoking the COPY command followed by selecting the objects to be copied, with **Noun/verb selection** checked, you can select the objects first and then invoke the COPY command, and AutoCAD skips the object selection prompt. When **Noun/verb selection** is checked, the cursor changes to resemble a running OSNAP (for "object snap") cursor. Whenever you want to use the **Noun/verb selection** feature, first create a selection set. Subsequent modify commands that you invoke execute by using the objects in the current selection set without prompting for object selection. To clear the current selection set, press ESC. This step clears the selection set, so that any subsequent editing command will once again prompt for object selection.

Selecting **Use Shift to add to selection** controls how you add objects to an existing selection set. When **Use Shift to add to selection** is checked, it activates an additive selection mode in which the SHIFT key must be held down while adding more objects to the selection set. For example, when you first pick an object, it is highlighted. If you pick another object, it is highlighted, and the first object is no longer highlighted. The only way to add objects to the selection set is to select objects by holding down the SHIFT key. Similarly, to remove objects from the selection set, select the objects by holding down the SHIFT key. When **Use Shift to add to selection** is not checked, as is the default, objects are added to the selection set by just picking them individually or by using one of the selection options; AutoCAD adds the objects to the selection set.

Selecting **Object grouping** controls the automatic group selection. If **Object grouping** is checked, selecting an object that is a member of a group selects the whole group. Refer to Chapter 6 for a detailed description of how to create groups.

Selecting **Associative Hatch** controls which objects will be selected when you select an associative hatch. If **Associative Hatch** is checked, selecting an associative hatch also selects the boundary objects. Refer to Chapter 9 for a detailed description of hatching.

Selecting **Implied windowing** allows you to create a selection window automatically when the "Select objects:" prompt appears. When **Implied windowing** is checked (which is the default) it works like the box option explained earlier. If **Implied windowing** is not checked, you can create a selection window by using the WINDOW or CROSSING selection-set methods.

Selecting **Allow press and draw on object** controls the manner by which you draw the selection window with your pointing device. When **Allow press and draw on object** is checked, you can create a selection window by holding down the pick button and dragging the cursor diagonally while you create the window. When **Allow press and draw on object** is not checked (which is the default) you need to use two separate picks of the pointing device to create the selection window. In other words, you need to pick once to define one corner of the selection window and then pick a second time to define its diagonal corner.

The **Window selection method** controls the method of drawing a selection window. The Click and Click selection creates a selection window using two points. The Press and drag selection creates a selection window clicking and dragging. Release the mouse button to complete the selection. The Both—Automatic detection selection creates a selection window using either of the methods.

> Another way to control how selection windows are drawn is to set the PICKDRAG system variable set to 0, to draw the selection window using two points, set to 1, to draw the selection window using dragging and set to 3, for either of the two methods.

The **Object limit for Properties palette** determines the limit of the number of objects that can be changed at one time with the Properties and Quick Properties palettes.

SELECT SIMILAR TOOL

The Select Similar tool enables you to select an object and automatically include all other objects of the same type and with the same properties, in the selection set. Select an object (with PICKFIRST system variable set to ON) and then from the right-click menu select *Select Similar* from the shortcut menu (see Figure 5–12). AutoCAD selects all the objects based on the properties to filter set in the Select Similar settings box. A Settings option (accessible when you invoke the SELECTSIMILAR command) enables you to specify which properties to filter. If only the Layer property is enabled when you select an arc, for example, AutoCAD automatically selects all arcs on the same layer as the one you selected. If both the Layer and Linetype properties are enabled, however, AutoCAD only selects the arcs on the same layer and with the same linetype as the selected one.

FIGURE 5–12 *Selection of Select Similar from the shortcut menu*

You can also select one or more objects with the Select Similar tool to create the matching selection set accordingly.

Invoke the Select Similar tool from the on-screen prompt by typing **selectsimilar,** AutoCAD displays the following prompt:

    Select objects or [Settings]: (select one or more objects to
    create a matching selection or select settings to change the
    controls which properties must match for an object of the
    same type to be selected)

For example, if the Layer filter is enabled and you select two arcs, each on different layers, AutoCAD selects all the arcs on both layers. If, instead, you select an arc and a line, AutoCAD selects all the arcs on the same layer as the selected arc and all the lines on the same layer as the selected line.

In addition to general object properties, you can filter selections based on object-specific properties including object style and reference name. Object Style properties apply to text and mtext, leaders and mleaders, dimensions and tolerances, as well as tables and multilines. Reference names apply to blocks and externally references files including xrefs and images, as well as *PDF*, *DWG*, or *DGN* files.

Choosing *Settings* causes AutoCAD to open the Select Similar Settings dialog box (see Figure 5–13) to set the control properties to match for the selected object(s).

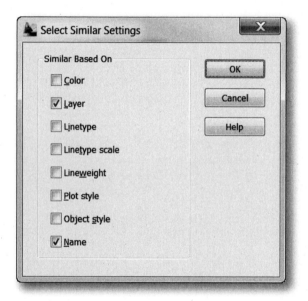

FIGURE 5–13 *Select Similar Settings dialog box*

The **Color** selection selects objects with matching colors to be similar.

The **Layer** selection selects objects on matching layers to be similar.

The **Linetype** selection selects objects with matching linetypes to be similar.

The **Linetype Scale** selection selects objects with matching linetype scales to be similar.

The **Lineweight** selection selects objects with matching lineweights to be similar.

The **Plot Scale** selection selects objects with matching plot styles to be similar.

The **Object Style** selection selects objects with matching styles (such as text styles, dimension styles, and table styles) to be similar.

The **Name** selection selects referenced objects (such as blocks, xrefs, and images) with matching names to be similar.

After making the necessary changes, choose OK to close the dialog box and prompt returns to select objects for matching selection.

OBJECT CREATION FROM AN EXISTING OBJECT

The Add Selected tool enables you to quickly create a new object in your drawing based on the properties of an existing object. Select an object (with PICKFIRST system variable set to ON) and then from the right-click menu select *Add Selected* from the shortcut menu (see Figure 5–14).

FIGURE 5–14 *Selection of Add Selected from the shortcut menu*

AutoCAD invokes the command of the selected object and prompts to draw additional objects. For example, if you select an arc and invoke the *Add Selected* tool from the shortcut menu, AutoCAD automatically launches the ARC command with basic object properties including color, layer, linetype, linetype scale, plotstyle, lineweight, transparency, and material preset to match the selected object.

You can also select an object with the Add Selected tool to invoke the selected object command. Invoke the Add Selected tool from the on-screen prompt by typing **addselected,** AutoCAD displays the following prompt:

> Select objects: *(select an object to invoke the command related to the selected object)*

AutoCAD invokes the command of the selected object and prompts to draw additional objects.

MODIFYING OBJECTS

This section describes seven modify commands: LENGTHEN, STRETCH, ROTATE, SCALE, PEDIT, JOIN, and MATCHPROP. Additional modify commands were explained in Chapters 2 and 4.

Lengthening Objects

The LENGTHEN command is used to increase or decrease the length of line objects or the included angle of an arc.

Invoke the LENGTHEN command from the Modify panel (Figure 5–15) located in the Home tab, and AutoCAD displays the following prompt:

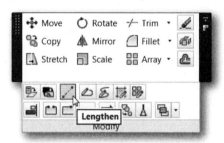

FIGURE 5–15 *Invoking the LENGTHEN command from the Modify panel*

> Select an object or ⬇ *(select an object or right-click and choose one of the options)*

Available options include *Delta*, *Percent*, *Total*, and *Dynamic*.

Choosing *Delta* allows you to change the length or, where applicable, the included angle from the end point of the selected object closest to the pick point. A positive value results in an increase in extension; a negative value results in a trim. When you select the *Delta* option, AutoCAD displays the following prompts:

> Enter delta length or ⬇ *(specify positive or negative value)*
>
> Select an object to change or ⬇ *(select an object, and its length is changed on the end nearest the selection point)*
>
> Select an object to change or ⬇ *(select additional objects; when done, press* **Enter** *to exit the command sequence)*

If you prefer to specify the included angle instead of specifying the delta length, then choose the *Angle* option. AutoCAD displays the following prompts:

> Enter delta angle or ⬇ *(specify positive or negative angle)*
>
> Select an object to change or ⬇ *(select an object, and its included angle is changed on the end nearest the selection point)*
>
> Select an object to change or ⬇ *(select additional objects; when done, press* **Enter** *to exit the command sequence)*

Choosing *Percent* allows you to set the length of an object by a specified percentage of its total length. It will increase the length/angle for values greater than 100 and decrease them for values less than 100. For example, a 12-unit-long line will be changed to 15 units by using a value of 125. A 12-unit-long line will be changed to 9 units by using a value of 75. When you select the *Percent* option, AutoCAD displays the following prompts:

> Enter percentage length: *(specify positive nonzero value and press* **Enter** *)*
>
> Select an object to change or ⬇ *(select an object, and its length is changed on the end nearest the selection point)*

```
Select an object to change or ⊡ (select additional objects;
when done, press [Enter] to exit the command sequence)
```

Choosing *Total* allows you to change the length/angle of an object to the value specified. When you select the *Total* option, AutoCAD displays the following prompts:

```
Specify total length or ⊡ (specify distance, right-click and
choose angle, and then specify an angle for change)

Select an object to change or ⊡ (select an object, and its
length is changed to the specified distance on the end
nearest the selection point)

Select an object to change or ⊡ (select additional objects;
when done, press [Enter] to exit the command sequence)
```

Choosing *Dynamic* allows you to change the length/angle of an object in response to the cursor's final location, relative to the end point nearest to where the object is selected. When you select the *Dynamic* option, AutoCAD displays the following prompts:

```
Select an object to change or ⊡ (select an object to change
the endpoint)

Specify new endpoint: (specify new endpoint for the selected
object)

Select an object to change or ⊡ (select additional objects;
when done, press [Enter] to exit the command sequence)
```

Choosing *Undo* reverses the most recent change made by the LENGTHEN command.

Stretching Objects

The STRETCH command allows you to stretch the shape of an object without affecting other crucial parameters. A common example is stretching a square into a rectangle. The length is changed while the width remains the same.

AutoCAD stretches lines, polyline segments, rays, arcs, elliptical arcs, and splines that cross the selection window. The STRETCH command moves the end points that lie inside the window, leaving those outside the window unchanged. If the entire object is inside the selection window, the STRETCH command operates like the MOVE command.

It is possible to make multiple selection sets using the crossing method, and AutoCAD will modify all of the selection sets with a single STRETCH command operation.

Invoke the STRETCH command from the Modify panel (Figure 5–16) located in the Home tab, and AutoCAD displays the following prompts:

```
Select objects: (select the first corner of a crossing window
or polygon)

Specify opposite corner: (select the opposite corner)

Select objects: (select additional objects; when done, press
[Enter] to exit the command sequence)

Specify base point or ⊡ (specify a base point or press [Enter])

Specify second point <or use first point as displacement>:
(specify the second point of displacement or press [Enter] to
use first point as displacement)
```

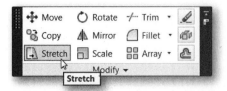

FIGURE 5–16 *Invoking the* STRETCH *command from the Modify panel*

If you provide the base point and second point of displacement, AutoCAD stretches the selected objects the vector distance from the base point to the second point. If you press Enter at the prompt for the second point of displacement, AutoCAD assigns the first point as the X,Y displacement value.

Figure 5–17 shows two examples of using the STRETCH command.

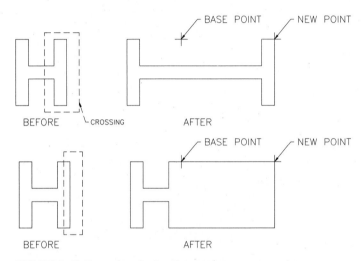

FIGURE 5–17 *Examples of using the* STRETCH *command*

Rotating Objects

The ROTATE command changes the orientation of existing objects by rotating them about a specified point, labeled as the base point. Design changes often require that an object, feature, or view be rotated. By default, a positive angle rotates the object in the counterclockwise direction, and a negative angle rotates in the clockwise direction.

Invoke the ROTATE command from the Modify panel (Figure 5–18) located in the Home tab, and AutoCAD displays the following prompts:

Select objects: *(select the objects to rotate and press* Enter *to complete the selection)*

Specify base point: *(specify a base point about which selected objects are to be rotated)*

Specify rotation angle or ⊡ *(specify a positive or negative rotation angle, or choose an option from the shortcut menu)*

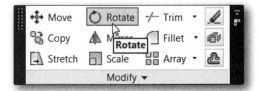

FIGURE 5–18 *Invoking the ROTATE command from the Modify panel*

The base point can be anywhere in the drawing. If the base point selected is on the selected object itself, the selected base point becomes an anchor point for rotation. The rotation angle determines how far an object rotates around the base point.

The following command sequence shows an example of rotating a group of objects selected by the *Window* option (Figure 5–19).

> rotate `Enter`
>
> Select objects: *(select the objects to rotate and press* `Enter` *to complete the selection)*
>
> Specify base point: *(pick the base point)*
>
> Specify rotation angle or ⬇ 45 `Enter`

AutoCAD rotates the selected objects by 45° (Figure 5–16).

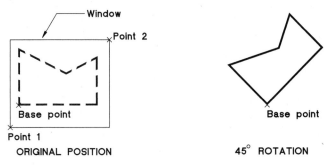

FIGURE 5–19 *Rotating a group of objects by means of the ROTATE command*

Choosing the *Reference* option allows you to rotate objects from a specified angle to a new, absolute angle. Specify the current orientation as referenced by the angle, or provide the angle by selecting two end points of a line to be rotated and specifying the desired rotation angle. AutoCAD automatically calculates the rotation angle and rotates the object appropriately. This method of rotation is very useful when you want to straighten an object or align it with other features in a drawing.

The following command sequence shows an example of rotating a group of objects, selected by the *Window* option, in reference to the current orientation (Figure 5–20).

> rotate
>
> Select objects: *(select the objects to rotate and press* `Enter` *to complete the selection)*
>
> Specify base point: *(pick the base point)*
>
> Specify rotation angle or ⬇ *(choose Reference from the shortcut menu)*

```
Specify the reference angle <current>: 60 Enter
Specify the new angle or ⊞ 30 Enter
```

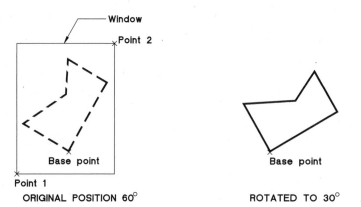

FIGURE 5-20 *Rotating a group of objects by means of the ROTATE command in reference to the current orientation*

Choosing the *Copy* option creates a copy of the selected objects for rotation.

Scaling Objects

The SCALE command lets you change the size of selected objects or the complete drawing. Objects are made larger or smaller, and the same scale factor is applied to the X, Y, and Z directions. To enlarge an object, specify a scale factor greater than 1. For example, a scale factor of 3 makes the selected objects three times larger. To reduce the size of an object, use a scale factor between 0 and 1. Do not specify a negative scale factor. For example, a scale factor of 0.75 would reduce the selected objects to three-quarters of their current size.

Invoke the SCALE command from the Modify panel (Figure 5–21) located in the Home tab, and AutoCAD displays the following prompts:

```
Select objects: (select the objects to scale and press Enter
to complete the selection)
Specify base point: (specify a base point about which
selected objects are to be scaled)
Specify scale factor or ⊞ (specify a scale factor, or choose
an option from the shortcut menu)
```

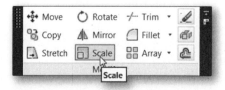

FIGURE 5-21 *Invoking the SCALE command from the Modify panel*

The base point can be anywhere in the drawing. If the base point selected is on the selected object itself, the selected base point becomes an anchor point for scaling. The scale factor multiplies the dimensions of the selected objects by the specified scale.

The following command sequence shows an example of enlarging a group of objects by a scale factor of three.

> scale
>
> Select objects: *(select the objects to scale, and press* [Enter] *to complete the selection)*
>
> Specify base point: *(pick the base point)*
>
> Specify scale factor or [⊡] 3 [Enter]

AutoCAD enlarges the selected objects by a factor of three (Figure 5–22).

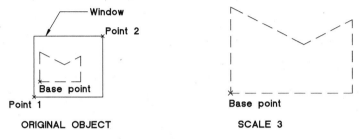

FIGURE 5–22 *Enlarging a group of objects by means of the* SCALE *command*

Choosing *Reference* causes the selected objects to be scaled based on a reference length and a specified new length. Specify the current dimension as a reference length, or select two end points of a line to be scaled, and specify the desired new length. AutoCAD will automatically calculate the scale factor and enlarge or shrink the object appropriately.

The following command sequence shows an example of using the SCALE command to enlarge a group of objects selected by means of the WINDOW option in reference to the current dimension (Figure 5–23).

> scale
>
> Select objects: *(select the objects to scale and press* [Enter] *to complete the selection)*
>
> Specify base point: *(pick the base point)*
>
> Specify scale factor or [⊡] *(choose Reference from the shortcut menu)*
>
> Specify reference length <1> 3.8 [Enter]
>
> Specify new length or [⊡] 4.8 [Enter]

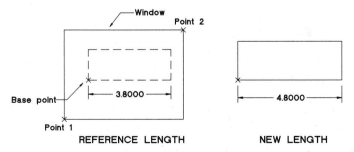

FIGURE 5–23 *Enlarging a group of objects by means of the* SCALE *command in reference to the current dimension*

Choosing the Copy option creates a copy of the selected objects for scaling.

Modifying Polylines

The `PEDIT` command allows you to modify polylines. In addition to using modify commands such as `MOVE`, `COPY`, `BREAK`, `TRIM`, and `EXTEND`, you can use the `PEDIT` command to modify polylines. The `PEDIT` command has special editing features for dealing with the unique properties of polylines and is perhaps the most complex AutoCAD command, with several multioption submenus, totaling some 70 command options.

Invoke the `PEDIT` command from the Modify panel (Figure 5–24) located in the Home tab, and AutoCAD displays the following prompt:

```
Select polyline or ⏎ (select polyline, line, or arc, or
    select Multiple from the shortcut menu)
```

If you select a line or an arc instead of a polyline, you are prompted as follows:

```
Do you want it to turn into one? <Y>
```

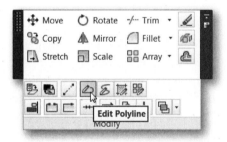

FIGURE 5–24 *Invoking the* `PEDIT` *command from the Modify panel*

Responding **Y** or pressing Enter turns the selected line or arc into a single-segment polyline that can then be edited. Normally this is done in order to use the *Join* option to add other connected segments that, if not already polylines, will be transformed into polylines. It should be emphasized that in order to join segments together into a polyline, their end points must coincide. This occurs during Line-Line, Line-Arc, Arc-Line, and Arc-Arc Continuation methods or by using the Endpoint Object Snap mode.

The second prompt does not appear if the first segment selected is already a polyline. It may even be a multisegment polyline. If you select the *Multiple* option, AutoCAD allows you to select more than one polyline to modify. After the object selection process, you will be returned to the multioption prompt as follows:

```
Enter an option (select one of the available options)
```

Choosing *Close* operates in a manner similar to the *Close* option of the `LINE` command. If, however, the last segment was a polyline arc, the next segment will be similar to the Arc-Arc Continuation, using the direction of the last polyarc as the starting direction and drawing another polyarc, with the first point of the first segment as the ending point of the closing polyarc.

Figure 5–25 and Figure 5–26 show examples of applying the *Close* option.

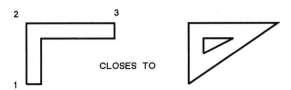

FIGURE 5–25 *Using the* `PEDIT` *command Close option with polylines*

FIGURE 5–26 *Using the* `PEDIT` *command Close option with polyarcs*

Choosing *Open* deletes the segment that was drawn with the *Close* option. If the polyline was closed by drawing the last segment to the first point of the first segment without using the *Close* option, the *Open* option will not have a visible effect.

Choosing *Join* takes selected lines, arcs, or polylines and combines them with a previously selected polyline into a single polyline if all segments are connected at sequential and coincidental end points.

Choosing *Width* permits uniform or varying widths to be specified for polyline segments.

Choosing *Edit Vertex* allows you to edit the vertices of the polyline. A vertex is the point where two segments join. The visible vertices are marked with an X to indicate which one is to be modified. You can modify vertices of polylines in several ways. When you select the *Edit Vertex* option, AutoCAD prompts you with this suboption:

> `Enter a vertex editing option`

Choose *Next* or *Previous* when you wish to move the mark to the next or previous vertex, whether or not you have modified the marked vertex.

Choosing *Break* establishes the marked vertex as one vertex for the *Break* option. AutoCAD then displays the following prompt:

> `Enter an option`

The choices of the *Break* option permit you to switch to another vertex for the second break point, to initialize the break, or to exit the option. If two vertices are selected, you may use the *Go* option to remove the segment(s) between the vertices; if you select the end points of a polyline, this option will not work. If you select the *Go* option immediately after the *Break* option, the polyline will be divided into two separate polylines. If it is a closed polyline, it will be opened at that point.

Choosing *Insert* allows you to specify a point and have the segment between the marked vertex and the next vertex become two segments that meet at the specified point. The selected point does not have to be on the polyline segment.

For example, the following command sequence shows the application of the *Insert* option (Figure 5–27).

> pedit ⏎Enter
> Select polyline or ⏎ *(select the polyline)*
> Enter an option: *(choose Edit Vertex)*
> Enter a vertex editing option: *(choose Insert)*
> Specify location for new vertex: *(specify a new vertex)*

BEFORE   AFTER

FIGURE 5–27 *Using the* PEDIT *command Insert option*

Choosing *Move* allows you to specify a point and relocate the marked vertex to the selected point.

For example, the following command sequence shows the application of the *Move* option (Figure 5–28).

> pedit ⏎Enter
> Select polyline or ⏎ *(select the polyline)*
> Enter an option: *(choose Edit Vertex)*
> Enter a vertex editing option: *(choose Move)*
> Specify new location for marked vertex: *(specify the new location)*

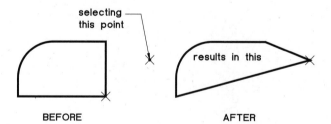

BEFORE   AFTER

FIGURE 5–28 *Using the* PEDIT *command Move option*

Choosing *Regen* regenerates the polyline without having to cancel the PEDIT command.

Choosing *Straighten* establishes the marked vertex as one vertex for the STRAIGHTEN option. AutoCAD then displays the following prompt:

> Enter an option

When the two vertices are selected, you may use the *Go* option to replace the segment(s) between the vertices with a single straight-line segment.

For example, the following command sequence shows the application of the *Straighten* option (Figure 5–29).

```
pedit  Enter
Select polyline or ⬇ (select the polyline)
Enter an option: (choose Edit Vertex)
Enter a vertex editing option: (choose Straighten)
Enter an option: (choose Next)
Enter an option: (choose Go)
Enter a vertex editing option: (choose Exit)
```

BEFORE   results in   AFTER

FIGURE 5–29 *Using the* PEDIT *command Straighten option*

Choosing *Tangent* permits you to assign a tangent direction to the marked vertex that can be used for the curve-fitting option. AutoCAD displays the following prompt:

```
Specify direction of vertex tangent:
```

You can either specify the direction with a point or type the coordinates at the keyboard.

Choosing *Width* permits you to specify the starting and ending widths of the segment between the marked vertex and the next vertex. AutoCAD displays the following prompt:

```
Specify new width for all segments:
```

For example, the following command sequence shows the application of the *Width* option (Figure 5–30).

```
pedit  Enter
Select polyline or ⬇ (select a polyline)
Enter an option: (select Edit Vertex)
Enter a vertex editing option: (choose Edit Vertex)
Enter an option: (choose Width)
Specify new width for all segments: 0.25  Enter
```

Choosing *Exit* exits the *Vertex Editing* option and returns you to the PEDIT multi-option prompt.

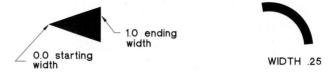

FIGURE 5–30 *Using the* PEDIT *command Width option*

Choosing *Fit* draws a smooth curve through the vertices, using any specified tangents.

Choosing *Spline* provides several ways to draw a curve, based on the polyline being edited. These include Quadratic B-spline and Cubic B-spline curves.

Choosing *Decurve* returns the polyline to the way it was drawn originally. Figure 5–31 illustrates the differences between *Fit, Spline*, and *Decurve*.

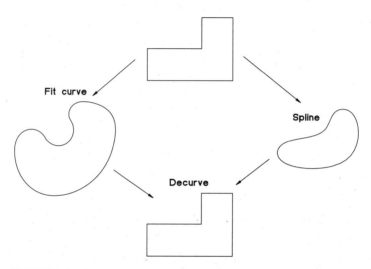

FIGURE 5–31 *Comparing* PEDIT *command Fit curve, Spline curve, and Decurve options*

Choosing *Ltype Gen* controls the display of the linetype at vertices. When it is set to ON, AutoCAD generates the linetype in a continuous pattern through the vertices of the polyline. And when it is set to OFF, AutoCAD generates the linetype starting and ending points with a dash at each vertex. The *Ltype Gen* option does not apply to polylines with tapered segments.

Choosing *Reverse* will reverse the order of vertices of the polyline.

Choosing *Undo* reverses the latest PEDIT operation.

Joining Similar Objects

The JOIN command allows you to create one line, arc, elliptical arc, polyline, or spline from multiple like objects. Invoke the JOIN command from the Modify panel (Figure 5–32) located in the Home tab, and AutoCAD displays the following prompt:

```
Select source object or multiple objects to join at once:
(select a line, polyline, arc, elliptical arc, or spline)
```

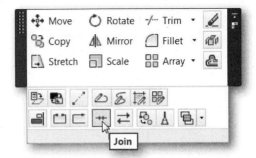

FIGURE 5–32 *Invoking the* JOIN *command from the Modify panel*

AutoCAD displays the following prompt:

> Select objects to join to source: *(select one or more objects and press* Enter *)*

The objects can be lines, polylines, or arcs. The objects cannot have gaps between them, and they must lie on the same plane parallel to the UCS XY plane.

Reversing the Direction of Objects

The REVERSE command reverses the vertices of selected lines, polylines, splines, and helixes, which is useful for linetypes with included text, or wide polylines with differing beginning and ending widths.

Invoke the REVERSE command from the Modify panel located in the Home tab, and AutoCAD displays the following prompt:

> Select objects: *(select objects to change the direction)*

Vertices of selected objects are reversed.

Matching Properties

The MATCHPROP command allows you to copy selected properties from an object to one or more other objects in the current drawing or any other drawing currently open. Properties that can be copied include color, layer, linetype, linetype scale, lineweight, transparency, thickness, plot style, and, in some cases, dimension, text, polyline, table, material, viewport, and hatch.

Invoke the MATCHPROP command by selecting Match Properties from the Clipboard panel (Figure 5–33) located in the Home tab, and AutoCAD displays the following prompts:

> Select source object: *(select the object whose properties you want to copy)*
>
> Select destination object(s) or ⬇ *(select the destination objects and press* Enter *to terminate the selection, or choose Settings from the shortcut menu to display the Property Settings dialog box)*

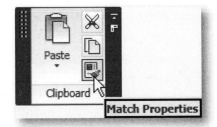

FIGURE 5–33 *Invoking the* MATCHPROP *command from the Clipboard panel*

Image(s) © Cengage Learning 2013

Selecting the *Settings* option displays the Property Settings dialog box (Figure 5–34). Use the *Settings* option to control which object properties are copied. By default, all object properties in the Property Settings dialog box are set to ON for copying.

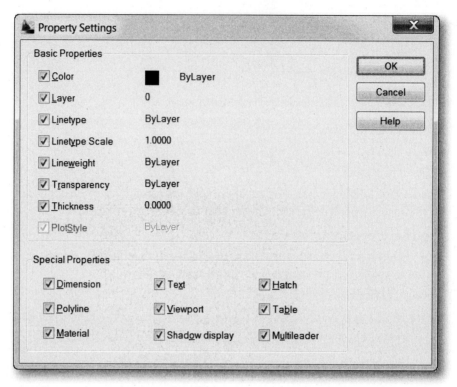

FIGURE 5–34 *Property Settings dialog box*

Choosing **Color** changes the color of the destination object to that of the source object.

Choosing **Layer** changes the layer of the destination object to that of the source object.

Choosing **Linetype** changes the linetype of the destination object to that of the source object. Available for all objects except attributes, hatches, multiline text, points, and viewports.

Choosing **Linetype Scale** changes the linetype scale factor of the destination object to that of the source object. Available for all objects except attributes, hatches, multiline text, points, and viewports.

Choosing **Lineweight** changes the lineweight of the destination object to that of the source object.

Choosing **Transparency** changes the transparency of the destination object to that of the source object.

Choosing **Thickness** changes the thickness of the destination object to that of the source object. Available only for arcs, attributes, circles, lines, points, 2D polylines, regions, text, and traces.

Choosing **PlotStyle** changes the plot style of the destination object to that of the source object. If you are working in color-dependent plot style mode (`PSTYLE-POLICY` is set to 1), this option is unavailable.

Choosing **Dimension** changes the dimension style of the destination object to that of the source object. Available only for dimension, leader, and tolerance objects.

Choosing **Polyline** changes the width and linetype generation properties of the destination polyline to those of the source polyline. The fit/smooth property and the elevation of the source polyline are not transferred to the destination polyline. If the source polyline has variable widths, the width property is not transferred to the destination polyline.

Choosing **Material** changes the material applied to the object. If the source object does not have a material assigned and the destination object does, the material is removed from the destination object.

Choosing **Text** changes the text style of the destination object to that of the source object.

Choosing **Viewport** changes the following properties of the destination paper space viewport to match those of the source viewport: ON/OFF, display locking, standard or custom scale, shade plot, snap, grid, and UCS icon visibility and location. The settings for clipping and for UCS per viewport and the freeze/thaw state of the layer are not transferred to the destination object.

Choosing **Shadow display** changes the shadow display. The object can cast shadows, receive shadows, or both, or it can ignore shadows.

Choosing **Hatch** changes the hatch pattern of the destination object to that of the source object.

Choosing **Table** changes the table style of the destination object to that of the source object.

Choosing **Multileader** changes the multileader style and annotative properties of the destination object to that of the source object. Available only for multileader objects.

After making the necessary changes, choose **OK** to close the Property Settings dialog box. AutoCAD displays the "Select destination object(s):" prompt. Press Enter to complete the object selection.

Delete Duplicate Objects

The `OVERKILL` command removes duplicate or overlapping lines, arcs, and polylines. In addition, the `OVERKILL` command can also combine partially overlapping or contiguous ones.

Invoke the `OVERKILL` command from the Modify panel (Figure 5–35) located in the Home tab, and AutoCAD displays the following prompts:

> `Select objects:` *(select the objects to delete the duplicates or removing overlapping objects and press Enter to complete the selection)*

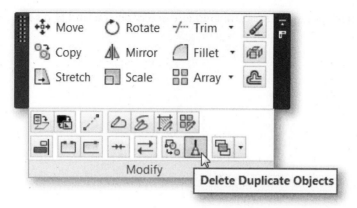

FIGURE 5–35 *Invoking the* OVERKILL *command from the Modify panel*

AutoCAD displays the Delete Duplicate Objects dialog box similar to shown in Figure 5-36.

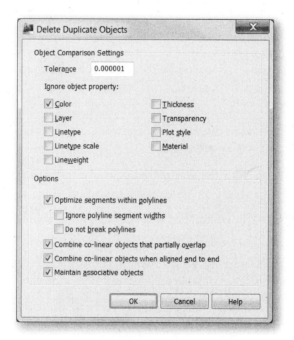

FIGURE 5–36 *Delete Duplicate Objects dialog box*

The **Tolerance** edit field controls the precision with which OVERKILL command makes numeric comparisons. If this value is 0, the two objects being compared must match before OVERKILL modifies or deletes one of them.

The available options under **Ignore object property** section determines which object properties are ignored during comparison.

The **Optimize segments within polylines** when selected causes individual line and arc segments within selected polylines to be examined. Duplicate vertices and segments are removed.

In addition, OVERKILL command compares individual polyline segments with completely separate line and arc segments. If a polyline segment duplicates a line or arc object, one of them is deleted.

If this option is not selected, polylines are compared as discreet objects and two sub-options are not selectable.

The **Ignore polyline segment width** when selected, ignores segment width, while optimizing polyline segments.

When **Do not break polylines** is selected Polyline objects are unchanged.

When **Combine colinear objects that partially overlap** is selected overlapping objects are combined into single objects.

When **Combine colinear objects when aligned end to end** is selected objects that have common endpoints are combined into single objects.

When **Maintain associative objects** is selected associative objects are not deleted or modified.

Open the Exercise Manual PDF for Chapter 5 for discipline-specific exercises. Related files are downloaded from the student companion site mentioned in the Introduction (refer to page number xii for instructions).

REVIEW QUESTIONS

1. The command that allows you to draw freehand lines is:
   a. SKETCH
   b. DRAW
   c. PLINE
   d. FREE

2. To create a six-sided area that would select all the objects completely within it, you should respond to the "Select Objects:" prompt with:
   a. WP
   b. Addition
   c. W
   d. Fence

3. The MATCHPROP command does not allow you to modify an object's:
   a. Linetype
   b. Fillmode
   c. Color
   d. Layer

4. By default, most selection set prompts default to which option:
   a. ALL
   b. AUTO
   c. BOX
   d. WINDOW
   e. CROSSING

5. The MULTIPLE option selection sets allow you to:
   a. Select multiple objects that lie on top of each other
   b. Scan the database only once to find multiple objects
   c. Use the Window or Crossing options
   d. Both A and B
   e. Both B and C

6. In regard to using the SOLID command, which of the following statements is true?
   a. The order of point selection is unimportant
   b. FILL must be turned ON in order to use the SOLID command
   c. The points must be selected on existing objects
   d. The points must be selected in a clockwise order
   e. None of the above

7. What command is commonly used to create a filled rectangle?
   a. FILL-ON
   b. PLINE
   c. RECTANGLE
   d. LINE
   e. SOLID

8. To turn a series of line segments into a polyline, which option of the PEDIT command would you use?
   a. JOIN
   b. FIT CURVE
   c. SPLINE
   d. CONNECT
   e. Line segments cannot be connected.

9. Which command allows you to change the size of an object, where the X and Y scale factors are changed equally?
   a. ROTATE
   b. SCALE
   c. SHRINK
   d. MODIFY
   e. MAGNIFY

10. The BOX option for creating selection sets is most useful for:
   a. Employing a menu macro
   b. Creating rectangular selection areas
   c. Extending a selection into 3D
   d. Nothing—It is not a valid option

11. Using the SCALE command, what number would you enter to enlarge an object by **50**%?

    a. 0.5

    b. 50

    c. 3

    d. 1.5

12. Scale factors smaller that **1** increase the size.

    a. True

    b. False

13. Which panel can you find the Move Command?

    a. Modify Panel

    b. Move Panel

    c. Draw Panel

    d. Edit Panel

14. To avoid changing the location of an object when using the SCALE command:

    a. The reference length should be less than the limits

    b. The scale factor should be less than one

    c. The base point should be on the object

    d. The base point should be at the origin

15. The ROTATE command is used to rotate objects around:

    a. Any specified point

    b. The point -1,-1 only

    c. The origin

    d. It is usable only in 3D drawings

    e. None of the above

16. The REMOVE option for forming selection sets deletes objects from the drawing in much the same manner as the ERASE command.

    a. True

    b. False

17. Which of the following is not supported by the Noun/verb selection feature?

    a. COPY

    b. MOVE

    c. TRIM

    d. ROTATE

    e. ERASE

18. When Shift to add selection is set to ON, you may add objects to the selection set by:

    a. Picking the objects

    b. Windowing the objects

    c. Using SHIFT to add

    d. Using CTRL to add

Directions: Use Figure 5–37 to answer the following questions:

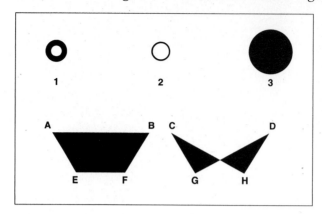

FIGURE 5–37

19. Which of the figures were drawn with the DONUT command?
    a. 1
    b. 2
    c. 3
    d. All of the objects
    e. 1, 2, and 3

20. Which of the donut objects has the smallest inside diameter?
    a. 1
    b. 2
    c. 3

21. Which of the options below is the correct sequence to create the solid object at the lower left?
    a. A, B, F, E
    b. A, E, F, B
    c. A, B, E, F
    d. F, E, A, B
    e. B, F, A, E

22. Which of the options below is the correct sequence to create the solid object at the lower right?
    a. H, G, C, D
    b. G, H, D, C
    c. G, D, H, C
    d. D, C, H, G
    e. C, G, H, D

23. If the space between the two solid shapes were to be filled, what sequence of points would completely fill this space?
    a. G, B, F, C
    b. B, G, F, C
    c. C, B, F, G
    d. F, G, C, B
    e. B, F, C, G

OBJECTIVES

After completing this chapter, you will be able to do the following:

- Construct geometric figures by means of the MULTILINE (MLINE), SPLINE, WIPEOUT, and REVCLOUD commands

- Modify multilines and splines

- Create or modify multiline styles

- Use grips to modify, group, and select objects by Quick Select for modification

- Use inquiry commands

- Change the settings of system variables

MULTILINES

The MLINE command allows you to draw multiple parallel line segments called multilines. You can modify the intersection of two or more multilines or cut gaps in them with the MLEDIT command. AutoCAD also allows you to create new multiline styles or edit existing styles composed of up to 16 lines, called elements, with the MLSTYLE command.

Drawing Multiple Parallel Lines

Multilines are multiple parallel line segments, similar to polyline segments that have been offset one or more times. Examples of multilines are shown in Figure 6–1.

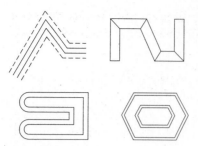

FIGURE 6–1 *Examples of multilines*

The properties of each element of a multiline are determined by the current style when the multiline is drawn. The properties of multilines controlled by the style include whether to display the line at the joints (miters), display of the end caps, and background fill. In addition, the style controls element properties such as color, linetype, and offset distance between two parallel lines and defines additional parallel lines (limited to a maximum of 16 parallel lines).

Invoke the `MLINE` command by typing **mline** at the on-screen prompt, and AutoCAD displays the following prompt:

> Specify start point or ⏎ *(specify a point or right click and select one of the options)*

Specify the starting point of the multiline, known as its origin. Once you specify the origin for a multiline, AutoCAD displays the following prompt:

> Specify next point:

When you respond by selecting a point, the first multiline segment is drawn according to the current style. AutoCAD displays the following prompt:

> Specify next point or ⏎

If you specify a point, the next segment is drawn. After two segments have been drawn, the shortcut menu will include the *Close* option. Choosing *Close* causes the next segment to join the origin of the multiline, fillets all elements, and exits the command.

Choosing *Undo* after any segment is drawn, and before the `MLINE` command has been terminated, causes the last segment to be erased, and you are prompted again for the next point.

After invoking the `MLINE` command, you can right-click and choose one of these options: *Justification*, *Scale*, and *Style*.

Choosing *Justification* determines the relationship between the elements of the multiline and the line you specify by way of the placement of the points. The justification is set by selecting one of the three available suboptions.

Choosing *Top* causes the element with the most positive offset value to be drawn aligned with the specified points. All other elements will be below or to the right of the element with the most positive offset value. In other words, if the line is drawn left to right, the element with the most positive offset value will be above all the other elements.

Choosing *Zero* causes the elements to center between the specified points. Elements with positive offsets will be to the right or above, and those with negative offsets will be to the left or below the selected points.

Choosing *Bottom* causes the element with the most negative offset value to be drawn aligned with the specified points. All other elements will be to the left or above the element with the most negative offset value. In other words, if the line is drawn left to right, the element with the most negative value will be below all the other elements.

Figure 6–2 shows the location for various justifications.

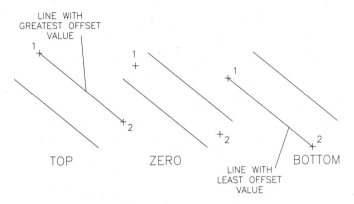

FIGURE 6–2 *Location of various justifications*

Choosing *Scale* determines the value used for offsetting elements relative to the values assigned in the style. For instance, if the scale is changed to 3.0, elements that are assigned 0.5 and −1.5 will be drawn with offsets of 1.5 and −4.5, respectively. If a negative value is given for the scale, the signs of the values assigned to them in the style will be inverted: positive to negative and negative to positive. The value can be entered in decimal form or as a fraction. Also, a zero scale value produces a single line.

Choosing *Style* allows you to choose the current multiline style from the available styles. A detailed explanation for creating or modifying multiline styles is provided later in this chapter.

Editing Multiple Parallel Lines

The MLEDIT command allows you to modify the intersections of two or more multilines or cut gaps in the lines of one multiline. The tools available depend on the type of intersection to be modified, such as cross, tee, or vertex, and if one or more elements need to be cut or welded.

Invoke **Multiline** edit by typing **mledit** from the on-screen prompt, and AutoCAD displays the Multilines Edit Tools dialog box, as shown in Figure 6–3.

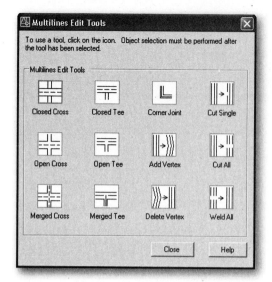

FIGURE 6–3 *Multilines Edit Tools dialog box*

To choose one of the available tools, select the appropriate image tile. AutoCAD then displays prompts for the required information.

The first column in the Multilines Edit Tools dialog box contains tools for multilines that cross, the second for multilines that form a tee, the third for corner joints and vertices, and the fourth for multilines to be cut or welded.

Choose **Closed Cross** (Figure 6–3) to cut all lines that make up the first multiline you select at the point where it crosses the second multiline (Figure 6–4).

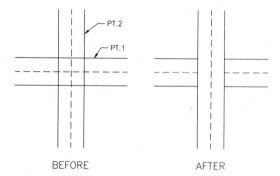

FIGURE 6–4 *An example of a closed cross*

Choose **Open Cross** (Figure 6–3) to cut all lines that make up the first multiline you select and cut only the outside line of the second multiline (Figure 6–5).

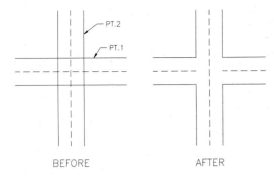

FIGURE 6–5 *An example of an open cross*

Choose **Merged Cross** (Figure 6–3) to cut all lines that make up the intersecting multiline that you select except the centerlines (Figure 6–6).

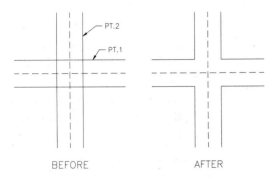

FIGURE 6–6 *An example of a merged cross*

Choose **Closed Tee** (Figure 6–3) to extend or shorten the first multiline you select to its intersection with the second multiline (Figure 6–7).

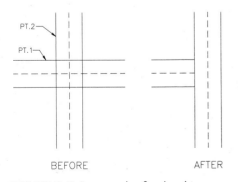

FIGURE 6–7 *An example of a closed tee*

Choose **Open Tee** (Figure 6–3) to create an open-tee intersection between two multilines (Figure 6–8). The first multiline is trimmed or extended to its intersection with the second multiline.

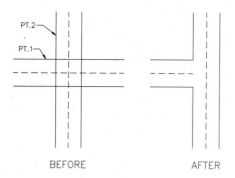

FIGURE 6–8 *An example of an open tee*

Choose **Merged Tee** (Figure 6–3) to create a merged-tee intersection between two multilines (Figure 6–9). The multiline selected first is trimmed or extended to its intersection with the second multiline.

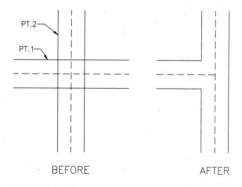

FIGURE 6–9 *An example of a merged tee*

Choose **Corner Joint** (Figure 6–3) to lengthen or shorten each of the two multilines you select as necessary to create a clean intersection forming a corner (Figure 6–10).

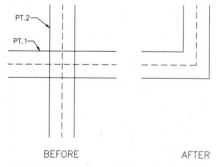

FIGURE 6–10 *An example of a corner joint*

Choose **Add Vertex** (Figure 6–3) to add a vertex to a multiline (Figure 6–11).

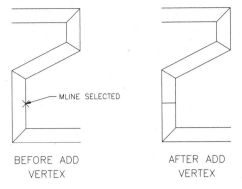

FIGURE 6–11 *An example of adding a vertex*

Choose **Delete Vertex** (Figure 6–3) to delete a vertex from a multiline (Figure 6–12).

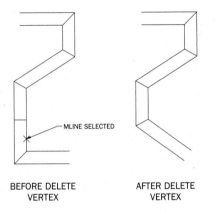

FIGURE 6–12 *An example of deleting a vertex*

Choose **Cut Single** (Figure 6–3) to cut a selected element of a multiline between two cut points (Figure 6–13).

FIGURE 6–13 *An example of removing a selected element of a multiline between two cut points*

Choose **Cut All** (Figure 6–3) to remove a portion of a multiline between two selected cut points (Figure 6–14).

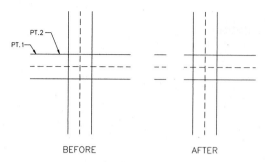

FIGURE 6–14 *An example of removing a portion of a multiline between two cut points*

Choose **Weld All** (Figure 6–3) to rejoin multiline segments that have been cut (Figure 6–15).

FIGURE 6–15 *An example of rejoining multiline segments that have been cut*

AutoCAD displays prompts to select additional multilines to continue with the selected multiline tool. Choosing *Undo* undoes the edited intersection, and AutoCAD continues with the "Select first mline or ⬐" prompt. To terminate the command sequence, press ⌈Enter⌋.

AutoCAD allows you to trim and extend multilines in the same manner as other objects such as lines, arcs, and polylines.

Creating and Modifying Multiline Styles

The MLSTYLE command is used to create a new Multiline Style or edit an existing one. You can define a multiline style consisting of up to 16 lines called elements. The style controls the number of elements and the properties of each element. In addition, you can specify the background color and the end caps of each multiline.

Invoke the Multiline Style by typing **mlstyle** from the on-screen prompt, and AutoCAD displays the Multiline Style dialog box (Figure 6–16).

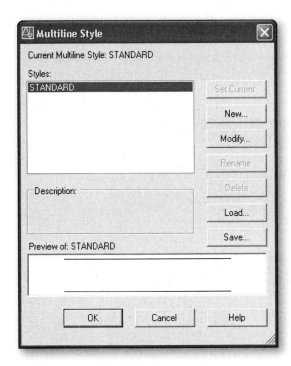

FIGURE 6–16 *Multiline Style dialog box*

Next to the label **Current Multiline Style** is the name of the multiline style that is current and that will be used for all new multilines.

The **Styles** list box allows you to choose from the available multiline styles loaded in the current drawing.

The **Description** area displays a description, if available, of the selected multiline style.

The **Preview of** area displays the name and an image of the selected multiline style.

Set Current is used to make the selected style current.

Selecting **New** causes the Create New Multiline Style dialog box to be displayed (Figure 6–17).

FIGURE 6–17 *Create New Multiline Style dialog box*

To create a new Multiline Style, enter a style name in the **New Style Name** text box. The **Start With** list box lets you choose an existing multiline style to use as a basis for the new style. After selecting the base style and naming the new one, choose **Continue**, and AutoCAD displays the New Multiline Style dialog box (Figure 6–18).

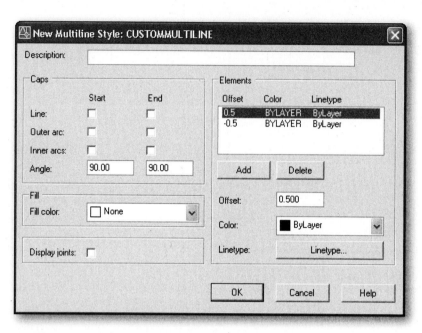

FIGURE 6–18 *New Multiline Style dialog box*

The **Description** text box allows you to add a description of up to 255 characters, including spaces. The description entered will be displayed in the **Description** text box when this style is selected in the Multiline Styles dialog box.

The **Elements** section sets element properties, such as the offset, color, and linetype, of new and existing multiline elements. AutoCAD lists all the elements in the current multiline style. Choose **Add** to add a new element to the multiline style. Each new element in the style is defined by its offset, its color, for which the default is set to ByLayer, and its linetype, for which the default is also set to ByLayer. The **Offset, Color,** and **Linetype** fields allow you to specify an element's offset distance from the middle of the multiline, color, and linetype. The offset distance is measured from a reference line that is assigned a value of 0. If the offset distance is set to a positive value, it will be drawn above or left of the reference line; a negative value will be drawn below or right of the reference line. Elements are always displayed in descending order of their offsets. To delete the selected element, choose **Delete**.

The **Caps** section allows you to specify the appearance of multiline start and end caps. The **Line** checkboxes control the display of the start and end caps by adding a straight line to the start or end of a multiline, as shown in Figure 6–19. The **Outer arc** checkboxes control the display of the start and end caps by connecting the ends of the outermost elements with a semicircular arc, as shown in Figure 6–20. The **Inner arc** checkboxes control the display of the start and end caps, by connecting the ends of the innermost elements with a semicircular arc, as shown in Figure 6–21. For a multiline with an odd number of elements, the center element is not connected. For an even number of elements, connected elements are paired with elements that are the same number from each edge. For example, the second element from the outer-left will be connected to the second element from the outer-right, the third to the third, and so on. The **Angle** edit field sets the angle of end caps. Figure 6–22 shows the display of the end caps with an angular cap.

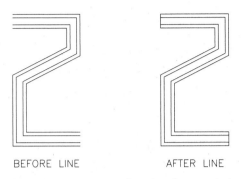

BEFORE LINE   AFTER LINE

FIGURE 6–19 *Display of the line for start and end caps*

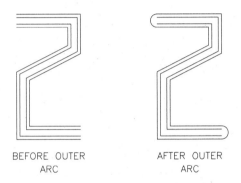

BEFORE OUTER   AFTER OUTER
ARC   ARC

FIGURE 6–20 *Display of the outer arc for start and end caps*

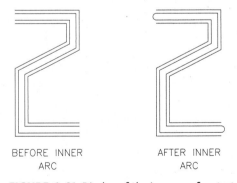

BEFORE INNER   AFTER INNER
ARC   ARC

FIGURE 6–21 *Display of the inner arc for start and end caps*

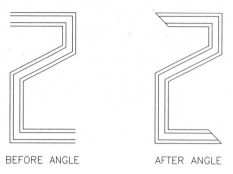

BEFORE ANGLE   AFTER ANGLE

FIGURE 6-22 *Display of the end caps with an angular cap*

The **Fill** section contains a **Fill color** list box that allows you to include a specified background color for the multiline style. Choosing **Select Color** displays the Select Color dialog box from which you can specify a nonstandard color for the background fill.

Selecting **Display joints** causes a line, or miter, to be drawn at the vertices of each multiline segment (Figure 6–23).

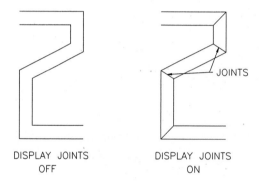

JOINTS

DISPLAY JOINTS   DISPLAY JOINTS
OFF   ON

FIGURE 6-23 *Display of the joints*

Choose **OK** to create the selected multiline style.

Choose **Modify** to change the element properties of an existing multiline style. AutoCAD displays the Modify Multiline Style dialog box, which is similar to the Create New Multiline Style dialog box (Figure 6–17).

Choose **Rename** to change the name of the selected multiline style.

Choose **Delete** to delete the selected multiline style.

Choose **Load** to load multiline style from a specified *.mln* file.

Choose **Save** to save multiline style to a multiline library (*.mln*) file. If you specify an *.mln* file that already exists, the new style definition is added to the file, and existing definitions are not erased. The default file name is *acad.mln*.

Choose **OK** to close the New Multiline Style dialog box to create your new multiline style.

SPLINE CURVES

The SPLINE command is used to draw a curve through specified points. The curve drawn is a nonuniform rational B-spline (NURBS). This type of curve has irregularly varying radii and is used to draw objects such as topographical contour lines. There are two methods for creating splines in AutoCAD: with fit points or with control vertices. Each method has different options.

Invoke the SPLINE command from the Draw panel (Figure 6–24) located in the Home tab, and AutoCAD displays the following prompt:

> Specify first point or ⊕ (*specify a point or choose one of the available options from the shortcut menu*)

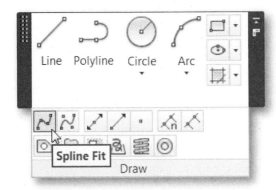

FIGURE 6–24 *Invoking* SPLINE *command from the Draw panel*

The default selection method is set for Fit and lets you specify the point from which the spline starts. After you enter the first point, a preview line appears. AutoCAD displays the following prompt:

> Enter next point or ⊕

When you respond by specifying a point, the spline segments are displayed as a preview spline, curving from the first point, through the second point, and ending at the cursor. AutoCAD displays the following prompt:

> Enter next point or ⊕

If you specify a point, the next segment is added to the spline. This will occur with each additional point specified until you press ⟨Enter⟩ to terminate the command sequence. If you select the *Close* option you are prompted for a tangentency determining point as follows:

> Specify tangent:

You can specify a point to determine the tangency at the connection of the first and last segments. If you press ⟨Enter⟩, AutoCAD calculates the tangency and draws the spline accordingly. You can also use the perpendicular or tangent object snap modes to cause the tangency of the spline to be perpendicular or tangent to a selected object.

The *End Tangency* option selection lets you specify the last tangent of the last entered fit point.

The *Fit Tolerance* option lets you vary how the spline is drawn relative to the selected points. AutoCAD displays the following prompt:

```
Specify fit tolerance <current>:
```

Entering zero causes the spline to pass through the specified points. A positive value causes the spline to pass within the specified value of the points.

Choosing *Undo* erases the most recent line without terminating the SPLINE command.

Instead of specifying the first point with the default Fit method, if you switch to Control Vertices method from the Method option, AutoCAD prompts:

```
Specify first point or ⊡ (specify a point or choose one of the
available options from the shortcut menu)
```

After you enter the first point, a preview line appears. AutoCAD displays the following prompt:

```
Enter next point
```

When you respond by specifying a point, the spline segments are displayed as a preview spline, curving from the first point, through the second point, and ending at the cursor. AutoCAD displays the following prompt:

```
Enter next point or ⊡
```

If you specify a point, the next segment is added to the spline. This will occur with each additional point specified until you press [Enter] to terminate the command sequence. Instead if you select the *Close* option [Enter], AutoCAD closes the last segment to beginning the spline curve.

Object Selection

The *Object Selection* option prompts to select an object. AutoCAD converts selected 2D or 3D quadratic or cubic spline-fit polylines to equivalent splines and, depending on the setting of the DELOBJ system variable, deletes the polylines.

Degree Selection

The *Degree* selection sets the maximum number of "bends" you can get in each span; the degree can be 1, 2, or 3. There will be one more control vertex than the number of degrees, so a degree 3 spline has four control vertices.

Editing Spline Curves

Splines created by means of the SPLINE command have numerous characteristics that can be changed with the SPLINEDIT command. These include quantity and location of fit points, end characteristics such as opened or closed, tangency, and tolerance of the spline, or how near the spline is drawn to fit points.

The SPLINEDIT command operates on control points, which are different from fit points, of the selected spline. Features include adding control points and changing the weight of individual control points, which determines how close the spline is drawn to the individual control points.

Invoke the SPLINEDIT command from the Modify panel (Figure 6–25) located in the Home tab, and AutoCAD displays the following prompts:

> Select Spline: *(select a spline curve)*
>
> Enter an option: *(choose one of the available options from the shortcut menu)*

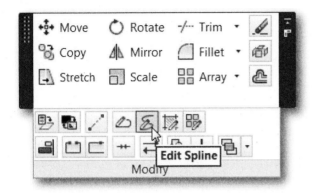

FIGURE 6–25 *Invoking Edit Spline from the Modify panel*

Control points appear in the grip color, and, if the spline has fit data, fit points also appear in the grip color. If you select a spline whose fit data is deleted, the *Fit Data* option is not available. A spline can lose its fit data if you use the *Purge* option while editing fit data, refining the spline, moving its control vertices, fitting the spline to a tolerance, or opening or closing the spline.

The *Open* option will replace *Close* if you select a closed spline and vice versa.

Choosing *Fit Data* allows you to edit the spline by providing the following sub-options: *Add, Open, Delete, Move, Purge, Tangents, Tolerance,* and *Exit*.

Choosing *Add* allows you to add fit points to the selected spline. AutoCAD displays the following prompt:

> Specify control point <exit>: *(select a fit point)*

After you select one of the fit points, AutoCAD highlights it and displays the following prompt for the next point:

> Specify new point or ⬇ *(specify a point)*
>
> Specify new point or ⬇ *(specify another point or press* Enter *)*

Selecting a point places a new fit point between the highlighted ones.

Choosing *Close* closes an open spline smoothly with a segment or smoothes a spline with coincidental starting and ending points.

Choosing *Open* opens a closed spline, disconnecting it and changing the starting and ending points.

Choosing *Delete* deletes a selected fit point.

Choosing *Move* moves fit options to a new location by prompting:

> Specify new location or ⬇ *(choose one of the available options to navigate through fit point)*

Choosing *Purge* deletes fit data for the selected spline.

Choosing *Tangents* edits the start and end tangents of a spline.

Choosing *Tolerance* refits the spline to the existing points with new tolerance value.

Choosing *Exit* exits the fit data option and returns to the main prompt.

Choosing *Close* causes the spline to be joined smoothly at its start point.

Choosing *Open* opens a closed spline. Previously open splines with coincidental starting and ending points will lose their tangency. Others will be restored to a previous state.

Choosing *Move Vertex* relocates a spline's control vertices, by providing the following suboptions: *Next*, *Previous*, *Select Point*, and *Exit*.

Choosing *Refine* allows you to fine-tune a spline definition, by providing the following suboptions: *Add Control Point*, *Elevate Order*, *Weight*, and *Exit*.

Choosing *Add Control Point* increases the number of control points that control a portion of a spline.

Choosing *Elevate Order* increases the order of the spline. You can increase the current order of a spline up to 26 (the default is 4), causing an increase in the number of control points.

Choosing *Weight* changes the weight at various spline control points, by providing the following suboptions: *Next*, *Previous*, *Select Point*, and *Exit*.

The default weight value for a control point is 1.0. Increasing it causes the spline to be drawn closer to the selected point. A negative or zero value is not valid.

From the refine menu, the *Exit* suboption returns you to the main prompt.

Choosing *Reverse* reverses the direction of the spline. Reversing the spline does not delete the fit data.

Choosing *Convert to Polyline* converts the spline to a polyline.

Choosing *Undo* undoes the effects of the last subcommand.

You can also edit splines using multifunctional grips. Multifunctional grips provide options that include adding control vertices and changing the tangent direction of the spline at its endpoints. Menu options can be dispalyed by hovering over a grip. The editing options available with multifunctional grips differ depending on whether the spline is set to display control vertices or fit points.

WIPEOUT

The WIPEOUT command allows you to create an area on the screen that obscures previously drawn objects within its boundary. These objects can be displayed with or without a visible boundary, called a frame.

Invoke the WIPEOUT command from the Draw panel (Figure 6–26) located in the Home tab, and AutoCAD displays the following prompts:

```
Specify start point or ⊡ (specify the start point of the
wipeout area, or right-click and select one of the options)
Specify next point: (specify the second point)
Specify next point or ⊡ (specify the third point, or right-
click and select one of the options)
```

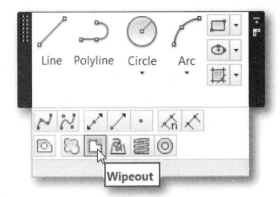

FIGURE 6–26 *Invoking* WIPEOUT *command from the Draw panel*

After drawing a connected series of lines, press Enter to terminate the WIPEOUT command. The polygonal boundary of the wipeout object is determined from the series of specified points. The shortcut menu displayed when you right-click at the prompt for the first point includes the *Frames* and *Polyline* options.

Selecting *Frames* determines whether the edges of all wipeout objects are displayed or hidden. Choose *On* to display all wipeout frames, and choose *Off* to suppress the display of all wipeout frames. When the frames are turned off, the wipeout object cannot be selected on the screen with the pointing device.

Selecting *Polyline* allows you to select a polyline, which determines the polygonal boundary of the wipeout area.

After selecting *Polyline*, AutoCAD displays the following prompts:

Select a closed polyline: *(use an object selection method to select a closed polyline)*
Erase polyline? *(enter yes or no)*

Enter **y** to erase the polyline that was used to create the wipeout object. Enter **n** to retain the polyline. The shortcut menu displayed when you right-click at the prompt for the third or subsequent point includes *Close* and *Undo* options. These operate in the same manner as in the LINE and POLYLINE commands. The Enter and *Cancel* options are also available.

The wipeout object created by the WIPEOUT command will cover existing objects. Objects drawn on top of the wipeout object will not be hidden. If one or more of the objects being covered is modified, by the MOVE command for example, then it will no longer be obscured. Likewise, if the wipeout area is modified, it will then cover all objects that overlap it. You can use the DRAWORDER command to change the effect of the wipeout object on other objects.

NOTE

REVISION CLOUD

The REVCLOUD command allows you to draw a connected series of arcs encircling objects in a drawing to signify an area on the drawing that has been revised (Figure 6–27).

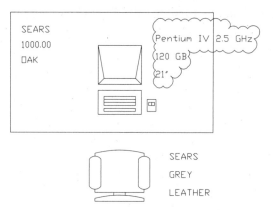

FIGURE 6–27 *An example of using the* REVCLOUD *command*

Invoke the REVCLOUD command from the Draw tab (Figure 6–28), and AutoCAD displays the following prompts:

> Specify start point or ⬇ *(specify the start point of the revision cloud)*
>
> Guide crosshairs along cloud path... (move the cursor crosshairs along the path of the desired revision cloud)

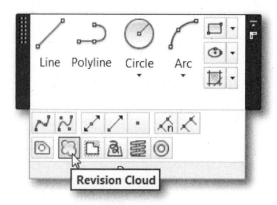

FIGURE 6–28 *Invoking Revision Cloud from the Draw panel*

When the crosshairs cursor approaches the starting point, the cloud is automatically closed without requiring any additional action or input.

At the "Specify start point or ⬇" prompt, right-click for the shortcut menu, which includes the options *Arc Length*, *Object*, and *Style*.

Choose *Arc Length* to specify the size range of the arcs when prompted for minimum and maximum arc lengths. These values are then multiplied by the value of the dimension variable DIMSCALE to compensate for drawings with different scale factors.

Choose *Object* to select a closed shape, such as a polyline, rectangle, circle, and so on, from which AutoCAD creates a revision cloud. AutoCAD displays the following prompt:

> Reverse direction *(choose* YES *or* NO*)*

Choose *Yes* to cause the arcs to be redrawn on the opposite side of the line or arc defining the cloud. Choose *No* to cause the revision cloud to remain as drawn. To reverse the bulges of an existing revision cloud, use the *Object* option to select it and then choose the *Yes* option at the "Reverse direction" prompt.

Choose *Style* to select between *Normal* and *Calligraphy*. The *Normal* selection causes AutoCAD to draw the revision cloud in the current linetype and lineweight. The *Calligraphy* selection causes the revision cloud to be drawn in bulges of tapered line width, producing an artistic effect. To change the style of an existing revision cloud, invoke the REVCLOUD command, select *Style* from the shortcut menu, change the style, select *Object* from the shortcut menu, and select the revision cloud.

EDITING WITH GRIPS

Grips allow you to edit AutoCAD drawings in an entirely different way than using the traditional AutoCAD modify commands. With grips you can move, stretch, rotate, copy, scale, and mirror selected objects without invoking one of the regular AutoCAD modify commands. For many objects, you can also hover over a grip to access a menu with object-specific, and sometimes grip-specific, editing options. Press Ctrl to cycle through the grip menu options. When you select an object with grips, small squares appear at strategic points on the object, enabling you to edit the selected objects.

To change the grip options, choose *Options* from the shortcut menu, and AutoCAD displays the Options dialog box (Figure 6–29).

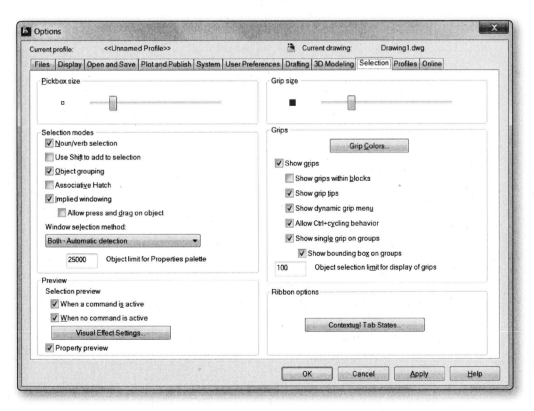

FIGURE 6–29 *Options dialog box*

Choose the **Selection** tab, and AutoCAD displays options related to Grips (Figure 6–29).

The **Grip Size** slider bar allows you to change the size of the grips. To adjust the size of the grips, move the slider box left or right. As you move the slider, the size is illustrated to the right of the slider.

Selecting **Show Grips** controls the display of grips on selected objects. If it is set to ON, the grips display is enabled; if it is set to OFF, the grips display is disabled. You can edit an object with grips by selecting a grip and using the shortcut menu. Displaying grips in a drawing significantly affects performance. Set to OFF to optimize performance. You can also enable the grips feature by setting the GRIPS system variable to one.

Selecting **Show grips within blocks** controls the display of grips on objects within blocks. If it is enabled, the grips are displayed on all objects within the block; if it is disabled, a grip is displayed only on the insertion point of the block.

Selecting **Show grip tips** causes a grip-specific tip to be displayed when the cursor hovers over a grip on an object that supports grip tips. Standard AutoCAD objects are not affected by this option.

Selecting **Show dynamic grip menu** controls the display of the dynamic menu when hovering over a multi-functional grip.

Selecting **Allow Ctrl-cycling Behavior** allows the Ctrl-cycling behavior for multi-functional grips.

Selecting **Show Single Grip on Groups** displays a single grip for an object group.

Selecting **Show Bounding Box on Groups** displays a bounding box around the extents of grouped objects.

The **Object selection limit for display of grips** textbox allows you to specify a number that limits the display of grips when the initial selection set includes more than the specified number of objects. The valid range is 1 to 32,767. The default setting is 100.

Choose the **Grip Colors...** button to display the Grip Colors dialog box (see Figure 6–30) where you can specify the colors for different grip status and elements.

FIGURE 6–30 *Grip Colors dialog box*

The **Unselected grip color** list box allows you to change the color of the unselected grips.

The **Selected grip color** list box allows you to change the color of the selected grips.

The **Hover grip color** list box allows you to change the color of the grips in hover mode.

The **Grip Contour Color** allows to you to change the color of the grip contour.

After making changes, choose **OK** to accept the changes and close the Options dialog box.

AutoCAD gives you a visual cue when grips are enabled by displaying a pickbox at the intersection of the crosshairs, even when you have not invoked an AutoCAD command (Figure 6–31).

FIGURE 6–31 *The pickbox displayed at the intersection of the crosshairs*

NOTE

The pickbox is also displayed on the crosshairs when the PICKFIRST (Noun/Verb selection) system variable is set to ON. Grips appear on the endpoints and midpoint of lines and arcs; on the vertices and endpoints of polylines, on quadrants and the center of circles; on dimensions, text, solids, 3D faces, 3D meshes, and viewports; and on the insertion point of a block. Figure 6–32 shows location of the grips on some of the commonly used objects.

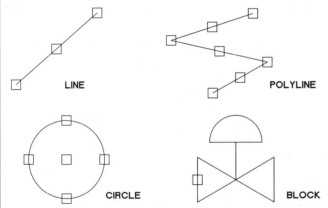

FIGURE 6–32 *Locations of grips on commonly used objects*

Using Grips

This section explains how to use grips to modify your drawing. Learning to use grips speeds up the editing of your drawing while at the same time maintaining the accuracy of your work.

When you move your cursor over a grip, it automatically snaps to the grip point. This allows you to specify exact locations in the drawing without having to use grid, snap, ortho, object snap, or coordinate entry tools.

To clear a selection set, press ESC. The grips on the selected objects will disappear. When you invoke a nonmodifying AutoCAD command, such as LINE or CIRCLE, AutoCAD clears the selection set.

To edit an object using grips, select the object and then select a grip to act as the base point for the editing operation. Place the cursor over the grip and press the pick button. You can use multiple grips to keep the shape of the object intact. Hold down SHIFT as you select the grips. Selecting a grip starts the Grip modes, which includes *Stretch*, *Move*, *Rotate*, *Scale*, and *Mirror*. You can cycle through the grip modes by pressing SPACEBAR, pressing Enter, entering a keyboard shortcut, or selecting from the right-click shortcut menu (Figure 6–33). To cancel a grip mode, enter **X** (for the mode's EXIT option). You can also use a combination of the current grip mode and a copy operation on the selection set.

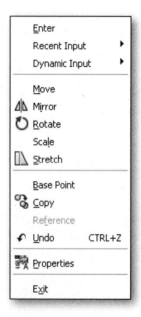

FIGURE 6–33 *Shortcut menu displaying the grip modes*

Stretch

The *Stretch* selection works like the STRETCH command. It allows you to stretch the shape of an object without affecting other parts. When you are in the stretch mode, AutoCAD displays the following prompt:

Specify stretch point or ⊡

Stretch, the default, refers to the stretch displacement point. As you move the cursor, you see that the shape of the object is stretched dynamically from the base point. You can specify the new point with the cursor or by entering coordinates.

If necessary, you can change the base point to be a point other than the base grip, by selecting *Base Point* from the shortcut menu. Then specify the new base point with the cursor, or enter the coordinates. You can also select multiple base points by holding SHIFT and choosing additional grips.

To make multiple copies while stretching objects, choose *Copy* from the shortcut menu. Specify destination copy points with the cursor, or enter their coordinates.

Move

The *Move* selection works like the MOVE command. It allows you to move one or more objects from their present location to a new one without changing their orientation or size. In addition, you can make copies of the selected objects at the specified displacement, leaving the original objects intact. To invoke the move mode, cycle through the modes by pressing [Enter] until it takes you to the move mode, or choose *Move* from the shortcut menu.

At the "Specify move point or [⊻]" prompt, enter a displacement point. As you move the cursor, AutoCAD moves all the objects in the current selection set to a new point relative to the base point. You can specify the new point with the cursor or by entering the coordinates.

If necessary, you can change the base point to be a point other than the base grip by selecting *Base Point* from the shortcut menu. Specify the new base point with the cursor, or enter the coordinates. You can also select multiple base points by holding [SHIFT] and choosing additional grips.

To make multiple copies while stretching objects, choose Copy from the shortcut menu. Specify destination copy points with the cursor, or enter their coordinates.

Rotate

The *Rotate* selection works like the ROTATE command. It allows you to change the orientation of objects by rotating them about a specified base point. In addition, you can make copies of the selected objects and at the same time rotate them about a specified base point.

To invoke the Rotate mode, cycle through the modes by pressing [Enter] until it takes you to the rotate mode, or choose *Rotate* from the shortcut menu.

At the "Specify rotation angle or [⊻]" prompt, enter the rotation angle to which objects are to be rotated. As you move the cursor, AutoCAD allows you to specify the rotation angle, relative to the base point, with your cursor. You can specify the new orientation with the cursor or from the keyboard. If you specify an angle by entering a value from the keyboard, this is taken as the degree that the objects should be rotated based on their current orientation. A positive value rotates counterclockwise, and a negative value rotates clockwise. Similar to the ROTATE command, the *Reference* option allows you to specify the desired new rotation.

If necessary, you can change the base point to be a point other than the base grip by selecting *Base Point* from the shortcut menu. Specify the new base point with the cursor, or enter the coordinates. You can also select multiple base points by holding [SHIFT] and choosing additional grips.

To make multiple copies while rotating objects, choose *Copy* from the shortcut menu. Specify destination copy points with the cursor, or enter their coordinates.

Scale

The *Scale* selection works like the SCALE command. It allows you to change the size of objects. In addition, you can make copies of the selected. To invoke the Scale mode, cycle through the modes by pressing [Enter] until it takes you to the scale mode, or choose *Scale* from the shortcut menu.

At the "Specify scale factor or [⊻]" prompt, enter the scale factor by which objects should be scaled. As you move the cursor, AutoCAD allows you to specify the scale factor, in relation to the base point, by moving your cursor closer to or farther from the base point. You can specify the new scale factor with the cursor or from the keyboard. If you specify the scale factor by entering a value from the keyboard, this is taken as a relative scale factor by which all dimensions of the objects in the current selection set are to be multiplied. To enlarge an object, enter a scale factor greater than 1; to reduce an object, use a scale factor between 0 and 1. Similar to the SCALE command, you can use the *Scale* option to specify the current length and the desired new length.

If necessary, you can change the base point to be a point other than the base grip by selecting Base Point from the shortcut menu. Specify the new base point with the cursor, or enter the coordinates. You can also select multiple base points by holding SHIFT and choosing additional grips.

To make multiple copies while scaling objects, choose *Copy* from the shortcut menu. Specify destination copy points with the cursor, or enter their coordinates.

Mirror

The *Mirror* selection works like the MIRROR command. It allows you to make mirror images of existing objects. To invoke the Mirror mode, cycle through the modes by pressing Enter until it takes you to the mirror mode, or choose *Mirror* from the shortcut menu.

Two points are required to define a line about which the selected objects are mirrored. AutoCAD considers the base grip point as the first point; the second point is the one you specify or enter in response to the "Specify second point or ⬇" prompt.

If necessary, you can change the base point to be a point other than the base grip by selecting *Base Point* from the shortcut menu. Specify the new base point with the cursor, or enter the coordinates. You can also select multiple base points by holding SHIFT and choosing additional grips.

To make multiple copies while mirroring objects, choose *Copy* from the shortcut menu. Specify destination copy points with the cursor, or enter their coordinates.

For quadrant grips on circles and ellipses, distance is measured from the center point, not the selected grip. For example, in Stretch mode, you can select a quadrant grip to stretch a circle and then specify a distance at the Command prompt for the new radius. The distance is measured from the center of the circle, not the selected quadrant. If you select the center point to stretch the circle, the circle moves.

You can also modify polylines, splines, and non-associative polyline hatch objects with multi-functional grips. With multi-functional grips, you can modify the position, size, and orientation of objects. You can also edit vertices, fit points, control points, segment types, and tangent directions. To activate a multi-functional grip, select the grip, or hover over it and choose a multi-functional grip-editing option from the dynamic menu. Once a grip is active, change the grip behavior with the Hot Grip shortcut menu, or by cycling through pressing Enter or Spacebar to cycle through the Grip modes.

For many objects (lines, polylines, arcs, elliptical arcs, splines, Dimension objects, multileaders, 3D faces, edges, and vertices) you can also hover over a grip to access a menu with object-specific, and grip-specific, editing options. Press Ctrl to cycle through the grip menu options.

SELECTING OBJECTS BY QUICK SELECT

The `QSELECT` command is used to create selection sets based on filters that select objects with similar characteristics or properties. For example, you can create a selection set of all lines that are equal to or less than 2.5 units long. Or, you can create a selection set of all objects that are not text objects on a specific layer. The combinations of possible filters are almost limitless.

To create a filtered selection set, invoke **Quick Select** (`QSELECT`) from the Utilities panel (Figure 6–34) located in the Home tab, and AutoCAD displays the Quick Select dialog box (Figure 6–35).

FIGURE 6–34 *Invoking* `QSELECT` *command from the Utilities panel*

FIGURE 6–35 *Quick Select dialog box*

The **Apply to** list box lets you select whether to apply the specified filters to the Current selection or to the entire drawing. If there is a current selection, then Current selection is the default; otherwise, Entire drawing is the default. If the **Append to**

current selection set checkbox is checked, then Current selection is not an option. If you wish to create a selection set, choose the **Select Objects** icon, located next to the **Apply to** list box. **Select Objects** is available only when **Append to current selection set** is not checked.

Choosing **Select Objects** returns you to the drawing screen so that you can select objects. After selecting the objects to be included in the selection set, press Enter.

The **Object type** drop-down list lets you select whether to include certain types of objects or multiple objects.

The **Properties** list box lists the properties that can be used for filters for objects specified in the **Object type** drop-down list box.

The **Operator** drop-down list box lets you apply logical operators to values. These include Equals, Not Equal, Greater than, Less than, and Select All. Greater than and Less than apply primarily to numeric values, and Select All applies to text strings.

The **Value** field is dynamic, based on the **Properties** selection, and it lets you specify a value to which the operator applies. If you specify the Greater than operator and the value of 1.0 for the length of lines, lines with lengths greater than 1.0 will be filtered to be included in or excluded from the selection set, depending on which radio button in the **How to apply** section has been chosen.

The **How to apply** section lets you specify whether the filtered objects will be included in or excluded from the selection set.

The **Append to current selection set** checkbox lets you specify whether the filtered selection set replaces the current selection or is appended to it.

Once you have all the required selection criteria set, choose **OK** to close the dialog box. AutoCAD displays grips on the selected objects. You can proceed with the appropriate modification for the selected objects.

SELECTION SET BY FILTER TOOL

The FILTER command displays the Object Selection Filters dialog box, which lets you create filter lists that you can apply to a selection set. This is another method of selecting objects. With the FILTER command, you can select objects based on object properties, such as location, object type, color, linetype, layer, block name, text style, and thickness. For example, you could use the FILTER command to select all the blue lines and arcs with a radius of 2.0 units. You can even name filter lists and save them. The selection set created by the FILTER command can be reselected by using *Previous* option at the next "Select object" prompt.

Invoke the FILTER command by typing **filter** at the on-screen prompt, and AutoCAD displays the Object Selection Filters dialog box (Figure 6–36).

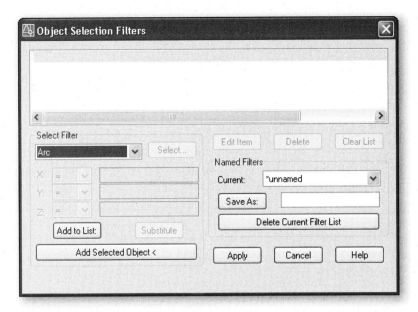

FIGURE 6–36 *Object Selection Filters dialog box*

The filter list box at the top of the dialog box displays the filters being used. The first time you use the FILTER command, the list box is empty.

The **Select Filter** section lets you add filters to the filters list box based on object properties. Your selection filters may be based on an object or on multiple objects compared by Boolean operators, and both options are found in the **Select Filter** drop-down list box. The Boolean operators must be paired and balanced correctly in the filter list. For example, each Begin OR operator must have a matching End OR operator. If you select more than one filter object, AutoCAD, by default, uses AND as a grouping operator between each filter.

Choosing **Select** displays a dialog box that lists all items of the specified type within the drawing. From the list, you can select as many items as you want to filter. This process is more accurate than typing the specific filter parameters.

Choosing **Add to List** adds the filter in the **Select Filter** section to the filter list box.

Choosing **Add Selected Object** allows you to select an object from the drawing and add it to the filter list box.

Choosing **Substitute** replaces the selected filter with the one in the **Select Filter** section.

Choosing **Edit Item** allows you to edit a selected filter from the filter list box by moving the filter into the **Select Filter** section for editing. Select the filter from the filter list box and then choose the **Edit Item** button. When finished making changes in the **Select Filter** section, choose the **Substitute** button. The edited filter replaces the selected filter.

Choosing **Delete** deletes the selected filter in the filter list box.

Choosing **Clear List** deletes all of the filters from the filter list box.

To save the filter list, name the filter list in the **Save As** field, and choose **Save As**.

The **Current** list box displays saved filter lists by name. Select a list to make it current.

Choosing **Delete Current Filter List** deletes the named filter lists from the filter file.

Choosing **Apply** closes the dialog box, and AutoCAD displays the "Select Objects" prompt. AutoCAD applies the filter to the objects you select.

CHANGING PROPERTIES OF SELECTED OBJECTS

The Quick Properties panel lists the most commonly used properties for selected objects (see Figure 6–37). AutoCAD allows you to change the properties for the selected object(s) from the Quick Properties panel.

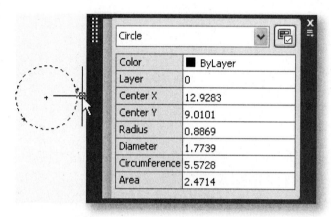

FIGURE 6–37 *Quick Properties panel for selected object*

By default, when a group of object(s) is selected, AutoCAD displays the Quick Properties panel listing the most commonly used properties. You can turn the Quick Properties panel display for selected object(s) on or off by clicking the **Quick Properties** button on the status bar. You can customize the quick properties panel for any object in the Customize User Interface (CUI) editor. For detailed explanation on customization, refer to Chapter 18.

The **Quick Properties** Tab of the **Drafting Settings** Dialog Box (see Figure 6–38) specifies the settings for displaying the Quick Properties panel.

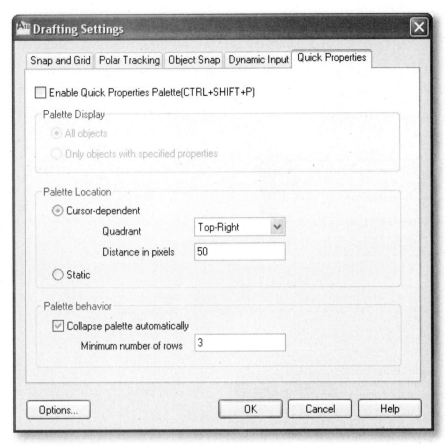

FIGURE 6–38 *Drafting Settings Dialog box with Quick Properties Tab Selection*

The **Enable Quick Properties Palette** setting turns the display of the Quick Properties panel on or off depending on the object type. You can also turn the Quick Properties panel on or off by clicking **Quick Properties** on the status bar.

The **Palette Display** section assigns the display settings of the Quick Properties palette. The **All Objects** selection displays the Quick Properties panel for any selection of objects. It doesn't matter what object type is selected, the Quick Properties panel will be displayed with appropriate information. If **Only Objects with Specified Properties** is selected, then the Quick Properties panel is displayed only for objects that are defined in the CUI editor.

The **Palette Location** section sets the display position of the Quick Properties palette. In **Cursor-Dependent** mode selection, the Quick Properties palette is displayed relative to Quadrant selection to the specified Distance setting. The **Quadrant** selection specifies the relative location to display the Quick Properties panel. You can select one of the four quadrants: top-right, top-left, bottom-right, or bottom-left. The **Distance in pixels** text box sets the distance in pixels and the values can be set in the range of 0 to 400 (only integer values). In **Static** mode selection, the Quick Properties panel is displayed in the same position unless you relocate the panel manually.

The **Palette Behavior** section sets the behavior of the Quick Properties palette. The **Collapse Palette Automatically** selection displays the Quick Properties panel listing the number of properties set in the **Minimum Number of Rows** edit field. You can specify default number of rows in the range of 1 to 30 (only integer values).

Choose **OK** to close and save the changes to the Drafting Settings dialog box.

In addition, the PROPERTIES command can be used to manage and change the properties of selected objects by means of the Properties palette.

To change the objects in a selection set, invoke PROPERTIES by choosing the down arrow from the Properties panel (Figure 6–39) located in the Home tab, and AutoCAD displays the Properties palette (Figure 6–40).

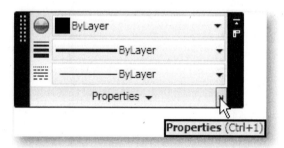

FIGURE 6–39 *Invoking the* PROPERTIES *command from the Properties panel*

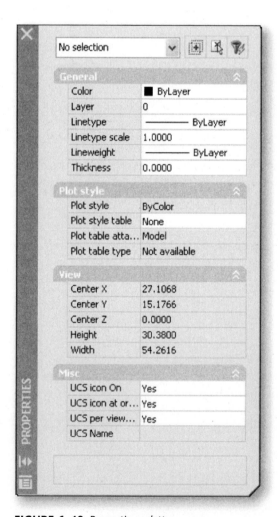

FIGURE 6–40 *Properties palette*

The default position for the Properties palette is floating on the left side of the screen. The Properties palette can be docked by double-clicking in the title bar or by placing the cursor over the title bar and dragging the window to the side at which you wish to dock it. When docked, it can be undocked by placing the cursor over the double-line bar at the top of the window and either double-clicking or dragging the window into the screen area. You can also drag the window across to a docking position on the right side of the screen. Double-clicking causes the Properties palette to become undocked and to float in the drawing area.

When the Properties palette is displayed and an object, say a circle, is selected, the Properties palette lists all the properties of that circle. Make sure that the PICK-FIRST and/or GRIPS system variable is set to ON. Not only are properties such as color, linetype, and layer listed, but the circle's X, Y, and Z coordinates, radius, diameter, circumference, and area are also listed. However, if two circles are selected, only the values of properties that are the same for both circles are listed. Properties that are not the same, but are common properties, show as *VARIES*. If you enter a value for an uncommon value, the new value will be applied to both circles. For example, if the centers of the two circles are different and you enter X and Y coordinates in the **Center X** and **Center Y** edit fields, respectively, both circles will be moved to have their centers coincide with the specified coordinates. If different types of objects are in the selection set, the properties listed in the Properties palette will include only those common to the selected objects such as color, layer, and linetype. Whenever you change the properties in the Properties palette, AutoCAD reflects the changes immediately in the drawing window.

You can also open the Properties palette by double-clicking on an object when the PICKFIRST system variable is set to ON. You can also open the Quick Select dialog box to select objects by choosing the **Quick Select** icon, located at the top-right side of the Properties palette. If you need to select objects by the traditional method, that is, by window and/or crossing, choose the **Select Objects** icon located on the top-right side of the Properties palette. You can change the PICKADD system variable value, by choosing the toggle value of the PICKADD system variable located on the top-right side of the Properties palette. The PICKADD system variable setting controls how you add objects to an existing selection set.

You can also change the properties of the selected objects by invoking the DDCHPROP or CHANGE commands.

The DDCHPROP command also displays the Properties palette, as shown in Figure 6–39.

The CHANGE command also allows you to change properties for selected objects such as their color, lineweight, linetype, layer, elevation, thickness, and plotstyle. The CHANGE command also lets you modify some of the characteristics of lines, circles, text, and blocks. Invoke the CHANGE command by typing **change** at the on-screen prompt, and AutoCAD displays the following prompts:

> Select objects: *(select objects and press* Enter *to complete the selection)*
>
> Specify change point or ⬇ *(specify change point for the selected objects, or choose properties from the shortcut menu to change one of the properties of the selected object)*

CHANGE allows you to modify some of the characteristics of lines, circles, text, and blocks. If you select one or more lines, the closest endpoint(s) are moved to the new change point. If you select a circle, CHANGE allows you to change the radius of the circle. If you selected more than one circle, AutoCAD moves on to the next circle and repeats the prompt. If the selected object is a text string, AutoCAD allows you to change one or more parameters such as text height, text style, text rotation angle, and text string. If the selected object is a block, specifying a new location repositions the block.

Instead of changing a point, you can change the properties of selected objects by right-clicking and choosing the PROPERTIES option from the shortcut menu. Properties that can be changed include color, lineweight, linetype, layer, elevation, thickness, and plot style.

GROUPING OBJECTS

The GROUP command adds flexibility when modifying a group of objects. It allows you to name a selection set, which combines two powerful AutoCAD drawing features. One is being able to modify a group of unrelated objects as a group. It is similar to using the *Previous* to select the last selection set when prompted to "Select object" for a modify command. The advantage of using GROUP instead of *Previous* is that you are not restricted to only the last selection set. The other feature combined in the GROUP command is that of giving a name to a selected group of objects for later recall by using the name of the group, similar to the BLOCK command. The advantage of using GROUP instead of BLOCK is that the GROUP command's "selectable" switch can be set to OFF, allowing for modification of an individual member without losing its "membership" in the group. Named groups, like blocks, are saved with the drawing.

A named group can be selected for modification as a group only when its "selectable" switch is set to ON. Figure 6–41 shows the result of trimming an object with the selectable switch set to ON or to OFF. Modifying objects (such as MOVE or COPY command) that belong to a group can be accomplished by two methods. One is to select one of its members. The other method is to select the GROUP option by typing **g** at the "Select Objects" prompt, and AutoCAD prompts for the group's name. Enter the group name, and press [Enter] or [SPACEBAR]. AutoCAD highlights the objects that belong to the specified group.

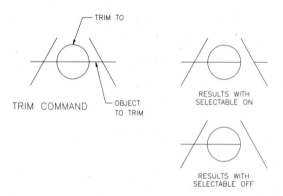

FIGURE 6–41 *Trimming an object with the selectable switch set to ON or to OFF*

To create a new group, invoke GROUP command from the Groups panel (Figure 6–42) located in the Home tab, and AutoCAD displays the following prompt:

Select objects or ⊡ (select objects to group together or choose one of the available options from the shortcut menu)

FIGURE 6–42 *Invoking* GROUP *command from the Groups panel*

The selected objects are grouped together into an unnamed group, which is assigned a default name such as *A1.

Unnamed groups are not displayed in the Object Grouping dialog box unless Include Unnamed Groups is selected.

Choosing **Name** from the shortcut menu allows you to assign a name to the newly created group.

Choosing **Description** from the short cut menu allows you to assign a description to the selected group.

Choosing **Group Selection On/Off** (see Figure 6–43) controls whether or not the selected group is selectable.

FIGURE 6–43 *Invoking Group Selection On/Off from the Groups panel*

To add and/or remove objects from a group or rename a group name, invoke the GROUPEDIT command from the Groups panel (see Figure 6–44). AutoCAD prompts:

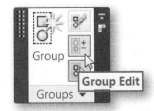

FIGURE 6–44 *Invoking Group Selection On/Off from the Groups panel*

> *Select group or* 🔽 *(select the group by selecting one of the object that belong to the group or choose name from the shortcut menu to specify the name of the group to edit)*

To add objects to the selected group, choose **Add objects** and select objects to add to the group. To remove objects from the selected group, choose **Remove objects** and select objects that have to be removed from the group. To rename the selected group, choose Rename, and specify the name for the group.

To remove all objects from the selected group, invoke the UNGROUP command from the Groups panel (see Figure 6–45). AutoCAD prompts:

> *Select group or* 🔽 *(select the group by selecting one of the object that belong to the group or choose name from the shortcut menu to specify the name of the group)*

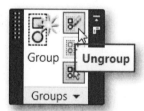

FIGURE 6–45 *Invoking* UNGROUP *command from the Groups panel*

The selected group no longer exists as a group. The members remain in the drawing and in any other group(s) of which they are members.

To manage groups in the existing drawing, invoke the CLASSICGROUP command from the Groups panel (see Figure 6–46) located in the Home tab, and AutoCAD displays the **Object Grouping** dialog box (Figure 6–47).

FIGURE 6–46 *Invoking the* CLASSICGROUP *command from Groups panel*

FIGURE 6–47 *Object Grouping dialog box*

The **Group Name** list box lists the names of existing groups defined in the current drawing. The **Selectable** column indicates whether a group is selectable. If it is listed as selectable, selecting a single group member selects all the members except those on locked layers. If it is listed as unselectable, selecting a single group member selects only that object.

In the **Group Identification** section, AutoCAD displays the group name and its description when a group is selected in the **Group Name** list.

Choosing **Find Name** lists the groups to which an object belongs. AutoCAD prompts for the selection of an object and displays the **Group Member List** dialog box, which lists the group or groups to which the selected object belongs.

Choosing **Highlight** lets you see the members of the group selected from the **Group Name** list box.

Choosing **Include Unnamed** controls the listing of the unnamed groups in the **Group Name** list box.

The **Create Group** section is used for creating a new group. In addition, you can set whether or not the new group is initially selectable and if it will have a name.

To create a new group, enter the group name and description in the **Group Name** and **Description** text boxes. Group names can include letters, numbers, and the special characters $ and _. To create an unnamed group, check **Unnamed**. AutoCAD assigns a default name, *An, to unnamed groups. The n represents a number that increases with each new group.

To create a selectable group, check **Selectable**, and then choose **New**. AutoCAD prompts for the selection of objects. Select all the objects to be included in the new group, and press $\boxed{\text{Enter}}$ to complete the selection.

The **Change Group** section is for making changes to individual members of a group or to the group itself. The buttons are disabled until a group name is selected in the **Group Name** list box.

Remove lets you remove objects from the selected group. To remove objects from the selected group, choose **Remove**, and AutoCAD displays the following prompt:

```
Remove objects: (select objects that are to be removed from
the selected group and press Enter)
```

AutoCAD then redisplays the Object Grouping dialog box.

Choosing **Add** allows you to add objects to the selected group. To add objects to the selected group, choose **Add**, and AutoCAD displays the following prompt:

```
Select objects: (select objects that are to be added to the
selected group and press Enter)
```

AutoCAD then redisplays the Object Grouping dialog box.

Choosing **Rename** allows you to change the name of the selected group to the name entered in the **Group Name** text box in the **Group Identification** section.

Re-Order allows you to change the numerical order of objects within the selected group. Initially, the objects are numbered in the order in which they were selected. Re-ordering is useful when creating tool paths. Choose **Re-Order**, and AutoCAD displays the Order Group dialog box, as shown in Figure 6–48.

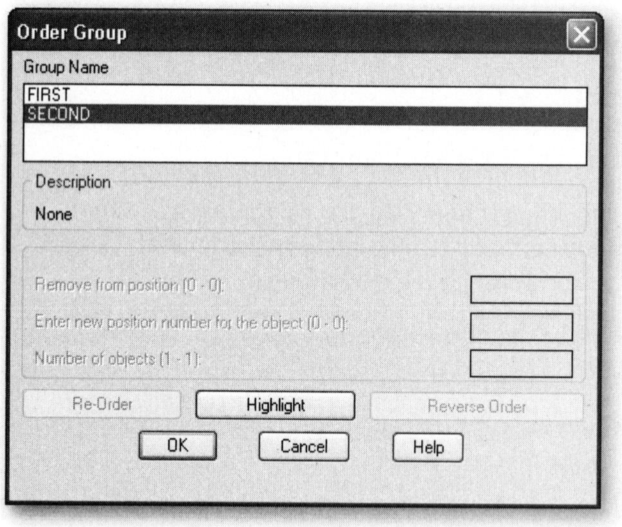

FIGURE 6–48 *Order Group dialog box*

The **Group Name** list box gives the names of the groups defined in the current drawing. Members of a group are numbered sequentially starting with 0 (zero).

Remove from position (0-0) identifies the position number of an object.

Enter new position number for the object (0-0) identifies the new position number of the object.

Number of objects (1-1) identifies the number/range of objects to reorder.

Re-Order and **Reverse Order** allow you to change the numerical order of objects as specified and reverses the order of all members, respectively.

Highlight allows AutoCAD to display the members of the selected group in the graphics area.

Choosing **Description** allows you to change the description of the selected group.

Choosing **Explode** in the **Change Group** section of the Object Grouping dialog box deletes the selected group from the current drawing. Thus, the group no longer exists as a group. The members remain in the drawing and in any other group(s) of which they are members.

Choosing **Selectable** controls whether or not the selected group is selectable.

After making changes in the Object Grouping dialog box, choose **OK** to accept the changes and close the dialog box.

INFORMATION ABOUT OBJECTS

AutoCAD provides several commands for displaying information about the objects in the drawing. These commands do not create anything, nor do they modify or affect the drawing or objects therein. The only effect is that on single-screen systems, the screen switches to the AutoCAD Text window—not to be confused with the TEXT command—and the information requested by the particular inquiry command is then displayed on the screen. If you are new to AutoCAD, it is helpful to know that you can switch between the text and graphics screens so that you can continue with your drawing. On most systems, this is accomplished with the F2 function key. You can also toggle between graphic and text screens with the GRAPHSCR and TEXTSCR commands. The inquiry commands include LIST, AREA, ID, DBLIST, and DIST.

List Command

The LIST command displays information about individual objects stored by AutoCAD in the drawing database. The location, layer, object type, and space (model or paper) of the selected object as well as the color, lineweight, and linetype, if not set to BYLAYER or BYBLOCK, is listed. The distance of the main axes between the endpoints of a line, that is the delta-X, delta-Y, and delta-Z, are also listed, as well as the area and circumference of a circle or the area of a closed polyline. The insertion point, height, angle of rotation, style, font, obliquing angle, width factor, and actual character string of a selected text object are also listed. The object handle, reported in hexadecimal, is listed.

Invoke **List** (LIST) from the Properties panel (Figure 6–49) located in the Home tab, and AutoCAD displays the following prompt:

```
Select objects: (select the objects and press Enter to
terminate object selection)
```

AutoCAD lists the information about the selected objects in the AutoCAD Text Window.

FIGURE 6–49 *Invoking List from the Properties panel*

DBLIST Command

The DBLIST command lists the data about all of the objects in the drawing. It can take a long time to scroll through all the data in a large drawing. DBLIST can, like other commands, be terminated by pressing ESC.

Invoke the DBLIST command by typing **dblist** at the on-screen prompt, and Auto-CAD lists the information about all the objects in the drawing. To view the results in AutoCAD Text Window, press F2.

Measure Geometry

The MEASUREGEOM command measures the distance, radius, angle, area, and volume of the selected objects or sequence of points. Invoke the MEASUREGEOM command by typing **measuregeom** at the on-screen prompt. AutoCAD displays the following prompt:

```
Enter an option.[Distance/Radius/Angle/Area/Volume]
(select one of the available options)
```

The *Distance* option displays the distance, in the current units, between two points selected on the screen or keyed in from the keyboard. The information includes the horizontal and vertical distances (delta-X and delta-Y, respectively) between the points and the angles in and from the XY plane.

The *Distance* option can also be invoked from the Utilities panel (Figure 6–50) located in the Home tab, and AutoCAD displays the following prompts:

```
Specify first point: (specify the first point to measure
from)
Specify second point: (specify the endpoint to measure to)
```

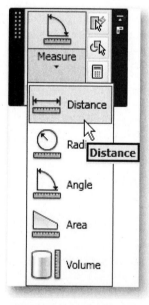

FIGURE 6–50 *Invoking the Distance option from the Utilities panel*

AutoCAD displays the distance between the two selected points.

The *Radius* option measures the radius and diameter of a specified arc or circle. The radius and diameter of the specified arc or circle display at the Command prompt and in the tooltip.

The *Radius* option can also be invoked from the Utilities panel (Figure 6–51) located in the Home tab, and AutoCAD displays the following prompts:

> Select Arc or Circle: *(select an arc or circle)*

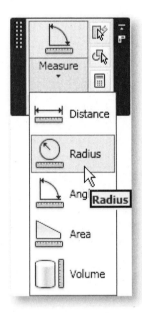

FIGURE 6–51 *Invoking the Radius option from the Utilities panel*

AutoCAD displays the radius of the selected arc or circle.

The *Angle* option measures the angle of a specified arc, circle, line, or vertex. The *Angle* option can be invoked from the Utilities panel (Figure 6–52) located in the Home tab, and AutoCAD displays the following prompts:

> Select Arc, Circle, Line, or <Specify Vertex>: (select an
> arc, circle, or line, or enter s to specify vertex)

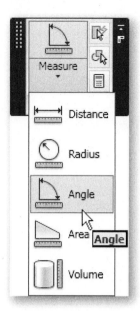

FIGURE 6–52 *Invoking the Angle option from the Utilities panel*

The angle of the specified object or vertex displays at the Command prompt and in the tooltip.

The *Area* option is used to report the area, in square units, and perimeter of a selected, closed geometric figure such as a circle, polygon, or polyline. You may also specify a series of points that AutoCAD considers to be a closed polygon.

The *Area* option can also be invoked from the Utilities panel (Figure 6–53) located in the Home tab, and AutoCAD displays the following prompt:

```
Specify first corner point or ☒ (specify a point or select
one of the available options from the shortcut menu)
```

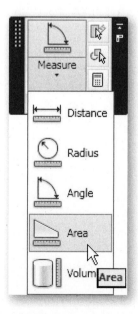

FIGURE 6–53 *Invoking Area from the Utilities panel*

The default actions calculate the area when you select the vertices of the objects. If you want to know the area of a specific object such as a circle, polygon, or closed polyline, select *Object* from the shortcut menu.

The following command sequence is an example of finding the area of a polygon using the *Object* option (Figure 6–54).

```
Specify first corner point or ⊡ (select object from the
shortcut menu)
Select objects: (select an object)
Area = 12.21, Perimeter = 13.79
```

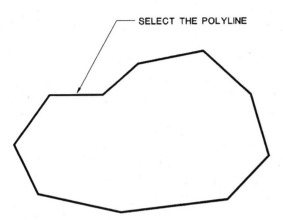
SELECT THE POLYLINE

FIGURE 6–54 *Finding the area of a polygon using the Area command object option*

The *Add* option allows you to add selected objects to form a total area. You can then use the *Subtract* option to remove selected objects from the running total.

The following example demonstrates the application of the *Add* and *Subtract* options. In this example, the area is determined for the closed shape after subtracting the area of the four circles shown in Figure 6–55. When using the *Add* and *Subtract* options of the AREA command, the results are not shown at the on-screen prompt. Therefore, you should press F2 to toggle to AutoCAD Text Window to view the results after each addition or subtraction.

```
Specify first corner point or ⊡ (select Add, from the
shortcut menu)
Specify first corner point or ⊡ (select Object, from the
shortcut menu)
  (ADD mode) Select objects: (select the polyline)
Result: Area = 12.9096, Perimeter = 15.1486
Result: Total area = 12.9096
  (ADD mode) Select objects: [Enter]
Specify first corner point or ⊡ (select Subtract, from the
shortcut menu)
Specify first corner point or ⊡ (select Object, from the
shortcut menu)
  (SUBTRACT mode) Select objects: (select circle A)
```

```
Result: Area = 0.7125, Circumference = 2.9992
Result: Total area = 12.1971
   (SUBTRACT mode) Select objects: (select circle B)
Result: Area = 0.5452, Circumference = 2.6175
Result: Total area = 11.6179
   (SUBTRACT mode) Select objects: (select circle C)
Result: Area = 0.7125, Circumference = 2.9922
Result: Total area = 10.9394
   (SUBTRACT mode) Select objects: (select circle D)
Result: Area = 0.5452, Circumference = 2.6175
Result: Total area = 10.3942
   (SUBTRACT mode) Select objects: ⊡
Specify first corner point or ⊡ Enter
```

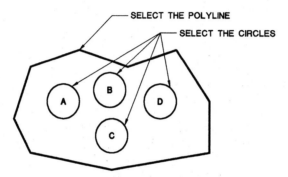

SELECT THE POLYLINE

SELECT THE CIRCLES

FIGURE 6–55 *Using the Add and Subtract options of the area command*

The *Volume* option measures the volume of an object or a defined area. The *Volume* option can also be invoked from the Utilities panel (Figure 6–56) located in the Home tab, and AutoCAD displays the following prompts:

```
Specify first corner point or [Object/Add volume/Subtract
volume/eXit] <Object>: Specify a point or enter an option
```

The available options for the *Volume* selection are similar to those for the *Area* selection explained earlier in this section.

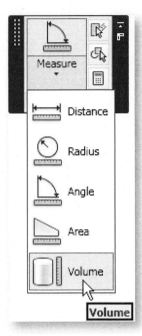

FIGURE 6–56 *Invoking the Volume option from the Utilities panel*

ID Command

The ID command is used to obtain the coordinates of a selected point. If you do not use an Object Snap mode to select a point that is not in the current construction plane, AutoCAD assigns the current elevation as the Z coordinate of the selected point.

Invoke the ID command by typing **id** at the on-screen prompt, and AutoCAD displays the following prompt:

```
Specify point: (select a point)
```

AutoCAD then displays the information about the selected point.

If the BLIPMODE system variable is set to ON, a blip appears on the screen at the specified point, if it is in the viewing area.

SYSTEM VARIABLES

AutoCAD stores the settings, or values, for its operating environment and some of its commands in system variables. Each system variable has an associated type: integer, for switching or for numerical value, real, point, or text string. Unless they are read-only, you can examine and change these variables at the on-screen prompt by typing the name of the system variable, or you can change them by using the SETVAR command.

Integers are used for switching system variables that have limited non-numerical settings and can be switched by setting them to the appropriate integer value. For example, the snap can be either ON or OFF. The purpose of the SNAPMODE system variable is to turn the snap ON or OFF by using the AutoCAD SETVAR command or the AutoLISP (SETVAR) function.

Turning the snap ON or OFF is demonstrated in the following example by changing the value of its SNAPMODE system variable. First, its current value is set to "0," which is OFF. Invoke the `SETVAR` command by typing **setvar** at the on-screen prompt, and AutoCAD displays the following prompts:

```
Enter variable name or ⊡ snapmode  [Enter]
Enter new value for SNAPMODE <0>: 1  [Enter]
```

For any system variable whose status is associated with an integer, the method of changing the status follows the preceding example. In the case of `SNAPMODE`, "0" turns it OFF and "1" turns it ON. In a similar manner, you can use `SNAPISOPAIR` to switch from one isoplane to another by setting the system variable to one of three integers: 0 is the left isoplane, 1 is the top isoplane, and 2 is the right isoplane.

It should be noted that the settings for the OSNAP system variable named `OSMODE` are members of the binomial sequence. The integers are 1, 2, 4, …, 512, 1024, 2048, 4096, 8192. See Table 6.1 for the meaning of `OSMODE` values. While the settings are switches, they are more than just ON and OFF settings. Several OSNAP modes may be active at one time. It is important to note that the value of an integer (switching) has nothing to do with its numerical value.

TABLE 6–1 *Values for the OSMODE System Variable*

NONe	0
ENDpoint	1
MIDpoint	2
CENter	4
NODe	8
QUAdrant	16
INTersection	32
INSertion	64
PERpendicular	128
TANgent	256
NEArest	512
QUIck	1024
APPint	2048
EXTension	4096
PARallel	8192

Integers (for Numeric Value)—System variables such as `APERTURE` and `AUPREC` are changed by using an integer whose value is applied numerically in some way to the setting rather than just as a switch. For instance, the size of the aperture, or the target box that appears for selecting OSNAP points, is set in pixels according to the integer value entered in the `SETVAR` command. For example, setting the value of `APERTURE` to 9 should render a target box that is three times larger than setting it to 3.

`AUPREC` is the variable that sets the precision of the angular units in decimal places. The value of the setting is the number of decimal places; therefore, it is considered a numeric integer setting.

Real—System variables that have a real number for a setting, such as `VIEWSIZE`, are called real.

Point (X Coordinate, Y Coordinate)—LIMMIN, LIMMAX, and VIEWCTR are examples of system variables whose settings are points in the form of the X coordinate and the Y coordinate.

Point (Distance, Distance)—Some system variables, whose type is point, are primarily used for setting spaces rather than a particular point in the coordinate system. For instance, the SNAPUNIT system variable, though called a point type, uses its X and Y distances from (0,0) to establish the snap X and Y resolution, respectively.

String—These variables have names such as CLAYER, for the current layer name, and DWGNAME, for the drawing name.

Open the Exercise Manual PDF for Chapter 6 for discipline-specific exercises. Related files are downloaded from the student companion site mentioned in the Introduction (refer to page number xii for instructions).

REVIEW QUESTIONS

1. The elements of a multiline can have different colors.
    a. True
    b. False

2. A SPLINE object:
    a. Does not actually pass through the control points
    b. Is always shown as a continuous line
    c. Requires you to specify tangent information for each control point
    d. Requires you to specify tangent information for the first and last points only

3. Once objects are grouped together, they can be ungrouped by:
    a. Using the EXPLODE command
    b. Using the UNGROUP command
    c. Using the GROUP command
    d. They cannot be ungrouped

4. The options dialog box allows you to change all of the following, except:
    a. Grip size
    b. Grip color
    c. Toggle grips system variable ON/OFF
    d. Specify the location of the grips

5. Grips do not allow you to _____ an object.
    a. Trim
    b. Erase
    c. Mirror
    d. Move
    e. Stretch

6. When a grip has been made a hot grip you are automatically placed in:

   a. Move mode

   b. Stretch mode

   c. Rotate mode

   d. Scale mode

   e. None of the above

7. Hovering the cursor over the quadrant grip on a circle will display:

   a. The Diameter of the circle

   b. The Radius of the circle

   c. The Circumference of the circle

   d. All of the above

8. A hot grip is:

   a. A grip that has been selected

   b. A grip that has been selected

   c. A grip that is blue

   d. A grip the can only be used to move the object

9. Multiline styles can be saved to an external file, thus allowing their use in multiple drawings.

   a. True

   b. False

10. The ID command will:

    a. Display the serial number of the AutoCAD program

    b. Display the X, Y, and Z coordinates of a selected point

    c. Allow you to password protect a drawing

    d. None of the above

11. Which command offers a quick way to find the length of a line?

    a. LIST

    b. INQUIRY

    c. DIMENSION

    d. LENGTH

    e. COORDINATES

12. To obtain a full listing of all the objects contained in the current drawing, you should use:

    a. LIST

    b. DBLIST

    c. LISTALL

    d. DBDUMP

    e. DUMP

13. You can change the setting of a system variable by using the command:

   a. SYSVAR

   b. SETVAR

   c. VARSET

   d. VARSYS

   e. None of the above

14. Which of the following is not a valid type of system variable?

   a. Real

   b. Integer

   c. Point

   d. String

   e. Double

Directions: Use the following X and Y locations to construct Figure 6–57. Your completed drawing will be used to answer the remaining questions.

Point	X value	Y value
A	1.5000	1.0000
B	2.1717	1.0000
C	2.7074	1.7500
D	4.5000	1.7500
E	4.8200	1.0000
F	5.5000	1.0000
G	4.1838	4.0711
H	3.5071	3.8100
J	3.7974	3.3861
K	3.6104	3.3139
L	2.8125	2.2500
M	4.2843	2.2500

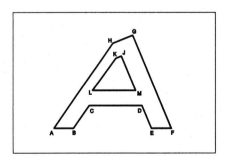

FIGURE 6–57

15. Using the appropriate AutoCAD command, what is the length of Line AH?

   a. 3.5432

   b. 3.4523

   c. 3.4532

16. Using the appropriate AutoCAD command, what is the angle of Line AH?

    a. 324 degrees

    b. 234 degrees

    c. 432 degrees

17. Using the appropriate AutoCAD command, what is the length of Line CD?

    a. 1.7962

    b. 1.7926

    c. 1.7629

18. Using the appropriate AutoCAD command, what is the length of Line FG?

    a. 3.3143

    b. 3.3314

    c. 3.3413

19. Using the appropriate AutoCAD command, what is the angle of Line FG?

    a. 131 degrees

    b. 311 degrees

    c. 113 degrees

20. Using the appropriate AutoCAD command, what is the length of Line GH?

    a. 0.7253

    b. 0.7532

    c. 0.7325

21. Using the appropriate AutoCAD command, what is the angle of Line GH?

    a. 210 degrees

    b. 102 degrees

    c. 201 degrees

22. What is the AREA of shape JKLM?

    a. 0.8833

    b. 0.9067

    c. 0.8838

23. What is the perimeter of shape JKLM?

    a. 4.1281

    b. 4.2382

    c. 4.1182

24. What is the linear distance from point E to point J?

    a. 2.2589

    b. 2.5960

    c. 2.5958

Dimensioning

INTRODUCTION

AutoCAD provides a full range of dimensioning commands and utilities. These enable the drafter to comply with the conventions of most disciplines, including the architectural, civil, electrical, and mechanical engineering fields.

OBJECTIVES

After completing this chapter, you will be able to do the following:

- Draw linear dimensioning

- Draw aligned dimensioning

- Draw angular dimensioning

- Draw diameter and radius dimensioning for arcs and circles

- Draw arc length dimensioning

- Draw leaders with annotation and geometric tolerance

- Draw ordinate dimensioning

- Draw baseline and continue dimensioning

- Use Quick dimensioning

- Disassociate and reassociate dimensioning

- Draw center marks for circles and arcs

- Edit dimension text

- Create and modify dimension styles

Figure 7–1 is a three-dimensional drawing of a lever arm. This "pictorial" rendering gives a viewer a good idea of what the object looks like, especially if the viewer has trouble understanding engineering drawings. Figure 7–2 is a typical engineering drawing of the same object. Because of the symmetry of the object, two views are sufficient. In order to communicate the size and location information necessary to make the object, the drawing in Figure 7–3 has the vital data that the manufacturer needs. It is possible that other information, such as material and finish specifications, and tolerances could also be included. This section covers how to use AutoCAD to produce dimensions and notes for this type of drawing as well as drawings for other disciplines such as architecture, civil engineering, and surveying.

FIGURE 7–1 *Pictorial view of lever arm*

AutoCAD provides commands to draw the full range of dimension types: Linear, Angular, Arc Length, Diameter and Radius, and Ordinate. Each type includes primary and secondary commands. For example, Linear dimensioning includes Horizontal, Vertical, Angle, and Rotated options. Figure 7–3 shows both horizontal and vertical formats of the Linear dimension type. AutoCAD also provides other general utility, editing, and style-related commands and subcommands that assist you in drawing the correct dimensions quickly and accurately.

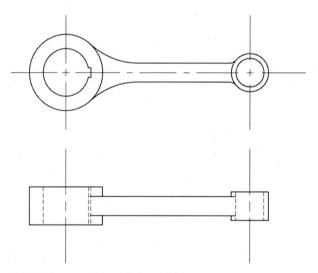

FIGURE 7–2 *Orthogonal views of lever arm*

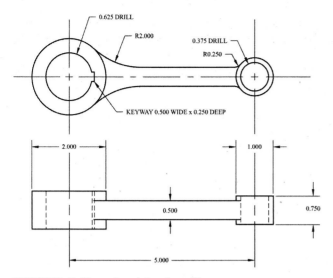

FIGURE 7–3 *Dimensioned drawing of lever arm*

AutoCAD makes drawing dimensions easy. For example, the width of the rectangle shown in Figure 7–4 can be dimensioned by selecting the two endpoints of the top corners, using an OSNAP mode such as *Endpoint* or *Intersection*, and then specifying a point to determine the location of the dimension line. AutoCAD provides a preview image of the dimension to indicate how it will look while you move the cursor to specify the location of the dimension line.

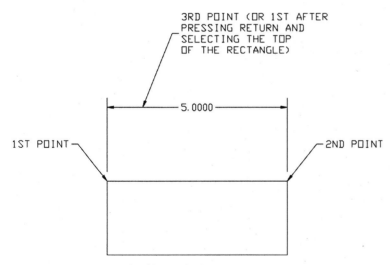

FIGURE 7–4 *An example of linear dimensioning*

The Linear dimensioning options include *Horizontal, Vertical, Angle,* and *Rotated.* Angular dimensioning is covered in the section on "Angular Dimensioning." Diameter, radius, arc length, and ordinate dimensioning are also covered later, under their respective sections.

Approximately 60 dimensioning system variables are available. Most of these have names that begin with "DIM." They are used for such purposes as determining the size of the gap between the extension line and the point specified on an object or whether one or both of the extension lines will be drawn or suppressed. Combinations of these variable settings can be named and saved as dimension styles and later recalled for applying where needed. See Appendix C for the listing of available dimensioning system variables.

Dimension variables change when their associated settings are changed in the Modify Dimension Style dialog box. For example, the value of the `DIMEXO`, the extension line offset, dimensioning system variable is established by the number in the **Offset from origin** text box in the Modify Dimension Style dialog box. See the section on "Dimension Styles" later in this chapter.

Dimension utilities include Override, Center, Leader, Baseline, Initial Length, Dimscale, Oblique, Linetype, Continue, and Feature Control Frames for adding tolerancing information.

The dimension text editing command options include *Home, New, Oblique, Edit,* and *Rotate*.

DIMENSION TERMINOLOGY

Following are the terms for the different parts that affect dimensions in AutoCAD.

Figure 7–5 shows the different components of a typical dimension.

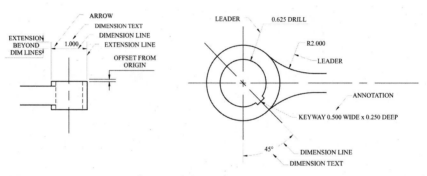

FIGURE 7–5 *Components of a typical dimension*

Dimension Line

The dimension line is offset from the measured feature. Sometimes drawn as two segments outside of the extension lines if a single line with its related text will not fit between the extension lines, the dimension line indicates the direction and length of the measured distance. It is normally offset for visual clarity. If the dimension is measured between the parallel lines of one or two objects, it may not be offset, but it is drawn on the object or between the two objects. Dimension lines are usually terminated with markers such as arrows or ticks, which are short slanted lines. Angular dimension lines become arcs whose centers are at the vertex of the angle.

Arrowhead

The arrowhead is a mark at the end of a dimension line to indicate its termination. Shapes other than arrows are used in some styles.

Extension Line

When a dimension line is offset from the measured feature, the extension lines, sometimes referred to as *witness lines*, indicate such offset. Unless you have invoked the OBLIQUE option, the extension lines will be perpendicular to the direction of the measurement.

Dimension Text

The dimension text consists of numbers, words, characters, and symbols used to indicate the measured value and type of dimension. Unless the text settings have been altered in the Standard dimension style, the number/symbol format is decimal. This conforms to the same linear and angular units as the default of the drawing. The text style conforms to that of the Current text style.

Leader

The leader is a radial line used to point from the dimension text to the circle or arc whose diameter or radius is being dimensioned. A leader can also be used for general annotation.

Center Mark

The center mark is made up of lines or a series of lines that cross in the center of a circle for the purpose of marking its center.

ASSOCIATIVE, NONASSOCIATIVE, AND EXPLODED DIMENSIONS

Dimensions in AutoCAD can be drawn as associative, nonassociative, or exploded, depending on the setting of the DIMASSOC dimensioning system variable. The variable setting for associative is 2 (default), nonassociative is 1, and exploded is 0.

Associative Dimensions

An associative dimension becomes associated with an object by selecting points on the object—using an Object Snap mode—when prompted to do so during a dimensioning command. If the object is subsequently modified in a manner that changes the location of one or both of the selected points, the associated dimension is automatically updated to correctly indicate the new distance or angle. Associative dimensioning does not support multilines. The associativity between a dimension and a block is lost when the block is redefined.

An associative dimension drawn with the DIMASSOC dimensioning system variable set to 2 will have all of its separate parts become members of a single object. Therefore, if any one of its members is selected for modifying, all members are highlighted and subject to being modified. This is similar to the manner in which member objects of a block reference are treated. If you have dimensioned the width of a rectangle with an associative dimension and then specified one end of the rectangle to stretch, the dimension will also be stretched, and the dimension text will be changed to correspond to the new measurement. In addition to the customary visible parts, AutoCAD draws point objects at the ends, where the measurement actually occurs on the object.

Figure 7–6 shows an example of a dimensioned object that has been revised. If the second extension line origin for the dimension on the right is moved horizontally with the STRETCH command, the associative dimension and the dimension text will reflect the new location.

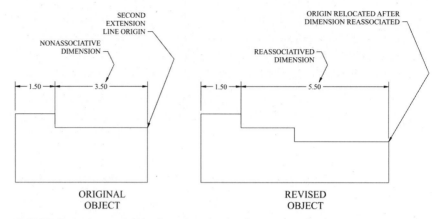

FIGURE 7–6 *Nonassociative dimension that has been reassociated*

Nonassociative Dimensions

Nonassociative dimensions are drawn while the DIMASSOC dimensioning system variable is set to 1. The separate parts of the nonassociative dimension are, like those of the associative dimensions, considered members of a single object when any one of them is selected for modification. However, if the object that was dimensioned is selected for modification, without selecting the dimension, the dimension itself will remain unchanged.

Exploded Dimensions

Exploded dimensions are drawn while the DIMASSOC dimensioning system variable is set to 0, and the members are drawn as separate objects. If one of the components of the dimension is selected for modification, that component will be the only one modified.

NOTE   An associative dimension can be converted to an exploded dimension with the EXPLODE command. Once the dimension is exploded, you cannot recombine the separate parts back into the associative dimension from which they were exploded except by means of the UNDO command, if feasible.

DIMENSIONING COMMANDS

The dimensioning commands are accessible from the Dimensions panel on the Annotate tab. You can also access them from the Dimension menu in the Menu Bar or by entering their names directly from the keyboard at the on-screen prompt. Dimension types include Linear, Aligned, Ordinate, Radius, Diameter, Arc Length, Angular, Baseline, Continue, Quick, Qleader, Leader, with Tolerances, Cross Marks for circles and arcs, and Oblique.

Linear Dimensioning

Invoke **Linear** (DIMLINEAR) from the Dimensions panel (Figure 7–7) located in the Annotate tab, and AutoCAD displays the following prompt:

> Specify first extension line origin or <select object>: (specify a point)

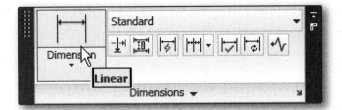

FIGURE 7-7 *Invoking the* `DIMLINEAR` *command by selecting Linear from the Dimension panel*

AutoCAD uses the first point specified as the origin for the first extension line. This point can be the endpoint of a line, the intersection of objects, the center point of a circle, the insertion point of a text object, or a point on the object itself. Auto-CAD provides a gap between the object and the extension line that is equal to the value of the `DIMEXO` dimensioning system variable. After you specify the origin, AutoCAD displays the following prompt:

> `Specify second extension line origin:` *(specify a point at which the second extension line should start)*

Dynamic Horizontal and Vertical Dimensioning

Dynamic horizontal and vertical dimensioning is the default mode after you have specified the second point, in response to the `DIMLINEAR` command. If you specify two points on the same horizontal line, moving the cursor above or below the line causes a preview image of the dimension to appear. AutoCAD assumes you wish to draw a horizontal dimension. Likewise, AutoCAD assumes a vertical dimension if the specified points are on the same vertical line.

> Horizontal and vertical formats of Linear dimensions refer to their orientation in the drawing, not their orientation on the object or in real space. Horizontal dimensions are aligned with the X axis, orthogonal left to right, in the drawing coordinate system. Vertical dimensions are aligned with the Y axis, orthogonal up and down.

NOTE

Dynamically switching between horizontal and vertical dimensioning is more applicable when the two points specified are not on the same horizontal or vertical line. That is, they can be considered diagonally opposite corners of an imaginary rectangle with both width and height. After the two points are specified, you are prompted to specify the location of the dimension line. You are also shown a preview image of the dimension on the screen, as you move the cursor. If you position the cursor above the top line, or below the bottom line of the rectangle, the dimension will be horizontal. If the cursor moves to the right of the right side, or to the left of the left side of the rectangle, the dimension will be vertical. If the cursor is dragged to one of the outside quadrants or inside of the rectangle, it will maintain the orientation in effect before the cursor was moved.

After two points have been specified, AutoCAD displays the following prompt:

> `Specify dimension line location or` ⬇ *(specify location for dimension line or select one of the options from the shortcut menu)*

Drawing the Dimension Line

After specifying the two points, or using the shortcut menu to change the dimension text or type, specify a point through which the dimension line is to be drawn. When AutoCAD draws the dimension line, if there is enough room between the extension lines, the dimension text will be centered in line with or above this line. If the dimension text is to be in line with the dimension line, the dimension line will be broken to allow room for the text. However, if the dimension line, arrows, and text do not fit between the extension lines, they are drawn outside and the text will be drawn near the second extension line.

Changing the Dimension Text

The *Mtext* and *Text* options allow you to change the measured dimension text. The *Angle* option allows you to change the rotation angle of the dimension text. After responding appropriately for *Text* or *Angle*, AutoCAD repeats the prompts for the dimension line location.

Changing Dimension Type

The *Horizontal* option forces a horizontal dimension to be drawn, even when dynamic dimensioning calls for a vertical dimension. Likewise, the *Vertical* option forces a vertical dimension to be drawn, even when dynamic dimensioning calls for a horizontal dimension.

Rotated Dimension

The *Rotated* option allows you to draw the dimension at a specified angle. Figure 7–8 shows a situation where Rotated dimensions are useful. Here is a case where the desired angle of dimensioning is the angle created by a line from point A to point B. The dimension from point A to point C can be properly oriented by choosing the Rotated option, after specifying points A and C. Points A and B are then specified to determine the angle.

Choose Rotated from the shortcut menu, and AutoCAD displays the following prompt:

```
Specify angle of dimension line <0>: (specify the dimension
line angle or specify points A and B to determine the angle)
```

The points specified do not have to be parallel to the dimensioned object.

The following command sequence is an example of drawing linear dimensioning for a horizontal line by providing data points for the first and second line origins, respectively, as shown in Figure 7–9.

```
dimlinear Enter
Specify first extension line origin or <select object>:
(specify the origin of the first extension line)
Specify second extension line origin: (specify the origin of
the second extension line)
Specify dimension line location or ⊡ (specify the location
for the dimension line)
```

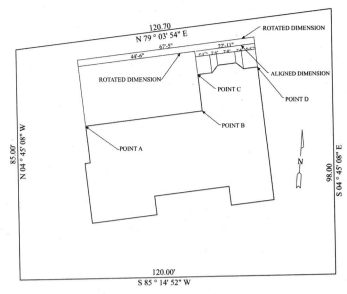

FIGURE 7–8 *Example of using rotated dimensioning*

The following command sequence is an example of drawing a linear dimension for a vertical line by providing data points for the first and second line origins, respectively, as shown in Figure 7–10.

```
dimlinear [Enter]
Specify first extension line origin or <select object>:
(specify the origin of the first extension line)

Specify second extension line origin: (specify the origin of
the second extension line)

Specify dimension line location or [↧] (specify the location
for the dimension line)
```

> **NOTE**
>
> You can dynamically switch dimensions between parallel and perpendicular to the angle between the endpoints. This is similar to dynamically switching between horizontal and vertical dimensioning for normal linear dimensioning.

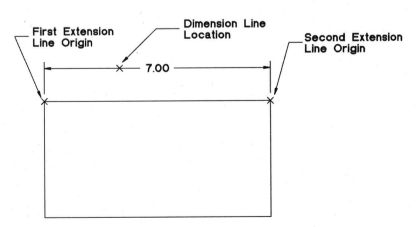

FIGURE 7–9 *Drawing a linear dimension for a horizontal line*

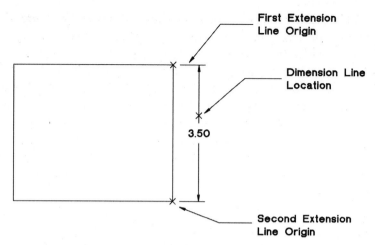

FIGURE 7–10 *Drawing linear dimensioning for a vertical line*

Jogged Linear Dimension

The DIMJOGLINE command allows you to draw a jog in the dimension line of a linear dimension (Figure 7–11). This is the convention used to convey that the dimension and object have foreshortened.

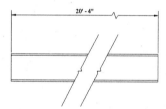

FIGURE 7–11 *Jog added to a linear dimensioning*

To add a jog to a linear dimension, select Dimjogline from the Dimension panel, and then select the dimension line. The jog can be moved along the dimension line by selecting the dimension and using the grip attached to the jog.

Linear Dimensioning by Selecting an Object

If the dimension you wish to draw is between endpoints of a line, arc, or circle diameter, AutoCAD allows you to bypass specifying those endpoints separately. This action speeds up drawing the dimension, especially if you have to invoke the endpoint object snap modes. When you have invoked the DIMLINEAR command and you press Enter in response to the "Specify first extension line origin or <select object>:" prompt, AutoCAD displays the following prompts:

> Select object to dimension: *(select line, polyline segment, arc, or circle object)*
>
> Specify dimension line location or ⊡ *(specify location for dimension line or select one of the options from the shortcut menu)*

If you select a line object, AutoCAD automatically uses the line's endpoints as the first and second points to determine the distance to measure. You will be prompted to specify the location of the dimension line. If you right-click and choose *Horizontal*, a horizontal dimension is drawn accordingly. In a similar manner, if you choose *Vertical*, a vertical dimension is drawn. If you do not choose the type of dimension to be drawn, a horizontal dimension is drawn if the specified point for the dimension line is above or below the line object. Similarly a vertical dimension is drawn if the specified point for the dimension line is to the right or left of the line object. If you right-click and choose *Rotated* from the shortcut menu, AutoCAD will use the endpoints of the line for the distance reference points and then use the next two points specified to determine the direction of the dimension and measure the distance between the two endpoints of the line object in that direction. The dimension line passes through the last point specified.

The following command sequence shows an example of drawing a linear dimension applied to a single object (Figure 7–12).

```
dimlinear Enter
Specify first extension line origin or <select object>: Enter
Select object to dimension: (select Line A)
Specify dimension line location or ⬇ (specify the location
for the dimension line location)
```

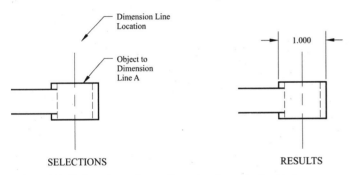

SELECTIONS   RESULTS

FIGURE 7–12 *Drawing a linear dimension for a single object*

If the object is a circle, AutoCAD automatically uses the diameter of the circle as the distance to dimension. If you choose the *Horizontal* option, a horizontal dimension is drawn using the endpoints of the horizontal diameter. Likewise, with the *Vertical* option the endpoints of the vertical diameter are used. If you choose the *Rotated* option, AutoCAD uses the first two points specified to determine the direction of the dimension and to measure the distance between the two endpoints of a diameter in that direction. The dimension line passes through the last point specified.

If the object is an arc, AutoCAD automatically uses the arc's endpoints as the distance to dimension. You will be prompted to specify the location of the dimension line. If you choose the *Horizontal* option, a horizontal dimension is drawn accordingly. If you choose the *Vertical* option, a vertical dimension is drawn accordingly. If you have not yet specified the orientation of the dimension to be drawn, a horizontal dimension will be drawn if the specified point for the dimension line is above or below the arc, and a vertical dimension will be drawn if the specified point for the dimension line is to the right or left of the arc. If you choose the Aligned option, AutoCAD uses the endpoints of the selected arc as the first and second points and the direction from one endpoint to the other endpoint as the direction to measure. If you choose the *Rotated* option, AutoCAD uses the first two points specified to determine the dimension's direction and the distance to dimension. The dimension line passes through the last point specified.

Aligned Dimensioning

Invoke **Aligned** (DIMALIGNED) from the Dimensions panel (Figure 7–13) located in the Annotate tab, and AutoCAD displays the following prompts:

> Specify first extension line origin or <select object>: *(specify the origin of the first extension line)*
>
> Specify second extension line origin: *(specify the origin of the second extension line)*

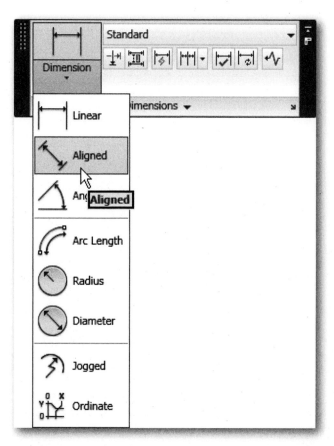

FIGURE 7–13 *Invoking the DIMALIGNED command by selecting Aligned from the Dimensions panel*

After two points have been specified, AutoCAD displays the following prompt:

> Specify dimension line location or ⊡ *(specify the location for the dimension line or select one of the options from the shortcut menu)*

Specify a point where the dimension line is to be drawn or choose one of the available options. The *Mtext, Text,* and *Angle* options are the same as in Linear dimensioning, explained earlier in this section.

The DIMALIGNED command creates an aligned linear dimension using the three familiar points on the object (Figure 7–14). To draw the proper dimension, provide "first extension line origin," "second extension line origin," and "dimension line location" as prompted in the following command sequence:

```
dimaligned
Specify first extension line origin or <select object>:
(specify Point A)
Specify second extension line origin: (specify Point B)
Specify dimension line location or ⤓ (specify Point C)
```

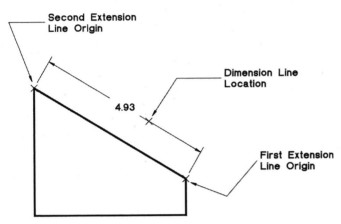

FIGURE 7–14 *Drawing aligned dimensioning for a line drawn at an angle*

Ordinate Dimensioning

AutoCAD uses the mutually perpendicular X and Y axes of the world coordinate system (WCS) or current user coordinate system (UCS) as the reference lines from which to base the X or Y coordinate displayed in an ordinate dimension, sometimes referred to as a *datum dimension*. In the following examples, Figure 7–16 is valid when the base of the rectangle lies on the X axis, giving it a Y value of 0.0000, and Figure 7–17 is valid when the left side of the rectangle lies on the Y axis, giving it an X value of 0.0000. If this is not where the objects are located in the drawing and you still wished to have their values to be 0.0000, then a new coordinate system would have to be created, even if temporarily for drawing the dimensions. See Chapter 16 for a detailed discussion on creating a UCS.

Invoke **Ordinate** (DIMORDINATE) from the Dimensions panel (Figure 7–15) located in the Annotate tab, and AutoCAD displays the following prompt:

```
Specify feature location: (specify a point)
```

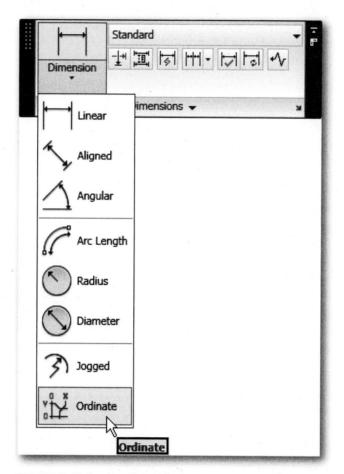

FIGURE 7–15 *Invoking the* DIMORDINATE *command by selecting Ordinate from the Dimensions panel*

Although the default prompt is "Specify feature location," AutoCAD is actually looking for a point that is significant in locating a feature point on an object, such as the endpoint or an intersection where planes meet, or the center of a circle representing a hole or shaft. Therefore, an object snap mode, such as an endpoint, intersection, quadrant, or center, will need to be invoked when responding to the "Specify feature location" prompt. Specifying a point determines the origin of a single orthogonal leader that will point to the feature when the dimension is drawn. AutoCAD displays the following prompt:

> Specify leader endpoint or ⊡ (specify a point or select one of the options from the shortcut menu)

If the Ortho mode is ON, the leader will be a single horizontal line for a Ydatum ordinate dimension (Figure 7–16) or a single vertical line for an Xdatum ordinate dimension (Figure 7–17).

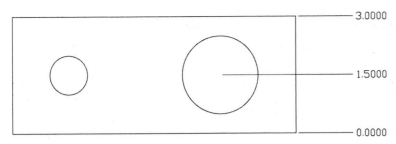

FIGURE 7–16 *Ydatum dimension*

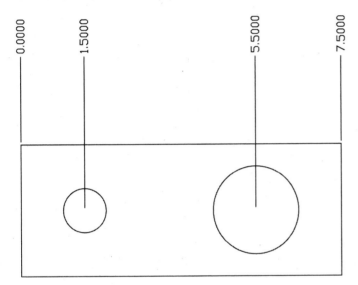

FIGURE 7–17 *Xdatum dimension*

If the Ortho mode is OFF, the leader will be a three-part line consisting of orthogonal lines on each end and joined by a diagonal line in the middle. It may be necessary to use the nonorthogonal leader if the text has to be offset to avoid interfering with other objects in the drawing. The type of dimension drawn, Ydatum or Xdatum, depends on the direction of the second endpoint from the first endpoint. A preview image of the dimension is displayed during specification of the "leader endpoint."

You can right-click and choose *Xdatum* or *Ydatum* from the shortcut menu, and AutoCAD draws an Xdatum dimension or Ydatum dimension regardless of the location of the "leader endpoint" point relative to the "feature location" point.

Radius Dimensioning

Radius dimensioning allows you to create radius dimensions (Figure 7–18) for circles and arcs. The type of dimensions that AutoCAD uses depends on the dimensioning system variable settings. See the section on "Dimension Styles," later in this chapter, for instructions on how to change the dimensioning system variable settings to draw appropriate radius dimensioning.

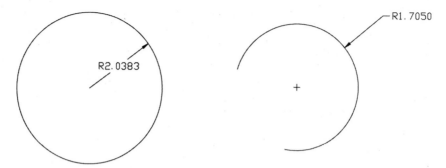

FIGURE 7–18 *Radius dimensioning of a circle and an arc*

Invoke **Radius** (DIMRADIUS) from the Dimensions panel (Figure 7–19) located in the Annotate tab, and AutoCAD displays the following prompts:

Select arc or circle: *(select an arc or a circle to dimension)*

Specify dimension line location or ⊡ *(specify the location for the dimension leader line or select one of the options from the shortcut menu)*

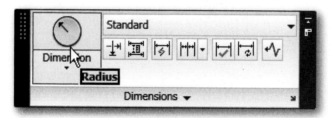

FIGURE 7–19 *Invoking the* DIMRADIUS *command by selecting Radius from the Dimensions panel*

The *Mtext, Text,* and *Angle* options are the same as those in linear dimensioning, explained earlier in this section. Dimension text for radius dimensioning is preceded by the letter R.

The following command sequence shows an example of drawing radius dimensioning for a circle (Figure 7–20).

dimradius

Select arc or circle: *(select the circle object)*

Specify dimension line location or ⊡ *(specify a point to draw the dimension leader line or select one of the options from the shortcut menu)*

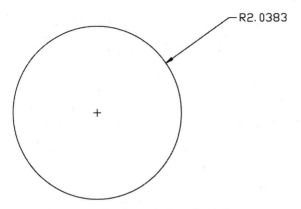

FIGURE 7–20 *Radius dimensioning of a circle*

Diameter Dimensioning

Diameter dimensioning allows you to create diameter dimensions for circles and arcs (Figure 7–21). The type of dimensions that AutoCAD uses depends on the dimensioning system variable settings. See the section on "Dimension Styles," later in this chapter, for instructions on how to change the dimensioning system variable settings to draw an appropriate diameter dimensioning.

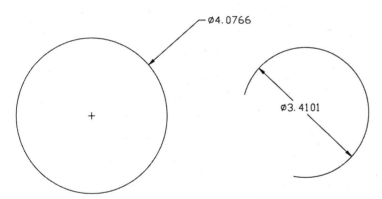

FIGURE 7–21 *Diameter dimensioning of a circle and an arc*

Invoke **Diameter** (DIMDIAMETER) from the Dimensions panel (Figure 7–22) located in the Annotate tab, and AutoCAD displays the following prompts:

> Select arc or circle: *(select an arc or a circle to dimension)*
>
> Specify dimension line location or ⊡ *(specify the location for the dimension line or select one of the options from the shortcut menu)*

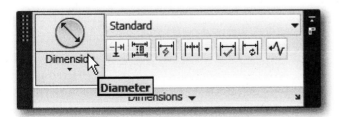

FIGURE 7–22 *Invoking the DIMDIAMETER command by selecting Diameter from the Dimensions panel*

The *Mtext*, *Text*, and *Angle* options are the same as those in linear dimensioning, explained earlier in this section.

Specifying a point determines the location of the diameter dimension. The following command sequence shows an example of drawing diameter dimensioning for a circle (Figure 7–23).

```
dimdiameter
Select arc or circle: (select an arc or a circle to dimension)
Specify dimension line location or ⊥ (specify a point to draw
the dimension)
```

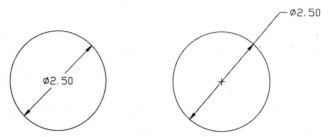

FIGURE 7–23 *Diameter dimensioning of a circle*

Arc Length Dimensioning

The Arc Length command allows you to create length dimensions for an arc (Figure 7–24).

Invoke **Arc Length** (DIMARC) from the Dimensions panel (Figure 7–25) located in the Annotate tab, and AutoCAD displays the following prompts:

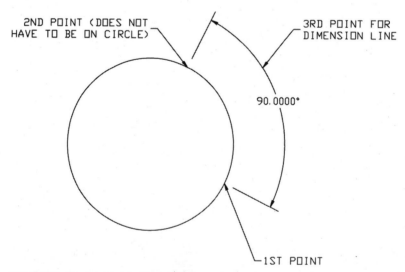

FIGURE 7–24 *Arc Length dimensioning*

```
Select arc or polyline arc segment: (select an arc or poly arc
to dimension)
Specify arc length dimension location, or ⊥ (specify the
location for the dimension line or select one of the options
from the shortcut menu)
```

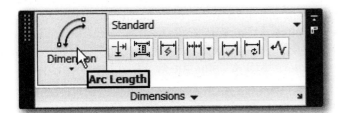

FIGURE 7–25 *Invoking the* `DIMARC` *command by selecting Arc Length from the Dimensions panel*

The *Mtext, Text,* and *Angle* options are the same as those in linear dimensioning, explained earlier in this section.

Selecting *Leader* adds a leader object. This option is displayed only if the arc (or arc segment) is greater than 90 degrees. The leader is drawn radially, pointing towards the center of the arc being dimensioned.

Selecting *Partial* allows you to specify a point other than the second endpoint of the arc, and AutoCAD will dimension the length of the portion of the arc between the first endpoint of the arc and the point specified (Figure 7–26).

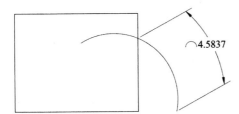

FIGURE 7–26 *Partial Arc Length dimensioning of an arc*

Angular Dimensioning

Angular Dimensioning allows you to draw angular dimensions using various methods: three points (Vertex/Point/Point), between two nonparallel lines, on an arc, or between the two endpoints of the arc, with the center as the vertex, and on a circle, or between two points on the circle, with the center as the vertex.

Invoke **Angular** (`DIMANGULAR`) from the Dimensions panel (Figure 7–27) located in the Annotate tab, and AutoCAD displays the following prompt:

```
Select arc, circle, line, or <specify vertex>: (select an
object, or press Enter to specify a vertex)
```

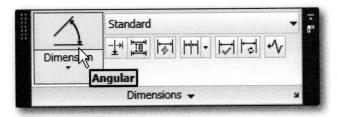

FIGURE 7–27 *Invoking the* `DIMANGULAR` *command by selecting Angular from the Dimensions panel*

The default method of angular dimensioning is to select an object. If the object selected is an arc (Figure 7–28), AutoCAD automatically uses the arc's center as the vertex, its endpoints for the first angle endpoint, and second angle endpoint to determine the three points of a Vertex/Endpoint/Endpoint angular dimension. Auto-CAD displays the following prompt:

> Specify dimension arc line location or ⊡ *(specify the location for the dimension arc line, or right-click and select one of the options from the shortcut menu)*

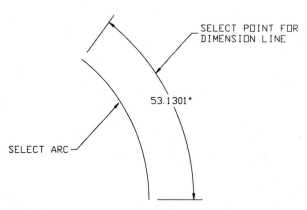

FIGURE 7–28 *Angular dimensioning of an arc*

The *Mtext, Text,* and *Angle* options are the same as those in linear dimensioning, explained earlier in this section. Specify a point for the location of the dimension and AutoCAD automatically draws radial extension lines.

Selection of the *Quadrant* option specifies the quadrant that the dimension should be locked to. When quadrant behavior is on, the dimension line is extended past the extension line when the dimension text is positioned outside of the angular dimension.

If the object selected is a circle (Figure 7–29), AutoCAD automatically uses the circle's center as the vertex and the point at which you select the circle as the endpoint for the first angle endpoint. AutoCAD displays the following prompt:

> Specify second angle endpoint: *(specify a point to determine the second angle endpoint)*

Specify a point, and AutoCAD makes this point the second angle endpoint to use along with the previous two points as the three points of a Vertex/Endpoint/Endpoint angular dimension. Note that the last point does not have to be on the circle, but it does determine the origin for the second extension line. AutoCAD displays the following prompt:

> Specify dimension arc line location or ⊡ *(specify the location for the dimension arc line, or select one of the options from the shortcut menu)*

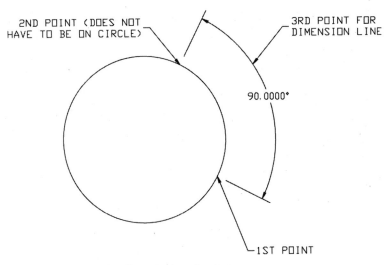

FIGURE 7-29 *Angular dimensioning of a circle*

The *Mtext, Text, Quadrant,* and *Angle* options are the same as those in linear dimensioning, explained earlier in this section.

When you specify a point for the location of the dimension arc line, AutoCAD automatically draws radial extension lines and draws either a minor or a major angular dimension, depending on whether the point used to specify the location of the dimension arc line is in the minor or major projected sector.

If the object selected is a line (Figure 7-30), AutoCAD displays the following prompt:

Select second line: *(select a line object)*

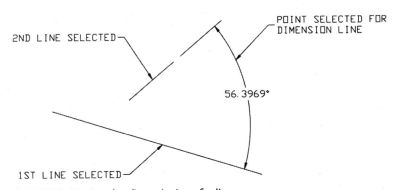

FIGURE 7-30 *Angular dimensioning of a line*

Select another line, and AutoCAD uses the apparent intersection of the two lines as the vertex for drawing a Vertex/Vector/Vector angular dimension. You are then prompted to specify the location of the dimension arc, which will always be less than 180°.

If the dimension arc is beyond the end of either line, AutoCAD adds the necessary radial extension lines. AutoCAD displays the following prompt:

Specify dimension arc line location or ⏎ *(specify the location for the dimension arc line or select one of the options from the shortcut menu)*

The *Mtext, Text,* and *Angle* options are the same as those in linear dimensioning, explained earlier in this section.

If you specify a point for the location of the dimension arc, AutoCAD automatically draws extension lines and draws the dimension text.

If instead of selecting an arc, a circle, or two lines for angular dimensioning you press Enter , AutoCAD allows you to do three-point angular dimensioning. The following command sequence shows an example of creating an angular dimension by providing three data points (Figure 7–31).

```
dimangular Enter
Select arc, circle, line, or <specify vertex>: Enter
Specify angle vertex: (specify point 1)
Specify first angle endpoint: (specify point 2)
Specify second angle endpoint: (specify point 3)
Specify dimension arc line location or ⬇ (specify the
dimension arc line location or select one of the options from
the shortcut menu)
```

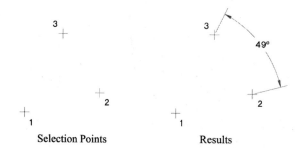

Selection Points   Results

FIGURE 7–31 *Using three points for angular dimensioning*

Baseline Dimensioning

Baseline dimensioning, sometimes referred to as *parallel dimensioning,* is used to draw dimensions to multiple points from a single datum baseline (Figure 7–32), the first extension line origin of the initial dimension. It can be a linear, angular, or ordinate dimension, and it establishes the base from which the subsequent baseline dimensions are drawn. That is, all of the dimensions in the series of baseline dimensions will share a common first extension line origin. AutoCAD automatically draws a dimension line/arc beyond the initial, or previous baseline, dimension line/arc. The location of the new dimension line/arc is an offset distance established by the DIMDLI, for dimension line increment, dimensioning variable.

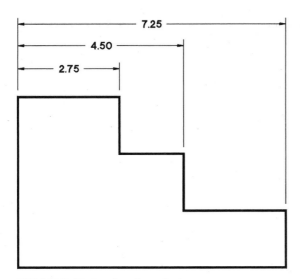

FIGURE 7–32 *Baseline dimensioning*

Invoke **Baseline** (DIMBASELINE) command from the Dimensions panel (Figure 7–33) located in the Annotate tab and AutoCAD displays the following prompt:

Specify a second extension line origin or ⊡ *(specify a point or select one of the options from the shortcut menu)*

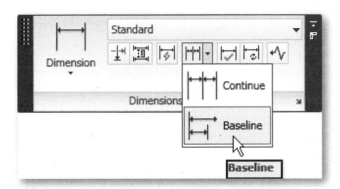

FIGURE 7–33 *Invoking the DIMBASELINE command by selecting Baseline from the Dimensions panel*

After you specify a point for the second extension line origin, AutoCAD will use the first extension line origin of the previous linear, angular, or ordinate dimension as the first extension line origin for the new dimension, and the prompt is repeated. To exit the command, press the ESC key, right-click and choose Cancel from the shortcut menu.

For the DIMBASELINE command to be valid, there must be an existing linear, angular, or ordinate dimension. If the previous dimension was not a linear, angular, or ordinate dimension, or if you press Enter without providing the second extension line origin, AutoCAD displays the following prompt:

Select base dimension: *(select a dimension object)*

You may select the base dimension, with the baseline being the extension line nearest to where you select the dimension.

The following command sequence shows an example of drawing baseline dimensioning for a circular object to an existing angular dimension (Figure 7–34).

```
dimbaseline Enter
Second extension line origin or ⬇ (specify a point)
Second extension line origin or ⬇ (specify a point)
Second extension line origin or ⬇ (press ESC or right-click
and choose Enter from the shortcut menu)
```

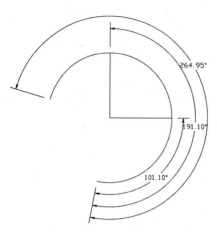

FIGURE 7–34 *Baseline dimensioning of a circular object*

Dimension Space

The DIMSPACE command allows you to specify the space between the dimension lines of parallel linear or concentric arc dimensions. Invoke **Adjust Space** (DIM-SPACE) from the Dimensions panel (Figure 7–35) located in the Annotate tab, and AutoCAD displays the following prompts:

```
Select base dimension: (select a dimension)
Select dimensions to space: (select one or more parallel or
concentric dimensions and right-click to terminate the
selection of dimensions)
Enter Value or ⬇ (specify a distance for spaces between
dimension lines)
```

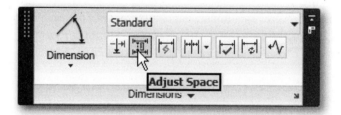

FIGURE 7–35 *Invoking the DIMSPACE command by selecting Adjust Space from the Dimensions panel*

AutoCAD will move the dimension lines so that they are spaced using the distance you specify. The first dimension line selected is the baseline for the second. The second dimension line selected is the baseline for the third, and so on. If a stack of dimension lines is selected in sequence from top to bottom, bottom to top, or side to side, the Dimension Space feature will maintain their order with the new spacing. Be careful: if you randomly select the dimension lines in a group, the spacing will be applied in the order you selected, and each space change will be in the direction of each pair of selections. The results could be unpredictable.

Dimension Break

The DIMBREAK command allows you to create a break in a dimension extension line where it crosses another line. Invoke **Break** (DIMBREAK) from the Dimensions panel (Figure 7–36) located in the Annotate tab, and AutoCAD displays the following prompts:

        Select a dimension or [Multiple]: *(select the dimension)*
        Select object to break dimension or [Auto/Restore/Manual]
        <Auto>: *(select line that dimension extention line crosses)*
        Select object to break dimension:

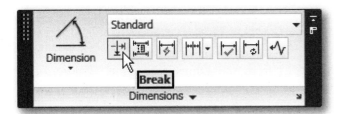

FIGURE 7–36 *Invoking the* DIMBREAK *command by selecting Break from the Dimensions panel*

Figure 7–37 shows the dimension being selected.

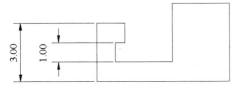

FIGURE 7–37 *Selecting Dimension*

Figure 7–38 shows the breaks in the dimension extension lines after the line on the object is selected.

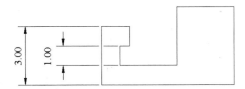

FIGURE 7–38 *Dimension extension line breaks*

Options to the DIMBREAK command include *Auto, Restore,* and *Manual.* Choosing *Auto* places dimension breaks automatically at all the intersection points of the objects that intersect the selected dimension. Any dimension break created using this option is updated automatically when the dimension or an intersecting object is modified.

Choosing *Restore* removes all dimension breaks from the selected dimensions. Choosing *Manual* places a dimension break manually. You specify two points on the dimension or extension lines for the location of the break. Any dimension break that is created using this option is not updated if the dimension or intersecting objects are modified.

Continue Dimensioning

The DIMCONTINUE command, shown in Figure 7–39, is used for drawing a string of dimensions, each of whose second extension line origin coincides with the next dimension's first extension line origin.

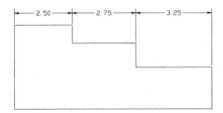

FIGURE 7–39 *Continue dimensioning*

Invoke **Continue** (DIMCONTINUE) from the Dimensions panel (Figure 7–40) located in the Annotate tab, and AutoCAD displays the following prompt.

> Specify a second extension line origin or ⊡ *(specify a point or select one of the options from the shortcut menu)*

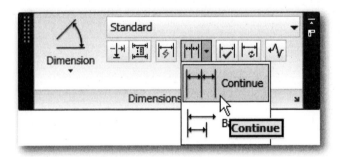

FIGURE 7–40 *Invoking the* DIMCONTINUE *command by selecting Continue from the Dimensions panel*

After you specify a point for the second extension line origin, AutoCAD will use the second extension line origin of the previous linear, angular, or ordinate dimension as the first extension line origin for the new dimension, and the prompt is repeated. To exit the command, right-click and select Enter from the shortcut menu.

For the DIMCONTINUE command to be valid, there must be an existing linear, angular, or ordinate dimension. If the previous dimension was not a linear, angular, or ordinate dimension, or if you pressed Enter without providing the second extension line origin, AutoCAD displays the following prompt:

> Select continued dimension: *(select a dimension object)*

You may select the continued dimension, with the coincidental extension line origin being the one nearest to where the existing dimension is selected.

The following command sequence shows an example of drawing continue dimensioning for a linear object to an existing linear dimension (Figure 7–41).

> dimcontinue [Enter]
>
> Specify a second extension line origin or ⬇ *(specify the right end of the 3.60-unit line)*
>
> Specify a second extension line origin or ⬇ *(specify the right end of the right 1.80-unit line)*
>
> Specify a second extension line origin or ⬇ *(press ESC or right-click and choose enter from the shortcut menu)*

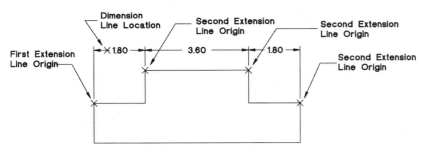

FIGURE 7–41 *Continue dimensioning of a linear object*

Quick Dimensioning

Quick Dimension is used to draw a string of dimensions between all of the end and center points of the selected object(s).

Invoke **Quick Dimension** (QDIM) from the Dimensions panel (Figure 7–42) located in the Annotate tab, and AutoCAD displays the following prompts:

> Select geometry to dimension: *(select one or more objects and then press* [Enter] *)*
>
> Specify dimension line position, or ⬇ *(specify location for dimension line or select one of the options from the shortcut menu)*

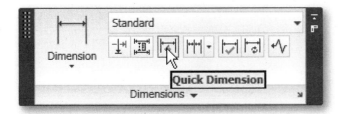

FIGURE 7–42 *Invoking the QDIM command by selecting Quick Dimension from the Dimensions panel*

If you specify a location for a dimension line, AutoCAD will draw continuous dimensioning between all endpoints or centerpoints of the objects selected horizontally or vertically, depending on where the dimension line location is specified. See Figure 7–43 for an example. You can right-click and select one of the options that will draw the type of dimension chosen.

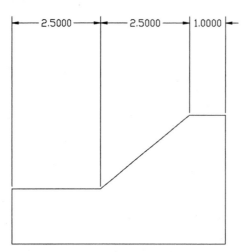

FIGURE 7–43 *An example of Quick Dimensioning*

Multileader

The Multileader, or MLEADER, command is used to connect annotation to objects or other features in the drawing. You can draw leaders, multiple leaders that connect one annotation to multiple objects, align multiple leaders, space them equally, and collect multiple leaders. You can draw the leader arrowhead first, the leader landing first, or the content first. The leader can be a line or curve, and the content can be either text or a block; blocks are discussed in Chapter 11. Leaders can be annotative; annotative objects are discussed in Chapter 8.

Invoke Multileader from the Multileaders panel (Figure 7–44) located in the Annotate tab, and AutoCAD displays the following prompts:

```
Specify leader arrowhead location or [leader Landing first/
Content first/Options] <Options>: (specify location for
arrowhead or right-click and choose an option from the
shortcut menu)

Specify leader landing location: (specify another point for
the leader landing)
```

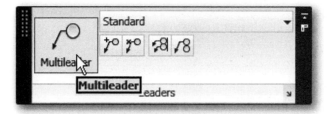

FIGURE 7–44 *Invoking Multileader from the Multileaders panel*

After specifying the arrow location and leader landing location, the Text formatting dialog box is displayed for you to enter the content. If you choose Landing first from the shortcut menu, you can specify the landing, then the arrowhead, and then the content. If you choose Content first, you specify a location for the text and enter the text. AutoCAD uses the location of the text to specify the location of the landing. Then you specify the arrowhead location.

Figure 7–45 shows a typical leader with the content "SEE DETAIL A."

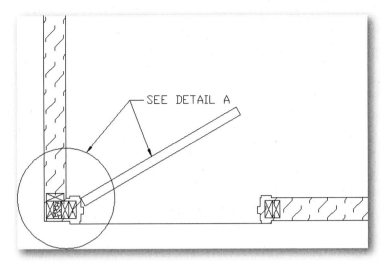

FIGURE 7–45 *Typical leader*

Options to the Multileader command include *Leader Type, Leader Landing, Content Type, Maxpoints, First Angle,* and *Second Angle.*

Choosing *Leader Type* allows you to select Straight, Spline, or None.

Choosing *Leader Landing* changes the distance of the horizontal landing line and the available options include *Yes* or *No.* If you choose *No* at this point, no landing line is associated with the multileader object.

Choosing *Content Type* allows you to select *Block, Mtext,* or *None.*

> Annotative blocks cannot be used as either content or arrowheads in multileader objects. Blocks are discussed in Chapter 11.

NOTE

Choosing *Maxpoints* allows you to specify the maximum allowable points used from the arrowhead to the landing in a multileader. The number of points will be one more than the number of lines. For example, selecting 5 points will allow a maximum of four lines.

Choosing *First Angle* allows you to specify a constraining angle for the first leader line. The angle from the arrowhead to the landing of a single leader line multileader will be constrained to a multiple of this angle. The angle from the arrowhead to the second point of a multiple leader line multileader will be constrained to a multiple of this angle.

Choosing *Second Angle* allows you to specify a constraining angle of the second leader line. The angle from the second point to the landing of a two-leader line multileader will be constrained to a multiple of this angle.

Adding a Leader Line

You can add leader line and arrow combinations to an existing multileader. Invoke Add Leader from the Multileaders panel (Figure 7–46) located in the Annotate tab, and AutoCAD displays the following prompts:

```
Select a multileader: (select an existing multileader)
Specify leader arrowhead location: (select a location for
the arrowhead)
```

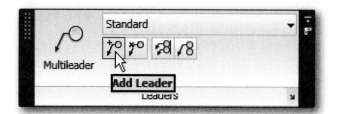

FIGURE 7–46 *Invoking Add Leader from the Multileaders panel*

After selecting the multileader, select a location for the arrowhead (Figure 7–47). AutoCAD adds the leader line and continues to prompt for additional leaders to add to the multileader until you press Enter to terminate the command.

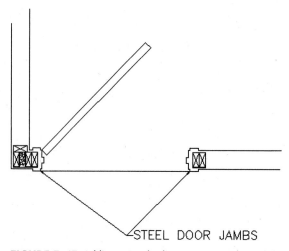

STEEL DOOR JAMBS

FIGURE 7–47 *Adding a Leader line to an existing multileader*

Removing a Leader Line

You can delete one of the leader lines of a multileader by choosing the Remove Leader command from the Multileaders panel located in the Annotate tab. Auto-CAD displays the following prompts:

> Select a multileader: *(select an existing multileader)*
>
> Specify leaders to remove: *(select leader lines for removal)*

Aligning Multileaders

If you have drawn several multileaders in an area and wish to align them, you can invoke the Align Multileaders command from the Multileaders panel located in the Annotate tab. AutoCAD displays the following prompt:

> Select multileaders: *(select an existing multileader)*

AutoCAD continues to prompt until you press enter to terminate the selecting process. AutoCAD displays the following prompt:

> Select multileader to align to or [Options]: *(select multileader to which subsequent leader contents will be aligned)*
>
> Specify direction: *(specify second point to establish alignment)*

Figure 7–48 shows three multileaders with randomly placed contents. After invoking the Align Multileader command, the three multileaders are selected for alignment.

The "JAMB" multileader is then selected for basing the alignment for the other two. Figure 7–49 shows the three multileaders with their content aligned.

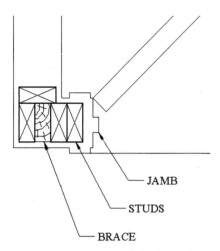

FIGURE 7–48 *Three multileaders to be aligned*

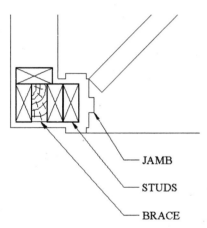

FIGURE 7–49 *Three multileaders aligned*

Options for the Align Multileader include *Distribute, Make Leader Segments Parallel, Specify Spacing,* and *Use Current.* Choosing *Distribute* spaces content evenly between two selected points. Choosing *Make Leader Segments Parallel* places content so that each of the last leader segments in the selected multileaders is parallel. Choosing *Specify Spacing* specifies spacing between the extents of the content of selected multileaders. Choosing *Use Current* uses the current spacing between multileader content.

Collecting Multileaders

Disconnected multileaders with blocks as content can be collected with a single line/ arrow. Invoke the Collect Multileaders command from the Multileaders panel located in the Annotate tab, and AutoCAD displays the following prompt:

```
Select multileaders: (select multileaders whose contents
will be aligned)
```

AutoCAD continues to prompt until you press enter to terminate the selecting process:

```
Specify collected multileader location or [Vertical/
Horizontal/Wrap]
```

```
<Horizontal>: (specify location for collecting the
collected content)
```

Figure 7–50 shows three multileaders with blocks as content. After invoking the Collect Multileader command, the contents of the three multileaders are collected, and the location is specified. Figure 7–51 shows the three multileaders with their content collected.

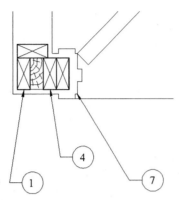

FIGURE 7–50 *Three multileaders with blocks as content*

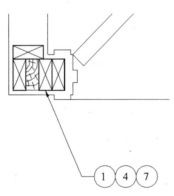

FIGURE 7–51 *Collected multileaders*

Options for the Collected Multileader include *Vertical, Horizontal,* and *Wrap.* Choosing *Vertical* places the multileader collection in a vertical orientation. Choosing *Horizontal* places the multileader collection in a horizontal orientation. Choosing *Wrap* specifies a width for a wrapped multileader collection. AutoCAD prompts for a number, which specifies a maximum number of blocks per row in the multileader collection.

Multileader Styles

The appearance of a leader can be modified and saved in a multileader style. You can use the Standard (default) style, or create new multileader styles. With a multileader style, you can specify formatting for landing lines, leader lines, arrowheads, and content. The Standard multileader style uses a straight leader line with a closed, filled arrowhead and multiline text content. You can change it to a curved line with an open arrowhead. You can also use a block for the content and then save this configuration for repeated use. It can be set as the current multileader style to be used when the MLEADER command is invoked.

Invoke the **Multileader Style** (MLEADERSTYLE) command by choosing the down arrow from the Multileaders panel (Figure 7–52) located in the Annotate tab, and AutoCAD displays the Multileader Style Manager dialog box, as shown in Figure 7–53.

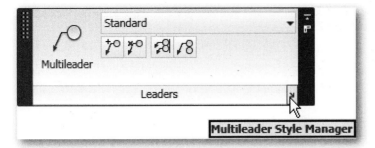

FIGURE 7–52 *Invoking the Multileader Style command from the Multileaders panel*

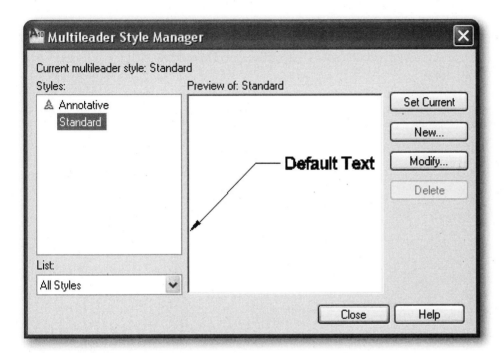

FIGURE 7–53 *Multileader Style Manager dialog box*

Styles displays a list of multileader styles with the current style highlighted. From the **List** text box, you can control the content of the Styles list. Choose **All Styles** to display all available multileader styles in the drawing. Choose **Styles in Use** to display only the multileader styles that are referenced by multileaders in the current drawing. The **Preview of** window displays a preview image of the style that is selected in the Styles list. Choose **Set Current** to set the multileader style selected in the Styles list as the current style. All new multileaders are created using this multileader style.

Choose **New,** and AutoCAD displays the Create New Multileader Style dialog box, where you can define new multileader styles (Figure 7–54). In the **New style name** text box, you can name the style. The default is "Copy of Standard." In the **Start with** text box, choose the style to use as a template for the new style. Select **Annotative** to make the multileaders that will be created with this style annotative.

FIGURE 7–54 *Create New Multileader Style dialog box*

Choose **Continue** in the Create New Multileader Style dialog box, and AutoCAD displays the Modify Multileader Style dialog box (Figure 7–55). This dialog box is also displayed when you choose **Modify** from the Multileader Style Manager dialog box. Here you can modify multileader styles. Choosing **Delete** deletes the multileader style selected in the Styles list. A style that is being used in the drawing cannot be deleted.

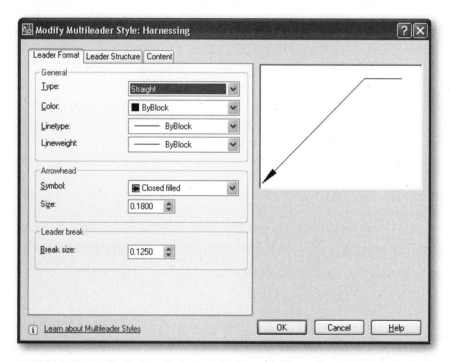

FIGURE 7–55 *Modify Multileader Style dialog box with the Leader Format tab displayed*

The **General** section (Figure 7–55) controls the general appearance of the multileader. The **Type** option determines the type of leader line; you can choose from straight leader, spline, or no leader. The Color, Linetype, and Lineweight features determine the color, linetype, and lineweight of the leader.

The **Arrowhead** section controls the appearance of the multileader arrowheads. Symbol sets the arrowhead symbol for the multileader, and Size displays and sets the size of arrowheads.

The **Leader break** section controls the settings used when adding a dimension break to a multileader. The Break size displays and sets the break size used for the DIMBREAK command when the multileader is selected.

The preview window to the right of the tabs shows how the leader will appear with the current formatting.

On the Leader Structure tab (Figure 7–56), you can specify physical constraints and dimensions for the leader and landing. The options in the Constraints section are **Maximum leader points, First segment angle,** and **Second segment angle.** A detailed explanation of these features is provided earlier in this chapter.

The options in the Landing settings section are **Automatically include landing,** which attaches a horizontal landing line to the multileader content, and **Set landing distance,** which determines the fixed distance for the multileader landing line.

In the Scale section, you can control how the leader is scaled. Options include **Annotative,** which specifies that the multileader is annotative. If the multileader is not annotative, options available include **Scale Multileaders to Layout,** which determines a scaling factor for the multileader based on the scaling in the model space and paper space viewports, and **Specify Scale,** which allows you to specify the scale for the multileader.

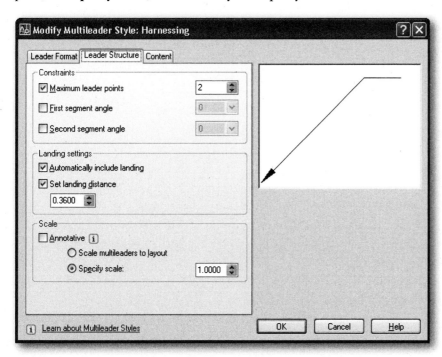

FIGURE 7–56 *Leader Structure tab of the Modify Multileader Style dialog box*

On the Content tab (Figure 7–57), you can specify the format for the content. The **Multileader type** text box allows you to choose **Mtext, Block,** or **None.**

The options in the Text options section are **Default text, Text angle, Text color, Text height, Always left justify,** and **Frame text.**

Choosing **Default Text** sets the default text for the multileader content. Choose the [...] button, and AutoCAD displays the text Editor.

Choosing **Text Style** allows you to specify a text style for attribute text.

Choosing **Text Angle** allows you to specify the rotation angle of the multileader text.

Choosing **Text Color** allows you to specify the color of the multileader text.

Choosing **Text Height** allows you to specify the height of the multileader text.

Choosing **Always Left Justify** causes the multileader text to always be left-justified.

Choosing **Frame Text Check Box** causes a text box to frame the multileader text.

The options in the Leader connection section control the leader connection settings of the multileader. The **Horizontal attachment** selection inserts the leader to the left or right of the text content. The **Vertical attachment** selection inserts the leader at the top or bottom of the text content. Choosing **Left Attachment** allows you to control the attachment of the landing line to the multileader text when the text is to the left of the leader. Choosing **Right Attachment** allows you to control the attachment of the landing line to the multileader text when the text is to the right of the leader. Choosing **Landing Gap** allows you to specify the distance between the landing line and the multileader text.

After making the necessary changes to the Multileader style, choose OK to close the dialog box and save the changes.

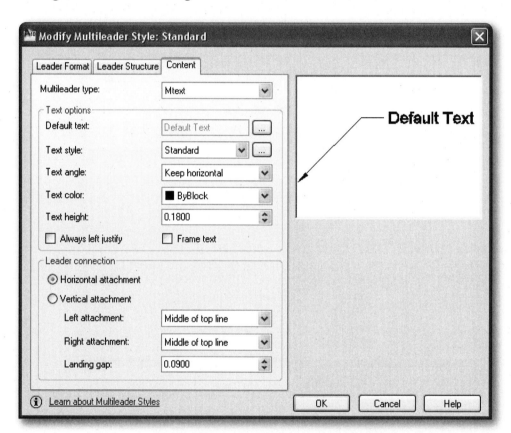

FIGURE 7–57 *Content tab of the Modify Multileader Style: Standard dialog box*

Tolerances

Tolerance symbols and text can be included with the dimensions that you draw in AutoCAD. AutoCAD provides a special set of subcommands for the two major methods of specifying tolerances. One set is for lateral tolerances; the other is for geometric tolerances.

Lateral tolerances are the traditional tolerances. Though they are often easier to apply, they do not always satisfy tolerancing in all directions, circularly and cylindrically, and they are more subject to misinterpretations, especially in the international community.

Geometric tolerance values are the maximum allowable distances that a form or its position may vary from the stated dimension(s).

NOTE

> Lateral tolerance symbols and text are accessible from the Dimension Style dialog box. Geometric tolerance symbols and text are accessible from the Dimensions panel located in the Annotate tab.

Lateral Tolerance

Lateral tolerancing draws the traditional symbols and text for Limit, Plus or Minus, or unilateral and bilateral, Single Limit, and Angular tolerance dimensioning. Lateral tolerances will appear with the dimension text in accordance with how they are set in the **Tolerances** tab of the Modify Dimension Style dialog box.

Lateral tolerance is the range from the smallest to the greatest that a dimension is allowed to deviate and still be acceptable. For example, if a dimension is called out as 2.50 ± 0.05, the tolerance is 0.1, and the feature being dimensioned may be anywhere between 2.45 and 2.55. This is the symmetrical plus-or-minus convention. If the dimension is called out as $2.50^{+0.10}_{-0.00}$, the feature may be 0.1 greater than 2.50, but it may not be smaller than 2.50. This is referred to as *unilateral*. The Limits tolerance dimension may also be shown as $\frac{2.55}{2.45}$.

To set the tolerance method, first open the Dimension Style Manager dialog box by typing `DIMSTYLE`. For a discussion of the other features of the Dimension Style manager, see the section on "Dimension Style Manager Dialog Box." Choose **Modify**, and AutoCAD displays the Modify Dimension Style dialog box. Select the **Tolerances** tab and select the **Tolerance** method from the **Method** option menu. Figure 7–58 shows the options available in the **Tolerance Format** section and the

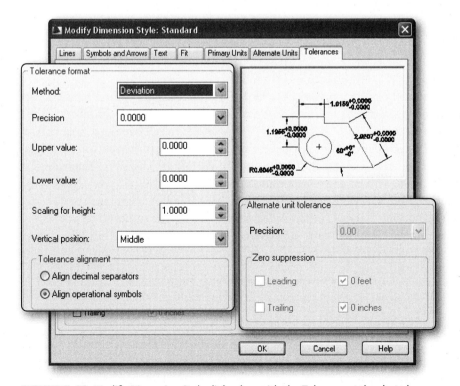

FIGURE 7–58 *Modify Dimension Style dialog box with the Tolerances tab selected*

Alternate Unit Tolerance section of the **Tolerances** tab. Selecting a method other than **None** in the **Method** text box applies only to lateral tolerancing.

Choosing the **Method** list box in the **Tolerance Format** section of the Tolerances tab displays the following options: **None, Symmetrical, Deviation, Limits,** and **Basic.** See the following table for examples of the various options.

EXAMPLES OF LATERAL TOLERANCE METHODS

Tolerance method	Description	Example
Symmetrical	Only the Upper Value box is usable. Only one value is required.	1.00 ± 0.05
Deviation	Both Upper Value and Lower Value boxes are active. A 0.00 value may be entered in either box, indicating that the variation is allowed in only one direction.	$1.00^{+0.03}_{-0.00}$
Limits	Both Upper and Lower Value boxes are active. In this case, there is no base dimension. The example at the right is using equal values of 0.07 for both Upper and Lower Values. Using an Upper value of 0.0500 and a Lower value of 0.0250 will cause annotation of a 1.000 basic dimension to be $^{1.0500}_{0.9750}$.	1.070 0.930
Basic	The dimension value is drawn in a box, indicating that it is the base value from which a general tolerance is allowed. The general tolerance is usually given in other notes or specifications on the drawing or other documents.	100

NOTE   To change how your tolerance will appear in relation to the dimension, the format is set in the *Primary Units* tab of the Modify Dimension Style dialog box.

In the **Tolerance Format** and **Alternate Unit Tolerance** sections, the **Precision** text box allows you to determine how many decimals will be shown in decimal units. This value is recorded in DIMDEC. The **Upper Value** and **Lower Value** text boxes allow you to preset tolerance values. The **Scaling for height** text box allows you to set the height of the tolerance value text. The **Vertical position** selection box determines if the tolerance value will be at the top, middle, or bottom of the text space. Choosing **Align Decimal Separators** causes the values to be stacked by aligning their decimal separators. Choosing **Align Operational Symbols** causes the values to be stacked by aligning their operational symbols. The **Zero Suppression** section controls suppression of leading and trailing zeros. See the following explanation.

In the **Alternate Unit Tolerance** section, the **Zero suppression** subsection allows you to specify whether or not zeros are displayed, as follows:

Selecting the **Leading** check box causes zeros ahead of the decimal point to be suppressed. For example, .700 is displayed instead of 0.700.

Selecting the **Trailing** check box causes zeros behind the decimal point to be suppressed. For example, 7 or 7.25 is displayed instead of 7.000 or 7.250, respectively.

Selecting the **0 Feet** check box causes zeros representing feet to be suppressed if the dimension text represents inches and/or fractions only. For example, 7 or 7 1/4 is displayed instead of 0'-7 or 0'-7 1/4.

Selecting the **0 Inches** check box causes zeros representing inches to be suppressed if the dimension text represents feet only. For example, 7' is displayed instead of 7'-0.

Geometric Tolerance

Geometric tolerancing draws a Feature Control Frame for use in describing standard tolerances according to the geometric tolerance conventions. Geometric tolerancing is applied to forms, profiles, orientations, locations, and runouts. Forms include squares, polygons, planes, cylinders, and cones.

Invoke **Tolerance** (TOLERANCE) from the Dimensions panel (Figure 7–59) located in the Annotate tab, and AutoCAD displays the Geometric Tolerance dialog box (Figure 7–60).

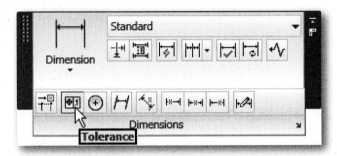

FIGURE 7–59 *Invoking the Tolerance from the Dimensions panel*

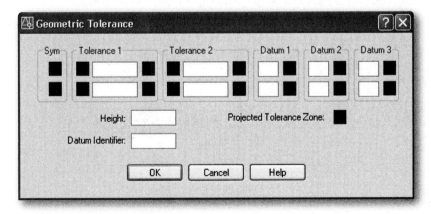

FIGURE 7–60 *Geometric Tolerance dialog box*

The conventional method of expressing a geometric tolerance for a single dimensioned feature is in a Feature Control Frame, which includes all necessary tolerance information for a particular dimension. A Feature Control Frame has the Geometric Characteristic Symbol box and a Tolerance Value box. Datum reference and/or material condition datum boxes may be added where needed. The Feature Control

Frame is shown in Figure 7–61. An explanation of the Characteristic Symbols is given in Figure 7–62. The supplementary material conditions of datum symbols are shown in Figure 7–63.

Once the symbol button located in the **Sym** column in the Geometric Tolerance dialog box is chosen, AutoCAD displays the Symbol dialog box (Figure 7–64). Select one of the available symbols, and specify appropriate tolerance values in a Tolerance field.

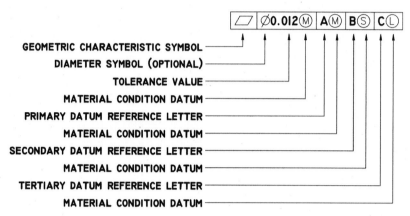

GEOMETRIC CHARACTERISTIC SYMBOL
DIAMETER SYMBOL (OPTIONAL)
TOLERANCE VALUE
MATERIAL CONDITION DATUM
PRIMARY DATUM REFERENCE LETTER
MATERIAL CONDITION DATUM
SECONDARY DATUM REFERENCE LETTER
MATERIAL CONDITION DATUM
TERTIARY DATUM REFERENCE LETTER
MATERIAL CONDITION DATUM

FIGURE 7–61 *Feature Control Frame*

Maximum Materials Condition (MMC) means that a feature contains the maximum material permitted within the tolerance dimension, such as the minimum hole size or the maximum shaft size.

Least Material Condition (LMC) means that a feature contains the least material permitted within the tolerance dimension, such as the maximum hole size or the minimum shaft size.

Regardless of Feature Size (RFS) applies to any size of the feature within its tolerance. This is more restrictive. For example, RFS does not allow the tolerance of the center-to-center dimension of a pair of pegs fitting into a pair of holes greater leeway if the peg diameters are smaller or the holes are bigger, whereas MMC does.

The diameter symbol, Ø, is used in lieu of the abbreviation DIA.

CHARACTERISTIC SYMBOLS

FORM	▱	FLATNESS
	—	STRAIGHTNESS
	○	ROUNDNESS
	⌀	CYLINDRICITY
	⌒	LINE PROFILE
	⌓	SURFACE PROFILE
	∠	ANGULARITY
	//	PARALLELISM
	⟂	PERPENDICULARITY
LOCATION	◎	CONCENTRICITY
	⊕	POSITION
	⌸	SYMMETRY
RUNOUT	↗	CIRCULAR RUNOUT
	↗↗	TOTAL RUNOUT

FIGURE 7–62 *Geometric characteristics symbols*

MATERIAL CONDITIONS OF DATUM SYMBOLS

Ⓜ	MAXIMUM MATERIAL CONDITIONS (MMC)
Ⓢ	REGARDLESS OF FEATURE SIZE
∅	DIAMETER

FIGURE 7–63 *Material conditions of datum symbols*

FIGURE 7–64 *Symbol dialog box*

Geometric tolerancing is becoming widely accepted. It is highly recommended that you study the latest drafting texts concerning the significance of the various symbols, so that you will be able to apply geometric tolerancing properly.

Drawing Cross Marks for Arcs or Circles

The DIMCENTER command is used to draw the cross marks that indicate the center of an arc or circle (Figure 7–65).

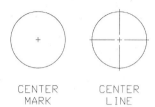

CENTER   CENTER
MARK   LINE

FIGURE 7–65 *Circles with center cross marks*

Invoke **Center Mark** (DIMCENTER) from the Dimensions panel (Figure 7–66) located in the Annotate tab, and AutoCAD displays the following prompt:

    Select arc or circle:

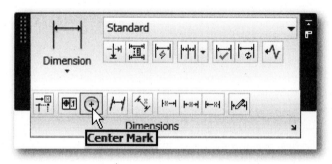

FIGURE 7–66 *Invoking Center Mark from the Dimensions panel*

Select an arc or circle, and AutoCAD draws the cross marks in accordance with the setting of the DIMCEN dimensioning variable.

Oblique Dimensioning

The OBLIQUE command allows you to slant the extension lines of a linear dimension to a specified angle. The dimension line will follow the extension lines, retaining its original direction. This feature is useful for ensuring that the dimension stays clear of other dimensions or objects in your drawing. It is also a conventional method of dimensioning isometric drawings.

Invoke the OBLIQUE command from the Dimensions panel (Figure 7–67) located in the Annotate tab, and AutoCAD displays the following prompt:

    Select objects: *(select the dimension(s) for obliquing,*
    press Enter *)*
    Enter obliquing angle (press Enter for none): *(specify an*
    angle or press Enter *)*

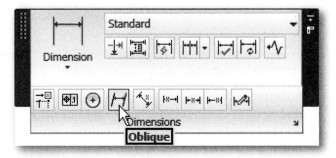

FIGURE 7–67 *Invoking Oblique from the Dimensions panel*

The OBLIQUE command is actually the DIMEDIT command in which AutoCAD automatically chooses the OBLIQUE option for you. All you have to do is select a dimension.

The extension lines of the selected dimension(s) will be slanted at the specified angle.

Dimension Jogline

The DIMJOGGED command creates jogged dimensions for circles and arcs. It measures the radius of the selected object and displays the dimension text with a radius symbol in front of it. The origin point of the dimension line can be specified at any convenient location.

Invoke the DIMJOGGED command from the Dimensions panel located in the Annotate tab, and AutoCAD displays the following prompt:

> Select arc or circle: *(select an arc, polyline arc, or circle)*
>
> Specify center location override: *(specify a point for new center point for a jogged radius dimension)*
>
> Specify dimension line location or ⊡ *(specify dimension location)*

The *Mtext, Text,* and *Angle* options available from the shortcut menu are the same as those in linear dimensioning, explained earlier in this section.

> Specify the jog location: *(specify the middle point of the jog)*

The transverse angle of the jog is determined by the Dimension Style Manager.

EDITING DIMENSIONS AND DIMENSION TEXT

AutoCAD allows you to edit dimensions with modification commands and grip editing options. AutoCAD also provides two additional Modify commands specifically designed to work on dimension text objects: DIMEDIT and DIMTEDIT.

DIMEDIT Command

The options available with the DIMEDIT command allow you to replace the dimension text with new text, rotate the existing text, move the text to a new location, and, if necessary, restore the text to its home position, which is the position defined by the current style. In addition, these options allow you to change the angle of the extension lines, normally perpendicular, relative to the direction of the dimension line by means of the *Oblique* option.

Invoke Dimension Edit (DIMEDIT) from the on-screen prompt, and AutoCAD displays the following prompt:

Enter type of dimension editing (press [Enter] for the HOME option, or select one of the options)

Dimension Editing Options

Choosing *Home* returns the dimension text to its default position, and AutoCAD displays the following prompt:

Select objects: (select the dimension objects and press [Enter])

Choosing *New* allows you to change the original dimension text to the new text. AutoCAD opens the MTEXT editor. Enter the new text, choose **OK,** and AutoCAD displays the following prompt:

Select objects: (select the dimension objects for which the existing text will be replaced by the new text)

Choosing *Rotate* allows you to change the angle of the dimension text. AutoCAD displays the following prompts:

Specify angle for dimension text: (specify the rotation angle for text)

Select objects: (select the dimension objects for which the dimension text has to be rotated)

Choosing *Oblique* adjusts the obliquing angle of the extension lines for linear dimensions. This is useful to keep the dimension parts from interfering with other objects in the drawing. It is also an easy method by which to generate the slanted dimensions used in isometric drawings. AutoCAD displays the following prompts:

Select objects: (select the dimension objects, and press [Enter])

Enter obliquing angle (press [Enter] for none): (specify the angle, or press [Enter])

DIMTEDIT Command

The DIMTEDIT command is used to change the location of dimension text, with the LEFT, RIGHT, CENTER, and HOME options, along the dimension line and its angle, with the ROTATE option.

Invoke Dimension Text Edit (DIMTEDIT) command from the on-screen prompt, and AutoCAD displays the following prompt:

Select dimension: (select the dimension object to modify)

A preview image of the dimension selected is displayed near the cursor. You will be prompted:

Specify new location for dimension text or ⬇ (specify a new location for the dimension text, or right-click for the shortcut menu and select one of the available options)

By default, AutoCAD allows you to position the dimension text with the cursor, and the dimension updates dynamically as it is moved.

Dimension Text Editing Options

Choosing *Center* will cause the text to be drawn at the center of the dimension line.

Choosing *Left* will cause the text to be drawn toward the left extension line.

Choosing *Right* will cause the text to be drawn toward the right extension line.

Choosing *Home* returns the dimension text to its default position.

Choosing *Angle* changes the angle of the dimension text. AutoCAD displays the following prompt:

> Specify angle for dimension text: *(specify the angle)*

The angle specified becomes the new angle for the dimension text.

Editing Dimensions with Grips

If Grips are enabled, you can select an associative dimension object, and its grips will be displayed at strategic points. The grips will be located at the object ends of the extension lines, at the intersections of the dimension and extension lines, and at the insertion point of the dimension text. In addition to normal grip editing of the dimension objects as a group, using ROTATE, MOVE, COPY, and so on, each grip can be selected for editing the dimension. Moving the object end grip of an extension line will move that specified point, making the value change accordingly. Horizontal and vertical dimensions remain horizontal and vertical. Aligned dimensions follow the alignment of the relocated point. Moving the grip at the intersection of the dimension line and one of the extension lines causes the dimension line to be nearer to or farther from the object dimensioned. Moving the grip at the insertion point of the text does the same as the intersection grip and also permits you to move the text back and forth along the dimension line.

DIMDISASSOCIATE Command

The DIMDISASSOCIATE command removes associativity from selected associative dimensions. Invoke the DIMDISASSOCIATE command by typing at the on-screen prompt. AutoCAD displays the following prompt:

> Select objects: *(select the dimension objects, and press* Enter *to complete the selection)*

AutoCAD removes the associativity of the selected dimensions.

DIMREASSOCIATE Command

The DIMREASSOCIATE command associates selected dimensions to geometric objects. When prompted, select the dimension to be reassociated to an object and then step through the point selection process, using Object Snap mode, when each extension line point is marked with an X. If an X appears in a box, then that point is already associated with a point on an object. You may skip this point by pressing Enter. Invoke the DIMREASSOCIATE command from the Dimensions panel (Figure 7–68) located in the Annotate tab, and AutoCAD displays the following prompt:

> Select objects: (select the dimension objects, and press Enter to complete the selection)

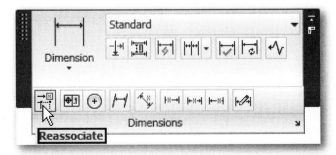

FIGURE 7–68 *Invoking* REASSOCIATE *command from the Dimensions panel*

Follow through the prompts by selecting the points to reassociate the dimensions.

DIMREGEN **Command**

The DIMREGEN command updates the locations of all associative dimensions. This action is sometimes necessary after panning or zooming with a wheel mouse, after opening a drawing that was modified in an earlier version of AutoCAD, or after opening a drawing that contains external references (xrefs) that have been modified. Invoke the DIMREGEN command by typing at the on-screen prompt.

Editing Dimensions Using the Shortcut Menu

Dimensions already drawn can be easily edited. Select them when the Noun/Verb selection, or the PICKFIRST system variable, is enabled, and choose the appropriate available options from the shortcut menu (Figure 7–69).

The *Dim Text Position* flyout includes options for *Above dim line, Centered, Home text, Move text alone, Move with leader,* and *Move with dim line.* Choosing one of these options causes the text in the selected dimension to conform to the position option chosen.

The *Flip Arrow* (AIDIMFLIPARROW) command allows you to reverse the position of a selected arrow relative to the extension line.

The *Precision* flyout includes a range of 0.0 to 0.000000 for Decimal dimensions and a range of 0'-0 to 0'-0 1/128 for Architectural and Fractional dimensions. Choosing one of the decimal or fractional options causes the distances or angles in the selected dimension to conform to the level of precision chosen.

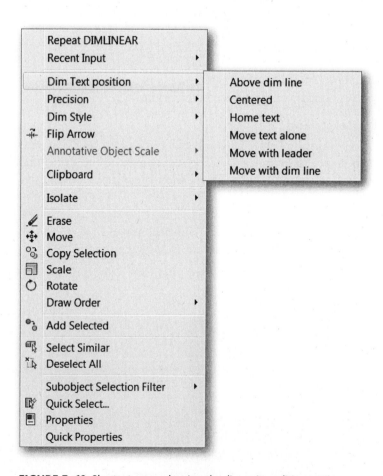

FIGURE 7–69 *Shortcut menu showing the dimension editing section*

> If the dimensions selected include styles with both decimal and fractional units, the *Precision* option is not available in the shortcut menu.

The *Dim Style* flyout includes an option for *Save as New Style*, and a list of available dimension styles (dimstyles) from which to choose. If you have made changes to settings of individual variables in a particular dimension, you can choose *Save as New Style* to create and name a new, unique dimstyle conforming to the selected dimension's variable settings.

DIMENSION STYLES

Each time a dimension is drawn, it conforms to the settings of the dimensioning variables in effect at the time. The entire set of dimension variables settings can be saved in their respective states as a dimension style. Named dimstyles can be recalled for application to a dimension later in the drawing session or in a subsequent session. Some dimensioning system variables affect every dimension. For example, every time a dimension is drawn, the DIMSCALE setting determines the relative size of the dimension. DIMDLI, the variable that determines the offsets for baseline dimensions, comes into effect only when a baseline dimension is drawn. However, when a dimension style is created and named, all dimensioning system variable settings are recorded in that dimension style. The associativity of dimensions is not controlled by dimstyles. The DIMASSOC dimensioning variable settings are saved separately from the dimension styles. As stated at the beginning of this chapter, "Dimension variables change when their associated settings are changed in the Modify Dimension Style dialog box. For example, the value of the DIMEXO, the extension line offset, dimensioning system variable is established by the number in the **Offset from origin** text box in the Modify Dimension Style dialog box.

Dimension Style Manager Dialog Box

AutoCAD provides a comprehensive set of dialog boxes, accessible through the Dimension Style Manager dialog box, for creating new dimension styles and managing existing ones. In turn, these dialog boxes compile and store dimensioning system variable settings. Creating dimension styles through use of the DIMSTYLE command's dialog boxes allows you to make the desired changes to the appearance of dimensions without having to search for or memorize the names of the dimensioning system variables in order to change the settings directly.

Invoke **Dimension Style** (DIMSTYLE) command by choosing the down arrow from the Dimensions panel (Figure 7–70) located in the Annotate tab, and AutoCAD displays the Dimension Style Manager, as shown in Figure 7–71.

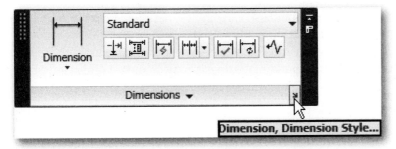

FIGURE 7–70 *Invoking the* DIMSTYLE *command by selecting Dimension Style from the Dimensions panel*

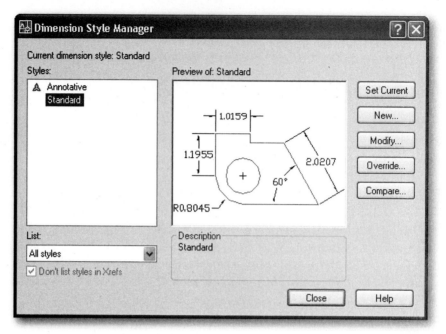

FIGURE 7–71 *Dimension Style Manager dialog box*

The **Current Dimstyle** label shows the name of the dimension style to which the next drawn dimension will conform. The name of the current dimstyle is recorded as the value of the DIMSTYLE dimensioning system variable.

The **Styles** list box displays the name(s) of the current style and any other style(s) available depending on the option chosen in the **List** selection box. If you choose one of the styles in the list, it is highlighted and will be the dimstyle acted upon when one of the buttons—**Set Current, New, Modify, Override,** or **Compare**—is chosen. It will also be the dimstyle whose appearance is shown in the **Preview of** viewing area.

The **List** text box allows you to list all of the styles available in the drawing displayed or only those in use by choosing the appropriate option from this selection box. The Standard style cannot be deleted and is always available.

The area below the **Preview of** heading shows how dimensions will appear when drawn using the current dimstyle.

The **Description** section shows the difference(s) between the dimstyle chosen in the Styles section and the current dimstyle. For example, if the current dimstyle is Standard and the dimstyle chosen in the Styles section is named Harnessing, the **Description** section might say "Standard + Angle format = 1, Fraction format = 1, Length units = 4 to indicate that the angle format is in Degrees Minutes Seconds, the fraction format is in diagonal, and the length units are architectural. These settings are different from the corresponding settings for the Standard dimstyle."

Choosing **Set Current** causes the dimstyle chosen in the **Styles** section to become the current dimstyle.

Choosing **New** causes the Create New Dimension Style dialog box to appear (Figure 7–72). The Create New Dimension Style dialog box allows you to create and name a new dimstyle.

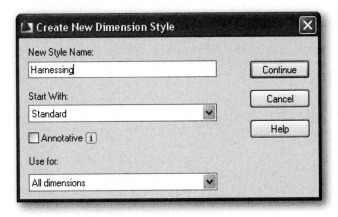

FIGURE 7–72 *Create New Dimension Style dialog box*

In the **New Style Name** text box, you can enter the name of the new dimstyle you wish to create.

The **Start With** list box allows you to choose an existing dimstyle that you would like to start with in creating your new dimstyle. The dimensioning system variables will be the same as those in the **Start With** list box until you change their settings. In many cases, you may wish to change only a few dimensioning system variable settings.

The **Annotative** check box specifies that the dimension style is annotative. For detailed explanation on creating annotative objects, refer to Chapter 8.

The **Use For** list box allows you to choose the type(s) of dimensions to which the new Dimstyle you will be creating will apply.

Choosing **Continue** causes the New Dimension Style dialog box to be displayed, from which the dimension variables of the style specified in the **Start With** text box can be modified and then saved with the new name specified. The New Dimension Style dialog box is similar to the Modify Dimension Style dialog box described in the next section.

Choosing **Modify** in the Dimension Style Manager dialog box causes the Modify Dimension Style dialog box to appear (Figure 7–73). Seven tabs are available: **Lines, Symbols and Arrows, Text, Fit, Primary Units, Alternate Units,** and **Tolerances.**

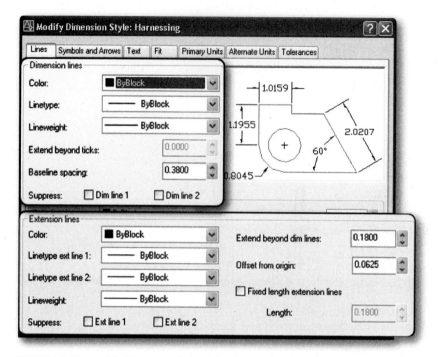

FIGURE 7–73 *Modify Dimension Style dialog box with Lines tab chosen*

The **Lines** tab allows you to modify the geometry of the various elements that comprise the dimensions, such as dimension lines and extension lines. As you change the dimensions, a preview displays, showing how the dimensions will appear when drawn using the settings.

The **Dimension Line** and **Extension Line** sections allow you to control the color, linetype, and lineweight of the dimension and extension lines.

The **Color** list boxes let you choose how the color of dimension lines and extension lines will be determined. **ByLayer** causes the color of the line to match that of its layer. **ByBlock** causes the color of the line to match that of its block reference if it is part of a block reference. You can choose one of the standard colors or another color for the line color to match. The value is recorded in the DIMCLRD system variable.

The **Linetype** list boxes let you choose how the linetype of Dimension and extension lines will be determined. **ByLayer** causes the linetype of the line to match that of its layer. **ByBlock** causes the linetype of the line to match that of its block reference if it is part of a block reference. You can also choose one of the standard linetypes or another linetype to load through the Select Linetype dialog box.

The **Lineweight** list boxes let you choose how the lineweight of Dimension and extension lines will be determined. **ByLayer** causes the lineweight of the line to match that of its layer. **ByBlock** causes the lineweight of the line to match that of its block reference if it is part of a block reference. You can also choose one of the standard lineweights or enter a value. The value is recorded in the DIMLWD system variable.

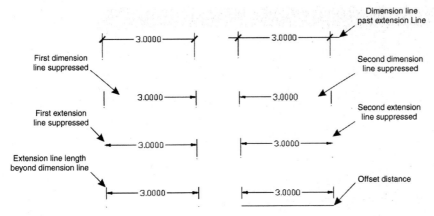

FIGURE 7–74 *Various examples with Dimension Line and Extension Line settings*

In the **Dimension Line** section only:

The **Extend beyond ticks** text box allows you to enter the distance that dimension lines will be drawn past extension lines (Figure 7–74) if the arrowhead type is a tick, that is, a small diagonal line. The value is recorded in the DIMDLE system variable.

The **Baseline spacing** text box allows you to enter the distance that AutoCAD uses between dimension lines when they are drawn using the BASELINE DIMENSION command. The value is recorded in the DIMDLI system variable.

Selecting **Suppress** let you determine if one or more of the dimension lines is suppressed when the dimension is drawn (Figure 7–74). The value is recorded in the DIMSD1 and DIMSD2 system variables.

In the **Extension Line** section only:

The **Extend beyond dim lines** text box allows you to enter the distance that extension lines will be drawn past dimension lines (Figure 7–74).

The **Offset from origin** text box allows you to enter the distance that AutoCAD uses between the extension line and the origin point specified when drawing an extension line (Figure 7–74). The value is recorded in the DIMEXO system variable. Selecting **Fixed length extension lines** causes dimension extension lines to be drawn to the length specified in the **Length** text box.

Selecting **Suppress** allows you to determine if one or more of the extension lines is suppressed when the dimension is drawn (Figure 7–74). The value is recorded in the DIMSE1 and DIMSE2 system variables.

The **Symbols and Arrows** tab allows you to specify the appearance, location, and display of symbols and arrowheads (Figure 7–75).

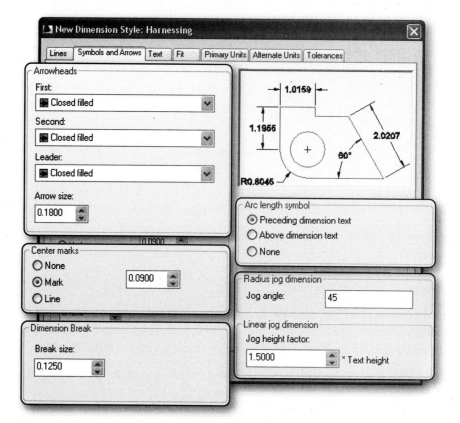

FIGURE 7–75 *Modify Dimension Style dialog box with the Symbols and Arrows tab chosen*

The **Arrowheads** section controls the appearance of the arrowheads.

The **First, Second,** and **Leader** list boxes allow you to determine whether and how arrowheads are drawn at the termini of the extension lines or the start point of a Leader. Unless you specify a different type for the second line terminus, it will be the same type as the first line terminus. The types of arrowheads are recorded in DIMBLK, if both the first and second are the same, or in DIMBLK1 and DIMBLK2, if they are different.

The **Arrow size** text box allows you to enter the distance that AutoCAD uses for the length of an arrowhead when drawn in a dimension or leader. The value is recorded in the DIMASZ system variable.

Some of the arrowhead types included in the standard library are **None, Closed, Dot, Closed filled, Oblique, Open, Origin Indication,** and **Right angle.** Figure 7–76 presents an example of each type.

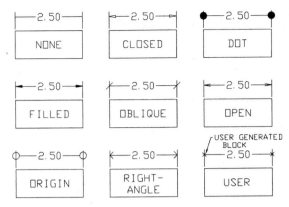

FIGURE 7–76 *Arrowhead types*

Choosing **User Arrow** allows you to use a saved block definition by entering its name. Figure 7–77 shows the Select Custom Arrow Block dialog box. The block definition should be created as though it were drawn for the right end of a horizontal dimension line, with the insertion point at the intersection of the extension line and the dimension line.

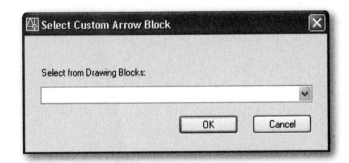

FIGURE 7–77 *Select Custom Arrow Block dialog box*

The **Center Marks** section controls the size and type of center marks drawn with the DIMCENTER command.

Selecting **None** suppresses the center mark; none will be drawn.

Selecting **Mark** causes cross marks to be drawn according the size specified in the **Size** text box.

Selecting **Line** causes cross lines to be drawn extending outside the circle in addition to the cross marks. Figure 7–78 presents an example of this function.

Center Mark set to None   Center Mark set to Mark   Center Mark set to Line

FIGURE 7–78 *Circles with various center marks*

The **Dimension Break** section controls the gap width of dimension breaks.

The **Break Size** text box sets the size of the gap used for dimension breaks.

The **Arc length symbol** section controls the appearance of the symbol used for the DIMARC command.

Selecting **Preceding dimension text** causes the arc length symbol to be drawn before the dimension text.

Selecting **Above dimension text** causes the arc length symbol to be drawn above the dimension text.

Selecting **None** suppresses the arc length symbol; none will be drawn.

The **Radius dimension jog** section controls the appearance of the line used to dimension long radius arcs whose center is located out of the drawing area.

The **Jog Angle** allows you to specify the angle that is drawn in the shortened dimension line for the long radius arc. Figure 7–79 shows an example of a line with a jog in it, representing a long radius arc being dimensioned.

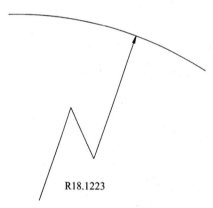

FIGURE 7–79 *A long radius arc with jog*

The **Linear jog dimension** section controls the display of the jog for linear dimensions. Jog lines are often added to linear dimensions when the actual measurement is not accurately represented by the dimension.

The **Linear jog dimension** text box sets the height of the jog, which is determined by the distance between the two vertices of the angles that make up the jog.

The **Text** tab allows you to modify the appearance, location, and alignment of dimension text that is included when a dimension is drawn. A preview displays, showing how the dimensions will appear when drawn using the settings as you change them (Figure 7–80).

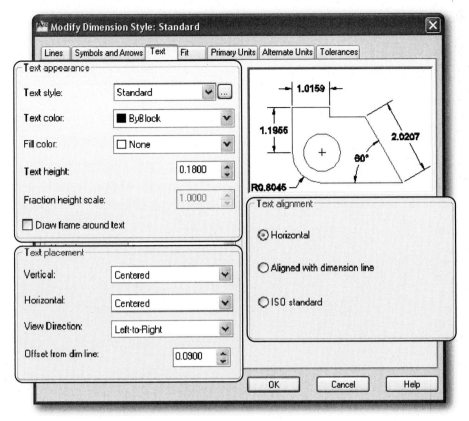

FIGURE 7–80 *Modify Dimension Style dialog box with the Text tab chosen*

The **Text Appearance** section controls the appearance of dimension text.

The **Text style** list box allows you to choose the text style to which the dimension text will conform. The value is recorded in the DIMTXSTY system variable.

> Do not confuse Text style with Dimension Style. Dimensions are drawn in accordance with the current Dimstyle, which has as part of its configuration a text style to which Dimension Text will conform.

The **Text color** list box allows you to choose how the color of dimension text will be determined. **ByLayer** causes the color of the text to match that of its layer. **ByBlock** causes the color of the text to match that of its block reference if it is part of a block reference. You can also choose one of the standard colors for the text color to match. The value is recorded in the DIMCLRT system variable.

The **Fill color** list box allows you to choose how the color for the text background in dimensions will be determined.

The **Text height** text box allows you to enter the distance that AutoCAD uses for text height when drawing dimension text as part of the dimension. The value is recorded in the DIMTXT system variable.

The **Fraction height scale** text box allows you to enter a scale that determines the distance that AutoCAD uses for text height when drawing dimension text for fractions as part of the dimension text. This scale is the ratio of the text height for normal dimension text to the height of the fraction text. The value is recorded in the DIMTFAC system variable.

Selecting **Draw frame around text** causes AutoCAD to draw the dimension text inside a rectangular frame. The value is recorded in the DIMGAP system variable as a negative value.

The **Text Placement** section controls how text is placed with relation to the dimension line.

The **Vertical** list box allows you to choose how the dimension text will be drawn in relation to the dimension line. The options include **Centered, Above, Outside,** and **JIS.** The value is recorded in the DIMTAD system variable. Figure 7–81 shows an example of each option.

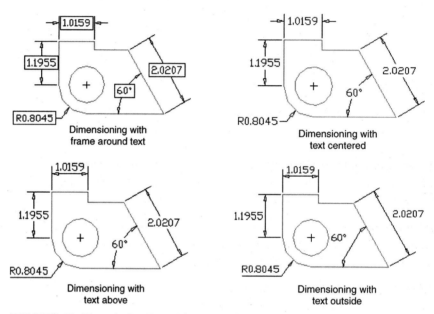

FIGURE 7–81 *Dimensioning Text with various Text Appearance selections*

The **Horizontal** list box allows you to choose how the dimension text will be drawn in relation to the extension lines. The options include **Centered, At Ext Line 1, At Ext Line 2, Over Ext Line 1,** and **Over Ext Line 2.** The value is recorded in the DIMJUST system variable. Figure 7–82 shows an example of each option.

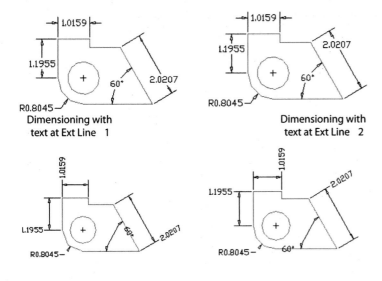

Dimensioning with
text at Ext Line   1

Dimensioning with
text at Ext Line   2

Dimensioning with
text over Ext Line   1

Dimensioning with
text over Ext Line   2

FIGURE 7–82 *Dimensioning Text with various Text Placement (horizontal) selections*

The **View Direction** list box allows you to choose the dimension text viewing direction. The Left-to-Right selection places the text to enable reading from left to right and the Right-to-Left selection places the text to enable reading from right to left.

The **Offset from dim line** text box allows you to enter the distance from the dimension line that AutoCAD uses when drawing dimension text as part of the dimension. The value is recorded in the DIMGAP system variable.

The **Text Alignment** section controls the alignment of the dimension text. The value is recorded in the DIMTIH and DIMTOH system variables.

Selecting **Horizontal** causes dimension text in nonhorizontal dimension lines to be drawn horizontally.

Selecting **Aligned with dimension line** causes dimension text in nonhorizontal dimension lines to be drawn aligned with the dimension line.

Selecting **ISO standard** causes dimension text to comply with ISO standards.

The **Fit** tab allows you to determine the arrangement of the various elements of dimensions. A preview displays, showing how dimensions will appear using the settings as you change them (Figure 7–83).

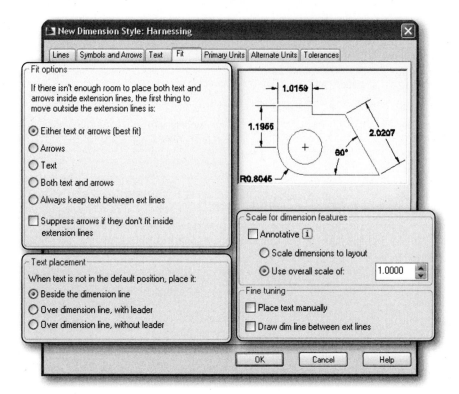

FIGURE 7–83 *Modify Dimension Style dialog box with the Fit tab chosen*

The **Fit Options** section allows you to choose which of the text or arrowheads will be drawn between the extension lines. The value is recorded in the DIMATFIT, DIMTIX, and DIMSOXD system variables.

Selecting **Either text or the arrows (best fit)** causes the text to be drawn outside and the arrows inside the extension lines if there is room only for the arrows.

Selecting **Arrows** causes the text to be drawn outside and the text inside the extension lines if there is room for the arrows only.

Selecting **Text** causes the arrows to be drawn outside and the text inside the extension lines if there is room only for the text.

Selecting **Both text and arrows** causes both the text and arrows to be drawn outside, if the dimension lines are forced outside (Figure 7–84), and both to be drawn inside, if space is available.

Selecting **Always keep text between ext lines** causes text to always be drawn between the extension lines (Figure 7–84). The value is recorded in the DIMTIX system variable.

Selecting **Suppress arrows if they don't fit inside the extension lines** causes AutoCAD to not draw the arrowheads between extension lines if they do not fit (Figure 7–84). The value is recorded in the DIMSOXD system variable.

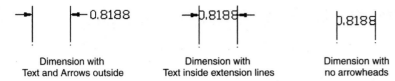

FIGURE 7–84 *Dimensioning examples with various Fit options*

The **Text Placement** section allows you to choose how the text will be placed when text is not in the default position. The value is recorded in the DIMTMOVE system variable.

Selecting **Beside the dimension line** causes text to be drawn beside the dimension lines (Figure 7–85).

Selecting **Over the dimension line, with leader** draws a leader line connecting the text to the dimension line (Figure 7–85). The leader line is omitted when text is too close to the dimension line. Selecting **Over the dimension line, without leader** draws the text without a leader line connecting the text to the dimension line (Figure 7–85).

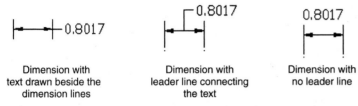

FIGURE 7–85 *Dimensioning example with various Text placement options*

The **Scale for Dimension Features** section sets the overall dimension scale value or the paper space scaling. This allows you to increase or decrease the size of all of the dimensioning components uniformly by the factor entered.

The **Annotative** check box specifies that the dimension style is annotative. For detailed explanation on creating annotative objects, refer to Chapter 8.

Selecting **Use overall scale of** activates the text box that allows you to set a scale for all dimension style settings that specify size, distance, or spacing, including text and arrowhead sizes. This scale does not change dimension measurement values. A scale factor greater than zero causes text sizes, arrowhead sizes, and other scaled distances to plot at their face values. The value is recorded in the DIMSCALE system variable. Selecting **Scale dimensions to layout (paper space)** determines a scale factor based on the scaling between the current model space viewport and paper space. When you work in paper space, but not in a model space viewport, or when TILEMODE is set to 1, the default scale factor of 1.0 is used or the DIMSCALE system variable.

The **Fine Tuning** section allows you to place dimension elements manually.

Selecting **Place text manually** allows you to dynamically specify where dimension text is placed. The value is recorded in the DIMUPT system variable.

Selecting **Draw dim line between ext lines** causes a dimension line to be drawn between the extension lines regardless of the distance between extension lines. The value is recorded in the DIMTOFL system variable.

The **Primary Units** tab allows you to determine the appearance and format of numerical values of distances and angles. A preview displays, showing how dimensions will appear when drawn using the settings as you change them (Figure 7–86).

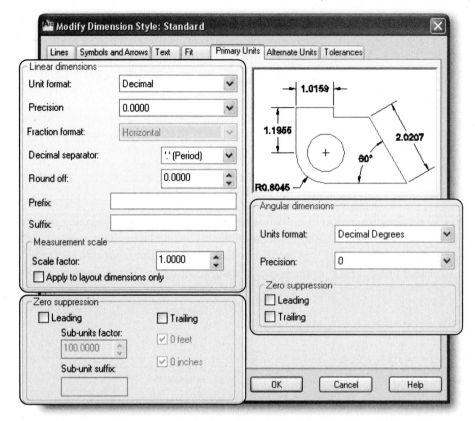

FIGURE 7–86 *Modify Dimension Style dialog box with the Primary Units tab chosen*

The **Linear Dimensions** section controls the type, format, and precision of the primary units used in the dimension text for calling out numeric values of distances and angles.

The **Unit format** list box allows you to choose the format of the units that AutoCAD uses when drawing dimension text as part of the dimension. The options include **Scientific, Decimal, Engineering, Architectural, Fractional,** and **Windows Desktop.** The value is recorded in the DIMLUNIT system variable. Figure 7–87 shows examples of these options.

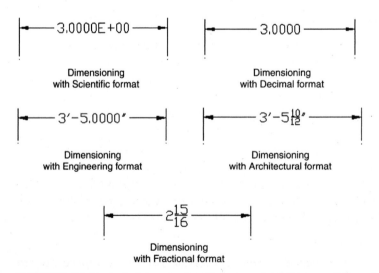

FIGURE 7–87 *Dimensioning example with various dimensioning unit formats*

The **Precision** list box allows you to determine how many decimal places the text will be shown if the format is in scientific, decimal, or Windows desktop units. It shows how many decimal places the inch measurements will have, if the format is in engineering units. It shows the precision fractions will have in architectural or fractional units. The value is recorded in the DIMDEC system variable.

The **Fractional format** text box shows whether the fractions will be shown horizontal, diagonal, or not stacked. The value is recorded in the DIMFRAC system variable.

The **Decimal separator** list box allows you to choose whether the decimal will be separated by a period, comma, or a space if the format is in scientific, decimal, or Windows desktop units. The value is recorded in the DIMDSEP system variable.

The **Round off** text box allows you to enter the value to which distances will be rounded off. For example, a value of 0.5 causes dimensions to be rounded to the nearest 0.5 units.

The **Prefix** text box allows you to include a prefix in the dimension text (Figure 7–88). The prefix text will override any default prefixes, such as those used in radius (R) dimensioning. The value is recorded in the DIMPOST system variable.

The **Suffix** text box allows you to include a suffix in the dimension text (Figure 7–88). If you specify tolerances, AutoCAD includes the suffix in the tolerances, as well as in the main dimension.

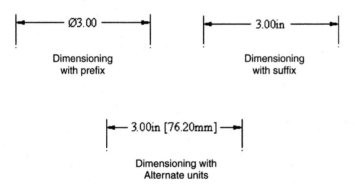

FIGURE 7–88 *Dimensioning example with prefix, suffix, and alternate units*

The **Measurement Scale** section controls the scale of the specified dimension values.

The **Scale factor** text box allows you to enter the number AutoCAD uses as a ratio of the true dimension distances, or the linear scale factor for linear measured distances of a dimension without affecting the components, angles, or tolerance values. For example, say that you are drawing with the intention of plotting at quarter-size scale, or 3 = 1'-0. You scale a detail by a factor of 4 so that it will plot to full scale. If you wish to dimension it after it has been enlarged, you can set the measurement scale factor in this section of the Primary Units dialog box to .25 so that dimensioned distances will represent the dimensions of the object before it was scaled up. This method keeps the components at the same size as dimensions created without a scale change. The value is recorded in DIMLFAC. Adjusting the scale factor is useful when you wish to create different parts of the drawing at different scales but have the dimension elements uniform in size.

Selecting **Apply to layout dimensions only** causes the ratio to be applied only to layout dimensions.

The **Zero Suppression** section controls whether or not leading and trailing zeros are displayed. The value is recorded in the DIMZIN system variable.

Selecting **Leading** causes zeros ahead of the decimal point to be suppressed. For example, .700 is displayed instead of 0.700.

The **Sub-units factor** sets the number of sub-units to a unit. It is used to calculate the dimension distance in a sub-unit when the distance is less than one unit. For example, specify **100** if the suffix is *m* and the sub-unit suffix is to display in *cm*.

The **Sub-unit suffix** specifies a suffix to the dimension value sub-unit. You can enter text or use control codes to display special symbols. For example, enter **cm** for .96 m to display as 96 cm.

Selecting **Trailing** causes zeros behind the decimal point to be suppressed. For example, 7 or 7.25 is displayed instead of 7.000 or 7.250.

Selecting **0 Feet** causes zeros representing feet to be suppressed if the dimension text represents inches and/or fractions only. For example, 7 or 7 1/4 is displayed instead of 0'-7 or 0'-7 1/4.

Selecting **0 Inches** causes zeros representing inches to be suppressed if the dimension text represents feet only. For example, 7' is displayed instead of 7'-0.

The **Angular Dimensions** section controls the format and precision of angular units. The value is recorded in the DIMAUNIT system variable.

The **Units format** list box allows you to choose the format of the units for the angular dimension text. Options include **Decimal Degrees, Degrees Minutes Seconds, Grads,** and **Radians.**

The **Precision** text box allows you to set the number of decimal places for angular dimension units corresponding to the specified angular format. The value is recorded in the DIMADEC system variable.

The **Zero Suppression** section controls whether or not leading or trailing zeros are displayed. The value is recorded in the DIMAZIN system variable.

Selecting **Leading** causes zeros ahead of the decimal point to be suppressed.

Selecting **Trailing** causes zeros behind the decimal point to be suppressed.

The **Alternate Units** tab (Figure 7–89) is similar to the **Primary Units** tab, except for the **Display alternate units** check box, which allows you to enable (DIMALT set to 1) or disable (DIMALT set to 0) alternate units. There is no **Angular Dimensions** section. See Figure 7–88 for an example of dimensioning with alternate units. A **Placement** section with **After primary value** and **Before primary value** lets you determine where to place the Alternate units.

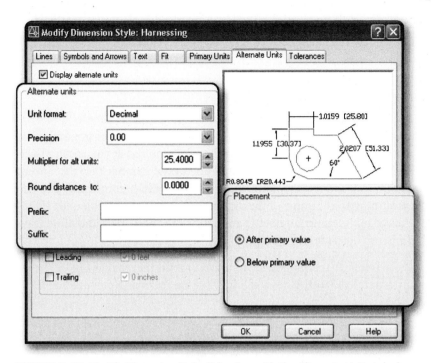

FIGURE 7–89 *Modify Dimension Style dialog box with the Alternate Units tab chosen*

The **Tolerances** tab (Figure 7–90) allows you to set the tolerance method and appropriate settings. A detailed explanation is provided earlier in this chapter in the "Tolerances" section.

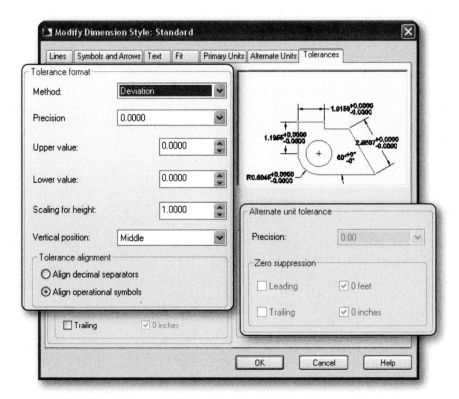

FIGURE 7–90 *Modify Dimension Style dialog box with the Tolerances tab chosen*

Once the necessary changes are made for the appropriate settings, choose **OK** to close the Modify Dimension Style dialog box. Choose **Close** to close the Dimension Style Manager dialog box.

OVERRIDING THE DIMENSION FEATURE

The DIMOVERRIDE command allows you to change one of the features in a dimension without having to change its dimension style or create a new dimension style. For example, you may wish to have one leader with the text centered at the ending horizontal line rather than over the ending horizontal line in the manner that the current dimension style may call for. By invoking the DIMOVERRIDE command, you can respond to the prompt with the name of the dimensioning system variable—DIMTAD in this case—and set the value to 0 rather than 1. You can then select the dimension leader that you wish to override.

Invoke the DIMOVERRIDE command from the Dimensions panel (Figure 7–91) located in the Annotate tab, and AutoCAD displays the following prompt:

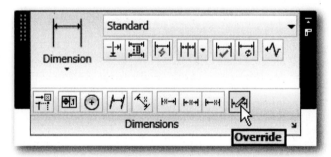

FIGURE 7–91 *Invoking the DIMOVERRIDE command by selecting Override from the Dimensions panel*

```
Enter dimension variable name to override or ⏎ (specify the
dimensioning system variable name or choose CLEAR from the
shortcut menu to clear overrides)
```

If you specify a dimensioning system variable, you are prompted:

```
Enter new value for dimension variable <current>: (specify
the new value, and press (Enter))
Enter dimension variable name to override: (specify another
dimension system variable or press (Enter))
```

If you press (Enter), you are prompted:

```
Select objects: (select the dimension objects)
```

The dimensions selected will have the specified dimensioning system variable settings overridden in accordance with the value specified.

If you choose CLEAR, you are prompted:

```
Select objects: (select the dimension to clear the
overrides)
```

The dimensions selected will have the dimensioning system variable setting overrides cleared.

UPDATING DIMENSIONS

Dimension Update permits you to make selected existing dimension(s) conform to the settings of the current dimension style.

Invoke Dimension Update from the Dimensions panel (Figure 7–92) located in the Annotate tab, and AutoCAD displays the following prompt:

> Select objects: *(select any dimension(s) whose settings you wish to have updated to conform to the current dimension style).*

FIGURE 7–92 *Invoking Dimension Update from the Dimensions panel*

> Dimension Update is actually the DIMSTYLE command, in which AutoCAD automatically chooses the APPLY option for you. All you have to do is select a dimension.

INSPECTING DIMENSIONS

The DIMINSPECT dimension adds or removes inspection information for a selected dimension. Inspection dimensions specify how frequently manufactured parts should be checked to ensure that the dimension value and tolerances of the parts are within the specified range.

Invoke DIMINSPECT command from the Dimensions panel (Figure 7–93) located in the Annotate tab, and AutoCAD displays the Inspection Dimension dialog box as shown in Figure 7–94.

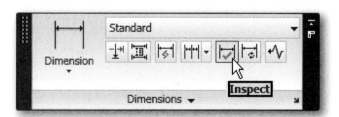

FIGURE 7–93 *Invoking the DIMINSPECT command from the Dimensions panel*

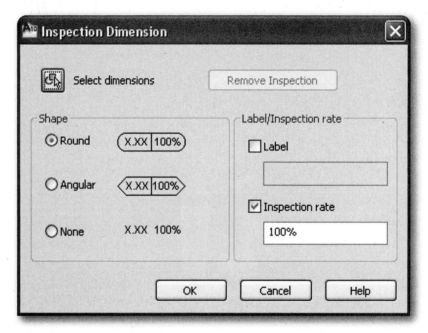

FIGURE 7–94 *Inspection Dimensions dialog box*

The **Shape and Inspection Label/Rate settings** section allows you to set the appearance of the frame of the inspection dimension and the inspection rate value.

Choose the **Select Dimensions** button to select dimensions to add or remove inspection information. Choose **OK** to apply the changes.

Open the Exercise Manual PDF for Chapter 7 for discipline-specific exercises. Related files are downloaded from the student companion site mentioned in the Introduction (refer to page number xii for instructions).

REVIEW QUESTIONS

1. Dimension types available in AutoCAD include:

   a. Linear

   b. Angular

   c. Diameter

   d. Radius

   e. All of the above

2. The associative dimension drawn with the DIMASSOC variable set to ON has all of its separate parts drawn as separate objects.

   a. True

   b. False

3. Linear (DIMLINEAR) allows you to draw horizontal, vertical, and aligned dimensions.

   a. True

   b. False

4. To draw a linear dimension you must (**1**) specify the first extension line origin, (**2**) locate the dimension line, and then (**4**) specify the second extension line.
   a. True
   b. False

5. Linear dimension are vertical rather than horizontal.
   a. True
   b. False

6. Angular (DIMANGULAR) allows you to draw angular dimensions between two parallel lines.
   a. True
   b. False

7. By default, the dimension text for a radius dimension is preceded by:
   a. Radius
   b. Rad
   c. R

8. Dimensions do not recalculate automatically when the object that they refer to are modified. For example, when you stretch a wall **2'-0"** to the right, you need to manually update the associated dimensions.
   a. True
   b. False

9. Using the PROPERTIES command will allow you to override the dimensioning system variable settings for a single dimension, without modifying the base dimension style.
   a. True
   b. False

10. Baseline (DIMBASELINE) is used to draw dimensions from a single datum baseline.
    a. True
    b. False

11. Center Mark (DIMCENTER) allows you to draw center cross marks or centerlines in a circle.
    a. True
    b. False

12. You must explode a dimension before you can use the DIMTEDIT command to edit the dimension text.
    a. True
    b. False

13. When you create stacked angular or linear dimensions of any type, they might be too close together or unevenly spaced. Instead of moving each dimension, you can modify the space between sets of dimensions using what command?
    a. Dimension Break
    b. Angular
    c. Adjust Space
    d. Jogged

14. The suppress option in the **Lines** tab of the Modify Dimension Style dialog box allows you to suppress only one extension line at a time.
   a. True
   b. False

15. By default, AutoCAD creates continuous dimensions if you select linear objects or more than one object.
   a. True
   b. False

16. The **Arrowhead** section in **Symbols and Arrows** tab of the Modify Dimension Style dialog box allows you to change the size and style of your arrowheads.
   a. True
   b. False

17. You must use the UNITS command to determine how many decimal places will be displayed in the dimension text.
   a. True
   b. False

Directions: For each of the following questions, select which of the dimensioning tabs the option can be found on in the following list.

- Symbols and Arrows tab
- Text tab
- Fit tab
- Primary Units tab
- Tolerances
- Cannot be found in the dimensioning dialog boxes

   Arrowhead size

   Associative/nonassociative setting

   Text height

   Arrowhead style

   Text location

   Suppressing of extension lines

   Linear tolerance settings

   Number of decimal places for a linear dimension

Directions: Answer the following questions based on Figure 7–95.

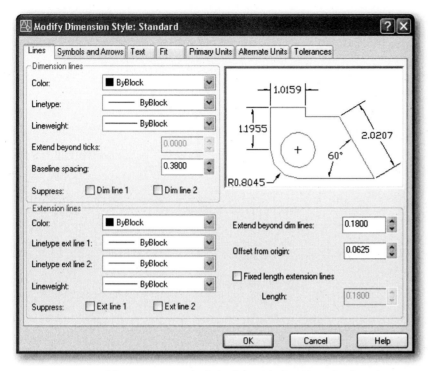

FIGURE 7–95 *Modify Dimension Style dialog box with the Lines tab selected*

18. Which dimension variable must be changed to modify the size of Arrowheads?

    a. DIMARO

    b. DIMASZ

    c. DIMTSZ

    d. DIMDLI

19. Which dimension variable must be changed to modify Arrowhead types?

    a. DIMASZ

    b. DIMTSZ

    c. DIMBLK

    d. DIMEXE

20. Which dimension variable must be modified to change the color of dimension lines?

    a. DIMCLRE

    b. DIMCLRT

    c. DIMCLRL

    d. DIMCLRD

21. Which dimension variable is used to hide the first extension line?

    a. DIMEXL

    b. DIMOFF

    c. DIMSE1

    d. DIMEL1

22. Which dimension variable establishes the distance from the object to the start of the extension line?

    a. DIMEXO

    b. DIMSEL

    c. DIMGAP

    d. DIMSPC

23. Which dimension variable establishes the distance from the first dimension line to the second dimension line when using baseline dimensions?

    a. DIMD12

    b. DIMSPC2

    c. DIMDLI

    d. DIMDSPC

24. Which dimension variable is used to modify tick terminators?

    a. DIMARC

    b. DIMART

    c. DIMTIC

    d. DIMTSZ

25. Which dimension variable is used to hide the second dimension line?

    a. DIMSDL

    b. DIMHD2

    c. DIM2OFF

    d. DIMSD2

26. When one desires for the extension lines to be drawn past the dimension lines which dimension variable is used to perform this function?

    a. DIMEXE

    b. DIMELD

    c. DIMEXT

    d. DIMDLE

27. Which dimension variable is used to have a second arrowhead that is different from the first?

    a. DIMARW2

    b. DIMABLK

    c. dimblk2

    d. dimaro2

Plotting and Layouts

INTRODUCTION

In the transition from board drafting to CAD, one task has not changed: obtaining a hard copy. The term "hard copy" describes a tangible reproduction of a screen image. The hard copy is usually a reproducible medium from which prints are made, and it can take many forms including slides, videotape, prints, or plots. This chapter describes the most commonly used processes for creating a hard copy: plotting or printing.

In manual drafting, if you need your drawing to be done in two different scales, you must physically draw it for two different scales. In CAD, with minor modifications, you plot or print the same drawing in different scale factors on different sizes of paper. In AutoCAD, you can compose your drawing in paper space (also referred as Layout) with limits that equal the sheet size and plot it at 1:1 scale.

OBJECTIVES

After completing this chapter, you will be able to do the following:

- Plan the plotted sheet
- Plot from model space
- Set up a layout
- Create and modify a layout
- Create floating viewports
- Scale views relative to paper space
- Control the visibility of layers within viewports
- Plot from layout: What You See Is What You Get (WYSIWYG)
- Use annotative scaling
- Create and modify plotstyle tables
- Change the plotstyle property for an object or layer
- Configure plotters

PLANNING THE PLOTTED SHEET

Planning ahead is required in laying out the objects to be drawn on the final sheet. The objects drawn on the plotted sheet must be arranged. In CAD, with its true-size capability, an object can be started without first laying out a plotted sheet, but eventually, limits, or at least a displayed area, must be determined. For schematics, diagrams, and graphs, plotted scale is of little concern, but for architectural, civil, and mechanical drawings, plotting to a conventional scale is a professionally accepted practice that should not be abandoned just because it can be circumvented.

When setting up the drawing limits, you must take the plotted sheet into consideration to get the entire view of the object(s) on the sheet, so even with all the power of the CAD system, some thought must be given to the concept of scale, which is the ratio of true size to the size plotted. In other words, before you start drawing, you should have an idea of the scale at which the final drawing will be plotted or printed on a given size of paper.

The limits should correspond to some factor of the plotted sheet. If the objects fit on a 24 × 18 sheet at full size with room for a border, title block, bill of materials, dimensioning, and general notes, then set up your limits to (0,0) the lower-left corner and (24,18) the upper-right corner. The drawing can be plotted or printed at 1:1 scale, that is, one object unit equals one plotted unit.

Plot scales can be expressed in several formats. Each of the following five plot scales is exactly the same; only the display formats differ.

$$1/4 = 1'\text{-}0''$$
$$1 = 4'$$
$$1 = 48$$
$$1:48$$
$$1/48$$

A plot scale of 1:48 means that a line 48 units long in AutoCAD will plot with a length of 1 unit. The units can be any measurement system, including inches, feet, millimeters, nautical miles, chains, angstroms, and light-years, but, by default, plotting units in AutoCAD are inches.

Four variables control the relationship between the size of objects in the AutoCAD drawing and their sizes on a sheet of paper produced by an AutoCAD plot:

- Size of the object in AutoCAD. For simplification, it will be referred to as the ACAD_size.
- Size of the object on the plot. For simplification, it will be referred to as the ACAD_plot.
- Maximum available plot area for a given sheet of paper. For simplification, it will be referred to as the ACAD_max_plot.
- Plot scale. For simplification, it will be referred as to the ACAD_scale.

The relationship between the variables can be described by the following three algebraic formulas:

$$\text{ACAD\_scale} = \text{ACAD\_plot}/\text{ACAD\_size}$$
$$\text{ACAD\_plot} = \text{ACAD\_size} \times \text{ACAD\_scale}$$
$$\text{ACAD\_size} = \text{ACAD\_plot}/\text{ACAD\_scale}$$

Example of Computing Plot Scale, Plot Size, and Limits

An architectural elevation of a building 48' wide and 24' high must be plotted on a 36 × 24 sheet. First you determine the plotter's maximum available plot area for the given sheet size, and this depends on the model of plotter you use.

In the case of an HP plotter, the available area for 36 × 24 is 33.5 × 21.5. Next you determine the area needed for the title block, general notes, and other items such as an area for revision notes and a list of reference drawings. For the given example, let's say that an area of 27 × 16 is available for the drawing.

The objective is to arrive at one of the standard architectural scales in the form of × in. = 1 ft. The usual range is from 1/16 = 1'-0" for plans of large structures to 3 = 1'-0" for small details. To determine the plot scale, substitute these values for the appropriate variables in the formula:

ACAD_scale = ACAD_plot/ACAD_size

ACAD_scale = 27/48' for X-axis

= 0.5625/1'-0" or 0.5625 = 1'-0"

The closest standard architectural scale that can be used in the given situation is 1/2 = 1'-0" (0.5 = 1'-0", 1/24 or 1:24).

To determine the size of the object on the plot, substitute these values for the appropriate variables in the formula:

ACAD_plot = ACAD_size × ACAD_scale

ACAD_plot = 48' × (0.5/1') for X-axis

= 24 (less than the 27 maximum allowable space on the paper)

ACAD_plot = 24' × (0.5/1') for Y-axis

= 12 (less than the 16 maximum allowable space on the paper)

If instead of 1/2 = 1'-0" scale you wish to use a scale of 3/4 = 1'-0", the size of the object on the plot will be 48' × (0.75/1') = 36 for the X-axis. This is larger than the available space on the given paper, the drawing will not fit on the given paper size. You must select a larger paper size.

Once the plot scale is determined, and you have verified that the drawing fits on the given paper size, you can then determine the drawing limits for the plotted sheet size of 33.5 × 21.5.

To determine the limits for the *X*- and *Y*-axes, substitute the appropriate values in the formula:

ACAD_limits (X-axis) = ACAD_max_plot/ACAD_scale

= 33.5/(0.5/1'-0")

= 67'

ACAD_limits (Y-axis) = 21.5/(0.5/1'-0")

= 43'

Appropriate limits settings in AutoCAD for a 36 × 24 sheet with a maximum available plot area of 33.5 × 21.5 at a plot scale of 0.5 = 1'-0" would be as follows:

Lower-left corner: 0,0

Upper-right corner: 67',43'

Another consideration in setting up a drawing for user convenience is to have the (0,0) coordinates at some point other than the lower-left corner of the drawing sheet. Many objects have a reference point from which other parts of the object are dimensioned. Being able to set that reference point to (0,0) is very helpful. In many cases, the location of (0,0) is optional. In other cases, the coordinates should coincide with real coordinates such as those on an industrial plant area block. In still other cases, only one set of coordinates might be a governing factor.

In this example, the 48' wide × 24' high front elevation of the building is to be plotted on a 36 × 24 sheet at a scale of 1/2 = 1'-0". It has been determined that (0,0) should be at the lower-left corner of the front elevation view, as shown in Figure 8–1.

Centering the view on the sheet requires a few minutes of layout time. Several approaches allow the drafter to arrive at the location of (0,0) relative to the lower-left corner of the plotted sheet or limits. Having computed the limits to be 67' wide × 43' high, the half-width and half-height, or the dimensions from the center, of the sheet are 33.5' and 21.5' to scale, respectively. Subtracting the half-width of the building from the half-width of the limits will set the X coordinate of the lower-left corner at –9.5' from the equation, 24' – 33.5'. The same is done for the Y coordinate –9.5' (12' – 21.5'). Therefore, the lower-left corner of the limits is at (–9.5',–9.5').

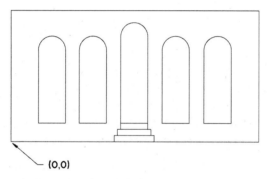

(0,0)

FIGURE 8–1 *Setting the reference point to the origin (0,0) in a location other than the lower-left corner*

Appropriate limits settings in AutoCAD for a 36 × 24 sheet with a maximum available plot area of 33.5 × 21.5 by centering the view at a plot scale of 0.5 = 1'-0" (Figure 8–2) are as follows:

Lower-left corner: −9.5',−9.5'

Upper-right corner: 57.5',33.5'

NOTE

The absolute X coordinates, when added (57.5' + 9.5'), equal 67', which is the width of the limits, and the absolute Y coordinates (33.5' + 9.5') equal 43', which is the height of the limits.

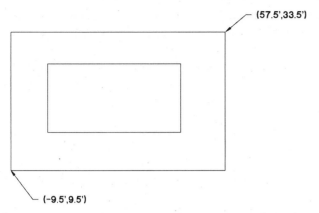

(57.5',33.5')

(-9.5',9.5')

FIGURE 8–2 *Setting the limits to the maximum available plot area by centering the view*

Setting for LTSCALE

As explained in Chapter 3, the LTSCALE system variable provides a method of adjusting the linetypes to a meaningful scale for the drawing. This action sets the length of dashes and spaces in linetypes. When the value of LTSCALE is set to the reciprocal of the plot scale, the linetypes provided with AutoCAD plot on paper at the sizes in which they are defined in *acad.lin*.

$$LTSCALE = 1/ACAD\_scale$$

Setting for DIMSCALE

As explained in Chapter 7, AutoCAD provides a set of dimensioning variables that control the way dimensions are drawn. The DIMSCALE dimension variable is applied globally to all dimensions that are applied to sizes or distances as an overall scaling factor. The default DIMSCALE value is set to 1. When DIMSCALE is set to the reciprocal of the plot scale, it applies globally to all dimension variables for the plot scale factor.

$$DIMSCALE = 1/ACAD\_scale$$

If necessary, you can set individual dimensioning variables to the size that you want the dimension to appear on the paper by substituting the appropriate values in the following formula:

size of the plotted dimvars_value=dimvars_value $\times$ ACAD_scale $\times$ -DIMSCALE

The following example shows you how to determine the arrow size DIMASZ for a plot scale of $1/2 = 1'\text{-}0''$, DIMSCALE set to 1, and default DIMASZ of 0.18:

size of the plotted arrow $= 0.18 \times 24 \times 1$

$= 4.32$

Scaling Text and Symbols by the Traditional Method

How can you determine the size at which text and symbols, or blocks, will plot? As mentioned earlier, you almost always draw objects in their actual size, that is, to real-world dimensions. Even in the case of text and blocks, you place them at the real-world dimensions. In the previous example, the architectural elevation of a building $48' \times 24'$ is drawn to actual size and plotted to a scale of $1/2 = 1'\text{-}0''$. Let's say that you wanted your text to plot at 1/4 high. If you were to create your text and annotations at 1/4, they would be so small relative to the elevation drawing itself that you could not read the characters.

Before you begin placing the text, you need to know the scale at which you will eventually plot the drawing. In the previous example of an architectural elevation, the plot scale is $1/2 = 1'\text{-}0''$, and you want the text to plot 1/4 high. You need to find a relationship between 1/4 on the paper and the size of the text for the real-world dimensions in the drawing. If 1/2 on the paper equals 12 in the model, 1/4-high text on the paper equals 6, so text and annotations should be drawn at 6 high in the drawing to plot at 1/4 high at this scale of $1/2 = 1'\text{-}0''$. Similarly, you can calculate the various text sizes for a given plot scale. The following table shows the model text size needed to achieve a specific plotted text height at some common scales (Table 8–1).

TABLE 8–1 *Text size corresponding to specific plotted text height at various scales*

Scale	Factor	Plotted Text Height								
		1/16	3/32	1/8	3/16	1/4	5/16	3/8	1/2	5/8
$1/16 = 1'\text{-}0''$	192	12	18	24	36	48	60	66	96	120
$1/8 = 1'\text{-}0''$	96	6	9	12	18	24	30	36	48	60
$3/16 = 1'\text{-}0''$	64	4	6	8	12	16	20	24	32	40
$1/4 = 1'\text{-}0''$	48	3	45	6	9	12	15	18	24	30
$3/8 = 1'\text{-}0''$	32	2	3	4	6	8	10	12	16	20
$1/2 = 1'\text{-}0''$	24	1.5	2.25	3	4.5	6	7.5	9	12	15
$3/4 = 1'\text{-}0''$	16	1	1.5	2	3	4	5	6	8	10
$1 = 1'\text{-}0''$	12	0.75	1.13	1.5	2.25	3	3.75	4.5	6	7.5
$1\ 1/2 = 1'\text{-}0''$	8	0.5	.75	1	1.5	2	2.5	3	4	5
$3 = 1'\text{-}0''$	4	0.25	.375	0.5	0.75	1	1.25	1.5	2	2.5
$1 = 10'$	120	7.5	11.25	15	22.5	30	37.5	45	60	75
$1 = 20'$	240	15	22.5	30	45	60	75	90	120	150
$1 = 30'$	360	22.5	33.75	45	67.5	90	112.5	135	180	225
$1 = 40'$	480	30	45	60	90	120	150	180	240	300
$1 = 50'$	600	37.5	56.25	75	112.5	150	187.5	225	300	375
$1 = 60'$	720	45	67.5	90	135	180	225	270	360	450
$1 = 70'$	840	52.5	78.75	105	157.5	210	262.5	315	420	525
$1 = 80'$	960	60	90	120	180	240	300	360	480	600
$1 = 90'$	1080	67.5	101.25	135	202.5	270	337.5	405	540	675
$1 = 100'$	1200	75	112.5	150	225	300	375	450	600	750

Instead of manually calculating the appropriate text size, you can use the annotative scaling feature, which will automatically scale, display, and plot at the correct size and in the desired views. For a detailed explanation refer to the section "Annotative Scaling" later in the chapter.

WORKING IN MODEL SPACE AND PAPER SPACE

One of AutoCAD's more useful features is the option to work on your drawing in two different environments. AutoCAD allows you to plot a drawing from model space as well as from paper space, also referred to as Layout. Like its name, model space is where you draw the model: the object that you are visually depicting. Paper space is for drawing the border, title block with its text, and general notes that you put on the plotted sheet—elements that are not part of the model itself. Most of the drafting and design work is created in the 3D environment of model space, even though your objects may have been drawn only in a 2D plane. You can plot the drawing from model space to any scale by specifying the plot scale in the plot dialog box.

Paper space is a 2D environment, like paper, used for arranging various views, or floating viewports, of what was drawn in model space. It represents the paper on which you arrange the drawing prior to plotting. With AutoCAD, single or multiple paper space environments, or layouts, can be easily set up and manipulated. Each layout represents an individual plot output sheet or an individual sheet in a drawing project. You can apply different scales and specify different visibility for layers in each floating viewport. After arranging the views and scaling appropriately, you can plot the drawing from layout at 1:1, or full scale, allowing each separate viewport to display the selected parts of model space at the scale determined by the factor you specify. Paper space allows you to plot the drawing in a mode called WYSIWYG, which stands for "What You See Is What You Get."

You can switch between model space and paper space by selecting the appropriate tab provided at the bottom of the drawing window. Model space can be accessed from the Model tab. Selecting one of the available Layout tabs will access paper space. You can also switch between model space and paper space by changing the value of the system variable TILEMODE. When TILEMODE is set to 1, the default, you will be working in model space; when it is set to 0, you will be working in paper space.

When you are working in a layout, you can access model space by double-clicking inside one of the floating viewports. If you double-click anywhere outside the viewport, AutoCAD will switch to paper space. You must have at least one floating viewport in paper space to see objects drawn in model space. You can also maximize the active floating viewport to fill the screen and switch to model space for editing.

By default, AutoCAD creates two layout tabs called Layout1 and Layout2. You can rename these and, if necessary, create additional layouts. A detailed explanation is provided later in this chapter for creating and modifying layouts.

PLOTTING FROM MODEL SPACE

To plot and print the current drawing from the model space, invoke the PLOT command from the Plot panel (Figure 8–3) on the Output tab, and AutoCAD displays the Plot dialog box (Figure 8–4).

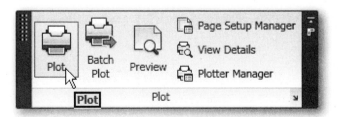

FIGURE 8–3 *Invoking the* PLOT *command from the Plot panel*

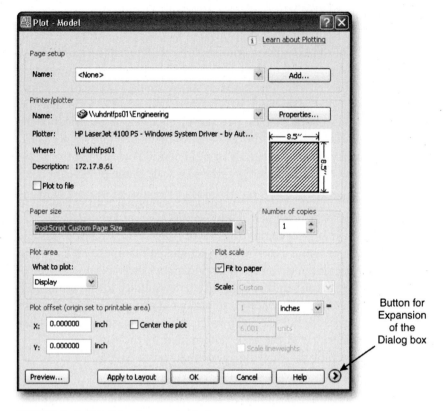

FIGURE 8–4 *Plot dialog box, contracted*

Plot Settings

The Plot dialog box is almost identical to the Page Setup dialog box except for the title and the **Name** text box in the **Page setup** section. The Plot dialog box can be expanded or contracted in size by choosing the arrow at the lower-right corner of the dialog box. The first seven sections described in the following section on dialog box details are displayed when the dialog box is contracted. The last four are displayed only when the dialog box is expanded (Figure 8–5).

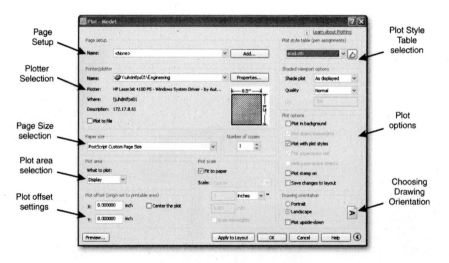

FIGURE 8–5 *Expanded Plot dialog box with additional options*

The **Page setup** section displays a list of any named and saved page setups in the drawing. You can base the current setup on a named page setup saved in the drawing, or you can create a new named page setup based on the current settings in the Plot dialog box by choosing **Add**.

The **Printer/plotter** section displays the currently configured plotting device.

The **Name** list box lists the system printers/plotters or PC3 files that are available to select for plotting. If a plotter is supported by AutoCAD but not by the Windows® operating system, you can use one of the nonsystem printer drivers. The available nonsystem driver includes methods for creating PostScript, raster, Design Web Format (DWF), and portable document format (PDF) files.

AutoCAD stores information about the media and plotting device in configured plot (PC3) files. Plot configurations are portable and can be shared in an office or on a project as long as they are for the same driver, model, and driver version. Shared plot configurations for Windows system printers may also need to be for the same version of Windows.

You can configure AutoCAD for many devices and store multiple configurations for a single device. Each plotter configuration (PC3) contains information such as the device driver and model, the output port to which the device is connected, and various device-specific settings. You can create several PC3 files with different output options for the same plotter. After you create a PC3 file, it's available in the list of plotter configuration names in the Plot dialog box.

The DWG To PDF configuration file selection creates a PDF file of the selected drawing file, and DWF6 ePlot configuration file creates a DWF file of the selected drawing file. For a detailed explanation of DWF file formats, refer to Chapter 15.

To create PC3 files, use the Add-a-Plotter wizard in the Autodesk Plotter Manager. For a detailed explanation, refer to the section "Configuring Plotters" later in this chapter.

There are a number of ways to modify the default settings for a Windows system printer without creating a PC3 file. For example, you can modify the properties across the system from the Control Panel. You can also choose Properties in the Plot dialog box and plot without saving the properties.

Plotter lists the currently selected plotter or plotter assigned in the currently selected page setup.

Where gives the port to which the selected plotter is connected or its network location.

Choosing **Properties** allows access to the Plotter Configuration of the currently configured plotting device. Refer to the section "Configuring Plotters" later in this chapter for a detailed explanation on plotter configuration.

Description lists information about the output device currently selected.

Selecting **Plot to file** causes plots to be output to a file rather than to a plotter or printer. If **Plot to file** is selected when you choose **OK** in the Plot dialog box, the Browse for Plot File dialog box is displayed. Specify the file name to save the plot file and press **Save**.

The **Paper size** section allows you to select the paper size to plot. Select the paper size to plot from the **Paper size** box for the selected plotting device. Actual paper sizes are indicated by the width (X-axis direction) and height (Y-axis direction). A default paper size is set for the plotting device when you create a plotter configuration file with the Add-a-Plotter Wizard.

The **Number of copies** text box lets you specify the number of copies to plot. This option is not available when you plot to file.

The **Plot area** section allows you to select the area to be plotted from the **What to plot** box.

Selecting **Display** causes the area displayed on the screen to be plotted.

Selecting **Limits** causes everything within the drawing limits to be plotted to the specified scale when the PLOT command is invoked from the **Model** tab.

Selecting **Layout** causes everything within the area of the paper size specified by the layout to be plotted when the PLOT command is invoked from paper space in a layout tab.

Selecting **Extents** causes the portion of the current space of the drawing that contains objects to be plotted. All visible geometry in the current space is plotted.

Selecting **View** plots a named view that was previously saved with the VIEW command. You can select a named view from the list. If there are no saved views in the drawing, this option is unavailable.

Selecting **Window** allows you to specify a window on the screen and plot the area covered by the window. The lower-left corner of the window becomes the origin of the plot.

The **Plot scale** section controls the plot area.

Selecting **Fit to paper** causes the plot to fit within the selected paper size. To specify the exact scale for the plot, clear the **Fit to paper** checkbox and select the plot scale from the **Scale** list box. You can create a custom scale by entering the number of inches or millimeters equal to the number of drawing units in the appropriate text fields. When you plot from a layout, the default setting is 1:1 or full scale.

Selecting **Scale lineweights** causes AutoCAD to plot lineweights in proportion to the plot scale. Otherwise AutoCAD will plot the objects with assigned lineweights. Lineweights normally specify the linewidth of printed objects and are plotted with the linewidth size regardless of the plot scale.

The **Plot offset** section specifies an offset of the plotting area from the lower-left corner of the paper. In a layout, the lower-left corner of a specified plot area is positioned at the lower-left margin of the plotting area. You can offset the origin by entering a positive or negative value. Select **Center the plot** to automatically center the plot on the paper.

The following sections of the Plot dialog box (Figure 8–5) are displayed only when the dialog box is expanded by choosing the arrow at the lower-right corner of the dialog box.

The **Plot style table** section allows you to set the plotstyle table, edit a plotstyle table, or create a new plotstyle table. Plotstyle tables are settings that give you control over how objects in your drawing are plotted into hardcopy plots. By modifying an object's plotstyle, you override that object's color, linetype, and lineweight. You can also specify end, join, and fill styles as well as output effects such as dithering, grayscale, pen assignment, and screening. You can use plotstyles if you need to plot the same drawing in different ways. This section displays the plotstyle table that is assigned to the current Model tab or Layout tab and provides a list of the currently available plotstyle tables. If you select **New** from the **Plot style table** list, the Add Color-Dependent Plotstyle Table Wizard is displayed, which you can use to create a new plotstyle table. Choose **Edit** to modify the currently assigned plotstyle table. Refer to the section "Creating a Plotstyle Table" later in this chapter for a detailed explanation about creating and modifying a plotstyle table.

The **Shaded viewport options** section specifies how shaded and rendered viewports are plotted and determines their resolution level with corresponding dpi.

The **Shade plot** text box displays how views are plotted, and it is specified through the Properties dialog box for the selected viewport.

The **Quality** text box specifies the resolution at which shaded and rendered viewports are plotted. Selecting Draft sets rendered and shaded model space views to plot as wireframe. Selecting Preview sets rendered and shaded model space views to plot at a maximum of 150 dpi. Selecting Normal sets rendered and shaded model space views to plot at a maximum of 300 dpi. Selecting Presentation sets rendered and shaded model space views to plot at the current device resolution to a maximum of 600 dpi. DPI values may vary with different computer or plotter configurations. Selecting Maximum sets rendered and shaded model space views to plot at the current device resolution with no maximum. Selecting Custom sets rendered and shaded model space views to plot at the resolution setting you specify in the DPI box up to the current device resolution.

The **Plot options** section specifies the following options for lineweights, plotstyles, and the current plotstyle table:

Selecting **Plot in background** causes the plot to be processed in the background while you perform other tasks on the computer.

Selecting **Plot object lineweights** plots the objects with assigned lineweights. Otherwise AutoCAD will plot with the default lineweight.

Selecting **Plot transparency** specifies whether object transparency is plotted. This option should only be used when plotting drawings with transparent objects.

Selecting **Plot with plotstyles** plots using the object plotstyles that are assigned to the geometry, as defined by the selected plotstyle table.

Selecting **Plot paperspace last** plots model space geometry before paper space objects are plotted.

Selecting **Hide paperspace objects** plots layouts (paper space) with hidden lines removed from objects.

Selecting **Plot stamp on** includes a plot stamp on the plotted sheet. Refer to the section "Plot Stamp Settings" for a detailed explanation of modifying plot stamp settings.

Select **Save changes to layout** to save the changes that you make in the Plot dialog box to the layout.

The **Drawing orientation** section specifies the orientation of the drawing on the paper for plotters that support landscape or portrait orientation.

Selecting **Portrait** orients and plots the drawing so that the short edge of the paper represents the top of the page.

Selecting **Landscape** orients and plots the drawing so that the long edge of the paper represents the top of the page.

Selecting **Plot upside-down** orients and plots the drawing upside down.

Choosing **Preview** displays the drawing on the screen as it would appear when plotted. AutoCAD temporarily hides the plotting dialog boxes, draws an outline of the paper size, and displays the drawing as it would appear, using all the current settings, when it is plotted (Figure 8–6).

The cursor changes to a magnifying glass with plus and minus signs. Holding the pick button and dragging the cursor toward the top of the screen enlarges the preview image. Dragging it toward the bottom of the screen reduces the preview image. Right-click, and AutoCAD displays a shortcut menu offering additional preview options: *Pan, Zoom, Zoom Window, Zoom Original, Plot,* and *Exit*. To end the full preview, choose *Exit* from the shortcut menu or press Enter. AutoCAD returns to the Plot dialog box.

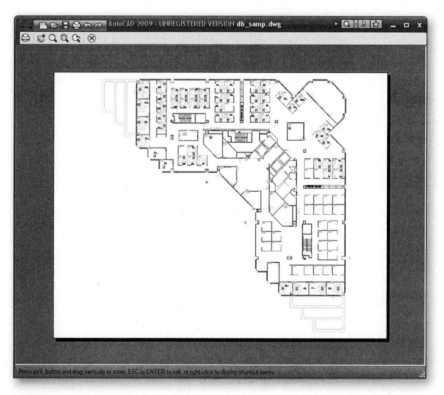

FIGURE 8–6 *Plot preview*

TIP You can also access a full preview by invoking the Preview command from the Plot panel.

After making the necessary changes in the plot settings, choose **OK**. AutoCAD starts plotting and reports its progress as it converts the drawing to the plotter's graphics language by displaying the number of vectors processed.

If something goes wrong or if you want to stop immediately, choose **Cancel** at any time. AutoCAD cancels the plotting.

Plot Stamp Settings

The Plot Stamp dialog box allows you to specify the information for the plot stamp that can be placed on a specified corner of each drawing and, if necessary, logs the information to a file (Figure 8–7). Choose **Plot Stamp Settings** from the Plot dialog box to open the Plot Stamp dialog box. This feature appears only when the **Plot stamp on** checkbox is set to ON.

Plot stamp information includes drawing name, layout name, date and time, login name, plot device name, paper size, plot scale, and user-defined fields, if any. Once you check the **Plot stamp on** checkbox in the **Plot options** section of the Plot dialog box, it remains active with whatever settings have been most recently entered until you specifically clear the checkbox. AutoCAD creates a plot stamp at the time the drawing is being plotted, and the stamp is not saved with the drawing. Before you plot the drawing, you can preview the position of the plot stamp, but not the contents, in the Plot Stamp dialog box. The plot stamp can be set to plot at one of the four drawing corners and can print up to two lines.

> Plot stamp information is plotted with pen number 7 or the highest numbered available pen if the plotter doesn't hold seven pens. If you are using a non-pen (raster) device, color 7 is always used for plot stamping.

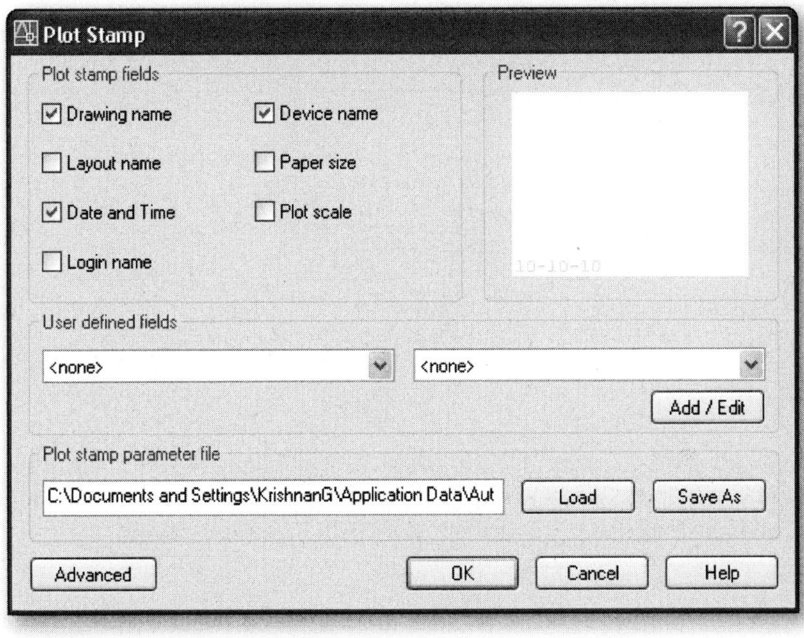

FIGURE 8–7 *Plot Stamp dialog box*

The **Plot stamp fields** section specifies the drawing information you want applied to the plot stamp. You can include seven items in the plot stamp by selecting them by their field name as follows: **Drawing name**, Layout name, Date and Time, Login name, Device name, Paper size, and Plot scale.

The **Preview** section of the Plot Stamp dialog box provides a visual display of the plot stamp location based on the location and rotation values specified in the Advanced Options dialog box.

The **User defined fields** section provides text that can optionally be plotted. You can choose one or both user-defined fields for the plot stamp information. If the user-defined value is set to <none>, no user-defined information is plotted. To add, edit, or delete user-defined fields, choose **Add/Edit**. AutoCAD displays the User Defined Fields dialog box (Figure 8–8).

FIGURE 8–8 *User Defined Fields dialog box*

Choose **Add** to add an editable user-defined field to the bottom of the list; choose **Edit** to edit the selected user-defined field; and choose **Delete** to delete the selected user-defined field. Choose **OK** to save the changes and close the User Defined Fields dialog box. Choose **Cancel** to discard the changes and close the User Defined Fields dialog box.

The **Plot stamp parameter file** section displays the name of the file in which the plot stamp settings are stored. If necessary, you can save the current plot stamp settings to a new file by choosing **Save As** and providing an appropriate file name. AutoCAD stores plot stamp information in a file with a *.pss* extension. If you need to load a different parameter file, choose **Load**. AutoCAD displays a standard file selection dialog box, in which you can specify the location of the parameter file you want to use.

To set the location, text properties, and units of the plot stamp, choose **Advanced**. AutoCAD displays the Advanced Options dialog box (Figure 8–9).

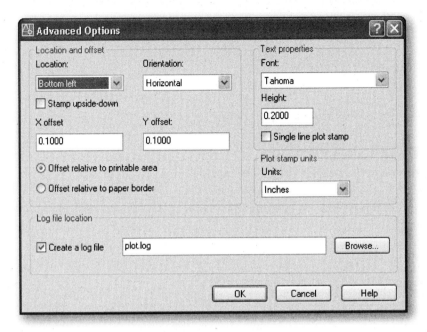

FIGURE 8–9 *Advanced Options (Plot Stamp) dialog box*

Location allows you to select the area where you want to place the plot stamp. The options include Top Left, Bottom Left (default), Bottom Right, and Top Right. The location is relative to the image orientation of the drawing on the page.

Orientation allows you to select the rotation of the plot stamp in relation to the specified page. The options include Horizontal and Vertical for each location.

Stamp upside-down controls whether or not to rotate the plot stamp upside down.

X offset and **Y offset** determine the offset distance calculated from either the corner of the paper or the corner of the printable area, depending on which setting you specify. Select one of the two options: **Offset relative to printable area** or **Offset relative to paper border** to set the reference point from which to measure the offset distance.

The **Text properties** section determines the font, height, and number of lines you want to apply to the plot stamp text. **Font** specifies the font you want to apply to the text used for the plot stamp information. **Height** specifies the text height you want to apply to the plot stamp information. **Single line plot stamp** controls whether or not to place the plot stamp information in a single line of text. The plot stamp information can consist of up to two lines of text, but the placement and offset values you specify must accommodate text wrapping and text height. If the plot stamp contains text that is longer than the printable area, the plot stamp text will be truncated. If **Single line plot stamp** is set to OFF, the plot stamp text is wrapped after the third field.

The **Plot stamp units** section allows you to specify the units used to measure the X offset, Y offset, and height. From the **Units** box, you can select one of the available units: inches, millimeters, or pixels.

Log file location specifies the name of the file to which the plot stamp information is saved instead of, or in addition to, stamping the current plot. The default log file name is *plot.log*, and it is located in the AutoCAD folder. Choose **Browse** to specify a different file name and path. After the initial plot log file is created, the plot stamp information in each succeeding plotted drawing is added to this file. Each drawing's plot stamp information is a single line of text. If necessary the log file can be placed on a network drive and shared by multiple users.

Choose **OK** to save the changes and close the Advanced Options dialog box. Choose **OK** to save the changes and close the Plot Stamp dialog box.

PLOTTING FROM LAYOUTS

As mentioned earlier, AutoCAD allows you to plot a drawing from layouts (paper space) as well as from model space. The layout (paper space) environment allows you to set up multiple layouts. You can have as many layouts as you like, and each one can be set up for a different type of output. A layout is used to compose or lay out your model drawing for plotting. A layout may consist of a title block, one or more viewports, and annotations. As you create a layout, you can design floating viewport configurations to display different details in your drawing. You can apply different scales to each viewport within the layout and specify different visibility for layers in the viewport. Layouts are accessible by choosing the Layout tab at the bottom of the drawing area. Using a paper space layout with viewports makes it easy to produce plotted sheets with AutoCAD, rearrange and scale objects that are drawn in model space, and add non-object (usually annotative) elements in paper space.

NOTE Reference to viewports in layouts means floating viewports. The distinction between floating viewports and model space viewports is explained in the section on viewports in this chapter.

The combination of paper space and viewports is a special and powerful application for producing the most commonly used communication tool in the architect or engineer's repertoire: a set of paper drawings, traditionally referred to as "blueprints."

AutoCAD allows you to perform the following:

Create or import a page setup, applying it to a layout that is the exact width and height of the desired plotted paper sheet.

Create, attach, import, or otherwise position a border, title block, symbol, table, and any other annotation or AutoCAD object in full size ($12 = 1'\text{-}0''$) on the sheet.

Open windows of specified sizes and locations, referred to as viewports, at specified scales on the sheet for viewing desired parts of objects drawn in model space.

Plot the layout to 1:1 scale.

When configured, the layout will be like a separate drawing sheet that represents the final printed sheet with a border, title block, and note information, making it easy to arrange objects to their desired scale on the sheet and in their viewports. This makes plotting a simple task.

Planning to Plot

Figure 8–10 shows a model space drawing of a residential elevation and floor plan without a border, title block, or any notation except dimensions in the plan view. In Figure 8–11, by using the Layout feature, the views are juxtaposed for conventional arrangement, and a border, title block, view titles, and text tables are added in paper space. The two figures illustrate the application of model space for creating the design objects and paper space for arranging the design objects on a plot-friendly sheet with border, title block, view titles, and annotation, accomplished in a single drawing file.

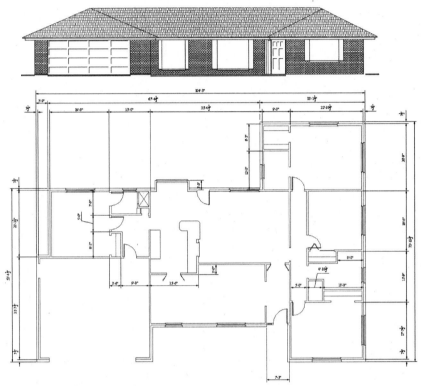

FIGURE 8–10 *Architectural floor plan and elevation drawn in model space*

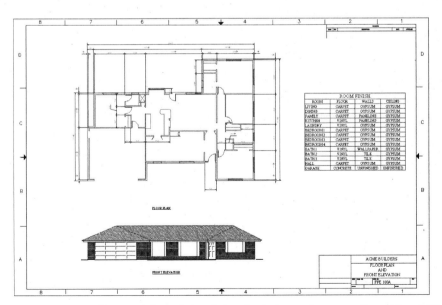

FIGURE 8–11 *Architectural floor plan and elevation in layout format configured for plotting*

Image(s) © Cengage Learning 2013

The earlier part of this chapter gave an example of setting up a drawing for plotting from model space (Figures 8–1 and 8–2). This section describes procedures for setting up a drawing for plotting and printing from layouts. The problem with plotting from model space is that the objects must be located and sized in the coordinate system according to how they will be arranged for plotting. Even with AutoCAD's true-size capability, plotting from model space requires parts that need to be plotted at different scales be drawn at different scales. This is not necessary when the Layout/Paper Space feature is utilized.

Setting up the drawing for plotting from layouts is not as restrictive as it is from model space. However, some thought must still be given to the concept of scale, which is the ratio of true size to the size plotted. In other words, before you start drawing, you should have an idea of which parts of the drawing must be plotted at what scale. The example earlier in this chapter showed how to determine the appropriate scale so that objects will fit on the desired sheet size. That example can be used as a guide for plotting to scale in layouts, except that each view of the objects drawn in model space can be treated as a separate drawing on the paper space sheet, like pictures on a page in a scrapbook.

Setting Up Layouts

By default, a new drawing using the *acad.dwt* file as a template starts with two layouts: Layout1 and Layout2. A drawing started with another template might have only one layout, for example, the *ANSI B-Named Plot Styles.dwt* that is configured with an ANSI B Title Block on an 11 × 17 sheet. However, each of the two default layouts in the drawing started with *acad.dwt* represents a sheet in landscape mode 11 units by 8.5 units with a dashed rectangle 10.5 units by 8 units outlining the expected printable area.

The following examples of setting up and using Layouts are shown with the Model and Layout tabs displayed at the bottom of the Graphic Area. To display these tabs, right-click on the Model or Layout icon on the Status bar and select **Display Model and Layout Tabs**.

Figure 8–12 shows how views of a design object drawn in model space will appear with the Model tab selected, and with the Layout1 tab selected. Layout1 has the basic elements of a layout: a paper space drawing sheet with specific width and height and a viewport through which you can view design objects that are drawn in model space. If you erase the viewport, model space objects will not be visible.

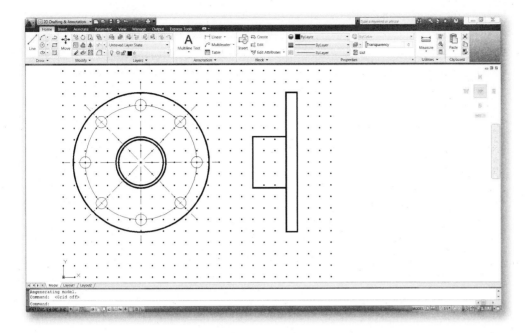

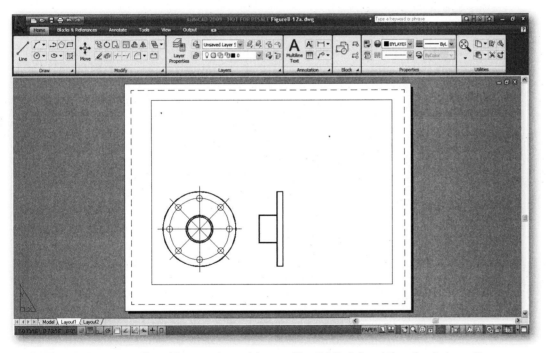

FIGURE 8–12 *Two views of an object, one in model space (11 × 8.5 limits) and the other in Layout1*

Figure 8–13 shows the basic parts of the default layout with only one viewport and nothing drawn in paper space.

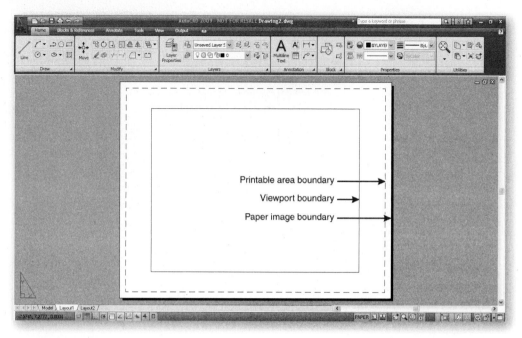

FIGURE 8–13 *Parts of a Layout tab*

By default, every initialized layout has an unnamed page setup associated with it. If necessary, you can change the layout's page setup through the Page Setup Manager, explained in detail in the section on "Reconfiguring the Layout with Page Setup" later in this chapter. You can modify the settings of a page setup at any time.

Layout(s) by Way of a Template

The most common procedure for creating a new drawing is to use a template file that has already been configured for the application or discipline in effect. The template will contain one or more layouts, each created at the desired sheet size and normally having a border, title block, revision history table, or other non-object elements drawn in paper space on it. Non-object elements are elements such as borders, title blocks, dimensions, and callouts versus object elements (lines, circles, arcs, etc.) that represent real objects such as walls, pipes, switches, and streets. A template drawing can also have layers, system variables, and styles for text, dimensions, and other features configured that conform to the standards of the proposed set of drawings.

The model space objects shown in Figure 8–12 will be used as an example of how to use a layout contained in an existing template drawing file, ANSI-A Color Dependent Plot Style template, to produce the desired plot. When the new drawing is created, the starting view will be of the layout named ANSI A Title Block, as shown in Figure 8–14a. After you switch to model space by selecting **Model** in the Status bar and draw the objects shown in Figure 8–12, you switch back to the ANSI A Title Block Layout tab. The objects will appear in the one viewport as shown in Figure 8–14b, and you can plot the drawing from paper space at 1:1 scale. The outline of the single viewport in this layout is not very distinguishable because it coincides with the inside lines of the border and title block. If you double-click inside the viewport, you will be switched to model space in the viewport while still in layout mode. This is different from switching to model space by using the Model tab. When you do this, the outline of the viewport becomes a heavy line and is more visible, as shown in Figure 8–14b.

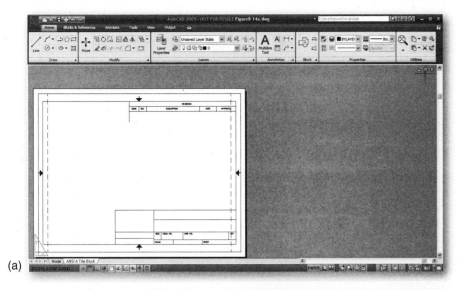

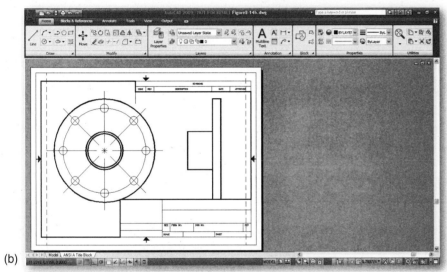

FIGURE 8–14 *Layout with the objects: (a) Title block and (b) title block with the objects*

Creating a New Layout

To create a new layout, right-click on the Layout tab and select *New Layout* from the shortcut menu (Figure 8–15). Up to 255 layouts can be created in a single drawing. AutoCAD creates a new layout. If necessary, you can rename the layout by double-clicking on the name and typing a new name.

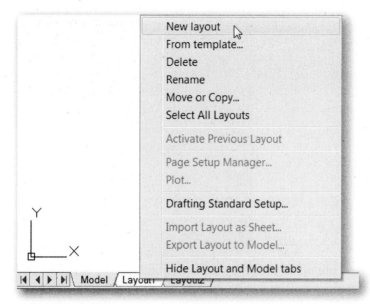

FIGURE 8–15 *Choosing New Layout from shortcut menu*

Layout names must be unique. Layout names can be up to 255 characters long and are not case sensitive. Only the first 31 characters are displayed on the tab.

By default, every initialized layout has an unnamed page setup associated with it. Once you create a layout, you can change the settings for the layout's page setup with the help of the Page Setup Manager dialog box, described later in this chapter, which includes the plot device settings and other settings that affect the appearance and format of the output. The settings you specify in the page setup are stored in the drawing file with the layout.

 NOTE If you want the Page Setup Manager to be displayed each time you begin a new layout, select the **Show Page Setup Manager for new layouts** option on the *Display* tab in the Options dialog box. If you don't want a viewport to be automatically created for each new layout, clear the **Create Viewport in new layouts** option on the Display tab in the Options dialog box.

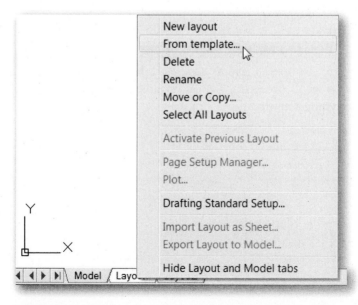

FIGURE 8–16 *Choosing Layout From template from shortcut menu*

Choosing *From template* from the shortcut menu (Figure 8–16) creates a new layout tab based on an existing layout in a template (*.dwt*), drawing (*.dwg*), or drawing interchange (*.dxf*) file. AutoCAD displays a standard file selection dialog box to select a file. Once you select a file, AutoCAD displays the Insert Layout(s) dialog box (Figure 8–17), which displays the layouts saved in the selected file. After you select a layout, the layout and all objects from the specified template or drawing file are inserted into the current drawing.

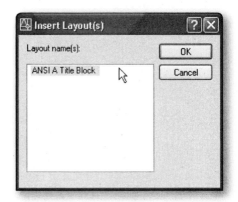

FIGURE 8–17 *Insert Layout(s) dialog box*

Additional options are available by invoking the LAYOUT command. The *Copy* selection creates a new layout by copying an existing layout, the *Delete* selection deletes an existing layout, the *Rename* selection renames an existing layout, and the *SaveAs* selection saves a layout as a drawing template (*.dwt*) file without saving any unreferenced symbol table and block definition information. You can access all of the available options from the shortcut menu that appears when you right-click the name of the layout tab.

Working with Floating Viewports

As mentioned earlier, in a layout, you can create multiple, overlapping, contiguous, or separated floating viewports, as shown in Figure 8–18. To repeat, viewports are configurable windows from paper space, or a layout, with a view into the model space design that you can move and resize. You can use any of the standard AutoCAD modify commands such as MOVE, COPY, STRETCH, SCALE, and ERASE to manipulate the floating viewports. For example, you can use the MOVE command to grab one viewport and move it around the screen without affecting other viewports. A viewport can be of any size and can be located anywhere in the layout. You must have at least one floating viewport to view the objects drawn in model space.

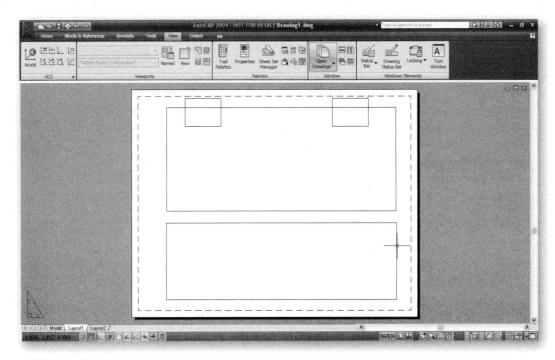

FIGURE 8–18 *Floating viewports in a layout*

You can create or manipulate floating viewports only while in paper space. While working in a layout, you can switch back and forth between model space and paper space. When you make a floating viewport current in a layout by double-clicking inside of it, you are then working in model space in that floating viewport. Any modification to objects in the drawing in model space is reflected in all paper space viewports in which those objects are visible as well as tiled viewports. When you double-click outside of a floating viewport, AutoCAD switches the mode to paper space. In paper space, you can add annotations or other graphical object such as a title block. You can even dimension model space objects while in paper space. Objects you add in paper space do not change the model or other layouts.

NOTE

While working in paper space, you can use an object snap mode (OSNAP) to snap to a point on an object in model space but not vice versa.

Figure 8–19 shows the model that is shown in Figure 8–14b in the desired plot-ready form. It has two floating viewports with dimensions and an annotation added in paper space. The topics that follow explain the steps required to proceed from Figure 8–14b to Figure 8–19.

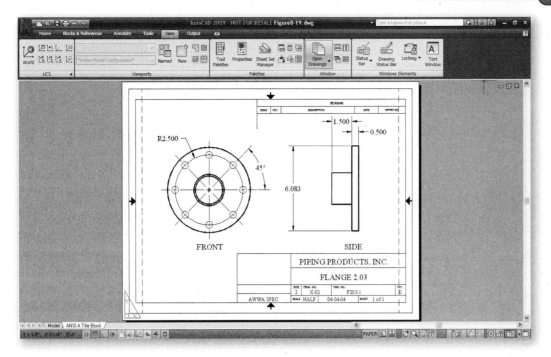

FIGURE 8–19 *Completed plot-ready drawing*

Almost all of the plotted sheets in a set of project drawings are the same size and share the same border/title block configuration, with repeated title block information such as the name of the project, client, and architect or engineer. The content, arrangement, and scale of the design objects are naturally different from one drawing sheet to the next. Each layout needs to have its own individualized group of floating viewports. Therefore, when a new drawing is created using a template file, first get rid of any unusable floating viewports in the layout that came with the template drawing by selecting and erasing them.

The example in Figure 8–14b shows the model space design in a single floating viewport. It has been decided that the front view and side view are too close together to allow for dimensioning. Rather than relocate the views in model space, we will create two viewports, one for each view of the object, and separate them on the layout to allow for the dimensioning. The existing viewport needs to be erased. To accomplish this, the existing viewport must be selected. Because it coincides with lines in the border/title block, it may be difficult to select by the pick method. The selection Window option of the object selection function can be employed. Figure 8–20 shows where the window needs to be in order to select only the viewport without including the border/title block. When the ERASE command is invoked, the drawing will appear as it is in Figure 8–14a. The model space design objects will not be visible.

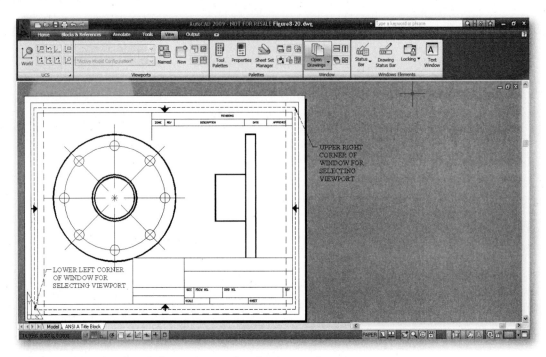

FIGURE 8–20 *Using the Window option to select the viewport for erasure*

Creating Floating Viewports

You can create a single floating viewport that fits the entire layout or create multiple floating viewports in the layout. Once you create a viewport, you can change its size and properties and move it as needed.

NOTE

It is important to create layout viewports on their own separate layer. When you are ready to plot, you can turn OFF the layer and plot the layout without plotting the boundaries of the layout viewports. Turning OFF the layer on which a viewport has been created turns off the border of that viewport only. The window to model space is not affected, and model space objects remain visible.

To create a single rectangular floating viewport, invoke the MVIEW command at the on-screen prompt, and AutoCAD displays the following prompts:

> Specify corner of viewport or ⊡ *(specify the first corner of viewport)*
>
> Specify opposite corner: *(specify the opposite corner to create the floating viewport)*

The viewport will be created on the active layer and, if necessary, you can move, copy, rotate, scale, or stretch the viewport or place it on a different layer, just as you would modify any other object.

To create an irregularly shaped floating viewport, choose Create Polygonal from the Viewports panel (Figure 8–21) located in the View tab, and AutoCAD displays the following prompts:

> Specify start point: *(specify first point to create an irregularly shaped floating viewport)*
>
> Specify next point or ⊡ *(specify next point or select one of the options from the shortcut menu)*

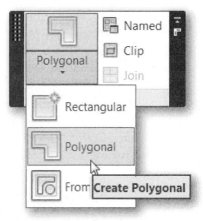

FIGURE 8–21 *Choosing Polygonal Viewport from the Viewports panel*

The *Arc* option adds arc segments to the polygonal viewport.

The *Length* option draws a line segment of a specified length at the same angle as the previous segment. If the previous segment is an arc, AutoCAD draws the new line segment tangent to that arc segment.

The *Undo* option removes the most recent line or arc segment added to the polygonal viewport.

The *Close* option closes the polygon to create the polygonal viewport.

To create a floating viewport from a closed polyline, ellipse, spline, region, or circle, choose Convert Object to Viewport from the Viewports panel (Figure 8–22) located in the View tab, and AutoCAD displays the following prompt:

```
Select object to clip viewport: (Select an object)
```

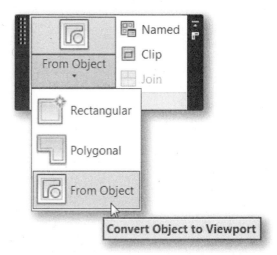

FIGURE 8–22 *Choosing Object from the Viewports panel*

The polyline you specify must be closed and must contain at least three vertices. It can be self-intersecting, and it can contain an arc as well as line segments.

Choose Named from the Viewports panel (Figure 8–23) on the View tab to create multiple floating viewports. This action opens the Viewports dialog box (Figure 8–24) and select New Viewports tab. AutoCAD lists standard viewport configurations. Choose the name of the configuration you want to use from the **Standard viewports** list. AutoCAD displays how the corresponding configuration will look in the **Preview** window. The **Setup** list specifies either a 2D or a 3D setup. When you select 2D, the new viewport configuration is initially created with the current view in all of the viewports. When you select 3D, a set of standard orthogonal 3D views is applied to the viewports in the configuration. The **Preview** section displays a preview of the viewport configuration that you select and the default views assigned to each individual viewport in the configuration. Choosing **Change view to** lets you replace the view in the selected viewport with the view you select from the list. You can choose a named view, or, if you have selected 3D setup, you can select from the list of standard views. Use the **Preview** area to see the choices. After choosing the viewport configuration and setting corresponding values, choose **OK** to close the dialog box, and AutoCAD displays the following prompts:

Specify first corner or ⬇ *(specify the first corner to define selected viewport configuration, or right-click and select fit to create the selected viewport configuration to fit the paper size)*

Specify opposite corner: *(specify opposite corner to define selected viewport configuration)*

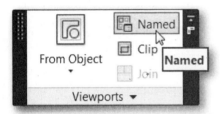

FIGURE 8–23 *Choosing New from the Viewports panel*

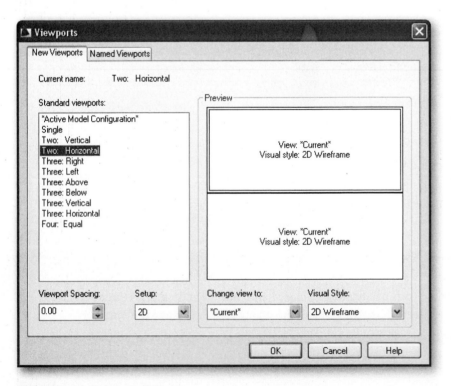

FIGURE 8–24 *Viewports dialog box*

Using the Viewports dialog box (Figure 8–24) for creating new viewports will be the method used in the example of transforming the layout shown in Figure 8–14b to the plot-friendly layout of Figure 8–19. Select the points shown in Figure 8–25 when prompted to specify the corners to determine the rectangle outlining the two vertical viewports. The model space design objects will be displayed similarly in both viewports (Figure 8–26).

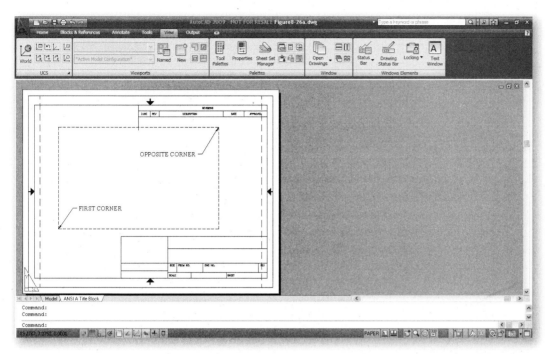

FIGURE 8–25 *Selection of points to create two vertical viewports*

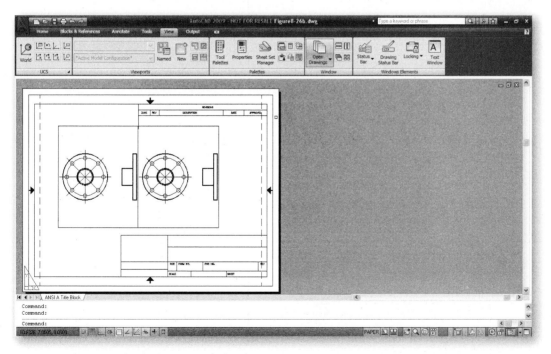

FIGURE 8–26 *Two vertical viewports with the objects*

Modifying Floating Viewports

As mentioned earlier, once you create a viewport, you can change its size and properties, and you can reposition it as needed. If you want to change the shape or size of a layout viewport, use its grips to edit the vertices just as you would edit any object with grips.

By using grips and/or the STRETCH and MOVE commands, the left viewport is widened and the right viewport is moved to the right and made narrower, as shown in Figure 8–27. The model space design objects are visible and will be arranged properly within each viewport, as described in later steps.

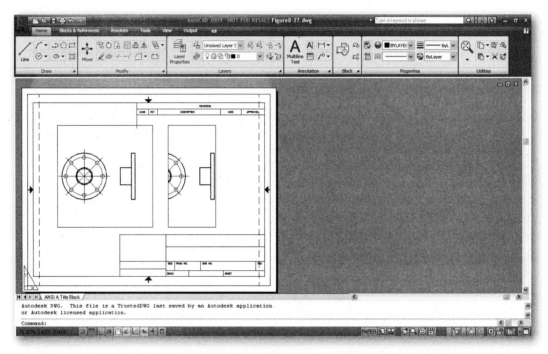

FIGURE 8–27 *Resizing and rearranging the two viewports*

You can also redefine the boundary of a layout viewport by using the VPCLIP command and maximize the viewport by using the VPMAX command. When you right-click on an active viewport, you can control whether or not the objects display in a viewport by changing the *Display Viewport Objects* setting. You can also control the setting of the locking feature, which prevents the zoom-scale factor in the selected viewport from being changed when working in model space.

The VPCLIP command allows you to clip a floating viewport to a user-drawn boundary. AutoCAD reshapes the viewport border to conform to a user-drawn boundary. To clip a viewport, you can select an existing closed object or specify the points of a new boundary. Invoke the CLIP command from Viewports panel (Figure 8–28) located in the View tab. AutoCAD displays the following prompts:

> Select viewport to clip: *(select viewport to clip)*
>
> Select clipping object or ⊙ *(select clipping object or select one of the options from the shortcut menu)*

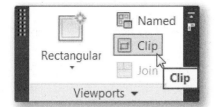

FIGURE 8–28 *Choosing Clip existing viewport from the Viewports panel*

If you select an object for clipping, AutoCAD converts the object to a clipping boundary. Objects that are valid as clipping boundaries include closed polylines, circles, ellipses, closed splines, and regions.

The *Polygonal* option allows you to draw line segments or arc segments by specifying points to create a polygonal clipping boundary.

The *Delete* option deletes the clipping boundary of a selected viewport. This option is available only if the selected viewport has already been clipped. If you clip a viewport that has been previously clipped, the original clipping boundary is deleted, and the new clipping boundary is applied.

Choosing *Maximize Viewport* from the shortcut menu when a viewport is selected causes the selected viewport in the current layout to fill the screen drawing area, making the entire drawing area accessible for viewing and editing. The size of the area displayed depends on the zoom factor in effect. When the viewport has been maximized, choosing the MINIMIZE VIEWPORT option in the shortcut menu returns the display to the previous layout state. You can also maximize or minimize viewports from the button located on the Status bar.

Selecting *Display Viewport Objects* from the shortcut menu when a viewport is selected controls the display of objects in the selected viewport. When OFF is selected, objects in the selected viewport are not visible, and the viewport cannot be selected when switching viewports in model space in the current layout. When ON (default) is selected, AutoCAD turns on a viewport, making it active and making its objects visible.

Selecting *Display Locked* from the shortcut menu when a viewport is selected prevents or enables the zoom-scale factor in the selected viewport from being changed when working in model space.

Scaling Views Relative to Paper Space

AutoCAD allows you to scale viewport objects relative to paper space, which establishes a consistent scale for each displayed view. To accurately scale the plotted drawing, you must scale each viewport relative to paper space. The layout is usually plotted at a 1:1 ratio. The ratio is determined by dividing the paper space units by the model space units. The scale factor of model space design objects in a viewport can be set with the XP option of the ZOOM command while model space is active in that viewport. For example, entering **1/24xp** or **0.04167xp** (1/24 = 0.01467) in response to the ZOOM command prompt will display an image to a scale of 1/2 = 1'-0", which is the same as 1:24 or 1/24. You can also change the plot scale of the viewport using the Viewport Scale Control available in the status bar (Figure 8–29). You can also change the viewport scale from Properties palette.

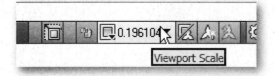

FIGURE 8-29 *Viewport Scale Control on the Status Bar*

In the case of the viewports shown in Figure 8–27, the views need to be set to half scale. First double-click in one of the viewports to make it active in model space, and then enter a scale factor in the Viewport Scale Control box of the Viewports toolbar of 0.5 or choose 1:2 from the drop-down list. AutoCAD displays $6 = 1'$, and the objects are rescaled in the selected viewport. Repeat the procedure for the second viewport and the result will appear, as shown in Figure 8–30.

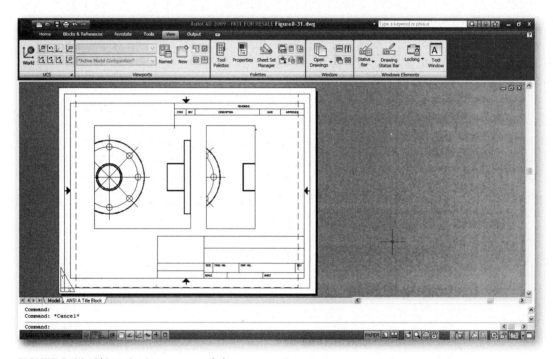

FIGURE 8-30 *Objects in viewports rescaled*

Centering Model Space Objects in a Viewport

In order to center the front view of the object in the left viewport, respond to the zoom Center option with the coordinates **3.5,5.0**, which is the center of the circle. Repeat the procedure in the right viewport using the coordinates **9.5,5.0**. Specifying the same Y coordinate for centering the model spaces in both viewports ensures that the objects will line up horizontally. Some practice, as well as some trial and error, is needed to size the viewports and center the model space design objects to ensure that only the desired object views are visible in the appropriate viewports. The resulting image will appear as shown in Figure 8–31.

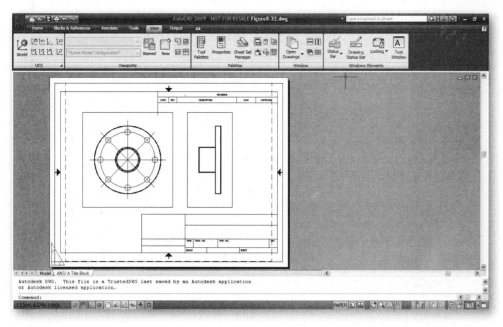

FIGURE 8–31 *Objects in viewports centered*

Rotate Views in Layout Viewports

When you rotate the viewport with the ROTATE command, AutoCAD also rotates the view if the VPROTAATEASSOC system variable is set to 1 (default setting). If the VPROTAATEASSOC system variable is set to 0, the view is not rotated when the viewport is rotated.

Hiding Viewport Borders

After the viewports are scaled and the objects are centered, double-click outside the viewports to return to paper space. Two viewports were initially created; turn the layer named Viewports OFF, and the result will appear as shown in Figure 8–32.

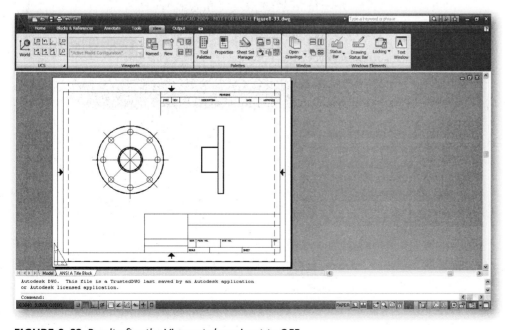

FIGURE 8–32 *Result after the Viewports layer is set to OFF*

While in paper space, the drawing name and other information can be entered in the title block as appropriate. The views can be named, and the objects can also be dimensioned, as shown in Figure 8–33.

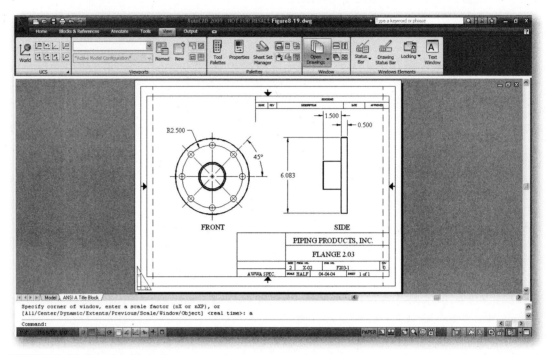

FIGURE 8–33 *Completed layout*

Controlling the Visibility of Layers within Viewports

The Layer Properties Manager dialog box controls the visibility of layers in a single viewport or in a set of viewports (Figure 8–34). This feature enables you to select a viewport and freeze a layer in it while still allowing the contents of that layer to appear in another viewport. Figure 8–35 shows two viewports containing the same view of the drawing. In one viewport, the layer containing the dimensioning, or the DIMLAYER, is set to ON, and in the other, the dimensioning layer is set to OFF.

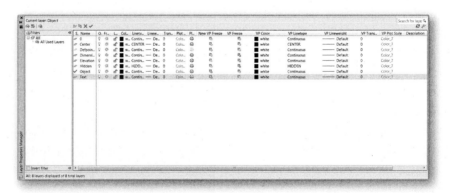

FIGURE 8–34 *Layer Properties Manager dialog box*

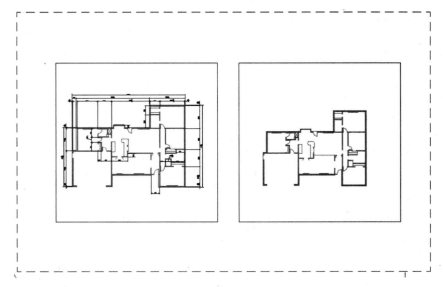

FIGURE 8–35 *One viewport with* DIMLAYER *ON and the other with* DIMLAYER *OFF*

The **VP Freeze** column is available only from a layout tab, the 12th column from the left in Figure 8–34. It freezes selected layers in the current layout viewport. You can freeze or thaw layers in the current viewport without affecting layer visibility in other viewports. **VP Freeze** is an override to the Thaw setting in the drawing. In other words, you can freeze a layer in the current viewport if it is thawed in the drawing, but you can't thaw a layer in the current viewport if it is frozen or OFF in the drawing. A layer is not visible when it is set to OFF or frozen in the drawing.

The **New VP Freeze** column is available only from a layout tab, the eleventh column from the left in Figure 8–34. It freezes selected layers in new layout viewports. For example, freezing the Text layer in all new viewports restricts the display of text on that layer in any newly created layout viewports but does not affect the Text layer in existing viewports.

The **Transparency** column lets you specify how transparent a layer is in the selected Viewport. The range is from 0 (not transparent) to 90 (almost invisible).

In Figure 8–34, the layers Dimension, Elevation, and Hidden are frozen in the current viewport, and Object and Text are frozen in all the new viewports.

Plotting from Layout

Invoke the PLOT command to plot the drawing from the selected layout. Before you plot the drawing from the layout, make sure you complete the following tasks:

- Create a model drawing.
- Create or activate a layout.
- Open the Page Setup dialog box, and set settings such as plotting device, paper size, plot area, plot scale, and drawing orientation. If necessary, configure the plotting device.
- If necessary, insert a title block or attach a title block as a reference file.
- Create and position floating viewport(s) in the layout.
- Annotate or create geometry in the layout as needed.
- Set the view scale of the floating viewport(s).
- Plot the layout.

To plot the current drawing from the layout, invoke the PLOT command from the Quick toolbar.

AutoCAD displays the Plot dialog box, as shown in Figure 8–36. This dialog box is similar to one displayed when you plot from model space except that by default Layout is selected in the **What to plot** list box in the **Plot area** section. For a detailed explanation, refer to the "Plotting from Model Space" section.

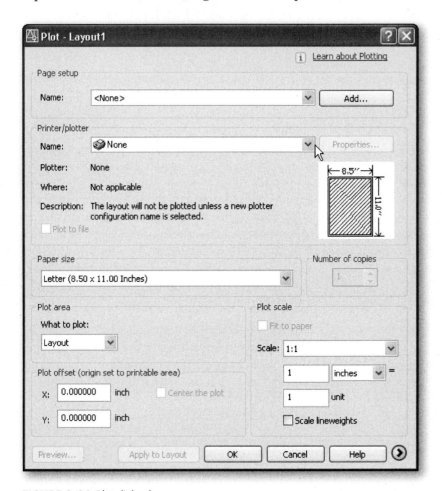

FIGURE 8–36 *Plot dialog box*

After making the necessary changes in the plot settings, choose **OK**. AutoCAD starts plotting the current layout.

RECONFIGURING THE LAYOUT WITH PAGE SETUP

As mentioned earlier, by default, every initialized layout has an unnamed page setup associated with it. You can modify the settings for the layout's page setup with the help of the Page Setup Manager dialog box. Choose Page Setup Manager from the Plot panel (Figure 8–37) located in the Output tab, and AutoCAD displays the Page Setup Manager dialog box (Figure 8–38).

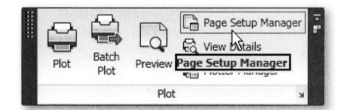

FIGURE 8–37 *Choosing Page Setup Manager from the Layouts panel*

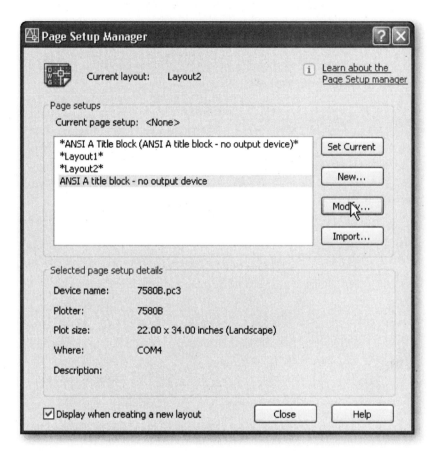

FIGURE 8–38 *Page Setup Manager dialog box*

AutoCAD displays the current layout name in the **Current layout** box. In the **Page setups** section of the Page Setup Manager dialog box, **Current page setup** displays the name of the page setup that is applied to the current layout. If the name is displayed as <None>, an unnamed page setup is assigned to the current layout. The **Page setups** section lists the page setups that are available to apply to the current layout. If the Page Setup Manager is opened from a layout, the current page setup is selected by default. The list includes the named page setups and layouts that are available in the drawing. Layouts that have a named page setup applied to them are enclosed in asterisks, with the named page setup in parentheses; for example, *ANSI A Title Block (portrait)*. You can double-click a page setup or a layout name that has an unnamed page setup associated in this list to set it as the current page setup for the current layout.

Figure 8–38 lists three layouts—*ANSI A Title Block (portrait)*, *Layout1*, and *Layout2*—and one page setup, ANSI A Title Block (portrait). **Current layout** is listed as Layout2. Layout1 and Layout2 have unnamed page setups assigned to them.

Changing the Current Page Setup

To change the page setup for the current layout, select the named page setup or a layout that has an unnamed page setup associated, and choose **Set Current** to set the selected page setup as the current page setup.

Modifying the Page Setup

To modify the page setup assigned to the current layout, choose **Modify**. AutoCAD displays the Page Setup dialog box, as shown in Figure 8–39.

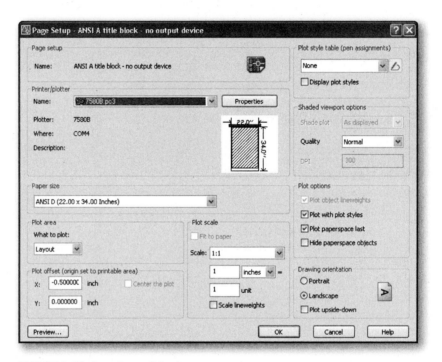

FIGURE 8–39 *Page Setup dialog box*

The Page Setup dialog box specifies page layout and plotting device settings. The Page Setup dialog box is similar to the expanded Plot dialog box (Figure 8–5) except it cannot be reduced in size and it does not contain some of the features that apply primarily to plotting. Items that do not appear are **Add** in the **Page setup** section, and **Plot in background, Plot stamp on**, and **Save changes to layout** in the **Plot options** section. Also, **Apply to layout** at the bottom of the Plot dialog box is not on the Page Setup dialog box. All of the other options and features are explained in the "Plot Settings" section, discussed earlier in this chapter.

To create a new page setup that can be assigned to any of the layouts, choose **New** in the Page Setup Manager dialog box. AutoCAD displays the New Page Setup dialog box, as shown in Figure 8–40.

FIGURE 8–40 *New Page Setup dialog box*

Specify the name of the new page setup in the **New page setup name** box. Select one of the available page setups to use as a starting point for the new page setup. <None> specifies that no page setup is used as a starting point. <Default output device> specifies the default output device. Choose **OK** to close the dialog box, and AutoCAD displays the Page Setup dialog box with the settings of the selected page setup, which you can modify as necessary.

To import a page setup from a drawing template or drawing file, choose **Import**. AutoCAD displays the Select Page Setup From File dialog box, a standard file selection dialog box, in which you can select a drawing format (*.dwg*), or drawing template (*.dwt*) file from which to import one or more page setups. After selecting the appropriate file, choose **Open**, and AutoCAD displays the Import Page Setups dialog box. Choose one of the available page setups to import to the current drawing, and choose **OK** to close the dialog box.

In the **Selected page setup details** section of the Page Setup Manager dialog box (Figure 8–38), AutoCAD displays information relative to the selected page setup. **Device name** displays the name of the plot device, **Plotter** displays the type of plot device, **Plot size** displays the plot size and orientation, **Where** displays the physical location of the output device, and **Description** displays descriptive text about the output device.

The **Display when creating a new layout** box specifies that the Page Setup Manager dialog box is displayed when a new layout tab is selected or when a new layout is created.

After making necessary changes in the Page Setup Manager dialog box, choose **Close** to close the dialog box.

CREATING A LAYOUT BY LAYOUT WIZARD

Once you have mastered the concepts of layouts and viewports, you can capitalize on the time-saving features in the Layout Wizard for creating new layouts. The Layout Wizard lets you create a new layout (paper space) for plotting. Each wizard page instructs you to specify layout and plot settings for the new layout you are creating. Once the layout is created using the wizard, you can modify layout settings using the Page Setup dialog box.

Open the Layout Wizard by typing **layoutwizard** at the on-screen prompt. AutoCAD displays the Begin page of the Layout Wizard, as shown at in Figure 8–41(a).

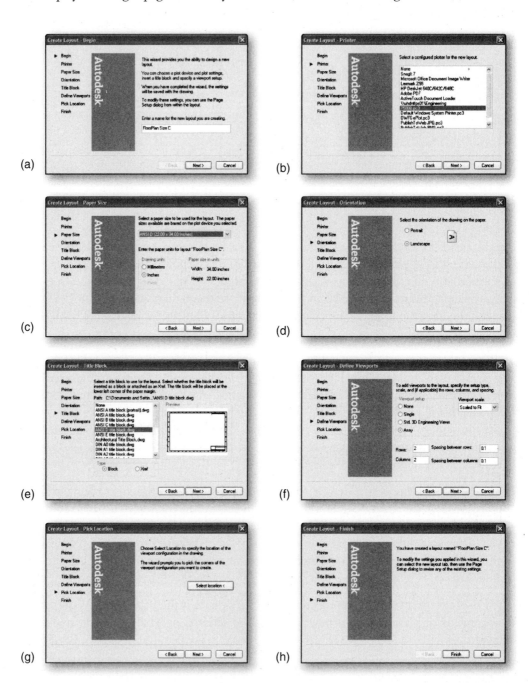

FIGURE 8–41 *Layout Wizard pages: (a) Create Layout – Begin page, (b) Create Layout – Printer page, (c) Create Layout – Paper Size page, (d) Create Layout – Orientation page, (e) Create Layout – Title Block page, (f) Create Layout – Define Viewports page, (g) Create Layout – Pick Location page, (h) Create Layout – Finish page*

Specify the name of the layout in the **Enter a name for the new layout you are creating** text field. Choose **Next**. AutoCAD displays the Create Layout – Printer page of the Layout Wizard, as shown in Figure 8–41(b).

Select a configured plotter for the new layout from the list box. If you do not see the name of the plotter to which you want to plot, refer to the section on "Configuring Plotters" to configure a plotter. Once you have selected the plotter configuration, choose **Next**. AutoCAD displays the Create Layout – Paper Size page of the Layout Wizard, as shown in Figure 8–41(c).

Select a paper size to be used for the layout from the list box. The paper sizes available are based on the plot device you selected. Select drawing units from one of the two radio buttons located in the **Drawing units** section. Choose **Next**, and AutoCAD displays the Create Layout – Orientation page of the Layout Wizard, as shown in Figure 8–41(d).

Select the orientation of the drawing on the paper from one of the two radio buttons: **Portrait** or **Landscape**. Choose **Next,** and AutoCAD displays the Create Layout – Title Block page of the Layout Wizard, as shown in Figure 8–41(e).

Select a title block from the list box to use for the layout. Select whether the title block will be inserted as a block or attached as an external reference. Choose **Next**, and AutoCAD displays the Create Layout – Define Viewports page of the Layout Wizard, as shown in Figure 8–41(f). Choose one of the following four available options in the Viewport setup section.

- **None**—if you do not need any floating viewports.
- **Single**—to create one floating viewport.
- **Std. 3D Engineering Views**—to create four viewports with top left set for top view, top right for isometric view, bottom left for front view, and bottom right for right side view. If necessary, you can specify the distance between the viewports in the **Spacing Between rows** and **Spacing between columns** text fields.
- **Array**—to create an array of viewports. Specify the number of viewports in rows and columns in the **Rows** and **Columns** text fields. If necessary, you can specify the distance between the viewports in the **Spacing between rows** and **Spacing between columns** text fields.

Choose **Next** and AutoCAD displays the Create Layout – Pick Location page of the Layout Wizard, as shown in Figure 8–41(g).

Choose **Select location** to specify the location of the viewport configuration in the drawing. The wizard prompts you to specify the corners of the viewport configuration that you want to create. After you specify the location, AutoCAD displays the Create Layout – Finish page of the Layout Wizard, as shown in Figure 8–41(h).

Choose **Finish** to create the layout. If necessary, you can make any changes to the newly created layout by using the Page Setup dialog box.

ANNOTATIVE SCALING

The real object that your drawing is depicting is called the model, and your drawing provides a picture of something to be built, manufactured, or otherwise visualized and understood by the viewer. Words and symbols are necessary to complete the picture. These words and symbols, referred to as "annotations" describe the model, tell its size, and are used to point out parts of the model or to label the views. The size of annotation does not depend on the size of the model except for dimension lines. Annotations are drawn in a view and at a size that makes them readable.

Annotative objects include hatches, text (single line and multiline), dimensions, tolerances, leaders and multileaders (created with MLEADER), blocks, and attributes. Blocks and attributes might or might not be used annotatively. They are discussed in Chapter 11.

AutoCAD 2008 introduced a new and easy method to display and plot such elements as text, symbols, and dimensions to the desired size: annotative scaling. Without annotative scaling, if you wanted to plot objects at one-quarter scale ($3 = 1'-0''$) and make the text 3/16 high, you would have to draw the text with a text height of 3/4, as described earlier in this chapter. With annotative scaling, you can create annotative objects such as text and dimensions and have them be automatically scaled, displayed, and plotted at the correct size and in the desired views. To achieve the proper effects, it takes a little planning. The feature works only on the annotative objects such as those listed in the previous paragraph, those that are normally used to describe and tell the size of the real object being drawn.

Space Relationships

To fully utilize annotative scaling, it is important to have a good grasp of the uses of model space and paper space. Annotative scaling is specifically designed to be used on annotative objects that have been drawn in model space so that, when they are displayed in a Layout viewport (paper space), they will be the correct size. Figure 8–33 in the previous discussion provides an example of how you can add dimensions and text while still in paper space mode. In that example, however, the dimensions of the model and the text were drawn in paper space. It illustrates the flexibility of the program and reminds us that most actions in AutoCAD can be accomplished by more than one method. Annotative scaling lessens the need for dimensioning the model from paper space or drawing dimensions at a calculated size so that they will plot at the correct scale. Some of the important concepts regarding model space and paper space that apply to annotative scaling are highlighted below.

- Model space is where the model is drawn, usually to full scale.
- Paper space is intended for use as a 2D plane in a layout that represents one drawing sheet at the full scale width and height of the final plotted sheet.
- Floating viewports are areas in the layout that are windows to the model or parts of it.
- Paper space is made available when a layout is created.
- In a layout, objects drawn in model space are visible only through a viewport. You can view the objects through a viewport, but you can only add, erase, or otherwise modify model space objects while that viewport is active.
- A layout can have multiple viewports, each displaying a different part of the model from different angles and at different scales.

Applying Annotative Scaling

Annotative scaling is used to draw objects that have been defined at a specific size and then display and plot them at that size, primarily from viewports whose scale is not 1:1. For example, 3/16 is a common height for plotting text for many drawings. This is the default text height for the standard text style and the height at which text will be plotted when the drawing is plotted at a scale of 1:1. However, when the scale at which you wish to plot changes, say to $3 = 1'-0''$ (1:4), the text must be drawn four times larger, or 3/4 high, in order to still be plotted at a height of 3/16. The Annotative scaling feature does this automatically.

To work as intended, annotative scaling must be applied to both the annotative object and to the Layout viewport in which it will be displayed. When an annotative scale is applied to an object such as text, it is said to support that scale.

The Model tab and the Layout viewports automatically have an annotation scale applied to them. When you start AutoCAD for the first time, in the default drawing, the Model tab that is displayed has its annotative scale set at 1:1. When you select Layout1 or Layout2 and double-click inside the viewport to make it active, each of them also has its annotative scale set at 1:1. The Model tab or Layout viewport, whichever is active, has its annotative scale displayed on right end of the Status bar.

> **NOTE**
>
> In addition to the annotation scale, the Layout viewport will have its VP scale displayed on the Status bar.

For an annotative object to be displayed and automatically sized in a Layout viewport, one of the annotative scales that the object supports must be the same as the annotation scale of the selected viewport. When the object and viewport scale match, the object will be scaled correctly so it will be plotted at the appropriate size. In the Model tab, all annotative objects are displayed regardless of annotative scale settings.

The following example demonstrates how annotative scaling is used for text with a specified height of 3/16. It is plotted at 3/16 high even though it has been displayed in a layout viewport at one-quarter scale $(3 = 1'\text{-}0'')$. Instead of manually drawing the text four times larger to compensate for the reduced scale, AutoCAD automates the process with annotative scaling.

> **NOTE**
>
> This example shows the text being drawn on the Model tab with annotative scaling set for both the text and model space. You will learn later how you can draw the text in the destination viewport with similar results. You will also learn how to change non-annotative text to annotative text and to change or add to the annotative scales that an annotative object supports.

Figure 8–42 is a drawing of a simple rectangular building with a door in one corner. The annotation scale is set at 1:1 (default) and displayed in the Status bar.

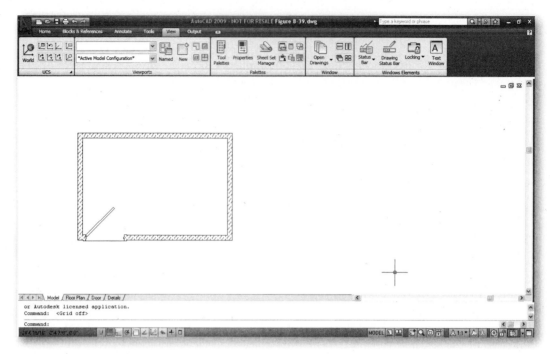

FIGURE 8–42 *Simple building plan*

Figure 8–43 shows that the annotation scale for the Model tab has been changed to 3" = 1'–0", and the ZOOM command was used to display the lower-left corner of the building. The annotative text is drawn with the text size set to 3/16. AutoCAD automatically displays the text at a size of 3/4" (four times the size of 3/16").

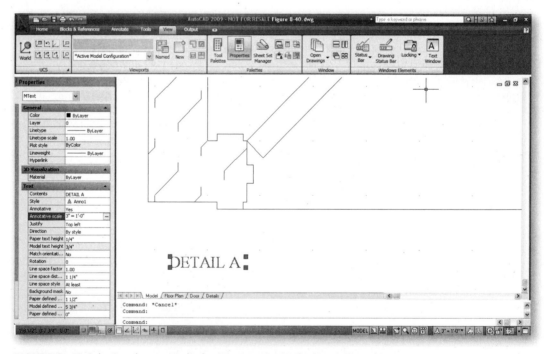

FIGURE 8–43 *Selecting the text to display its properties in the Properties palette*

In Figure 8–43, the Properties palette lists the properties related to the annotation of the selected text DETAIL A. The properties listed include the following: the annotative scale is set to $3 = 1'\text{-}0''$, the annotative property is set to Yes, the model text size is set to $3/4''$, and the paper text size is set to $3/16''$.

> Zooming in and out on the Model tab does not affect the annotation scale, and changing the annotation scale does not affect the zoom factor. You will learn, however, that changing the annotation scale of a viewport on one of the Layout tabs will reset the zoom factor to match.

NOTE

Figure 8–44 shows how the text is displayed in the viewport on the Details Layout that has had its annotative scale set to $3'' = 1'\text{-}0''$. When the cursor hovers over an annotative object, a special icon is displayed to report that the object's annotative property has been set to Yes.

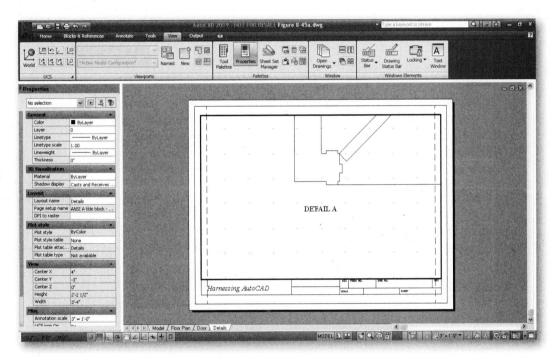

FIGURE 8–44 *A Layout with its Annotation Scale set to 3 = 1'–0''*

Displaying Annotative Objects

While the primary purpose of annotative scaling is to make it easier to draw annotative objects at the desired size, it includes another significant feature: displaying or hiding objects in the desired viewports.

Annotative objects are displayed or hidden in a viewport depending upon the annotative scales they support as being or not being the same as the annotation scale of the viewport. This function also depends on whether or not the object's annotative property is set to Yes.

In model space or a Layout viewport, you can display all the annotative objects or only those that support the current annotation scale. This reduces the need for multiple layers to manage the visibility of your annotations. You can use the **Annotation Visibility** button on the Status bar to choose the display setting for annotative objects.

By default, the **Annotation visibility** is turned ON. When annotation visibility is turned on, all annotative objects are displayed. When annotation visibility is turned OFF, only annotative objects for the current scale are displayed. If an object supports more than one annotation scale, the object will display at the current scale.

Annotation Scale Setting

Annotative scaling can be applied to objects, to styles, and to Layout viewports. When an object has had its annotative scale set to a certain scale, say $\frac{1}{4} = 1'-0''$, it will automatically be displayed at the correct size when it is displayed in a viewport whose scale is the same, in this case $\frac{1}{4} = 1'-0''$. This is provided that the layer on which it is drawn is turned ON in the viewport and the part of model space where the object is drawn is visible in the viewport. If an annotative object does not support the annotative scale of the current viewport, it will not be displayed in that viewport.

> If an object supports more than one annotative scale, it will be displayed in each of those viewports whose annotative scale it supports.

Annotative objects can be created to support one or more annotative scales in several ways. As described above, the annotative scale of the current space is set to the desired value before creating the object. Objects created will automatically support that scale if the current text style is annotative. If the annotative scale of the space is subsequently changed, the annotative scale(s) that the previously created objects support remain unchanged. Objects created after the change will now support the annotative scale equal to that of the new annotative scale of the model space.

The Annotation Scale option is available while you are working in the Model tab or in an active Layout viewport. Select the arrow on the Status bar (Figure 8–45), and AutoCAD displays the shortcut menu with the list for standard scales (Figure 8–46). Choose the desired scale to set it as the current annotative scale.

FIGURE 8–45 *Choosing the Annotation Scale arrow*

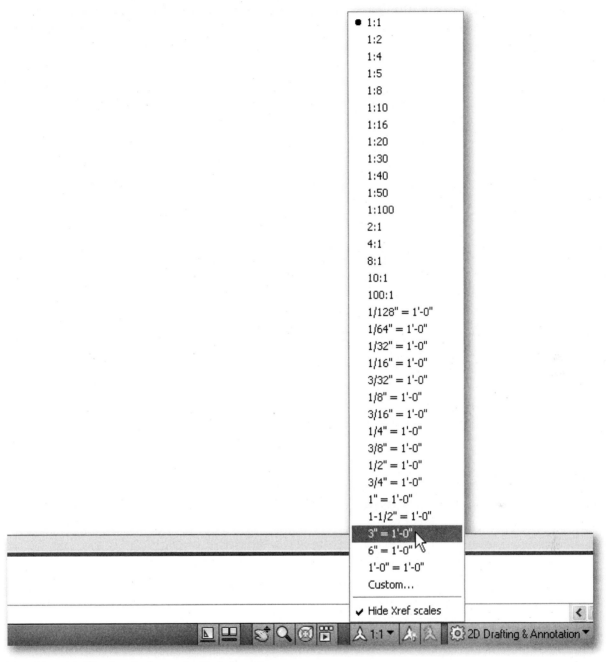

FIGURE 8–46 *Choosing the Annotation Scale from the menu*

If you change the annotation scale in a Layout viewport, the zoom factor in the view will adjust to conform to that scale, changing the VP scale to match. If you change the VP scale, AutoCAD will change the annotation scale to match it. For example, if you change the annotation scale in a viewport from 1:1 to $3 = 1'-0''$, object will appear four times smaller. If you zoom in or out with the ZOOM command in the active viewport, the VP scale will change but not the annotation scale.

Creating Annotative Objects through Styles and Dialog Boxes

Many of the dialog boxes that are used to create styles for annotative objects have an annotative checkbox for making objects that are drawn using that style annotative. When you create a new text style or modify an existing one, you can select **Annotative** to make text drawn with this style annotative (Figure 8–47). When a style is set to annotative, it displays a special triangle icon in the styles list.

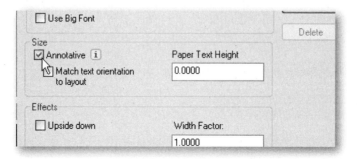

FIGURE 8–47 *Selecting Annotative in the text style dialog box*

In a similar manner, when you create a new dimension style or modify an existing one, you can select **Annotative** on the Fit tab in the Dimension Style dialog box to make dimensions drawn with this style annotative (Figure 8–48).

FIGURE 8–48 *Selecting Annotative on the Fit tab of the Dimension Style dialog box*

When you create a new hatch pattern or gradient or modify an existing one, you can select **Annotative** on the Hatch and Gradient dialog box to make patterns drawn with this setup annotative (Figure 8–49).

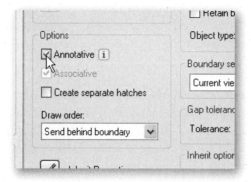

FIGURE 8–49 *Selecting Annotative in the Hatch and Gradient dialog box*

When you create a new multileader style or modify an existing one, you can select **Annotative** on the Leader Structure tab in the Multileader Style dialog box to make multileaders drawn with this style annotative (Figure 8–50).

FIGURE 8–50 *Selecting Annotative on the Leader Structure tab of the Multileader Style dialog box*

Making Objects Annotative with the Properties Palette

Existing objects can be made annotative by changing their annotative property to **Yes** in the Properties palette. From the same palette, you can also change the existing annotative scale that an object supports or add annotative scales that the object will support.

To add annotative scales to the selected object, select the ellipsis button on the right of the Annotative scale text box (Figure 8–51).

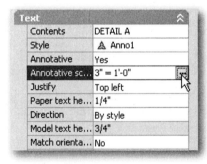

FIGURE 8–51 *Choosing the ellipsis button on the right of the Annotative scale text box in the Properties palette*

AutoCAD displays the Annotation Object Scale dialog box (Figure 8–52). Choose the **Add** button to add annotative scales.

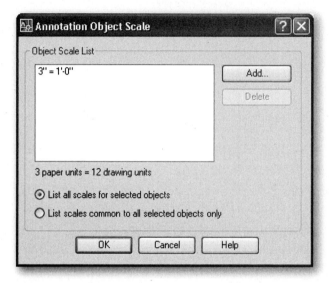

FIGURE 8–52 *Choosing the Add button in the Annotation Object Scale dialog box*

AutoCAD displays the Add Scales to Object dialog box (Figure 8–53). Select the desired scale, and AutoCAD adds the selected scale to the Object Scale list in the Annotation Object Scale dialog box.

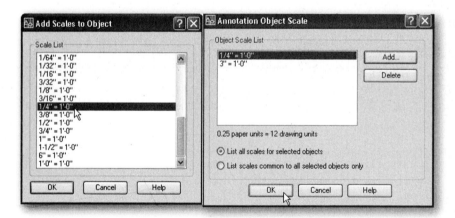

FIGURE 8–53 *Choosing a scale from the Add Scales to Object Scale dialog box and adding it to the list in the Annotation Object Scale dialog box*

You can also add annotative scales when the **Automatically add the scales to annotative objects when the annotation scale changes** button (on the Status bar) is set to ON. AutoCAD will add the newly set annotation scale to all annotative objects that are displayed.

When you select the annotation object that has multiple annotation scales, Auto-CAD shows the sizes that represent the annotative scales that the object supports (Figure 8–54).

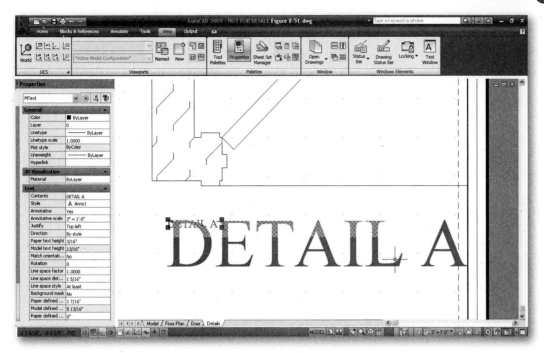

FIGURE 8–54 *Selecting text object to display its image in multiple annotative scale representations*

Multiple Viewports in the Same Layout with Different Scales

Annotative objects that support different scales can all be created in model space and be visible at the same time (Figure 8–55). In a Layout, when multiple viewports are set to different scales, only the annotative objects that support a particular viewport's scale will be visible in that viewport (Figure 8–56).

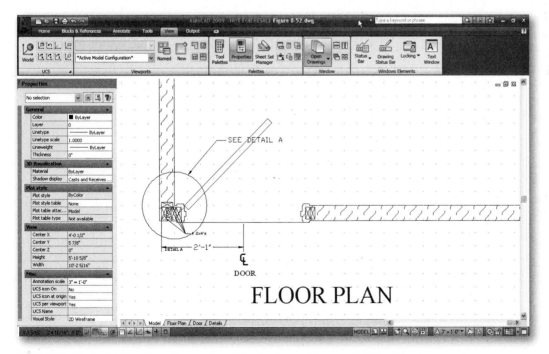

FIGURE 8–55 *Annotative dimensions and multileaders in model space at different scales*

In Figure 8–56, the $2' - 1''$ dimension, the text DOOR, and the leader with the callout SEE DETAIL A are drawn as annotative objects supporting the annotative scale of $1\frac{1}{2} = 1' - 0''$. The multileader with the callout "$42 \times 4's$" and the text DETAIL A are drawn as annotative objects supporting the annotative scale of $3 = 1' - 0''$. These were all drawn using the same basic styles having the same text heights, arrow sizes, and so on. They were resized to accommodate the scale they supported. In this example, the text style called for the text height to be $3/16''$.

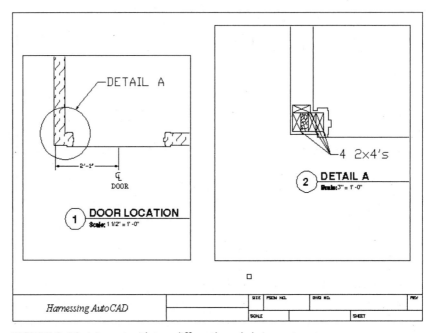

FIGURE 8–56 *A Layout with two differently scaled viewports*

Relocating Representations of an Object

When an annotative object supports more than one annotative scale, multiple representations are displayed when the object is selected. The representation that matches the annotative scale of the current viewport is the one that can be selected with the cursor. You can manipulate the annotative object by using the grips as you would with any other object. Representations for other scales will be displayed in gray. This allows the object to be displayed at one location and orientation in one viewport, whose annotative scale matches the image selected, and at another location and orientation in another viewport.

Figure 8–57, on the left, shows a viewport with the annotation scale set at $1\frac{1}{2} = 1'-0''$ and the one on the right set at $3 = 1'-0''$. The annotative text 3068 DOOR supports both annotative scales and is therefore visible in both viewports. However, its location is different in the two viewports. On the left viewport, the text is centered in the doorway. On the right, it is closer to the left door jamb so that it will be visible in the displayed area. This step is accomplished by selecting the object in the active viewport in which you wish to manipulate it using its grips.

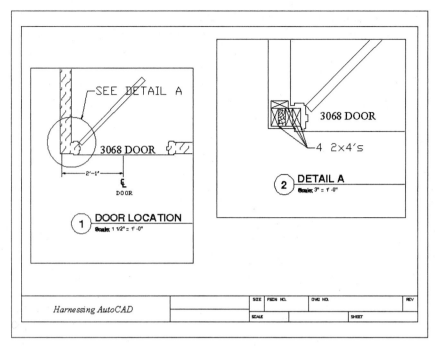

FIGURE 8–57 *Annotative text at different locations in different viewports*

Annotative Orientation

You can set the orientation of annotative blocks and text objects so that it matches the orientation of the layout. This action can be accomplished even when the view in the layout viewport is not planar or is twisted. The orientation of blocks and text can be set either through the Block Definition or Style dialog boxes or through the Properties palette.

Figure 8–58 shows a floor plan with the text SALTILLO TILE, which is drawn using an annotative style.

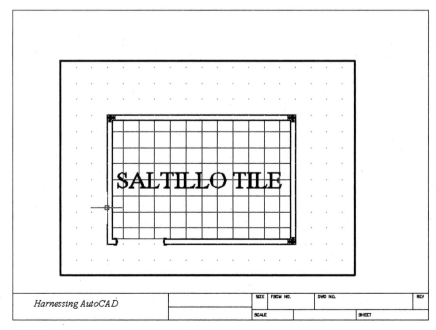

FIGURE 8–58 *Text in a planar view in a viewport*

Select the text and, in the Properties palette, set the Match Orientation to Layout option to Yes. Figure 8–59 shows how the text remains planar even when the viewport is changed to a view that is not planar.

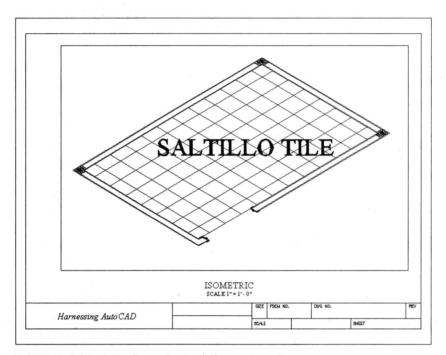

FIGURE 8–59 *Text that has maintained planar orientation*

MAKING THINGS LOOK RIGHT FOR PLOTTING

Once you have mastered the concepts of layouts and viewports, you can utilize the following features that will enhance the appearance of your plots.

Setting Paper Space Linetype Scaling

Linetype dash lengths and the space lengths between dots or dashes are based on the drawing units of the model or paper space in which the objects were created. They can be scaled globally by setting the value of the system variable LTSCALE, as explained in Chapter 3. If you want to display objects in viewports at different scales in layout, the linetype objects would be scaled to model space rather than paper space by default. However, by setting paper space linetype scaling with the system variable PSLTSCALE set to 1 (default), dash and space lengths are based on paper space drawing units, including the linetype objects that are drawn in model space. For example, a single linetype definition with a dash length of 0.30, displayed in several viewports with different zoom factors, would be displayed in paper space with dashes of length 0.30, regardless of the scale of the viewpoint in which it is being displayed (PSLTSCALE set to 1).

NOTE When you change the PSLTSCALE value to 1, the linetype objects in the viewport are not automatically regenerated. Use the REGEN or REGENALL command to update the linetypes in the viewports.

Dimensioning in Model Space and Paper Space

Dimensioning can be done in both model space and paper space. There are no restrictions placed on the dimensioning commands by the current mode. For dimensioning in model space, the DIMSCALE factor should be set to 0.0. This setting causes AutoCAD to compute a scale factor based on the scaling between paper space and the current model space viewport.

Figure 8–60 shows dimensions and view labels that have been drawn in paper space of model space objects.

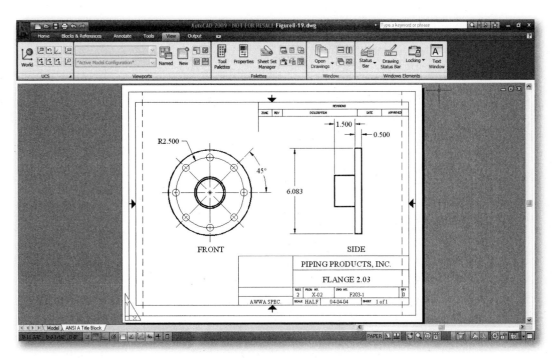

FIGURE 8–60 *Model space objects dimensioned in paper space*

CREATING A PLOTSTYLE TABLE

Plotstyle tables are settings that give you control over how objects in your drawing are plotted into hardcopy plots. By modifying an object's plotstyle, you can override that object's color, linetype, and lineweight. You can also specify end, join, and fill styles as well as output effects such as dithering, grayscale, pen assignment, and screening. You can use plotstyles if you need to plot the same drawing in different ways.

By default, every object and layer has a plotstyle property. The actual characteristics of plotstyles are defined in plotstyle tables that you can attach to a Model tab and layouts within drawings. If you assign a plotstyle to an object and then detach or delete the plotstyle table that defines the plotstyle, the plotstyle will have no effect on the object.

AutoCAD provides two plotstyle modes: color dependent and named.

The 255 color-dependent plotstyles are based on object color. You cannot add, delete, or rename color-dependent plotstyles. You can control the way all objects of the same color plot in color-dependent mode by adjusting the plotstyle that corresponds to that object color. Color-dependent plotstyle tables are stored in files with the extension *.ctb*.

Named plotstyles work independently of an object's properties. You can assign any plotstyle to any object regardless of that object's color. Named plotstyle tables are stored in files with the extension *.stb*.

By default, all the plotstyle table files are saved in the path that is listed in the **Files** section of the Options dialog box.

The default plotstyle mode is set in the Plot Style Table Settings dialog box (Figure 8–61), which can be opened by choosing **Plot Style Table settings** in the Plot and Publish tab of the Options dialog box.

FIGURE 8–61 *Plot Style Table Settings dialog box*

Every time you start a new drawing in AutoCAD, the plotstyle mode set in the Options dialog box is applied. Whenever you change the mode, it is applied only to the new drawings or to an open drawing that has not yet been saved in AutoCAD.

The CONVERTPSTYLES command converts a currently open drawing from color-dependent plotstyles to named plotstyles or from named plotstyles to color-dependent plotstyles, depending on which plotstyle method the drawing is currently using.

Creating a New Plot Style Table

AutoCAD allows you to create a named plotstyle table to use all the flexibility of named plotstyles or a color-dependent plotstyle table to work in a color-based mode. The Add Plot Style Table Wizard allows you to create a new plotstyle, modify an existing plotstyle table, import plotstyle properties from an *acadr14.cfg* file, or import plotstyle properties from an existing *.pcp* or *.pc2* file. After you invoke the Add Plot Style Table option from the Wizards flyout of the Tools menu (while in the

AutoCAD Classic workspace), AutoCAD displays the introductory text of the Add Plot Style Table Wizard, as shown in Figure 8–62(a).

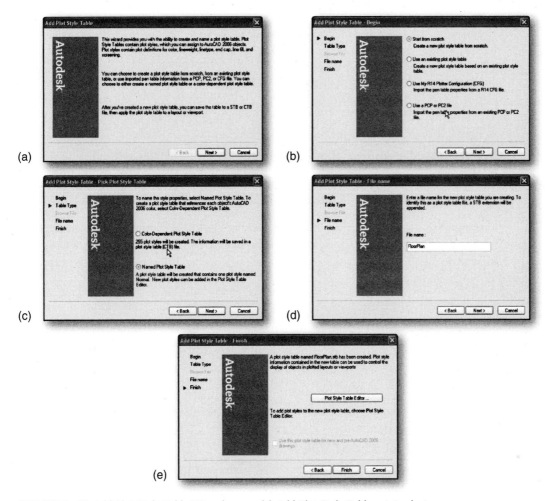

(a) (b) (c) (d) (e)

FIGURE 8–62 *Add Plot Style Table Wizard pages: (a) Add Plot Style Table – Introductory page, (b) Add Plot Style Table – Begin page, (c) Add Plot Style Table – Pick Plot Style Table page, (d) Add Plot Style Table – File name page, (e) Add Plot Style Table – Finish Page*

Choose **Next**, and AutoCAD displays the Add Plot Style Table – Begin page, as shown in Figure 8–62(b). The following four options are available:

Selecting **Start from scratch** allows you to create a new plotstyle.

Selecting **Use an existing plotstyle table** creates a new plotstyle using an existing plotstyle table.

Selecting **Use My R14 Plotter Configuration (CFG)** creates a new plotstyle table using the pen assignments stored in the *acadr14.cfg* file. Select this option if you do not have an equivalent *.pcp* or *.pc2* file.

Selecting **Use a PCP or PC2 file** creates a new plotstyle table using pen assignments stored in a *.pcp* or *.pc2* file.

To create a new pen table, select **Start from scratch** and choose **Next**. AutoCAD displays the Add Plot Style Table – Pick Plot Style Table page, as shown in Figure 8–62(c). Select one of the following options:

Select **Color-Dependent Plot Style Table** to create a plotstyle table with 255 plotstyles.

Select **Named Plot Style Table** to create a named plotstyle table.

Choose **Next**, and AutoCAD displays the Add Plot Style Table – File name page, as shown in Figure 8–62(d). Specify the file name in the **File name** box. By default, the new style table is saved in the path that is listed in the **Files** section of the Options dialog box.

Choose **Next**, and AutoCAD displays the Add Plot Style Table – Finish page, as shown in Figure 8–62(e).

Set the **Use this plotstyle table for new and pre-AutoCAD 2006 drawings** to ON to attach this plotstyle table to all new drawings and pre-AutoCAD 2006 drawings by default.

Choose **Finish** to create the plotstyle table and close the wizard.

Modifying a Plot Style Table

AutoCAD allows you to add, delete, copy, paste, and modify plotstyles in a plotstyle table by using the Plot Style Table Editor. You can open more than one instance of the Plot Style Table Editor at a time, and you can copy and paste plotstyles between the tables. Open the Plot Style Table Editor using any of the following methods:

- Choose the **Plot Style Table Editor** button from the Finish screen in the Add Plot Style Table Wizard.
- Open the Plot Style Manager from the File menu, right-click a .ctb or .stb file, and choose OPEN from the shortcut menu.
- In the **Plot style table (pen assignment)** section of the Plot dialog box or Page Setup dialog box, select the plotstyle table that you want to edit from the **Plot Style Table** list, and choose the **Edit** button.
- On the Plot and Publish tab of the Options dialog box, choose the **Add or Edit Plot Style Tables Settings** button.

Figure 8–63 shows an example of the Plot Style Table Editor for a named plotstyle table, and Figure 8–64 shows an example of the Plot Style Table Editor for a color-dependent plotstyle table.

FIGURE 8–63 *Plot Style Table Editor dialog box for a named plotstyle table: General tab selection, Table View tab selection, Form View tab selection*

FIGURE 8–64 *Plot Style Table Editor dialog box for a Color-Dependent Style table: General tab selection, Table View tab selection, Form View tab selection*

The following three tabs are available in the Plot Style Table Editor:

- **General**—Displays the name of the plotstyle table, description (if any), location of the file, and version number; see Figure 8–63 (left) and Figure 8–64 (left). You can modify the description and apply scaling to non-ISO linetypes (those not found in the *acad.lin* file) and to fill patterns.

- **Table View**—Lists entire plotstyles in the plotstyle table and their settings in tabular form; see Figure 8–63 (middle) and Figure 8–64 (middle). The styles are displayed in columns from left to right. The setting names of each row appear at the left of the tab. By default, in the case of a named plotstyle table, AutoCAD sets up a style named Normal and represents an object's default properties. You cannot modify or delete the Normal style. In the case of a color-dependent plotstyle table, AutoCAD lists all the 255 color styles in tabular form. In general, this is convenient if you have a small number of plotstyles to view.

- **Form View**—The plotstyle names are listed under the **Plot styles** list box and the settings for the selected plotstyle are displayed at the right side of the dialog box; see Figure 8–63 (right) and Figure 8–64 (right).

To create a new plotstyle, choose **Add Style** from the Plot Style Table Editor. AutoCAD adds a new style. You can change the name to a descriptive name, if necessary, though it cannot exceed 255 characters. You cannot duplicate names within the same plotstyle table.

To delete a pen style, click the gray area above the plotstyle name in the Table View. The entire column will be highlighted. Choose **Delete Style**. In the Form View, select the style name from the **Plot styles** list box, and choose **Delete Style**.

The settings on the **Form View** tab include the following:

Description field allows you to specify a description for plotstyles and modify an existing description for a plotstyle if necessary. The description cannot exceed 255 characters.

The **Color** list box allows you to assign a plotstyle color. If you assign a color from one of the available colors, AutoCAD overrides the object's color at plot time. By default, all of the plotstyles are set to Use object color.

Set **Dither** to ON for the plotter to approximate colors with dot patterns, giving the impression of plotting with more colors than the number of inks available in the plotter. If you set **Dither** to OFF, AutoCAD maps colors to the nearest color, which limits the range of colors used for plotting. The most common reason for turning off dithering is to avoid false linetyping from dithering of thin vectors and to make dim colors more visible. If the plotter does not support dithering, the dithering setting is ignored. The default setting is ON.

Set **Grayscale** to ON for AutoCAD to convert the object's colors to grayscale if the plotter supports grayscale. If you set **Grayscale** to OFF, AutoCAD uses the RGB (red, green, blue) values for the object's colors. The default setting is OFF.

The **Pen** ⊡ setting in the Plot Style Table Editor specifies which pen to use for each plotstyle. You can specify a pen to use in the plotstyle by selecting from a range of pen numbers from 1 to 32. By using the BACKSPACE or DELETE keys, you can set the field to read Automatic. AutoCAD uses the information you provide under Physical Pen Configuration in the Plotter Configuration Editor to select the pen closest in color to the object you are plotting. The default setting is Automatic.

Specify a virtual pen number in the **Virtual pen** ⊡ edit field for plotters that do not use pens but can simulate the performance of a pen plotter by using virtual pens. The default is set to Automatic to specify that AutoCAD should make the virtual pen assignment from the AutoCAD Color Index. You can specify a virtual pen number between 1 and 255. The virtual pen number setting in a plotstyle is used only by plotters without pens and only if they are configured for virtual pens. If this is the case, all other style settings are ignored and only the virtual pen is used. The default setting is Automatic.

The **Screening** text field sets a color intensity setting that determines the amount of ink AutoCAD places on the paper while plotting. The valid range is 0 through 100. Selecting 0 reduces the color to white. Selecting 100 (default) displays the color at its full intensity. The default setting is 100%.

The **Linetype** list box allows you to assign a plotstyle linetype. If you assign a linetype from one of the available linetypes, AutoCAD overrides the object's linetype at plot time. By default, all the plotstyles are set to Use object linetype.

The **Adaptive** toggle adjusts the scale of the linetype to complete the linetype pattern. Set **Adaptive** to ON if it is more important to have complete linetype patterns than correct linetype scaling. Set **Adaptive** to OFF if linetype scale is more important. The default setting is ON.

The **Lineweight** list box allows you to assign a plotstyle lineweight. If you assign a lineweight from one of the available lineweights, AutoCAD overrides the object's lineweight at plot time. By default, all the plotstyles are set to Use object lineweight.

The **Line end style** list box allows you to assign a line end style. The line end style options include Butt, Square, Round, and Diamond. If you assign a line end style from one of the available line end styles, AutoCAD overrides the object's line end style at plot time. By default, all the plotstyles are set to Use object end style.

The **Line join style** list box allows you to assign a line end style. The line join style options include Miter, Bevel, Round, and Diamond. If you assign a line join style from one of the available line join styles, AutoCAD overrides the object's line join style at plot time. By default, all the plotstyles are set to Use object line join style.

The **Fill style** list box allows you to assign a fill style. The fill style options include Solid, Checkerboard, Crosshatch, Diamonds, Horizontal Bars, Slant Left, Slant Right, Square Dots, and Vertical Bars. The fill style applies only to solids, splines, donuts, and 3D faces. If you assign a fill style from one of the available fill styles, AutoCAD overrides the object's fill style at plot time. By default, all the plotstyles are set to Use object fill style.

AutoCAD allows you to edit the available lineweights by choosing **Edit Lineweights**. You cannot add or delete lineweights from the list.

To save the changes and close the Plot Style Table Editor, choose **Save & Close**. To save the changes to another plotstyle table, choose **Save As**. AutoCAD displays the Save As dialog box. Specify the file name in the **File name** text field, and choose **Save** to save and close the Save As dialog box.

Changing Plot Style Property for an Object or Layer

As mentioned earlier, every object that is created in AutoCAD has a plotstyle property in addition to its color, linetype, and lineweight. Similarly, every layer has a color, linetype, and lineweight, in addition to a plotstyle property. The default settings for plotstyles for objects and layers are set in the Plot Style Table Settings dialog box (Figure 8–65), which can be opened from the Options dialog box.

FIGURE 8–65 *Plot Style Table Settings dialog box*

The default plotstyle for objects can be any of the following:

- **Normal**—Uses the object's default properties.
- **ByLayer**—Uses the properties of the layer that contains the object.
- **ByBlock**—Uses the properties of the block that contains the object.
- **Named plotstyle**—Uses the properties of the specific named plotstyle defined in the plotstyle table.

The default plot setting for an object is ByLayer, and the initial plotstyle setting for a layer is Normal. When the object is plotted, it retains its original properties.

If you are working in a Named plotstyle mode, you can change the plotstyle for an object or layer at any time. If you are working in a color-dependent plotstyle mode, you cannot change the plotstyle for objects or layers—by default, they are set to ByColor.

To change the plotstyle for one or more objects, first select the objects, with the system variable PICKFIRST set to ON, and select the plotstyle from the Plot style control list box on the Properties toolbar, as shown in Figure 8–66. If the list does not include the style you wish to select, choose **Other**, and AutoCAD displays the Current Plot Style dialog box, as shown in Figure 8–67.

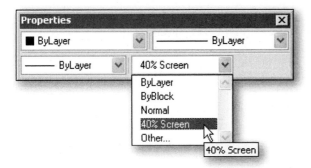

FIGURE 8–66 *Properties toolbar, Plot Style control*

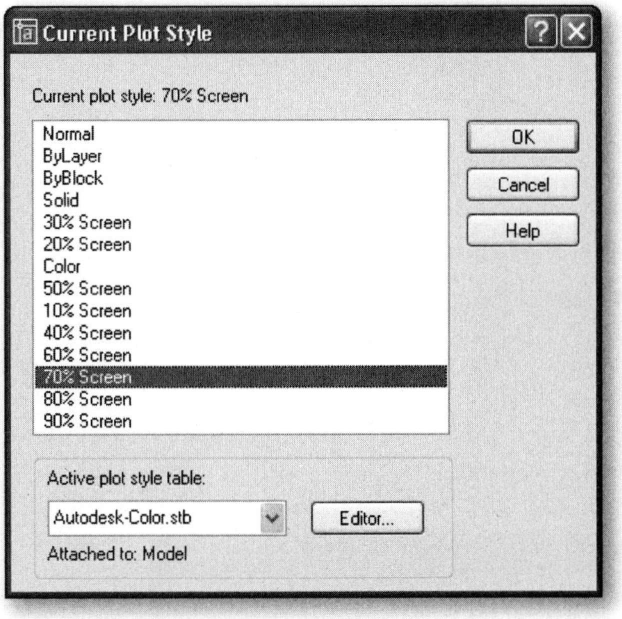

FIGURE 8–67 *Current Plot Style dialog box*

Select the plotstyle you want to apply to the selected object(s) from the **Current plot style** list box. If you need to select a plotstyle from a different plotstyle table, select the plotstyle table from the **Active plot style table** list box. AutoCAD lists all the available plotstyles in the **Current plot style** list box from where you can select the one you want to apply to the select object(s). Choose **OK** to close the dialog box. You can also change the plotstyle of the selected object(s) from the Properties dialog box.

To change the plotstyle for a layer, open the Layer Properties Manager dialog box. Select the layer you want to change, and select a plotstyle for the selected layer similar to changing color or linetype.

CONFIGURING PLOTTERS

Autodesk Plotter Manager allows you to configure a local or network nonsystem plotter. In addition, you can configure a Windows system printer with nondefault settings. AutoCAD stores information about the media and plotting device in configured plot (PC3) files. The PC3 files are stored in the path that is listed in the **Files** section of the Options dialog box. Plot configurations are therefore portable and can be shared in an office or on a project. If you calibrate a plotter, the calibration information is stored in a plot model parameter (PMP) file that you can attach to any PC3 files you create for the calibrated plotter.

AutoCAD allows you to configure plotters for many devices and store multiple configurations for a single device. You can create several PC3 files with different output options for the same plotter. After you create a PC3 file, it is available in the list of plotter configuration names in the **Printer/plotter** section of the Plot dialog box.

Open the Autodesk Plotter Manager from the Plot panel, and AutoCAD displays the Plotters window explorer (Figure 8–68), which lists all the configured plotters.

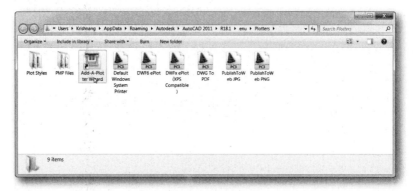

FIGURE 8–68 *Plotters window explorer (Windows 7 version)*

Double-click the Add-A-Plotter Wizard, and AutoCAD displays the Add Plotter – Introduction Page, as shown in Figure 8–69.

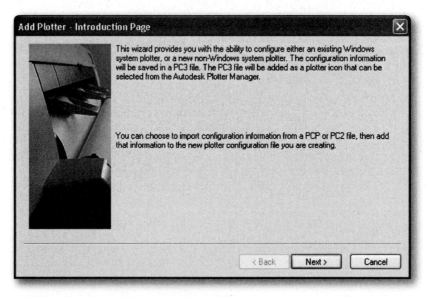

FIGURE 8–69 *Add Plotter – Introduction Page*

Choose **Next**, and AutoCAD displays the Add Plotter – Begin page, as shown in Figure 8–70.

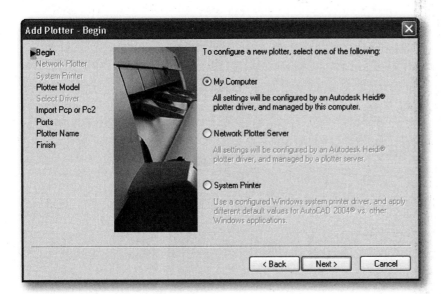

FIGURE 8–70 *Add Plotter – Begin page*

Choose one of these three radio buttons:

- **My Computer**—To configure a local nonsystem plotter.
- **Network Plotter Server**—To configure a plotter that is on the network.
- **System Printer**—To configure a Windows system printer. If you want to connect to a printer that is not in the list, you must first add the printer using the Windows Add Printer wizard in the Control Panel.

If you select the **My Computer** option, the wizard prompts you to select a plotter manufacturer and model number, identify the port to which the plotter is connected, specify a unique plotter name, and choose **Finish** to close the wizard.

If you select the **Network Plotter Server** option, the wizard prompts you to identify the network server, select the plotter manufacturer and model number, specify a unique plotter name, and choose **Finish** to close the wizard.

If you select the **System Printer** option, the wizard prompts you to select one of the printers configured in the Windows operating system, specify a unique plotter name, and choose **Finish** to close the wizard.

AutoCAD saves the configuration file in PC3 file format with a unique given name in the path that is listed in the Files section of the Options dialog box.

If necessary, you can edit the PC3 file using the Plotter Configuration Editor. The Plotter Configuration Editor provides options for modifying a plotter's port connections and output settings including media, graphics, physical pen configuration, custom properties, initialization strings, calibration, and user-defined paper sizes. You can drag these options from one PC3 file to another.

You can open the Plotter Configuration Editor using one of the following methods:

- From the File menu, choose **Page Setup**. Choose **Properties**.
- From the File menu, choose **Plot**. Choose **Properties**.
- Double-click a PC3 file from Windows Explorer, right-click the file, and choose **Open**.
- Choose **Edit Plotter Configuration...** on the Add Plotter – Finish page in the Add-A-Plotter Wizard.

Figure 8–71 shows the Plotter Configuration Editor for an HP7580B plotter.

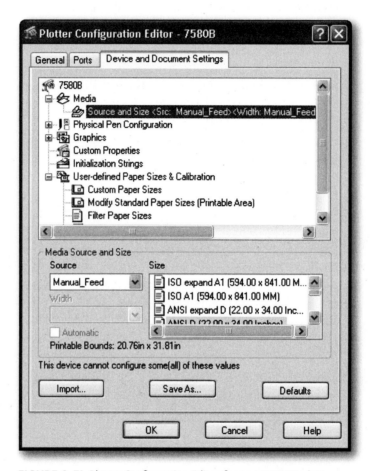

FIGURE 8–71 *Plotter Configuration Editor for an HP7580B plotter*

The Plotter Configuration Editor contains three tabs:

- General tab—Contains basic information about the configured plotter.
- Ports tab—Contains information about the communication between the plotting device and your computer.
- Device and Document Settings tab—Contains plotting options.

In the Device and Document Settings tab, you can change many of the settings in the configured plot (PC3) file. You can make changes in the following six areas:

- **Media**—Specifies a paper source, size, type, and destination.
- **Physical Pen Configuration**—Specifies settings for pen plotters.
- **Graphics**—Specifies settings for printing vector graphics, raster graphics, and TrueType® fonts.

- **Custom Properties**—Displays settings related to the device driver.
- **Initialization Strings**—Sets pre-initialization, post-initialization, and termination printer strings.
- **User-defined Paper Sizes & Calibration**—Attaches a PMP file to the PC3 file, calibrates the plotter, and adds, deletes, or revises custom or standard paper sizes.

The areas correspond to the categories of settings in the PC3 file that you can edit. Double-click any of the six categories to view and change the specific settings. When you change a setting, your changes appear in angle brackets (<>) next to the setting name unless there is too much information to display. To save the changes to another PC3 file, choose **Save As**. AutoCAD displays the Save As dialog box. Specify the file name in the **File name** text field, and choose **Save**. To save the changes to the PC3 file and close the Plotter Configuration Editor, choose the **OK** button.

Open the Exercise Manual PDF for Chapter 8 for discipline-specific exercises. Related files are downloaded from the student companion site mentioned in the Introduction (refer to page number xii for instructions).

REVIEW QUESTIONS

1. If you were to plot a drawing at a scale of **1 = 60'**, what LTSCALE setting should you specify?
   a. 60
   b. 1/60
   c. 720
   d. 1/720

2. If you want to plot a drawing requiring multiple pens and you are using a single pen plotter, AutoCAD will:
   a. Not plot the drawing at all
   b. Pause when necessary to allow you to change pens
   c. Invoke an error message
   d. Plot all of the drawing using the single pen
   e. None of the above

3. A drawing created at a scale of **1:1** and plotted to "Scaled to Fit" is plotted:
   a. To fit the specified paper size
   b. At the prototype scale
   c. None of the above

4. To plot a full-scale drawing at a scale of **1/4 = 1'**, use a plot scale of:
   a. 0.25 = 1
   b. 1 = 48
   c. 12 = 0.25
   d. 1 = 24

5. What is the file extension assigned to all files created when plotting to a file?

   a. DRW

   b. DRK

   c. PLO

   d. PLT

6. When plotting, pen numbers are assigned to:

   a. Layers

   b. Thickness

   c. Linetypes

   d. None of the above

You need to draw three orthographic views of an airplane whose dimensions are as follows: wingspan of 102 feet, a total length of 118 feet, and a height of 39 feet. The drawing has to be plotted on a standard 12" by 9" sheet of paper. No dimensions will be added, so you will need only 1" between the views. Answer the following five questions using the information from this drawing:

7. What would be a reasonable scale for the paper plot?

   a. 1=5'

   b. 1=15'

   c. 1=25'

   d. 1=40'

8. What would be a reasonable setting of LTSCALE?

   a. 1

   b. 5

   c. 60

   d. 25

   e. 300

   f. 480

9. If you were plotting from paper space, what ZOOM-scale factor would you use?

   a. 1/5X

   b. 1/25X

   c. 1/60X

   d. 1/300X

   e. 1/5XP

   f. 1/25XP

   g. 1/60XP

   h. 1/300XP

10. When inserting your border in paper space, what scale factor should you use?

    a. 1

    b. 5

    c. 25

    d. 60

    e. 300

11. Which of the following options will the plot preview give you?

    a. Seeing what portion of your drawing will be plotted

    b. Seeing the plotted size of your drawing

    c. Seeing the plotted drawing relative to the page size

    d. All of the above

12. Which of the following determine the relationship between the size of the objects in a drawing and their sizes on a plotted copy?

    a. Size of the object in the AutoCAD drawing

    b. Size of the object on the plot

    c. Maximum available plot area

    d. Plot scale

    e. All of the above

13. Once you have set the scale, you can pan in the viewport without changing the scale. However, using the Zoom command modifies the scale.

    a. True

    b. False

14. AutoCAD permits plotting in which of the following environment modes?

    a. Model space

    b. Paper space

    c. Layout

    d. All of the above

15. Any changes you make to the model object through a viewport are reflected in the drawing and in all other viewports.

    a. True

    b. False

16. Which TILEMODE system variable setting corresponds to model space?

    a. 0

    b. 1

    c. Either A or B

    d. None of the above

17. When starting a new drawing, how many default plotting layouts does AutoCAD create?

    a. 0

    b. 1

    c. 2

    d. Unlimited

18. How many viewports can be active at any one time in Layout?

    a. ONE

    b. TWO

    c. THREE

    d. Multiple

19. Within multiple floating viewports, you can establish various scale and layer visibility settings for each individual viewport.

    a. True

    b. False

20. Paper sizes are indicated by an X-axis direction (drawing length) and a Y-axis direction (drawing width).

    a. True

    b. False

21. In Paper Space Layout Printing there is only one model per drawing. But how many layouts can you have?

    a. Multiple

    b. ONE

    c. TWO

    d. None

22. Floating viewports such as lines, arcs, and text can be manipulated using AutoCAD commands such as move, copy, stretch, scale, or erase.

    a. True

    b. False

23. While in paper space, both the floating viewports and the 3D model can be modified or edited.

    a. True

    b. False

24. Which of the following can be converted into a viewport?

    a. Ellipses

    b. Splines

    c. Circles

    d. All of the above

25. Which of the following commands allows for the control of layer visibility in a specific viewport?

    a. VPORTS

    b. VPLAYER

    c. VIEWLAYER

    d. LAYERVIS

Hatching, Gradients, and Boundaries

OBJECTIVES

After completing this chapter, you will be able to do the following:

- Create hatch and gradient patterns using the HATCH command
- Modify hatch patterns via the HATCHEDIT command
- Control the visibility of hatch and gradient patterns

WHAT IS HATCHING?

Drafters and designers use repeating patterns called hatching to fill regions in a drawing for various purposes (Figure 9–1). In a cutaway (cross-sectional) view, hatch patterns help the viewer differentiate between components of an assembly and indicate the material of each. In surface views, hatch patterns depict material and add to the readability of the view. In general, hatch patterns greatly help drafters and designers achieve their purpose: communicating information. Because drawing hatch patterns is a repetitive task, it is an ideal application of computer-aided drafting.

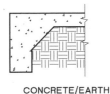

CONCRETE/EARTH

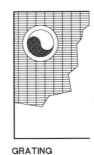

GRATING

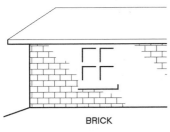

BRICK

FIGURE 9–1 *Examples of hatch patterns*

Image(s) © Cengage Learning 2013

You can use patterns that are supplied in an AutoCAD support file called *acad.pat*, patterns in files available from third-party custom developers, or you can create your own custom hatch patterns. See Appendix E for the list of patterns supplied with *acad.pat*.

AutoCAD allows you to fill objects with a solid color in addition to a hatch pattern. AutoCAD creates an associative hatch, which updates when its boundaries are modified, or a nonassociative hatch, which is independent of its boundaries. Before AutoCAD draws the hatch pattern, it allows you to preview the hatching and to adjust the definition if necessary.

Examples of Hatch Patterns

Hatch patterns are considered separate drawing objects, and the hatch pattern operates as one object. You can separate it into individual objects with the EXPLODE command, but once separated into individual objects, the hatch pattern will no longer be associated with the boundary object.

Hatch patterns are stored with the drawing, so they can be updated even if the pattern file containing the hatch is not available. You can control the display of the hatch pattern with the FILLMODE system variable. If FILLMODE is set to OFF, the patterns are not displayed, and regeneration calculates only the hatch boundaries. By default, FILLMODE is set to ON.

The hatch pattern is drawn with respect to the current coordinate system, elevation, layer, color, linetype, and snap origin.

WHAT IS GRADIENT FILL?

A gradient fill is a solid hatch fill that gives the blended-color effect of a surface with light on it. You can use gradient fills to suggest a curved surface in two-dimensional drawings. The color in a gradient fill makes a smooth transition from light to dark or from dark to light. You may select a predefined pattern such as a linear, spherical, or radial sweep and specify an angle for the pattern. In a two-color gradient fill, the transition is from light to dark and from the first color to the second.

Gradient fills are applied to objects in the same way that solid fills are, and they can be associated with their boundaries. An associated fill is automatically updated when the boundary changes. The Hatch and Gradient dialog box allows you to modify the settings for both hatch and gradient patterns.

NOTE | You cannot use plotstyles to control the plotted color of gradient fills.

HATCH AND GRADIENT FILL WITH THE HATCH COMMAND

Invoke the HATCH command from the Draw panel (Figure 9–2) located in the Home tab, and AutoCAD displays the Hatch Creation Contextual tab is displayed (Figure 9–3).

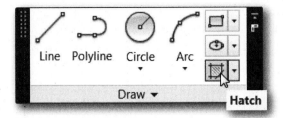

FIGURE 9–2 *Invoking the HATCH command from the Draw panel*

FIGURE 9–3 *Hatch Creation Contextual tab*

The majority of the options available are applicable to both hatch and gradient patterns such as the sections on **boundaries, properties, origin,** and **options.**

AutoCAD prompts:

> Pick internal point or ⊡ *(specify a point within the area to be hatched or select one of the available options)*

When you move the cursor inside a closed area, AutoCAD displays the preview of the hatch pattern with the current setting.

Hatch and Gradient Related Settings

The Hatch Creation Contextual tab provides settings related to Hatch and Gradient Fill.

The **Pattern** panel (see Figure 9–4) displays preview images for all predefined and custom patterns and allows you to select the type of pattern to be drawn.

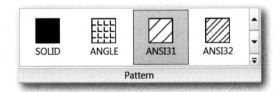

FIGURE 9–4 *Pattern panel*

The **Properties** panel (see Figure 9–5) allows you to set the properties of the selected pattern. The **Hatch Type** selection specifies whether to create a solid fill, a gradient fill, a predefined hatch pattern, or a user-defined hatch pattern. Predefined patterns are stored in the acad.pat or acadiso.pat files supplied with the program. User-defined patterns are based on the current linetype in your drawing. A custom pattern is a pattern that is defined in any custom PAT files that you have added to the search path.

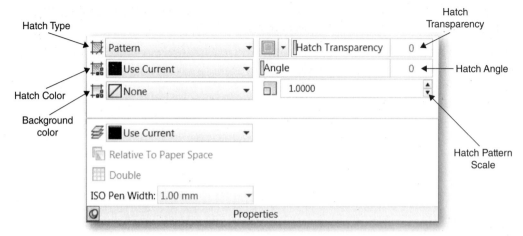

FIGURE 9–5 *Properties panel*

The **Hatch Color** selection overrides the current color for solid fills and hatch patterns, or specifies the first of two gradient colors.

The **Background Color** specifies the color for hatch pattern backgrounds, or the second gradient color. When Hatch Type is set to Solid, **Background Color** selection is not available.

The **Hatch Transparency** sets the transparency level for new hatches or fills, overriding the current object transparency. Select Use Current to use the current object transparency setting.

The **Hatch Angle** specifies an angle for the hatch or fill relative to the X-axis of the current UCS. Valid values are from 0 to 359.

The **Hatch Scale** expands or contracts a predefined or custom hatch pattern. This option is available only when Hatch Type is set to Pattern.

The Hatch Angle and Hatch Scale settings can be changed to suit the desired appearance, as shown in Figure 9–6.

FIGURE 9–6 *Hatch pattern with different scale and angle values*

The **Hatch Spacing** specifies the spacing of lines in a user-defined pattern. This option is available only when Hatch Type is set to User Defined.

The **Tint Slider** specifies the tint (the selected color mixed with white) or shade (the selected color mixed with black) of a color to be used for a gradient fill of one color. This option is available only when Hatch Type is set to Gradient.

The **Layer Name** assigns new hatch objects to the specified layer, overriding the current layer. Select Use Current to use the current layer.

The **Relative to Paper Space** setting scales the hatch pattern relative to paper space units. This allows you to display hatch patterns at a scale that is appropriate for your layout. This option is available only from a layout.

The **Double** selection draws a second set of lines at 90 degrees to the original lines, creating a crosshatch for a user-defined pattern. This option is available only when Hatch Type is set to User Defined.

The **ISO Pen Width** scales an ISO predefined pattern based on the selected pen width. This option is available only when an ISO pattern has been specified.

The **Shade and Tint** slider lets you specify the tint, or the amount of white mixed in of the selected color or the shade, or the amount of black mixed in of the selected color for a one-color gradient fill. This option is available only when Hatch Type is set to Gradient.

The **Hatch origin** panel (see Figure 9–7) controls the origin of the pattern drawn. Changing the origin causes the lines in the hatch pattern to be offset by the distance between 0,0 and the specified origin. This step is sometimes necessary when you wish to offset the whole pattern of lines for visual effects. For example, if you wish to use the same hatch pattern in adjacent boundaries but do not want their lines to coincide, you can use different origins for the two patterns. Or, if a hatch pattern, such as those for brick and masonry, needs to begin at a certain point, you can specify that point as the origin.

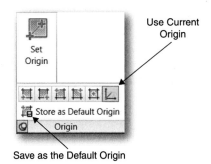

FIGURE 9–7 *Hatch origin panel*

Choosing **Set origin** lets you specify an origin different than the current one with the pointing device or by entering the new coordinates. You can also set origin by choosing one of the five available options: Bottom Left, Bottom Right, Top Left, Top Right, and Center.

Selecting **Use current origin** causes the hatch pattern to use the current setting of the hatch origin that is stored in the HPORIGIN system variable. By default, it is set to 0,0 in the current UCS. If it has been specified as another point, selecting **Use current origin** uses the current setting.

Choosing **Store as default origin** lets you store the newly specified origin in the HPORIGIN system variable.

The Options panel (See Figure 9–8) controls several commonly used hatch or fill options.

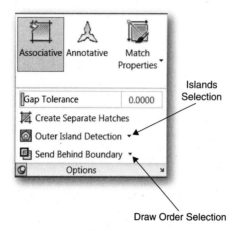

FIGURE 9–8 *Options panel*

Associative controls whether the hatch or fill is associative or nonassociative. Choosing **Associative** causes the hatch pattern elements to be associated with the objects that make up the boundary. For example, if the object is stretched, the hatch pattern expands to fill the new size. Figure 9–9 shows examples of associative and nonassociative hatch patterns.

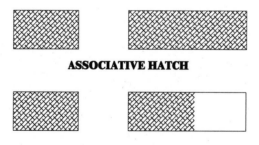

FIGURE 9–9 *Examples of associative and nonassociative hatch pattern when an object is stretched*

Choosing **Annotative** toggles Annotative ON or OFF for the selected hatch or fill pattern. The Annotative property allows you to automate the process of scaling the hatch or fill pattern. Refer to Chapter 8 for detailed explanation on Annotative scaling.

Match Properties allows you to apply the hatch pattern settings, such as pattern type, pattern angle, and pattern scale, from an existing pattern to another area to be hatched. Choosing **Use Current Origin** sets the properties of a hatch with a selected hatch object, except the hatch origin. Choosing **Use Source Hatch Origin** sets the properties of a hatch with a selected hatch object, including the hatch origin.

The **Gap tolerance** textbox lets you specify a value, in drawing units, from 0 to 5,000 to set the maximum size of gaps that can be ignored, when the objects serve as a hatch boundary. Any gaps equal to or smaller than the value you specify are ignored, and the boundary is treated as closed. The default value, 0, specifies that the objects enclose the area with no gaps.

The **Create Separate Hatches** controls whether a single hatch object or multiple hatch objects are created when several separate closed boundaries are specified.

The **Islands** selection specifies the method used to hatch or fill objects within the outermost boundary. If no internal boundaries exist, specifying an island detection style has no effect. Choosing **Normal Island Detection** hatches or fills between alternate areas, starting with the outermost area. Choosing **Outer Island Detection** hatches or fills only the outermost area and leaves the internal structure blank. Choosing **Ignore Island Detection** hatches or fills the entire area enclosed by the outermost boundary, regardless of how you select the object, as long as its outermost objects comprise a closed polygon and are joined at their endpoints.

In Figure 9–10, specifying the point shown in the upper-left image in response to the **Add Pick points** option results in hatching for the **Normal** style, as shown in the upper right. **Outer** style is shown at the lower left, and **Ignore** style is shown at the lower right.

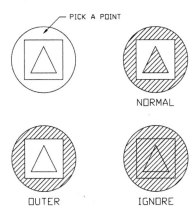

FIGURE 9–10 *Examples of hatching by specifying a point for Normal, Outer, and Ignore styles*

The **Draw Order** selection lets you assign an order to a hatch or fill. You can place a hatch or fill behind all other objects, in front of all other objects, behind the hatch boundary, or in front of the hatch boundary.

Defining the Hatch or Gradient Boundary

A region of a drawing may be filled with a hatch pattern or gradient fill if it is enclosed by a boundary of connecting lines, circles, or arc objects. Overlapping boundary objects can be considered as terminating at their intersections with other boundary objects. There must not be any gaps between boundary objects. Figure 9–11 illustrates various objects and the potential boundaries that might be established from them.

Note in Figure 9–11 how the enclosed regions are defined by their respective boundaries. A boundary might include all or part of one or more objects. In addition to lines, circles, and arcs, boundary objects can include 2D and 3D polylines, 3D faces, and viewports. Boundary objects should be parallel to the current UCS. You can also hatch Block References that have been inserted with unequal X- and Y-scale factors.

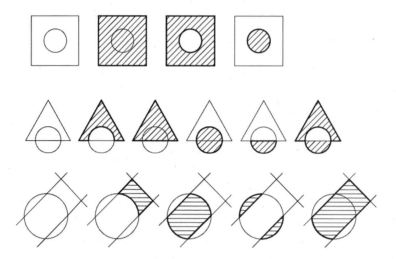

FIGURE 9–11 *Allowed hatching boundaries made from different objects*

AutoCAD provides two methods of determining the area to be filled with a hatch or gradient pattern. Using the select object method (Figure 9–12), select the four lines via the Window option. These four lines compose the hatching boundary. The four objects are valid boundary segments, if they connect at their endpoints and do not overlap. Instead of using the Window option of the object selection process, you can select the four lines individually, which may be desirable if there are unwanted objects within the window used to select them.

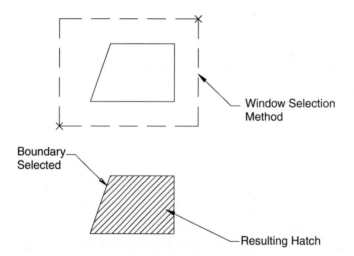

FIGURE 9–12 *Using the select objects method to define the boundary*

In Figure 9–13, using the pick points method, select a point in the region enclosed by the four lines. AutoCAD creates a polyline with vertices that coincide with the intersections of the lines. An option is available that allows you to retain or discard the boundary when the hatching is complete.

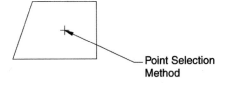

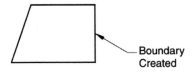

FIGURE 9–13 *Using the pick points method to create the boundary*

If the four lines shown in Figure 9–13 had been segments of a closed polyline, you could have selected that polyline by picking it with the cursor. Otherwise, all objects enclosing the region to be hatched must be selected, and those objects must be connected at their endpoints. For example, to use the select objects method for the region shown in Figure 9–14, you would need to draw three lines—from 1 to 2, 2 to 3, and 3 to 4—and an arc from 4 to 1, select the four objects or connect them into a polyline, and select them to be the boundary.

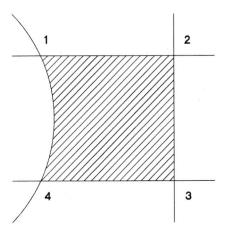

FIGURE 9–14 *A region bounded by three lines and an arc*

The Select Points method permits you to select a point in the region and have AutoCAD automatically create the needed polyline boundary. The ease of use and automation of the select points method almost eliminate the need for the select objects method except for rare, specialized applications. When a hatch boundary area is not found, AutoCAD attempts to show you where the problem may have occurred. Red circles appear around endpoints near where any gap in geometry is estimated to be. You can close the gap or increase the **Gap Tolerance** setting in the Hatch and Gradient settings box to a value that is more than the gap size and proceed with hatching.

The **Boundaries** panel (see Figure 9–15) provides various methods and related settings by which you can select the objects that determine the boundary for drawing hatch patterns.

FIGURE 9–15 *Boundaries panel*

Choosing **Pick points** from the Boundaries panel lets you determine a boundary from existing objects that form an enclosed area around the specified point. AutoCAD draws an imaginary line from the selected point to the nearest object and traces the boundary in the counterclockwise direction. If it cannot trace a closed boundary, AutoCAD returns to the drawing without hatching the object. How HATCH detects objects using this option depends on which island detection method is selected in the Options panel: **Normal Island Detection**, **Outer Island Detection**, or **Ignore Island Detection**.

For example, in Figure 9–16, point A is valid and point B is not when the **No Island Detection** is selected. The object nearest to point A is the line that is part of a potential boundary, the square, of which point A is inside, and AutoCAD considers the square as the hatch boundary. Conversely, point B is nearest a line that is part of a potential boundary, the triangle, of which point B is outside, and AutoCAD displays an error message noting that the selected point is outside the boundary.

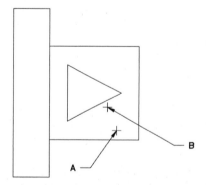

FIGURE 9–16 *Selecting points for hatching*

Figure 9–17 shows an example of hatching by specifying a point inside a boundary.

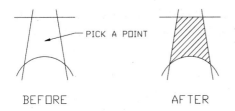

FIGURE 9–17 *Hatching by specifying a point*

Choosing **Add Select objects** lets you select specific objects for hatching that determine a boundary to form an enclosed area.

When you use the **Add Select objects** option, HATCH does not detect interior objects automatically. You must select the objects within the selected boundary to hatch or fill those objects according to the current island detection style.

Use caution when hatching over dimensioning. Dimensions are not affected by hatching as long as the DIMASSOC, short for "associative dimensioning," dimension variable is set to ON when the hatching is created and the dimension has not been exploded. The DIMASSOC system variable toggles between associative and nonassociative dimensioning. If the dimensions are drawn with DIMASSOC set to OFF or exploded into individual objects, the lines, dimension, and extension have an unpredictable and undesirable effect on the hatching pattern. In this case, selecting should be done by specifying the individual objects on the screen.

Blocks are hatched as though they are separate objects. Note, however, that when you select a block, all objects that make up the block are selected as part of the group to be considered for hatching.

If the selected items include text objects, shapes, or attribute objects, AutoCAD does not hatch through these items if identified in the selection process. AutoCAD leaves an unhatched area around the text objects so that they can be clearly viewed, as shown in Figure 9–18. Using the **Ignore** style will negate this feature so that the hatching is not interrupted when passing through text, shape, and attribute objects.

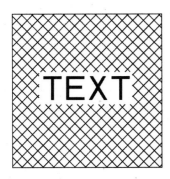

FIGURE 9–18 *Hatching in an area that includes text*

When you select objects individually, after selecting the *Select Objects* option, AutoCAD no longer creates a closed border automatically. Therefore, any objects selected that will be part of the desired border must be either connected at their endpoints or a closed polyline.

NOTE

When a filled solid or trace with width is selected in a group to be hatched, AutoCAD does not hatch inside that solid or trace. However, the hatching stops at the outline of the filled object, leaving no clear space around the object as it does around text objects, shapes, and attributes.

The **Remove boundaries** and **Recreate boundary** options are not available during creation of a new hatch. They are used on existing hatches and fills with internal boundaries. See their explanation in the section on the Hatch Edit dialog box.

The **Boundary retention** selection specifies whether to retain boundaries as objects and allows you to select the object type AutoCAD applies to those objects. You can choose the object type to a polyline or a region.

The **Boundary set** allows you to select a set of objects, called a boundary set, that AutoCAD analyzes when defining a boundary from a specified point. The selected boundary set applies only when you use the **Add Pick points** selection to create a boundary to draw hatch patterns. By default, when you use **Add Pick points** to define a boundary, AutoCAD analyzes all objects visible in the current viewport.

The **Boundary set** list box lets you select whether the boundary set will be selected from the Current Viewport or Existing Set. Selecting **Use Current Viewport** causes the boundary set to be defined from everything in the current viewport extents. Selecting this option discards any current boundary set. Selecting **Existing Set** causes the boundary set to be defined from the objects that you selected with **New**. If you have not created a boundary set with **New,** the **Existing Set** option is not available. Choosing **New** causes AutoCAD to clear the dialog box and return you to the drawing area to select objects from which a new boundary set will be defined. AutoCAD creates a new boundary set from those objects selected that are hatchable; existing boundary sets are abandoned. If hatchable objects are selected, they remain as a boundary set until you define a new one or exit the HATCH command. Defining a boundary set will be helpful when you are working on a drawing that has too many objects to analyze to create a boundary for hatching.

HATCH PATTERNS USING TOOL PALETTES

You can also hatch a closed shape by dragging a hatch pattern from the tool palette (Figure 9–19).

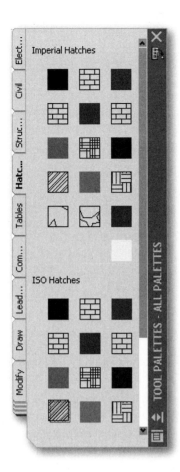

FIGURE 9–19 *Tool Palettes window with the Hatches tab selected*

Tool palettes are tabbed areas within the Tool Palettes window, and hatches that reside on a tool palette are called tools. Several tool properties including scale, rotation, and layer can be set for each tool individually. To change the tool properties, right-click a tool and select *Properties* from the shortcut menu. You can change the tool's properties in the Tool Properties dialog box, as shown in Figure 9–20. The Tool Properties dialog box has two categories of properties: the Pattern properties category, which controls object-specific properties such as scale, rotation, and angle, and the General properties category, which overrides the current drawing property settings such as layer, color, and linetype.

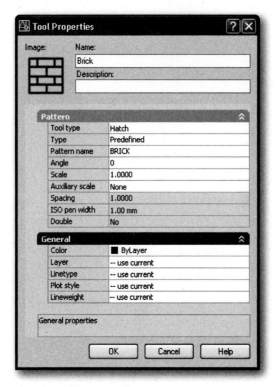

FIGURE 9–20 *Tool Properties of a selected tool*

You can place hatches that you use often on a tool palette by dragging hatch patterns from the DesignCenter by opening the *acad.pat* file. The *acad.pat* file can be found on the following path: *C:\Program Files\Autodesk\AutoCAD 2011\Support\acad.pat*. When this file is selected in the DesignCenter window, available patterns are displayed in the Content panel. See Chapter 13 for a detailed explanation on using the DesignCenter.

EDITING HATCHES AND GRADIENTS

Select the hatch object to be edited and AutoCAD displays Hatch Editor Contextual tab. You can modify hatch-specific properties, such as pattern, scale, and angle for the selected hatch or fill pattern.

The HATCHEDIT command also allows you to modify hatch patterns and gradient fills or choose a new pattern for an existing hatch. It also allows you to change the pattern style of an existing pattern.

Invoke the HATCHEDIT command from the Modify panel (Figure 9–21) located in the Home tab, and AutoCAD displays the following prompt:

```
Select hatch object: (select an associative hatch object)
```

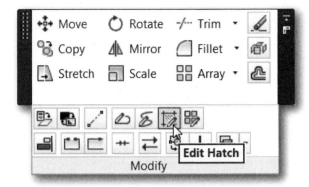

FIGURE 9–21 *Invoking the* `HATCHEDIT` *command from the Modify panel*

AutoCAD displays the Hatch Edit dialog box (Figure 9–22).

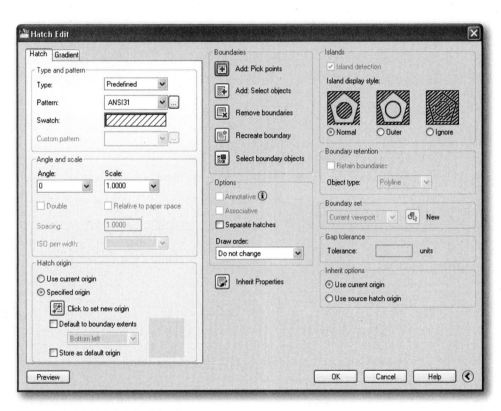

FIGURE 9–22 *Hatch Edit dialog box*

The Hatch Edit dialog box has the same options that are available in the Hatch Editor Contextual tab.

If a hatch or gradient is created using the normal island display style, you can make inner objects that are used as boundaries can be removed with the **Remove boundaries** option. This action does not physically remove the objects; it only prevents them from being used as boundaries. The **Recreate boundary** option can be applied only to boundaries that have been removed with the **Remove boundary** option.

Choosing **Remove boundaries** causes AutoCAD to prompt you to select a boundary set to be removed from the defined boundary set. You cannot remove the outermost boundary.

Choosing **Recreate boundary** creates a polyline or region around the selected hatch or fill, with the option to associate the hatch object with it. Choosing **Recreate Boundary** causes the dialog box to close temporarily, and AutoCAD displays the following prompts:

> Select objects: *(select the object(s) by one of the standard methods, and press* Enter *)*
>
> Enter type of boundary object: *(Enter r to create a region or p to create a polyline)*
>
> Reassociate hatch with new boundary? ⊡ *(Enter y or n)*

You can also display and change the current properties for hatch or fill objects with the **Properties** palette. Open the **Properties** palette and view and change the settings for all properties of the selected hatch or fill objects (Figure 9–23). The **Properties** palette lists the current settings for properties of the selected objects. You can modify any property that can be changed by specifying a new value. The **Properties** palette also enables you to view the area of a hatch (Figure 9–23). If you select multiple hatch objects, you can now view their cumulative area.

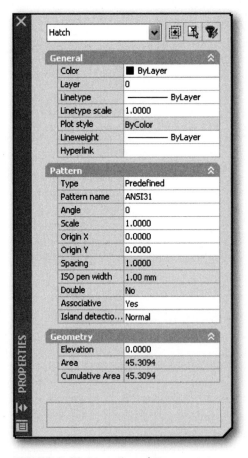

FIGURE 9–23 *Properties palette*

CONTROLLING THE VISIBILITY OF HATCH PATTERNS

The FILL command controls the visibility of hatch patterns in addition to the fill of multilines, traces, solids, and wide polylines. Invoke the FILL command by typing **fill** and pressing Enter at the on-screen prompt. AutoCAD displays the following prompt:

> Enter mode *(choose on to display the hatch pattern and off to turn off the display of hatch pattern)*

Invoke the REGEN command after changing the setting of FILL to see the effect.

Open the Exercise Manual PDF for Chapter 9 for discipline-specific exercises. Related files can be downloaded from the student companion site mentioned in the Introduction (refer to page number xii for instructions).

REVIEW QUESTIONS

1. AutoCAD will ignore text within a crosshatching boundary.
   a. True
   b. False

2. The HATCH command allows you to create an associative hatch pattern that updates when its boundaries are modified.
   a. True
   b. False

3. By default, hatch patterns are drawn at a 45-degree angle.
   a. True
   b. False

4. All of the following may be used as boundaries of the HATCH command except:
   a. ARC
   b. LINE
   c. BLOCK
   d. CIRCLE
   e. PLINE

5. The following are all valid AutoCAD commands except:
   a. ANGLE
   b. POLYGON
   c. HATCH
   d. ELLIPSE
   e. MULTIPLE

6. When using the HATCH command with a named hatch pattern, you can change:

    a. The color and scale of the pattern

    b. The angle and scale of the pattern

    c. The angle and linetype of the pattern

    d. The color and linetype of the pattern

    e. The color and angle of the pattern

7. Boundary hatch patterns inserted with an asterisk "*" preceding the name of the pattern will:

    a. Exclude inside objects

    b. Ignore inside objects

    c. Be inserted as individual objects

    d. Be inserted on layer 0

    e. None of the above

8. The type of hatch can be Predefined, User-defined, or Custom.

    a. True

    b. False

9. The AutoCAD hatch feature:

    a. Provides a selection of numerous hatch patterns

    b. Allows you to change the color and linetype

    c. Hatches over the top of text when the text is contained inside the boundary

    d. All of the above

10. The HATCH command will allow you to create a polyline around the area being hatched and to retain that polyline upon completion of the command.

    a. True

    b. False

11. Hatch patterns created with the HATCH command can be nonassociative or associative.

    a. True

    b. False

12. You can hatch any closed area by selecting a point outside it.

    a. True

    b. False

13. The HATCH command will place a hatch pattern over any text contained within the hatch boundary.

    a. True

    b. False

14. Which command automatically defines the nearest boundary surrounding a point you have specified.

    a. HATCH

    b. BHATCH

    c. BOUNDARY

    d. PTHATCH

15. When hatching, what would you press to select objects in the drawing to remove them from the selection?

    a. Remove Boundaries

    b. Remove Hatching

    c. Remove Selection

    d. Remove Objects

16. The Hatch and Gradient and Hatch Edit dialog boxes look the same.

    a. True

    b. False

17. When associative hatches are selected, what displays at the centroid of the hatch?

    a. A single grip

    b. The remove option

    c. A select Boundary Objects option

    d. A dialog box displaying the properties of the hatches selected

18. What default file does AutoCAD use to load hatch patterns from?

    a. ACAD.mnu

    b. ACAD.pat

    c. ACAD.dwg

    d. ACAD.hat

19. You can terminate the HATCH command before applying the hatch pattern by pressing ESC.

    a. True

    b. False

20. AutoCAD allows the properties of a nonassociative hatch pattern to be inherited.

    a. True

    b. False

10

Parametric Drawing Using Constraints

INTRODUCTION

The Parametric Drawing feature, introduced in AutoCAD 2010, allows you to apply geometric and dimensional constraints to objects. Applying constraints makes design and drafting easier and more accurate by assuring that objects maintain certain specified relationships with other objects, or with one of the coordinate axes, during the design process. Constraints can also be used to ensure that sizes of objects, measured distances, or angles between points or lines on objects can be fixed so that they will not change when the objects are modified. Inferred Constraints, introduced in AutoCAD 2011, allows you to create and modify objects and have constraints applied automatically as long as the constraints conform to the constraint conditions. You can also define parameter groups and filters in the Parameters Manager.

OBJECTIVES

After completing this chapter, you will be able to do the following:

- Apply geometric constraints to objects
- Apply dimensional constraints to objects
- Apply constraints using the AutoConstrain functionality
- Provide a name to a dimensional parameter and then assign a numeric value or formula as its expression
- Set dimensional parameters to reference other parameters for automatically updating when those other parameters are changed
- Define Parameter Groups and Filters
- Use Inferred Constraints

CONSTRAINTS

Geometric constraints allow you to control how objects react when the objects they are linked to are modified. For example, you can constrain two circles to always be concentric or two lines to always be parallel. Then if one of the pair of linked objects is moved, the other is automatically moved the same displacement. In another example, if the endpoint of a line is constrained to be coincident with the center of a circle, then when the circle is moved the endpoint of the line follows the center of the circle. Conversely, if the endpoint of the line is moved, the center of the circle will follow the endpoint of the line. In addition, vertical and horizontal constraints applied to pairs of points on objects will force the specified points on the constrained objects to maintain alignment with the Y or X axis, respectively. In most cases, key points on the objects are used. These are usually the same points used by Object Snap, except they are limited to endpoints, midpoints, center points, and insertion points. When a key point is involved, the nearest applicable point will be highlighted with the constraint key point marker.

Dimensional constraints can be applied to two points on an object so that the distance between those points (the length of the object) will remain unchanged during any modification to the object. Likewise, if the distance between points on two different objects is dimensionally constrained, that distance will not change during any modification. Dimensional constraints can be applied to the angle between objects so it cannot be changed.

Geometric Constraints

There are 12 geometric constraints available for use in an AutoCAD drawing. They are *Coincident, Colinear, Concentric, Fix, Parallel, Perpendicular, Horizontal, Vertical, Tangent, Smooth, Symmetric,* and *Equal,* and can be invoked from the Geometric panel (see Figure 10–1) available on the Parametric tab.

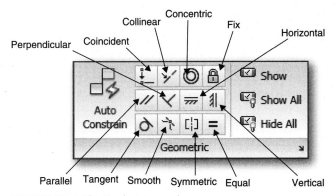

FIGURE 10–1 *The Geometric panel*

Table 10.1 lists the geometric constraints with a brief description of the function of each.

TABLE 10–1 *Geometric Constraints*

Button	Tool	Function
	Coincident	Two selected points are constrained to have the same location.
	Colinear	Two selected lines are constrained to be on the same infinite line, whether they overlap or not.
	Concentric	The centers of two selected circle/arcs have the same location.
	Fix	The selected point is constrained to its location at the time it is selected.
	Parallel	Two selected lines are constrained to always be parallel.
	Perpendicular	Two selected lines are constrained to be perpendicular to each other.
	Horizontal	The selected line is constrained to be parallel to the X axis.
	Vertical	The selected line is constrained to be parallel to the Y axis.
	Tangent	A selected line or circle/arc is constrained to be tangent to a selected circle/arc.
	Smooth	Two selected splines are constrained to a curvature continuous (G2) condition at their shared coincident endpoints.
	Symmetric	A second selected object is constrained to be symmetric to the first selected object about a selected line.
	Equal	Two selected objects are constrained to be equal in length or radius.

Coincident Constraint

The *Coincident* constraint causes selected points on two separate objects to maintain the same location. If the two points selected are not at the same location, then the object selected second will move or be modified for this to occur. You can apply the *Coincident* constraint to any point on one object to be coincident with any point on another object unless there are other constraints in effect that would cause the geometry to become over-constrained. Objects that can be selected for *Coincident* constraint include lines, polylines, circles, arcs, ellipses, and splines.

Invoke the *Coincident* constraint from the Geometric panel (Figure 10–2) located in the Parametric tab, and AutoCAD displays the following prompts:

```
Select first point or ⊡: (select a point on an object, or
right-click and select Object from the shortcut menu)
```

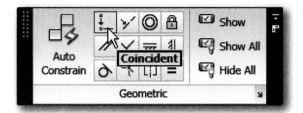

FIGURE 10–2 *Invoking the Coincident constraint from the Parametric panel*

By default, AutoCAD prompts you to select a point. Move the cursor over LINE A (Figure 10–3) and select near point A1 in the same manner as you would specify a point using Object Snap. AutoCAD displays a symbol near the cursor indicating a valid point. AutoCAD prompts:

> Select second point or ⊡: *(select a point on a different object or right-click and select OBJECT from the shortcut menu)*

When you apply the *Coincident* constraint to points on two objects, the resulting action depends on which object's point is selected first. Figure 10–3 shows the available constraint points for LINE A (A1, A2, and A3), LINE B (B1, B2, and B3), CIRCLE C (C1) and CIRCLE D (D1).

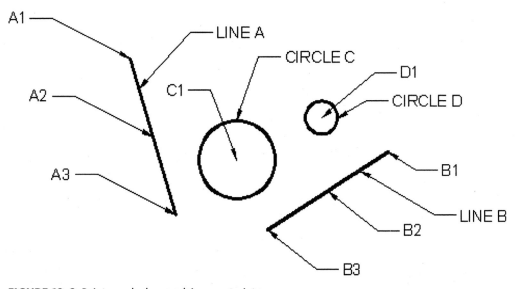

FIGURE 10–3 *Points used when applying constraints*

If you apply the *Coincident* constraint by selecting point A1 first and B1 second, then LINE A will remain where it is and LINE B will be moved so that point B1 is coincident with point A1 (Figure 10–4). If LINE B has no other constraints, it will retain its original length and angle after relocation.

Invoke the *Coincident* constraint and AutoCAD displays the following prompts:

> Select first point or ⊡: *(select LINE A near point A1)*
> Select second point or ⊡: *(select LINE B near point B1)*

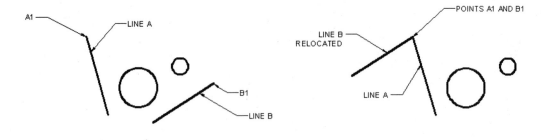

BEFORE   AFTER

FIGURE 10–4 *Applying the Coincident constraint by selecting point A1 first and B1 second*

If you apply the *Coincident* constraint by selecting point B1 first and A1 second, then LINE B will remain where it is and LINE A will be moved so that point A1 is coincident with point B1 (Figure 10–5).

Invoke the *Coincident* constraint and AutoCAD displays the following prompts:

> Select first point or ⬇: *(select LINE B near point B1)*
> Select second point or ⬇: *(select LINE A near point A1)*

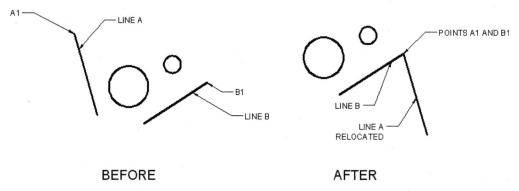

BEFORE   AFTER

FIGURE 10–5 *Applying the Coincident constraint by selecting point B1 first and A1 second*

Object Option to the Coincident Constraint. Using the *Object* option of the *Coincident* constraint causes the second selected object to be moved in a direction perpendicular to the first selected object until that the key point of the second object meets the first object. In the example shown in Figure 10–6, CIRCLE C is the second selected object. Its center is the key point. If you draw a line from its center perpendicular to LINE A (the first selected object), then that line is the direction in which the circle is moved.

Invoke the *Coincident* constraint from the Geometric panel and right-click and choose ENTER or *Object* from the shortcut menu. AutoCAD displays the following prompts:

> Select object: *(select LINE A)*
> Select point or ⬇: *(select CIRCLE C)*

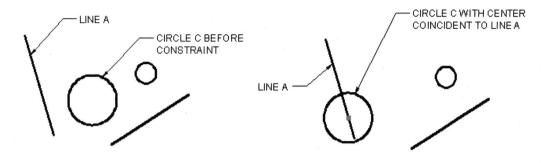

BEFORE   **AFTER**

FIGURE 10–6 *Applying the Coincident constraint by selecting LINE A first and CIRCLE C second*

When a Circle/Arc is the Object in the Object Option. If a circle or an arc is the first object selected when using the *Object* option of the *Coincident* constraint, the second selected object is moved (using the key point of the selection process) toward the center of the first selected circle/arc until the key point is on the circle, arc, or extension of the arc. In the example above, the key point of the circle is the center. But if the second object selected is a line, then the key point will be the one nearest to where the object is selected. It may be the middle point or one of the endpoints. For example, if you select CIRCLE C as the first object and then select LINE A near point A1 in response to the second prompt, the result will be as shown in Figure 10–7.

> Select object: *(select CIRCLE C)*
>
> Select point or 🔽: *(select LINE A near point A1)*

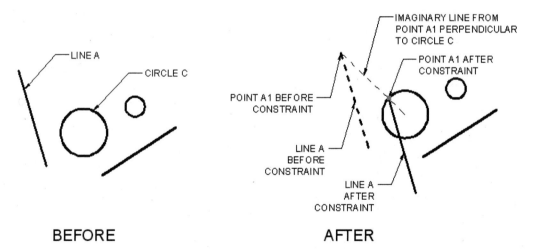

BEFORE   **AFTER**

FIGURE 10–7 *Applying the Coincident constraint by selecting CIRCLE C first and LINE A second*

The imaginary line is shown to illustrate how the second selected object is moved to the first selected object. It is in a direction determined by a line from the nearest key point perpendicular to the first selected object. In this case the perpendicular extends to the center of the circle.

Colinear Constraint

The *Colinear* constraint causes the second selected line to lie in the same infinite line as the first line selected (Figure 10–8). The second line, when relocated and rotated, might or might not overlap the first line. AutoCAD moves the second selected line until its nearest endpoint is on the infinite line and then it is rotated to the new alignment in the direction of the smallest angle.

Invoke the *Colinear* constraint from the Geometric panel and AutoCAD displays the following prompts:

```
Select first object or ⬇: (select LINE A)
Select second object: (select LINE B near point B3)
```

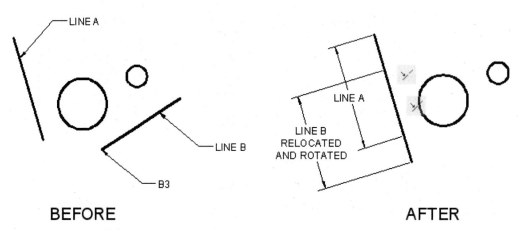

BEFORE   AFTER

FIGURE 10–8 *Applying the Colinear constraint for LINE B to be colinear with LINE A*

LINE B is relocated and rotated so that it lies on the same infinite line on which LINE A lies. When LINE B is moved perpendicular to the infinite line that LINE A is on, it may be at a far enough distance from LINE A that it does not even touch LINE A when rotated. It would not overlap LINE A but it would still be on the same infinite line.

The Multiple Option to the Colinear Constraint. The Multiple option to the *Colinear* constraint allows you to select more than one object to be colinear to the first line selected. Invoke the Colinear constraint, then right-click and select *Multiple* from the shortcut menu. After selecting the first line, AutoCAD repeats the "Select object to make colinear to first:" prompt until you press ENTER. Lines selected after selecting the first object will become colinear with the first.

Concentric Constraint

The *Concentric* constraint constrains the second selected circle/arc to be concentric with the first circle/arc (Figure 10–9).

Invoke the *Concentric* constraint from the Geometric panel and AutoCAD displays the following prompts:

```
Select first object: (select CIRCLE C)
Select second object: (select CIRCLE D)
```

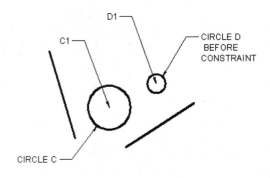 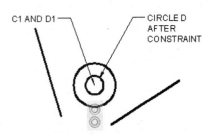

BEFORE   **AFTER**

FIGURE 10–9 *Applying the Concentric constraint for CIRCLE D to be concentric with CIRCLE C*

The second selected circle (CIRCLE D) will be moved so that its center will coincide with the center of the first selected circle (CIRCLE C). Once the circles have had the *Concentric* constraint applied and one of them is moved, the other will move to maintain concentricity.

Fix Constraint

The *Fix* constraint constrains the selected point on an object to maintain its location.

Invoke the *Fix* constraint from the Geometric panel and AutoCAD displays the following prompts:

> Select point or ⊡: (select point on object)

The point selected will be fixed to its current location. The remainder of the object may be stretched from or rotated about the fixed point if no other constraints prevent it.

The Object Option to the Fix Constraint. The Object option to the *Fix* constraint is accessible by choosing ENTER or *Object* from the shortcut menu. When a line or straight segment of a polyline is the object selected, it is fixed on the infinite line on which it lies. You can edit the endpoint grips only by moving them along the line to lengthen or shorten the selected line/pline object.

When a circle is the object selected in response to the *Object* option to the *Fix* constraint, it cannot be moved nor can its radius be changed. If the object selected is an arc, its center location cannot be changed but you can edit the endpoint grips only by moving them along the circle described by the center and radius to lengthen or shorten the arc.

Multiple Constraints and Over-constraining

AutoCAD will not allow you to over-constrain objects. If you attempt to apply a constraint that conflicts with an existing one AutoCAD displays the message shown in Figure 10–10.

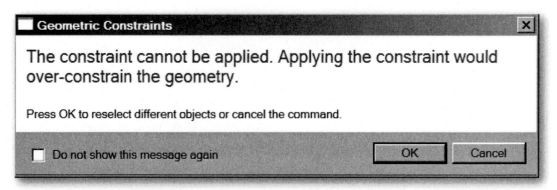

FIGURE 10–10 *Message displayed when attempting to over-constrain the geometry*

For example, if the endpoints of two skew lines have been fixed, then you cannot apply the *Parallel* constraint to them.

If you apply the *Coincident* constraint as shown in Figure 10–4 where you select point A1 and then select B1, but point B3 has been fixed, the result will be as shown in Figure 10–11.

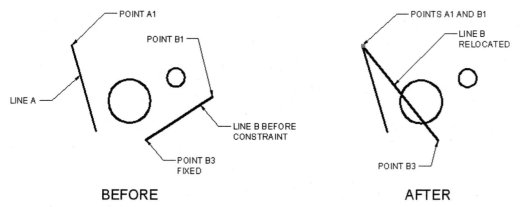

FIGURE 10–11 *Applying the Coincident constraint by selecting point A1 first and B2 second when point B3 is fixed*

Multiple constraints can be applied to objects and to points on objects. The more constraints there are the less freedom there is to manipulate. This can be very useful when understood and applied properly.

Parallel Constraint

The *Parallel* constraint causes the second selected line to rotate until it is parallel to the first line selected. The procedure involves rotating the second selected line about the endpoint nearest to the point where the line is selected.

Invoke the *Parallel* constraint from the Geometric panel and AutoCAD displays the following prompts:

        Select first object: *(select LINE A)*
        Select second object: *(select LINE B)*

Line B is rotated about point B3 if the point you select is nearest to B3 (Figure 10–12).

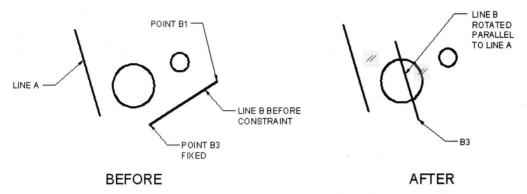

FIGURE 10–12 *Applying the Parallel constraint by selecting LINE A first and LINE B second near point B3 for parallel constraint*

LINE B is rotated counterclockwise about point B3 until it is parallel to LINE A. This is because LINE B was selected at a point nearest to point B3 and counterclockwise is the smaller angle of rotation. If LINE B had been selected at a point nearer to point B1, then LINE B would have been rotated about point B1. It would have still been rotated counterclockwise.

Perpendicular Constraint

The *Perpendicular* constraint causes the second selected line to rotate until it is perpendicular to the first line selected. The procedure involves rotating the second selected line about the endpoint nearest to the point where the line is selected.

Invoke the *Perpendicular* constraint from the Geometric panel and AutoCAD displays the following prompts:

> Select first object: *(select LINE A)*
> Select second object: *(select LINE B)*

LINE B is rotated about point B3 if the point you select is nearest to B3 (Figure 10–13).

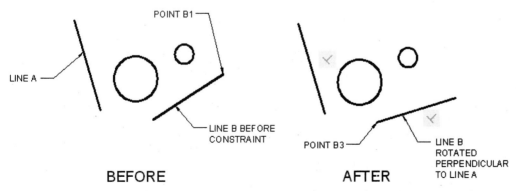

FIGURE 10–13 *Applying the Perpendicular constraint by selecting LINE A first and LINE B second near point B3*

LINE B is rotated clockwise about point B3 until it is perpendicular to LINE A. This is because LINE B was selected at a point nearest to point B3 and clockwise is the smaller angle of rotation. If LINE B had been selected at a point nearer to point B1, then LINE B would have been rotated about point B1. It would have still been rotated clockwise.

Horizontal Constraint

The *Horizontal* constraint causes the selected line to rotate until it is parallel to the X-axis of the current coordinate system. The procedure involves rotating the selected line about the endpoint nearest to the point where the line is selected.

Invoke the *Horizontal* constraint from the Geometric panel and AutoCAD displays the following prompts:

Select an object or ⊡: (select LINE A near point A1)

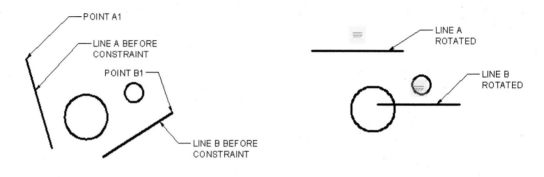

BEFORE   AFTER

FIGURE 10–14 *Applying the Horizontal constraint by selecting LINE A near point A1 and LINE B near point B1.*

LINE A is rotated about point A1 if the point you select is nearest to A1 (Figure 10–14). It is made parallel to the X-axis.

Repeat the procedure for LINE B. LINE B is rotated about point B1 if the point you select is nearest to B1 (Figure 10–14).

The 2Points Option to the Horizontal Constraint. The *2Points* option to the *Horizontal* constraint allows you to select two points on an object to be parallel to the X-axis of the current coordinate system (Figure 10–15). Invoke the *Horizontal* constraint, then right-click and select *2Points* from the shortcut menu and AutoCAD displays the following prompts:

Select first point: *(select LINE A near the midpoint)*
Select second point: *(select LINE A near an endpoint)*

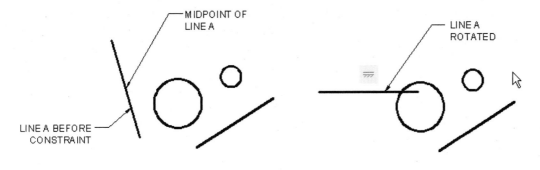

BEFORE **AFTER**

FIGURE 10–15 *Applying the Points option to the Horizontal constraint*

LINE A is rotated to horizontal about the first selected point, which is the midpoint of LINE A in this example.

Vertical Constraint

The *Vertical* constraint causes the selected line to rotate until it is parallel to the Y-axis of the current coordinate system. The procedure involves rotating the selected line about the endpoint nearest to the point where the line is selected.

Invoke the *Vertical* constraint from the Geometric panel and AutoCAD displays the following prompts:

Select an object or ⊡: (select LINE A near point A3)

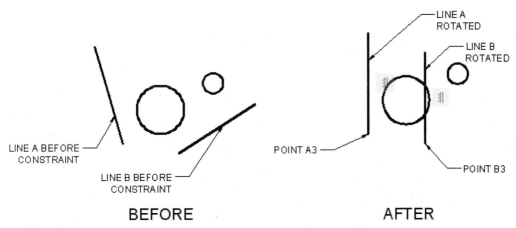

BEFORE **AFTER**

FIGURE 10–16 *Applying the Vertical constraint by selecting LINE A near point A3 and LINE B near point B3*

LINE A is rotated about point A3 if the point you select is nearer to A3 (Figure 10–16). It is made parallel to the Y-axis.

Repeat the procedure for LINE B. Line B is rotated about point B3 if the point you select is nearer to B3 (Figure 10–16).

The 2Points Option to the Vertical Constraint. The *2Points* option to the *Vertical* constraint allows you to select two points on an object to be parallel to the Y-axis of the current coordinate system (Figure 10–17). Invoke the *Vertical* constraint, then right-click and select *2Points* from the shortcut menu and AutoCAD displays the following prompts:

```
Select first point: (select LINE B near the midpoint)
Select second point: (select LINE B near an endpoint)
```

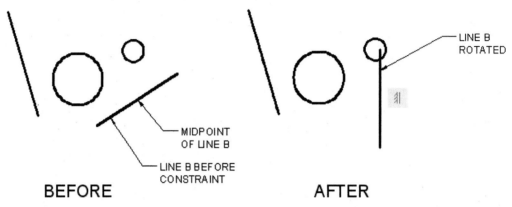

FIGURE 10–17 *Applying the Vertical constraint with the 2Points option*

LINE B is rotated vertically about the first selected point which is the midpoint of LINE B in this example.

Tangent Constraint

The *Tangent* constraint causes the second selected object to move until it is tangent to the first selected object (Figure 10–18).

Invoke the *Tangent* constraint from the Geometric panel and AutoCAD displays the following prompts:

```
Select first object: (select CIRCLE C)
Select second object: (select LINE A)
```

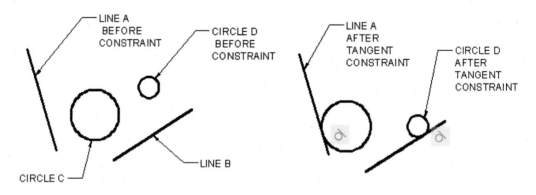

FIGURE 10–18 *Applying the Tangent constraint by selecting CIRCLE C first and then LINE A to be tangent to CIRCLE C and selecting LINE B first and then CIRCLE D to be tangent to LINE B*

LINE A is moved toward the circle until it becomes tangent to CIRCLE C (Figure 10–18). LINE A maintains its length and angle of rotation.

Repeat the procedure for LINE B and CIRCLE D, except select LINE B first. In this case, CIRCLE D is moved toward LINE B until it is tangent to LINE B (Figure 10–18).

Smooth Constraint

The *Smooth* constraint forces a spline to maintain geometric continuity with another spline, line, arc, or polyline. Invoke the *Smooth* constraint from the Geometric panel and AutoCAD displays the following prompts:

> Select first spline curve: *(select a spline curve)*
>
> Select second curve: *(select spline, line, arc or polyline)*

The first selected spline becomes contiguous and maintains G2 continuity with the second selected spline, line, arc, or polyline. G2 continuity means the splines touch at the join point where they share a common tangent direction and also share a common center of curvature. Endpoints of the curves to which you apply the smooth constraints are made coincident.

Symmetric Constraint

The *Symmetric* constraint causes selected objects to become symmetrically constrained about a selected line. For lines, the line's angle is made symmetric (but not the endpoints). For arcs and circles, the center and radius are made symmetric (but not the endpoints of the arc).

> There must be an axis about which you constrain the objects or points to be symmetrical. This is referred to as the symmetry line.

The Symmetric Constraint Applied to Lines. Figure 10–19 shows an example of applying the *Symmetric* constraint to a line. Invoke the *Symmetric* constraint from the Geometric panel and AutoCAD displays the following prompts:

> Select first object or ⬇: *(select LINE A)*
>
> Select second object: *(select LINE B)*
>
> Select symmetry line: *(select LINE E)*

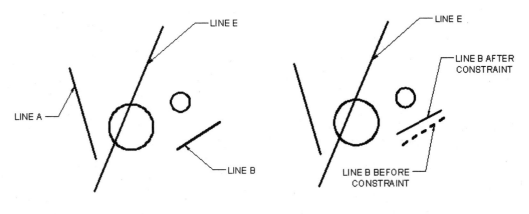

FIGURE 10–19 *Applying the Symmetric constraint by selecting LINE A first, LINE B second and then LINE E for the symmetry line*

LINE B is moved and rotated so that it mirrors the angle that LINE A makes with LINE E. LINE A maintains its length and angle of rotation. The angle of LINE B in relation to LINE E is now symmetrical to that of LINE A. If you perform a fillet between lines A and B with a zero radius, they will intersect on LINE E.

The Symmetric Constraint Applied to Circles and Arcs. In the case of circles and arcs, the *Symmetric* constraint causes the second selected circle to be mirrored about the line of symmetry and have the same radius as the first selected circle (Figure 10–20).

Invoke the *Symmetric* constraint from the Geometric panel and AutoCAD displays the following prompts:

```
Select first object or ⊡: (select CIRCLE D)
Select second object: (select CIRCLE C)
Select symmetry line: (select LINE B)
```

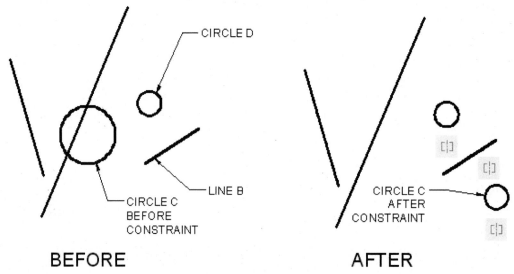

BEFORE   **AFTER**

FIGURE 10–20 *Applying the Symmetric constraint by selecting CIRCLE D, CIRCLE C, and LINE B as the symmetry line*

The result is CIRCLE C mirrors CIRCLE D about LINE B and has the same radius as CIRCLE D.

The 2Points Option to the Symmetric Constraint. The *2Points* option to the *Symmetric* constraint handles second selected objects in a different manner than the default option. In the case of lines, the point selected on the second object mirrors the point selected on the first object but the second object is not rotated to mirror the first line's angle. In the case of circles and arcs, the center of the second selected circle/arc mirrors the center of the first selected circle/arc but the second selected circle/arc retains its original radius.

Equal Constraint

The *Equal* constraint causes the second selected objects to become the same length for lines and the same radius for circles and arcs as the first selected object. For arcs and circles, the center and radius are made symmetric (not the endpoints of the arc).

Deleting Constraints

To delete a constraint, move the cursor over a constraint icon and choose *Delete* from the shortcut menu.

Autoconstrain

Choosing AUTOCONSTRAIN lets you apply geometric constraints to a selection set of objects based on orientation of the objects relative to each other. The constraints that are applied to the selection set depend on the configuration of the **AutoConstrain** tab of the Constraint Settings dialog box (Figure 10–21) which can be displayed by choosing the arrow in the lower-right corner of the Geometric panel of the Parametric tab.

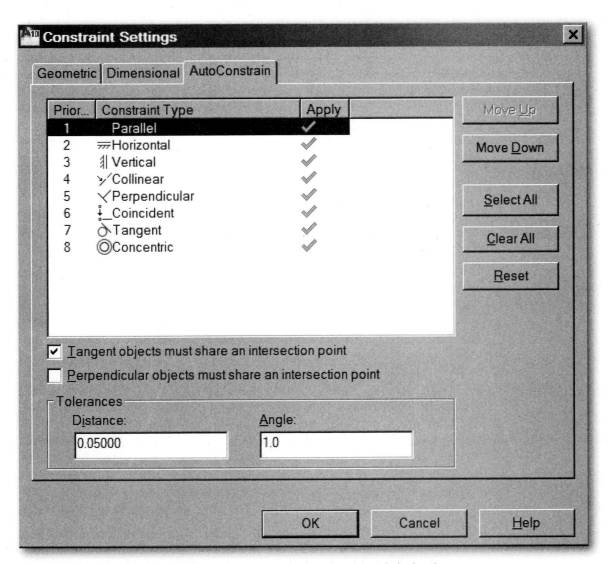

FIGURE 10–21 *The Constraint Settings dialog box with the AutoConstrain tab displayed*

The constraints will be applied in the order of priority of those that are checked in the list. To assign a higher priority to a constraint, move it up in the list. When two constraints conflict, the one with the higher priority will be applied.

Constraints will only be applied to geometry that will accept them without moving or resizing objects. For example, the *Coincident* constraint will be applied to key points that are at the same location. The *Parallel* constraint will be applied only to objects that already parallel, the *Horizontal* and *Vertical* constraints only to existing horizontal and vertical objects respectively.

Choosing **Move Up** or **Move Down** changes the order of the selected item by moving it up or down respectively in the list.

Selecting **Select All** causes all of the geometric constraint types to be checked.

Selecting **Clear All** causes all of the geometric constraint types to be unchecked.

Choosing **Reset** resets AutoConstrain settings to default values.

Choosing **Tangent Objects Must Share an Intersection Point** specifies that two curves must share a common point (as specified within the distance tolerance) for the tangent constraint to be applied.

Choosing **Perpendicular Objects Must Share an Intersection Point** specifies that lines must intersect or the endpoint of one line must be coincident with the other line or endpoint of the line as specified within the distance tolerance.

The **Tolerances** section lets you set the acceptable tolerance values to determine whether a constraint can be applied. The **Distance** box sets the distance tolerance for determining whether constraints are applied. The **Angle** box sets the angle tolerance for determining whether constraints are applied.

Geometric Constraint Settings

The Geometric tab of the Constraint Settings dialog box (Figure 10–22) lets you control the settings that affect the display of constraint bars or constraint point markers for objects in the drawing editor.

For example, you can hide the display of constraint bars for *Horizontal* and *Vertical* constraints.

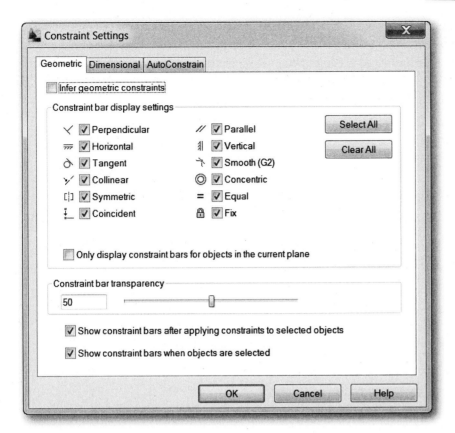

FIGURE 10–22 *The Constraint Settings dialog box with the Geometric tab displayed*

The **Constraint bar settings** section controls the parameters of the AutoCAD drawing window.

Selecting **Select All** causes all of the geometric constraint types to be checked.

Selecting **Clear All** causes all of the geometric constraint types to be unchecked.

Choosing **Only display constraint bars for objects in the current plane** only displays constraint bars for geometrically constrained objects on the current plane.

The **Constraint bar transparency** section controls the transparency of the constraint bars by adjusting the slider bar between the values of 10 and 90.

Choosing **Show constraint bars after applying constraints to selected objects** displays relevant constraint bars after you apply a constraint manually or when you use the AUTOCONSTRAIN command.

Choosing **Show constraint bars when objects are selected objects** displays relevant constraint bars of the objects selected.

Combining Geometric Constraints

The following example uses a combination of *Coincident, Horizontal, Vertical, Fix,* and *Equal* constraints to accurately draw a simple flight of stairs (Figure 10–23).

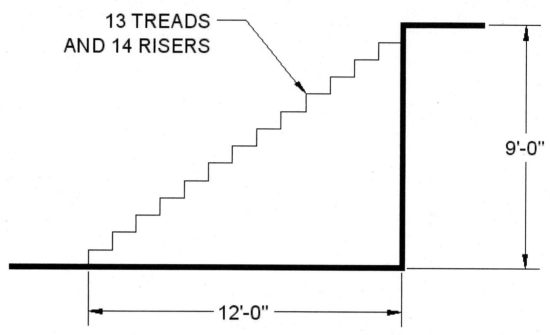

FIGURE 10–23 *Example of a flight of stairs for the application of Coincident, Horizontal, Vertical, Fix and Equal constraints*

It is a tedious task to divide the rise by 14 and the run by 13 to calculate the dimension of each riser and tread respectively. The left side of Figure 10–24 shows a unit module with each riser and tread one unit long. Detail A shows the first several risers and treads with the constraint bars being displayed. Note that the bottom of the first riser is locked into its location with the *Fix* constraint. Every riser will maintain the same length as every other riser by using the *Equal* constraint and likewise every tread will be constrained equal to every other tread. Every riser has the *Vertical* constraint applied to it and every tread the *Horizontal* constraint. To cause the risers and treads to stay connected when manipulated, every intersection of a tread and riser is tied together with the *Coincident* constraint.

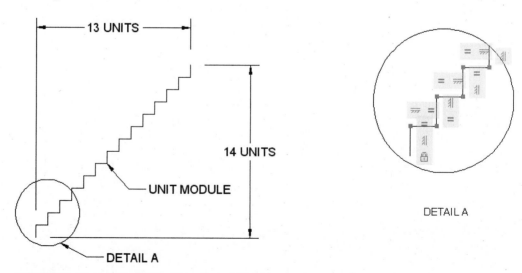

FIGURE 10–24 *Unit Module for 13 treads and 14 risers*

This Unit Module can be saved in a drawing and then inserted in a drawing of the cross section of the proposed stairway with the bottom of the lowest riser at the base of the stairs. Then it can be manipulated by selecting the top riser and then moving it by the highest grip to the upper landing point. The desired total run and total rise is achieved. When you stretch the module and lock the highest grip of the top riser onto the top landing of the proposed staircase in this manner, all risers maintain equal length as do all treads.

NOTE

> The above example is in architectural units with the rise and run being 9'-0" and 12'-0" respectively. Making the unit module risers and treads 1 unit (1 inch in this case) in length causes the module to be so small relative to its intended size that it is difficult to see in a full size cross section. To make it easier to work with, you can create the unit module with all risers 6 inches long and all treads 12 inches. These are arbitrary suggested values. Before it is manipulated, this module will be 13 feet long and 7 feet high which will make it more visible in a cross section. The risers do not need to be equal to the treads in the unit module, as long as all risers are constrained to be equal to other risers and all treads are constrained to be equal to other treads.

Flexibility Using Constraints

The stairway example explained earlier is a simple, straightforward application of the power of constraints. The risers are 7 11/16" high and the treads are 11 1/16" long. These dimensions are a result of stretching the unit module until the upper riser coincides with the upper landing point.

Normally, the rise of the staircase is fixed by the floor to floor change in elevation, so it doesn't change. But if you wished to have the run different from 12'-0", it is easy to change. After the top riser has been located at the top landing, delete the *Fix* constraint from the bottom riser and then apply the *Fix* constraint to the top of the top riser. Then the bottom of the bottom riser can be moved left to make the run longer or to the right to make it shorter. If you move it 6" to the left the total run will be 12'-6" long and the length of the treads is changed to 11 9/16".

If you wished to change the number of risers and treads from 14 and 13 respectively to 13 and 12, it is a simple matter of using the top grip of the 13th riser to lock onto the upper landing when manipulating the module. The bottom riser must still be fixed during this modification. After making the necessary changes, you can erase the risers and treads that extend above the landing. So, it is advisable to create a unit module with enough risers (and treads) to accommodate the maximum number anticipated. It is easy to erase the excess. It is not as easy to add more risers and treads with all the necessary constraints.

Dimensional Constraints

There are 7 dimensional constraints available for use in an AutoCAD drawing. They are *Linear, Horizontal, Vertical, Aligned, Angular, Radial,* and *Diameter*. Dimensional constraints maintain specified distances and angles between geometric objects or points on objects. For example, you can specify that the length of a line should always remain at 10.00 units, that the vertical distance between two points be maintained at 5.00 units, and that the diameter of a circle should always remain at 3.00 units.

Figure 10–25 shows the Dimensional panel, which can be accessed from the Parametric tab, with the Linear option expanded to show the Horizontal and Vertical options.

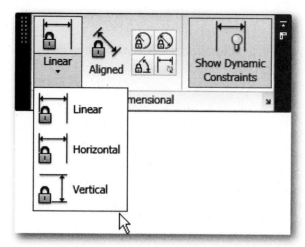

FIGURE 10–25 *The Dimensional panel expanded to show Horizontal and Vertical*

The Dimensional panel allows you to invoke dimensional constraints and control the display of dimensional constraints. You can also convert existing normal dimensions to be dimensional constraints.

Forms of Dimensional Constraints

When a dimensional constraint is applied to an object, a constraint variable is created which maintains the constraint value. Names such as *d1* or *rad1* are assigned by default, but they can be renamed as they are created or can be renamed in the Parameters Manager (explained later). Dimensional constraints can be created as either dynamic constraints or annotational constraints, each of which has a different purpose. In addition, any dynamic or annotational constraint can be converted to a reference constraint (explained later).

Dynamic Constraints. By default, dimensional constraints are *dynamic* and are used for normal parametric drawing and design tasks.

Dynamic constraints keep the same size when zooming in or out and can be globally turned on or off in the drawing. Dynamic constraints are displayed using a fixed, predefined dimension style and the textual information is positioned automatically. They have triangle grips for changing the value of a dimensional constraint and are not displayed when plotted. For controlling the dimension style of dynamic constraints or plotting them, you can change dynamic constraints to annotational constraints in the Properties palette.

Annotational Constraints. Annotational constraints are used when you wish to change the size of constraints when zooming in or out, or to display them individually with layers or display them using the *current* dimension style. Annotational constraints have grip capabilities similar to those on dimensions, and display when the drawing is plotted. You can convert the dynamic constraints to annotational constraints for plotting purposes and then convert them back to dynamic constraints in the Properties palette after plotting.

Linear, Horizontal, and Vertical Dimensional Constraints

Applying dimensional constraints is similar to creating normal dimensions except for being limited to the key constraint points on objects. When the *Horizontal* or *Vertical* dimensional constraint is applied to two points on a line, arc, or polyline segment, then those points will maintain their horizontal or vertical dimensional value respectively during any modification. Figure 10–26 shows the dimensional constraints applied to LINE A (vertical) and LINE B (horizontal) and to the distance between circles C and D (horizontal). Invoke the *Linear* or V*ertical* dimensional constraint from the Dimensional panel and AutoCAD prompts:

> Select first constraint point or ⊡: *(select point A1 on LINE A)*
>
> Select second constraint point: *(select point A3 on LINE A)*
>
> Specify dimension line location: *(specify a point for the dimension line)*

Invoke the *Linear* or *Horizontal* dimensional constraint from the Dimensional panel and AutoCAD prompts:

> Select first constraint point or ⊡: *(select point B1 on LINE B)*
>
> Select second constraint point: *(select point B3 on LINE B)*
>
> Specify dimension line location: *(specify a point for the dimension line)*

Invoke the *Linear* or *Horizontal* dimensional constraint from the Dimensional panel and AutoCAD prompts:

> Select first constraint point or ⊡: *(select CIRCLE C)*
>
> Select second constraint point: *(select CIRCLE D)*
>
> Specify dimension line location: (specify a point for the dimension line)

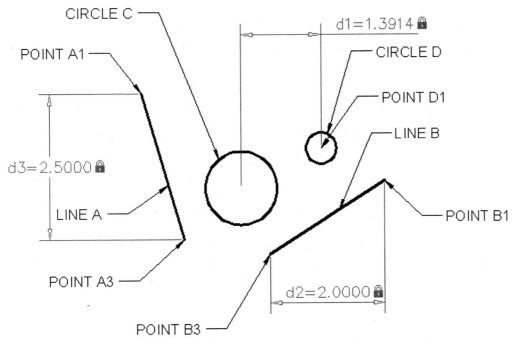

FIGURE 10–26 *The Vertical dimensional constraint applied to LINE A and the Horizontal dimensional constraint applied to LINE B and between circles C and D*

After specifying the location for the dimension line, the name and value displayed are highlighted. It can be modified at this time if desired or you can press the pick button on the mouse or press ENTER to accept the displayed value. The name is automatically created for labeling that constraint value, distinguishing it from other existing dimensional constraints. By default, they are assigned names such as *d1* or *dia1*. You can also rename them in the Parameters Manager palette, discussed later. In Figure 10–26 the *Vertical* dimensional constraint name and value is *d3=2.5000* for line A, and for line B the *Horizontal* dimensional constraint name and value is *d2=2.000*. The *Horizontal* dimensional constraint name and value for the distance between the circles is *d1=1.3914*.

Object Option to the Linear, Horizontal, or Vertical Dimensional Constraint.

Using the *Object* option to the *Linear, Horizontal,* or *Vertical* dimensional constraint causes the two endpoints of the selected object to be used for the particular dimensional constraint in effect. The *Object* option to the *Linear, Horizontal,* or *Vertical* dimensional constraint is accessible by choosing ENTER or *Object* from the shortcut menu.

Invoke the *Object* option and AutoCAD displays the following prompts:

```
Select object: (select a line, arc or a polyline segment)
Specify dimension line location: (specify a point for the
dimension line)
```

How Dimensional Constraints Constrain.

Figure 10–27 shows the result of modifying the objects as shown in Figure 10–26 after they have had dimensional constraints applied to them.

LINE A has a *Vertical* dimensional constraint applied to it. When the grip at point A3 is selected for relocating the object, point A1 remains fixed (even without using the *Fix* constraint at point A1). The vertical component of the length of LINE A is constrained to the distance of 2.5000 and A3 is constrained to being on an imaginary horizontal line that passes through its original location. This means that even though the final selection point S1 is beyond that imaginary horizontal line, Point A3 will remain on it.

LINE B has a *Horizontal* dimensional constraint applied to it. When the grip at point B1 is selected for editing, point B3 remains fixed (even without using the *Fix* constraint at point B3). The horizontal component of the length of LINE B is constrained to the distance of 2.0000 and B1 is constrained to being on an imaginary vertical line that passes through its original location. This means that even though the final selection point S2 does not extend to that imaginary vertical line, Point B1 will remain on it.

NOTE

In the case of a line having a horizontal or vertical constraint applied to it, the effect of modifying it by selecting one of the end grips is that the location of the cursor only determines the direction of the modified line. The length of the modified line is such that it will have the horizontal or vertical component remain unchanged, depending on the constraint applied.

The distance between CIRCLE C and CIRCLE D has a *Horizontal* dimensional constraint applied to it. When points on two separate objects have a linear constraint, that distance will remain the same when either is moved. CIRCLE D has been moved but the horizontal component of the distance between it and CIRCLE C remains the same. If CIRCLE D is moved either left or right, CIRCLE C will move the same horizontal displacement as CIRCLE D moves.

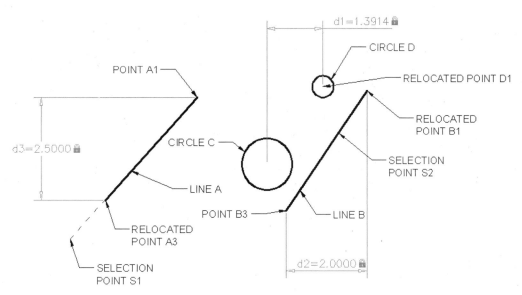

FIGURE 10–27 *Modifying lines A and B*

The vertical dimensional component of the length of LINE A and the horizontal dimensional component of the length of LINE B and distance between circles C and D are now constrained and will not change during drawing modifications.

Programmable Constraint Values. The name in a dimensional constrain is like an algebraic variable that can be referenced by another dimensional constraint in order to establish its value. The value may be in the form of an equation, function, or formula which can, in turn, reference the name of other dimensional constraint values. This feature adds powerful control over the dimensions of objects and their relationships in distances and angles. The geometry that is associated with the dimensional constraint is automatically resized when the value in a dimensional constraint is edited.

Figure 10–28 shows how variable names in dimensional constraints are referenced to values in other dimensional constraints. In Figure 10–28 (left side) LINE A has a *Vertical* dimensional constraint with the variable name *d1* shown equal to the numeric value of 2.500. This is the default format. LINE B has a *Horizontal* constraint with the variable name *d2* which has been edited to be equal to 0.8 times the value of *d1*. This makes it a function of *d1*. So, when normal dimensions are applied, the value will be 2.00 (2.50 × 0.8). The distance between the centers of the two circles is a *Horizontal* dimensional constraint with the variable name *d3*. It has been edited to be equal to 0.5 times *d2*. It will evaluate to 1.00 (2.00 × 0.5). In Figure 10–28 (right side), the *Vertical* dimensional constraint of line A (the value of *d1*) has been changed to 7.00. This causes *d2* to evaluate to 5.6 (7.00 times 0.8) and *d3* to 2.80 (5.60 times 0.5).

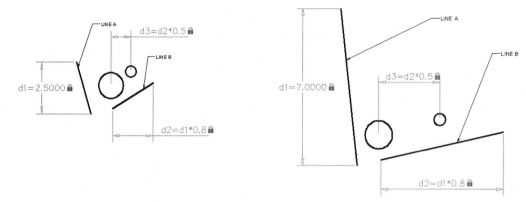

FIGURE 10–28 *Example for Variable names for dimensional constraint values*

Dimensional Constraint Key Points. Dimensional constraints are intended to constrain lengths of objects, and angles and distances between objects. For this reason, only certain key constraint points are used. Therefore, some points used by the Osnap feature are not valid constraint points, such as node, quadrant, intersection, and nearest. If you try to use an invalid point, AutoCAD displays the message "Invalid Point". Table 10.2 lists the valid constraint points for each type of object.

TABLE 10–2 *Dimensional Constraints*

Objects	Valid Constraint Points
Line	Endpoints, Midpoint
Arc	Center, Endpoints, Midpoint
Spline	Endpoints
Ellipse, Circle	Center
Polyline	Endpoints, Midpoint of line and arc subobjects, Center of arc subobjects
Block, Xref, Text, Mtext, Attribute, Table	Insertion point

Geometric Constraints and Dimensional Constraints Together. Dimensional constraints can work in conjunction with geometric constraints to satisfy certain design requirements. Earlier in this chapter an example of a flight of stairs was used to show how applying several geometric constraints to create a unit module can make it easier to keep all of the treads the same length and all of the risers the same length while stretching the top riser to reach the upper landing. This was facilitated by using the *Fix* geometric constraint at the bottom to lock it in place while fitting the unit module into position. The result was a stair case with a total run of 12'-0" and a rise of 9'-0" (Figure 10–29). The treads were 11 1/16" each and the risers were 7 11/16". Figure 10–31 shows the staircase from the previous example with the upper landing cut away for an unobstructed look at the top riser. For the next example, the *Fix* geometric constraint that had been applied in the previous example is deleted from the lower riser and one has been applied to the top of the upper riser.

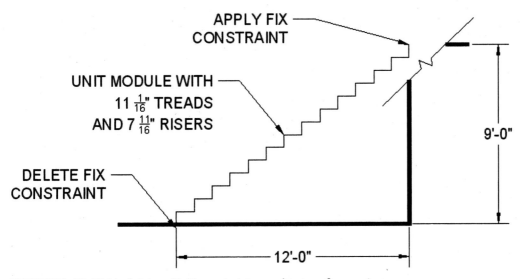

FIGURE 10–29 *Flight of stairs with Fix constraint moved to top of upper riser*

With the top of the unit module fixed at the upper landing, a *Horizontal* dimensional constraint is applied to the top tread (Figure 10–30). When the dimension is located the value is changed from 11 1/16″ to 11 1/2″.

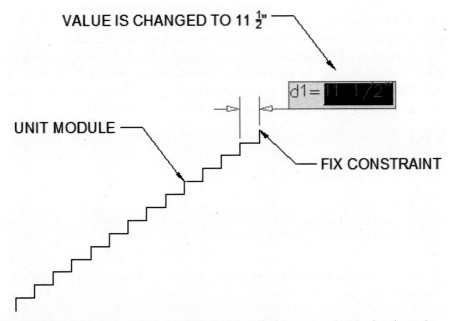

FIGURE 10–30 *Horizontal dimensional constraint applied to top tread and value changed*

When the value is changed to 11 1/2″, all of the other treads are changed to the same length because they have had the *Equal* geometric constraint applied to them. The result is a new total run length (Figure 10–31).

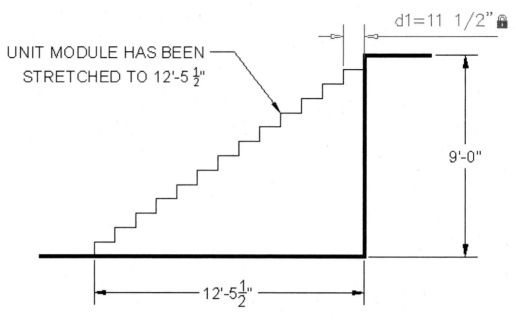

FIGURE 10–31 *Unit module stretched to new length with all treads equal*

Aligned Dimensional Constraints

Aligned dimensional constraints are similar to *Linear* dimensional constraints. They are applied to two points on a line, arc, or polyline segment or to the distance between objects. *Aligned* dimensional constraints, however, align with the selected points, either on an object or on two separate objects. *Aligned* dimensional constraints assure that the distance between the two selected points remains the same during any changes in the geometry. Figure 10–32 shows the *Aligned* dimensional constraints applied to lines A and B and to the distance between circles C and D.

Invoke the *Aligned* dimensional constraint from the Dimensional panel and AutoCAD prompts:

> Select first constraint point or ⊻: *(select point A1 on LINE A)*
>
> Select second constraint point: *(select point A3 on LINE A)*
>
> Specify dimension line location: *(specify a point for the dimension line)*

Invoke the *Aligned* dimensional constraint from the Dimensional panel and AutoCAD prompts:

> Select first constraint point or ⊻: *(select point B1 on LINE B)*
>
> Select second constraint point: *(select point B3 on LINE B)*
>
> Specify dimension line location: *(specify a point for the dimension line)*

Invoke the *Aligned* dimensional constraint from the Dimensional panel and AutoCAD prompts:

> Select first constraint point or ⊻: *(select CIRCLE C)*
>
> Select second constraint point: *(select CIRCLE D)*

Specify dimension line location: (specify a point for the
dimension line)

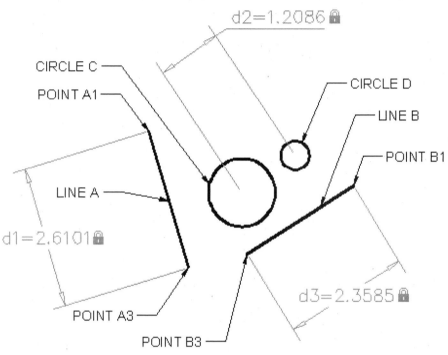

FIGURE 10–32 *The Aligned dimensional constraint applied to lines A and B and between circles C and D*

The lengths of lines A and B and the distance between circles C and D are now constrained and will not change during drawing modifications.

Modifying Dimensionally Constrained Objects. Figure 10–33 shows the result of modifying LINE A by moving it using the midpoint grip or the MOVE command.

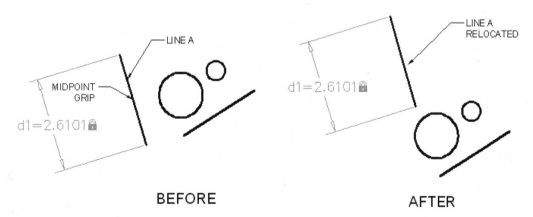

FIGURE 10–33 *Example of modifying dimensional constrained object LINE A*

LINE A has been moved. Its length and orientation is unchanged.

Figure 10–34 shows the result of modifying LINE B by moving the grip at point B3.

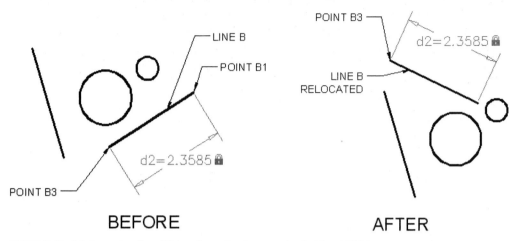

BEFORE　　　　　　　　　　　　　　**AFTER**

FIGURE 10–34 *Example of modifying dimensional constrained object LINE B*

LINE B has been moved. Its length is unchanged. However, its orientation has changed such that the new direction from point B3 to point B1 points to the original location of point B1.

Figure 10–35 shows the result of moving CIRCLE C.

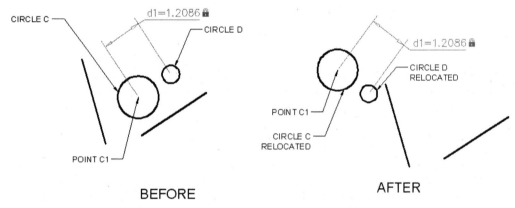

BEFORE　　　　　　　　　　　　　　**AFTER**

FIGURE 10–35 *Example of moving the dimensional constrained object CIRCLE C*

CIRCLE D "follows" CIRCLE C's movement so that the distance between CIRCLE C and CIRCLE D is unchanged. However, the orientation of the distance between the circles has changed.

NOTE

In the examples above only the first one, which involved moving LINE A by its midpoint grip or using the MOVE command, caused the object to maintain both its length and orientation. If you wished to have both the distance and orientation remain unchanged, you can apply a second dimensional constraint. If you add either a *Horizontal* or *Vertical* dimensional constraint to the *Aligned* dimensional constraint, the distance and the orientation will not be changed by moving the endpoint grip, as illustrated in Figure 10–34, or a point on one of the objects, as illustrated in Figure 10–35.

Angular Dimensional Constraints

Angular dimensional constraints are applied to objects so that the angle between line or polyline segments, the angle swept out by an arc or a polyline arc segment, or the angle between three points does not change during the design process. Figure 10–36 shows the *Angular* dimensional constraint applied to lines A and B.

Invoke the *Angular* dimensional constraint from the Dimensional panel and AutoCAD prompts:

> Select first line or arc or ⊡: *(select LINE A)*
>
> Select second line: *(select LINE B)*
>
> Specify dimension line location: (specify a point for the dimension line)

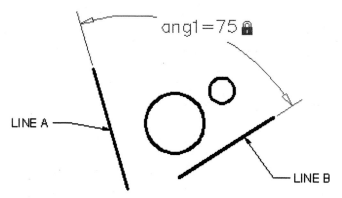

FIGURE 10–36 *The Angular dimensional constraints applied to lines A and B*

The angle between lines A and B is now constrained and will not change during drawing modifications. If LINE A is rotated LINE B will rotate to maintain the constrained angle, and if LINE B is rotated LINE A will rotate.

Radial and Diameter Dimensional Constraints

Radial and *Diameter* dimensional constraints are applied to a circle or an arc so that its radius/diameter does not change during the design changes. Figure 10–37 shows the *Diameter* dimensional constraint applied to circle C and the *Radial* dimensional constraint applied to circle D.

Invoke the *Diameter* dimensional constraint from the Dimensional panel and AutoCAD prompts:

> Select arc or circle: *(select CIRCLE C)*

Invoke the *Radial* dimensional constraint from the Dimensional panel and AutoCAD prompts:

> Select arc or circle: *(select CIRCLE D)*

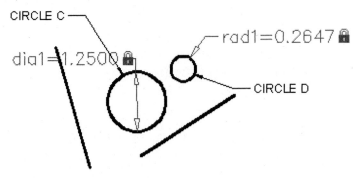

FIGURE 10–37 *The Diameter dimensional constraint applied to CIRCLE C and the Radial dimensional constraint applied to CIRCLE D*

The diameter of CIRCLE C and the radius of CIRCLE D are now constrained and will not change during drawing modifications.

Reference Constraints. *Reference* constraints do not control geometry but only report distances. They can be either dynamic or annotational. Reference constraints are a convenient way to display measurements that you would otherwise have to calculate. For example, LINE A and LINE B in Figure 10–38 are constrained by the *Vertical* dimensional constraint *d1* and the *Horizontal* dimensional constraint *d2* respectively. The reference constraint, *d3*, displays the total width but does not constrain it. The textual information in reference constraints is always displayed within parentheses. You can set the Reference property in the Properties palette to convert a dynamic or annotational constraint to a reference constraint. You cannot change a reference constraint back to a dimensional constraint if doing so would over-constrain the geometry.

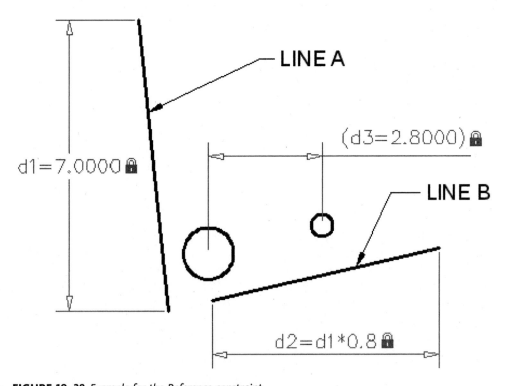

FIGURE 10–38 *Example for the Reference constraint*

Parameters Manager Palette

To manage the parameters for the dimensional constraints choose **Parameters Manager** from the Manager panel of the Parametric tab. AutoCAD displays the Parameters Manager palette (Figure 10–39).

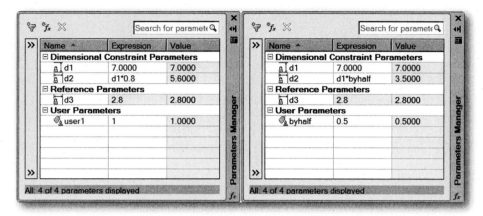

FIGURE 10–39 *Parameters Manager palette*

The **Name** list shows the names of the existing dimensional constraints in the drawing.

The **Expression** list shows the expressions of the dimensional constraints.

The **Value** list shows the value that the dimensional constraints evaluate to.

You can add two columns by right-clicking on the header bar and selecting **Type** and **Description** from the shortcut menu. All of the columns, except the **Name** column can be switched on and off in this manner.

Choosing **fx** from the toolbar opens the **User Variables** list below the **Dimensional Constraints** list and adds a variable named *user1*. The name can be changed if desired. In the palette on the right in Figure 10–39 the user variable has been renamed *byhalf* and given the value of 0.5 and its name has been inserted in the expression for constraint *d2* in place of the value of 0.8. The result is a constraint (*d2*) whose value is a function of a dimensional constraint (*d1*) and a user variable (*byhalf*).

Choosing **X** from the toolbar causes the selected constraint or variable to be deleted from the list.

The Search feature allows you to display one of the parameters by entering its name.

The **Filters** panel can be opened by selecting one of the double arrows on the top or bottom of the bar to the left of the tree view panel as shown in Figure 10–39. AutoCAD displays the Filter panel as shown in Figure 10–40. The default hierarchy has **All,** which lists all of the parameters, and the branch labeled **All Used in Expressions,** which lists only parameters that are associated with functions.

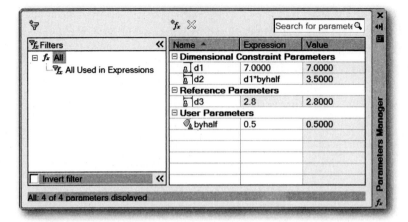

FIGURE 10–40 *Filter Panel*

Choosing the funnel icon from the toolbar creates a new filter group with the default name **Group Filter 1** as shown in Figure 10–41.

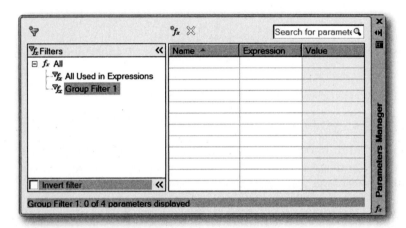

FIGURE 10–41 *New Filter Group*

Figure 10–42 shows how, while the **All** filter is selected, you can select one of the parameters in the right panel and drag it into the new filter group that has been renamed **d1-d2 Link**.

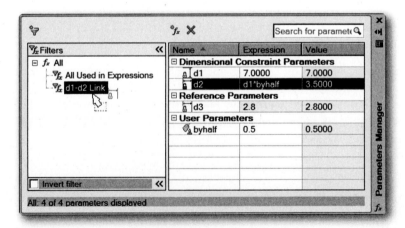

FIGURE 10–42 *Renaming of the filter group*

After dragging parameters **d2** and **byhalf** into filter group **d1-d2 Link,** they will be displayed when **d1-d2 Link** is selected from the Filters panel as shown in Figure 10–43.

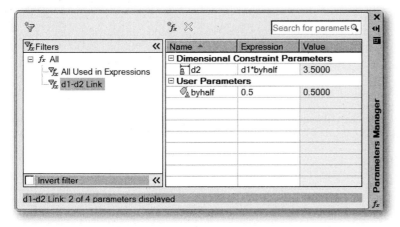

FIGURE 10–43 *Listing of parameters for the selected group*

Dimensional Constraint Settings. The Dimensional tab of the Constraint Settings dialog box (Figure 10–44) allows you to set preferences in behavior when displaying dimensional constraints.

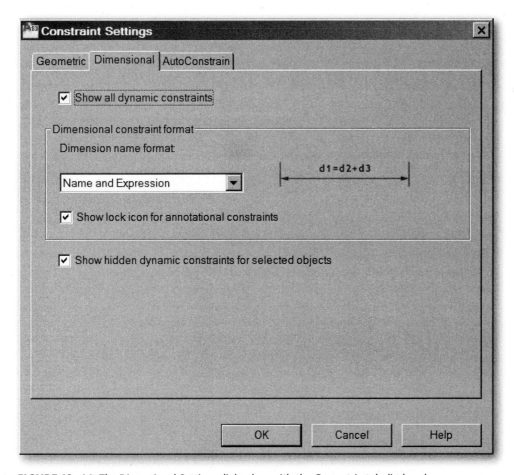

FIGURE 10–44 *The Dimensional Settings dialog box with the Geometric tab displayed*

Choosing **Show All Dynamic Constraints** causes all dynamic dimensional constraints to be displayed by default.

The **Dimensional Constraint Format** section allows you to set the display of the dimensional name format and lock icon. The **Dimension Name Format** text box allows you to specify the format for the text displayed when dimensional constraints are applied. Formats you can choose from are **Name, Value,** or **Name and Expression.** An expression might be in the form of Length=Width *3

Choosing **Show Lock Icon for Annotational Constraints** causes a lock icon to be displayed against an object that has an annotational constraint applied.

Choosing **Show Hidden Dynamic Constraints for Selected Objects** causes dynamic constraints that have been set to hide when selected to be displayed.

Inferred Geometric Constraints

With the use of Inferred Geometric Constraints certain constraints can be automatically applied as you draw and edit objects. The *Infer Constraints* feature is accessible through the Infer Constraints button on the Status Bar at the bottom left of the drawing area. Switching it on and off is similar to using the Grid, Ortho, and Osnap buttons to switch their respective features on and off.

Figure 10–45 shows selecting the Infer Constraints button on the Status Bar.

FIGURE 10–45 *Selecting the Infer Constraints button on the Status Bar*

The Inferred Constraints feature is similar to the AUTOCONSTRAIN command except Inferred Constraints is initiated before objects are drawn or edited. Once Inferred Constraints is turned on, when you draw certain objects whose key points satisfy the conditions of certain constraints, those constraints are automatically applied. In most cases this requires using one or more of the OSNAP modes to make this occur. For example, if you draw multiple segments with one LINE command then a Coincident Constraint will be applied at each common point of connecting segments. If, however, you draw one segment, end the LINE command and then reinitiate the LINE command to draw a second connecting segment and wish to have a Coincident Constraint applied where they connect, you will need to use the Endpoint Osnap mode. Line-Line continuation (launched when you press ENTER in response to the "Specify start point:" prompt of the second LINE command) does not support Inferred Constraints. If you draw a circle and then use the Center Osnap mode to start or end a line or arc at the center of the circle, a Coincident Constraint will be applied at that point.

Table 10.3 shows the OSNAP modes that can be used with Inferred Constraints and how they are applied between the created or edited object and the snapped object.

TABLE 10–3 *Dimensional Constraints*

Object SNAP	Description
Endpoint, Midpoint, Center, Node, Insertion	Applies a Coincident Constraint
Nearest	Applies a Coincident Constraint
Perpendicular	Applies a Perpendicular or Coincident Constraint
Tangent	Applies a Tangent Constraint
Parallel	Applies a Parallel Constraint

OSNAP modes that do not infer constraints include the Intersection, Apparent Intersection, Extension, and Quadrant.

Constraints that are not inferred include Fix, Smooth, Symmetric, Concentric, Equal, and Collinear.

Horizontal and Vertical Constraints conform to the settings of the Polar and Ortho.

As mentioned earlier, segments drawn using the LINE command in a repeating fashion have constraints applied. However, segments drawn using the PLINE command do not. But, if you join a polyline object to another object (line, circle, arc, or another polyline) using one of the valid OSNAP modes, the appropriate constraint will be applied.

The RECTANG command will apply Coincident constraint to the corners of the rectangle.

The FILLET command will apply Tangent and Coincident constraint between the arc created and the filleted lines.

The CHAMFER command will apply Coincident constraint between the chamfer line created and the chamfered lines.

Commands that are not affected by Inferred Constraints include ARRAY, BREAK, EXTEND, MIRROR, OFFSET, SCALE, and TRIM.

Open the Exercise Manual PDF for Chapter 10 for discipline-specific exercises. Related files are downloaded from the student companion site mentioned in the Introduction (refer to page number xii for instructions).

REVIEW QUESTIONS

1. The number of points that can be constrained with one Coincident constraint:
   a. 1
   b. 2
   c. 3
   d. 4
   e. 255

2. The number of lines that can be constrained with one Colinear constraint:

   a. 1

   b. 2

   c. 3

   d. 4

   e. unlimited

3. The objects that can be constrained with the Concentric constraint:

   a. Circle

   b. Arc

   c. Ellipse

   d. All of above

4. What can be used to modify a geometrically constrained drawing?

   a. Constraint Bar

   b. Delete

   c. Grips

   d. Fix command

5. Only points and not objects can be constrained with the Fix constraint:

   a. True

   b. False

6. The objects that can be constrained with the Colinear constraint:

   a. Circles

   b. Arcs

   c. Ellipses

   d. Lines

   e. All of above

7. What command sets up the parallel, perpendicular, and vertical/horizontal relationships?

   a. Fix

   b. Symmetrical

   c. Tangent

   d. Auto-Constrain

8. The objects that can be constrained with the Parallel constraint:

   a. Circles

   b. Arcs

   c. Ellipses

   d. Lines

   e. All of the above

9. What kind of object does the Tangent command constrain drawing elements to?

    a. Linear

    b. Spline

    c. Radial

    d. Parallel

10. The objects that can be constrained with the Perpendicular constraint:

    a. Circles

    b. Arcs

    c. Ellipses

    d. Lines

    e. All of the above

11. The Horizontal constraint can be applied to the endpoint of a line and the center of a circle:

    a. True

    b. False

12. The Vertical constraint can only be applied to a line:

    a. True

    b. False

13. Aligned Dimensional constraints can only be applied to lines:

    a. True

    b. False

14. The Parallel command constrains two linear objects to take the angle of the_____.

    a. First Linear Object

    b. Spline

    c. Perpendicular

    d. Radii

15. You can convert normal dimensions to Dimensional constraints:

    a. True

    b. False

CHAPTER
11

Block References
and Attributes

INTRODUCTION

The AutoCAD BLOCK command feature is a powerful design/drafting tool. The BLOCK command enables a designer to create an object from one or more objects, save it under a user-specified name, and later place it back into the drawing. When block references are inserted in the drawing, they can be scaled up or down in either of the X, Y, and Z axes. They can also be rotated as they are inserted in the drawing. Block references can best be compared with their manual drafting counterpart, the template. Even though an inserted block reference can be created from more than one object, the block reference acts as a single unit when operated on by certain modify commands such as MOVE, COPY, ERASE, ROTATE, ARRAY, and MIRROR. You can export a block reference to become a drawing file outside the current drawing and create a symbol library from which block references are inserted into other drawings. Like the plastic template, block references greatly reduce repetitious work.

The Dynamic Block feature, explained in this chapter, uses parametric data technology to make the use of blocks much more powerful. After you have mastered the concepts for creating and using the BLOCK command, you can apply the power of the dynamic blocks to provide drawings of objects and parts that have similar geometry with more flexibility, ease, and accuracy.

The BLOCK command can save time because you don't have to draw the same object(s) more than once. Block references save computer storage because the computer only needs to store the object descriptions once. When inserting block references, you can change the scale and/or proportions of the original object(s).

OBJECTIVES

After completing this chapter, you will be able to do the following:

- Create and insert block references in a drawing
- Convert individual block references into drawing files
- Define attributes, edit attributes, and control the display of attributes
- Use the DIVIDE and MEASURE commands

- Create Dynamic Blocks

- Use dynamic geometric and dimensional constraints in Dynamic Blocks

- Use parametric constraints in Dynamic Blocks

CREATING BLOCKS

When you invoke the BLOCK command to create a block, AutoCAD refers to this as defining the block. The resulting definition is stored in the drawing database. The same block can be inserted as a block reference as many times as needed.

Blocks may comprise one or more objects. The first step in creating blocks is to create a block definition. In order to do this, the objects that make up the block must be visible on the screen. That is, the objects that will make up the block definition must have already been drawn so that you can select them when prompted to do so during the BLOCK command.

It is very important to take into consideration the layer on which the objects are drawn that comprises the block. Objects that are on Layer 0 when the block is created will assume the color, linetype, and lineweight of the layer on which the block reference is inserted. Objects on any layer other than 0 when included in the block definition will retain the characteristics of that layer, even when the block reference is inserted on a different layer (Figure 11–1).

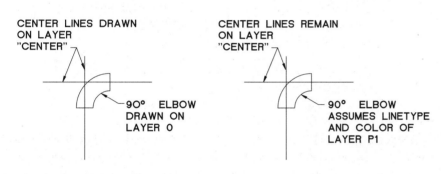

FIGURE 11–1 *Example of inserting block references drawn in different layers*

You should be careful when invoking the PROPERTIES command to change the color, linetype, or lineweight of elements of a block reference. It is best to keep the color, linetype, and lineweight of block references and the objects that comprise them in the BYLAYER state.

Examples of some common uses of blocks in various disciplines are shown in Figure 11–2.

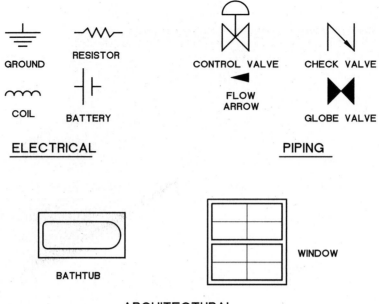

FIGURE 11–2 *Examples of common uses of blocks in various disciplines*

Creating a Block Definition

The BLOCK command creates a block definition for selected objects.

Invoke the BLOCK command from the Block panel (Figure 11–3) located in the Home tab, and AutoCAD displays the Block Definition dialog box (Figure 11–4).

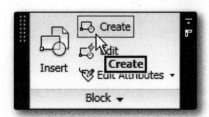

FIGURE 11–3 *Invoking the BLOCK command from the Block panel*

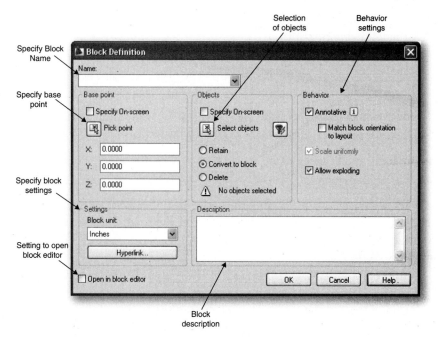

FIGURE 11–4 *Block Definition dialog box*

The **Name** text box lets you specify the block name. The block name can be up to 255 characters long and may contain letters, numbers, and any special character not used by Microsoft® Windows® and AutoCAD for other purposes, if the system variable EXTNAMES is set to 1. DIRECT, LIGHT, AVE_RENDER, RM_SDB, SH_SPOT, and OVERHEAD cannot be used as block names. To list the block names in the current drawing, click the down-arrow to the right of the **Name** text box. AutoCAD lists the blocks in the current drawing.

The Base point section lets you specify the insertion point for the block. The insertion point specified during the creation of the block becomes the basepoint for future insertions of this block as a block reference. It is also the point about which the block reference can be rotated or scaled during insertion. When determining where to locate the base insertion point, it is important to consider what will be on the drawing before you insert the block reference. Therefore, you must anticipate this preinsertion state of the drawing. It is sometimes more advantageous for the insertion point to be somewhere off the object than on it.

Choose **Pick point** to specify the basepoint on the screen, and AutoCAD displays the following prompt:

    Specify insertion point: *(specify the insertion point)*

The **X**, **Y**, and **Z** text boxes let you specify the X, Y, and Z coordinates of the insertion point. Once you have specified the insertion point, the Block Definition dialog box reappears.

The Objects section lets you select objects to be included in the block and determine how the objects will be treated once the block is created.

Choose **Select Objects** to select objects to include in the block definition, and Auto-CAD displays the following prompt:

```
Select objects: (select objects using one of the AutoCAD
object selection methods, and press [Enter] to complete object
selection)
```

Choosing **Quick Select** lets you use the QSELECT command, explained in Chapter 6, to select the object to include in the block definition.

Choose **Retain** for the objects selected to be included in the block definition to remain in place as separate objects.

Choose **Convert to Block** for the block definition created from the selected objects to become a block reference inserted into the drawing at the location where the block definition was created.

Choose **Delete** for the objects selected to be included in the block definition to be deleted from the drawing after the block definition is created.

The Behavior section lets you determine the behavior of the block.

Choosing **Annotative** specifies that the block is annotative. For additional information on Annotation, refer to Chapter 8.

Choosing **Match Block Orientation to Layout** specifies that the orientation of the block references in paper space viewports matches the orientation of the layout. This option is available only when the **Annotative** option is selected.

Choosing **Scale Uniformly** specifies whether or not block reference is prevented from being scaled differently in the X, Y, and Z axes. This option is available only when the **Annotative** option is not selected.

Choosing **Allow exploding** specifies whether or not the block reference can be exploded. The Settings section controls units, scale, explode options, description, and hyperlink options.

The **Block unit** text box lets you specify the insertion units for the block reference.

The **Description** text box can be used to enter a description of the block if desired.

Choosing **Hyperlink** causes AutoCAD to display the Insert Hyperlink dialog box (Figure 11–5).

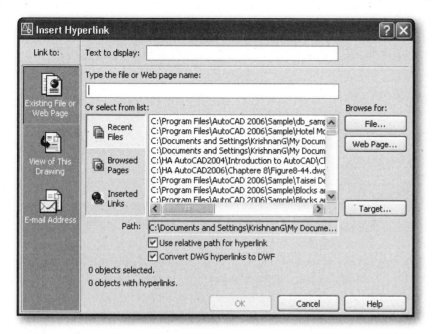

FIGURE 11–5 *Insert Hyperlink dialog box*

Hyperlinks are created in AutoCAD drawings as pointers to associated files. Hyperlinks can launch a word processing program and open a specific file and even point to a named location in a file. Hyperlinks can activate your Web browser and load a specified HTML page. You can specify a view in AutoCAD or a bookmark in a word processing file. Hyperlinks can be attached to a graphical object in an AutoCAD drawing.

Hyperlinks can be either absolute or relative. Absolute hyperlinks have the full path to a file location stored in them. Relative hyperlinks have only a partial path to a file location stored in them, relative to a default URL or directory you have specified by setting the HYPERLINKBASE system variable.

Hyperlinks can point to locally stored files, files on a network drive, or files on the Internet. Cursor feedback is automatically provided to indicate when the crosshairs are over a graphical object that has an attached hyperlink. You can then select the object and use the Hyperlink shortcut menu to open the file associated with the hyperlink. This hyperlink cursor and shortcut menu display can be turned off in the Options dialog box.

When a hyperlink to an AutoCAD drawing that has a named view is opened, that view is restored. This also applies to a hyperlink created with a named layout. AutoCAD opens that drawing in that layout.

If a hyperlink is created that points to an AutoCAD drawing template (DWT) file, AutoCAD will create a new drawing file based on the template. This prevents overwriting the original template. For additional information on hyperlinks and how to create and edit them, refer to Chapter 14.

Choosing **Open in block editor** opens the current block definition in the Block Editor when you click OK. For details on the Block Editor, refer to the section on Dynamic Blocks.

After making necessary changes to Block Definition dialog box, choose **OK** to create the block definition with the given name. If the given name is the same as an existing block in the current drawing, AutoCAD displays a warning (Figure 11–6).

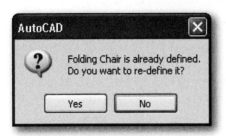

FIGURE 11–6 *Warning dialog box regarding block definition*

To redefine the block, choose **Yes** in the Warning dialog box. The block with that same name is then redefined. Once the drawing is regenerated, any insertion of this block reference already inserted in the drawing is redefined to the new block definition with this name.

Choose **No** in the Warning dialog box to cancel the block definition. To create a new block definition, specify a different block name in the **Name** text box of the Block Definition dialog box, and choose **OK**.

If you create a block without selecting objects, AutoCAD displays a warning that nothing has been selected, and provides an opportunity to select objects before the named block is created.

AutoCAD creates the block from the selected objects that make up the definition from the screen using the specified name.

INSERTING BLOCK REFERENCES

You can insert previously defined blocks into the current drawing by invoking the INSERT command. If there is no block definition with the specified name in the current drawing, AutoCAD searches the drives and folders on the path for a drawing of that name and inserts it instead.

NOTE

If blocks were created and stored in a template drawing, and you make your new drawing equal to the template, those blocks will be in the new drawing ready to insert. Any drawing inserted into the current drawing will bring with it all of its block definitions, whether they have been inserted or are only stored as definitions.

Invoke the INSERT command from the Block panel (Figure 11–7) located in the Home tab, and AutoCAD displays the Insert dialog box (Figure 11–8).

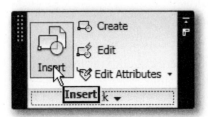

FIGURE 11–7 *Invoking the* INSERT *command from the Block panel*

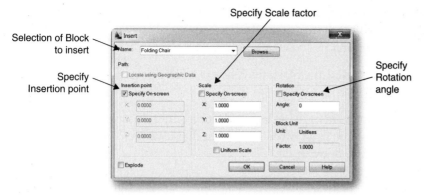

FIGURE 11–8 *Insert dialog box*

Specify a block name in the **Name** text box. Or, you can choose the down-arrow to display a list of blocks defined in the current drawing, and select the block you wish to insert.

Choosing **Locate Using Geographic Data** inserts drawing using geographic data as the reference. This option is valid only if the current and inserted block/drawing contain geographic data.

The Insertion point section of the Insert dialog box allows you to specify the insertion point for inserting a copy of the block definition.

The **X**, **Y**, and **Z** text boxes let you specify the X, Y, and Z coordinates of the insertion point by entering them from the keyboard when **Specify On-screen** is not checked. Set **Specify On-screen** to ON if you prefer to specify the insertion point on screen with your pointing device.

The Scale section of the Insert dialog box allows you to specify the scale for the inserted block. The default scale factor is set to 1 (full scale). You can specify a scale factor between 0 and 1 to insert the block reference smaller than the original size of the block and specify more than 1 to increase the size from the original size.

The **X**, **Y**, and **Z** text boxes let you specify the X, Y, and Z scales by entering them from the keyboard. If necessary, you can specify different X, Y, and Z scale factors for the block reference. If you specify a negative scale factor, AutoCAD inserts a mirror image of the block about the insertion point. If −1 were used for both X- and Y-scale factors, it would "double-mirror" the object, the equivalent of rotating it 180°.

Set **Specify On-screen** to ON if you prefer to specify the scale factor on screen with your pointing device.

Select **Uniform Scale** and then you can enter a value only in the **X** text box. The Y and Z scales will be the same as that entered for the X scale.

The Rotation section of the Insert dialog box allows you to specify the rotation angle for the inserted block.

Set **Specify On-screen** to ON if you prefer to specify the rotation angle on screen with your pointing device.

The **Angle** text box lets you specify the angle to rotate the block reference by entering the value from the keyboard.

The Block Unit section displays the units, such as inches or millimeters, that were used when the selected block was created and the unit scale factor, calculated based on the block units value and the drawing units.

Set **Explode** to ON to insert the block reference as a set of individual objects rather than as a single unit.

To specify a drawing file to insert as a block definition, enter the drawing file name in the **Name** text box. Or, you can choose **Browse** to display a standard file dialog box and select the appropriate drawing file.

The name of the last block reference inserted during the current drawing session is remembered by AutoCAD. The name becomes the default for subsequent use of the INSERT command.

Choose **OK** to insert the selected block.

NESTED BLOCKS

Blocks can contain other blocks. That is, when you are using the BLOCK command to combine objects into a single object, one or more of the selected objects can themselves be blocks, and the blocks selected can have blocks nested within them. There is no limitation to the depth of nesting. You may not, however, use the name of any of the nested blocks as the name of the block being defined. This would mean that you were trying to redefine a block, using its old definition in the new.

Any objects within blocks, such as nested blocks, that were on Layer 0 when made into a block will assume the color, linetype, and lineweight of the layer on which the block reference is inserted. If an object originally on Layer 0 when included in a block definition is in a block reference that has been inserted on a layer other than Layer 0, it will retain the color, linetype, and lineweight of the layer it was on when its block was included in a higher-level block. For example, say that you draw a circle on Layer 0 and include it in a block named Z1. You then insert Z1 on Layer R, whose color is red. The circle would then assume the color of Layer R; in this case it will be red. Create another block called Y3 by including the block Z1. If you insert block reference Y3 on a layer whose color is blue, the block reference Y3 will retain the current color of layer R—in this case it will be red—instead of taking up the color of blue.

EXPLODE COMMAND

The EXPLODE command causes block references, hatch patterns, and associative dimensioning to be turned into the separate objects from which they were created. It also causes polylines/polyarcs and multilines to separate into individual simple line and arc objects. The EXPLODE command causes 3D polygon meshes to become 3Dfaces, and 3D polyface meshes to become 3Dfaces and simple line and point objects. When an object is exploded, the new, separate objects are created in the model or paper space of the exploded objects.

Invoke the EXPLODE command from the Modify panel (Figure 11–9) in the Home tab, and AutoCAD displays the following prompt:

```
Select objects: (select objects to explode, and press Enter
to complete object selection)
```

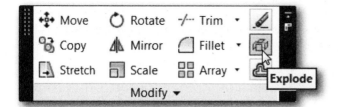

FIGURE 11–9 *Invoking the EXPLODE command from the Modify panel*

You can use one or more object selection methods. The object selected must be eligible for exploding, or an error message will appear. An eligible object may or may not change its appearance when exploded.

Possible Changes Caused by the EXPLODE Command

A polyline segment having width will revert to a zero-width line and/or arc. Tangent information associated with individual segments is lost. If the polyline segments have width or tangent information, the EXPLODE command will be followed by this message:

```
Exploding this polyline has lost (width/tangent)
information.
The undo command will restore it.
```

Individual elements within blocks that were on Layer 0 when the block was created, and whose color was BYLAYER, but were inserted on a layer with a color different than that of Layer 0, will revert to the color of Layer 0.

Attributes are special text objects that, when included in a block definition, take on the values, or names and numbers, specified at the time the block reference is inserted. The power and usage of attributes are discussed later in this chapter. To understand the effect of the EXPLODE command on block references that include attributes, it is sufficient to know that the object from which an attribute is created is called an attribute definition. It is displayed in the form of an attribute tag before it is included in the block.

An attribute within a block will revert to the attribute definition when the block reference is exploded and will be represented on the screen by its tag. The value of the attribute specified at the time of insertion is lost. The group will revert to those elements created by the ATTDEF command prior to combining them into a block via the BLOCK command.

In brief, an attribute definition is turned into an attribute when the block in which it is a part is inserted; conversely, an attribute is turned back into an attribute definition when the block reference is exploded.

Exploding Block References with Nested Elements

Block references containing other blocks and/or polylines are separated for one level only. That is, the highest-level block reference will be exploded, but any nested blocks or polylines will remain block references or polylines. They in turn can be exploded when they come to the highest level.

Viewport objects in a block definition cannot be turned on after being exploded unless they were inserted in paper space.

Block references with equal X, Y, and Z scales explode into their component objects. Block references with unequal X, Y, and Z scales, or nonuniformly scaled block references, might explode into unexpected objects.

> **NOTE** Block references inserted via the MINSERT command or external references and their dependent blocks cannot be exploded.

BASE COMMAND

The BASE command allows you to establish a base insertion point for the whole drawing in the same manner that you specify a base insertion point when using the BLOCK command to combine elements into a block. The purpose of establishing this basepoint is primarily so that the drawing can be inserted into another drawing by way of the INSERT command and having the specified basepoint coincide with the specified insertion point. The default basepoint is the origin (0,0,0). You can specify a 2D point, and AutoCAD will use the current elevation as the base Z coordinate, or you can specify the full 3D point.

Invoke the BASE command from the Block panel (Figure 11–10) in the Home tab, and AutoCAD displays the following prompt:

    Enter base point: (*specify a point, or press* Enter *to accept the default*)

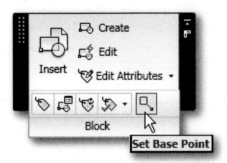

FIGURE 11–10 *Invoking the* BASE *command from the Block panel*

ATTRIBUTES

Attributes can be used for automatic annotation during insertion of a block reference. Attributes are special text objects that can be included in a block definition and must be defined beforehand and then selected when you are creating a block definition.

Attributes have two primary purposes:

The first use of attributes is to permit annotation during insertion of the block reference to which the attributes are attached. Depending on how you define an attribute, it either appears automatically with a preset (constant) text string or it prompts you (or other users) for a string to be written as the block reference is inserted. This feature permits you to insert each block reference with a string of preset text or with its own unique string.

The second, and perhaps the more important purpose of attaching attributes to a block reference is to have extractable data about each block reference stored in the drawing database file. When the drawing is complete, or even before, you can invoke the ATTEXT—short for "attribute extract"—command to have attribute data extracted from the drawing and written to a file in a form that database-handling programs can use. You can have as many attributes attached to a block reference as you wish. As just mentioned, the text string that makes up an attribute can be either constant or user-specified at the time of insertion.

A Definition within a Definition

When creating a block, you select objects to be included. Objects such as lines, circles, and arcs are drawn by means of their respective commands. Normal text is drawn with the TEXT command or the MTEXT command.

As with drawing objects, attributes must be drawn before they can be included in the block. This step is complicated, and it requires additional steps to place them in the drawing; AutoCAD calls this procedure defining the attribute. Therefore, an attribute definition is simply the result of defining an attribute by means of the ATTDEF command. The attribute definition is the object that is selected during the BLOCK command. Later, when the block reference is inserted, the attributes that are attached to it, and the manner in which they become a part of the drawing is a result of how you created (defined) the attribute definition.

Visibility and Plotting

If an attribute is to be used only to store information, then you can, as part of the definition of the attribute, specify whether or not it will be visible. If you plan to use an attribute with a block as a note, label, or callout, you should consider the effect of scaling, whether equal or unequal XY factors, on the text that will be displayed. The scaling factor(s) on the attribute will be the same as on the block reference. Therefore, be sure that they will result in the size and proportions desired. You should also be aware of the effect of rotation on visible attribute text. Attribute text that is defined as horizontal in a block will be displayed vertically when that block reference is inserted with a 90° angle of rotation.

Like any other object in the drawing, the attribute must be visible on the screen, or would be if the plotted view were the current display, for that object to be eligible for plotting.

Attribute Components

Four components associated with attributes should be understood before attempting a definition: tag, value, prompt, and default. The purpose of each is described in the following sections.

Tag

An attribute definition has a tag, just as a layer or a linetype has a name. The tag is the identifier of the attribute definition and is displayed where this attribute definition is located, depicting text size, style, and angle of rotation. The tag cannot contain spaces. Two attribute definitions with the same tag should not be included in the same block. Tags appear in the block definition only, not after the block reference is inserted. However, if you explode a block reference, the attribute value, described herein, changes back into the tag.

If multiple attributes are used in one block, each must have a unique tag in that block. This restriction is similar to each layer, linetype, block, and other named object having a unique name within one drawing. An attribute's tag is its identifier at the time that attribute is being defined, before it is combined with other objects and attributes by the BLOCK command.

Value

The value of an attribute is the actual string of text that appears, if the visibility mode is set to ON, when the block reference of which it is a part is inserted. Whether visible or not, the value is tied directly to the attribute, which, in turn, associates it with the block reference. It is this value that is written to the database file. It might be a door or window size or, in a piping drawing, the flange rating, weight, or cost of a valve or fitting. In an architectural drawing, the value might represent the manufacturer, size, color, cost, or other pertinent information attached to a block representing a desk.

NOTE

When an extraction of attribute data is performed, it is the value of an attribute that is written to a file, but it is the tag that directs the extraction operation to that value. This will be described in detail in the later section on "Extracting Attributes."

Prompt

The prompt is what you see when inserting a block reference with an attribute whose value is not constant or preset. During the definition of an attribute, you can specify a string of characters that will appear in the prompt area during the insertion of the block reference to prompt you to enter the appropriate value. The message of the prompt during insertion is the message provided when you defined the attribute.

Default

You can specify a default value for the attribute during the definition procedure. During insertion of the block reference, it will appear behind the prompt in brackets: for example, <default>. It will automatically become the value if you press Enter in response to the prompt for a value.

Attribute Commands

The four primary commands to manage attributes are as follows:

ATTDEF—attribute definition

ATTDISP—attribute display

ATTEDIT—attribute edit

ATTEXT—attribute extract

As explained previously, the ATTDEF command is used to create an attribute definition. The attribute definition is the object that is selected during the BLOCK command.

The ATTDISP command controls the visibility of the attributes.

The ATTEDIT command provides a variety of ways to edit without exploding the block reference.

The ATTEXT command allows you to extract the data from the drawing and have it written to a file in a form that database-handling programs can use, as shown in Tables 11.1 and 11.2.

TABLE 11–1 *Doors*

Size	THKNS	CORE	FINISH	LOCKSET	HINGES	INSET
3070	1.750	SOLID	PAINT	PASSAGE	4×4	3/4
3070	1.750	SOLID	VARNISH	KEYED	4×4	20×20
2868	1.375	HOLLOW	PAINT	PRIVACY	3×3	3/4

TABLE 11–2 *Room finishes*

NAME	WALL	CEILING	FLOOR	BASE	REMARKS
LIVING	GYPSUM	GYPSUM	CARPET	NONE	PAINT
FAMILY	PANEL	ACOUSTICAL	TILE	STAIN	STAIN
BATH	PAPER	GYPSUM	TILE	COVE	4'_CERAMIC_TILE
GARAGE	GYPSUM	GYPSUM	CONCRETE	NONE	TAPE_FLOAT_ONLY

Let's look at an example to include an attribute to a block named WDW to record the size of the window. A suggested procedure would be to zoom in near the insertion point and create an attribute definition with a tag that reads WDW-SIZE, as shown in Figure 11–11.

If, during the insertion of the WDW block reference, you respond to the prompt for the SIZE attribute with 2054 for a 2'-0-wide×5'-4-high window, the resulting block reference object would be as shown in Figure 11–12, with the normally invisible attribute value shown here for illustration purposes. Figure 11–13 shows the result of the attribute being inserted with unequal scale factors.

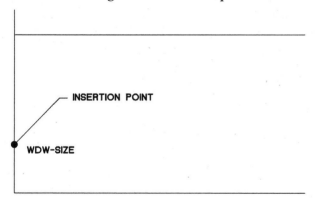

FIGURE 11–11 *Create the attribute definition before defining the block*

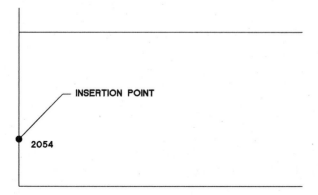

FIGURE 11–12 *The attribute value (2054) is visible with the inserted block*

Even though the value displayed is distorted, the string is not affected when extracted to a database file for a bill of materials.

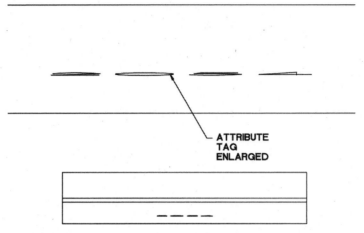

FIGURE 11–13 *The attribute in a block reference with exaggerated, unequal scale factors*

To solve the distortion and rotation problem, if you wish to have an attribute displayed for rotational purposes, you can create a block that contains attributes only or only one attribute. It can then be inserted at the desired location and rotated for readability to produce the results shown in Figure 11–14.

Attribute definitions would be created, as shown in Figure 11–14, and inserted as shown in Figure 11–15.

FIGURE 11–14 *Attributes created as separate blocks*

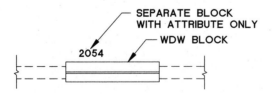

FIGURE 11–15 *Attribute inserted separately for the window*

The insertion points selected would correspond to the midpoint of the outside line that would result from the insertion of the WDW block reference. The SIZE attributes could be defined into blocks called WDWSIZE and inserted separately with each WDW block reference, thereby providing both annotation and data extraction.

> The main caution in having an attribute block reference separate from the symbol (WDW) block reference is in editing. Erasing, copying, and moving the symbol block reference without the attribute block reference could mean that the data extraction results in the wrong quantity.

There is another solution to the problem of a visible attribute not being located or rotated properly in the inserted block reference. If the attribute is not constant, you can edit it independently after the block reference has been inserted by way of the ATTEDIT command. It permits changing an attribute's height, position, angle, value, and other properties. Although the ATTEDIT command is covered in detail in the later section on "Editing Attributes," it should be noted here that the height editing option applies to the X and Y scales of the text. Therefore, for text in the definition of a block that was inserted with unequal X- and Y-scale factors, you will not be able to edit its proportions back to equal X- and Y-scale factors.

Creating an Attribute Definition

The ATTDEF command allows you to create an attribute definition through a dialog box.

Invoke the DEFINE ATTRIBUTES (ATTDEF) command from the Block panel (Figure 11–16) in the Home tab, and AutoCAD displays the Attribute Definition dialog box (Figure 11–17).

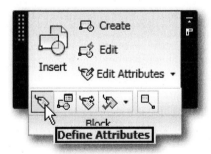

FIGURE 11–16 *Invoking the ATTDEF command from the Block panel*

Set one or more of the available modes to ON in the Mode section of the Attribute Definition dialog box.

Selecting **Invisible** causes the attribute value not to be displayed when the block reference is completed. Even if visible, the value will not appear until the insertion is completed. Attributes needed only for data extraction should be invisible, to quicken regeneration and to avoid cluttering your drawing. You can use the ATTDISP command to override the Invisible mode setting. Setting the Invisible mode to Y (Yes, or ON) does not affect the visibility of the tag in the attribute definition.

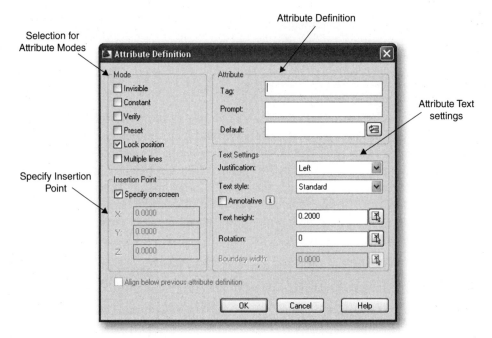

FIGURE 11–17 *Attribute Definition dialog box*

Selecting **Constant** requires that you must enter the value of the attribute while defining it. That value will be used for that attribute every time the block reference to which it is attached is inserted. There will be no prompt for the value during insertion, and you cannot change the value.

Selecting **Verify** means that you will be able to verify its value when the block reference is inserted. For example, if a block reference with three (nonconstant value) attributes is inserted, once you have completed all prompt/value sequences that have displayed the original defaults, you will be prompted again, with the latest values as new defaults, giving you a second chance to be sure the values are correct before the INSERT command is completed. Even if you press Enter to accept an original default value, it also appears as the second-chance default. If, however, you make a change during the verify sequence, you will *not* get a third chance, that is, a second verify sequence.

Selecting **Preset** causes the attribute to automatically take the value of the default that was specified at the time of defining the attribute. During a normal insertion of the block reference, you will not be prompted for the value. You must be careful to specify a default during the ATTDEF command or the attribute value will be blank. A block consisting only of attributes whose defaults were blank when Preset modes were set to ON could be inserted, but it would not display anything and cannot be purged from the drawing. The only adverse effect would be that of adding to the space taken in memory. One way to get rid of a nondisplayable block reference like this is to use a visible entity to create a block with the same name, thereby redefining it to something that can be edited, that is, erased and subsequently purged.

You can duplicate an attribute definition with the COPY command and use it for more than one block. You can also explode a block reference and retain one or more of its attribute definitions for use in subsequent blocks.

Selecting **Lock Position** locks the location of the attribute within the block reference. If it is unlocked, the attribute can be moved relative to the rest of the block using grip editing, and multiline attributes can be resized.

Selecting **Multiple Lines** lets you specify that the attribute value can contain multiple lines of text. When selected, you can define a boundary width for the attribute. In a dynamic block, an attribute's position must be locked for it to be included in an action's selection set.

The Attribute section of the Attribute Definition dialog box allows you to set attribute data. Enter the attribute's tag, prompt, and default value in the text boxes.

The attribute's tag identifies each occurrence of an attribute in the drawing. The tag can contain any characters except spaces. AutoCAD changes lowercase letters to uppercase.

The attribute's prompt appears when you insert a block reference containing the attribute definition. If you do not specify the prompt, AutoCAD uses the attribute tag as the prompt. If you turn on the Constant mode, the Prompt field is disabled.

The default Value specifies the default attribute value. This is optional, except if you turn on the Constant mode, for which the default value needs to be specified.

The Insertion Point section of the dialog box allows you to specify a coordinate location for the attribute in the drawing. Either choose **Pick Point** to specify the location on the screen or enter coordinates in the text boxes provided.

The Text Settings section of the dialog box allows you to set the justification, text style, height, and rotation of the attribute text.

Choosing **Annotative** in the Text Settings section specifies that the attribute is annotative. If the block is annotative, the attribute will match the orientation of the block. For additional information on Annotation, refer to Chapter 8.

Selecting **Align below previous attribute definition** allows you to place the attribute tag directly below the previously defined attribute. If you haven't previously defined an attribute definition, this option is unavailable.

Choose **OK** to define the attribute definition.

After you close the Attribute Definition dialog box, the attribute tag appears in the drawing. Repeat the procedure to create additional attribute definitions.

Inserting a Block Reference with Attributes

Blocks with attributes may be inserted in a manner similar to that for inserting regular block references. If there are any nonconstant attributes, you will be prompted to enter the value for each. You may set the system variable called ATTREQ to zero, thereby suppressing the prompts for attribute values. In this case, the values will either be blank or set to the default values, if they exist. You can later use either the DDATTE or ATTEDIT command to establish or change values.

Controlling the Display of Attributes

The ATTDISP command controls the visibility of attributes. Attributes will normally be visible if the Invisible mode is set to N (normal) when they are defined.

Invoke the ATTDISP command from the Block panel in the Home tab.

Choosing *On* makes all attributes visible.

Choosing *Off* makes all attributes invisible.

Choosing *Normal* displays the attributes as you created them.

The ATTMODE system variable is affected by the ATTDISP setting. If REGENAUTO is set to ON, changing the ATTDISP setting causes drawing regeneration.

Editing Attribute Values

Unlike other objects in an inserted block reference, attributes can be edited independently of the block reference. The EATTEDIT command allows you to change the value of attributes in blocks that have been inserted. This action permits you to insert a block reference with generic attributes; that is, the default values can be used in anticipation of changing them to the desired values later. Or, you can copy an existing block reference that may need only one or two attributes changed to make it correct for its new location. There is always the chance that an error was made in entering the value or that design changes necessitate subsequent changes. The ATTEDIT command can be used to change only the value of the attributes. In order to change other characteristics of attributes such as text size, font, and visibility, the Block Attribute Manager, described later in this chapter, must be used.

Editing an attribute is accomplished by invoking the EATTEDIT command. The EATTEDIT command edits individual, nonconstant attribute values associated with a specific block reference.

Invoke the EDIT ATTRIBUTE (EATTEDIT) command from the Block panel (Figure 11–18) in the Home tab, and AutoCAD displays the following prompt:

```
Select block reference: (select the block reference)
```

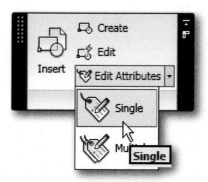

FIGURE 11–18 *Invoking the* EDIT ATTRIBUTE *command from the Block panel on the Home tab*

AutoCAD displays the Enhanced Attribute Editor dialog box (Figure 11–19). Selecting objects that are not block references or block references that contain no attributes will cause an error message to appear.

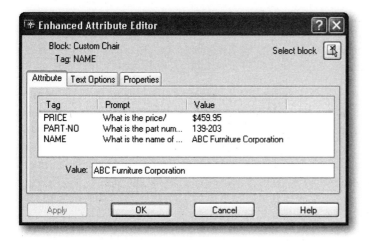

FIGURE 11-19 *Enhanced Attribute Editor dialog box*

The dialog box lists all the attributes defined with values for the selected block reference. **Block** and **Tag** display the tag identification of the selected attribute and name of the block whose attributes you are editing, respectively. Using the pointing device, you can select values to be changed in the dialog box. In addition, you can also change the text attributes and object properties of the selected attribute tag. Choose **Apply** to update the drawing with the attribute changes you have made, and leave the Enhanced Attribute Editor open. Choose **Select Block** to temporarily close the dialog box while you select a block for modifying its attribute definitions. If you modify attributes of a block and then select a new block before you save the attribute changes you made, you are prompted to save the changes before selecting another block. After making the necessary changes, choose **OK** to accept the changes, and close the Enhanced Attribute Editor dialog box.

Extracting Attributes

Extracting data from a drawing is one of the most innovative features in CAD. Paper copies of drawings have long been used to communicate more than just how objects look. In addition to dimensions, drawings tell builders or fabricators what materials to use, quantities of objects to make, manufacturers' names, models of parts in an assembly, coordinate locations of objects in a general area, and what types of finishes to apply to surfaces. Until computers came into the picture—or pictures came into the computer—extracting data from manual drawings involved making lists, usually by hand, while studying the drawing and often checking off the data with a marker.

The AutoCAD attribute feature and the EATTEXT command allows complete, fast, and accurate extraction of data of all objects such as block references, in this case.

The CAD drawing in Figure 11-20 shows a piping control set. The 17 valves and fittings are a fraction of those that might be on a large drawing. Each symbol is a block with attributes attached to it. Values that have been assigned to each attribute tag record the type (TEE, ELL, REDUCER, FLANGE, GATE VALVE, or CONTROL VALVE), size (3, 4, or 6), rating (STD or 150#), weld (length of weld), and many other vital bits of specific data. Keeping track of hundreds of valves, fittings, and even cut lengths of pipe is a time-consuming task subject to omissions and errors if done manually, even if the drawing is plotted from CAD. Just as important as extracting data from the original drawing is the need to update a list of data when the drawing is changed. Few drawings, if any, remain unchanged. The AutoCAD attribute feature makes the job fast, thorough, and accurate. Examples of some of the block references with attribute definition are shown in Figure 11–21 through Figure 11–23.

These are three examples of block references as defined and inserted, each attribute having been given a value during the INSERT command. Remember that the value that will be extracted is in accordance with how the template specifies the tags. Figure 11–24 shows the three block references inserted, with corresponding attribute values in tabular form. By using the ATTEXT command, you can write a complete listing of all valves and fittings in the drawing to a file, as shown in Table 11.3.

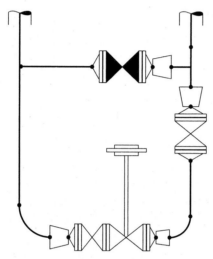

FIGURE 11–20 *Piping control set*

ENTITIES THAT WILL BE
INCLUDED IN EACH BLOCK

RED

TEXT ITEMS REPRESENT
ATTRIBUTE TAGS

TYPE
SIZE
RATING
BLTQUAN
BLTSIZE
GSKETMATL
GSKETSIZE
WELD

FIGURE 11–21 *Reducer with attribute definition*

ENTITIES THAT WILL BE
INCLUDED IN EACH BLOCK

FLANGE

TEXT ITEMS REPRESENT
ATTRIBUTE TAGS

TYPE
SIZE
RATING
BLTQUAN
BLTSIZE
GSKETMATL
GSKETSIZE
WELD

FIGURE 11–22 *Flange with attribute definition*

ENTITIES THAT WILL BE
INCLUDED IN EACH BLOCK

VALVE

TEXT ITEMS REPRESENT
ATTRIBUTE TAGS

TYPE
SIZE
RATING
BLTQUAN
BLTSIZE
GSKETMATL
GSKETSIZE
WELD

FIGURE 11–23 *Valve with attribute definition*

BLOCKS INSERTED WITH
VALUES ASSIGNED TO
ATTRIBUTES

TEXT ITEMS REPRESENT
ATTRIBUTE TAGS

REDUCER	RFWN FLG	GT VLV
6X4	4"	4"
STD	150#	150#
-	8	8
-	5/8"x3"	5/8"x3"
-	COMP	COMP
-	1/8"	1/8"
34.95	14.14	-

FIGURE 11–24 *Inserted block references with corresponding attribute values*

TABLE 11–3 *A list of inventory as specified by the ATTEXT command and tags*

Type	Size	Rating	Bltquan	Bltsize	Gsketmatl	Gsketsize	Weld
TEE	6	STD	4	5/8×3	COMP	1/8	62.44
ELL	6	STD	4	5/8×3	COMP	1/8	41.63
ELL	4	STD	4	5/8×3	COMP	1/8	14.14
RED	6×4	STD	8	5/8×3	COMP	1/8	34.95
RED	6×4	STD	8	5/8×3	COMP	1/8	34.95
RED	6×3	STD	4	5/8×3	COMP	1/8	31.81
RED	4×3	STD	8	5/8×3	COMP	1/8	25.13
FLG	4	150#	8	5/8×	COMP	1/8	14.14
FLG	4	150#	8	5/8×3	COMP	1/8	14.14
FLG	4	150#	8	5/8×3	COMP	1/8	14.14
FLG	4	150#	8	5/8×3	COMP	1/8	14.14
FLG	3	150#	4	5/8×3	COMP	1/8	11.00
FLG	3	150#	4	5/8×3	COMP	1/8	11.00
GVL	4	150#	16	5/8×3	COMP	1/8	11.00
GVL	4	150#	16	5/8×3	COMP	1/8	11.00
GVL	3	150#	8	5/8×3	COMP	1/8	11.00
CVL	3	150#	8	5/8×3	COMP	1/8	11.00

The headings above each column in this table are for your information only. They will not be written to the extract file by the EATTEXT command, and they signify the tags whose corresponding values will be extracted.

When operated on by a database program, this file can be used to sort valves and fittings by type, size, or other value. The scope of this book is too limited to cover database applications. However, generating a file similar to this inventory table that a database program can use is the important linkage between computer drafting and computer management of data for inventory control, material takeoff, flow analysis, cost, maintenance, and many other applications. The CAD drafter can apply the ATTEXT feature to perform this task.

The EATTEXT command uses the Attribute Extraction wizard to facilitate selecting objects and/or drawings from which to extract attribute data from inserted block references. It allows you to place the data in tabular format for use on the drawing and/or write the data to a file and in a format for use by a program, such as Microsoft Excel® or Access®, or other file for handling the resulting database.

Extracted data can be manipulated by a database application program. The telephone directory, a database, is an alphabetical listing of names, each followed by a first name (or initial), an address, and a phone number. A listing of pipe, valves, and fittings in a piping system can be a database if each item has essential data associated with it: its size, flange rating, weight, material of manufacture, product that it handles, cost, and location within the system, among many others.

The two elementary terms used in a database are the record and the field. A record is like a single listing in the phone book; it consists of a name and its associated first name, address, and phone number. The name Jones with its data is one record. Another Jones with a different first name (or initials) is another record. Another Jones with the same first name or initials at a different address is still another record. Each listing is a record. The types of data that may be in a record come under the heading of a field. Name is a field. All the names in the list come under the name field. Address is a field. Phone number is a field. Even though some names may have first names and some may not, first name is a field. It is possible to take the telephone directory that is listed alphabetically by name, input it into a computer database program, and generate the same list in numerical order by phone number. You can generate a partial list of all the Joneses sorted alphabetically by the first name, or you can generate a list of everyone who lives on Elm Street. The primary purpose of the EATTEXT command is to generate the main list that includes all of the desired objects to which the database program manipulations can be applied.

Invoke the ATTRIBUTE EXTRACTION (EATTEXT) command from the Linking and Extraction panel located in the Insert tab, and AutoCAD displays the Attribute Extraction wizard (Figure 11–25, first row, left).

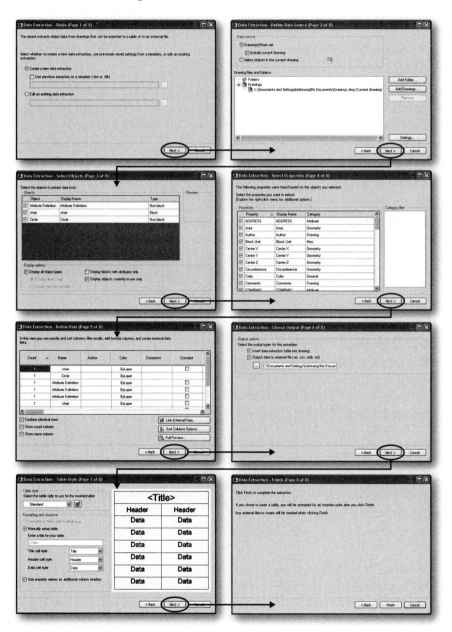

FIGURE 11–25 *Attribute Extraction wizard page*

Image(s) © Cengage Learning 2013

The **Attributes Extraction - Begin (Page 1 of 8)** page lets you extract block attribute data to a table in the current drawing or to an external file.

Choose **Create a new data extraction** to use the settings you specify as you proceed through the wizard to create a table or save data in an external file. To use the previous extraction file as a template for creating the new data extraction file, set the toggle button to ON for **Use previous extraction as a template**.

Choose **Edit an existing data extraction** to use settings previously saved in a data extraction (*DXE*) file or an attribute extraction template (*BLK*) file. As you move through the wizard, each page is already filled in with the settings in the template file. If necessary, you can change these settings.

Choose **Next** and AutoCAD displays the **Data Extraction - Define Data Source (Page 2 of 8)** page (Figure 11–25, first row, right). This page specifies the drawing files, including folders from which to extract data. Allows selection of objects in the current drawing from which to extract information. The Data source section lets you select objects in the current drawing, the current drawing itself, or drawings/sheet sets to search for blocks from which to extract attribute data.

Choose **Drawings/sheet sets** to select drawing/sheet sets from which to select blocks from which to extract attribute data. Choose **Browse** to the right of **Select drawings/ sheet sets**, and AutoCAD displays the Select Files dialog box to select the drawings and sheet sets. Select **Include current drawing** to select all blocks and xrefs from the current drawing for extracting attribute data. The name of the current drawing is listed in the Drawing Files section.

Choose **Select Objects in the current drawing** to select specific blocks and external references (xrefs) in the current drawing for extracting attribute data. The name of the current drawing is listed in the Drawing Files section. Choose **Select blocks** to select blocks in the current drawing, and AutoCAD temporarily returns to the drawing display so that you can select blocks for extracting attributes.

Choose **Settings** to set options for nested and external reference blocks and options that specify which blocks to count.

After selecting the desired objects or drawing(s), choose **Next**, and AutoCAD displays the **Data Extraction - Select Objects (Page 3 of 8)** page (Figure 11–25, second row, left). This page specifies the types of objects (blocks and nonblocks) and drawing information to be extracted. Valid objects are checked by default. The blocks are listed by block name, and nonblocks are listed by their object name. The Display options section provides the controls to display the type objects to be displayed.

After selecting the type of objects for data extraction, choose **Next**, and AutoCAD displays the **Data Extraction - Select Properties (Page 4 of 8)** page (Figure 11–25, second row, right). This page specifies the object, block, and drawing properties to extract. Each row displays a property name, its display name, and category. The **Category filter** displays a list of categories that are extracted from the property list. To exclude from the list uncheck the categories from the filter list. Available categories include 3D Visualization, Attribute, Drawing, Dynamic Block, General, Geometry, Misc, Pattern, Table, and Text.

After selecting the properties for data extraction, choose **Next**, and AutoCAD displays the **Data Extraction - Refine data (Page 5 of 8)** page (Figure 11–25, third row, left). This page allows you to modify the structure of the data extraction table. You can reorder and sort columns, filter results, add formula columns and footer rows, and create a link to data in a Microsoft Excel spreadsheet. You can display a full preview by choosing **Full Preview** of the final output, including linked external data, in the text window. The preview is for viewing only.

After refining the data, choose **Next,** and AutoCAD displays the **Data Extraction - Choose Output (Page 6 of 8)** page (Figure 11–25, third row, right). This page specifies the type of output to which the data is extracted. Choose **Insert Data Extraction Table into Drawing** to create a table that is populated with extracted data. You are prompted to insert the table into the current drawing when you click Finish on the Finish page. Choose **Output Data to External File** to create a data extraction file. Available file formats include Microsoft Excel (XLS), comma-separated file format (CSV), Microsoft Access (MDB), and tab-separated file format (TXT). The external file is created when you click Finish on the Finish page.

After choosing the output, choose **Next,** and AutoCAD displays the **Data Extraction - Table Style (Page 7 of 8)** page (Figure 11–25, fourth row, left). This page controls the appearance of the data extraction table. This page is displayed only if **Insert Data Extraction Table into Drawing** is selected. The **Select the table style to use for the inserted table** list box lets you select the style for the table from the available styles. If necessary, make changes to the formatting and structure of the table.

Choose **Next,** and AutoCAD displays the **Data Extraction - Finish (Page 8 of 8)** page (Figure 11–25, fourth row, right) which finishes the process of extracting attributes. This completes the process of extracting object property data that was specified in the wizard and creates the output type that was specified on the page. If data linking and column matching to an Excel spreadsheet was defined in the Link External Data dialog box, the selected data in the spreadsheet is also extracted. If the Insert Data Extraction Table into Drawing option was selected on the **Choose Output** page, you are prompted to insert the table into the drawing when you click Finish. If the **Output Data to External File** option was selected, the extracted data is saved to the specified file type.

Choose **Finish** to close the wizard, and AutoCAD returns to the drawing screen, prompting you to specify the location of the table if a table has been created during the attribute extraction process.

Redefining a Block and Its Associated Attributes

The ATTREDEF command allows you to redefine a block reference and updates associated attributes. Invoke the ATTREDEF command by typing **attredef** at the on-screen prompt, and AutoCAD displays the following prompts:

```
Enter name of the block you wish to redefine: (specify the
block name to redefine)
Select objects for new Block...
Select objects: (select objects for the block to redefine,
and press Enter)
Insertion base point of new Block: (specify the insertion
basepoint of the new block)
```

New attributes assigned to existing block references are given their default values. Old attributes in the new block definition retain their old values. Old attributes not included in the new block definition are deleted from the old block references.

Block Attribute Manager

The BATTMAN command provides a means of managing blocks that contain attributes.

Invoke the BATTMAN command from the Block panel (Figure 11–26) in the Home tab. AutoCAD displays the Block Attribute Manager dialog box, similar to Figure 11–27.

FIGURE 11–26 *Invoking the BATTMAN command from the Block panel*

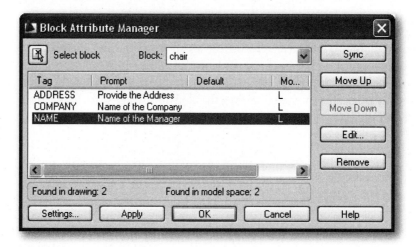

FIGURE 11–27 *Block Attribute Manager dialog box*

The BATTMAN command lets you edit the attribute definitions in blocks, change the order in which you are prompted for attribute values when inserting a block, and remove attributes from blocks. AutoCAD displays attributes of the selected block in the attribute list of the Block Attribute Manager dialog box. By default, the Tag, Prompt, Default, Modes, and Annotative attribute properties are shown in the attribute list. You can specify which attribute properties you want displayed in the list by choosing Settings. The number of instances of the selected block is shown in a description below the attribute list.

The **Block** list box lists all block definitions in the current drawing that have attributes. Select the block whose attributes you want to modify.

Instead of selecting a block from the **Block** list box, choose **Select Block** to use your pointing device to select a block from the drawing area. When you choose **Select Block**, the dialog box closes until you select a block from the drawing or cancel by pressing ESC. If you modify attributes of a block and then select a new block but have not saved the attribute changes you made, you are prompted to save the changes before selecting another block.

Choosing **Sync** causes AutoCAD to update all instances of the selected block with the attribute properties currently defined.

Selecting **Move Up** causes the selected attribute tag to move up in the prompt sequence when the block contains multiple attributes.

Selecting **Move Down** causes the selected attribute tag to move down in the prompt sequence when the block contains multiple attributes.

Selecting **Edit** causes AutoCAD to display the Edit Attribute dialog box (Figure 11–28), where you can modify selected attribute properties. Choose from three tabs in the Edit Attribute dialog box: Attribute, Text Options, or Properties.

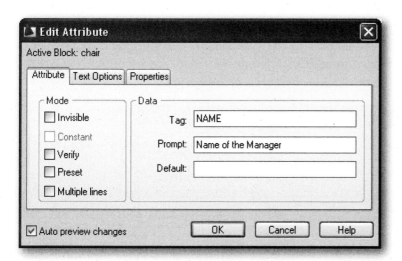

FIGURE 11–28 *Edit Attribute dialog box with the Attribute tab selected*

The Attribute tab controls the modes and data tag, prompt, and default of the selected attribute. The Mode section of the Attribute tab has checkboxes for setting the Invisible, Constant, Verify, Preset, and Multiple Lines modes. The Data section has text boxes in which you can set the tag name, prompt, and default. Selecting **Auto preview changes** causes changes to attributes to be immediately visible.

The **Text Options** tab lets you modify attribute text (Figure 11–29). It includes text boxes to set the Text Style, Justification, Height, Rotation, Width Factor, and Oblique Angle of the Attribute text. There are checkboxes for displaying the text Backwards or Upside down. Selecting **Annotative** specifies that the attribute is annotative.

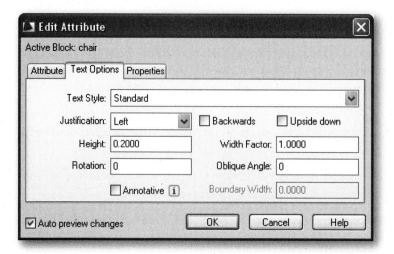

FIGURE 11–29 *Edit Attribute dialog box with the Text Options tab selected*

The **Properties** tab includes text boxes to set the layer that the attribute is on and the color, lineweight, and linetype for the selected attribute (Figure 11–30). If the drawing uses plot styles, you can assign a plot style to the attribute.

After making the necessary changes to the Edit Attribute dialog box, choose **OK** to save the changes and return to the Block Attribute Manager dialog box.

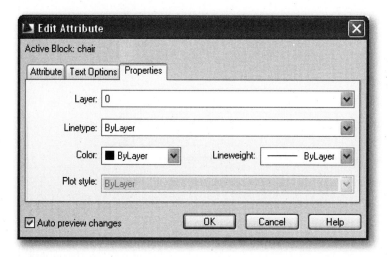

FIGURE 11–30 *Edit Attribute dialog box with the Properties tab selected*

Choosing **Remove** in the Block Attribute Manager dialog box removes the selected attribute from the block definition. **Remove** is not available for blocks with only one attribute.

Choosing **Settings** causes AutoCAD to display the Settings dialog box (Figure 11–31), where you can customize how attribute information is listed in the Block Attribute Manager.

FIGURE 11–31 *Settings dialog box*

The Display in List section specifies the properties to be displayed in the attribute list. Only the selected properties are displayed in the list. The Tag property is always selected.

Choose **Select All** to select all properties' checkboxes to be checked.

Choosing **Clear All** causes all properties' checkboxes to be cleared.

Selecting **Emphasize duplicate tags** causes duplicate attribute tags to be displayed in red type in the attribute list. If this option is cleared, duplicate tags are not emphasized in the attribute list.

Selecting **Apply changes to existing references** updates all instances of the block with the new attribute definitions. If this option is cleared, AutoCAD updates only new instances of the block with the new attribute definitions. You can choose **Sync** in the Block Attribute Manager to apply changes immediately to existing block instances. This action temporarily overrides the **Apply Changes to Existing References** option.

Choose **OK** to close the Settings dialog box and return to the Block Attribute Manager.

Choosing **Apply** in the Block Attribute Manager dialog box updates the drawing with the attribute changes you have made and leaves the Block Attribute Manager open. Choose **OK** to close the Attribute Manager dialog box.

DIVIDING OBJECTS

The DIVIDE command causes AutoCAD to divide an object into equal-length segments, placing markers at the dividing points. Objects eligible for application of the DIVIDE command are the line, arc, circle, ellipse, spline, and polyline. Selecting an object other than one of these will cause an error message to appear, and the command will terminate.

Invoke the DIVIDE command from the Draw panel located in the Home tab, and AutoCAD displays the following prompts:

```
Select the object to divide: (select a line, arc, circle,
ellipse, spline, or polyline)
Enter the number of segments or ⊕ (specify the number of
segments or select the BLOCK option from the shortcut menu)
```

You may respond with an integer from 2 to 32,767, causing points to be placed along the selected object at equal distances but not actually separating the object. The object snap (OSNAP) NODE can snap at the divided points. Logically, there will be one less point placed than the number entered, except in the case of a circle. The circle will have the first point placed at the angle from the center of the current snap rotation angle. A closed polyline will have the first point placed at the first point drawn in the polyline. The total length of the polyline will be divided into the number of segments entered without regard to the length of the individual segments that make up the polyline. An example of a closed polyline is shown in Figure 11–32.

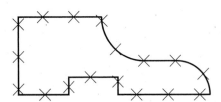

FIGURE 11–32 *The* DIVIDE *command as used with a closed polyline*

 NOTE   It is advisable to set the PDSIZE and PDMODE **system variables to values that will cause the points to be visible.**

The **Block** option allows a named block reference to be placed at the dividing points instead of a point. The sequence of prompts is as follows:

```
divide
Select object to divide: (select a line, arc, circle,
ellipse, spline, or polyline)
Enter the number of segments or ⊕ (choose BLOCK from the
shortcut menu)
Enter name of block to insert: (enter the name of the block)
Align block with object? or ⊕ (press [Enter] to align the block
reference with the object, or enter n, for not to align with
the block reference)
Enter the number of segments: (specify the number of
segments)
```

If you respond with No or N to the "Align block with object?" prompt, all of the block references inserted will have a 0 angle of rotation. If you choose Yes (default), the angle of rotation of each inserted block reference will correspond to the direction of the linear part of the object at its point of insertion or to the direction of a line tangent to a circular part of an object at the point of insertion.

MEASURING OBJECTS

The MEASURE command causes AutoCAD to divide an object into specified-length segments, placing markers at the measured points. Objects eligible for application of the MEASURE command are the line, arc, circle, ellipse, spline, and polyline. Selecting an object other than one of these will cause an error message to appear and the command will terminate.

Invoke the MEASURE command from the Draw panel located in the Home tab, and AutoCAD displays the following prompts:

> Select object to measure: (*select a line, arc, circle, ellipse, spline, or polyline*)
> Specify length of segment or ⬇ (*specify the length of the segment, or select the* BLOCK *option from the shortcut menu*)

If you reply with a distance, or show AutoCAD a distance by specifying two points, the object is measured into segments of the specified length, beginning with the closest endpoint from the selected point on the object. The **Block** option allows a named block reference to be placed at the measured point instead of a point. The sequence of prompts is as follows:

> measure
> Select object to measure: (*select a line, arc, circle, ellipse, spline, or polyline*)
> Specify length of segment or ⬇ (*choose* BLOCK *from the shortcut menu*)
> Enter name of block to insert: (*enter the name of the block*)
> Align block with object? or ⬇ (*press* Enter *to align the block reference with the object, or enter n, to not align with the block reference*)
> Specify length of segment: (*specify the length of segments*)

DYNAMIC BLOCKS

A dynamic block has flexibility and intelligence. Individual objects or groups of objects within a dynamic block reference can easily be changed in a drawing while you work. You can manipulate the geometry in a dynamic block reference through custom grips or custom properties. This allows you to adjust the block in place as necessary rather than searching for another block to insert or having to redefine the existing one.

For example, after inserting a block in a drawing representing a door, you might need to change the size of the door while you're editing the drawing. If the block is dynamic and defined to have an adjustable size, you can change the size of the door simply by dragging the custom grip or by specifying a different size in the Properties palette. You might also need to change the opening angle of the door. The door block might also contain an alignment grip, which allows you to align the door block reference easily to other geometry in the drawing.

Figure 11–33 shows three insertions of a block representing a simplified instrument panel. Figure 11–33 at the left shows the geometry as it was drawn when the block was defined with the ON/OFF switch in the center. In the middle and at the right of Figure 11–33 the same block is shown with the ON/OFF switch moved to the left and right, respectively. All three references are insertions of the same block definition. The dynamic capability of the block allows the user to change the size, shape, location, or orientation of preselected geometry in the reference, after it has been inserted. In Figure 11–33, all three references are drawn from a single block definition.

FIGURE 11–33 *Three block references of an instrument panel with the ON/OFF switch in the center (as defined), on the left, and on the right*

Three Steps to Create a Simple Dynamic Block

The first step to create a dynamic block is to create the geometry as shown in Figure 11–33 at the left. This action can be accomplished either in the normal AutoCAD drawing area or in the Block Editor drawing area. If the objects have been drawn in the normal drawing area, invoke the MAKE BLOCK command, and this causes the Block Definition dialog box to be displayed. While in the dialog box, you can name the block, specify the insertion point, and select the geometry to be included in the definition. Select the **Open in block editor** checkbox and choose **OK**. AutoCAD opens the Block Editor with the Authoring palette displayed.

You can go through the process of opening the Block Editor without selecting objects and then create the geometry in the Block Editor drawing area. You can also double-click an insertion (reference) of a nondynamic block. AutoCAD displays the Edit Block Definition dialog box with the name of the block selected in the **Block to create or edit** text box. Choose **OK** to select the block. Whether you arrive at the Edit Block Definition dialog box when first creating the block or you double-click an existing block, AutoCAD lets you open the Block Editor with the Authoring palette displayed. The Authoring palette has tabs from which you can add parameters, actions, and parameter sets to your block definition to make it dynamic.

The second step to create a dynamic block is to add a parameter. Parameters are used to establish points, distances, and angles on or near objects in the block so that actions can be applied to those objects, making that block dynamic. Parameters also allow you to control the visibility of objects in the reference and make use of the properties of the geometry in the reference. Parameters include *Point, Linear, Polar, XY, Rotation, Alignment, Flip, Visibility, Lookup,* and *Base Point*. For this example, *Point Parameter* is selected from the **Parameters** tab of the Block Authoring palette. AutoCAD displays the following prompt:

> Specify parameter location or ⬇ (*specify the center of the switch geometry*)

The point parameter is located on the center of the switch geometry (Figure 11–34).

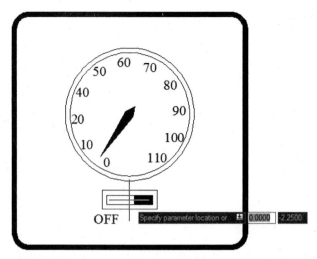

FIGURE 11–34 *Adding a point parameter while defining the dynamic block in the Block Editor*

AutoCAD prompts for the label location. You can locate the label for the point parameter, "Position" in this case, as shown in Figure 11–35. The label name "Position" is the default for a point parameter and can be changed if desired. The grip is displayed where you specified the location of the parameter with a leader to the label. An Alert icon, with the exclamation point, is displayed because no action is associated with the parameter yet.

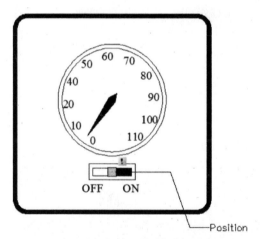

FIGURE 11–35 *The point parameter, labeled "Position," with Alert icon indicating that no action has been added*

The third step consists of adding an action, selecting objects, and closing the Block Editor. Actions are added to parameters and then associated with selected objects in the block. These objects are the selection set that will be affected by the subsequent selecting of the appropriate grip in a reference of the block and initiation of the action. The actions include *Move, Scale, Stretch, Polar Stretch, Rotate, Flip, Array,* and *Lookup.* Choose *Move* Action from the **Actions** tab (Figure 11–36), and AutoCAD displays the following prompt:

        Select parameter: *(select the "Position" label or other part*
        of the parameter)

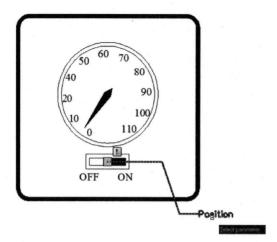

FIGURE 11–36 *A Move action is added by selecting the point parameter when prompted*

After you select the point parameter labeled "Position" (Figure 11–37), AutoCAD displays the following prompt:

Select objects: (*in this case, use the Window method to select the geometry that makes up the switch and the* OFF *and* ON *text*)

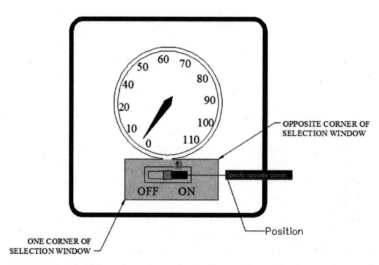

FIGURE 11–37 *Objects (geometry of switch) are selected for being associated with the Move action*

After you select the objects to be associated with the Move action (Figure 11–38), AutoCAD displays the following prompt:

Specify action location or ⊡ (*specify the symbol for the Move action near the label*)

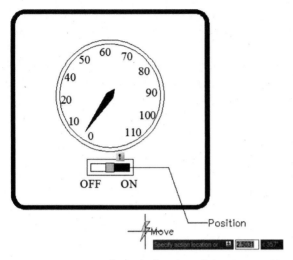

FIGURE 11–38 *Specifying the location of the Move action*

The location of the Action symbol does not affect how the action operates. Its purpose is to identify the type of action associated with the parameter only when you are in the Block Editor. When you use the window or crossing method to select the objects for adding them to the selection set or to define a stretch frame (explained later), the label, leader, grip, and other Block Editor symbols are not affected. The symbols, leaders, labels, and icons do not appear in a block reference. Only the grips appear when the block reference is selected for modifying.

After the parameter is added, the action is added, the objects associated with the action are selected, and the action symbol is located, you can exit the Block Editor by right-clicking and choosing CLOSE BLOCK EDITOR from the shortcut menu.

Using the Dynamics of a Dynamic Block

When a dynamic block is inserted and the reference is selected for modification, the geometry is highlighted, a grip will appear at the insertion point of the block as it usually does (Figure 11–37). A grip will also appear where you located the point parameter on the switch geometry. This grip is square for the Move action. You can select this grip and move the switch geometry and OFF and ON text with the cursor or enter new coordinates from the keyboard. Figure 11–39 at the left shows where grips are located. Figure 11–39 in the middle shows the switch being relocated by selecting the grip and initiating the Move action. Figure 11–39 at the right shows the result of applying the Move action in the dynamic block reference.

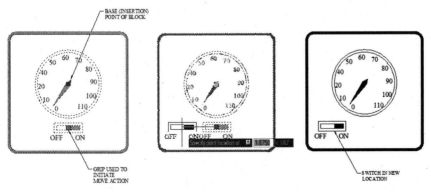

FIGURE 11–39 *The Move Action grip appears when the block reference is selected for modifying*

The Move action added to a point parameter in the block definition is the simplest form of using the dynamic block feature. Other, more powerful action/parameter combinations make dynamic blocks one of the most useful tools for the application of the computer to design/drafting. For example, the block in the example above can also have a Stretch action added to a linear parameter to allow the shape and/or size of the panel outline to be changed from the one in Figure 11–39 to the one shown in Figure 11–40. Geometry that was hidden in the block definition, such as the second dial, is made visible.

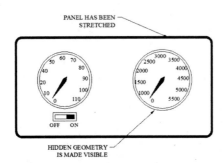

PANEL HAS BEEN
STRETCHED

HIDDEN GEOMETRY
IS MADE VISIBLE

FIGURE 11–40 *The Dynamic Block after using the Stretch action and making hidden geometry visible*

Commonly used parameter/action combinations can be quickly added by using the Parameter Sets available on the Parameter Sets tab of the Authoring palette. These include *Point Move, Linear Move, Linear Stretch, Linear Array, Linear Move Pair, Linear Stretch Pair, Polar Move, Polar Stretch, Polar Array,* and *Polar Move Pair.* Once you have learned how to use the actions and parameters individually, you can use these combinations to speed up dynamic block creation.

> **NOTE** Good planning is necessary in creating dynamic blocks that are flexible and useful. And a thorough understanding of how to apply the concepts of parameters, actions, and grips is necessary also. You are advised to begin with a simple geometric shape, perhaps a rectangle, add a parameter and action, save it in the Block Editor, insert it and then see how your parameter/action operates. Like many of the very powerful features of AutoCAD, applying the Dynamic Block feature requires study and practice. But once mastered, the savings in time and the flexibility are worth the effort.

The Block Editor—Making a Block Dynamic

The Block Editor is an authoring window dedicated to creating and editing dynamic blocks. You can use the Block Editor to make a dynamic block out of an existing non-dynamic block or while creating a new block. You can open the Block Editor by selecting **Open in block editor** in the Block Definition dialog box before choosing **OK** to create a new block (Figure 11–41).

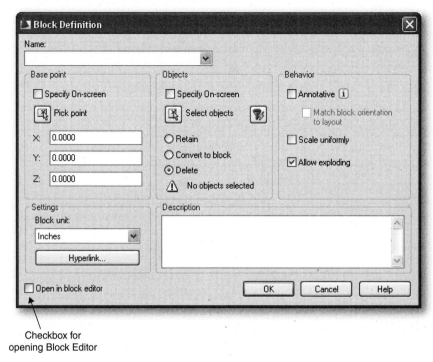

Checkbox for
opening Block Editor

FIGURE 11–41 *Block Definition dialog box with the Open in Block Editor checkbox*

You can also open the Block Editor by double-clicking on an inserted block reference on the screen. AutoCAD displays the Edit Block Definition dialog box from which you can select a block from the **Block to create or edit** list and choose **OK** (Figure 11–42). AutoCAD then opens the Block Editor, the selected block is displayed on the editing screen, and the Block Editor toolbar is displayed across the top of the drawing area.

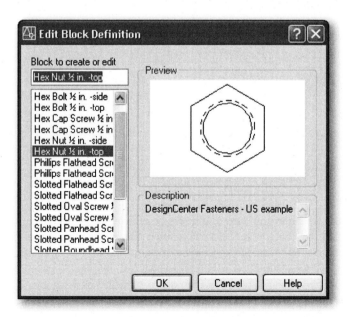

FIGURE 11–42 *Edit Block Definition dialog box*

AutoCAD displays the Block Editor (Figure 11–43 with the Parameter tab selected in the Authoring palette). The example shown in Figure 11–43 is for creating a dynamic block named RFWN-FLANGE, which indicates Raised-Face-Weld-Neck Flange.

The Block Editor provides a special Authoring palette. This palette provides quick access to block authoring tools. Figure 11–44 shows the three tabs available on the Authoring palette: Parameters, Actions, and Parameter Sets. In addition to the Block Authoring palettes, the Block Editor provides a drawing area in which you can draw and edit geometry as you would in the program's main drawing area. You can specify the background color for the Block Editor drawing area.

NOTE You can use most commands in the Block Editor. When you enter a command that is not allowed in the Block Editor, a message is displayed on the command line. It is recommended that the Command window be left open.

In the Block Editor, the Block Editor tab is displayed above the drawing area. The Block Editor tab includes the following panels: Open/Save, Geometric, Dimensional, Manage, Action Parameters, Visibility, and Close (Figure 11–43).

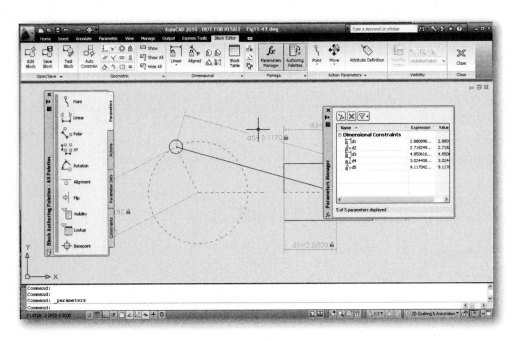

FIGURE 11–43 *Block Editor tab displayed*

NOTE The objects that you see on the screen of the Block Editor are all part of the geometry of the particular block being edited. Any objects you add will become part of the block when you accept the changes as you exit the Block Editor. Unlike the Help window, the Design-Center window, the Properties window, and other palettes, the Block Editor window must be closed before you can create and modify objects in the normal AutoCAD drawing area that are not associated with the selected block.

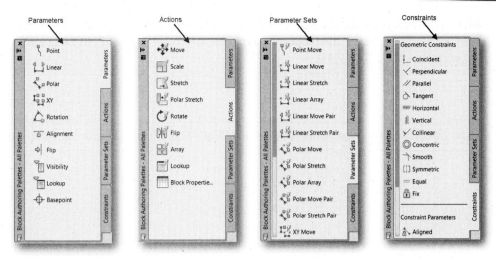

FIGURE 11–44 *Three tabs of the Block Editor Authoring palette: Parameters, Actions, Parameter Sets, and Constraints*

In the AutoCAD Classic Workspace there is a toolbar that is displayed above the drawing area when you are in the Block Editor. This toolbar shows the name of the block definition currently being edited and provides several tools (Figure 11–45).

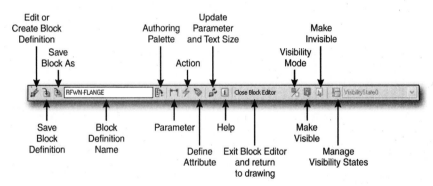

FIGURE 11–45 *Block Editor toolbar*

When the Block Editor is open for editing an existing named dynamic block or creating a dynamic block from the geometry that has just been defined as a named block, numerous special commands are used for this purpose, and they can only be used while the Block Editor is open.

Parameters

As mentioned previously, after the initial planning is done and the geometry has been created, the first step in making a block dynamic is to add parameter(s). Parameters include *Point, Linear, Polar, XY, Rotation, Alignment, Flip, Visibility, Lookup,* and *Base Point.* The type of parameter used depends on the type of action desired. Some parameters can have different types of actions added to them. Some are limited to only one type of action. The following table shows which actions can be associated with which parameters. Note that the *Alignment,*

Visibility, and *Base Point* parameters do not have actions associated with them, while the *Rotation, Flip,* and *Lookup* parameters can have only one action associated with them (Table 11.4).

TABLE 11–4 *Parameters*

Actions	Point	Linear	Polar	Xy	Rotation	Alignment	Flip	Visibility	Lookup	Base Pt
MOVE	X	X	X	X						
SCALE		X	X	X						
STRETCH	X	X	X	X						
POL. STR			X							
ROTATE					X					
FLIP							X			
ARRAY										
LOOKUP									X	

The parameter, in combination with the associated action, determines the location and appearance of the grips that appear when a reference of the block is selected for modifying.

Move, Scale, and *Stretch* actions can be associated with at least three parameters. Some of the reasons for associating an action with a particular parameter will become evident as you learn how the action operates with each parameter. For example, a *Move* action associated with a *Point* or *Polar* parameter allows the user to move the selection set of objects in any direction. If the *Move* action is associated with a *Linear* parameter, however, it restricts the movement to the direction, even if diagonal, that the *Linear* parameter was drawn.

Practice coordinating parameter geometry with the geometry of the objects in the selection set that are associated with the action added to the parameter. For example, if a rectangle in the block definition is 8 units wide and you plan to allow the user to stretch the rectangle to 10, 11, or 12 units in the block reference, a *Linear* parameter might be added from one end of the rectangle to the other, making its distance 8 units (Figure 11–46). For this example, the *Linear* parameter label has been changed to "8 UNITS DISTANCE" for illustrative purposes. When a *Stretch* action is added to this *Linear* parameter, the user will be able to resize the rectangle with a parameter whose base distance coincides with the original width of the rectangle. This is a logical approach to creating a dynamic block.

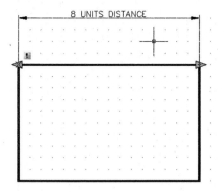

FIGURE 11–46 *Linear parameter added to rectangle*

It is possible to draw the *Linear* parameter with its endpoints located otherwise at a different distance and still enable the user to stretch the rectangle to the desired dimension, but unless there is a specific reason to do so, it is not recommended. The power of dynamic blocks can best be mastered when you understand the concept of how the parameters are manipulated by the actions that manipulate the objects in the selection set associated with the action.

Point Parameter

Figure 11–47 at the left shows a dynamic block with a Point parameter near the switch geometry and text. In this example, a Move action has been added to the Point parameter and associated with the objects that make up the switch and text. Figure 11–47 at the right shows how the switch and text were moved in the block reference by using the Move action.

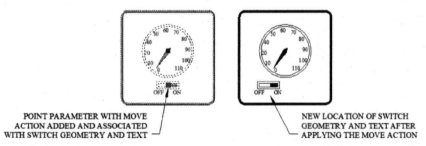

POINT PARAMETER WITH MOVE
ACTION ADDED AND ASSOCIATED
WITH SWITCH GEOMETRY AND TEXT

NEW LOCATION OF SWITCH
GEOMETRY AND TEXT AFTER
APPLYING THE MOVE ACTION

FIGURE 11–47 *Example of using the Move action with the point parameter*

The Point parameter establishes a point by setting user-defined X and Y properties for the block reference. A *Move* or *Stretch* action can be associated with a Point parameter. Invoke the *Point* parameter from the Authoring palette, and AutoCAD displays the following prompts:

> Specify parameter location or ⊡ (*specify a location, or right-click and select an option*)
> Specify label location: (specify a location for the label)

Options for the point parameter include *Name, Label, Chain, Description*, and *Palette*.

Choosing *Name* lets you specify the parameter name that is displayed in the Properties palette when you select the parameter in the Block Editor. When the parameter has been selected, AutoCAD displays the following prompt:

> Enter parameter name <default>: (*enter a name for the parameter, or press* Enter *to use the default name*)

Choosing *Label* lets you specify the parameter label, which identifies the custom property name added to the block. The label is displayed in the Properties palette as a Custom property when you select the block reference in a drawing. In the Block Editor, the parameter label is displayed next to the parameter. AutoCAD displays the following prompt:

> Enter position property label <default>: (*enter a label for the position property, or press* Enter *to use the default label*)

Choosing *Chain* lets you specify the Chain Actions property for the parameter. The point parameter may be included in the selection set of an action that is associated with a different parameter. When that parameter is edited in a block reference, its associated action may trigger a change in the values of other parameters included in the action's selection set. AutoCAD displays the following prompt:

> Evaluate associated actions when parameter is edited by another action? [Yes/No] <No>: (*enter y or press* Enter)

Setting the *Chain Actions* property to **Yes** triggers any actions associated with the Point parameter, just as if you had edited the parameter in the block reference through a grip or custom property.

Setting the *Chain Actions* property to **No** means that the Point parameter's associated actions are not triggered by the changes to the other parameter.

Choosing *Description* lets you specify the description for the custom property name (parameter label). This description is displayed in the Properties palette when you select the parameter in the Block Editor. In a drawing, when you select the custom property name (parameter label) for the block reference in the Properties palette, the description is displayed at the bottom of the Properties palette. AutoCAD displays the following prompt:

> Enter property description: (*enter a description for the parameter*)

Choosing *Palette* lets you specify whether or not the parameter label is displayed in the Properties palette when the block reference is selected in a drawing.

> Display property in Properties palette? [Yes/No] <Yes>: (*enter n or press* Enter)

 NOTE You can also specify and edit these properties in the Properties palette later, after you've added the parameter to the block definition.

A grip is displayed where you specified the location of the parameter with a leader to the label. An Alert icon, with the exclamation point, is displayed because no action is associated with the parameter yet. Refer to the section on "Using Actions" for a detailed explanation of adding actions.

Linear Parameter
Figure 11–48 at the left shows a dynamic block with a linear parameter added that is the width of the instrument panel. In this example, a Stretch action has been added to the linear parameter and associated with the objects that make up the outline of the

right end of the panel. Figure 11–48 in the middle shows the grip being moved to change the distance of the linear parameter and in turn stretch the width of the panel outline. Figure 11–48 at the right shows how the panel outline in the block reference was changed by using the Stretch action.

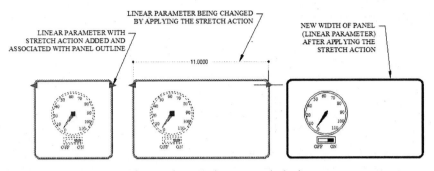

FIGURE 11–48 *Example of using the Stretch action with the linear parameter*

The Linear parameter establishes a vector by setting a user-defined distance property for the block reference. A Move, Scale, Stretch, or Array action can be associated with a Linear parameter. After you invoke the *Linear* parameter from the Authoring palette, AutoCAD displays the following prompts:

> Specify start point or ⬇ (*specify a start point for the parameter, or right-click and select an option*)
>
> Specify endpoint: (*specify an endpoint for the parameter*)
>
> Specify label location: (*specify a location for the label*)

Options for the linear parameter include *Name, Label, Chain, Description, Base, Palette,* and *Value set.*

Choosing *Name, Label, Chain, Description,* or *Palette* operates the same as described for the point parameter above.

Choosing *Base* lets you specify the Base Location property for the parameter. AutoCAD displays the following prompt:

> Enter base location [Startpoint/Midpoint]: (*specify an option*)

Choosing *Startpoint* specifies that the start point of the parameter remains fixed when the endpoint of the parameter is edited in the block reference.

Choosing *Midpoint* specifies a midpoint base location for the parameter. It is indicated by an X in the block definition. When you edit the linear parameter in the block reference, the midpoint of the parameter remains fixed, and the start point and endpoint of the parameter move simultaneously equal distances from the midpoint. For example, if you move the grip on the endpoint two units away from the midpoint, the start point simultaneously moves two units in the opposite direction.

Choosing *Value Set* lets you specify a value set for the parameter that limits the available values for the parameter in a block reference to the values specified in the set. AutoCAD displays the following prompt:

> Enter distance value set type [None/List/Increment] <None>: (*specify a value set type, or select one of the available option*)

Choosing *List* lets you specify a list of available values for the parameter in a block reference. AutoCAD displays the following prompts:

```
Enter list of distance values (separated by commas):
(specify a list of values separated by commas)
```

Choosing *Increment* lets you specify a value increment and minimum and maximum values for the parameter in the block reference.

AutoCAD displays the following prompts:

```
Enter distance increment: (specify an increment value for
the parameter)
Enter minimum distance: (specify a minimum distance value
for the parameter)
Enter maximum distance: (specify a maximum distance value
for the parameter)
```

Refer to the section on "Using Actions" for a detailed explanation of adding actions.

Polar Parameter

Figure 11–49 at the left shows a dynamic block with a polar parameter added that is the length and angle of the crane boom. In this example, a Polar Stretch action has been added to the polar parameter and associated with the objects that make up the outline of the boom. Figure 11–49 in the middle shows the grip being selected for modifying the geometry. Figure 11–49 at the right shows the grip being moved to change the distance and angle of the polar parameter and in turn stretching and relocating the end of the boom by using the Polar Stretch action.

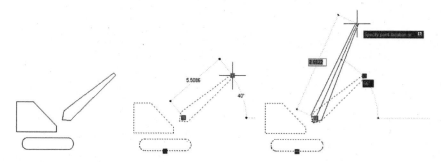

FIGURE 11–49 *Example of using the Polar Stretch action with the polar parameter*

The Polar parameter establishes a vector by setting user-defined distance and angle properties for the block reference. A *Move, Scale, Stretch, Polar Stretch,* or *Array* action is associated with a Polar parameter. After you invoke the *Polar* parameter from the Authoring palette, AutoCAD displays the following prompts:

```
Specify base point or 🡇 (specify a start point for the
parameter, or select an option from the shortcut menu)
Specify endpoint: (specify an endpoint for the parameter)
Specify label location: (specify a location for the label)
```

Options for the polar parameter include *Name, Label, Chain, Description, Palette,* and *Value set.*

Choosing *Name*, *Label*, *Chain*, *Description*, or *Palette* operates the same as described for the point parameter above. Choosing *Value Set* operates the same as described for the linear parameter above.

Refer to the section on "Using Actions" for a detailed explanation of adding actions.

XY Parameter

Figure 11–50 at the left shows a dynamic block representing a logo with an XY parameter added to it whose X distance is the width from one side to the other and the Y distance is the height from top to bottom. In this example, a Scale action has been added to the XY parameter and associated with the objects that make up the logo. Figure 11–50 in the middle shows using the Scale action to make the X distance of the logo fill the width of a rectangular shape. Figure 11–50 at the right shows using the Scale action to make the Y distance of the logo fill the height of a rectangular shape.

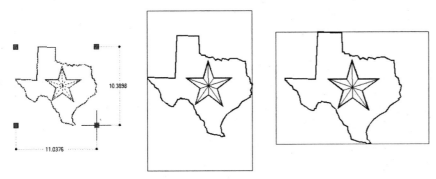

FIGURE 11–50 *Example of using the Scale action with the XY parameter*

The XY parameter establishes a pair of orthogonal vectors by setting user-defined horizontal and vertical distance properties for the block reference. A *Move*, *Scale*, *Stretch*, or *Array* action can be associated with an *XY* parameter. After you invoke the *XY* parameter from the Authoring palette, AutoCAD displays the following prompts:

```
Specify base point or ☑ (specify a base point, or select one
of the options from the shortcut menu)
Specify endpoint: (specify an endpoint for the parameter)
```

Options for the *XY* parameter include *Name*, *Label*, *Chain*, *Description*, *Palette*, and *Value set*.

Choosing *Name*, *Label*, *Chain*, *Description*, or *Palette* operates the same as described for the point parameter above. Choosing *Value set* operates the same as described for the linear parameter above.

Refer to the section on "Using Actions" for a detailed explanation of adding actions.

Rotation Parameter

Figure 11–51 at the left shows a dynamic block representing a door with jambs with a rotation parameter added to it. In this example, a Rotate action has been added to the rotation parameter and associated with the objects that make up the door. Figure 11–51 in the middle shows the grip being selected for modifying the geometry. Figure 11–51 at the right shows the result of using the Rotate action to change the door's opening angle from 45° to 90°.

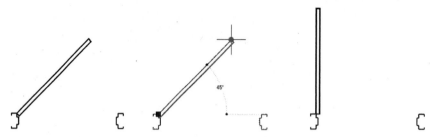

FIGURE 11–51 *Example of using the Rotate action with the rotation parameter*

The rotation parameter establishes a direction of rotation by setting a user-defined angle property for the block reference. Only a Rotate action can be associated with a *Rotation* parameter. After you invoke the *Rotation* parameter from the Authoring palette, AutoCAD displays the following prompts:

> Specify base point or ⊡ (*specify a base point for the parameter, or select an option from the shortcut menu*)
>
> Specify radius of parameter: (*specify a radius for the parameter*)
>
> Specify default rotation angle or [Base angle] <0>: (*specify a base angle for the parameter*)
>
> Specify label location: (*specify a location for the label*)

Choosing *Base Angle* lets you specify a base angle for the parameter and places the grip for the parameter at this angle. AutoCAD displays the following prompt:

> Specify base angle <0>: (*specify a base angle for the parameter, or press* Enter)

Options for the Rotation parameter include *Name, Label, Chain, Description, Palette,* and *Value set*.

Choosing *Name, Label, Chain, Description,* or *Palette* operates the same as described for the *Point* parameter above. Choosing *Value Set* operates the same as described for the *Linear* parameter above.

Refer to the section on "Using Actions" for a detailed explanation of adding a Rotation action.

Alignment Parameter

Figure 11–52 at the left shows a dynamic block with an alignment parameter at a point midway between the two jambs and aligned with where the door and jambs should be drawn on the hinge side of the wall where it will be installed. Figure 11–52 at the right shows how the block automatically aligns itself during the insertion process as the cursor nears the face of the wall. In this example, the door needed to be placed at the midpoint of the angled wall, so the OSNAP mode *Midpoint* was invoked.

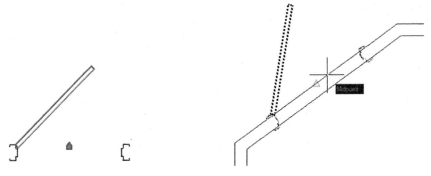

FIGURE 11–52 *Example of using the Alignment parameter*

The Alignment parameter establishes an X and Y location and an angle. It affects the entire block and does not require an action associated with it. An Alignment parameter allows the block reference to be automatically rotated around the base point to align with other objects in the drawing. It changes the angle property of the block reference. After you invoke the *Alignment* parameter from the Authoring palette, AutoCAD displays the following prompt:

```
Specify base point of alignment or [Name]: (Specify a point
or enter name)
```

When you specify the base point for the alignment parameter, an X is displayed in the Block Editor. When the command is completed, an alignment grip is added at this base point. The block reference automatically rotates about this point to align with another object in the drawing.

The parameter name is displayed in the Properties palette when you select the parameter in the Block Editor. AutoCAD displays the following prompt:

```
Specify alignment direction or alignment type or ⊡ (specify
an alignment direction, or select type from the shortcut
menu)
```

The alignment direction specifies the direction of the grip and the angle of alignment for the block reference. The *Type* selection allows you to select whether the parameter type is perpendicular or tangent, as shown in the following prompt:

```
Enter alignment type [Perpendicular/Tangent]
<Perpendicular>: (specify an alignment type)
```

Choosing *Perpendicular* lets you specify that the dynamic block reference aligns perpendicular to objects in a drawing.

Choosing *Tangent* lets you specify that the dynamic block reference aligns tangent to objects in a drawing.

Flip Parameter

Figure 11–53 at the left shows a dynamic block representing a logo with a Flip parameter added to it. In this example, a Flip action has been added to the Flip parameter and associated with the objects that make up the dolphin and the text. Figure 11–53 at the right shows the result of using the Flip action to flip the objects associated with it.

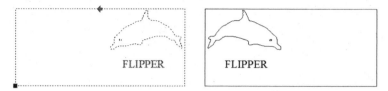

FIGURE 11–53 *Example of using the Flip action with the Flip parameter*

The Flip parameter establishes a user-defined flip property for the block reference. A Flip parameter flips objects. In the Block Editor, a Flip parameter is displayed as a reflection line. Objects can be flipped about this reflection line. A Flip parameter displays a value that shows whether or not the block reference has been flipped. You associate a Flip action with a Flip parameter. After you invoke the Flip parameter from the Authoring palette, AutoCAD displays the following prompts:

Specify base point of reflection line or ⊡ (*specify a base point for the reflection line, or select an option from the shortcut menu*)

Specify endpoint of reflection line: (*specify an endpoint for the reflection line*)

Specify label location: (*specify a location for the label*)

Options for the flip parameter include *Name*, *Label*, *Description*, and *Palette*.

Choosing *Name*, *Label*, *Description*, or *Palette* operates the same as described for the point parameter above.

Refer to the section on "Using Actions" for a detailed explanation of adding a Flip action.

Visibility Parameter

Figure 11–54 at the left shows a dynamic block representing an instrument panel with a Visibility parameter added to it. In this example, visibility states have been applied to enable the user to select between one set of objects to be visible in one state and another set to be visible in another state. Figure 11–54 at the right shows the result of using the visibility states to hide the objects associated with it and show others.

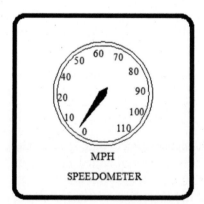

FIGURE 11–54 *Example of using the Visibility parameter*

The Visibility parameter establishes a user-defined visibility property for the block reference. A Visibility parameter allows you to create visibility states and to control the visibility of objects in the block. A Visibility parameter always applies to the entire block and needs no action associated with it. After you invoke the Visibility parameter from the Authoring palette, AutoCAD displays the following prompt:

> Specify parameter location or ⊡ (*specify a location for the parameter, or select an option from the shortcut menu*)

Options for the visibility parameter include *Name, Label, Description*, and *Palette*.

Choosing *Name, Label, Description*, or *Palette* operates the same as described for the point parameter above.

Lookup Parameter

Figure 11–55 shows four insertions of a dynamic block representing a raised-face-weld-neck piping flange in which dimensions of selected parts of the geometry are changed to suit different requirements. In this example, each of the four flange ratings—150#, 300#, 400#, and 600#—requires a unique combination of flange length and outside diameter.

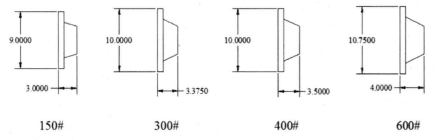

FIGURE 11–55 *Example of using the Lookup parameter*

The Lookup parameter establishes user-defined lookup properties for the block reference. A Lookup parameter defines a custom property that you can specify or set to evaluate to a value from a list or table you define. You associate a Lookup action with a Lookup parameter. Each Lookup parameter you add to the block definition can be added as a column in the Property Lookup Table dialog box. After you invoke the Lookup parameter from the Authoring palette, AutoCAD displays the following prompt:

> Specify parameter location or ⊡ (*specify a location for the parameter, or right-click and select an option*)

Options for the lookup parameter include *Name, Label, Description*, and *Palette*.

Choosing *Name, Label, Description*, or *Palette* operates the same as described for the point parameter above.

Refer to the section on "Using Actions" for a detailed explanation of adding a Lookup action.

Base Parameter

The Base point parameter defines the base point for the dynamic block reference in relation to the geometry in the block. This provides a way to control the location of the base point within the block reference when it is edited in a drawing. You do not associate any actions with a Base point parameter. The Base point parameter is generally included in a selection set of the block definition's actions. After you invoke the Base parameter from the Authoring palette, AutoCAD displays the following prompt:

Specify parameter location: *(specify a location)*

Only one Base point parameter is allowed in a dynamic block definition. Trying to add another base point parameter causes the following alert to be displayed:

Base point parameter already exists in block definition

In the Block Editor, when the Display UCS Icon option is turned ON, the icon is located at the base point of the block.

Action, Parameter, and Grip Properties

The font, size, and color of actions, parameters, and grip properties can be set in the Block Editor Settings dialog box (Figure 11–56) accessible from the Manage panel of the Block Editor tab.

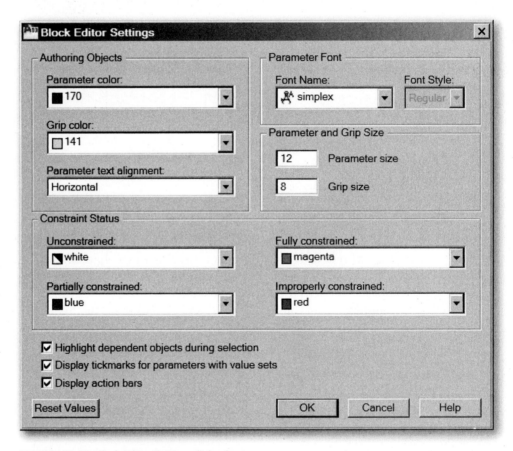

FIGURE 11–56 *Block Editor Settings dialog box*

The Authoring Objects section allows you to control the color of parameter and grip objects and specifies the orientation of the parameter text. The **Parameter Color** text box lets you specify the color of the parameter (BPARAMETERCOLOR system variable) You can choose ByLayer, ByBlock, a color by its integer from 1 to 255, or a true color specified by three integers each ranging from 1 to 255 in the format

RGB:000,000,000. The **Grip Color** text box lets you specify the color of the custom grip objects in the Block Editor (BGRIPOBJCOLOR system variable). The **Parameter Text Alignment** text box lets you specify the orientation of the parameter text (BPTEXTHORIZONTAL system variable).

The Parameter Font section allows you to set the font for the authoring objects. The **Font Name** text box lists the font family name for all registered TrueType fonts and all compiled shape (SHX) fonts in the *Fonts* folder (BPARAMETERFONT system variable). The **Font Style** text box lets you specify font character formatting for the authoring objects, such as italic, bold, or regular.

The Parameter and Grip Size section allows you to control the size of the parameter and grip objects. The **Parameter Size** text box lets you set the size of the parameter specified by an integer from 1 to 255 (pixels) (BPARAMETERSIZE system variable). The **Grip Size** text box lets you specify the size of grip objects in pixels (BGRIPOBJSIZE system variable).

The Constraint Status section allows you to specify the color overrides for objects in the Block Editor to show constraint status. The **Unconstrained** text box lets you set the color of the unconstrained objects. The **Partially Constrained** text box lets you set the color of the partially constrained objects. The **Fully Constrained** text box lets you set the color of the fully constrained objects. The **Improperly Constrained** text box lets you set the color of the over-constrained objects.

Choosing **Highlight Dependent Objects During Selection** automatically highlights all objects that are dependent on the currently selected authoring objects (BDEPENDENCYHIGHLIGHT system variable).

Choosing **Display Tickmarks for Parameters With Value Sets** allows you to control the display of value set markers for parameters (BTMARKDISPLAY system variable).

Choosing **Display Action Bars** allows you to control the display of action bars for the selected parameters (BACTIONBARMODE system variable).

Choosing **Reset Values** resets the Block Editor settings to default values.

Using Actions

Actions are associated with parameters to allow the user to manipulate and/or change the values of the geometric properties, specifically locations, sizes, distances, angles, and multiple instances. Actions are accessible while editing a dynamic block in the Block Editor either by invoking the BACTION command from the on-screen prompt or choosing a specific action from the Action tab of the Authoring palette. Actions include *Move, Scale, Stretch, Polar Stretch, Rotate, Flip, Array*, and *Lookup*.

Move Action

The Move action can be associated with a Point, Linear, Polar, or XY parameter. A Move action stipulates that a specified selection set of objects will move when the action is initiated in a dynamic block reference.

In the following example, a block named RECTANGLE is the current block being edited in the Block Editor. A Point parameter has already been added to a block definition (Figure 11–57), so you can choose it as the parameter for the Move action. Invoke the *Move* action from the Action tab of the Block Authoring palette and select the point parameter defined on the screen. AutoCAD displays the following prompts:

Select objects: (select the rectangle, and press Enter)

Specify action location (or ⊡): *(specify a location for the label, or select an option from the shortcut menu)*

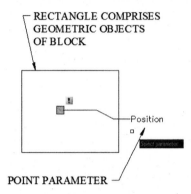

RECTANGLE COMPRISES
GEOMETRIC OBJECTS
OF BLOCK

Position

POINT PARAMETER

FIGURE 11–57 *Associating an action with a parameter*

Location is not significant, but when the block definition becomes crowded with other symbols for parameters, actions, and grips, it is advisable to locate the action symbol near the parameter with which it is associated (Figure 11–58).

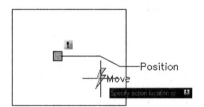

Position

Move

FIGURE 11–58 *Adding a Move action to a point parameter in the dynamic block definition*

NOTE

The names of the action and parameter are not visible in a block reference, or a block that has been inserted. Nor is the grip visible until the block has been selected for modifying. In the Block Editor, when selecting objects to include in an action, if a grip or the name of a parameter or action has been included in the window or crossing for determining the selection set, it has no effect on that grip, action, or parameter.

Instead of specifying an action location, you can choose one of the two options available from the shortcut menu: *Multiplier* or *Offset*.

Choose *Multiplier* to change the associated parameter value by a specified factor when the action is triggered. AutoCAD displays the following prompt:

Enter distance multiplier <current>: *(enter a numeric value)*

Specify a multiplier factor. For example, if you enter 10 for the multiplier and the user specifies a move that is 3.5 units in a specific direction, the objects will be moved 35 units in the specified direction. Or, if the multiplier factor is 0.25 and the user specifies a distance of 40 units, the selection set associated with that Move action will be moved 10 units.

Choose *Offset* to increase or decrease by a specified number the angle of the associated parameter when the action is triggered. AutoCAD displays the following prompt:

```
Enter angle offset <current>: (enter a numeric value)
```

Specify an angle offset. For example, if you specify 90 for the angle offset and the user specifies a move that is vertical, such as 90° in the current UCS units for a specific distance, the objects will be moved at a direction of 180° for the specified distance. If the angle offset is specified as −45° and the user specifies an angle of −45°, the selection set associated with that Move action will be moved in a direction of −90°.

The Multiple and Offset options can be applied together to the same Move action.

You can use the grips or custom property in the Properties palette to manipulate the block reference. By default, the color of the dynamic block grips will be different from the normal grip color. When you manipulate the block reference in a drawing, by moving a grip or changing the value of a custom property in the Properties palette, you change the value of the parameter that defines that custom property in the block. When you change the value of the parameter, it drives the action that is associated with that parameter, which changes the geometry, or a property, of the dynamic block reference.

Without applying the Multiple or Offset option to the Move action, it will operate like the Move option when selecting the normal grip associated with a block, except for one difference. When you use the Move action grip to move the selected geometry of the block, only that geometry moves. If you select a Linear or Polar parameter instead of a Point parameter (Figure 11–59), AutoCAD displays the following prompt:

```
Specify parameter point to associate with action or enter ⊡
(select one of the parameter points)
```

You can select a point or choose *Start Point* or *Second Point*. After you specify a point, AutoCAD displays the following prompts:

```
Select objects: (select one or more objects to be associated
with the Move action and then press Enter )
Specify action location or ⊡ (specify a location near the
selected point or choose Multiplier or Offset from the
shortcut menu)
```

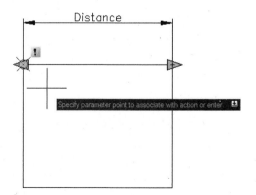

FIGURE 11–59 *Adding a Move action to a linear parameter in the dynamic block definition*

In Figure 11–59, a Linear parameter, which has two points, was selected. Whichever point is selected will become the location for the grip in the block reference that can be selected for initiating the associated Move action. Pressing [Enter] automatically associates the Move action with the second point. The grip for the other point will be visible when the block is selected, but it cannot be used to initiate this Move action. The primary difference in using a Linear parameter versus a Point parameter is that the move is restrained in the direction of the Linear parameter. In the example, the Linear parameter is horizontal; therefore, when the user applies the Move action associated with it, the selection set of objects can only be moved horizontally.

If a Polar parameter, which has two points, is selected with which to associate the Move action, AutoCAD prompts you to select one of the points in the same manner as the Linear parameter. Unlike the Linear parameter, when the user initiates the Move action by selecting that grip in the block reference, the selection set of objects associated with that Move action can be moved both specified distances and specified angles. The Move action associated with a point of a Polar parameter is practically the same as the Move action associated with a Point parameter.

When you are being prompted to "Specify parameter point to associate with action or enter 🔽," you can choose either *Start Point* or *Second Point* from the shortcut menu to specify the point with which to associate the Move action instead of selecting the point with the pointing device on the screen. If you add a Move action to an XY parameter (Figure 11–60), AutoCAD displays the following prompt:

```
Specify parameter point to associate with action or enter 🔽
(select one of the parameter points, or choose one of the
options from the shortcut menu)
```

Specify a point, and AutoCAD displays the following prompts:

```
Select objects: (select one or more objects to be associated
with the Move action, and then press [Enter])
Specify action location or 🔽 (specify a location near the
selected point, or choose one of the options from the
shortcut menu)
```

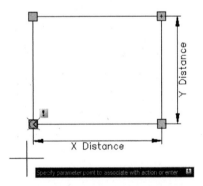

FIGURE 11–60 *Adding a Move action to an XY parameter in the dynamic block definition*

In Figure 11–60, an XY parameter, which has four points, was selected. The grip for the point on the same X axis as the primary point, referred to as the Xcorner, can be used to initiate a Move action in the direction of the Y axis, referred to as the Ycorner. The grip for the point on the same Y axis as the primary point can be used to initiate a Move action in the direction of the X axis. The grip diagonally opposite the primary point, referred to as the second point, cannot be used to initiate this Move action.

Instead of specifying the parameter point, you can select *Base Point, Second Point, XCorner,* or *YCorner* from the shortcut menu to specify the point with which to associate the Move action instead of selecting the point with the pointing device on the screen.

> Pressing enter automatically associates the Move action with the second point. However, with four points in the XY parameter, it is not always predictable which point AutoCAD might consider the second point. You should note which grip has an X placed on it during the selection process to be sure it is the one you wish to become the base point for the Move action to be able to work in any direction.

NOTE

Instead of specifying an action location, you can choose from the shortcut menu the *Multiple, Offset,* or *XY* option. Multiple and Offset behave similar to the point parameter explained earlier. If you choose the XY option, AutoCAD allows you to specify whether the distance that is applied to the action is the XY parameter's X distance, Y distance, or XY distance from the parameter's base point.

Scale Action

The Scale action can only be associated with a Linear, Polar, or XY parameter. A Scale action stipulates that a specified selection set of objects in a dynamic block reference will be scaled when the action is initiated in a dynamic block reference in a manner similar to the SCALE command. The Scale action causes the objects to scale when the associated parameter is edited by moving grips or by using the Properties palette.

In a dynamic block definition, you do not associate a Scale action with a key point on the parameter but with an entire parameter.

To add a Scale action, invoke the *Scale* action from the Action tab of the Block Authoring palette, and select the Polar parameter defined on the screen. AutoCAD displays the following prompts:

> Select objects: (*select one or more objects to be associated with the Scale action, and then press* Enter)
> Specify action location or ⬇ (*specify a location,* Figure 11-61)

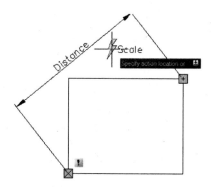

FIGURE 11–61 *Adding a Scale action to a polar parameter in the dynamic block definition*

You can instead choose the *Base* option from the shortcut menu. AutoCAD displays the following prompt:

```
Enter base point type [Dependent/Independent]: (select the
base point type)
```

Choosing *Dependent* causes the objects in the selection set to scale relative to the base point of the parameter with which the Scale action is associated. If the custom grip is used to scale the block, it scales relative to the lower left corner of the rectangle.

Choosing *Independent*, shown in the Block Editor as an X marker, lets you specify a base point independent of the parameter with which the Scale action is associated. The objects in the selection set will scale relative to this independent base point you specify.

When the block is inserted and the user selects it for modification, the grips for first and second points of the Polar parameter will appear. When the cursor is moved to the second point and polar tracking is ON, the distance and angle between the first and second points are displayed (Figure 11–62).

NOTE
The Scale action uses the distance between the first and second point as the base distance. If you enter a value, AutoCAD will scale the selected objects an amount determined by the ratio of the value entered divided by the distance between the first and second parameter points. In the example in Figure 11–62, that distance is 15.3832 units. If you enter a value of 20.0, the selected objects will be scaled up so that the new distance between the first and second points is 20.0. If you enter a value of 10.0, the selected objects will be scaled down so that the new distance between the first and second points is 10.0.

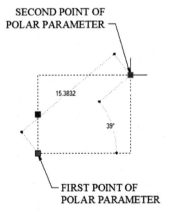

FIGURE 11–62 *Selecting the second point of a polar parameter to initiate the associated Scale action in a dynamic block reference*

After you have selected the grip at the second point of the polar parameter with which a Scale action has been associated, you can drag the cursor, and AutoCAD displays the distance and angle between the first point and the cursor (Figure 11–63). The angle has no effect on the Scale action, but if you specify the point as displayed, the distance, 21.0503 in the example, will be used as the new distance between the corresponding points on the scaled-up objects and all of the selected objects will be scaled accordingly. In this example, the base point of the Scale action is the first point specified when adding the Scale action to the polar parameter.

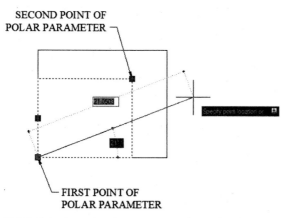

SECOND POINT OF
POLAR PARAMETER

FIRST POINT OF
POLAR PARAMETER

FIGURE 11–63 *Using the cursor to scale up the objects in a dynamic block reference associated with the Scale action that was added to a polar parameter*

Instead of using the cursor to determine the distance, you can enter a value from the keyboard. That value will be used as the new distance between the corresponding points on the scaled-up objects, and all of the selected objects will be scaled proportionately.

To add a Scale action to an XY parameter, invoke the *Scale* action from the Action tab of the Block Authoring palette, and select the XY parameter. The prompts are the same in the selection of Linear or Polar parameter, except that an additional XY option is provided to the base type. AutoCAD displays the following prompt when you choose the XY option:

```
Enter XY distance type [X/Y/XY] <XY>: (select X, Y, or XY
distance type)
```

Choosing X causes the proportion that the user scales the selected objects in a dynamic block reference to be determined by the movement of the cursor in direction of the X axis. Whichever grip the user selects to initiate the Scale action, the objects will be scaled according to the distance in the X direction that the cursor is from a base point. The base point will be the point opposite the selected grip in the X direction. The direction in which the objects are scaled depends on how the XY parameter was originally added to the object. For example, if the XY parameter was added by first selecting the lower-left corner and then the upper-right corner, the base point for certain Scale actions would be the lower left grip (Figure 11–64). If the XY parameter had been added by first selecting the upper-left corner and then the lower-right corner, the base point for certain Scale actions would be the upper-left grip (Figure 11–65).

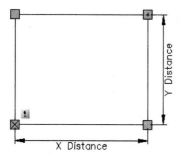

FIGURE 11–64 *Adding an XY parameter from the lower-left corner to the upper-right corner of the dynamic block definition*

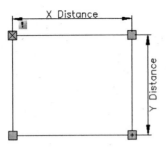

FIGURE 11–65 *Adding an XY parameter from the upper-left corner to the lower-right corner of the dynamic block definition*

Choosing Y causes the proportion that the user scales the selected objects in a dynamic block reference to be determined by the movement of the cursor in the direction of the Y axis (Figure 11–66). Whichever grip the user selects to initiate the Scale action, the objects will be scaled according to the distance in the Y direction that the cursor is from a base point. The base point will be the point opposite the selected grip in the Y direction. As in choosing X, in choosing Y the direction in which the objects are scaled depends on how the XY parameter was originally added to the object.

Choosing XY causes the proportion that the user scales the selected objects in a dynamic block reference to be determined by the movement of the cursor in any direction.

NOTE

In choosing X, Y, or XY, where the user moves the cursor after selecting a grip determines how the objects will be scaled based on the parameter's X, Y, or XY distance, respectively. When entering a value from the keyboard, the scale factor is based only on the X distance. No matter whether you choose X, Y, or XY, when the dynamic block reference is selected and highlighted, and a grip is selected, the only text box that is editable, or displayed as dynamic, is the X distance.

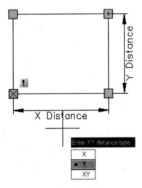

FIGURE 11–66 *Choosing the Y distance from which to base the Scale action added to an XY parameter*

Choosing the second point of the XY parameter of the dynamic block reference to initiate the Scale action lets you move the cursor up and down, or in the direction of the Y axis, to scale the selected objects larger or smaller in proportion to the Y distance (Figure 11–67). If you enter a value from the keyboard, however, it will be applied to the X distance. The X distance highlighted in an editable text box on the screen.

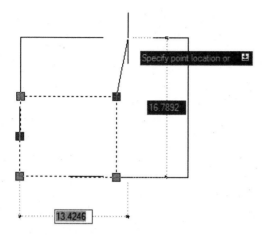

FIGURE 11–67 *Choosing the second point in the dynamic block reference to initiate the Scale action added to an XY parameter*

Stretch Action

The Stretch action can be associated with a Point, Linear, Polar, or XY parameter. A Stretch action stipulates that a specified selection set of objects in a dynamic block reference will be stretched when the action is initiated in a dynamic block reference in a manner similar to the STRETCH command. The Stretch action causes the objects to stretch when the associated parameter is edited by moving grips or by using the Properties palette.

Figure 11–68 shows a typical floor plan of a room with a symbol for a light in the ceiling connected to a wall switch by a dashed arc, indicating electrical wiring. Drawing the three entities of the light, switch, and arc would normally take three separate operations, even if the light and switch were inserted as blocks. However, with the power of the dynamic block, you can insert a single block and then use the STRETCH command to relocate the light symbol and have the dashed arc stretched to maintain the "connection."

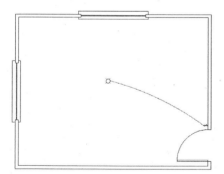

FIGURE 11–68 *Floor plan of a room with symbols for a light, switch, and a connecting dashed arc*

In the following example, a block named LT&SW, for "Light and Switch," is the block being currently edited in the Block Editor. A Point parameter has already been added to a block definition (Figure 11–69), so you can choose it as the parameter for the Stretch action. To add the Stretch action, invoke the *Stretch* action from the

Action tab of the Block Authoring palette and select the Point parameter, labeled "Position," defined on the screen. AutoCAD displays the following prompts:

Specify first corner of stretch frame or ⊡ (*specify the first corner*)

Specify opposite corner: (*specify the opposite corner*)

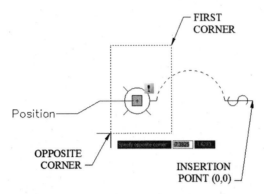

FIGURE 11–69 *Specifying the first point and opposite point of the stretch frame*

AutoCAD then prompts you to select the objects to be stretched:

Select objects: (*with the crossing selection mode, select the symbol for the light and one end of the dashed arc, and then press* Enter)

This action causes the light symbol to move and the arc to stretch when the Stretch action is used in a reference of this block in the drawing. AutoCAD then prompts you to locate the action label in the block definition:

Specify action location or ⊡ (*specify a location near the selected point*)

You can now stretch the dashed arc and move the light symbol by selecting the block reference, selecting the grip for the Stretch action, and then moving the cursor. Figure 11–70 (A) shows the dynamic block reference as it is inserted. Figure 11–70 (B) shows the grips highlighted when the reference is selected for modification. Figure 11–70 (C) shows the dynamic stretching of the dashed arc and moving of the light symbol. Figure 11–70 (D) shows the result of applying the Stretch action to a point parameter in the dynamic block.

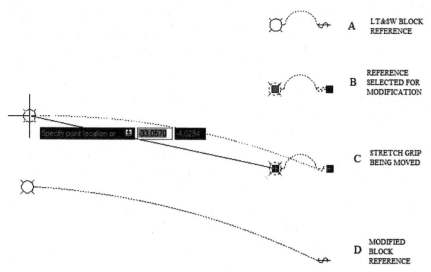

FIGURE 11–70 *Modifying a block reference with a Stretch action*

The following example of a weld-neck flange shows how to apply a Stretch action to a Linear parameter (Figure 11–71). A Linear parameter, with the label Distance, has been added that coincides with the length of the flange. In order for the parameter to align with the length of the flange, it was drawn from right (start point) to left (second point). This example shows how to add a Stretch action to enable the change in length.

To add the Stretch action, invoke the *Stretch* action from the Action tab of the Block Authoring palette, and select the Linear parameter (labeled "Distance") defined on the screen. AutoCAD prompts for the point to associate with the Stretch action:

> Specify parameter point to associate with action or enter ⊡ (*press* [Enter] *to accept the default second point to be associated with the Stretch action*)

If you want to specify the start point, select *Start Point* from the shortcut menu.

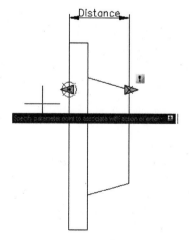

FIGURE 11–71 *Specifying the second point with the Stretch action added to a linear parameter*

AutoCAD prompts for the corners of the rectangle to determine the stretch frame:

```
Specify first corner of stretch frame or ⊡ (specify the first
corner)
Specify opposite corner: (specify the opposite corner)
```

The corners in this example need to include the rectangle representing the flange face and only the two lines representing the part of the weld neck that attach to the flange. The line representing the welded end of the weld neck should not be included; otherwise, it will be moved away from the insertion point when a Stretch action is initiated (Figure 11–72). AutoCAD then prompts you to select objects. In the example, specify the rectangle that includes the flange and two lines of the weld neck that are attached to the flange. After you select the objects, AutoCAD prompts you to place the action symbol, which can be placed at any convenient location.

FIGURE 11–72 *Specifying the first point and opposite point of the stretch frame*

When the block reference is selected for modifying in a drawing, the grips for the Stretch action are represented by arrows when associated with a linear parameter, and the distance that will be affected is highlighted in an editable text box (Figure 11–73). The cursor in Figure 11–73 is shown being moved upward and to the left, but because this Stretch action has been associated with a Linear parameter instead of a Point parameter, the only effect cursor movement will have is its change in the direction of the parameter: the block definition's X axis, in this case.

This is where the flange length can best be changed to the desired value. Once the associated grip has been selected, initiating the Stretch action, and the value of the Linear parameter is being highlighted, you can enter the desired value from the keyboard and have it applied in the correct direction without moving the insertion point.

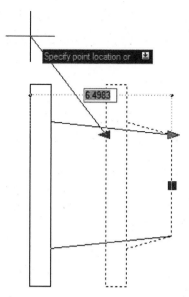

FIGURE 11–73 *Using the cursor to change the flange length by means of the Stretch action applied to a linear parameter*

Figure 11–74 shows the block reference that has been inserted on a 90° pipe elbow that is angled at 22.50° off horizontal. The block reference has been selected for modifying and the grip associated with the Stretch action has also been selected. When selected, it turned a specified color; the default is red. The Linear parameter that will be affected by the Stretch action rotates with the block when it is inserted, even though the text box remains horizontal for legibility.

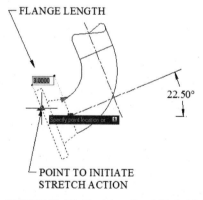

FIGURE 11–74 *Stretch action initiated in a block reference angled at 22.50°*

Figure 11–75 shows the block reference with its new flange length and the new location of the grip associated with the Stretch action. In the example, the value of 4.5 was entered from the keyboard when the text box was highlighted.

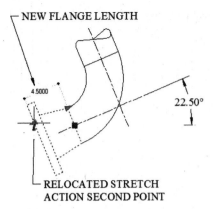

FIGURE 11–75 *Result of entering new value for flange length when text box was highlighted*

If, instead of a Point or Linear parameter, you choose an XY parameter for a Stretch action (Figure 11–76), AutoCAD prompts for the point to associate with the Stretch action:

> Specify parameter point to associate with action or enter ⊡ (*press* Enter *to choose the base point as the point associated with the Stretch action*)

You can choose *Start Point, Second Point, XCorner,* or *YCorner* from the shortcut menu to specify a point to associate with the XY parameter.

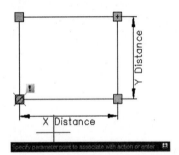

FIGURE 11–76 *Choosing the base point in the dynamic block reference to initiate the Stretch action added to an XY parameter*

AutoCAD prompts for the corners of the rectangle (see Figure 11–77) to determine the stretch frame:

> Specify first corner of stretch frame or ⊡ (*specify the first corner*)
>
> Specify opposite corner: (*specify the opposite corner*)

AutoCAD then prompts you to select objects. Select the rectangle in the given example. After you select the objects, AutoCAD prompts you to place the action symbol, which can be placed at any convenient location.

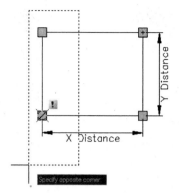

FIGURE 11–77 *Specifying the first point and opposite point of the stretch frame*

When the block reference has been selected for modifying in a drawing, the grips for the Stretch action are represented by squares when associated with an XY parameter (Figure 11–78). The cursor in Figure 11–78 is shown being moved downward and to the left and, because the base point had been selected, stretching is allowed in both axes. If the Xcorner or Ycorner is selected, this Stretch action is restricted to only the X or Y direction, respectively (Figure 11–79 and Figure 11–80).

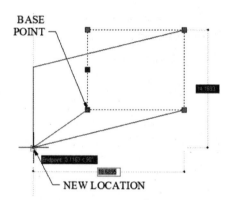

FIGURE 11–78 *The Stretch action applied to an XY parameter in the dynamic block reference using the base point*

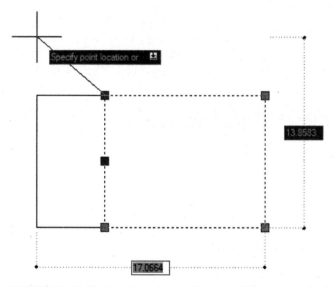

FIGURE 11–79 *The Stretch action applied to an XY parameter in the dynamic block reference using the Xcorner*

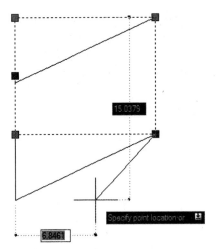

FIGURE 11–80 *The Stretch action applied to an XY parameter in the dynamic block reference using the Ycorner*

If, instead of a Point or Linear parameter, you choose a Polar parameter for a Stretch action, AutoCAD prompts for the point to associate with the Stretch action.

Applying a Stretch action to a polar parameter is similar to applying it to a linear parameter. When the grip associated with the Stretch action is selected, it acts in the same manner as selecting the base point when associated with an XY parameter; stretching is allowed in both axes.

Polar Stretch Action

The Polar Stretch action can be associated only with a polar parameter. A Polar Stretch action stipulates that a specified selection set of objects in a dynamic block reference will be stretched when the action is initiated in a dynamic block reference in a manner similar to the STRETCH command. The Stretch action causes the objects to stretch when the associated parameter is edited by moving grips or by using the Properties palette.

Figure 11–81 shows a rectangle as a block in the Block Editor with a Polar Stretch action added to a Polar parameter (labeled "Distance"). The stretch frame windowed the right end of the rectangle, and the rectangle was selected as the object associated with the Polar Stretch action. When this block is saved and inserted, the grip at the upper right of the polar parameter becomes the grip that will be used to initiate the Polar Stretch action.

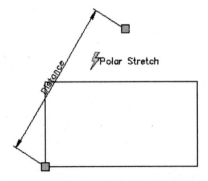

FIGURE 11–81 *The Polar Stretch action added to a polar parameter in the Block Editor*

Figure 11–82 shows how the rectangle will be changed by using the Polar Stretch action added to the Polar parameter. The shape and size on the left are affected by selecting the grip and moving the cursor on a line that is an extension of the Polar parameter. The changes on the right reflect moving the cursor at an angle, causing a rotation of the stretched rectangle.

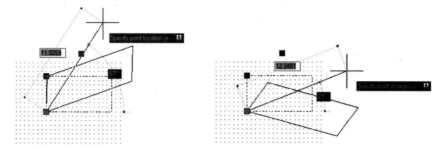

FIGURE 11–82 *The Polar Stretch action initiated in the dynamic block reference*

Rotate Action

The Rotate action can only be associated with a Rotation parameter. A Rotate action stipulates that a specified selection set of objects in a dynamic block reference will rotate when the action is initiated in a dynamic block reference in a manner similar to the ROTATE command. The Rotate action causes the objects to rotate when the associated parameter is edited by moving the cursor or by using the Properties palette.

Figure 11–83 shows a block representing a door with jambs. The block definition has been created with the door open at 45°.

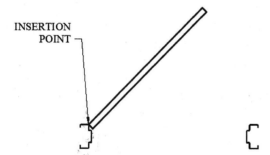

FIGURE 11–83 *Dynamic block reference showing door and jambs—door open at 45°*

In the Block Editor, a Rotation parameter is added to the definition with the base point located at the hinge point of the door, the insertion point of the block, the radius of the parameter set to the door width, and the default rotation angle set to 45° (Figure 11–84). This action places a round grip at the end of the door.

To add a Rotate action, invoke the *Rotate* action from the Action tab of the Block Authoring palette and select the Rotation parameter (labeled "Angle") defined on the screen (Figure 11-85). AutoCAD displays the following prompt:

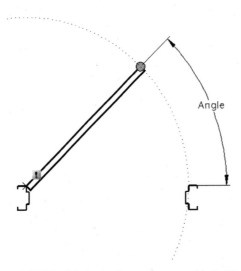

FIGURE 11–84 *A rotation parameter added with base point at hinge of door*

Select objects: (*select the rectangle that represents the door*)

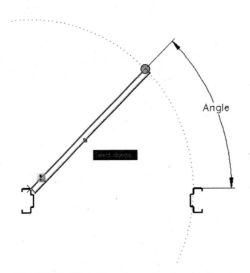

FIGURE 11–85 *A Rotation action is added and geometry for door selected to be associated with action*

Specify default rotation angle or ⊡ (*press* Enter *to accept the default*)

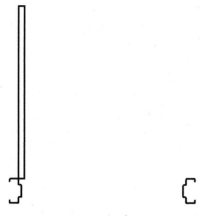

FIGURE 11–86 *Result of initiating Rotation action and changing property value from 45° to 90°*

Figure 11–86 shows the reference block that is modified with the Rotate action from the default angle of 45° to 90°.

Flip Action

The Flip action can only be associated with a flip parameter. A Flip action stipulates that a specified selection set of objects in a dynamic block reference will be flipped about a baseline when the action is initiated in a dynamic block reference in a manner similar to the ROTATE command. The Flip action causes the objects to flip when the associated parameter is edited by moving the cursor or by using the Properties palette.

Figure 11–87 shows a block with a rectangle, a logo (the dolphin), and text to which a flip parameter has been added in the Block Editor. Adding the Flip parameter establishes a baseline about which objects associated with a Flip action will be flipped. Note the label "Flip state" and the grip with the alert that an action has not been added.

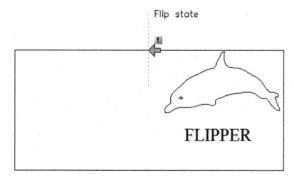

FIGURE 11–87 *A flip parameter added to a block in the Block Editor*

After the Flip action is added and associated with the logo and text, the block can be inserted. Figure 11–88 shows the result of the grip being selected for initiating the Flip action. Note that the logo graphic has been flipped about the base in both location and orientation, but the text is flipped in location only. The text maintains its original orientation so that it can be read. Having the name "FLIPPER" in the example of the Flip action is purely coincidental.

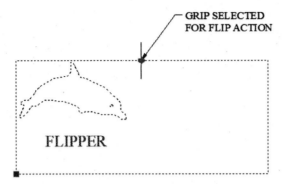

FIGURE 11–88 *Geometry location and orientation flipped—text location flipped and orientation unchanged*

Array Action

The Array action can be associated with a Linear, Polar, or XY parameter. An Array action stipulates that a specified selection set of objects in a dynamic block reference will be arrayed when the action is initiated in a dynamic block reference in a manner similar to the ARRAY command.

In Figure 11–89, a dynamic block has been created to represent a bolt with threads, a linear parameter has been added (labeled "Distance"), and two grips are displayed. To add the Array action, invoke the *Array* action from the Action tab of the Block Authoring palette, and select the Linear parameter (labeled "Distance") defined on the screen. AutoCAD prompts you to select the objects.

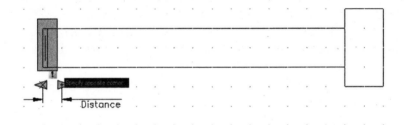

FIGURE 11–89 *Lines for bolt threads selected by window to be included in selection set of objects associated with an Array action*

Select the two lines near the end of the bolt, which represent the threads. Do not use crossing window to select the objects, because the lines for the outline of the bolt would be included in the selection set, which would cause them to be arrayed as well.

AutoCAD prompts you to specify the distance between columns. Because this is a Linear parameter, only the column value, not the row, is required. In Figure 11–90, the distance is specified as the pitch of the threads, which is 0.125.

```
Enter the distance between columns (&vert;&vert;&vert;):
(specify the distance between columns as 0.125)
Specify action location: (specify the action location in the
block definition)
```

The Array action symbol is placed in a convenient location.

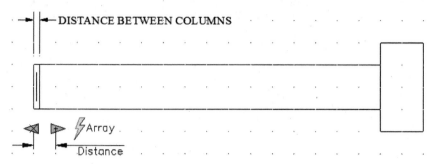

FIGURE 11–90 *Distance between columns being specified for an Array action*

The Linear parameter is the distance that the array will cover when the block is inserted as a reference. When the block is inserted and selected for modifying, the two grips of the Linear parameter are displayed (Figure 11–91). The base grip of the block is also displayed, as usual.

FIGURE 11–91 *Dynamic block reference selected for modification—grips for Array action displayed*

When the grip for the Array action is selected and moved to the right, the two lines associated with the action are arrayed in accordance with the distance between columns specified when the Array action was added to the block definition. Figure 11–92 shows the grip being moved to increase the Linear parameter distance to 4 units.

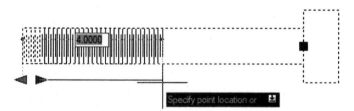

FIGURE 11–92 *Grip selected and moved to initiate Array action in block reference*

Figure 11–93 shows the block reference modified by using the Array action.

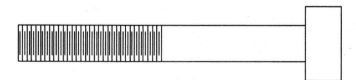

FIGURE 11–93 *Block reference after modifying with Array action*

If you select an XY parameter instead of a Linear or Polar parameter, AutoCAD prompts you to select the objects. AutoCAD then displays the following prompts:

```
Enter the distance between rows or specify unit cell (–):
(specify the distance between rows of arrayed objects, or
enter two values separated by a comma for each of the two
points for a unit cell for the arrayed objects)
Enter the distance between columns (&vert;&vert;&vert;):
(specify the distance between columns of arrayed object)
Specify action location: (specify the action location in the
block definition)
```

The Array action added to an XY parameter operates in both axes, that is, columns and rows, in the same manner that it acts in one axis for a linear or polar parameter.

Lookup Action

The Lookup action can only be associated with a lookup parameter. When you associate a Lookup action with a lookup parameter in a dynamic block definition, a lookup table is created. A lookup table can be used to assign custom properties and values to a dynamic block.

 Geometric and Dimensional constraints can be used in conjunction with Dynamic Blocks; however, constraint parameters cannot be added to a lookup table. Instead, you should use a Block Properties Table (explained later).

When you add a Lookup action to a lookup parameter, AutoCAD displays the following prompt:

```
Specify action location: (specify the action location in the
block definition)
```

No selection set is associated with the Lookup action. AutoCAD displays the Property Lookup Table dialog box (Figure 11–94).

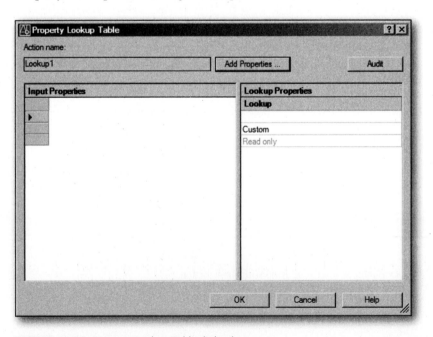

FIGURE 11–94 *Property Lookup Table dialog box*

The Property Lookup Table dialog box establishes a table for controlling parameter values (lookup properties) in the dynamic block definition. The lookup table also allows the values of lookup parameters to be controlled by the values of other parameters (input properties).

The lookup table assigns property values to the dynamic block reference based on how it is manipulated in a drawing. If Reverse Lookup, from the **Custom** list box, is selected for a lookup property, the block reference displays a lookup grip. When the lookup grip is clicked in a drawing, a list of lookup properties is displayed. Selecting an option from that list can be used to change the geometry of the dynamic block reference.

Figure 11–95 at the left shows a pictorial view of a piping make-up that includes a gate valve, two 90° ells, and two raised-face-weld-neck flanges. Figure 11–95 in the middle shows the two-dimensional drawing. The raised-face-weld-neck flange (RFWN) flange can be made into a block, as shown in Figure 11–95 at the right, for insertion into the drawing when required.

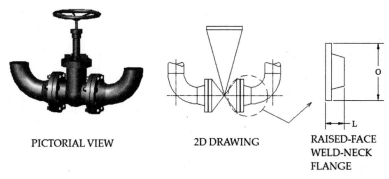

PICTORIAL VIEW   2D DRAWING   RAISED-FACE
WELD-NECK
FLANGE

FIGURE 11–95 *Piping make-up that includes two raised-face-weld-neck flanges*

The illustration of the block for the flange (Figure 11–95 at the right) shows two of the dimensions of the flanges that will vary depending on the flange rating (used to determine pressure rating). The flange length "L" and the outside diameter "O" are different for each flange rating. Table 11.5 shows the pipe size/flange ratings and their corresponding dimensions for the flange lengths and flange O.D.s (outside diameters).

TABLE 11–5

Pipe size	Flange rating	Flange length L	Flange dia O
4	150#	3.000	9.000
4	300#	3.375	10.000
4	400#	3.500	10.000
4	600#	4.000	10.750

Without using the dynamic block feature, it would be necessary to have a separate block drawn for each flange rating. This could require dozens of blocks for various fittings in a large piping drawing.

Figure 11–96 shows the block for the flange in the Block Editor with three linear parameters added to the definition. The parameters include FLANGE LENGTH, HALF FLANGE O.D. 1, and HALF FLANGE O.D. 2. The HALF FLANGE O.D. 1 and HALF FLANGE O.D. 2 are flange outside diameters that are to be stretched symmetrically about the center line of the flange. Stretch actions will be added to these three Linear parameters.

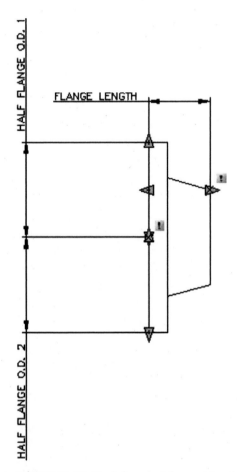

FIGURE 11–96 *Block for raised-face-weld-neck flange in Block Editor with three linear parameters added*

Figure 11–97 shows adding a Stretch action to the FLANGE LENGTH parameter. The rectangle represented by the dotted line is the stretch frame. All of the objects in the block will be associated with the Stretch action.

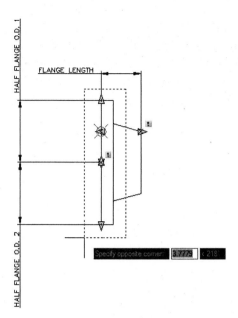

FIGURE 11–97 *Stretch frame defined for Stretch action added to FLANGE LENGTH linear parameter*

Figure 11–98 shows adding a Stretch action to the HALF FLANGE O.D. 1 parameter. The rectangle represented by the dotted line is the stretch frame. All of the objects in the block will be associated with the Stretch action. Similarly a Stretch action is added to HALF FLANGE O.D. 2.

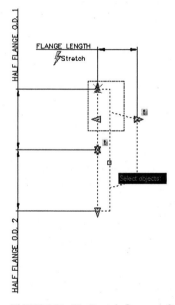

FIGURE 11–98 *Stretch frame defined and objects selected for Stretch action added to HALF FLANGE O.D. 1 linear parameter*

Figure 11–99 shows the three Linear parameters with three Stretch actions added in the Block Editor.

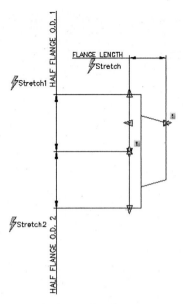

FIGURE 11–99 *Three linear parameters with Stretch actions added in Block Editor*

Figure 11–100 shows a Lookup action added to a lookup parameter. The location is not critical to any geometry. However, its location determines where it will appear when a reference of the block is selected for modifying.

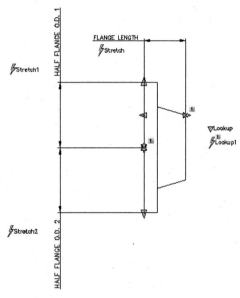

FIGURE 11–100 *Lookup action added to lookup parameter in Block Editor*

Figure 11–101 shows the Property Lookup Table dialog box being displayed when the Lookup action is added to the lookup parameter.

NOTE	If you have closed the Property Lookup Table dialog box and need to reopen it, double-click on the Lookup action.

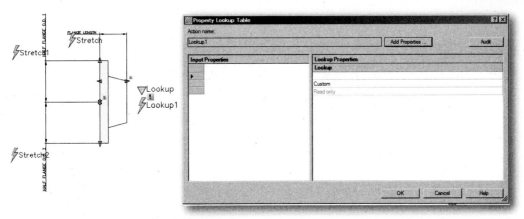

FIGURE 11–101 *Property Lookup Table dialog box*

Action Name displays the name of the Lookup action associated with the table.

To add properties to the Lookup Table, choose **Add Properties**. AutoCAD displays the Add Parameter Properties dialog box. Choose **Add input properties**, and Auto-CAD lists the available input properties (Figure 11–102).

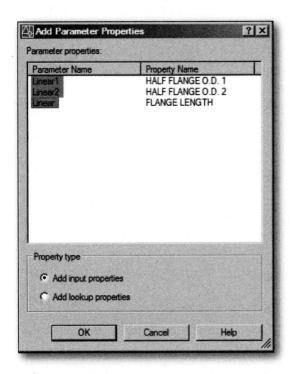

FIGURE 11–102 *Selecting parameters from the Property Lookup Table dialog box*

Select all three parameters listed, choose **OK**, and AutoCAD returns to the Property Lookup Table dialog box with the selected parameters listed under **Input Properties**. The values for the properties of each flange rating are entered under their respective headings in the Input Properties list and the corresponding flange rating is entered under the Lookup Properties list. Figure 11–103 shows Property Lookup Table with all the values entered in the Input Properties values and corresponding flange rating in the Lookup Properties. You can choose **Audit**, and AutoCAD checks that each record is unique; otherwise, a warning is displayed.

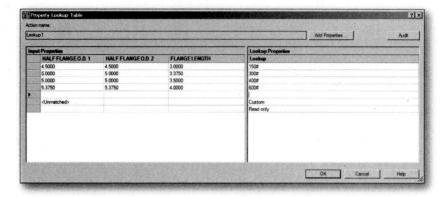

FIGURE 11–103 *Entering input properties for the parameters and corresponding lookup properties for each flange rating*

In the text box at the bottom of the Input Properties list, select **Allow reverse lookup** in the **Custom** list box (Figure 11–104). Selection of **Allow Reverse Lookup** enables the lookup property for a block reference to be set from a drop-down list that is displayed when the lookup grip is clicked in a drawing. Selecting an option from this list changes the block reference to match the corresponding input property values in the table.

Choose **OK** to accept the changes, and AutoCAD returns to the Block Editor.

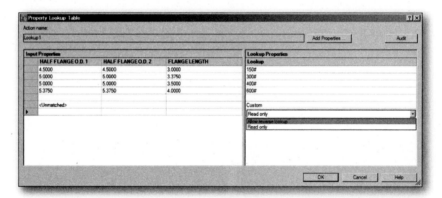

FIGURE 11–104 *Property Lookup Table with Allow reverse lookup selected in the Custom list box*

Close the Block Editor, and insert two references of the dynamic block. The dimensions are added to the figure for illustrative purposes. They will not be displayed unless added by using the DIMENSION command. The values for the flange length and flange O.D. have been changed in the block reference by selecting the lookup grip (Figure 11–105 at the left) and then choosing 600# in the list of Lookup Properties in the shortcut menu (Figure 11–105 at the right).

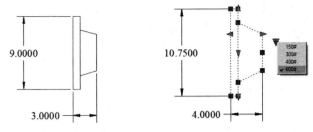

FIGURE 11–105 *Block reference with flange length and flange O.D. changed with Lookup action*

Adding Visibility States to Dynamic Blocks

By adding a Visibility parameter in your dynamic block definition, you can create and name visibility states and then specify which geometric objects are invisible for a given visibility state. One block can have multiple visibility states.

The Visibility parameter includes a lookup grip. A block reference that contains visibility states will always display the grip. When you select the grip in the block reference, a shortcut menu is displayed listing all the visibility states in the block reference. Selecting one of the states from the list causes the geometry that is visible for that state to be displayed.

An instrument panel is used to demonstrate the visibility feature in a dynamic block. In this case, a single dynamic block definition is created that represents two sets of geometry to suit two separate requirements. One visibility state will display the English version of a speedometer for reading miles per hour, as shown in Figure 11–106 at the left. The second visibility state will display the German version of a speedometer for reading kilometers per hour, as shown in Figure 11–106 at the right.

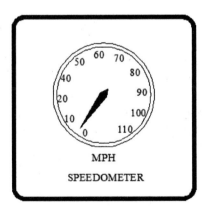 

FIGURE 11–106 *Two block references from the same dynamic block definition applying the visibility feature*

The fundamental geometry, including panel outline, dial outline, and indicating needle, is created as a block. It is opened in the Block Editor, and a visibility parameter is added to the definition. Double-click on the Visibility parameter, and AutoCAD displays the Visibility States dialog box (Figure 11–107).

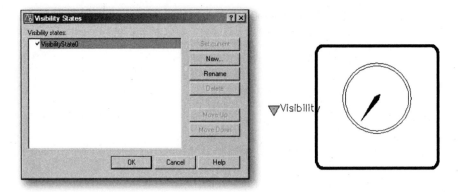

FIGURE 11–107 *Visibility States dialog box displayed in the Block Editor*

When the dialog box first appears, it contains one visibility state named Visibility-State0. Rename it to ENGLISH. Create a new visibility state by choosing **New**, and AutoCAD opens the New Visibility State dialog box (Figure 11–108). Rename it as GERMAN, and set it as current. Choose **OK** to close the dialog box.

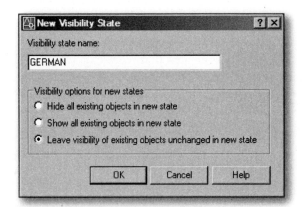

FIGURE 11–108 *Visibility States dialog box displayed in the Block Editor*

Draw numerals 0 through 110 and the text objects "MPH" and "SPEEDOMETER" (Figure 11–109).

To apply the visibility states, it is necessary to think in reverse. That is, a Visibility state is used primarily to make selected objects Invisible when that state is selected during the modifying of the block reference.

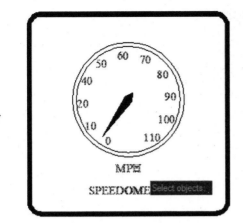

FIGURE 11–109 *Objects not made invisible in the English Visibility State*

Set English as the current visibility state by selecting English from the list box (Figure 11–110).

FIGURE 11–110 *Selecting English from the list box (Block Editor toolbar)*

Numerals 0 through 110 and the text objects "MPH" and "SPEEDOMETER" will disappear. Draw numerals 0 through 140 at the same location and the text objects "KPH" and "GESCHWINDIGKEITSMESSER" (Figure 11–111).

FIGURE 11–111 *Objects not made invisible in the German Visibility State*

By switching between English and German from the list box, you can see the corresponding objects disappear. Save the changes, and close the Block Editor. When the block reference is selected to change visibility, the grip for the visibility parameter is displayed. To change the geometry and text, select the grip, and then select *German* from the shortcut menu (Figure 11–112). Numerals 0 through 140 and the text objects "KPH" and "GESCHWINDIGKEITSMESSER" will appear. To change them back, select the grip and then select *English* from the shortcut menu. Numerals 0 through 110 and the text objects "MPH" and "SPEEDOMETER" will appear. From the shortcut menu of the selected reference block, you can control the visibility of objects.

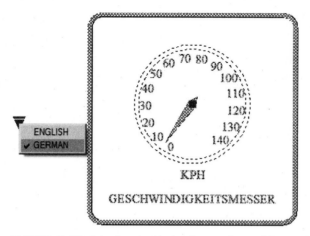

FIGURE 11–112 *Selecting the Block reference for modifying, selecting the visibility grip, and then selecting GERMAN from the shortcut menu*

Advanced Dynamic Block Utilities and Features

The fundamental elements that make the blocks dynamic are parameters, actions, and grips on blocks. This section covers the more refined and advanced features available when working with dynamic blocks.

Parameter Sets

The Parameter Sets tab of the Authoring palette contains 20 parameter sets (Figure 11–113). Parameter sets are combinations of parameters and actions with a preset number of grips that can be quickly applied to a dynamic block and associated with selected geometry without having to go through the process of adding them separately. For example, the Linear Move parameter set adds a Linear parameter with one grip and an associated Move action. The Linear Move Pair parameter set adds a Linear parameter with two grips and a Move action associated with each grip. Once the configuration is learned, using a parameter set can save time in adding the desired dynamics to a block.

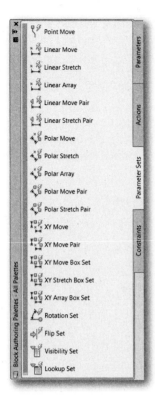

FIGURE 11–113 *Parameter Sets tab of the Authoring palette*

Property Values

The values of a selected object's properties are displayed in the Properties palette. The Block Editor lets you specify values for properties for a parameter in a dynamic block definition, which can be displayed under Custom properties for the dynamic block reference when it has been inserted into a drawing.

You can specify parameter labels while in the Block Editor so that when the dynamic block reference is selected in a drawing, these properties are listed by their label under Custom in the Properties palette. Parameter labels within the block should be unique.

Other parameter properties may be listed under Custom in the Properties palette when you select the dynamic block reference in a drawing depending on the parameters used. For example, a Polar parameter has an angle property that displays in the Properties palette.

Depending on how the dynamic block is defined, properties such as size, angle, and position might be displayed.

Geometric properties such as color, linetype, and lineweight can also be specified by using the Properties palette. They are listed in the Properties palette under Geometry when you select a parameter in the Block Editor.

How the block reference will function in a drawing is defined by parameter properties such as Value Set properties and Chain actions, whether or not a block can be exploded, and whether or not the block that is nonuniformly scaled can be specified in the Block Editor.

You can specify whether or not custom properties are displayed for the block reference when it is selected in a drawing.

Properties can be extracted using the Attribute Extraction wizard.

Changing the Insertion Point

The BCYCLEORDER command lets you specify certain grips in the dynamic block definition, which can be used as the insertion point when inserting the block. Figure 11–114 shows a dynamic block definition in the Block Editor with several parameter/action grips.

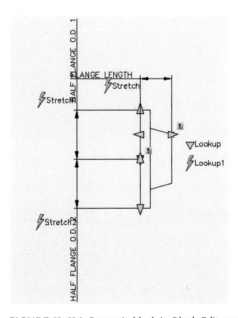

FIGURE 11–114 *Dynamic block in Block Editor with several action/parameter grips*

While in the Block Editor, type BCYCLEORDER and AutoCAD displays the Insertion Cycling Order dialog box (Figure 11–115). The **Grip cycling list** displays the grips, their parameter name, Type, and Location. If a grip has a check under the Cycling column, it will be available as an insertion point when the block is inserted.

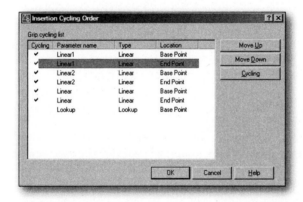

FIGURE 11–115 *Insertion Cycling Order dialog box showing action/parameter grips*

Choosing **Move Up** or **Move Down** causes the selected grip to move up or down respectively in the list.

Choosing **Cycling** causes the Cycling column for the selected grip to toggle between being checked or unchecked.

Choose **OK** to accept the selections and close the dialog box.

Figure 11–116 at the left shows the dynamic block as it is being inserted with the default insertion point. To cycle between selectable points, press CTRL. Figure 11–116 at the right shows the insertion point after cycling through the selectable points; press CTRL until the desired point is selected.

FIGURE 11–116 *Cycling from the default action/parameter grip to the desired one for the insertion point*

Using Annotative Blocks

Annotative Scaling can be used with blocks to annotate your drawing. See Chapter 8 for a complete explanation of annotative scaling. You can also use annotative attributes for nonannotative blocks. When you select a block as the content of a multileader, AutoCAD causes the geometry in the block to display on the paper based on the scale of the viewport, but the attribute text is displayed at the paper height defined for the attribute.

Annotative block references are created when you insert a block with an annotative block definition. The inserted block and attributes will initially support the current annotation scale at the time they are inserted. They should be inserted with a unit factor of 1. Annotative properties of individual block references cannot be changed. An annotative block's paper size can be set by defining the block in paper space or on the Model tab with the annotative scale set to 1:1. You can use annotative objects in an annotative block definition except for other annotative blocks. You should not manually scale blocks that contain annotative objects.

Figure 11–117 shows how a block has been used annotatively to label the two views, each in a separate viewport. This example illustrates the application of annotative scaling using a block. The block consists of a circle, a line, and three attributes. In this case, the three attributes are for the number of the view, seen in the circle, the title of the view, seen above the line, and the scale, seen below the line. The first insertion is in the viewport on the left, showing the door location. The viewport has an annotative scale of 1½ = 1' -0. The same block, inserted in the viewport on the right, has an annotative scale of $3 = 1'$ -0. Thus, nonannotative objects are displayed twice as large as the viewport on the left. However, because the block was annotative, it was automatically scaled so that it will display and plot at the same size in both viewports. You should not apply a scale factor when inserting an annotative block, as annotative scaling does this for you automatically. These blocks were inserted in the model space in the viewport.

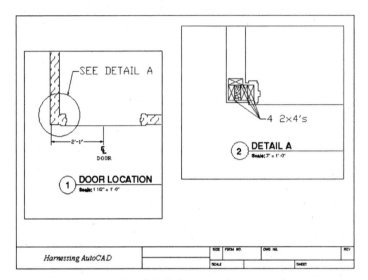

FIGURE 11–117 *Example of blocks being used annotatively*

Using Constraints and Parameters in Dynamic Blocks

When creating a block definition, you can apply Geometric constraints the same way that you would in a parametric drawing, as described in Chapter 10. Dimensional constraints can also be applied but have limited functionality when the block is inserted. Therefore, special dimensional-type constraints called Constraint parameters provide access to the parameter values after the block is inserted.

The Geometric constraints that you apply in the Block Editor include the *Vertical, Horizontal, Fix, Coincident, Parallel, Perpendicular,* and *Tangent* constraints discussed in Chapter 10.

Constraint parameters are dimensional-type constraints that behave like the Dimensional constraints discussed in Chapter 10. Constraint parameters can be invoked from the Dimensional panel of the Block Editor tab in the ribbon or with the BCPARAMETER command. Options include *Linear, Aligned, Horizontal, Vertical, Angular, Radial,* and *Diameter.*

NOTE

Dimensional constraints and Constraint parameters can both be used in a block definition, but only Constraint parameters will display editable custom properties for an inserted reference of that block. The Dimensional constraints (the DIMCONSTRAINT command) are invoked from the Dimensional panel of the Parametric tab and Constraint parameters are invoked from the Dimensional panel of the Block Editor tab as the BCPARAMETER command. While these two dimensioning methods look similar on their respective panels, you should be aware of the capabilities and limitations of the Dimensional constraints when applying dimensions in the Block Editor. It is possible to switch to the Parametric tab while still in the Block Editor and mistakenly use a Dimensional constraint when you need to use a Constraint parameter. Constraint parameters are accessible only from the Block Editor tab.

The Block Properties Table used in conjunction with Constraint parameters can achieve a similar functionality as the Lookup action and Lookup parameters described earlier in this chapter. However, once the proper constraint parameters have been applied, the Block Properties Table can be more easily incorporated.

This example applies Constraint parameters and a Block Table to a dynamic block. A raised-face-weld-neck piping flange (Figure 11–118) was created similar to the parametric drawing created earlier for use with a Lookup table (refer to Figure 11–95 and Table 11.5). The flange below was defined as a block with the geometry and dimensions as shown on the left. From this geometry the constraints, parameters, and attributes were added in the Block Editor and then saved with the block name "rfwn." On the right it is shown as it was inserted and then the Block Table was used to cause the Constraint parameters to comply with the dimensions of a 3" 150# flange. In the drawing below the dimensions were added for illustration purposes.

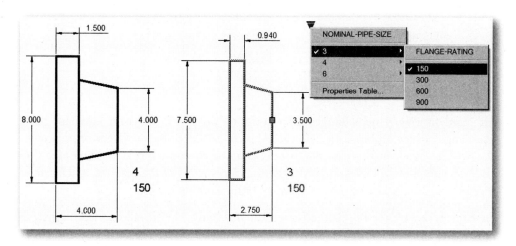

FIGURE 11–118 *Example of a Dimensioned Insertion of a Dynamic Block*

When drawing the geometry to use in the Block Editor the dimensions used are not critical, nor do they need to match any of the combinations of values for any of the standard flanges. The objects in this example are a rectangle and a three-segment polyline. Using polylines in this case insures that the corners stay connected. This means Coincident constraints are not needed for the corners. The attributes are included so they will be available in the Block Properties Table for selecting the pipe size/flange rating combination later.

Pipe size/Flange rating combinations with their dimensions for this example are shown in Table 11.6. These will be used in the Block Properties Table.

TABLE 11–6

Pipe size	Flange rating	Flange length L	Flange dia O	Pipe dia O	Flange tkns
3	150#	2.750	7.500	3.500	0.940
3	300#	3.125	8.250	3.500	1.250
3	600#	3.250	8.250	3.500	1.375
3	900#	4.000	9.500	3.500	1.625
4	150#	3.000	9.000	4.500	0.940
4	300#	3.375	10.000	4.500	1.440
4	600#	4.000	10.750	4.500	1.625
4	900#	4.500	11.500	4.500	1.875
6	150#	3.500	11.000	6.625	1.000
6	300#	3.875	12.500	6.625	1.810
6	600#	4.625	14.000	6.625	2.000
6	900#	5.500	15.000	6.625	2.250

All raised-face-weld-neck flanges share common geometry. To cover the many pipe size/flange rating combinations in a typical project might take hundreds of predrawn blocks. The intent in this example is to create one dynamic block that will take the place of many blocks by being able to set the dimensions of the inserted block reference to suit a multitude of pipe size/flange rating combinations.

First the geometry for the flange is created. In Figure 11–119 all of the Geometric constraints and Constraint parameters have been added to the geometry from the Block Editor. A Block Properties Table has also been included in the block definition along with two attribute definitions.

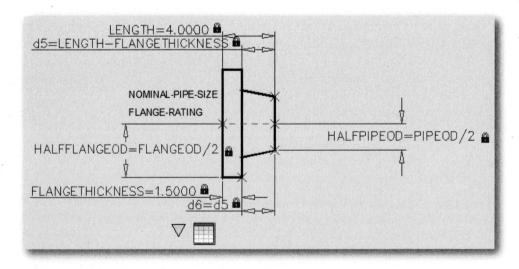

FIGURE 11–119 *Constraint Parameters applied to a Dynamic Block in the Block Editor*

Careful planning is necessary in order for the block reference to perform as required after being inserted. Figure 11–120 shows the initial Geometric constraints applied along with one Constraint parameter to establish control for the flange length. When applying a Constraint parameter it makes a difference in which sequence the points are selected. In the case of *d1* as shown in Figure 11–120, the pipe end of the weld neck part of the flange is selected first and then the flange face end second. This causes the geometry to change its length by moving the objects associated with the left extension line while leaving the right where it is.

When applying a constraint parameter or block properties table you will be prompted to specify number of grips. The Horizontal, Vertical, Aligned, Angular, and Diameter constraint parameters have the option of 0, 1, or 2 grips. When you apply the Radius constraint parameter or the block properties table you will be prompted to specify 0 or 1 grip. Specifying 0 grips means there will be no grips displayed when the inserted block reference is selected in the drawing. This means there will be no grips available to set dimensions or angles or have access to the block properties table from the drawing area. If you specify 1 grip there will be a grip available to set the value of a distance from one end of the parameter. Or, in the case of the block properties table, you will be able to display the shortcut menus from which to set parameter values. Specifying 2 grips allows you to change the value of a distance from either end of the parameter.

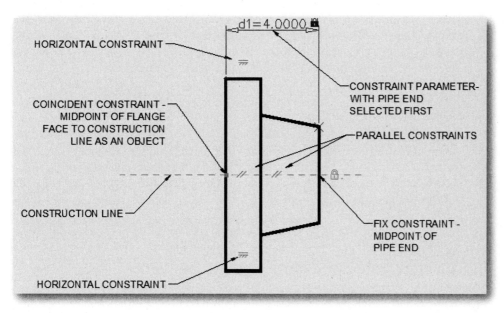

FIGURE 11–120 *Flange Geometry with Geometric Constraints, Construction Line, and one Constraint Parameter*

After the geometry has been created (Figure 11–120) and the Block Editor has been invoked, defining the dynamic block includes the following important steps:

1. Horizontal constraints are applied to the top and bottom horizontal lines of the rectangle so that they will maintain alignment when the flange face dimension is changed.

2. A Parallel constraint is applied to the front and back faces so they will maintain their alignment with each other.

3. A horizontal line is drawn through the midpoint of the flange face so that it (the midpoint of the flange face) will stay on the line as the face moves to adjust to the flange length.

4. The line added in step 3 is converted to construction geometry by using the BCONSTRUCTION command from the Manage panel of the Block Editor tab (Figure 11–121).

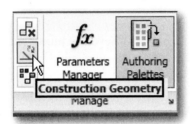

FIGURE 11–121 *Choosing the BCONSTRUCTION command from the Manage Panel*

5. A Coincident constraint is applied by first selecting the midpoint of the face of the flange and then invoking the OBJECT option and selecting the construction line.

6. A Fix constraint is applied to the midpoint of the pipe end of the weld neck. This anchors the block at the insertion point.

7. The first Constraint parameter, which has the equation *d1 = 4.000*, is applied. Before saving the block, the name *d1* is changed to *LENGTH* to make it easy to identify. Zero grips will be specified because there is no need to be able to set the length of the flange by using a grip after the block is inserted. Setting the length and other dimensions will be done automatically through the Block Properties Table.

Refer to Figure 11–119 for the remaining steps in defining the block:

8. A Horizontal Constraint parameter is applied to the rectangle and named *FLANGETHICKNESS*.

9. In order to keep the angled lines (representing the sides of the weld-neck) touching the back of the flange, a Horizontal Constraint parameter is applied. Here the value of the constrained horizontal dimension is made equal to the length of the flange minus the thickness of the flange. This is achieved by overwriting the value with the equation *d5 = LENGTH-FLANGETHICKNESS*. This value is not going to be used in the Block Properties Table so its name is left unchanged. Likewise a horizontal constraint parameter is applied to the opposite angled line and it value is replaced with the equation *d6 = d5* with 0 grips.

The next steps require that the dimensions for the flange face and the weld-neck pipe end be applied in a manner that keeps their midpoints on the center line (represented by the construction line created in steps 3 and 4). To accomplish this, a Constraint parameter is applied by selecting the midpoint as the first point and the end of the face as the second point. When applied in this manner, the value given for the "half-distance" will be doubled for the total dimension for the face. The adjustments in distance will be symmetrical about the midpoint of the line. That is, if you specify that the Constraint parameter be 6", the total length of the line will be 12" and its midpoint will not move from the center line. You can use the Parameters Manager to create a User Parameter as one of the parts of the equation that will be used in the value of the Constraint parameter. Invoke the PARAMETERS command from the Manage panel and AutoCAD displays the Parameters Manager palette (Figure 11–122).

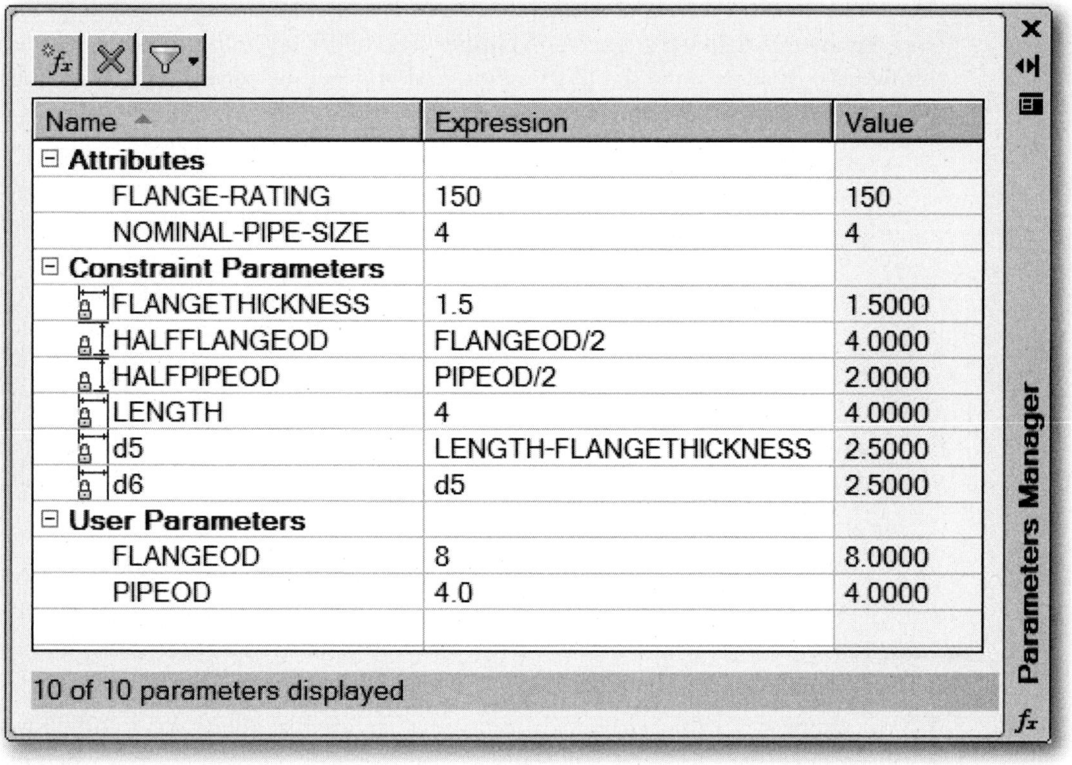

Name ▲	Expression	Value
⊟ **Attributes**		
FLANGE-RATING	150	150
NOMINAL-PIPE-SIZE	4	4
⊟ **Constraint Parameters**		
🔒 FLANGETHICKNESS	1.5	1.5000
🔒 HALFFLANGEOD	FLANGEOD/2	4.0000
🔒 HALFPIPEOD	PIPEOD/2	2.0000
🔒 LENGTH	4	4.0000
🔒 d5	LENGTH-FLANGETHICKNESS	2.5000
🔒 d6	d5	2.5000
⊟ **User Parameters**		
FLANGEOD	8	8.0000
PIPEOD	4.0	4.0000

10 of 10 parameters displayed

FIGURE 11–122 *Parameters Manager palette*

In the Parameters Manager palette the User Parameters *FLANGEOD* and *PIPEOD* (for flange outside diameter and pipe outside diameter) are created to receive the specified dimensions from the Block Properties Table when it is created. They must also exist before you can use their names in the equations for the *HALFFLANGEOD* and *HALFPIPEOD* or AutoCAD will display an error message (Figure 11–123).

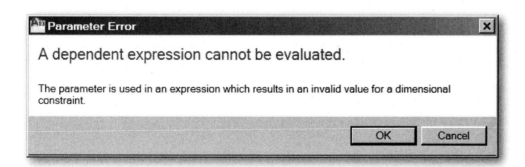

Parameter Error ✖

A dependent expression cannot be evaluated.

The parameter is used in an expression which results in an invalid value for a dimensional constraint.

OK   Cancel

FIGURE 11–123 *Parameter Error Message*

Remember that the values assigned to the Attributes, Constraint Parameters, and User Parameters in the Parameters Manager palette are arbitrary. These values are simply place holders until the block is inserted and will be set by using the Block Table to select the pipe size/flange rating combination.

10. A Vertical Constraint parameter is applied to the face of the flange. In order to have its value applied to the rectangle symmetrically about the center, the midpoint of the face is selected as the first point and one end as the second. The values will be half of the required outside diameter of the flange face and 0 grips is specified.

11. A Vertical Constraint parameter is applied to the pipe end of the weld-neck in the same manner as in step 10. The values will be half of the required outside diameter of the pipe with 0 grips specified.

Three final steps are needed to make it easy to interface with an inserted reference of the block so that it will automatically set dimensions to conform to the selected pipe size and flange rating.

12. An attribute is added with the value of the nominal pipe size.

13. An attribute is added with the value of the flange rating.

14. A Block Properties Table is created.

Block Properties Table

A Block Properties Table is a spreadsheet type of feature that allows you to store rows of values. These values can be applied to the parameters in a dynamic block when inserted so that a set of dimensions, angles, and other data can be changed by selecting the appropriate variable on the screen. To create a Block Properties Table invoke the command from the Dimensional panel (Figure 11–124).

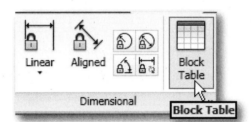

FIGURE 11–124 *Choosing the BTABLE command from the Dimensional Panel*

AutoCAD prompts for the location of the table:

```
Specify parameter location or ⊞ (specify a location for the
table)
Enter number of grips: (specify 1 grip which is the default)
```

Specify a location on the drawing that will display the grip when a reference of the block is selected. AutoCAD displays the Block Properties Table dialog box (Figure 11–125).

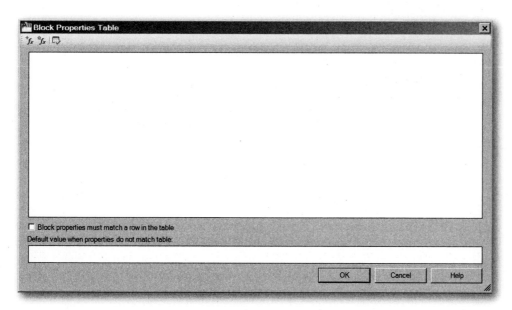

FIGURE 11–125 *Block Properties Table dialog box*

Choosing **Add Properties** (select the icon located on the top left side of the dialog box) displays the **Add Parameter Properties** dialog box where you can add parameters already defined in the Block Properties Table (Figure 11–126).

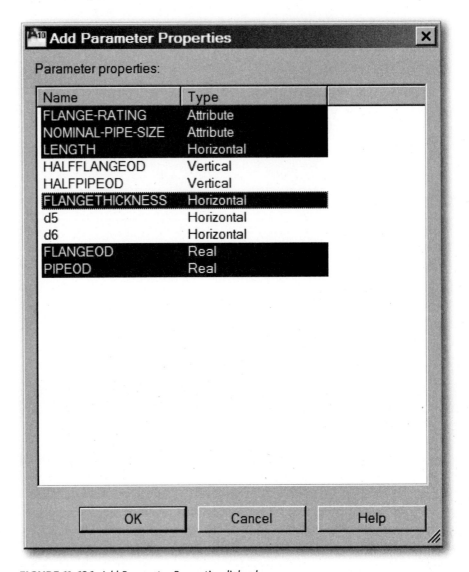

FIGURE 11–126 *Add Parameter Properties dialog box*

After selecting the 6 items whose values need to be determined from a spreadsheet, choose **OK** and AutoCAD displays the Block Properties Table (Figure 11–127) with these 6 items as column headings. Note that two of the items are Attributes, two are Horizontal Constraint parameters and two are Real (numbers) derived from User Parameters created in the Parameters Manager palette.

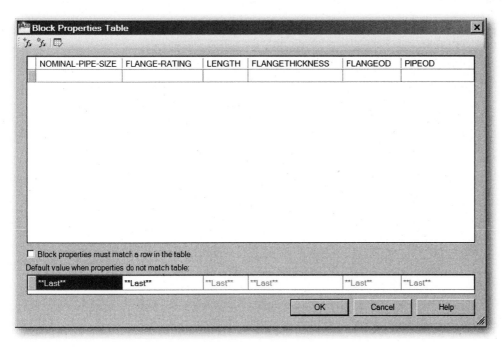

FIGURE 11–127 *Block Properties Table with column headings only*

Each row in the table will represent a pipe size/flange rating combination with corresponding dimensions for the flange length, flange thickness, flange outside diameter, and pipe outside diameter. The dimensions for these parameters will drive them when the pipe size and flange rating is selected after the block has been inserted and you select the grip for the Block Table. You can enter the values for each cell or you can cut and paste an array of rows and columns from a spreadsheet form. The result of entering the data for this example is shown in Figure 11–128.

NOMINAL-PIPE-SIZE	FLANGE-RATING	LENGTH	FLANGETHICKNESS	FLANGEOD	PIPEOD
3	150	2.7500	0.9400	7.5000	3.5000
3	300	3.1250	1.2500	8.2500	3.5000
3	600	3.2500	1.3750	8.2500	3.5000
3	900	4.0000	1.6250	9.5000	3.5000
4	150	3.0000	0.9400	9.0000	4.5000
4	300	3.3750	1.4400	10.0000	4.5000
4	600	4.0000	1.6250	10.7500	4.5000
4	900	4.5000	1.8800	11.5000	4.5000
6	150	3.5000	1.0000	11.0000	6.6250
6	300	3.8750	1.8100	12.5000	6.6250
6	600	4.6250	2.0000	14.0000	6.6250
6	900	5.5000	2.2500	15.0000	6.6250

FIGURE 11–128 *Block Properties Table with cells filled in*

Choosing **New Properties** (select the second icon located on the top left side of the Block Properties Table dialog box) displays the **New Parameters dialog box** (see Figure 11–129) where you can create and add new user parameters to the Block Properties Table.

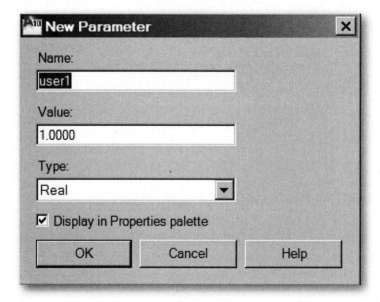

FIGURE 11–129 *New Parameter dialog box*

Choosing **Audit** (select the third icon located on the top left side of the Block Properties Table dialog box) checks the Block Properties Table for errors.

Choosing **Block Properties Must Match A Row In The Table** specifies whether the properties added to the grid control can be modified individually for a block reference.

The **Default Value When Entering Custom Values** box displays the default values when other properties are changed without matching a row.

You can press SHIFT+ENTER to add a new line for a multiline attribute value in the grid in the Block Properties Table dialog box.

Figure 11–130 shows how you can select the grip when an inserted block reference has been highlighted and then select the pipe size and flange rating from their respective shortcut menus. These menus are displayed when you select the grip. In the example below, the flange on the left was specified to be a 4"-150# flange. Then another block was inserted indicating that the pipe ends of the two weld necks were attached and the flange on the right has been specified to be a 4"-600# flange. This is a common occurrence when there needs to be a change in rating specifications in a pipeline.

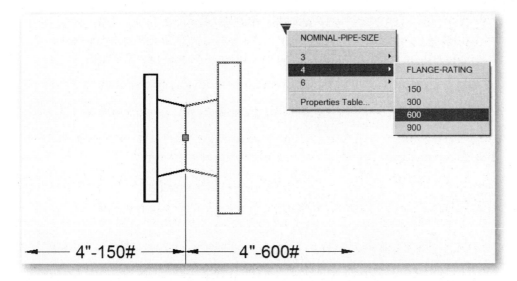

FIGURE 11–130 *Example of using the Block Table to set dimensions of a dynamic block*

Open the Exercise Manual PDF for Chapter 11 for discipline-specific exercises. Related files are downloaded from the student companion site mentioned in the Introduction (refer to page number xii for instructions).

REVIEW QUESTIONS

1. The maximum number of characters for a block name is:
   a. 8
   b. 16
   c. 23
   d. 31
   e. 255

2. A block is:
   a. A rectangular-shaped figure available for insertion into a drawing
   b. A single element found in a block formation of a building drawn with AutoCAD
   c. One or more objects stored as a single object for later retrieval and insertion
   d. None of the above

3. A block reference cannot be exploded if it:
   a. Consists of other blocks (nested)
   b. Has a negative scale factor
   c. Has been moved
   d. Has different X- and Y-scale factors
   e. None of the above

4. All of the following can be exploded except:

    a. Block references

    b. Associative dimensions

    c. Polylines

    d. Block references inserted with MINSERT command

    e. None of the above

5. To return a block reference back to its original objects, use:

    a. EXPLODE

    b. BREAK

    c. CHANGE

    d. UNDO

    e. STRETCH

6. To identify a new insertion point for a drawing file that will be inserted into another drawing, invoke the:

    a. BASE command

    b. INSERT command

    c. BLOCK command

    d. WBLOCK command

    e. DEFINE command

7. MINSERT places multiple copies of an existing block similar to the command:

    a. ARRAY

    b. MOVE

    c. INSERT

    d. COPY

    e. MIRROR

8. If one drawing is to be inserted into another drawing and editing operations are to be performed on the inserted drawing, you must first:

    a. Use the PEDIT command

    b. EXPLODE the inserted drawing

    c. UNDO the inserted drawing

    d. Nothing; it can be edited directly

    e. None of the above

9. The BASE command:

    a. Can be used to move a block reference

    b. Is a subcommand of PEDIT

    c. Will accept 3D coordinates

    d. Allows one to move a dimension baseline

    e. None of the above

10. Attributes are associated with:

    a. Objects

    b. Block references

    c. Text

    d. Layers

    e. Shapes

11. To merge two drawings, use:

    a. INSERT

    b. MERGE

    c. BIND

    d. BLOCK

    e. IGESIN

12. The DIVIDE command causes AutoCAD to:

    a. Divide an object into equal length segments

    b. Divide an object into two equal parts

    c. Break an object into two objects

    d. All of the above

13. One cannot explode:

    a. Polylines containing arcs

    b. Blocks containing polylines

    c. Dimensions incorporating leaders

    d. Block references inserted with different X-, Y-, and Z-scale factors

    e. None of the above

14. The DIVIDE command will:

    a. Place points along a line, arc, polyline, or circle

    b. Accept 1.5 as segment input

    c. Place markers on the selected object and separate it into different segments

    d. Divide any object into the equal number of segments

15. The MEASURE command causes AutoCAD to divide an object:

    a. Into specified length segments

    b. Into equal length segments

    c. Into two equal parts

    d. All of the above

16. A command used to edit attributes is:

    a. DDATTE

    b. EDIT

    c. EDITATT

    d. ATTFILE

    e. none of the above

17. Attributes are defined as the:
    a. Database information displayed as a result of entering the LIST command
    b. X and Y values that can be entered when inserting a block reference
    c. Coordinate information of each vertex found along a SPLINE object
    d. None of the above

18. It is possible to force invisible attributes to display on a drawing.
    a. True
    b. False

19. The first step in making a block dynamic is to add action.
    a. True
    b. False

20. The *Alignment, Visibility,* and *Base Point* parameters do not have actions associated with them.
    a. True
    b. False

21. The *Point* parameter can be associated with the:
    a. *Move* action
    b. *Stretch* action
    c. *Scale* action
    d. *Rotate* action
    e. a and b

22. The *Rotate* parameter associates only with the Rotate action.
    a. True
    b. False

23. A dynamic block cannot be associated with more than one action.
    a. True
    b. False

24. *Move* action to the dynamic block definition can be associated with:
    a. *Point* parameter
    b. *Linear* parameter
    c. *Polar* parameter
    d. *XY* parameter
    e. All of the above

25. Parameter sets are combinations of parameters and actions with a preset number of grips.
    a. True
    b. False

12

External References and Images

INTRODUCTION

One of the most powerful time-saving features of AutoCAD is that it allows one drawing, referred to as an external reference or xref, to become part of a second drawing while maintaining the integrity and independence of the first one. If the xref is changed, any changes will be reflected in the drawing in which it is referenced. This feature is provided by the XREF command.

OBJECTIVES

After completing this chapter, you will be able to do the following:

- Attach and detach reference files
- Change the path for reference files
- Load and unload reference files
- Decide whether to attach or overlay an xref file
- Clip xref files
- Control dependent symbols
- Edit xrefs
- Manage xrefs
- Use the BIND command to add dependent symbols to the current drawing
- Attach and detach DWF, DGN, and PDF underlays
- Attach and detach image files

EXTERNAL REFERENCES

When a drawing is externally referenced instead of being inserted as a block, the user can view and object snap to objects in the xref from the current drawing, but each drawing's data is still stored and maintained in a separate drawing file. The only information in the reference drawing that becomes a permanent part of the current drawing is the name of the reference drawing and its folder path. If necessary, externally referenced files can be scaled, moved, copied, mirrored, or rotated by using the AutoCAD modification commands. You can control the visibility, color, and linetype of the layers belonging to an external drawing file. These actions allow you to control which portions of the external drawing file are displayed and how. No matter how complex an xref drawing may be, it is treated as a single object by AutoCAD. If you invoke the MOVE command and select a line in the xref, for example, the entire xref moves, not just the line you have selected. You cannot explode the externally referenced drawing. A manipulation performed on an xref will not affect the original drawing file, because an xref is only an image, however scaled or rotated.

In addition to attaching an AutoCAD drawing as a reference file, you can also attach a raster image file.

You can also underlay 2D geometry objects stored in design files created with MicroStation. Underlays are similar to attached raster images in that they provide visual content, but they also provide some object snapping. Unlike xrefs, underlays cannot be bound into the drawing. DGN support is limited to V8 DGN files and 2D objects. The 2D objects will be imported or attached with full X, Y, Z coordinate information, as was present in the original file. If 3D solids, surfaces, or other 3D objects are encountered, a warning message will be displayed about the 3D content found in the file. If no 2D objects are encountered and only 3D objects are present in the file, a warning message will be displayed.

Similarly, you can also underlay DWF, DWFx, or PDF files to drawing files. Refer to Chapter 15 for details on DWF files. You can also change the position, scale, and orientation of DWF underlays.

Borders are excellent examples of drawing files that are useful as xref files. The objects that make up a border will use considerable space in a file, commonly around 20,000 bytes. If a border were drawn in each drawing file, this would waste a large amount of disk space—multiply 20,000 bytes by 100 drawing files. If xref files are used correctly, they would save 2 MB of disk space in this case.

Accuracy and efficient use of drawing time are other important design benefits that are enhanced through xref files. When an addition or change is made to a drawing file that is being used as an xref file, all drawings that use the file as an xref will reflect the modifications. For example, if you alter the title block of a border, all the drawing files that use that border as an xref file will automatically display the title block revisions. Can you imagine accessing 100 drawing files to change one small detail?

The number of xrefs that you can add to a drawing is limited to 32,000. In practice, this represents an unlimited number. If necessary, you can nest them so that loading one xref automatically causes another xref to be loaded. When you attach a drawing file as an xref file, it is permanently attached until it is detached or bound to the current drawing. When you load the drawing with xrefs, AutoCAD automatically reloads each xref drawing file; thus, each external drawing file reflects the latest state of the referenced drawing file.

Figure 12–1A through Figure 12–1F illustrate how xref files save time and ensure the drawing accuracy required to produce a set of drawings for a house. A logical sequence of design tells us that the foundation plan or roof plan cannot be drawn until the floor plan is completed. The layout and dimensions arrived in the design of the floor plan will be the basis of the other plans. The same border and title block can be used for all the final drawings.

Figure 12–1A is a separate drawing file containing the floor plan of the residence. Figure 12–1B is a separate drawing file containing a typical border/title block to be used for a set of drawings for the residence.

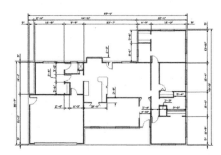

FIGURE 12–1A *Floor Plan*

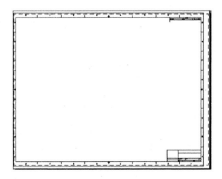

FIGURE 12–1B *Border and Title Block*

FIGURE 12–1C *Foundation Drawing*

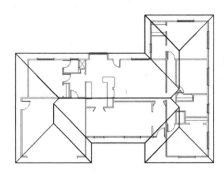

FIGURE 12–1D *Roof Drawing*

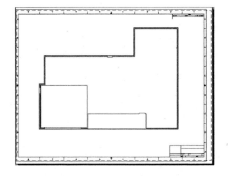

FIGURE 12–1E *Final Foundation Drawing*

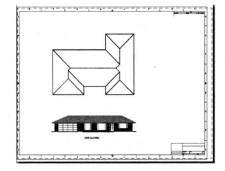

FIGURE 12–1F *Final Roof and Elevation Drawing*

Figure 12–1C and Figure 12–1D show the foundation and roof plans being drawn using the floor plan drawing as an xref to create the geometry based on the dimensions and details in the floor plan. When completed, these drawings are configured for plotting by turning OFF the layers of the floor plan xref. The xref is hidden but not removed from the drawing. If any changes are made to the original drawing of the floor plan, they will be reflected automatically in the drawing in which it is referenced. This ensures that corresponding changes will be made in the foundation and roof plans.

Figure 12–1E is the final foundation plan, and Figure 12–1F is the final roof plan and front elevation of the residence. Figure 12–1F also shows an xref of the drawing of the front elevation, which is included in the final plot. Figures 12–1E and 12–1F can now be completed by adding dimensions, filling in the title block, and adding notes and call-outs and labeling the views.

When combined with the networking capability of AutoCAD, the XREF command gives the project manager powerful tools for coping with the problems of file management. The project manager can instantaneously see the work of the departments and designers working on aspects of the contract. If necessary, you can overlay a drawing where appropriate, track the progress, and maintain document integrity. At the same time, departments need not lose control over individual designs and details.

If you need to make changes to an attached xref file while you are in the host drawing, you can do so by using the REFEDIT command. AutoCAD also allows you to open an attached xref in a separate window. With the XOPEN command, the xref opens immediately in a new window. You can make the necessary changes and save those changes. The changes will be reflected in the host drawing immediately.

You can also control the display of the xref file by means of clipping, so that you can display only a specific section of the reference file.

The INSERT command and the XREF command give users a choice of two methods for combining existing drawing files. The xref feature does not make the INSERT feature obsolete, as the user can decide which method is more appropriate for the current application.

EXTERNAL REFERENCES AND DEPENDENT SYMBOLS

The symbols that are carried into a drawing by an external reference (xref) drawing are called dependent symbols because they depend on the external file, not on the current drawing, for their characteristics. The symbols have arbitrary names and include blocks, layers, linetypes, text styles, and dimension styles.

When you attach an AutoCAD drawing as an external reference, AutoCAD automatically renames the xref's dependent symbols. AutoCAD forms a temporary name for each symbol by combining its original name with the name of the xref file name itself. The two names are separated by the vertical bar (|) character. Renaming the symbols prevents the xref's objects from taking on the characteristics of existing symbols in the drawing.

For example, say that you have created a drawing called PLAN1 with Layers 0, First-fl, Dim, and Text, in addition to blocks called Arrow and Monument. If you attach the PLAN1 drawing as an xref file, the layer First-fl will be renamed as PLAN1|First-fl, Dim as PLAN1|Dim, and Text as PLAN1|Text, as shown in Figure 12–2. Blocks Arrow and Monument will be renamed as PLAN1|Arrow and PLAN1|Monument, respectively. The only exceptions to renaming are unambiguous defaults such as Layer 0 and linetype continuous. The information on Layer 0 from the reference file will be placed on the active layer of the current drawing when the drawing is attached as an xref of the current drawing. It takes on the characteristics of the current drawing.

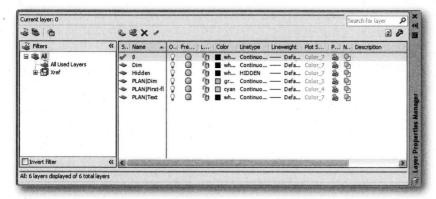

FIGURE 12-2 *Layer Properties Manager*

This prefixing is carried to nested xrefs. For example, if the external file PLAN1 included an xref named Title that has a layer Legend, it would get the symbol name PLAN1|Title|Legend, if PLAN1 were attached to another drawing.

This automatic renaming of an xref's dependent symbols has two benefits:

- It allows you to see at a glance which named objects belong to which xref file.
- It allows dependent symbols to have the same base name in both the current drawing and an xref and to coexist without conflict.

AutoCAD's commands and dialog boxes for manipulating named objects do not let you select an xref's dependent symbols. Usually, dialog boxes display these entries in lighter text.

For example, you cannot insert a block that belongs to an xref drawing in your current drawing, nor can you make a dependent layer the current layer and begin creating new objects.

You can control the visibility of the layers of an xref drawing using the ON/OFF and freeze/thaw commands. If necessary, you can change the color and linetype. When the VISRETAIN system variable is set to 0, any changes that you make to these settings apply only to the current drawing session. They are discarded when you end the drawing. If VISRETAIN is set to 1, the default, the current drawing visibility, color, and linetype for xref dependent layers take precedence. They are saved with the drawing and are preserved during xref reload operations.

There may be times when you want to make your xref data a permanent part of your current drawing. To accomplish this, use the *Bind* option of the XREF command. With the *Bind* option, all layers and other symbols, including the data, become part of the current drawing. This action is similar to inserting a drawing via the INSERT command.

If necessary, you can make dependent symbols such as layers, linetypes, text styles, and dim styles part of the current drawing by using the XBIND command instead of binding the whole drawing. This action allows you to work with the symbol just as if you had defined it in the current drawing.

ATTACHING AND MANIPULATING XREFS WITH THE EXTERNAL REFERENCES PALETTE

The External References palette lists the xrefs attached to the current or host drawing. You can also attach, detach, overlay, bind, reload, unload, rename, and modify paths to xrefs.

Choose down arrow from the References panel (Figure 12–3) located in the Insert tab, and AutoCAD displays the External Reference palette (Figure 12–4).

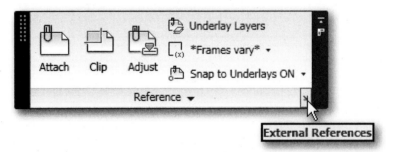

FIGURE 12–3 *Invoking the XREF command by selecting arrow from the References panel*

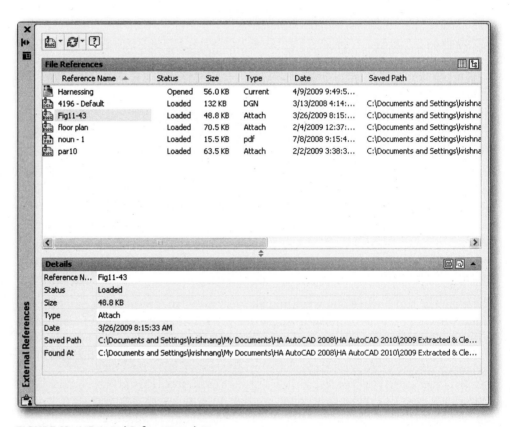

FIGURE 12–4 *External References palette*

The External References palette contains **a set of tool buttons**, two dual-mode data panes, and a messaging field. The upper data pane, **File References**, can be set to display file references in a list mode or tree mode. Shortcut menus and function keys provide options for working with the files. The lower data pane, **Details**, can display properties for the selected file references or it can display a thumbnail preview of the selected file reference. The messaging field at the bottom presents information about selected file references that is pertinent under current conditions.

The reference icons that precede the reference name indicate the type of reference as follows:

Icon	Description
	Indicates the current drawing icon. It represents the master drawing to which all xrefs are attached.
	Indicates a DWF underlay attachment.
	Indicates a DWG (xref) attachment.
	Indicates a DWG (xref) overlay.
	Indicates a raster image attachment.
	Indicates a DGN underlay attachment.
	Indicates a PDF underlay attachment.

When displaying the Details mode, properties for the selected file reference are reported. Each file reference has a core set of properties and some file references such as referenced images and display properties specific to the file type. The core set of details include the reference name, status, file size, file type, creation date, saved path, found at path, and file version if the Vault client is installed. Some of the properties can be edited.

The **Reference Name** column displays the file reference name. This property can be edited only if single file references are selected. The reference name shows *Varies* if multiple file references are selected. This property is editable for all the file references.

The **Status** column shows whether the file reference is loaded, unloaded, or not found. This property cannot be edited. A detailed discussion of these states is provided later in this chapter.

The **Size** column shows the file size of the selected file reference. The size is not displayed for file references that are unloaded or not found. This property cannot be edited.

The **Type** column indicates whether the file reference is an attachment or an overlay, the type of image file, or the DWF underlay. This property is editable for all the file references.

The **Date** column displays the last date the file reference was modified. This date is not displayed if the file reference is unloaded or not found. This property cannot be edited.

The **Saved Path** column shows the saved path of the selected file reference. This is not necessarily where the file reference is found. This property cannot be edited.

The **Found At** row in the **Details** pane displays the full path of the currently selected file reference. This is where the referenced file is actually found and is not necessarily the same as the saved path. Clicking the [...] button displays the Select Image File dialog box where you can select a different path or file name. You can also type directly into the path field. These changes are stored to the Saved Path property if the new path is valid.

The **File Version** property is defined by the Vault client. This property is displayed only when you are logged into the Vault.

You can also display the information as a Tree View. To do so, choose **Tree View** at the top right of the palette or press F4. To switch back to the List View, choose **List View** or press F3.

In a Tree View listing, AutoCAD displays a hierarchical representation of the xrefs in alphabetical order (Figure 12–5). Tree View shows the level of nesting relationship of the attached xrefs, whether they are attached or overlaid, whether they are loaded, unloaded, marked for reload or unload or not found, unresolved, or unreferenced.

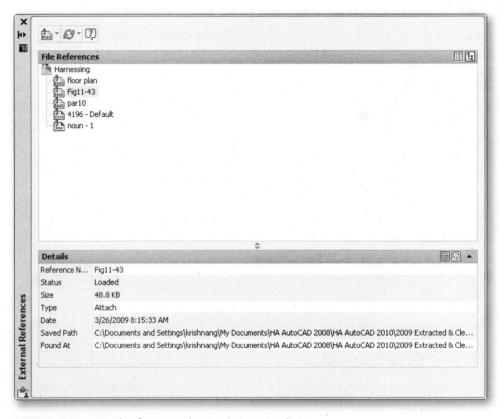

FIGURE 12–5 *External References palette with Tree View listing*

Attaching an AutoCAD Drawing as an Xref

To attach an AutoCAD drawing as an xref, choose **Attach DWG...** (Figure 12–6) in the External References palette, and AutoCAD displays the Select Reference File dialog box. You can also attach an AutoCAD drawing by invoking the ATTACH command from the Reference panel located in the Insert tab.

Select the drawing file from the appropriate folder, and choose **Open** to attach to the current drawing. AutoCAD displays the External Reference dialog box (Figure 12–7).

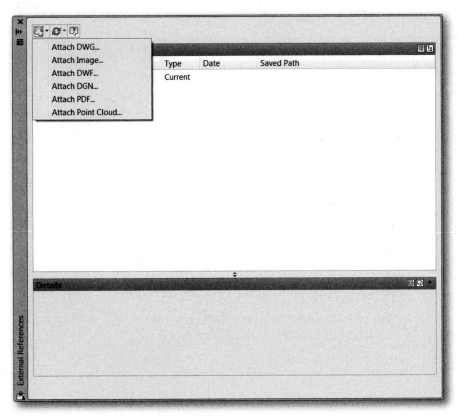

FIGURE 12–6 *Invoking Attach DWG... from the External References palette*

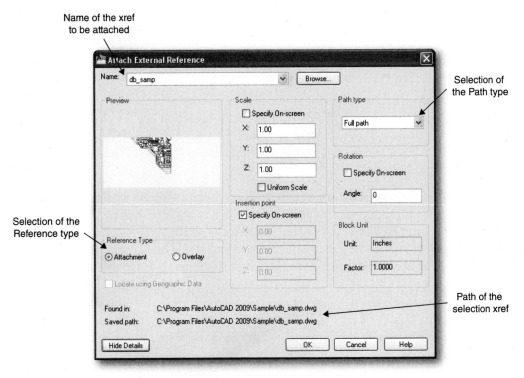

FIGURE 12–7 *External Reference dialog box*

The selected drawing name is added to the list box located next to the **Browse** button. Choose **Browse** to display the Select a Reference dialog box, in which you can select a different drawing file to attach to the current drawing. To add another instance of a drawing file that is already attached, select the drawing name from the list and choose **OK**.

The **Path type** list box specifies whether the saved path to the xref should be set to *No path*, *Full path*, or *Relative path*. When you specify *No path*, AutoCAD first looks for the xref file in the folder of the host drawing. This option is useful when the xref files are located in the same folder as the host drawing. If you specify *Full path*, AutoCAD saves the xref's precise location to the host drawing. This option is the most precise but the least flexible. If you move a project folder, AutoCAD cannot resolve any xrefs that are attached with a Full path. If you specify *Relative path*, AutoCAD saves the xrefs location relative to the host drawing. If you move a project folder, AutoCAD can resolve xrefs attached with a relative path as long as its location relative to the host drawing has not changed. You must save the current drawing before you can set the path type to Relative.

If *No Path* was saved for the xref, or if the xref is no longer located at the specified path, AutoCAD searches for the xref in the following order:

1. Current folder of the host drawing
2. Project search paths defined on the Files tab in the Options dialog box and in the PROJECTNAME system variable
3. Support search paths defined on the Files tab in the Options dialog box
4. Start-in folder specified in the Windows® application shortcut

In the Reference Type section of the dialog box, select one of two available options: Attachment or Overlay.

Selecting **Attachment** causes the xref to be included in the drawing when the drawing itself is attached as an xref to another drawing.

Selecting **Overlay** causes the xref to not be included in a drawing when the drawing itself is attached as an xref or overlaid xref to another drawing.

For example, PLAN-A drawing is attached as an overlaid xref to PLAN-B. PLAN-B is attached as an xref to PLAN-C. PLAN-A is not seen in PLAN-C because it is overlaid in PLAN-B, not attached as an xref. However, if PLAN-A is attached as an xref in PLAN-B, and in turn PLAN-B is attached in PLAN-C, both PLAN-A and PLAN-B will be seen in PLAN-C.

The only difference between overlays and attachments is how nested references are handled. Overlaid xrefs are designed for data sharing. If necessary, you can change the status from Attachment to Overlay, or vice versa, by double-clicking the field in the **Type** column in the External References palette or by selecting the appropriate radio button in the Reference Type section of the External Reference dialog box.

In the Insertion point section of the External Reference dialog box, you can specify the insertion point of the xref.

In the Scale section of the External Reference dialog box, you can specify the X, Y, and Z scale factor of the xref. Selecting **Uniform Scale** causes the Y and Z scale factors to be the same as the specified X scale factor.

In the Rotation section of the External Reference dialog box, you can specify the rotation angle of the xref.

The Insertion point, Scale, and Rotation features are similar to those in the insertion of a block, as explained in Chapter 11.

The Block Unit section displays the units such as inches or millimeters that were used when the selected reference drawing was created and the unit scale factor, calculated based on the reference drawing unit's value and the current drawing units.

The Found in section displays the path where the drawing to be attached was found.

The Saved path section displays the saved path, if any, that is used to locate the xref. This path can be a full or absolute path, a relative or partially specified path, or no path.

Choosing the **Hide Details** button hides the information displayed in the Found in section.

Choose **OK** to attach the selected xref drawing to the current drawing.

When you select the AutoCAD drawing that is attached as a reference file, a relevant contextual tab is automatically displayed in the ribbon as shown in Figure 12–8, providing easy access to reference edit tools.

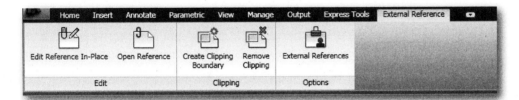

FIGURE 12–8 *External Reference contextual tab*

> Once a drawing is attached as a reference file, the Manage Xrefs icon is displayed in the status bar that allows you to open the Reference Manager.

Attaching a Raster Image as an Xref

To attach a raster image as an xref, choose **Attach Image...** in the External References palette, and AutoCAD displays the Select Image File dialog box. You can also attach a raster image by invoking the ATTACH command from the Reference panel located in the Insert tab. The image formats that can be inserted into AutoCAD include BMP, TIFF, RLE, DIB, JPG, PCX, FLIC, GEOSPOT, GIF, IG4, IGS, RLC, PCT, PCX, CALS1, PNG, and TGA. More than one image can be displayed in any viewport, and the number and size of images is not limited.

Select the image file from the appropriate directory and choose **Open**. AutoCAD displays the Image dialog box (Figure 12–9).

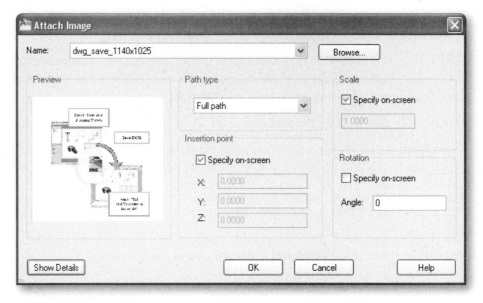

FIGURE 12–9 *Image dialog box*

The selected image name is added to the list box located next to the **Browse** button. Choose **Browse** to display the Select Image dialog box, from which you can select a different image file to attach to the current drawing. To add another instance of an image file that is already attached, select the image name from the list, and choose **OK**.

The **Path type** list box specifies whether the saved path to the xref should be set to *No path*, *Full path*, or *Relative path*.

The Insertion point section lets you specify the insertion point of the xref.

The Scale section lets you specify X, Y, and Z scale factor of the xref.

The Rotation section lets you specify the rotation angle of the xref.

The Insertion point, Scale, and Rotation features are similar to those in the insertion of a block, as explained in Chapter 11.

Choose **Details** to display the Image Information and saved path for the selected image file. The image formation includes image resolution in horizontal and vertical units, image size by width and height in pixels, and image size by width and height in the current selected units.

Choose **OK** to attach the selected xref image to the current drawing.

When you select the image that is attached as a reference file, a relevant contextual tab is automatically displayed in the ribbon as shown in Figure 12–10, providing easy access to reference edit tools.

FIGURE 12–10 *Image contextual tab*

Attaching a DWF as an Underlay Attachment

To attach a DWF file as an xref, choose **Attach DWF...** in the External References palette, and AutoCAD displays the Select DWF dialog box. You can also attach a DWF file by invoking the `ATTACH` command from the Reference panel located in the Insert tab. Select the DWF file from the appropriate directory, and choose **Open**. AutoCAD displays the Attach DWF Underlay dialog box (Figure 12–11).

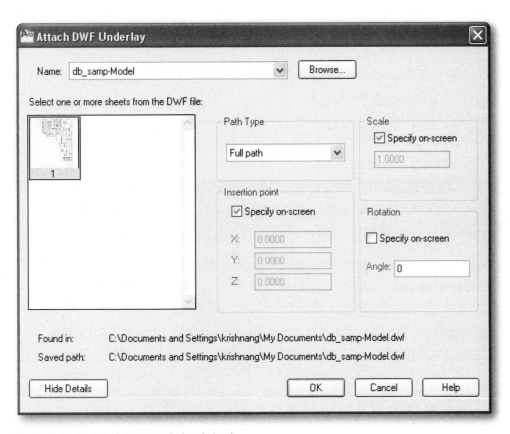

FIGURE 12–11 *Attach DWF Underlay dialog box*

The selected DWF file is added to the list box located next to the **Browse** button. Choose **Browse** to display the Select DWF File box, from which you can select a different DWF file to attach to the current drawing. To add another instance of a DWF file that is already attached, select the name from the list, and choose **OK**.

The **Path type** list box specifies whether the saved path to the xref should be set to *No path*, *Full path*, or *Relative path*.

AutoCAD displays all of the available sheets that are found in the selected DWF file in the Select a sheet from the DWF file section. Select the sheet from the list. If the DWF file only contains a single sheet, that sheet is listed. If the DWF file contains multiple sheets, only a single sheet can be selected for attachment. The first sheet in the list is selected by default.

The Insertion point section lets you specify the insertion point of the xref.

The Scale section lets you specify X, Y, and Z scale factor of the xref.

The Rotation section lets you specify the rotation angle of the xref.

The Insertion point, Scale, and Rotation features are similar to those in the insertion of a block, as explained in Chapter 11.

The Found in section displays the path where the drawing to be attached was found.

The Saved path section displays the saved path, if any, that is used to locate the xref. This path can be a full or absolute path, a relative or partially specified path, or no path.

Choosing the **Hide Details** button hides the information displayed in the Found in section.

Choose **OK** to attach the selected xref DWF file to the current drawing.

When you select the DWF that is attached as a reference file, a relevant contextual tab is automatically displayed in the ribbon as shown in Figure 12–12, providing easy access to reference edit tools.

FIGURE 12–12 *DWF Underlay contextual tab*

Attaching a MicroStation Design File as an Underlay Attachment

To attach a MicroStation Design file as an xref, choose **Attach DGN...** in the External References palette. You can also attach a MicroStation Design file by invoking the ATTACH command from the Reference panel located in the Insert tab. AutoCAD displays the Select DGN File dialog box. Select the DGN file from the appropriate directory, and choose **Open**. AutoCAD displays the Attach DGN underlay dialog box (Figure 12–13).

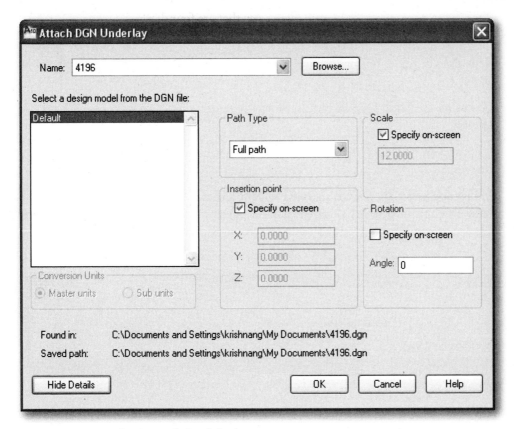

FIGURE 12–13 *Attach DGN Underlay dialog box*

The selected DGN file is added to the list box located next to the **Browse** button. Choose **Browse** to display the Select DGN File box, from which you can select a different DGN file to attach to the current drawing. To add another instance of a DGN file that is already attached, select the name from the list, and choose **OK**.

The **Path type** list box specifies whether the saved path to the xref should be set to *No path*, *Full path*, or *Relative path*.

AutoCAD displays all of the available design models that are found in the selected DGN file in the Select a design model from the DGN file section. Select the design model from the list. If the DGN file contains multiple models, only a single model can be selected for attachment. The first model in the list is selected by default.

Select the appropriate conversion units for the DGN underlay from the Conversion Units section. The DGN file contains working units, imperial or metric, called master units and subunits. The selected working units, master units or subunits, are converted one-for-one into DWG units.

The Insertion point section lets you specify the insertion point of the xref.

The Scale section lets you specify X, Y, and Z scale factor of the xref.

The Rotation section lets you specify the rotation angle of the xref.

The Insertion point, Scale, and Rotation features are similar to those in the insertion of a block, as explained in Chapter 11.

The Found in section displays the path where the drawing to be attached was found.

The Saved path section displays the saved path, if any, that is used to locate the xref. This path can be a full or absolute path, a relative or partially specified path, or no path.

Choosing the **Hide Details** button hides the information displayed in the Found in section.

Choose **OK** to attach the selected xref MicroStation Design file to the current drawing.

When you select the MicroStation DGN file that is attached as a reference file, a relevant contextual tab is automatically displayed in the ribbon as shown in Figure 12–14, providing easy access to reference edit tools.

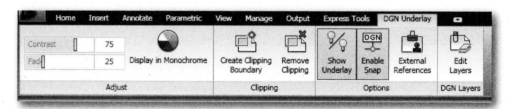

FIGURE 12–14 *DWF Underlay contextual tab*

Attaching a PDF File as an Underlay Attachment

To attach a PDF file as an xref, choose **Attach PDF...** in the External References palette. You can also attach a PDF file by invoking the ATTACH command from the Reference panel located in the Insert tab. AutoCAD displays the Select PDF File dialog box. Select the PDF file from the appropriate directory and choose **Open**. AutoCAD displays the Attach PDF underlay dialog box (Figure 12–15).

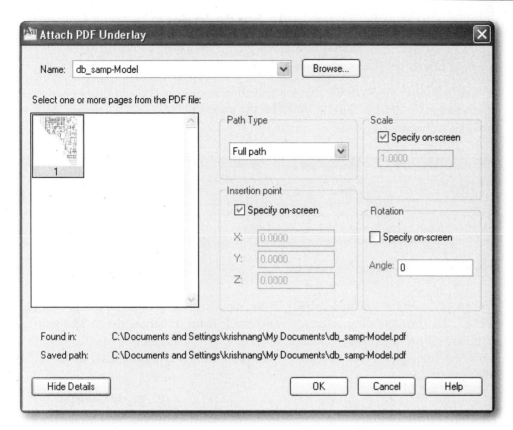

FIGURE 12–15 *Attach PDF Underlay dialog box*

The selected PDF file is added to the list box located next to the **Browse** button. Choose **Browse** to display the Select PDF File box, from which you can select a different PDF file to attach to the current drawing. To add another instance of a PDF file that is already attached, select the name from the list and choose **OK**.

The **Path type** list box specifies whether the saved path to the xref should be set to *No path, Full path,* or *Relative path*.

AutoCAD displays all of the available pages that are found in the PDF file in the Select one or more pages from the PDF file section. If the PDF file only contains a single page, that page is listed. You can select multiple pages by holding the Shift or CTRL key and selecting the pages to attach. PDF files with more than one page are attached one **page** at a time. Also, hypertext links from PDF files are converted to straight text and digital signatures are not supported.

The Insertion point section lets you specify the insertion point of the xref.

The Scale section lets you specify X, Y, and Z scale factors of the xref.

The Rotation section lets you specify the rotation angle of the xref.

The Insertion point, Scale, and Rotation features are similar to those in the insertion of a block, as explained in Chapter 11.

The Found in section displays the path where the drawing to be attached was found.

The Saved path section displays the saved path, if any, that is used to locate the xref. This path can be a full or absolute path, a relative or partially specified path, or no path.

Choosing the **Hide Details** button hides the information displayed in the Found in section.

Choose **OK** to attach the selected xref PDF file to the current drawing.

When you select the PDF file that is attached as a reference file, a relevant contextual tab is automatically displayed in the ribbon as shown in Figure 12–16, providing easy access to reference edit tools.

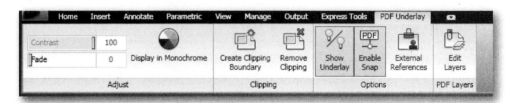

FIGURE 12–16 *PDF Underlay contextual tab*

When you select a reference object in the drawing (xrefs, images, DWF, DGN, PDF), the corresponding reference is selected in the External References palette. Similarly, if you select file references from the External References palette, the references are highlighted (but not selected) in the drawing, assuming they are visible in the current view.

If you pass the cursor over the edge of the attached image, DWF, or PDF file, a selection preview frame is displayed, even when frames are turned off. Similarly, passing the cursor over the geometry inside an attached DWF or PDF, DGN, or DWG file displays a selection preview frame. You can select the frame for any of the attached reference files to display a frame boundary with the ability to grip-edit a clipped boundary frame. With the selection of the reference file in the Reference Manager, AutoCAD automatically displays a frame boundary to the selected reference file.

Detaching Xrefs

In the External References palette, select the Reference Name that you wish to detach, right-click, and choose *Detach* from the shortcut menu. This action allows you to detach an xref file from the current drawing. Only the xref file that is attached or overlaid directly to the current drawing can be detached. You cannot detach an xref drawing referenced by another xref drawing. If the xref is currently being displayed as part of the current drawing, it disappears when you detach it.

Reloading Xrefs

In the External References palette, select the Reference Name that you wish to reload, right-click, and choose *Reload* from the shortcut menu. This action allows you to update the selected xref attached to the current drawing. When you open a drawing, it automatically reloads any xrefs attached. The *Reload* option will reread the external file in case it has been changed during the current AutoCAD session.

Unloading Xrefs

In the External References palette, select the Reference Name that you wish to unload, right-click, and choose *Unload* from the shortcut menu. This action allows you to unload the selected xref from the current drawing. Unlike the *Detach* option, the *Unload* option merely suppresses the display and regeneration of the xref definition to help current session editing and improve performance. This option can also be useful when a series of xrefs needs to be viewed during a project on an as-needed basis. Rather than have the referenced files displayed at all times, you can reload the drawing when you require the information.

The results of *Unload* and *Reload* take effect when you close the dialog box.

Binding Xrefs

In the External References palette, select the Reference Drawing Name that you wish to bind, right-click, and choose *Bind* from the shortcut menu. This action allows you to make your xref drawing data a permanent part of the current drawing. The *Bind* option is available only for the AutoCAD drawings that are attached as an xref. Auto-CAD displays the Bind Xrefs dialog box (Figure 12–17). Select one of two available bind types: **Bind** or **Insert**.

FIGURE 12–17 *Bind Xrefs dialog box*

Choose **Bind**, and the xref drawing becomes an ordinary block in your current drawing. It also adds the dependent symbols to your drawing, letting you use them as you would any other named objects. In the process of binding, AutoCAD renames the dependent symbols. The vertical bar symbol (|) is replaced with three new characters: a $, a number, and another $. The number is assigned by AutoCAD to ensure that the named object will have a unique name.

For example, if you bind an xref drawing named PLAN1 that has a dependent layer PLAN1|FIRST-FL, AutoCAD will try to rename the layer to PLAN1$0$FIRST-FL. If a layer by that name already exists in the current drawing, AutoCAD tries to rename the layer to PLAN1$1$FIRST-FL, incrementing the number until there is no duplicate.

If you do not want to bind the entire xref drawing but only specific dependent symbols, such as a layer, linetype, block, dimension style, or text style, you can use the XBIND command, explained later in this chapter in the section on "Adding Dependent Symbols to the Current Drawing."

Choose **Insert**, and the xref drawing is inserted in the current drawing similar to inserting a drawing with the INSERT command. AutoCAD adds the dependent symbols to the current drawing by stripping off the xref drawing name.

For example, if you insert an xref drawing named PLAN1 that has a dependent layer PLAN1|FIRST-FL, AutoCAD will rename the layer to FIRST-FL. If a layer by that name already exists in the current drawing, the layer FIRST-FL would assume the properties of the layer in the current drawing.

Opening the Xref

If you need to make changes to an attached xref drawing file while you are in the host drawing, choose *Open* from the shortcut menu, and AutoCAD opens the selected xref for editing in a new window. You can also open an xref for editing by invoking the XOPEN command. This option is available only for the AutoCAD drawings that are attached as an xref.

ADDING DEPENDENT SYMBOLS TO THE CURRENT DRAWING

The XBIND command lets you permanently add a selected subset of xref-dependent symbols to your current drawing. The dependent symbols include the block, layer, linetype, dimension style, and text style. Once the dependent symbol is added to the current drawing, it operates as if it was created in the current drawing and is saved with the drawing when you close the drawing session. While adding the dependent symbol to the current drawing, AutoCAD removes the vertical bar symbol (|) from each dependent symbol's name, replacing it with three new characters: a $, a number, and another $ symbol.

For example, you might want to use a block that is defined in an xref. Instead of binding the entire xref with the BIND option of the XREF command, it is advisable to use the XBIND command. With the XBIND command, the block and the layers associated with the block will be added to the current drawing. If the block's definition contains a reference to an xref, AutoCAD binds that xref and all its dependent symbols. After binding the necessary dependent symbols, you can detach the xref file.

Invoke the XBIND command by typing **Bind** from the on-screen prompt, and Auto-CAD displays the Xbind dialog box (Figure 12–18).

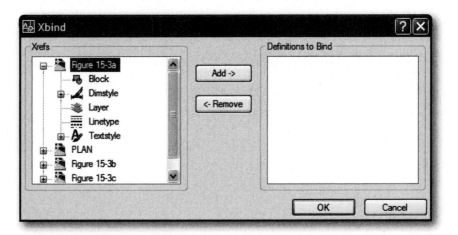

FIGURE 12–18 *Xbind dialog box*

On the left side of the Xbind dialog box, AutoCAD lists the xref files currently attached to the current drawing. Double-click on the name of the xref file, and Auto-CAD expands the list to show the dependent symbols. Select the dependent symbol from the list, and choose **Add**. AutoCAD moves the selected dependent symbol into the **Definitions to Bind** list. If necessary, return to the **Xrefs** list from the **Definitions to Bind** list by choosing **Remove** after selecting the appropriate dependent symbol.

Choose **OK** to bind the selected definitions to the current drawing.

CONTROLLING THE DISPLAY OF EXTERNAL REFERENCES

The CLIP command allows you to control the display of unwanted information by clipping the selected external reference, image, viewport, or underlay (DWF, DWFx, PDF, or DGN) to a specified boundary. Clipping does not edit or change the xref, it just prevents part of the object from being displayed. The defined clipping boundary can be visible or hidden. You can also define the front and back clipping planes.

The clipping boundary is created coincident with the polyline. Valid boundaries are 2D polylines with straight or spline-curved segments. Polylines with arc segments, or fit-curved polylines, can be used as the definition of the clip boundary, but the clip boundary will be created as a straight segment representation of that polyline. If the polyline has arcs, the clip boundary is created as if it had been decurved prior to being used as a clip boundary. An open polyline is treated as if it were closed.

If you set the clip boundary to *Off*, the entire xref or block is displayed. If you subsequently set the clip boundary to *On*, the clipped drawing is displayed again. If necessary, you can delete the clipping boundary; AutoCAD redisplays the entire xref. In addition, AutoCAD allows you to generate a polyline from the clipping boundary.

Invoke the CLIP command by choosing Clip from the Reference panel (Figure 12–19) of the Insert tab, and AutoCAD displays the following prompts:

    Select object to clip: *(select the xref to be included in the clipping)*

    Enter clipping option [ON/OFF/Clipdepth/Delete/generate Polyline/New boundary] *(choose one of the options from the shortcut menu)*

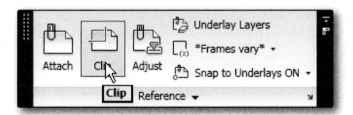

FIGURE 12–19 *Invoking the XREF CLIP command from the Reference panel*

Choose *New Boundary* to define a rectangular or polygonal clip boundary or generate a polygonal clipping boundary from a polyline. AutoCAD displays the following prompts:

    Specify clipping boundary:

    [Select polyline/Polygonal/Rectangular/Invert clip]: *(choose one of four options from the shortcut menu)*

Choosing *Rectangular* (default) allows you to define a rectangular boundary by specifying the opposite corners of a window. The clipping boundary is applied in the current UCS and is independent of the current view.

Choosing select *Polyline* defines the boundary by using a selected polyline. The polyline can be open or closed, and it can be made of straight-line segments, but it cannot intersect itself.

Choosing *Polygonal* defines a polygonal boundary by specifying points for the vertices of a polygon.

Choosing *Invert Clip* inverts the mode of the clipping boundary: either the objects outside the boundary, the default, or inside the boundary are hidden.

Once the clipping boundary is defined, AutoCAD displays only the portion of the drawing that is within the clipping boundary and then exits the command.

If you already have a clipping boundary of the selected xref drawing, and you choose the *New Boundary* option, AutoCAD displays the following prompt:

```
Delete old boundary(s)? [Yes/No]: (select one of two
   options)
```

If you choose Yes, the clipping boundary is deleted, the entire reference file is displayed and the command continues. If you choose *No*, the command sequence is terminated.

> **NOTE**
>
> The display of the boundary border is controlled by the XCLIPFRAME system variable. If it is set to 1 (ON), AutoCAD displays the boundary border. If it is set to 0 (OFF), AutoCAD does not display the boundary border.

The *On/Off* option controls the display of the clipped boundary. Choose *Off* to display all of the geometry of the xref or block, ignoring the clipping boundary. Choose *On* to display the clipped portion of the xref or block only.

Choosing *Clipdepth* sets the front and back clipping planes on an xref or block. Objects outside the volume defined by the boundary and the specified depth are not displayed.

Choosing *Delete* removes the clipping boundary for the selected xref or block. To temporarily turn the clipping boundary *Off*, use the *Off* option as explained earlier. The *Delete* option erases the clipping boundary and the clipdepth and displays the entire reference file.

> **NOTE**
>
> The ERASE command cannot be used to delete clipping boundaries.

AutoCAD draws a polyline coincident with the clipping boundary. The polyline assumes the current layer, linetype, and color settings. When you delete the clipping boundary, AutoCAD deletes the polyline. If you need to keep a copy of the polyline, choose the *Generate Polyline* option. AutoCAD makes a copy of the clipping boundary. You can use the PEDIT command to modify the generated polyline and then redefine the clipping boundary with the new polyline. To see the entire xref while redefining the boundary, choose OFF to turn the clipping boundary OFF.

If you select image, viewport, or underlay (DWF, DWFx, PDF, or DGN) to clip, *Clipdepth* and *Generate Polyline* options are not available.

EDITING REFERENCE FILES/XREF EDIT CONTROL

You can edit block references and xrefs while working in a drawing session by means of the REFEDIT command. This is referred to as in-place reference editing. If you select a reference for editing and it has attached xrefs or block definitions, the nested references and the reference are displayed and available for selection in the Reference Edit dialog box.

NOTE

> You can edit only one reference at a time. Block references inserted with the MINSERT command cannot be edited. You cannot edit a reference file if it is in use by someone else.

You can also display the attribute definitions for editing if the block reference contains attributes. The attributes become visible and their definitions can be edited along with the reference geometry. Attributes of the original reference remain unchanged when the changes are saved back to the block reference. Only subsequent insertions of the block will be affected by the changes.

As mentioned earlier, you can also edit an xref by opening it in a separate window with the XOPEN command.

Select the reference file for in-place editing and from the External Reference contextual tab choose Edit Reference In-Place button.

AutoCAD displays the Reference Edit dialog box (Figure 12–20).

FIGURE 12–20 *Reference Edit dialog box (Identify Reference tab)*

The Identify Reference tab provides visual aids for identifying the reference to edit and controls how the reference is selected. If you select an object that is part of one or more nested references, the nested references are displayed in the dialog box. Objects selected that belong to any nested references cause all the references to become candidates for editing. Select the specific reference you want to edit by choosing the name of the reference in the **Reference name** list box of the Reference Edit dialog box. This action will lock the reference file to prevent other users from opening the file. Only one reference can be edited in place at a time. The path of the selected reference is displayed at the bottom of the dialog box. If the selected reference is a block, no path is displayed.

The Preview section of the dialog box displays a preview image of the currently selected reference. The preview image displays the reference as it was last saved in the drawing. The reference preview image is not updated when changes are saved back to the reference.

Choosing **Automatically select all nested objects** controls whether or not nested objects are included automatically in the reference editing session. If this option is chosen, all the objects in the selected reference will be automatically included in the reference editing session, becoming part of the working set.

Choosing **Prompt to select nested objects** means that nested objects must be selected individually in the reference editing session. If this option is chosen, after you close the Reference Edit dialog box and enter the reference edit state, AutoCAD prompts you to select the specific objects in the reference that you want to edit.

The Settings tab (Figure 12–21) provides options for editing references.

Selecting **Create unique layer, style, and block names** controls whether or not layers and other named objects extracted from the reference are uniquely altered. If this option is set to ON, named objects in xrefs are altered. Names are prefixed with 0, similar to the way they are altered when you bind xrefs. If it is set to OFF, the names of layers and other named objects remain the same as in the reference drawing. Named objects that are not altered to make them unique assume the properties of those in the current host drawing that share the same name.

FIGURE 12–21 *Reference Edit dialog box (Settings tab)*

Selecting **Display attribute definitions for editing** controls whether or not all variable attribute definitions in block references are extracted and displayed during reference editing. If this option is set to ON, the attributes, except constant attributes are made invisible, and the attribute definitions are available for editing along with the selected reference geometry. When changes are saved back to the block reference, the attributes of the original reference remain unchanged. The new or altered attribute definitions affect only subsequent insertions of the block; the attributes in existing block instances are not affected. Xrefs and block references without definitions are not affected by this option.

Selecting **Lock objects not in working set** locks all objects not in the working set. If this option is set to ON, it will prevent you from accidentally selecting and editing objects in the host drawing while in a reference editing state. The action of locked objects is similar to objects on a locked layer. If you try to edit locked objects, they are filtered from the selection set.

Choose **OK** to close the Reference Edit dialog box. AutoCAD prompts to select objects if the Prompt to select nested objects option is selected.

The objects you choose are temporarily extracted for modification in the current drawing and become the working set. The working set objects are highlighted so that they can be distinguished from other objects. All objects not selected appear faded. You can now perform modifications on the working set objects.

Make sure that **Reference Edit fading intensity** is set to an appropriate setting in the Display tab of the Options dialog box.

NOTE

Adding/Removing Objects from the Working Set

If a new object is created while editing a reference, it is usually added to the working set automatically. However, if making changes to objects outside the working set causes a new object to be created, it will not be added to the working set.

Objects removed from the working set are added to the host drawing and are removed from the reference when the changes are saved back. Objects created or removed are automatically added to or deleted from the working set. You can tell whether or not an object is in the working set by the way it is displayed on the screen; a faded object is not in the working set. When a reference is being edited, the Edit Reference panel is displayed (Figure 12–22) in the External Reference contextual tab.

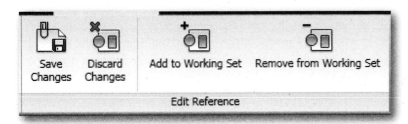

FIGURE 12–22 *The Edit Reference panel*

To add objects to the working set, select **Add to Working set** from the Edit Reference panel and select objects to be added. You can select items only when the type of space, Model or Paper, is in effect that was in effect when the REFEDIT command was initiated. To remove objects from the working set, select **Remove from Working set** from the Edit Reference panel, and select the objects to be removed. As mentioned earlier, if you remove objects from the working set and save the changes, the objects are removed from the reference and added to the current drawing. Any changes that you make to objects in the current drawing, but not in the xref or block, are not discarded. Once the modifications are complete, select **Save Changes** to save the changes to the reference file. To discard the changes, select **Discard Changes** from the Edit Reference panel.

MANAGING EXTERNAL REFERENCES

Several tools are available to help in the management and tracking of xrefs.

One tracking mechanism is an external ASCII log file that is maintained on each drawing that contains xrefs. This file, which AutoCAD generates and maintains automatically, has the same name as the current drawing with a file extension *.xlg*. You can examine the file with any text editor and/or print it. The log file registers each attach, bind, detach, and reload of each xref for the current drawing. AutoCAD writes a title block to the log file that contains the name of the current drawing, the date and time, and the operation being performed. Once a log file has been created for a drawing, AutoCAD continues to append it. The log file is always placed in the same folder as the current drawing. The log file is maintained only if the XREFCTL system variable is set to 1. The default setting for XREFCTL is 0.

Because of the xref feature, the contents of a drawing may now be stored in multiple drawing files; this means that new backup procedures are required to handle drawings linked in xref partnerships. Possible solutions include the following:

1. Make the xref drawing a permanent part of the current drawing prior to archiving with the BIND option of the XREF command.

2. Modify the path of the current drawing to the xref drawing so that they are both stored in the same folder, and archive them together.

3. Archive the folder location of the xref drawing with the drawing which references it. Tape backup machines do this automatically; you can use the ETRANSMIT command to automatically include all related dependent files such as xrefs and font files.

In AutoCAD, a combination of demand loading and saving drawings with indexes helps you to increase the performance of drawing with xrefs. In conjunction with the XLOADCTL and INDEXCTL system variables, demand loading provides a method of displaying only those parts of the referenced drawing that are necessary.

The XLOADCTL system variable controls whether demand loading is set to ON or OFF and whether it opens the original drawing or a copy. If XLOADCTL is set to 0, AutoCAD turns off demand loading, and the entire reference file is loaded. If XLOADCTL is set to 1, AutoCAD turns on the demand loading, and the reference file is kept open. AutoCAD loads only the objects that are necessary to display on the current drawing. AutoCAD places a lock on all reference drawings that are set for demand loading. Other users can open those reference drawings, but they cannot save changes to them. If XLOADCTL is set to 2, AutoCAD turns on demand loading and a copy of the reference file is opened. AutoCAD makes a temporary copy of the externally referenced file and demand-loads the temporary file. Other users are allowed to edit the original drawing. When you disable demand loading, AutoCAD reads in the entire reference drawing regardless of layer visibility or clip instances.

The INDEXCTL system variable determines whether or not layer, spatial, or layer and spatial indexes are created when a drawing file is saved. Using layer and spatial indexes increases performance when AutoCAD is demand-loading xrefs. If INDEXCTL is set to 0, indexes are not created. If INDEXCTL is set to 1, a layer index is created. The layer index maintains a list of objects that are on specific layers, and with demand-loading, it determines which objects need to be read in and displayed. If INDEXCTL is set to 2, a spatial index is created. The spatial index organizes lists of objects based on their location in 3D space, and it determines which objects lie within the clip boundary and reads only those objects into the current session. If INDEXCTL is set to 3, both layer and spatial indexes are created and saved with the drawing. If you intend to take full advantage of demand loading, INDEXCTL should be set to 3.

> If the drawing you are working on is not going to be referenced by another drawing, it is recommended that you set INDEXCTL to 0 (OFF).

NOTE

AutoCAD provides another system variable, VISRETAIN, to control the visibility of layers in the xref drawing. If the VISRETAIN system variable is set to 0 (OFF), any changes you make to settings of the xref drawing's layers, such as ON/OFF, freeze/thaw, color, and linetype, apply to the current drawing session only. If VISRETAIN is set to 1 (ON), any changes you make to settings of the xref drawing's layers take precedence over the xref layer definition.

Another tool that you can use to manage xrefs is Autodesk Reference Manager. It provides tools to list referenced files in selected drawings and to modify the saved reference paths without opening the drawing files in AutoCAD. With Reference Manager, drawings containing unresolved references can be easily identified and fixed.

Reference Manager is a stand-alone application that you can access from the Autodesk program group under Programs in the Start menu of Windows.

MANAGING IMAGES

AutoCAD has several tools to modify the settings related to images. When you select a raster image that is attached to a drawing, the Image Contextual tab appears on the ribbon.

The Adjust panel (see Figure 12–23) controls the brightness, contrast, and fade values of the selected image.

FIGURE 12–23 *Image Adjust panel*

You can adjust the **Brightness**, **Contrast**, and **Fade** within the range of 0 to 100.

The IMAGEQUALITY command controls the display quality of images. The quality setting affects display performance. A high-quality image takes longer to display. Changing the setting updates the display immediately without causing a regeneration. Images are always plotted using a high-quality display.

Invoke the IMAGEQUALITY command and AutoCAD displays the following prompt:

    Enter image quality setting [High/Draft]: (select one of the
    available options from the shortcut menu)

The High option produces a high-quality image on the screen, and the Draft option produces a lower-quality image on the screen.

The TRANSPARENCY command controls whether the background pixels in an image are transparent or opaque. Select the image and from the Image contextual tab choose the Transparency toggle button to turn ON or OFF.

Choosing ON turns transparency ON so that objects beneath the image are visible.

Choosing OFF turns transparency OFF so that objects beneath the image are not visible.

The IMAGEFRAME command controls whether image frames are displayed or hidden from view.

Invoke the IMAGEFRAME command, and AutoCAD displays the following prompt:

> Enter image frame setting [0/1/2]: *(select one of the available options from the shortcut menu)*

Choosing 0 turns display and plotting of frames around images OFF.

Choosing 1 turns display and plotting of frames around images ON.

Choosing 2 turns display of frames on but plotting of frames OFF.

Similar tools are available to adjust the settings for referenced DWF, DGN, and PDF files.

Open the Exercise Manual PDF for Chapter 12 for discipline-specific exercises. Related files are downloaded from the student companion site mentioned in the Introduction (refer to page number xii for instructions).

REVIEW QUESTIONS

1. If an externally referenced drawing named *FLOOR.dwg* contains a block called "TABLE" and is permanently bound to the current drawing, the new name of the block is:
   a. FLOOR0TABLE
   b. FLOOR|TABLE
   c. FLOOR$0TABLE
   d. FLOOR_TABLE
   e. FLOOR$|TABLE

2. If an externally referenced drawing named *FLOOR.dwg* contains a block called "TABLE," the name of the block is listed as:
   a. FLOOR0TABLE
   b. FLOOR|TABLE
   c. FLOOR$0TABLE
   d. FLOOR_TABLE
   e. FLOOR$|$TABLE

3. The maximum number of files that can be externally referenced into a drawing is:
   a. 32
   b. 1,024
   c. 8,000
   d. 32,000
   e. Only limited by memory

4. Can you snap to objects in a DWF underlay?

   a. Yes

   b. No

5. If you want to retain from one drawing session to another any changes you make to the color or visibility of layers in an externally referenced file, the system variable that controls this is:

   a. XREFRET

   b. RETXREF

   c. XREFLAYER

   d. VISRETAIN

   e. These changes cannot be saved from one session to another

6. XREFs are converted to blocks if you detach them.

   a. True

   b. False

7. When detaching xrefs from your drawing, it is acceptable to use wildcards to specify which xref should be detached.

   a. True

   b. False

8. Overlaying xrefs rather than attaching them causes AutoCAD to display the file as a bitmap image rather than a vector-based image.

   a. True

   b. False

9. When named objects are bound using Xbind, how is the reference prefixed?

   a. SOS

   b. 0

   c. %0%

   d. &X&

10. To make a reference file a permanent part of the current drawing database, use the XREF command with the:

    a. ATTACH option

    b. BIND option

    c. ? option

    d. RELOAD option

    e. PATH option

11. The XREF command is invoked from which panel?

    a. Home

    b. Modify

    c. External Reference

    d. Any of the above

    e. None of the above

12. What is the name of the line that surrounds clipped references?

    a. Boundary

    b. Print Variable

    c. Border

    d. Clip Frame

13. The ATTACH option of the XREF command is used to:

    a. Bind the external drawing to the current drawing

    b. Attach a new xref file to the current drawing

    c. Reload an xref drawing

    d. All of the above

14. Which command removes a reference file from a drawing temporarily?

    a. Unload

    b. Detach

    c. Dock

    d. Pin

15. The following are the dependent symbols that can be made a permanent part of your current drawing except:

    a. Blocks

    b. Dimstyles

    c. Text Styles

    d. Linetypes

    e. Grid and Snap

16. What type of file can be used as a reference file?

    a. PNG

    b. DWG

    c. DGN

    d. All of the above

17. Which of the following is not a valid file type to use with the IMAGE command?

    a. bmp

    b. tif

    c. wmf

    d. jpg

    e. gif

18. Which of the following parameters can be adjusted on a bitmapped image?

    a. Brightness

    b. Contrast

    c. Fade

    d. All of the above

    e. None of the above

AutoCAD DesignCenter and Content Explorer

INTRODUCTION

AutoCAD DesignCenter makes it much easier to manage content within your drawing. Content includes blocks, external references, layers, raster images, hatch and gradient fills, linetypes, layouts, text styles, dimension styles, and custom content created by third-party applications. You can now manage content between your drawing and sources such as other drawings whether currently open, stored on any drive, or even elsewhere on a network or somewhere on the Internet. DesignCenter provides a program window with a specialized drawing file-handling section. It allows you to drag and drop content and images into your current drawing or attach a drawing as an external reference. DesignCenter also gives immediate and direct access to thousands of symbols, manufacturers' product information, and content aggregators' sites through DC Online. Content in DesignCenter can be dragged onto a tool palette for use in the current drawing.

OBJECTIVES

After completing this chapter, you will be able to do the following:

- Open, undock, move, resize, dock, and close AutoCAD DesignCenter
- Locate drawings, files, and their content in a manner similar to Windows® Explorer
- Use the DesignCenter content area and Tree View
- Preview images, drawings, content, and their written descriptions
- Customize and use Autodesk Favorites folder
- Manage blocks, layers, xrefs, layouts, dimstyles, textstyles, and raster images
- Manage Web-based content and custom content from third-party applications
- Create shortcuts to drawings, folders, drives, the network, and Internet locations
- Browse sources for content and drag and drop, or copy, into drawings
- Access symbols and information directly over the Internet
- Using Content Explorer

THE DESIGNCENTER WINDOW

The DesignCenter window allows you to browse, find, preview, and insert content, which includes blocks, hatches, and external references or xrefs. Use the buttons in the toolbar at the top of DesignCenter for display and access options. You can control the size, location, and appearance of DesignCenter.

Content

As mentioned earlier, content includes block definitions, external references, layer names and compositions, raster images, linetypes, layouts, text styles, dimension styles, and custom content created by third-party applications. For example, a layer of one name may have the same properties as a layer of another name. However, because its name is part of its composition, it can be identified as a unique layer. Two items of content can have the same name if they are not the same type of content. For example, you can name both a text style and a dimension style "architectural."

NOTE

> You can only view the name of an item of content in the content area along with a raster image if it is a drawing, block, or image and a description, if one has been written. The item itself still resides in its container. Through DesignCenter, you can drag and drop copies of the item's definition into your current drawing, but you cannot edit the item itself from DesignCenter.

Content Type

Content types are only types or categories and not the items themselves. Lines, circles, and other objects are not included in what are considered content types, although a block may be made up of such objects. To view a particular item of content, you must first select the name of its content type, under a drawing name, in the Tree View or double-click the name of its content type in the content area. For example, if you were trying to locate an xref named "*1st floor plan*" you could search through drawing names and select the content type named "xref" under each drawing name. Under each drawing name, there will be a content type named "xref" but under each content type named "xref" there may or may not be any xrefs listed.

Container

The primary container is the drawing. It contains the blocks, images, linetypes, and other definitions that are most commonly sought to add to your current drawing. A folder can be considered a container because it contains files, and a drawing is a file. An image can also be a file contained in a folder. An image is normally not a container like a drawing. Drawings and images can be dragged and dropped into your current drawing from the content area. A drawing can be both content and a container.

Figure 13–1 shows an example of a content block named Full Wave Bridge a content type called Block, and a Container, a drawing named Basic Electronics.

Content

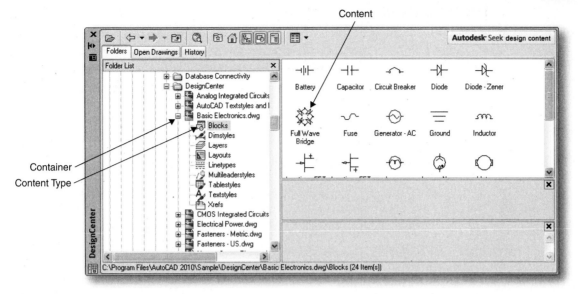

Container
Content Type

FIGURE 13–1 *An example showing a content, content type, and container from the DesignCenter window*

OPENING THE DESIGNCENTER WINDOW

DesignCenter is a window rather than a dialog box. It is like calling up a special program that runs alongside AutoCAD and expedites the tasks of managing files and handling drawing content.

Invoke the DesignCenter command from the Content panel (Figure 13–2) in the Insert tab, and AutoCAD displays the DesignCenter window (Figure 13–3).

FIGURE 13–2 *Invoking the DesignCenter from the Content panel*

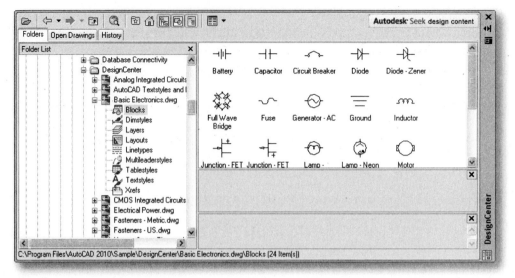

FIGURE 13–3 *DesignCenter window*

POSITIONING THE DESIGNCENTER WINDOW

The default docking position of DesignCenter is at the left side of the drawing area. This is where it will be located when you open the DesignCenter window for the first time, as shown in Figure 13–4. However, once DesignCenter has been repositioned and the drawing session is closed with the DesignCenter window open, the next time you open AutoCAD, it will appear at its relocated position.

FIGURE 13–4 *DesignCenter docked on the left side of the drawing area*

You can undock DesignCenter by double-clicking on its title bar or border. Or, you can drag the window into the drawing area until its drag preview image clears its present position (Figure 13–5).

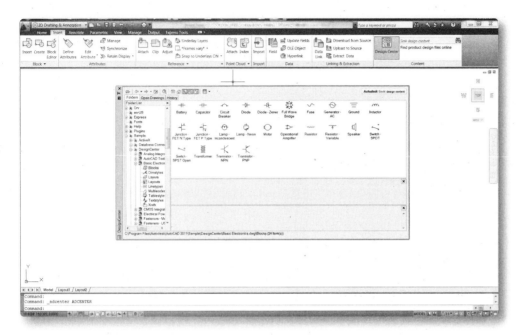

FIGURE 13–5 *DesignCenter floating (undocked) in the drawing area*

To redock DesignCenter, double-click the title bar or border or drag the Design-Center to the left or right side of the drawing area. It cannot be docked at the top or bottom of the screen. When undocked, DesignCenter can be resized by holding the cursor over an edge or corner of the window; when the cursor becomes a double arrow when holding down the pick button, drag the edge or corner with the pointing device.

VIEWING CONTENT

The names of content types, such as blocks, linetypes, textstyles, and so on, and containers such as drawings, image files, folders, drives, networks, and Internet locations, can be viewed in the Tree View as well as the content area. Names of content can be viewed in the content area but not in the Tree View pane. Either pane can be used to move up and down through the path from the drive to the item of content. It is usually quicker to navigate the path in the Tree View because of its ability to display multiple levels of hierarchy. When the container, a folder for drawings/images and a drawing/content type for blocks, images, and other items, appears in the Tree View, select it and then view the name of the items of content, if any, in the content area.

Using the Tree View

As mentioned earlier, in the Tree View you can view the content type and container but not the actual content. You can use the Tree View to display the icon for an item of content in the content area. Just doing this will not, however, cause a raster image to be displayed in the Preview panel, as shown in using the content area in this section.

The folder named Sample that installs with the AutoCAD program contains numerous drawings which will be used as examples in this section. First select **Up** as many times as necessary to display the level, in the Tree View pane, that includes the branch leading through the folder named AutoCAD 2011, or the folder in which AutoCAD has been installed, to the subfolder named Sample. Again, this action might require going all the way back to a particular drive. Click the plus sign (+) or double-click the folder name or drive name to display the folders within a folder. Figure 13–6 shows the folder named Sample highlighted in the Tree View. Note that the list of content folders is displayed in the content area.

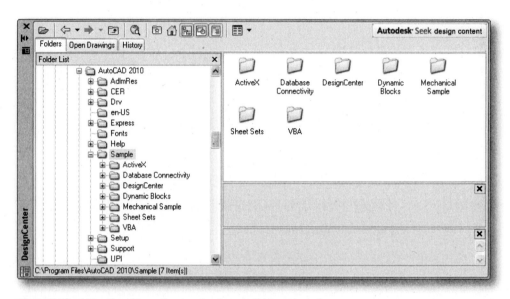

FIGURE 13–6 *The Tree View showing the Sample folder selected*

Next double-click on the drawing named *Mechanical – Text and Tables.dwg* located in the Mechanical Sample folder in the Tree View or select the box with the plus sign (+) at the left of the drawing name. The content types will be listed below the drawing name, as well as in the content area. To display the names of individual items of content, blocks for example, select the content type Blocks in the Tree View (Figure 13–7).

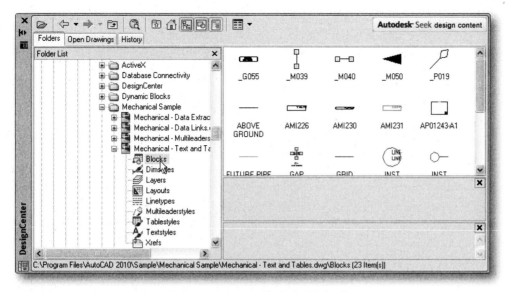

FIGURE 13–7 *The Tree View with the content type Blocks highlighted*

When one of the block icons displayed in the content area is chosen, AutoCAD displays the corresponding preview image and description, if any, in the Preview pane and Description pane, respectively. Figure 13–8 shows the selection of a block named "Pipe Support," and its corresponding preview image and description are shown in the Preview pane and in the Description pane.

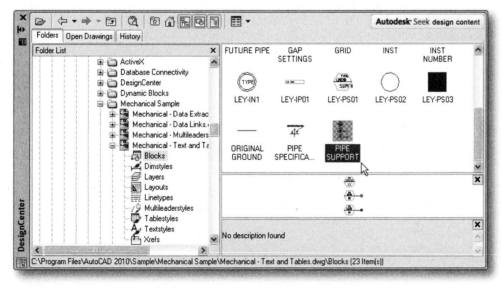

FIGURE 13–8 *Selection of a block named Pipe Support; the corresponding preview image and description are shown in the Preview pane and in the Description pane*

Using the Content Area

As mentioned earlier, the content area is used for displaying content. It can also display containers and content type. In the hierarchy of what can be displayed, containers are one step higher than content types. Content types are one step higher than content. From drawing files and other containers, the next step up is a folder. Progressing up through the folders, if there is a hierarchy of folders, the next level is the drive, which may lead to a network, a website on the Internet, a floppy disk, or a CD-ROM.

NOTE

> The content area displays only items of the same level that are members of a single container one level above them. The content area does not display more than one level at a time. For example, content types such as blocks, xrefs, and layers are members of one drawing. If there are any blocks in the drawing, they are listed when you have selected that drawing's content type called Blocks. Several drawings may be members of a particular folder. Folders are members of one drive or folders may actually be subfolders of a folder one level up. Remember, in the content area, only one level of the hierarchy is displayed, and all items shown are members of the same component/container one level above. For displaying more than one level at a time, use the Tree View.

You can navigate up and down through the hierarchy of drives, folders, drawings, content types, and content in the content area. It is not as easy as working in the Tree View, however, because only one level is displayed at a time. If the content is a drawing, block, or image, you can view a raster image of it in the Preview pane. Select the item in the content area. Remember that you can enlarge the Preview pane for viewing, even if only temporarily. As mentioned earlier, you can double-click the title bar to dock or undock DesignCenter.

For example, there is a block named "Gap Settings" in the drawing named *Mechanical – Text and Tables.dwg* in the folder named *Mechanical Sample*. To get to the block named "Gap Settings" in the content area, as described for the example in the Tree View above, select the Up button as many times as necessary to display the level that includes the branch leading to the drawing named *Mechanical – Text and Tables.dwg*. This might require going all the way back to a particular drive. Double-click the sequence of drive and folders on the path to the drawing named *Mechanical – Text and Tables.dwg* in the content area, and then double-click on *Mechanical – Text and Tables.dwg* icon. When you double-click the icon for *Mechanical – Text and Tables.dwg*, the icons for content types will be displayed including blocks, dimstyles, layers, layouts, linetypes, textstyles, and xrefs (Figure 13–9).

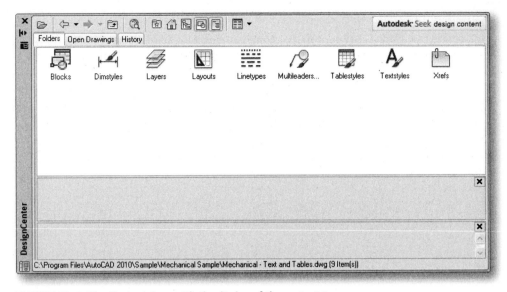

FIGURE 13–9 *The Content Area with the display of the content types*

AutoCAD also displays the path to the selected drawing at the bottom of the Design-Center window (Figure 13–9). Select the **Blocks** icon, and AutoCAD displays the icons when the display is set to Large icons representing all the blocks in the selected drawing. Select one of the block icons—do not double-click—and AutoCAD displays corresponding raster images and descriptions in the Preview pane and Description pane, respectively. Figure 13–10 shows the selection of the block named "Gap Settings" and the corresponding display of its raster image and description. You can enlarge the Preview pane to get a bigger picture. Note, however, that it is a raster image, and sharp details are not available in closeup views.

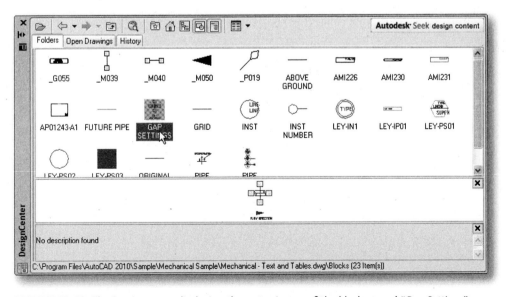

FIGURE 13–10 *The Preview pane displaying the raster image of the block named "Gap Settings"*

Viewing Images

If the item of content is a bitmap image, its name can be displayed in the content area by selecting the name of its container in the Tree View. For example, there is a sub-folder named *Sample* in the AutoCAD 2011 folder. In the subfolder named *Database Connectivity* is another folder named *CAO*. When *CAO* is highlighted in the Tree View, you can choose the bitmap file named *dbcm_query.eps* in the content area, and the icon resembling a question mark is displayed in the Preview pane (Figure 13–11). A description can be viewed in the Description pane. All the images in the subfolder have a file extension of *.eps*. Similarly, you can view any other graphic file type.

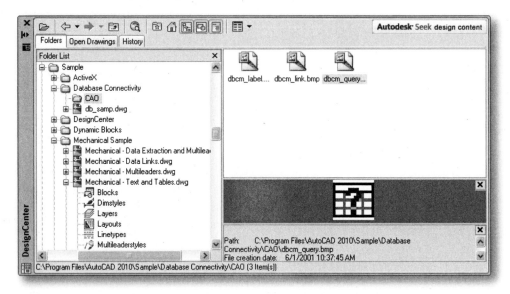

FIGURE 13–11 *Display of the image named dbcm_query.eps with its raster image in the Preview pane and its corresponding description in the Description pane*

 NOTE   When you have selected the name of a folder in the Tree View, if there are any viewable files in that folder, their names will appear in the content area. Some folders may have other files in them, but if they are not viewable, their names will not appear.

You can also load a drawing or folder from the Search dialog box by dragging and dropping it into the content area or Tree View pane. Or, you can right-click and select *Load into Content Area* from the shortcut menu.

 NOTE   The Load feature does not load the item selected into your drawing. It only loads its icon into the content area, as shown in the next section on adding content to drawings.

When file manipulations are made to the content area or Tree View, they do not always show up immediately on the screen. If this happens, you can right-click in one of the panes and choose *Refresh* from the shortcut menu. The views will be updated.

To open a drawing being displayed in the content area, right-click its icon, and choose *Open in Application Window* from the shortcut menu. You can also drag and drop the icon into the drawing area. Be sure to drop the icon in an area that is clear of another drawing. If necessary, resize or minimize any other drawing(s) first.

WORKING WITH DESIGNCENTER

DesignCenter has three tabs: **Folders, Open Drawings, and History.**

The Folders tab displays a hierarchy of navigational icons including Networks and computers, Web addresses (URLs), drives, folders, drawings and related support files, xrefs, layouts, hatch styles, and named objects, including blocks, layers, linetypes, text styles, dimension styles, and plotstyles within a drawing. The Open Drawings tab displays a list of the drawings that are currently open. The History tab displays a list of files opened previously with DesignCenter.

Folders

When the Folders tab is displayed, DesignCenter shows five major areas: a content area, a Tree View pane, a Preview pane, a Description pane, and a toolbar. The content area and toolbar are always displayed (Figure 13–12). The Tree View, Preview, and Description panes can be turned OFF and ON when desired.

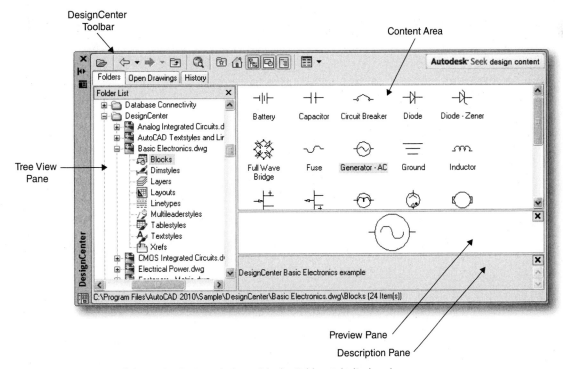

FIGURE 13–12 *Parts of the DesignCenter window with the Folders tab displayed*

Content Area

The content area, the right pane, is the primary area for displaying the names and icons representing content (Figure 13–12).

Tree View

The Tree View or navigation pane is an optional area on the left side of the Design-Center window for displaying files such as drawings and images and their locations. Choose the Tree View button on the toolbar (Figure 13–13), to display the Tree View.

FIGURE 13–13 *Invoking the Tree View from the DesignCenter toolbar*

The Tree View also displays content types. The Tree View displays multiple levels and operates in a manner similar to Windows Explorer. Figure 13–14 shows the DesignCenter window with the Tree View window display ON and OFF within the Folders tab.

NOTE

The Tree View can be used to navigate up and down through the hierarchy of networks, drives, folders, files, and content type. However, you cannot drag and drop to, from, or within the Tree View.

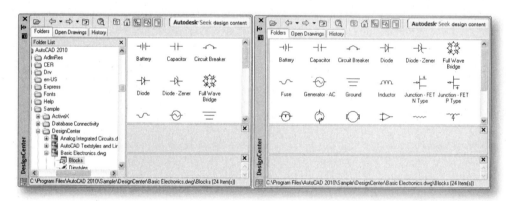

FIGURE 13–14 *DesignCenter window with Tree View window display ON (left) and OFF (right)*

Preview

The Preview pane displays a raster image of the selected item of content if the item is a drawing, block, or image (Figure 13–12).

Description

The Description pane displays a written description of the item of content selected on the content area, if a description has been written (Figure 13–12).

Toolbar

The buttons on the toolbar are used to manage the DesignCenter panes and what is being displayed in them (Figure 13–15).

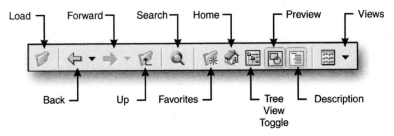

FIGURE 13–15 *DesignCenter toolbar*

Choosing **Load** causes the Load dialog box to be displayed (Figure 13–16). It is similar to the Windows File Manager dialog box. From it you can select a drawing file whose contents will be loaded into the content area.

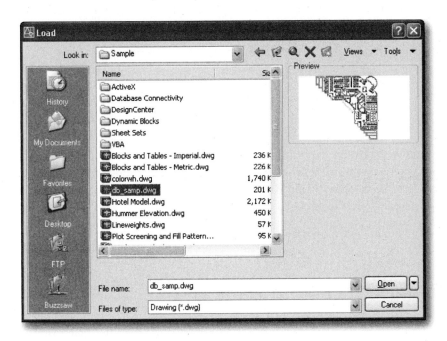

FIGURE 13–16 *Load dialog box*

The **Search** icon from the DesignCenter toolbar provides a means to locate containers, content type, and content in a manner similar to the Find feature of Windows. Choosing **Search** causes the Search dialog box to be displayed (Figure 13–17).

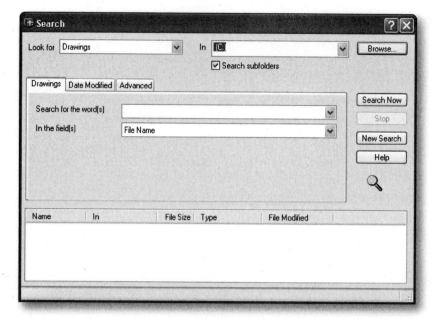

FIGURE 13–17 *Search dialog box with the Drawings tab displayed*

From the **Look for** list box in the Search dialog box, you can select the type of content you wish to find. Available options include **Blocks, Dimstyles, Drawings, Drawings and Blocks, Attach Pattern Files, Hatch Patterns Layers, Layouts, Linetypes, Tabstyles, Textstyles,** and **Xrefs.**

From the **In** list box, select the location for searching. Available options include My Computer, local hard drives (C:) and any others, 3 1/2 Floppy (A:), and any other drives.

Choosing **Browse** causes the Browse for Folder dialog box to be displayed, which has a file manager window (Figure 13–18). Here you can specify a path by stepping through the levels to the location which you wish to search. When the final location is highlighted, choose **OK**, and the path to this location is displayed in the **In** list box.

FIGURE 13–18 *Browse For Folder dialog box*

The number of tabs that will be displayed in the Search dialog box depends on the type of content selected in the **Look for** list box. Each of the content types will have a tab that corresponds to the type of content selected. If Drawings is selected, two additional tabs are available: **Date Modified** and **Advanced**.

On the Drawings tab (Figure 13–17), the **Search for the word(s)** textbox lets you enter the word(s), such as a drawing name or author, to determine what to search for. The **In the field(s)** textbox lets you select a field type to search. Fields include File Name, Title, Subject, Author, and Keywords.

On the Date Modified tab—available only when Drawings is selected in the **Look for** textbox of the Search dialog box—you can specify a search of drawing files by the date they were modified (Figure 13–19). The search will include all files that comply with other filters if **All files** is selected. If **Find all files created or modified** is selected, you can limit the search to the range of dates specified by selecting one of the following secondary radio buttons. Selecting **between and** allows you to limit the search to drawings modified between the dates entered. Selecting **during the previous months** allows you to limit the search to drawings modified during the number of previous months entered. Selecting **during the previous days** allows you to limit the search to drawings modified during the number of previous days entered.

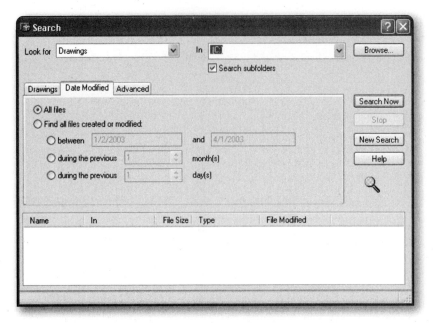

FIGURE 13–19 *Display of the Date Modified tab in the Search dialog box*

On the Advanced tab, which is available only when **Drawings** is selected in the **Look For** textbox of the Search dialog box, you can specify a search of files by additional parameters (Figure 13–20). In the **Containing** textbox, you can specify a file by one of four options: **Block name, Block and drawing description, Attribute tag,** and **Attribute value.** The search will then be limited to items containing the text entered in the **Containing** textbox. The **Size is** textboxes allow you to limit the search to drawings that are **At least** or **At most** in kilobytes, the size of the drawing file, as the number entered in the second textbox.

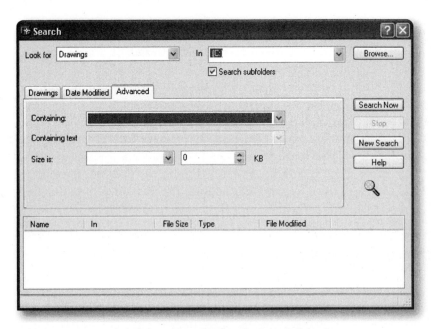

FIGURE 13–20 *Display of the Advanced tab in the Search dialog box*

All of the other tabs, **Blocks, Dimstyles, Drawings and Blocks, Hatch Patterns, Layers, Layouts, Linetypes, Textstyles,** and **Xrefs,** have a **Search for the name** textbox in which you can enter the name of the item you wish to find. This can be a block name, dimstyle, or one of the other content types.

To the right of the tab section are four buttons: **Search Now, Stop, New Search,** and **Help.**

Once the parameters have been specified such as the path, content type, and date modified, selecting **Search Now** initiates the search. Select **Stop** to terminate the search. Selecting **New Search** allows you to specify new parameters for another search. Selecting **Help** causes the AutoCAD Command Reference or Help dialog box to be displayed.

Each time a search is performed, the text entered in the **Search for the name** textbox or the **Search for the word(s)** textbox is saved. If you wish to repeat the same search, select the down-arrow next to the textbox, and select the text associated with the search you wish to repeat.

Choosing **Favorites** on the DesignCenter toolbar causes the Autodesk Favorites feature to display icons in the content area representing shortcuts to frequently used files. These are files for which you have previously set up shortcut icons to make them quickly accessible. The Tree View displays and highlights the subfolder Favorites of the folder Autodesk (Figure 13–21).

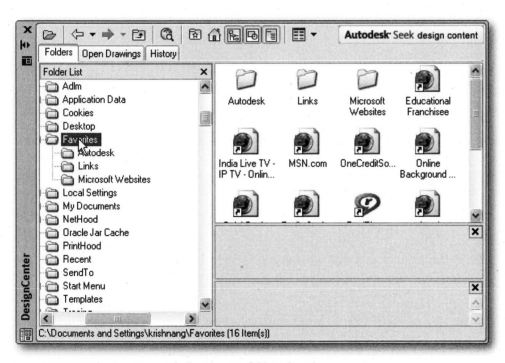

FIGURE 13–21 *The Tree View with the Favorites folder selected*

Only the shortcuts are in the Favorites folder. The files themselves remain in their original locations.

A drawing or image file or a folder can be added to the Favorites folder by right-clicking on the file name or folder name in the content area or Tree View and choosing *Add to Favorites* from the shortcut menu. To view and organize items in Favorites, choose *Organize Favorites* from the shortcut menu. An Autodesk window is displayed for managing items in Favorites.

When the Favorites folder is highlighted in the Tree View, the content area displays the icons, which are shortcuts to other files or folders. Double-click on the icon for the desired folder or file.

After you have double-clicked the desired icon in the content area, the selected file or folder will be displayed in the Tree View and will be highlighted.

In turn, the content area will display the contents of the highlighted folder or file. Figure 13–22 shows the selection of the Favorites folder and corresponding selection.

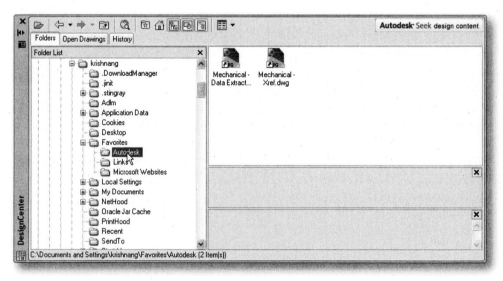

FIGURE 13–22 *Selection of a subfolder in the Favorites folder*

The buttons for Tree View, Preview, and Description cause their respective panes to be displayed or not displayed, depending on their current status.

Choosing **Home** from the DesignCenter toolbar causes the Tree View to display the drives, folders, subfolders, and files that are located on your computer's desktop. The hierarchy is expanded, and the DesignCenter folder is highlighted with its contents displayed in the content area (Figure 13–23). This is the type of view you normally see when you open Windows Explorer.

From there you can step through the path(s) necessary to the desired location. Select the **Folders** tab whenever you want to return from **Open Drawings** tree view and **History** view, explained later in this section, to display the drives, folders, subfolders, and files located on your computer's desktop.

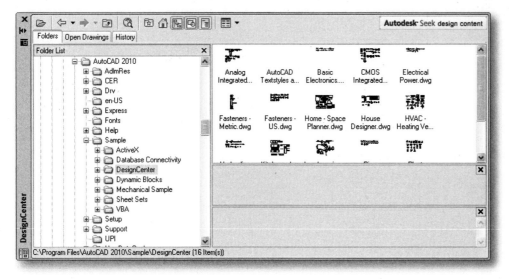

FIGURE 13–23 *DesignCenter displaying the Desktop with the DesignCenter folder highlighted*

Opening Drawings

The Open Drawings tab of the DesignCenter window displays a list in the Tree View of drawings that are currently open (Figure 13–24). When you select one of the drawings, its content types will be displayed in the content area. You can also double-click on the file name, and it will display the content types at one level below the name you double-clicked. Selecting one of the content types causes content of that type, if any exists in the drawing, to be listed in the content area.

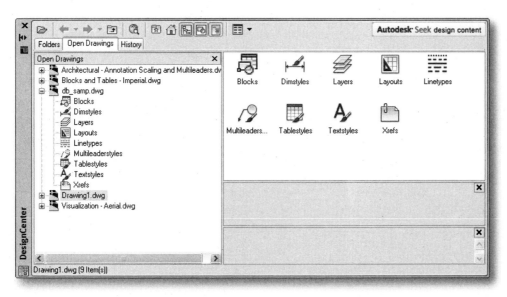

FIGURE 13–24 *The DesignCenter window with the Open Drawings tab displayed*

History

The History tab of the DesignCenter window displays the items accessed through DesignCenter including their paths (Figure 13–25).

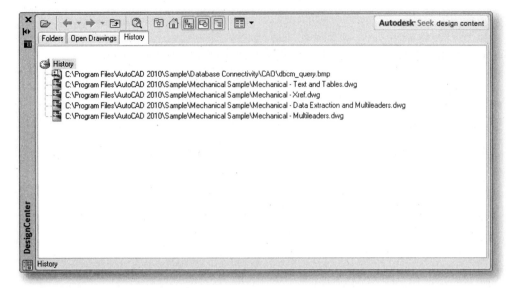

FIGURE 13–25 *The DesignCenter window with the History tab displayed*

DC Online

The DC Online tab of the DesignCenter window, when selected, causes AutoCAD to log on to the DesignCenter Online on the Internet, provided that your computer is connected to the Internet (Figure 13–26). DesignCenter Online provides access to pre-drawn content such as blocks, symbol libraries, manufacturers' content, and online catalogs. This content can be used in common design applications to assist you in creating your drawings.

NOTE

The DesignCenter Online (DC Online tab) is disabled by default. You can enable it from the CAD Manager Control utility. Install the utility from the AutoCAD1 2011 Installation Wizard and then you can enable/disable the DC Online tab in DesignCenter.

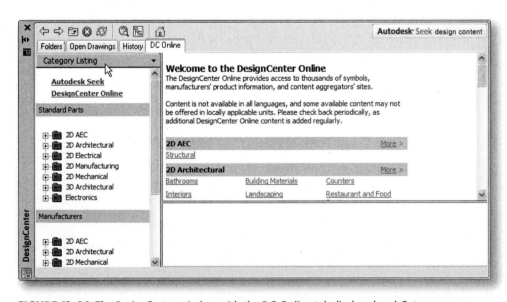

FIGURE 13–26 *The DesignCenter window with the DC Online tab displayed and Category Listing selected*

There are two panes in the DC Online tab of the DesignCenter window. The left pane has four views: **Category Listing, Search, Settings,** and **Collections** and the right pane displays the details of the selected view (see Figure 13–27).

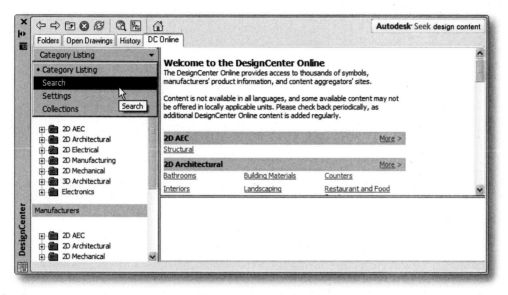

FIGURE 13–27 *Category Listings with 2D Architectural/Landscaping/Tables groups displayed*

Category Listing

The **Category Listing** view includes three categories: **Standard Parts, Manufacturers,** and **Aggregators** (Figure 13–26). The **Standard Parts** category includes groups of drawings and images that can be used in architectural and engineering design disciplines such as architecture, landscaping, mechanical, and GIS. For example, you can select the box to the left of the **2D Architectural** group and then select the box to the left of the **Landscaping** subgroup when expanded. This action expands to the list of types of content. If you select **Tables** from this list, the right pane will display thumbnail sketches of drawings or images that are available for download (Figure 13–28). When you select an image, additional links are displayed in the lower half of the right pane along with a larger sketch of the content.

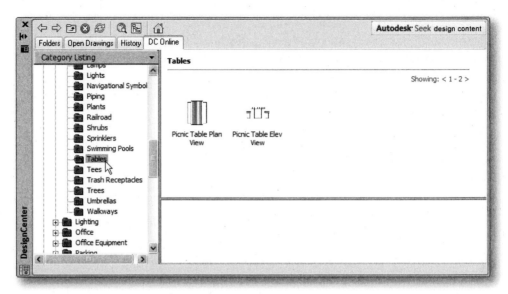

FIGURE 13–28 *Category Listings with 2D Architectural/Landscaping/Tables groups displayed*

NOTE

If you are looking for drawings, images, or links to drawings and images of a particular type of content, such as tables in the example above, there is usually more than one group or path to a group that includes such drawings, images, or links. See the explanation of using the Search view later in this section.

The **Manufacturers** category includes groups of websites or Internet addresses of manufacturers of products used in architectural and engineering construction. Through these websites, drawings and images can be downloaded where available. For example, in the **Manufacturers** group, similar to the selection in the **Category Listing** group, you can select the box to the left of the **2D Architectural** group and then select the box to the left of the **Landscaping** subgroup when expanded. This action expands to the list of types of content. If you select **Outdoor Furnishing** from this list, the right pane will display one or more Internet addresses of websites from which you can access drawings, images, or other content data that are available for download (Figure 13–29).

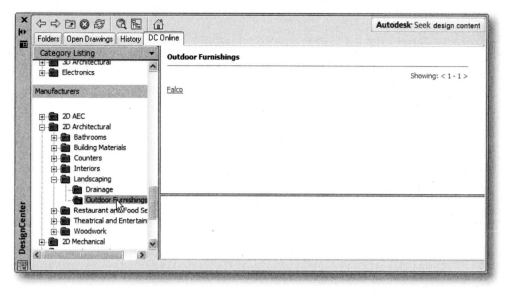

FIGURE 13–29 *Manufacturers with 2D Architectural/Landscaping/Outdoor Furnishings groups displayed*

The **Aggregators** category contains lists of libraries compiled by commercial catalog providers. Like the items in the **Manufacturers** category, you can access websites that contain or lead to drawings and blocks for use in architectural and engineering design and drafting. For example, in the **Aggregators** group, similar to the selection in the **Category Listing** and **Manufacturers** group, you can select the **AEC Aggregators** from this list, and the right pane will display one or more Internet addresses of websites from which you can access drawings, images, or other content data that are available for download (Figure 13–30).

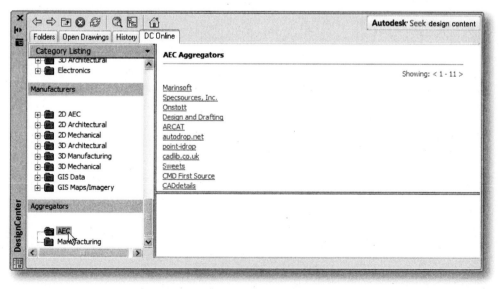

FIGURE 13–30 *Aggregators groups displayed*

Search

The Search view provides a textbox in which you can enter words or combinations of characters to tell AutoCAD what type of content to search for. You can display details for how to use Boolean and multiple-word search strings by selecting the **Need Help?** link. Figure 13–31 shows an example of entering "table" in the textbox and selecting the **Search** button. The right pane shows the result of the search. Selecting one of the items of content in the right pane causes additional links to be displayed in the lower half of the right pane along with a larger sketch of the content.

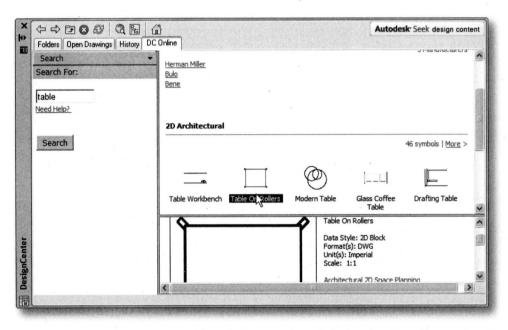

FIGURE 13–31 *The DesignCenter window with the DC Online tab displayed and Search view selected*

Settings

The Settings view provides two textboxes: **Number of Categories per page** and **Number of Items per page** (Figure 13–32). You can select from 5, 10, and 20 in the **Number of Categories per page** and from 50, 100, and 200 in the **Number of Items per page** textboxes to specify the **Max Search** numbers for categories and items, respectively.

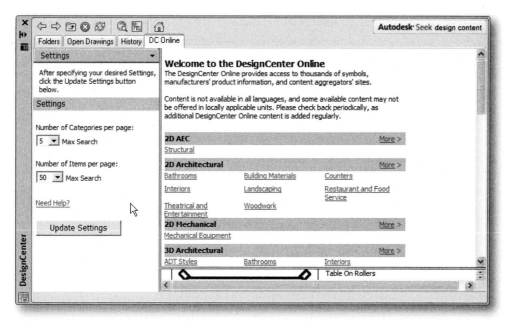

FIGURE 13–32 *The DesignCenter window with the DC Online tab displayed and Settings view selected*

Collections

The **Collections** view includes a list of collections for each of the three categories with a check box beside each collection (Figure 13–33). Check a particular collection's check box that you wish to be displayed in the **Category Listing** view. Once you have selected/deselected the desired collections, select **Update Collections**, and AutoCAD will return to the **Category Listing** view, displaying the list of collections whose boxes are checked.

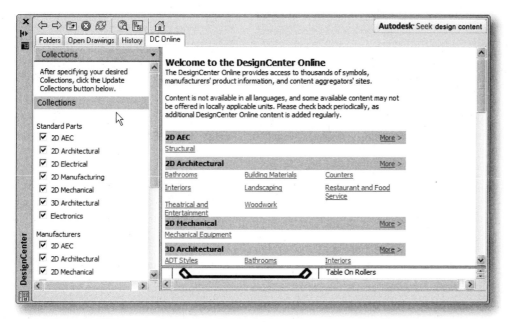

FIGURE 13–33 *The DesignCenter window with the DC Online tab displayed and Collections view selected*

Autodesk Seek Design Content

Autodesk Seek is a Web service that allows designers, architects, engineers, and students to search and find manufacturer-specific or generic building products and associated design content. This content could include 3D models, 2D drawings, and specifications.

To access the Autodesk Seek Web service, choose the Autodesk Seek design content button (see Figure 13–34) from DesignCenter and AutoCAD opens the default Web browser and displays the home page of the Autodesk seek website (see Figure 13–35). When you are working in a design program, you may want to include products that meet certain design standards. Autodesk Seek can help you locate such information and products and get them into your design. You can also upload AutoCAD drawing files and blocks to the Autodesk Seek service from the **Application** menu located in the **Publish** menu.

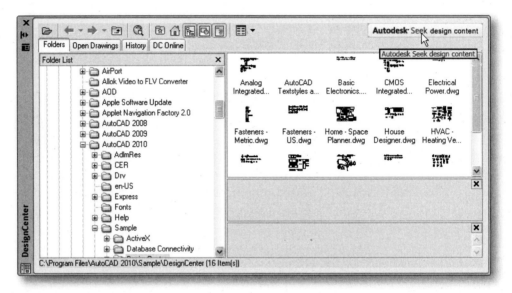

FIGURE 13–34 *The DesignCenter window with the Autodesk Seek design content selection*

FIGURE 13–35 *Autodesk Seek design content website*

You can also access the Autodesk Seek Web service by pointing your browser to **http:// seek.autodesk.com**

NOTE

ADDING CONTENT TO DRAWINGS

As discussed in the introduction to this chapter, blocks, external references, layers, raster images, linetypes, layouts, text styles, dimension styles, and custom content created by third-party applications are the content types that can be added to the current drawing session by using DesignCenter. Content in DesignCenter can be dragged into a tool palette for use in the current drawing.

Layers, Linetypes, Text Styles, and Dimension Styles

A definition of a layer or linetype or a style created for text or dimensions can be dragged into the current drawing from the content area. It will become part of the current drawing as if it had been created in that drawing.

Blocks

A block definition can be inserted into the current drawing by dragging its icon from the content area into the drawing area. You cannot, however, do this while a command is active. Only one block definition at a time can be inserted from the content area. Figure 13–36 shows a block in the content area of DesignCenter being dragged into the tool palette of the current drawing.

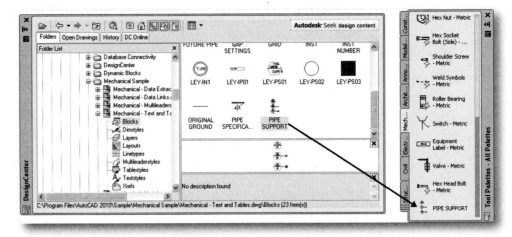

FIGURE 13–36 *The block named Pipe Support being dragged from the DesignCenter window into the tool palette of the current drawing*

NOTE

If you double-click a block that has nested blocks, the hierarchy is flattened.

There are two methods of inserting blocks from DesignCenter. One method uses Autoscaling, which scales the block reference as needed based on the comparison of the units in the source drawing to the units in the target drawing. Another method is to use the Insert dialog box to specify the insertion point, scale, and rotation.

NOTE

When dragging and dropping using the automatic scaling method, the dimension values inside the blocks will not be true.

When you drag a block definition from the content area or the Find dialog box into your drawing, you can release the button on the pointing device, or drop, when the block is at the desired location. This step is useful when the desired location can be specified with Running Osnap mode in effect. The block will be inserted with the default scale and rotation.

To invoke the Insert dialog box, double-click the block icon or right-click the block definition in the content area or Find dialog box, and then select *Insert Block* from the shortcut menu. In the Insert dialog box, specify the Insertion point, Scale, and Rotation, or select **Specify On-screen**. You can select **Explode** to have the block definition exploded on insertion.

Raster Images

A raster image such as a digital photo, print screen capture saved in a paint program as a bitmap, or a company logo can be copied into the current drawing by dragging its icon from the content area into the drawing area. Specify the Insertion point, Scale, and Rotation. You can also right-click the image icon and choose ATTACH IMAGE from the shortcut menu.

External References

To attach an external reference, or xref, to the current drawing, drag its icon from the content area or the Find dialog box with the right button on the pointing device into the drawing area. Release the button and select *Attach* from the shortcut menu. The Attach Xref dialog box is displayed, from which you can choose between **Attachment** or **Overlay** as the **Reference Type** option. From the Attach Xref dialog box, specify the Insertion point, Scale, and Rotation, or select **Specify On-screen**.

When you copy, insert, or attach content into a drawing that already has an item of the content type with the same name, AutoCAD will display a warning, and the item is not added to the drawing. If the item is a block or external reference, AutoCAD checks to determine if the name is already listed in the database. The warning "Duplicate definition of [object][name] ignored" is displayed and the item is not added to the drawing.

You can exit DesignCenter by either selecting the X in the upper-right corner of the DesignCenter window or entering **adcclose** at the AutoCAD on-screen prompt.

CONTENT EXPLORER, CONTENT SERVICE AND SEEK

Content Explorer collects and makes available for access the design content from different folders that have been called out to be watched by the Content Service. Content Service accompanies the Content Explorer search client installed with AutoCAD.

With Content Explorer you cause design content to be indexed for quick access, view the objects in each file, and search for content in folders (both local and network) and within the Autodesk Seek Library. Autodesk Seek is an online source for product information.

Using Content Explorer and Content Service

With Content Explorer, you can search for objects, all text, attributes (including block attributes), and files in specified local and network folders. You can also search for text contained in multiline text, tables, fields, multileaders, dimensions, and hyperlinks.

Content Service indexes, monitors and updates data associated with design files and design objects in watched folders. All computers on a network intended to be accessed must have Content Service installed on them.

Invoke Content Explorer (the `CONTENTEXPLORER` command) from the Content panel (Figure 13–37) in the Plug-ins tab and AutoCAD displays the Content Explorer window (Figure 13–38).

FIGURE 13–37 *Invoking the Content Explorer from the Content panel*

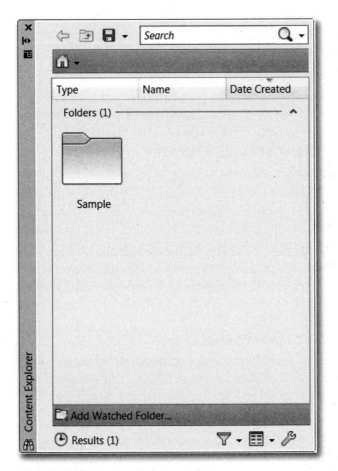

FIGURE 13–38 *Content Explorer window*

When the Content Explorer window initially appears, by default the Sample folder is added as shown in Figure 13–38. To add additional folders, click the **Add Watched Folder...** button located at the bottom left of the window. You can also drag and drop folders into the main view. Folders aren't copied; instead, their content will be added to the index. Double-clicking on the name of the folder or right-clicking on the folder and choosing Explore from the shortcut menu (Figure 13–39) allows you to navigate through the file structure. Selecting **Home** (see Figure 13–40) returns you to the top level of the listing of the folders that are currently watched and indexed.

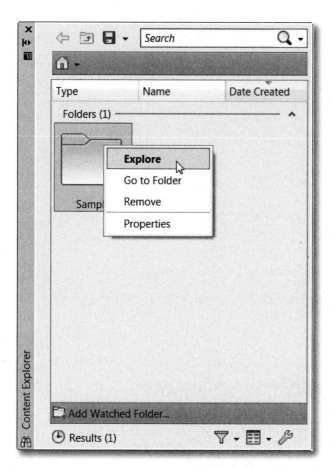

FIGURE 13–39 *Choosing Explore from the shortcut menu*

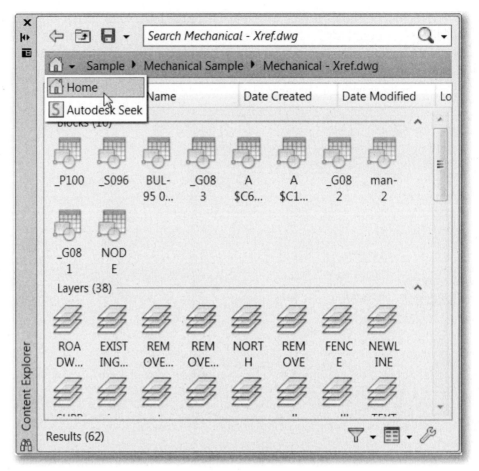

FIGURE 13–40 *Selecting Home to return to the top level of listing of folders*

The Sample folder contains numerous drawings which will be used as examples in this section. Double-click the Sample folder and then double-click the Mechanical Sample folder to access all of the drawing files. Next double-click on the drawing named *Mechanical – Text and Tables.dwg* located in the Mechanical Sample folder. AutoCAD displays all of the available blocks (19) in the selected drawing (see Figure 13–41). By scrolling you will see listing of layers, Layouts, Dimstyles, Linetypes, Tablestyles, and Textstyles. You can also navigate between the folders from the Breadcrumbs bar (Figure 13–41).

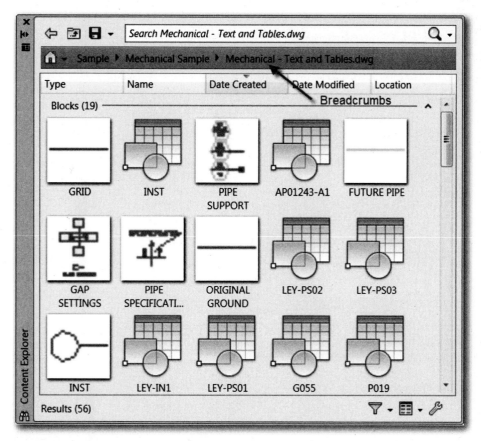

FIGURE 13–41 *Available blocks in the Mechanical – Text and Tables drawing*

The View Options drop-down arrow (Figure 13–42) allows you to configure the icon size, whether the folders are displayed in thumbnail or detailed view, which labels are displayed, and how the results are grouped.

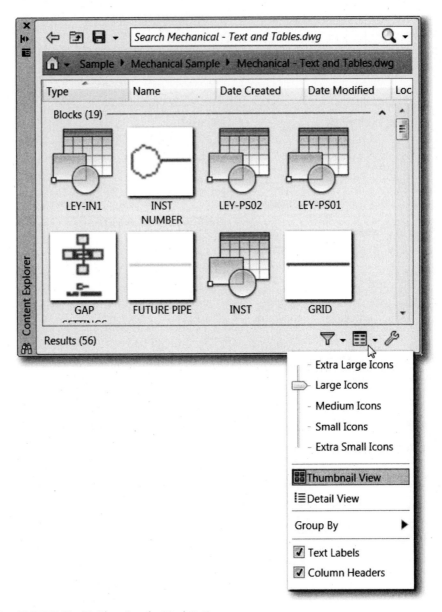

FIGURE 13–42 *Changing the View Options*

The Filter drop-down menu (Figure 13–43) allows you to specify which types of files and objects are displayed when you browse or search a folder. When a filter is active, the icon highlights to let you know that not everything is being displayed, based on your filter requisites.

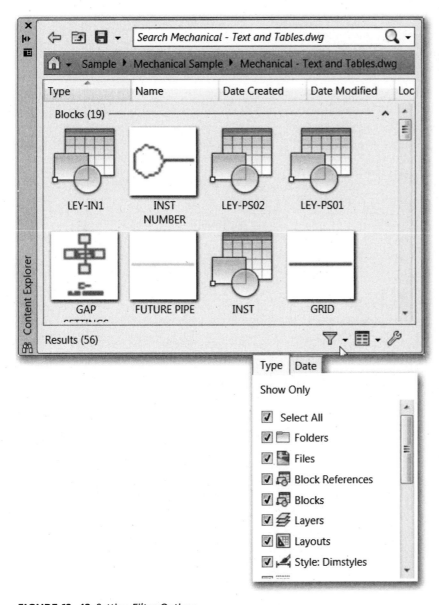

FIGURE 13–43 *Setting Filter Options*

Figure 13–44 shows the list of blocks available in the selected drawing as set in the Filter drop-down menu. You can toggle the filter ON and OFF as desired.

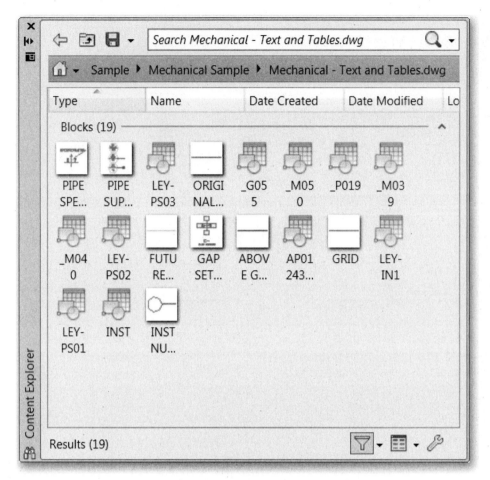

FIGURE 13–44 *Filter set to display only Blocks for the selected drawing*

The Search field allows you to locate the files and objects that meet the specified criteria. Only the current content source is searched and results are displayed based on the current filter settings. Figure 13–45 lists all the files and objects for the search word "ceiling" from the Sample folder.

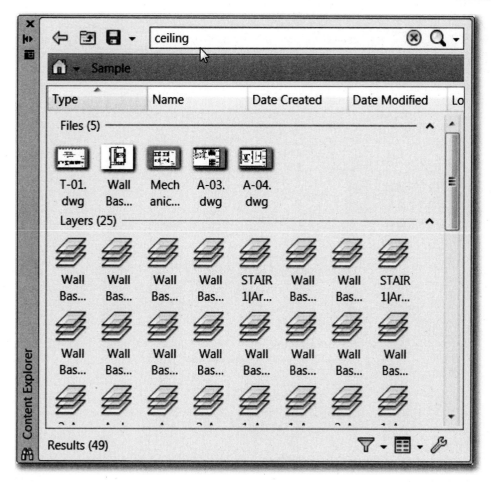

FIGURE 13–45 *Search result for "ceiling" categorized by Type*

You can drag and drop any of the listed objects in a similar manner to Design Center.

To locate the search string in the listed drawing, select **Open and Find text** from the shortcut menu (Figure 13–46). AutoCAD opens the drawing and if necessary, you can edit the text string.

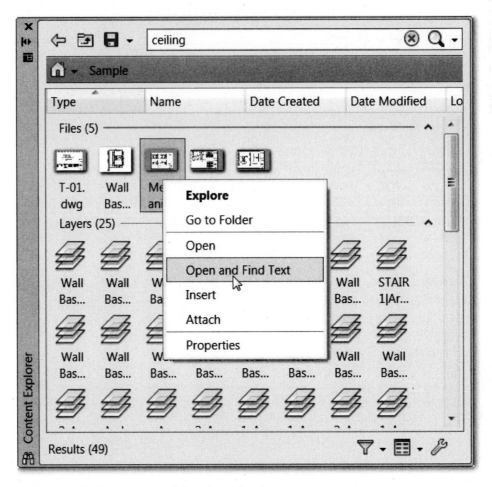

FIGURE 13–46 *Shortcut menu of the selected drawing*

Repeat the same thing for the other file(s) containing the string that was searched for.

Using Autodesk Seek

AutoCAD also allows you to search for objects, text strings, attributes (including block attributes), and files from products available on the Autodesk Seek web site.

Selection of the **Autodesk Seek** (see Figure 13–47) switches the search to the Autodesk Seek library.

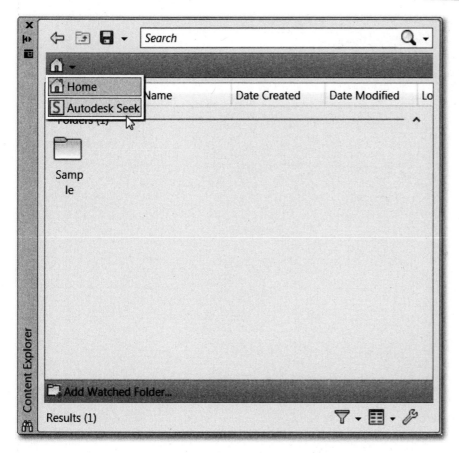

FIGURE 13–47 *Selecting Autodesk Seek to search the Autodesk Seek library*

The Autodesk Seek search box allows you to access and share product design information online. This includes generic or manufacturer-specific building products or components and associated design information. Blocks and drawings are uploaded with their attribute values to the Autodesk Seek web site. You can include products that meet design standards for the Americans with Disabilities Act (ADA), for example. The host drawing file uploaded to the Autodesk Seek web site does not include any external images, external references (xref), or underlay file references.

The search feature allows you to quickly locate files that meet your search criteria. When you enter a search string into the search field, Content Explorer examines the index and returns the files with a file name, author, keyword, comment, name, subject, or title that meets the search criteria. Any objects with a file name that meets the search string are also returned. Figure 13–48 shows searching to locate "chair" in Autodesk Seek.

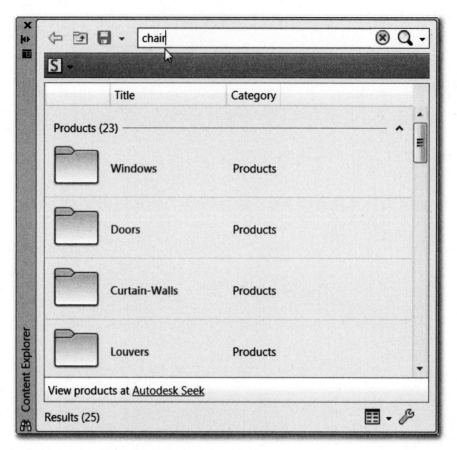

FIGURE 13–48 *Searching for "chair" in Autodesk Seek*

AutoCAD lists the links to content that match the search criteria as shown in Figure 13–49 at (A). By clicking on a link, AutoCAD provides Manufacturer's descriptions (Figure 13–49 (B)), Types/Specification (Figure 13–49 (C)), and any available files (Figure 13–49 (D)). To insert any of the available files, first select the file, then drag and drop it into the current drawing.

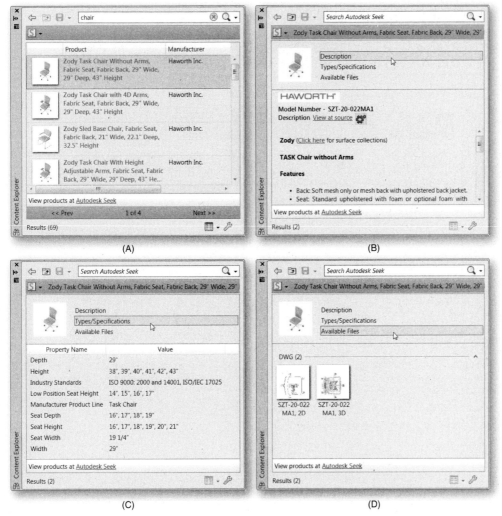

FIGURE 13–49 *(A) Listing of the links to content that match the criteria, (B) manufacturer's description of the selected link, (C) types/Specifications of the selected link, (D) available files of the selected link*

AutoCAD prompts for a new Block name as shown in Figure 13–50. Specify a block name and click the OK button. Then specify the insertion point, X and Y scale factor and rotation angle similar to inserting a block.

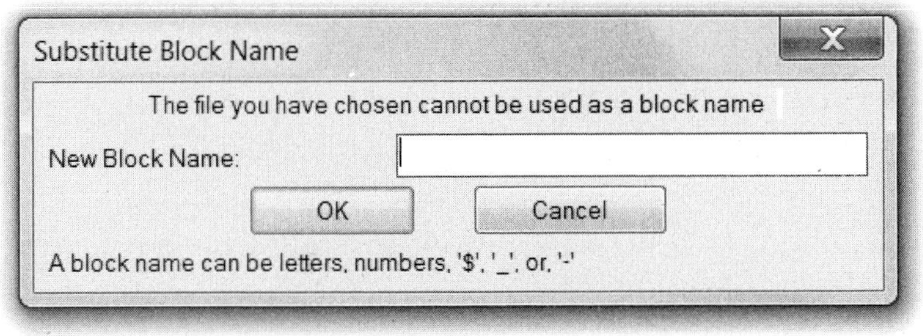

FIGURE 13–50 *Substitute Block Name dialog box*

Click Configure Settings as shown in Figure 13–51 to add, remove, enable, and disable content sources.

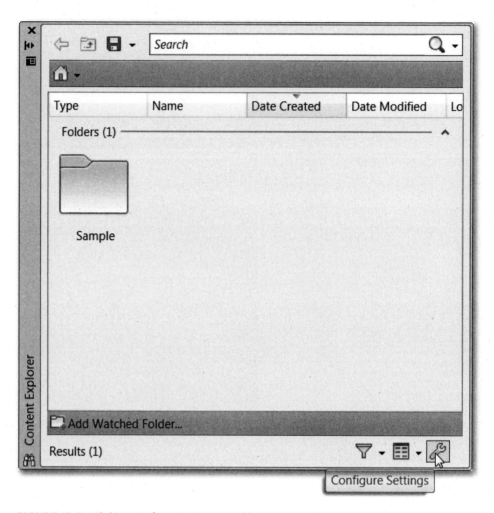

FIGURE 13–51 *Clicking Configure Settings to add, remove, enable, and disable content sources*

AutoCAD displays Configure Settings dialog box as shown in Figure 13–52.

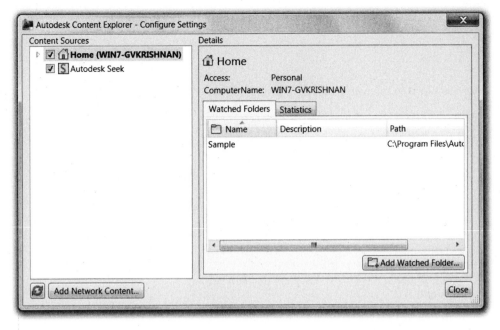

FIGURE 13–52 *Configure Settings dialog box*

To add a folder to the watched folder list, click **Add Watched Folder** and AutoCAD displays the Add Watched Folder dialog box. Navigate to the folder that you want to add, select it, and click OK. The folder is added to the watched folder list. To remove a folder, right-click on the folder in the Watch Folders list and select to remove. The folder is no longer available for searching or browsing with Content Explorer. Click the **Close** button to close the Configure Settings dialog box after making the necessary changes.

Folders on a network computer must be set as shared to be added as watched folders on that network computer. Only a network computer running Content Service can be added as a network content source in Content Explorer.

The use of the Content Explorer along with Autodesk Seek makes finding and manipulating objects quick and easy as long as the files and folders they are in are indexed.

REVIEW QUESTIONS

1. What command is used to invoke AutoCAD DesignCenter?

   a. DSNCEN

   b. DGNCEN

   c. ADCENTER

   d. DGNCTR

2. Which of the following are considered drawing content types?

   a. Blocks

   b. Layers

   c. Linetypes

   d. All the above

3. AutoCAD allows items to be directly edited from DesignCenter.
   a. True
   b. False

4. Which of the following are not drawing content types?
   a. Lines
   b. Circles
   c. Arcs
   d. All of the above

5. Within DesignCenter, a drawing is considered to be a _____.
   a. Content
   b. Container
   c. Folder
   d. A and B
   e. A and C

6. Invoking the DesignCenter command opens DesignCenter dialog box.
   a. True
   b. False

7. The default position for DesignCenter is in the lower-right corner.
   a. True
   b. False

8. Which of the following areas of DesignCenter displays the names and icons representing content?
   a. Toolbar
   b. Content area
   c. Tree
   d. Preview pane
   e. Description pane

9. Large Icons, Small Icons, List, and Details are four optional modes of displaying content using which of the following buttons?
   a. LOAD
   b. FIND
   c. UP
   d. VIEWS

10. Within the Tree View pane, which button can display the previous items and their paths accessed through DesignCenter?
    a. Open Drawings
    b. Desktop
    c. History
    d. None of the above

CHAPTER
14

Utility Commands

OBJECTIVES

After completing this chapter, you will be able to do the following:

- Create, customize, and use tool palettes
- Partial Load drawings
- Manage drawing properties
- Use the Calculator
- Manage named objects
- Delete unused named objects
- Use the utility display commands
- Use object properties
- Use X, Y, and Z filters
- Use the SHELL command
- Set up a drawing by means of the MVSETUP utility
- Use the Layer Translator
- Use the TIME and AUDIT commands
- Use Object Linking and Embedding
- Customize AutoCAD settings
- Export and import data
- Understand CAD Standards
- Use the Action Recorder
- Use script and slide commands
- Set up and use workspaces

TOOL PALETTES

Tool palettes are tabbed areas within the Tool Palettes window that provide an efficient method for organizing, sharing, placing blocks and hatches, executing a single command or a string of commands, and using custom tools provided by third-party developers. For example, one tool palette is named **Architectural**, with blocks representing doors, windows, and fixtures.

A special facility of tool palettes is the ability to pre-load a palette with objects with specific properties. When these objects are selected and dragged into the drawing area, they take their properties with them. For instance, if a green polyline on Layer 7 is selected and dragged onto a tool palette, dragging it back into the drawing area invokes the PLINE command, and the resulting polyline will be green and on Layer 7, regardless of the current layer or color control. The following objects can be dragged onto a tool palette:

- Blocks
- Dimensions
- External references
- Hatch patterns and gradient fills
- Objects such as arcs, circles, lines, and polylines
- Raster images
- AutoCAD commands or strings of commands

The TOOLPALETTES command causes the Tool Palettes window to be displayed. Invoke the TOOLPALETTES command from the Palettes panel located (Figure 14–1) in the View tab. AutoCAD displays the Tool Palettes window (Figure 14–2) with 29 tabs that include **Command Tools, Hatches, Civil, Structural, Electrical, Mechanical, Architectural**, and **Annotation**, and these contain commands, hatch patterns, icons representing blocks, and callout/bubble symbols.

FIGURE 14–1 *Invoking* TOOL PALETTES *the command from the Palettes panel*

NOTE	You can open or close the Tool Palettes window with the CTRL+3 key combination.

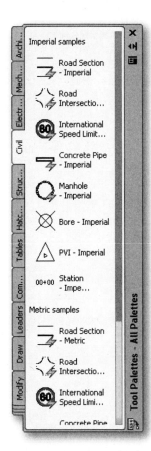

FIGURE 14–2 *The Tool Palettes window with the Civil tab displayed*

The default position for the Tool Palettes window is floating on the right side of the screen. When the Tool Palettes window is undocked, it can be docked by double-clicking in the title bar, which may be on the left or right side of the window (Figure 14–3), or by placing the cursor over the title bar and dragging the window all the way to the side on which you wish to dock it. Its position can be changed by placing the cursor over the double-line bar at the top of the window; either double-click or drag the window into the screen area or across to a docking position on the opposite side of the screen. Double-clicking causes the Tool Palettes window to become undocked and to float in the drawing area.

FIGURE 14–3 *The Tool Palettes window in the docked position*

Tool Palettes Window Shortcut Menus

Four different shortcut menus are available for managing the Tool Palettes window, its palettes, and elements on the palettes. The appearance of the shortcut menu depends on the location of the cursor (tab, tool palette element, open area in tool palette, and tool palette window title bar) when you right-click.

The shortcut menu displayed when you right-click the active tab in the Tool Palettes window (Figure 14–4) includes these options: *Move Up*, *Move Down*, *New Palette*, *Delete Palette*, *Rename Palette*, *View Options*, and *Paste*. The *Delete Palette*, *View Options*, and *Paste Function* are not available when an inactive tab is selected.

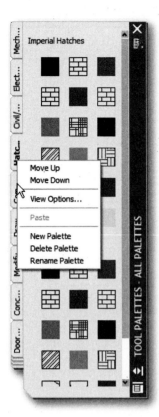

FIGURE 14–4 *The shortcut menu when right-clicking a Tool Palettes window tab*

Choose *Move Up* or *Move Down* to move the selected tab up or down one place, respectively, in the order of tabs.

Choose *New Palette* to create a new palette. AutoCAD will create a new tool palette to which you can add tools. Enter a name, or press [Enter] to use the default name.

> Tool palettes can be used only in the current or later version of AutoCAD in which they were created. For example, you cannot use a tool palette that was created in AutoCAD 2009 in AutoCAD 2006.

NOTE

Choose *Rename Palette* to rename the current tool palette.

Choose *Delete Palette* to remove the current tool palette.

Choose *View Options* to display the View Options dialog box (Figure 14–5) from which you can control the display of tools in the current tool palette or in all tool palettes. The **Image size** slider bar allows you to change the size of the images. The View style section allows you to set how the tool elements are displayed. Selecting **Icon only** causes the block/hatch pattern icon to be displayed as an image only without text. Selecting **Icon with text** causes the block/hatch pattern icon to be displayed as an image with the descriptive text below it. Selecting **List view** causes the block/hatch pattern icon to be displayed as an image with the descriptive text to its right, allowing for a more compressed listing of the symbols when used with a small image. The **Apply to** list box allows you to choose whether the changes are applied to the

Current Palette or to **All Palettes.** To exit the View Options dialog box and accept the changes, choose **OK.** To exit without accepting the changes, choose **Cancel.**

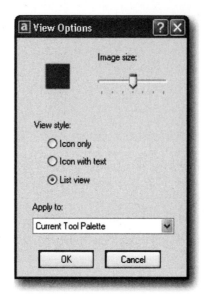

FIGURE 14–5 *The View Options dialog box*

Choose *Paste* to paste a block/hatch pattern shortcut that has been copied to the Clipboard on the tool palette. If there is nothing on the Clipboard, or if whatever is on the Clipboard is not a block/hatch pattern shortcut, the *Paste* option is not active and cannot be selected.

The shortcut menu displayed when you right-click one of the elements in a tool palettes window includes the options *Redefine, Block Editor, Cut, Copy, Delete, Rename, Update Tool Image, Specify Image,* and *Properties* (Figure 14–6). *Update Tool Image* and *Block Editor* are not available if the selected element is a command or hatch pattern.

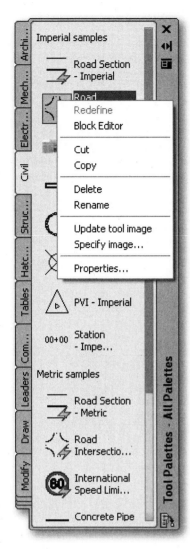

FIGURE 14–6 *The shortcut menu when right-clicking an element in the Tool Palettes window*

Choose *Redefine* to redefine the selected block and see Chapter 11 for details on redefining the block.

Choose *Block Editor* to invoke the BEDIT command, explained in Chapter 11, in the section that covers Dynamic Blocks.

Choose *Cut* to delete the selected element and place it on the Clipboard.

Choose *Copy* to copy the selected element and place it on the Clipboard.

Choose *Delete* to remove the selected element.

Choose *Rename* to rename the selected element.

Choose *Update Tool Image* to update the selected image of the tool when the definition for a block, xref, or raster image is changed. You must save the drawing before you can update the tool image. The icon for a block, xref, or raster image in a tool palette is not automatically updated if its definition changes.

Image(s) © Cengage Learning 2013

Choose *Specify Image* to replace the icon for a tool with an image that you specify. This action is useful when the automatically generated icon is too cluttered to be easily recognizable.

Choose *Properties* to display the Tool Properties dialog box showing the properties of the selected element. Tool properties are explained later in this chapter.

The shortcut menu that is displayed when you right-click in an open area of the Tool Palettes window includes the options *Allow Docking, Seek Design Content, Auto-Hide, Transparency, View Options, Sort By, Paste, Add Text, Add Separator, New Palette, Delete Palette, Rename Palette*, and *Customize* (Figure 14–7).

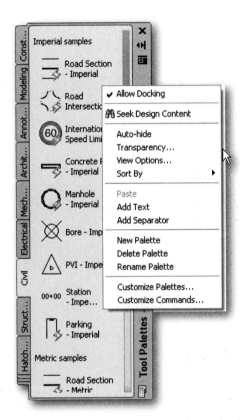

FIGURE 14–7 *The shortcut menu when right-clicking in an open area of the Tool Palettes window*

The *Allow Docking* option, when checked, allows you to drag the Tool Palettes window to one side of the screen and dock it. When *Allow Docking* is not checked, the window cannot be docked.

Choosing *Seek Design Content* causes the Autodesk Seek web page to open.

The *Auto-Hide* option, when checked, causes the Tool Palettes window, when floating, to be hidden, except for the title bar. To display the window, move the cursor over the title bar.

The *Transparency* option, when selected, causes the Transparency dialog box to be displayed (Figure 14–8). In the General section the **Clear-Solid** slider controls the degree of transparency when the dialog box is displayed. Sliding the bar to the left (Clear) causes the Tool Palettes window, only when floating, to be opaque. The closer the indicator is to the right (Solid), the more transparent the window will become. In the Rollover section the **Clear-Solid** slider controls the degree of transparency when the mouse is rolled over the dialog box. Checking **Apply these settings to all palettes** applies the settings to all palettes. Checking **Disable all window transparency (global)** disables the settings to all palettes.

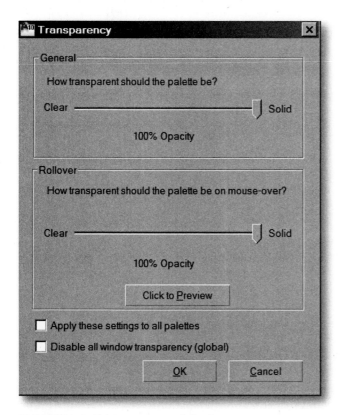

FIGURE 14–8 *The Transparency dialog box*

The *View Options* option, when selected, causes the View Options dialog box to be displayed.

Choosing *Sort By* causes a shortcut menu to be displayed from which you can sort the elements on the tool palette by *Name* or *Type*.

The *Paste* option functions similarly to the one explained earlier.

Choose *Add Text* to add descriptive text at the location of the cursor when right-clicked.

Choose *Add Separator* to create a separating line on the palette at the location of the cursor when right-clicked.

The *New Palette*, *Delete Palette*, and *Rename Palette* options function similarly to the ones explained earlier.

The *Customize Commands* option, when selected, causes the Customize User Interface dialog box to be displayed and which in turn allows you to customize the user interface.

The *Customize Palettes* option, when selected, causes the Customize dialog box to be displayed (Figure 14–9).

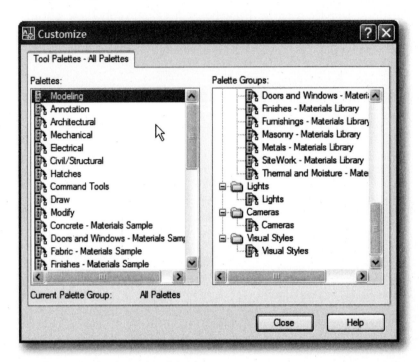

FIGURE 14–9 *The Customize dialog box*

The Customize dialog box allows you to manage the palettes and groups of palettes shown in the Tool Palettes window. It allows you to create, modify, and organize palettes in addition to importing and exporting palette files. You can organize tool palettes into groups and specify which group of tool palettes is displayed. For example, if you have several tool palettes that contain hatch patterns, you can create a group called Hatch Patterns. You can then add all your tool palettes that contain hatch patterns to the Hatch Pattern group. When you set the Hatch Pattern group as the current group, only those tool palettes you've added to the group are displayed.

The Palettes section lists all available palettes. Click and drag a palette to move it up or down in the list. From the shortcut menu that appears when you right-click a palette, you can rename, delete, or export the palette. When you export a palette, it is saved to a file with an *.xtp* extension. From the shortcut menu that appears when you right-click in the open area of the dialog box in the Palettes section, you can create a new palette or import a palette.

The Palettes Group section displays the organization of your palettes in a tree view. Click and drag a palette to move it into another group. From the shortcut menu that appears when you right-click a palette group, you can create, delete, rename, export, and import a group, remove a tool from a palette group, and set it current to display the selected group of palettes.

Choosing **Current Palette Group** displays the name of the palette group currently shown.

The shortcut menu displayed when you right-click on the Tool Palette window title bar includes the options *Move, Size, Close, Allow Docking, Auto-Hide, Transparency, New Palette, Rename, Customize, Dynamic Blocks, Samples,* and *All Palettes* (Figure 14–10).

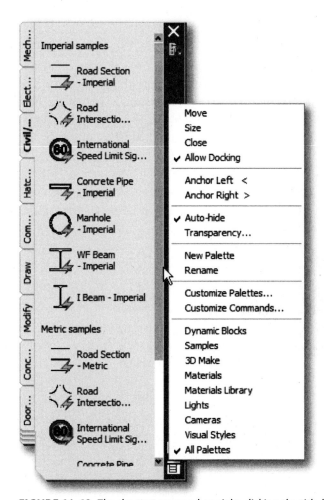

FIGURE 14–10 *The shortcut menu when right-clicking the title bar of the Tool Palettes window*

Choose *Move* to drag the Tool Palettes window to another location on the screen.

Choose *Size* to drag the title bar edge of the Tool Palettes window to make it wider or narrower. You can make the window longer or shorter (vertically) by placing the cursor on the top or bottom until the double arrow appears and then dragging the edge up or down.

Choose *Close* to close the Tool Palettes window.

Allow docking, auto-hide, transparency, new palette, rename, customize palettes, and customize are explained in the previous section.

The shortcut menu lists available palette groups, and you can switch between the groups.

Insert Blocks/Hatch Patterns from a Tool Palette

To insert a block from a tool palette, place the cursor on the block symbol in the tool palette, press the pick button and drag the symbol into the drawing area. The block will be inserted at the point where the cursor is located when the pick button is released. This procedure is best implemented by using the appropriate OSNAP mode. Another method of inserting a block from a tool palette is to select the block symbol in the tool palette and then select a point in the drawing area for the insertion point.

To draw a hatch pattern that is a tool in a tool palette, place the cursor on the hatch pattern symbol in the tool palette, press the pick button, drag the symbol into the boundary to receive the hatch pattern, and release the pick button. Another method of drawing a hatch pattern that is a tool in a tool palette is to select the hatch pattern symbol in the tool palette and then select a point within a boundary in the drawing area.

Block Tool Properties

Blocks whose symbols appear in a tool palette are not, as a rule, blocks defined in the current drawing. Usually, they reside as block definitions in another drawing or in some cases they might even be drawing files. As a tool in a tool palette, a block has tool properties. To access the block/drawing tool properties, right-click on the tool's symbol in the tool palette, and select *Properties* from the shortcut menu. The Tool Properties dialog box is displayed (Figure 14–11).

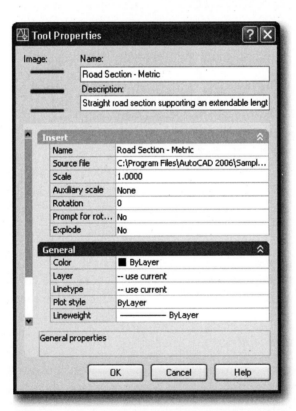

FIGURE 14–11 *The Tool Properties dialog box for block/drawing tools*

The Tool Properties dialog box controls the properties associated with the selected tool. The Insert properties of a block/drawing tool in a tool palette include **Name, Source file, Scale, Auxiliary Scale, Rotation, Prompt for rotation**, and **Explode.** The General properties include **Color, Layer, Linetype, Plot style**, and **Lineweight.**

The **Name** specifies the name of the block of the selected tool. The **Source file** property edit box lists the path to the drawing file where the block that is a tool in the tool palette resides. If the tool is a drawing file, the edit box lists the path to it.

> Any change to the path/file name in the *Source file* property text box will prevent the block/ drawing from being inserted.

NOTE

The **Scale** factor specifies the X, Y, and Z scale factor of the block. **Auxiliary scale** overrides the regular scale setting and multiplies your current scale setting by the plot scale or the dimension scale. The **Rotation** specifies the rotation angle of the block. If the **Prompt for rotation** list box is set to **No,** the block will be inserted with the default rotation angle. If it is set to **Yes,** you will be prompted for the angle of rotation. The **Explode** list box determines whether or not the block is exploded when inserted. If the **Explode** list box is set to **No,** the block will be inserted as an unexploded block. If it is set to **Yes,** the separate parts that make up the block will be drawn as separate entities, that is, as a block that is inserted and exploded. The **Color, Layer, Linetype, Plot Style**, and **Lineweight** list boxes specify the override for the color, layer, linetype, plot style, and lineweight, respectively.

Pattern Tool Properties

Hatch patterns whose symbols appear in a tool palette have tool properties. The Pattern properties of a hatch pattern tool in a tool palette include **Tool type, Type, Pattern name, Angle, Scale, Auxiliary scale, Spacing, ISO pen width**, and **Double.** The General properties include **Color, Layer, Linetype, Plot Style**, and **Lineweight.** To access the pattern tool properties, right-click on the tool's symbol in the tool palette, and select *Properties* from the shortcut menu. The Tool Properties dialog box is displayed (Figure 14–12).

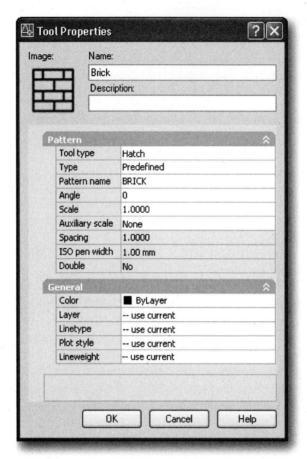

FIGURE 14–12 *The Tool Properties dialog box for pattern tools*

Tool type specifies whether the tool is a hatch or a gradient. **Type** specifies the pattern type of the hatch. The pattern types include **User-defined, Predefined** from one of the hatch patterns listed in the **Pattern** list box, or **Custom** from one of the hatch patterns in the drawing as listed in the **Custom Pattern** list box. **Pattern name** specifies the pattern name of hatch. **Angle, Scale**, and **Spacing** specify the angle, scale, and spacing of the hatch, respectively. **Auxiliary scale** overrides the regular scale setting and multiplies your current scale setting by the plot scale or the dimension scale. **ISO pen width** specifies the ISO pen width of an ISO hatch pattern. **Double** determines whether the hatch pattern is double. **Color, Layer, Linetype, Plot style**, and **Lineweight** specify the override for the color, layer, linetype, plot style, and lineweight, respectively.

Creating and Populating Tool Palettes

A new tool palette can be created from the Tool Palette shortcut menu by selecting the New Tool Palette option, as described earlier in this section.

You can also create a new tool palette from the Tree view or Content area of the DesignCenter. See Chapter 13 for an explanation of using the DesignCenter. From the Tree view or Content area, highlight an item, and then right-click to display the shortcut menu. If the selected item is a folder, one of the available options in the shortcut menu is *Create Tool Palette Of Blocks* (Figure 14–13). If there are no blocks in the folder, a message will be displayed stating "Folder does not contain any drawing files." If the selected item is a drawing, one of the available options in the shortcut menu is *Create Tool Palette*. If the drawing does not contain any blocks, a message will be displayed stating "Drawing does not contain any block definitions."

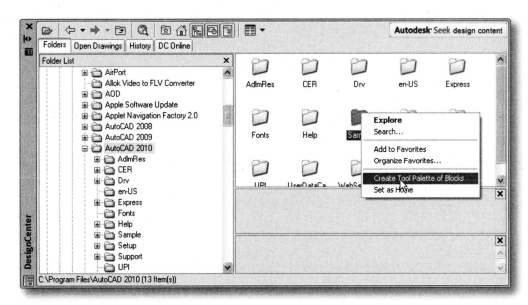

FIGURE 14–13 *Shortcut menu, DesignCenter content area*

When one of the options to create a tool palette is selected, AutoCAD creates a new tool palette, which will be populated with the drawings from the selected folder or blocks from the selected drawing.

From the Content area, in addition to creating tool palettes by right-clicking a folder or drawing, you can also right-click on a block and select the *Create Tool Palette* option from the shortcut menu. In this case, the new tool palette will contain only the selected block, and you will be prompted to name the tool palette.

You can also drag and drop a block from a drawing or a drawing from the Content area of the DesignCenter on to the one of the existing palettes.

PARTIAL LOAD

The PARTIALOAD command allows you to work with just part of a drawing by loading geometry only from specific views or layers. The command is available only when a drawing is partially open and when specified and named objects are loaded. It is recommended that you partially open a drawing by using the OPEN command and choosing **Partial Open** in the Select File dialog box to display the Partial Open dialog box. Named objects include blocks, layers, dimension styles, linetypes, layouts, text styles, viewport configurations, UCSs, and views. Invoke the PARTIALOAD command at the On-Screen prompt and AutoCAD displays the Partial Load dialog box (Figure 14–14).

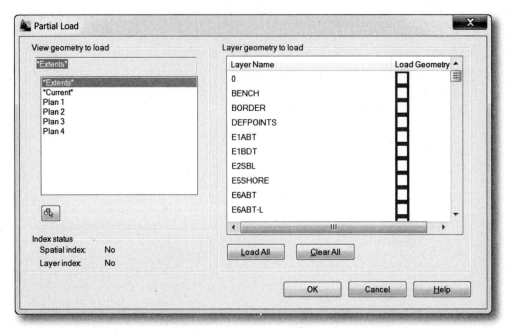

FIGURE 14–14 *The Partial Load dialog box*

In the Partial Load dialog box, select a view. The default view is Extents. Only geometry from model space views that are saved in the current drawing can be loaded. Select layer(s). No layer geometry is loaded into the drawing if you do not select any layers to load. However, all drawing layers will exist in the drawing. Even if the geometry from a view is specified to load, and no layer is specified to load, no geometry is loaded. When the view(s) and layer(s) have been specified, choose **OK** to partially load the geometry from the specified layers.

> **NOTE**
>
> If you invoke the PARTIALLOAD command in a drawing that has not been partially opened, AutoCAD will respond with the message "Command not allowed unless the drawing has been partially opened." This response is only displayed at the command line and not shown at the on-screen prompt.

DRAWING PROPERTIES

Drawing properties allow you to keep track of your drawings by having properties assigned to them. The properties you assign will identify the drawing by its title, author, subject, and keywords for the model or other data. Hyperlink addresses or paths can be stored along with custom properties.

Invoke **Drawing Properties** from the Application Menu under the Drawing Utilities submenu and AutoCAD displays the Drawing Properties dialog box (Figure 14–15).

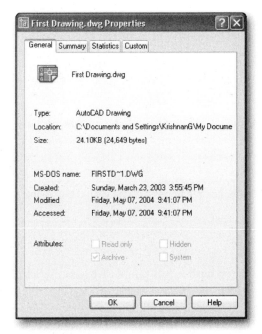

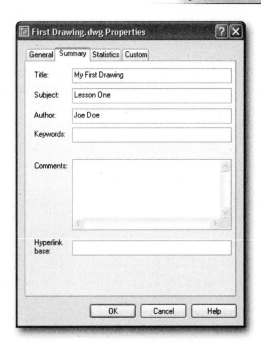

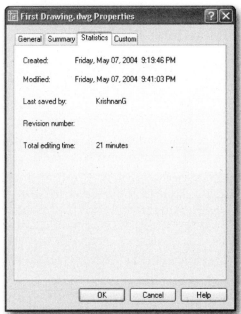

FIGURE 14–15 *The Drawing Properties dialog box: General tab (top left), Summary tab (top right), Statistics tab (bottom left), and Custom tab (bottom right)*

The Drawing Properties dialog box has four tabs: **General, Summary, Statistics,** and **Custom.**

The General tab (Figure 14–15, top left) displays the drawing type, location, size, and other information. These values come from the operating system (OS). These fields are read-only. However, the attributes options are made available by the OS if you access file properties through Windows® Explorer®.

The Summary tab (Figure 14–15, top right) allows you to enter a Title, Subject, Author, Keywords, Comments, and a Hyperlink base for the drawing. **Keywords** for drawings sharing a common property will help in your search. For a **Hyperlink base**, you can specify a path to a folder on a network drive or an Internet address.

The Statistics tab (Figure 14–15, bottom left) displays information such as the dates files were created and last modified. You can search for all files created at a certain time.

The Custom tab (Figure 14–15, bottom right) allows you to specify custom properties. Enter the names of the custom fields in the left column, and the value for each custom field in the right column.

NOTE

Properties entered in the Drawing Properties dialog box are not associated with the drawing until you save the drawing.

QUICKCALC

AutoCAD provides an "on demand" calculator with a full range of mathematical, scientific, and geometric calculations, which can be used to create and use variables and convert units of measurement. This QuickCalc calculator includes shortcut functions that are found in the Geometric Calculator but are more easily accessed in QuickCalc.

Invoke the QUICKCALC command from the Palettes panel (Figure 14–16) located in the View tab, and AutoCAD displays the QuickCalc palette (Figure 14–17).

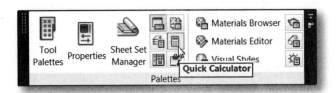

FIGURE 14–16 *Invoking the* QUICKCALC *command from the Palettes panel*

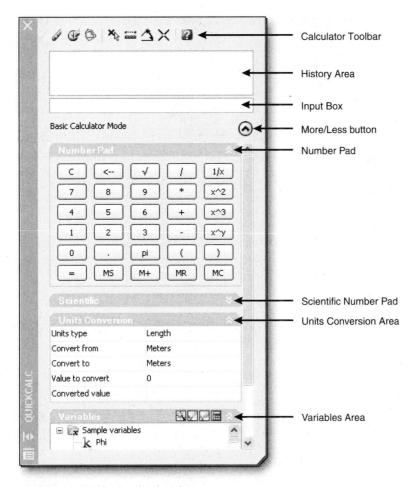

Calculator Toolbar

History Area

Input Box

More/Less button

Number Pad

Scientific Number Pad

Units Conversion Area

Variables Area

FIGURE 14–17 *The QuickCalc Palette*

The QuickCalc palette contains the following areas: Toolbar, History, Input Box, More/Less Button, Number Pad, Scientific, Units Conversion, and Variables.

The Toolbar allows you to perform quick calculations of common functions.

Choosing **Clear** clears the **Input** box.

Choosing **Clear History** clears the **History** area.

Choosing **Paste value to command line** pastes the value in the **Input** box to the command line. When QuickCalc is used transparently during a command, this button is replaced by **Apply.**

Choosing **Get Coordinates** allows you to calculate the coordinates of a point location that you have selected in the drawing.

Choosing **Distance Between Two Points** allows you to calculate the distance between two point locations that you have selected on an object.

Choosing **Angle of Line Defined by Two Points** allows you to calculate the angle of two point locations that you have selected on an object.

Choosing **Intersection of Two Lines Defined by Four Points** allows you to calculate the intersection of four point locations that you have selected on an object.

Choosing **Help** causes the QuickCalc calculator help to be displayed.

The **History** area shows the expressions that have previously been evaluated. From it you can copy expressions to the Clipboard.

The **Input** area is where expressions are entered and retrieved. Choosing equal (=) sign evaluates the expression and displays the results in the Input area.

Choosing the **More/Less** icon causes all function areas to be hidden or displayed. Right-clicking allows you to select the individual function areas to hide or display.

The **Number Pad** is a standard calculator keypad where numbers and symbols for arithmetic expressions can be entered. Enter a value or expression and then click the equal (=) sign to evaluate the expression.

The **Scientific** area allows you to evaluate trigonometric, logarithmic, exponential, and other expressions commonly associated with scientific and engineering applications.

The **Units Conversion** area allows you to convert units of measurement from one unit type to another unit type. Only decimal values without units are accepted.

The **Units type** box allows you to select length, area, volume, and angular values from a list.

The **Convert from** box lists the units of measurement from which to convert.

The **Convert to** box lists the units of measurement to which to convert.

The **Value to convert** box is where you enter the value to convert.

The **Converted value** box is where the converted value is displayed.

The **Variables** area allows you to access predefined constants and functions. You can use the **Variables** area to define and store additional constants and functions.

The **Variables Tree** has predefined shortcut functions and user-defined variables stored in tree format. The shortcut functions are common expressions that are a combination of a function and an object snap. Following are the available predefined shortcut functions:

Dee is short for **dist(end,end).** It allows you to specify the distance between two endpoints.

Ille is short for **ill(end,end,end,end).** It allows you to specify the intersection of two lines defined by four endpoints.

Mee is short for **(end + end)/2.** It allows you to specify the midpoint between two endpoints.

Nee is short for **nor(end,end).** It allows you to specify the unit vector in the XY plane and normal to two endpoints.

Rad allows you to specify the radius of a selected circle, arc, or polyline arc.

Vee is short for **vec(end,end).** It allows you to specify the vector from two endpoints.

Vee1 is short for **vec1(end,end).** It allows you to specify the unit vector from two endpoints.

Choosing the **Calculator** icon returns the converted value to the **Input** area.

Choosing **New Variable** opens the Variable Definition dialog box, where you can specify a new variable.

Choosing **Edit Variable** opens the Variable Definition dialog box, where you can make changes to the selected variable.

Choosing **Delete** variable deletes the selected variable.

Choosing **Calculator** returns the selected variable to the Input box.

MANAGING NAMED OBJECTS

The RENAME command allows you to change the names of blocks, dimension styles, layers, linetypes, text styles, views, User Coordinate Systems, or viewport configurations.

Invoke the RENAME command by entering **rename** at the on-screen prompt and AutoCAD displays the Rename dialog box (Figure 14–18).

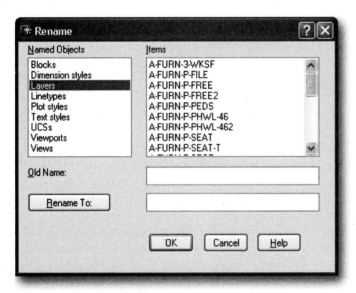

FIGURE 14–18 *Rename dialog box*

> Except for the layer named 0 and the linetype named Continuous, you can change the name of any of the named objects.

NOTE

In the **Named Objects** list box, select the object name you want to change. The **Items** list box displays the names of all objects that can be renamed. To change the object's name, pick the name in the **Items** list box or enter it into the **Old Name** text box. Enter the new name in the **Rename To** text box, and select **Rename To** to update the object's name in the **Items** list box. To close the dialog box, choose **OK.**

DELETING UNUSED NAMED OBJECTS

The PURGE command is used to selectively delete any unused named objects.

Invoke the PURGE command from the Application Menu and AutoCAD displays the Purge dialog box (Figure 14–19).

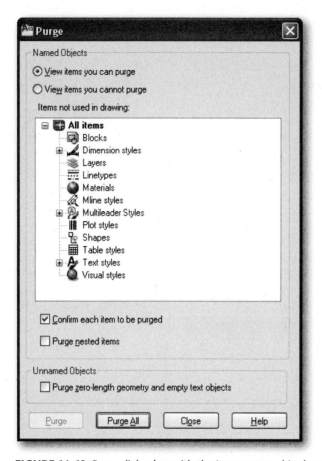

FIGURE 14–19 *Purge dialog box with the Items not used in drawing section displayed*

Choosing **View items you can purge** causes the **Items not used in drawing** list box to list categories of named items, under which are listed the individual named items that have been defined in the drawing but are not currently being used. For example, if a layer has been defined but has nothing drawn on it, and it is not the current layer, it can be purged. If the drawing contains a block definition but a reference to that block has not been inserted, it can be purged. Objects such as lines, circles, and other basic unnamed drawing elements cannot be purged.

Selecting **Confirm each item to be purged** causes AutoCAD to display the Confirm Purge dialog box and ask you to reply by selecting **Yes** or **No** before continuing. Selecting **Purge nested items** causes AutoCAD to purge nested items within any item selected.

In the Unnamed Objects section choosing **Purge zero-length geometry and empty text objects** deletes geometry of zero length (lines, arcs, polylines, and so on) in nonblock objects. It also deletes mtext and text that contains only spaces (no text) in nonblock objects.

Once an individual named item in the list has been selected or a category of items has been selected that has items that can be purged, the **Purge** and **Purge All** buttons at the bottom of the dialog box become operational. Choosing **Purge** purges the selected items, and **Purge All** purges all unused items. Choosing **View items you**

cannot purge causes the **Items currently used in drawing** list box to list categories of named items, under which are listed the individual named items that have been defined in the drawing but are currently being used (Figure 14–20). For example, if a layer has something drawn on it or it is the current layer, it cannot be purged. If a reference of a block has been inserted in the drawing, that block cannot be purged. The text area under the list of items informs you why a selected item cannot be purged. For example, if the Standard Text Style is selected, the message "The default text style, STANDARD, cannot be purged" will be displayed.

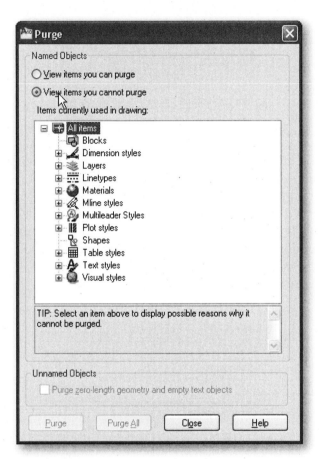

FIGURE 14–20 *Purge dialog box with the Items currently used in drawing: section displayed*

The PURGE command removes only one level of reference. For instance, if a block has nested blocks, the PURGE command removes only the most deeply nested block definition. To remove the second-, third-, or deeper-level blocks within blocks, you must repeat the PURGE command until there are no referenced objects. You can use the PURGE command at any time during a drawing session. Choosing **Purge nested items** removes all unused named objects from the drawing even if they are contained within or referenced by other unused named objects. The Confirm Purge dialog box is displayed, and you can cancel or confirm the items to be purged.

Individual shapes are part of a *.shx* file. They cannot be renamed, but references to those that are not being used can be purged. Views, user coordinate systems, and viewport configurations cannot be purged, but the commands that manage them provide options to delete those that are not being used.

COMMAND MODIFIER—MULTIPLE

MULTIPLE is not a command, but when used with another AutoCAD command, it causes automatic recalling of that command when it is completed. You must press ESC or right-click and select CANCEL from the short-cut menu to terminate this repeating process. Here is an example of using this modifier to cause automatic repeating of the ARC command:

> multiple (Enter)
>
> Enter command name to repeat: arc (Enter)

You can use the MULTIPLE command modifier with any of the draw, modify, and inquiry commands. PLOT, however, will ignore the MULTIPLE command modifier.

UTILITY DISPLAY COMMANDS

The utility display commands include VIEW, REGENAUTO, DRAGMODE, and BLIPMODE.

Saving Views

The VIEW command allows you to give a name to the display in the current viewport and have it saved as a view. You can recall a view later by using the VIEW command and responding with the name of the view desired. This is useful for moving back quickly to needed areas in the drawing without having to resort to zoom and pan. Naming and saving views is a significant part of creating a drawing set with the Sheet Set Manager, as discussed in Chapter 15.

Invoke the VIEW command by choosing **View Manager** from the **Views** panel located on the View tab and AutoCAD displays the View Manager dialog box (Figure 14–21).

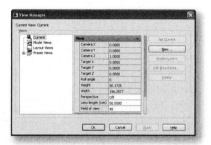

FIGURE 14–21 *View Manager and New View Shot Properties dialog boxes*

The View Manager dialog box is used to create named views and sets. In this box, you can also rename, modify, and delete named views including model named views, camera views, layout views, and preset views.

The **View** category lists available views. You can expand each node except for the Current node to display its views.

The **Current** selection displays the current view and its View and Clipping properties.

The **Model Views** selection displays a list of named views and cameras and lists General, View, and Clipping properties for a selected view.

The **Layout Views** selection displays a list of viewports on a layout that define a view and lists General and View properties for a selected view.

The **Preset Views** selection displays a list of orthogonal and isometric views and lists the General properties for a selected view.

To create a new view, choose **New.** AutoCAD displays the New View / Shot Properties dialog box with the View Properties tab current (Figure 14–21).

On the View Properties tab of the New View / Shot Properties dialog box, specify a name for the new view in the **View name** box. Specify a category for the named view in the **View category** box. You can select a view category from the list, enter a new category, or leave this option empty. Grouping by category will be helpful in creating drawing sheet sets.

The Boundary section allows you to select one of the two options to specify the area for the new view.

Choosing **Current display** creates the current display as the new view.

Choosing **Define window** allows you to specify diagonally opposite corners of a window that will define the area for the new view. AutoCAD temporarily closes the dialog boxes so that you can use the pointing device to define the opposite corners of the new view window.

The Settings section allows you to select where to save the new view.

Choosing **Save Layer Snapshot with View** saves the current layer visibility settings with the new named view.

Select **UCS** from the **UCS** menu for model and layout views to save with the new view. Select **Live Section** from the **Live Section** menu, which is available only for model views and which specifies the live section applied when the view is restored.

Select **Visual Style** from the Visual Style menu, which is available only for model views and which specifies a visual style to save with the view.

The Background section allows you to select the background type to apply to the selected view. Options are Solid, Gradient, Image, or Sun & Sky. AutoCAD opens the appropriate dialog box to adjust the settings.

The Save sun properties with view Current override toggle controls whether "Sun & Sky" data are saved with the named view. The option is automatically selected when choosing Sun & Sky for the background type. Saving "Sun & Sky" data to a named view is optional when using a background type other than Sun & Sky.

Choose **OK** to create a new view and close the New View / Shot Properties dialog box.

On the Shot Properties tab of the New View / Shot Properties dialog box, you can define the transition and motion to use for a view when the view is played with ShowMotion.

In the Transition section you can define the transition to use when playing back a view by choosing from the options in the text box. You can set the time of the transition in seconds in the **Transition duration** text box.

In the Motion section you can define the behavior of the motion to use when playing back a view. **Movement type** options include Zoom In, Zoom Out, Track Left, Track Right, Crank Up, Crank Down, Look, and Orbit. Selecting **Preview** displays a preview of the transition and motion assigned to the named view. Selecting **Loop** continuously plays back the transition and motion assigned to the named view.

To restore the selected named view, choose **Set Current** in the View Manager dialog box. You can also restore a named view by double-clicking its name in the list, or by right-clicking its name and choosing *Set Current* on the shortcut menu.

To update the layer information saved with the selected named view to match the layer visibility in the current model space or layout viewport, choose **Update Layers.**

To edit boundaries of the selected view, choose **Edit Boundaries.** AutoCAD displays the selected named view centered and zoomed out and with the rest of the drawing area in a lighter color to show the boundaries of the named view. You can specify opposite corners of a new boundary repeatedly until you press [Enter] to accept the results.

To delete the selected named view, choose **Delete**.

Choose **OK** to close the dialog box and save the changes.

Controlling the Regeneration

The REGENAUTO command controls automatic regeneration. When REGENAUTO is set to ON, AutoCAD drawings regenerate automatically. When it is set to OFF, you may have to regenerate the drawing manually to see the current status of the drawing.

Invoke the REGENAUTO command by entering **regenauto** at the on-screen prompt, and AutoCAD displays the following prompt:

    Enter mode *(select an option)*

When you set REGENAUTO to OFF, what you see on the screen may not always represent the current state of the drawing. When changes are made by certain commands, the display will be updated only after you invoke the REGEN command. The delay caused by constant regenerating can be avoided as long as you are aware of the changing status of the geometry versus what is being displayed. Returning the REGENAUTO setting to ON will cause a regeneration. If a command should require regeneration while REGENAUTO is set to OFF, you will be prompted:

    About to regen, proceed?

Responding with **n** (No) will abort the command.

Regeneration during a transparent command will be delayed until a regeneration is performed at the end of that transparent command. The following message will appear in the text screen window:

    REGEN QUEUED

Controlling the Dragging of Objects

The DRAGMODE command controls the way dragged objects are displayed. Certain draw and modify commands display highlighted dynamic (cursor-following) representations of the objects being drawn or edited. This can slow down the drawing process if the objects are very complex. Setting DRAGMODE to OFF turns off this dynamic display while dragging.

Invoke the DRAGMODE command by entering **dragmode** at the on-screen prompt and AutoCAD displays the following prompt:

Enter new value *(select one of the three available options)*

When DRAGMODE is set to OFF, all calls for dragging are ignored. Setting DRAGMODE to ON allows dragging by use of the DRAG command modifier. Setting DRAGMODE to AUTO (default) causes dragging wherever possible.

Controlling the Display of Marker Blips

The BLIPMODE command controls the display of marker blips. When BLIPMODE is set to ON, a small cross mark is displayed when points on the screen are specified with the cursor or by entering their coordinates. After you edit for a while, the drawing can become cluttered with these blips. They have no effect other than visual reference and can be removed at any time by using the REDRAW, REGEN, ZOOM, or PAN commands. Any other command requiring regeneration causes the blips to be removed. When BLIPMODE is set to OFF, the blip marks are not displaced.

Invoke the BLIPMODE command by entering **blipmode** at the on-screen prompt, and AutoCAD displays the following prompt:

Enter mode *(select one of the two options)*

CHANGING THE DISPLAY ORDER OF OBJECTS

The DRAWORDER command allows you to change the display order of objects as well as images. This will ensure proper display and plotting output when two or more objects overlay one another. For instance, when a raster image is attached over an existing object, AutoCAD obscures them from view. With the use of the DRAWORDER command, you can make the existing object display over the raster image.

Invoke the DRAWORDER command from the Modify panel located in the Home tab by selecting **Draw Order** (Figure 14–22) and then choosing from the options: *Bring To Front, Send To Back, Bring Above Objects*, or *Send Under Objects*. AutoCAD displays the following prompts:

Select objects: *(select the objects for which you want to change the display order, and press* Enter *to complete object selection)*

Enter object ordering option *(select one of the options)*

Select reference object: *(select the reference object for changing the order of display)*

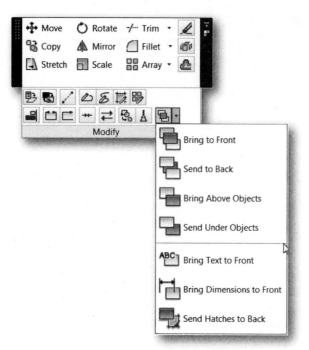

FIGURE 14–22 *Invoking the DRAWORDER command from the Modify panel*

When multiple objects are selected for reordering, the relative display order of the objects selected is maintained.

Choosing *Bring Above Objects* moves the selected object(s) above a specified reference object.

Choosing *Send Under Objects* moves selected object(s) below a specified reference object.

Choosing *Bring To Front* moves selected object(s) to the front of the drawing order.

Choosing *Send To Back* moves selected object(s) to the back of the drawing order.

Choosing *Bring Text To Front* moves all text in front of all other objects in the drawing.

Choosing *Bring Dimensions To Front* moves all dimensions in front of all other objects in the drawing.

Choosing *Send Hatches To Back* moves all hatch patterns, solid fills, and gradient fills (including hatch objects on locked layers) to be behind all other objects in the drawing.

NOTE

The DRAWORDER command terminates when selected object(s) are reordered. The command does not continue to prompt for additional objects to reorder.

CHANGING THE DIRECTION OF OBJECTS

The REVERSE command reverses the order of vertices of selected lines, polylines, splines, and helixes. This command is useful for reversing the directions of objects that use linetypes which include text. Invoke the REVERSE command from the Modify panel located in the Home tab. AutoCAD displays the following prompts:

> Select line, polyline, spline, or helix to reverse direction: *(select one or more objects to change direction, and press* Enter *to complete object selection)*

AutoCAD changes the direction of the selected objects.

OBJECT PROPERTIES

Three important properties control the appearance of objects: color, linetype, and lineweight. You can specify the color, linetype, and lineweight for the objects to be drawn with the help of the LAYER command, as explained in Chapter 3. You can perform the same actions from their respective list boxes in the Properties toolbar or by invoking the COLOR, LINETYPE, or LINEWEIGHT commands from the **Format** menu.

Setting an Object's Color

The Color option menu of the Properties panel in the Home tab (Figure 14–23) allows you to specify a color for the objects to be drawn, separate from the assigned color for the layer. You can select **ByLayer** or **ByBlock**, select one of the primary colors or the **Select Color** options to open the Select Color dialog box (Figure 14–24) to define the color of objects selecting from the 255 AutoCAD Color Index (ACI) colors, True Colors, and Color Book colors.

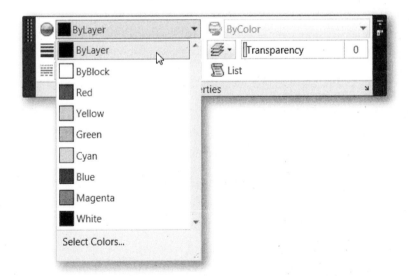

FIGURE 14–23 *Selecting from the Color menu of the Properties panel*

The default is set to ByLayer, which causes the objects drawn to assume the color of the layer on which they are drawn. You can also select one of the colors from the chart or one of the color bars below the chart on the Index Color tab (Figure 14–24, left) of the Select Color dialog box. All new objects you create are drawn with this color, regardless of the color of the current layer, until you again set the color to **ByLayer** or **ByBlock. ByBlock** causes objects to be drawn in white, or black against a white background, until selected for inclusion in a block definition. Subsequent insertion of a block reference that contains objects drawn under the **ByBlock** option causes those objects to assume the color of the current setting of the COLOR command.

Instead of choosing from 256 standard colors, you can also choose colors from the True Color graphic interface located on the True Color tab (Figure 14–24, center) with its controls for **Hue, Saturation, Luminance**, and **Color Model** or from standard Color Books, such as PANTONE®, located on the Color Books tab (Figure 14–24, right). True Color and Color Books options make it easier to match colors in your drawing with colors of actual materials.

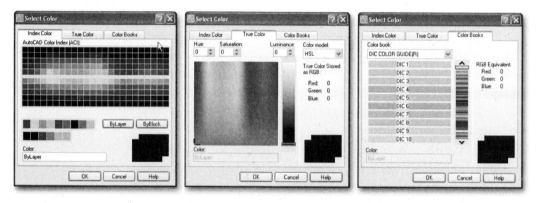

FIGURE 14–24 *Select Color dialog box: Index Color tab (left), True Color tab (center), and Color Books tab (right)*

NOTE

As noted in Chapter 3, the options to specify colors by both layer and the COLOR command can cause confusion in a large drawing, especially one containing blocks and nested blocks. You are advised not to mix the two methods of specifying colors in the same drawing.

Setting an Object's Linetype

The Linetype option menu of the Properties panel in the Home tab (Figure 14–25) allows you to specify a linetype for the objects to be drawn, separate from the assigned linetype for the layer. You can select **ByLayer** or **ByBlock**, select one of the linetypes loaded in the current drawing, or choose the **Other** option to open the Linetype Manager dialog box to load and set the current linetype (Figure 14–26).

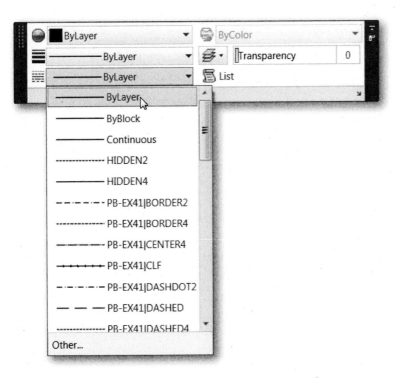

FIGURE 14–25 *Linetype option menu from the Properties panel*

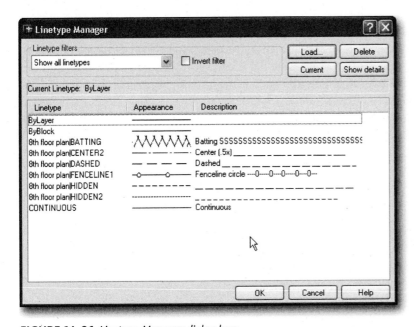

FIGURE 14–26 *Linetype Manager dialog box*

A linetype must exist in a library file and be loaded before you can apply it to an object or layer. Standard linetypes are in the library file called *acad.lin* and are not loaded with the LAYER command. You must load the linetype before you assign it to a specific layer.

Linetypes are combinations of dashes, dots, and spaces. Customized linetypes permit "out of line" objects in a linetype such as circles, wavy lines, blocks, and skew segments.

Dash, dot, and space combinations eventually repeat themselves. For example, a six-unit-long dash, followed by a dot between two one-unit-long spaces, repeats itself according to the overall length of the line drawn and the LTSCALE setting.

Lines with dashes that are not all dots usually have dashes at both ends. AutoCAD automatically adjusts the lengths of end dashes to reach the endpoints of the adjoining line. Intermediate dashes will be the lengths specified in the definition. If the overall length of the line is not long enough to permit the breaks, the line is drawn continuous.

There is no guarantee that any segments of the line fall at a particular location. For example, when placing a centerline through circle centers, you cannot be sure that the short dashes will be centered on the circle centers as most conventions call for. To achieve this effect, the short and long dashes have to be created by either drawing them individually or by breaking a continuous line to create the spaces between the dashes. This also creates multiple in-line lines instead of one line of a particular linetype. Or, you can use the CENTER option of the DIMENSION command to place the desired mark.

Individual linetype names and definitions are stored in one or more files whose extension is *.lin*. The same name may be defined differently in two different files. Selecting the desired one requires proper responses to the prompts in the **Load** option of the LINETYPE command. If you redefine a linetype, loading it with the LINETYPE command will cause objects drawn on layers assigned to that linetype to assume the new definition.

AutoCAD lists the available linetypes for the current drawing and displays the current linetype setting in the Linetype Manager dialog box (Figure 14–26). By default, it is set to ByLayer. To change the current linetype setting, double-click its name in the list. All new objects you create will be drawn with the selected linetype, regardless of the layer you are working with, until you again set the linetype to ByLayer or ByBlock. ByLayer causes the object drawn to assume the linetypes of the layer on which it is drawn. ByBlock causes objects to be drawn in the Continuous linetype until selected for inclusion in a block definition. Subsequent insertion of a block that contains objects drawn under the ByBlock option will cause those objects to assume the linetype of the block.

To load a linetype explicitly into your drawing, choose **Load**, and AutoCAD displays the Load or Reload Linetypes dialog box (Figure 14–27).

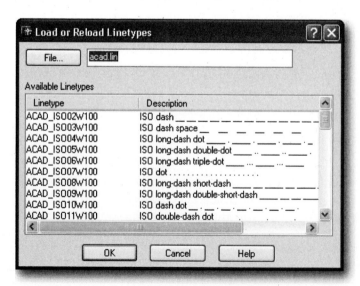

FIGURE 14–27 *Load or Reload Linetypes dialog box*

By default, AutoCAD lists the available linetypes from the *acad.lin* file. Select the linetype to load from the Available Linetypes list box, and choose **OK**.

If you need to load linetypes from a different file, choose **File** in the Load or Reload Linetypes dialog box. AutoCAD displays the Select Linetype File dialog box. Select the appropriate linetype file and choose **OK**. In turn, AutoCAD lists the available linetypes from the selected linetype file in the Load or Reload Linetypes dialog box. Select the appropriate linetype to load from the **Available Linetypes** list box, and choose **OK**.

To delete a linetype that is currently loaded in the drawing, first select the linetype from the list box in the Linetype Manager dialog box, and then choose **Delete**. You can delete only linetypes that are not referenced in the current drawing. You cannot delete the linetype Continuous, ByLayer, or ByBlock. Deleting is the same as using the PURGE command to purge unused linetypes from the current drawing.

To display additional information of a specific linetype, first select the linetype from the Linetype list box in the Linetype Manager dialog box, and then choose **Show details**. AutoCAD displays an extension of the dialog box, listing additional settings (Figure 14–28).

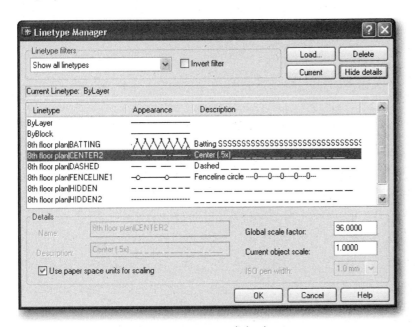

FIGURE 14–28 *Extended Linetype Manager dialog box*

The **Name** and **Description** edit fields display the selected linetype name and description, respectively.

The **Global scale factor** text field displays the current setting of the LTSCALE factor. The **Current object scale** text field displays the current setting of the CELTSCALE factor. If necessary, you can change the values of the LTSCALE and CELTSCALE system variables.

The **ISO pen width** box sets the linetype scale to one of a list of standard ISO values. The resulting scale is the global scale factor multiplied by the object's scale factor.

Selecting **Use paper space units for scaling** scales linetypes in paper space and model space identically.

After making the necessary changes, choose **OK** to keep the changes, and close the Linetype Manager dialog box.

NOTE As noted in Chapter 3, the options to specify linetypes by both the LAYER and the LINE-TYPE commands can cause confusion in a large drawing, especially one containing blocks and nested blocks. You are advised not to mix the two methods of specifying linetypes in the same drawing.

Setting an Object's Lineweight

The Lineweight menu of the Properties panel in the Home tab allows you to specify a lineweight for the objects to be drawn, separate from the assigned lineweight for the layer (Figure 14–29). You can select **ByLayer** or **ByBlock**, or one of the lineweights listed, or invoke the LINEWEIGHT command and select a lineweight from the listed **Lineweights** in the Lineweight Settings dialog box (Figure 14–30).

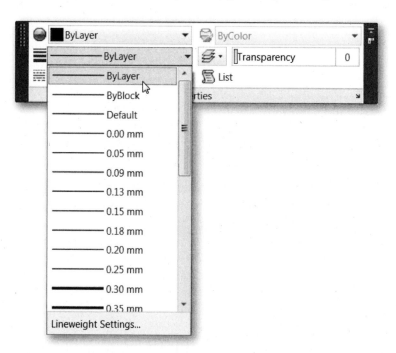

FIGURE 14–29 *Selecting one of the available lineweights from the menu in the Properties panel*

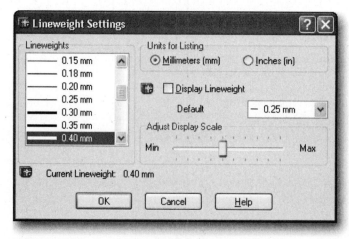

FIGURE 14–30 *Lineweight Settings dialog box*

AutoCAD displays the current lineweight at the bottom of the dialog box. To change the current lineweight, select one of the available lineweights from the **Lineweights** list box. ByLayer is the default. All new objects you create are drawn with the current lineweight, regardless of which layer is current, until you again set the lineweight to ByLayer, ByBlock, or Default. ByLayer causes the objects drawn to assume the lineweight of the layer on which it is drawn. ByBlock causes objects to be drawn in the default lineweight until selected for inclusion in a block definition. Subsequent insertion of a block reference that contains objects drawn under the ByBlock option causes those objects to assume the lineweight of the current setting of the LINEWEIGHT command. The Default selection causes the objects to be drawn to the default value as set by the LWDEFAULT system variable and defaults to a value of 0.01 inches or 0.25 mm. You can also set the default value from the Default option menu located on the right side of the dialog box. The lineweight value of 0 plots at the thinnest lineweight available on the specified plotting device and is displayed at 1 pixel wide in model space. You can use the PROPERTIES command to change the lineweight of the existing objects.

The Units for Listing section specifies whether lineweights are displayed in millimeters or inches.

The **Display Lineweight** check box controls whether lineweights are displayed in the current drawing. If it is enabled, lineweights are displayed in model space and paper space. AutoCAD regeneration time increases with lineweights that are represented by more than 1 pixel. If it is disabled, AutoCAD performance improves. Performance slows down when working with lineweights enabled in a drawing.

The **Adjust Display Scale** controls the display scale of lineweights on the Model tab. On the Model tab, lineweights are displayed in pixels. Lineweights are displayed using a pixel width in proportion to the real-world unit value at which they plot. If you are using a high-resolution monitor, you can adjust the lineweight display scale to better display different lineweight widths. The lineweight list reflects the current display scale. Objects with lineweight that are displayed with a width of more than 1 pixel may increase AutoCAD regeneration time. If you want to optimize AutoCAD performance when working in the Model tab, set the lineweight display scale to the minimum value or turn off lineweight display altogether.

Choose **OK** to close the dialog box and keep the changes in the settings.

> As noted in Chapter 3, the options to specify lineweights by both the LAYER and the LINEWEIGHT commands can cause confusion in a large drawing, especially one containing blocks and nested blocks. You are advised not to mix the two methods of specifying lineweights in the same drawing.

NOTE

Setting an Object's Transparency

The Transparency option of the Properties panel in the Home tab allows you to control the transparency level of objects and layers. Transparency can be set to ByLayer, ByBlock, or to a specified value. First select an object whose transparency you wish to change. Right-click the object and from the shortcut menu select Properties. From the Transparency option select from ByLayer, ByBlock, or enter the desired value for transparency.

COORDINATE FILTERS—OBJECT SNAP ENHANCEMENT

The AutoCAD filters feature allows you to establish a 2D point by specifying the individual X and Y coordinates one at a time in separate steps. In the case of a 3D point, you can specify the individual X, Y, and Z coordinates in three steps. You can also specify one of the three coordinate values in one step and a point in another step, from which AutoCAD extracts the other two coordinate values for use in the point being established.

The filters feature is used when you are prompted to establish a point, as in the starting point of a line, the center of a circle, drawing a node with the POINT command, or specifying a base point or second point in displacement for the MOVE or COPY command, to mention just a few.

> **NOTE**
>
> During the application of the filters feature you can input either single coordinate values or points, and you can input only points. It is necessary to understand these restrictions and options and when one type of input is more desirable than the other.

When selecting points during the use of filters, you need to know which coordinates of the specified point are going to be used for the point being established. It is also essential to know how to combine object snap (OSNAP) modes with those steps that use point input.

The filters feature is actually an enhancement to either the object snap or the @ (last point) feature. AutoCAD's ability to establish a point by snapping to a point on an existing object is one of its most powerful features, and being able to have Auto-CAD snap to such an existing point and then filter out selected coordinates for use in establishing a new point adds to that power. Therefore, in most cases you will not use the filters feature if it is practical to enter in all of the coordinates from the keyboard, because entering in all the coordinates can be done in a single step. The filters feature is a multistep process, and each step might include substeps, one to specify the coordinate(s) to be filtered out and another to designate the OSNAP mode involved.

> **NOTE**
>
> The filter prompts will be seen only when you invoke the appropriate commands in the Command window.

Filters with @

When AutoCAD is prompting for a point, the filters feature is initiated by entering a period followed by the letter designation for the coordinate(s) to be filtered out. For example, if you draw a point starting at (0,0) and use the relative polar coordinate response @3 < 45 to determine the endpoint, you can use filters to establish another point whose X coordinate is the same as the X coordinate of the end of the line just drawn. It also works for Y and Z coordinates and combinations of XY, XZ, and YZ coordinates. The following command sequence shows how to apply a filter to a line that needs to be started at a point whose X coordinate is the same as that of the end of the previous line and whose Y coordinate is 1.25. The line will be drawn horizontally

3 units long. The following command sequence is shown when it is invoked in the command window:

```
line
Specify first point: 0,0
Specify next point or ⬇: @3<45
Specify next point or ⬇: ( Enter )
line (or Enter )
Specify first point: (hold SHIFT and right-click, and from
the Point Filters flyout menu, choose .x)
X of end (use the endpoint snap mode to select the end of the
line)
of (need YZ): 0,1.25
Specify next point or ⬇: @3<0
Specify next point or ⬇:
```

Entering **.x** initiates the filters feature. AutoCAD then prompts you to specify a point from which it can extract the X coordinate. The @ (last point) does this. The new line has a starting point whose X coordinate is the same as that of the last point drawn. By using the filters feature to extract the X coordinate, that starting point will be on an imaginary vertical line through the point specified by @ in response to the "of" prompt.

When you initiate filters with a single coordinate (.x, in our example) and respond with a point (@), the prompt that follows also asks for a point. From this second point, AutoCAD extracts the other two coordinates for the new point.

Even though the prompt is for "YZ," the point may be specified in 2D format as 0,1.25, representing the X and Y coordinates, from which AutoCAD takes the second value as the needed Y coordinate. The Z coordinate is assumed to be the elevation of the current coordinate system.

You can use the two-coordinate response to initiate filters. Then specify a point, and all that AutoCAD requires is a single value for the final coordinate. An example of this process follows.

```
line
Specify first point: 0,0
Specify next point or ⬇: @3<45
Specify next point or ⬇: ( Enter )
line (or Enter )
Specify first point: (hold SHIFT and right-click, and from
the Point Filters flyout menu choose xz)
XZ of end (use the endpoint snap mode to select the end of the
line)
of (need Y): 1.25
Specify next point or ⬇: @3<0
Specify next point or ⬇:
```

You can also specify a point in response to the "of (need Y):" prompt:

```
of (need Y): 0,1.25 (or pick a point on the screen)
```

In this case, AutoCAD uses the Y coordinate of the point specified as the Y coordinate of the new point.

Remember, it is an individual coordinate in 2D—one or two coordinates in 3D—of an existing point that you wish AutoCAD to extract and use for the new point. In most cases, you will be object snapping to a point for the response. Otherwise, if you knew the value of the coordinate needed, you would probably enter it at the keyboard.

Filters with Object Snap

Without filters, an OSNAP mode establishes a new point to coincide with one on an existing object. With filters, an OSNAP mode establishes the selected coordinate(s) of a new point to coincide with the corresponding coordinate(s) of one or two of the coordinates, depending on the filter, on an existing object.

Extracting one or more coordinate values to be applied to corresponding coordinate values of a point that you are being prompted to establish is shown in the following example.

In Figure 14–31, a 2.75 × 7.1875 rectangle has a 0.875-diameter hole in its center. A board drafter would determine the center of a square or rectangle by drawing diagonals and centering the circle at their intersection. AutoCAD drafters (without filters) could do the same, or they might draw orthogonal lines from the midpoint of a horizontal line and from the midpoint of one of the vertical lines to establish a centering intersection. The following command sequence shows steps in drawing a rectangle with a circle in the center using filters.

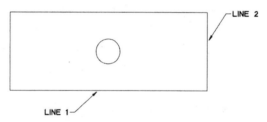

FIGURE 14–31 *Extracting coordinate values to be applied to corresponding coordinate values*

```
line
Specify first point: (select a point)
Specify next point or ⬇: @2.75<90
Specify next point or ⬇: @7.1875<0
Specify next point or ⬇: @2.75<270
Specify next point or ⬇: c
circle
Specify center point for circle or ⬇: (hold SHIFT and right-
click, and from the Point Filters flyout menu choose .x)
of mid (use the midpoint snap mode to select a horizontal line)
of (need YZ): (hold SHIFT and right-click, and from the Point
Filters flyout menu choose .y)
of mid (use the midpoint snap mode to select a vertical line)
of (need Z): (hold SHIFT and right-click, and from the Point
Filters flyout menu choose .z and click any line to place the
circle on the same plane)
```

Specify radius of circle or ⊡: d *(choose diameter)*
Specify diameter of circle: .875

SHELL COMMAND

The SHELL command allows you to execute OS programs without leaving AutoCAD. You can execute any operating system program as long as there is sufficient memory to execute.

Invoke the SHELL command by entering **shell** at the on-screen prompt, and AutoCAD displays the following prompt:

OS Command: *(invoke one of the available operating system's utility programs)*

When the utility program is finished, AutoCAD takes you back to the on-screen prompt. If you need to execute more than one OS program, press [Enter] at the "OS command" prompt. AutoCAD responds with the appropriate operating system prompt. You can now enter as many operating system commands as you wish. When you are finished, return to AutoCAD by entering EXIT. It will take you back to the on-screen prompt.

> Do not delete the AutoCAD lock files or temporary files created for the current drawing when you are at the operating system prompt.

NOTE

SETTING UP A DRAWING

The MVSETUP command is used to control and set up the view(s) of a drawing, including the choice of standard plotted sheet sizes with a border, the scale for plotting on the selected sheet size, and multiple viewports. AutoLISP is AutoCAD's embedded programming language, and an MVSETUP routine can be customized to insert any type of border and title block.

Options and associated prompts depend on whether the TILEMODE system variable is set to ON (1) or OFF (0). When TILEMODE is set to ON, Tiled Viewports is enabled. When TILEMODE is set to OFF, the Floating Viewports menu item is enabled. Other paper space-related drawing setup options are available.

Invoke the MVSETUP command by entering **mvsetup** at the on-screen prompt, and the AutoCAD prompts that appear depend on whether you are working in model space (Model tab) or paper space (Layout tab). If you are working in model space, AutoCAD displays the following prompt:

Enable paper space? *(press [Enter] to enable paper space, and AutoCAD changes the TILEMODE setting to 0, or choose no to stay in the TILEMODE setting of 1)*

If you are working in paper space (Layout tab), AutoCAD displays the following prompt:

Enter an option *(select one of the options from the shortcut menu)*

> If you invoke the command at the on-screen prompt, make sure to display the Text window to see the prompt sequence.

NOTE

The following is the procedure for setting up the drawing when you are working in model space. AutoCAD displays the following prompts:

Enable paper space? *(choose no from the shortcut menu)*

Enter units type *(Select a unit type. Depending on the units selected, AutoCAD lists the available scales. Select one of the available scales, or you can even specify a custom scale factor.)*

Enter the scale factor: *(specify a scale factor)*

Enter the paper width: *(specify the paper width at which the drawing will be plotted)*

Enter the paper height: *(specify the paper height at which the drawing will be plotted)*

AutoCAD sets up the appropriate limits to allow you to draw to full scale and also draws a bounding box enclosing the limits. Draw the drawing to full scale; when you are ready to plot, specify the scale mentioned earlier to plot the drawing.

Following is the procedure for setting up the drawing when you are working in paper space. AutoCAD displays the following prompt:

Enter an option

The options are explained here in the order that is logical to complete the drawing setup.

Choosing *Title Block* allows you to select an appropriate title block. AutoCAD displays the following prompt (Figure 14–32):

Enter title block option *(select one of the options from the shortcut menu)*

Choosing *Insert* (default) allows you to insert one of the available standard title blocks. AutoCAD lists the available title blocks, in the text screen, and prompts you to select one of the available title blocks as follows:

0. None
1. ISO A4 Size(mm)
2. ISO A3 Size(mm)
3. ISO A2 Size(mm)
4. ISO A1 Size(mm)
5. ISO A0 Size(mm)
6. ANSI-V Size(in)
7. ANSI-A Size(in)
8. ANSI-B Size(in)
9. ANSI-C Size(in)
10. ANSI-D Size(in)
11. ANSI-E Size(in)
12. Arch/Engineering (24 × 36in)
13. Generic D size Sheet (24 × 36in)

Enter number of title block to load or [Add/Delete/Redisplay]: *(select one of the available title blocks and press* Enter *, and AutoCAD inserts a border and title block)*

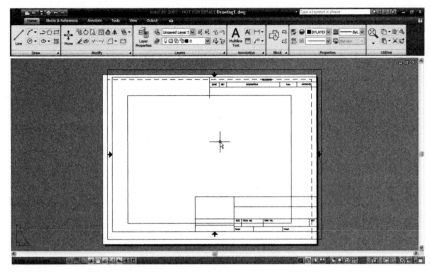

FIGURE 14–32 *Title block and border for ANSI-B Size*

Choosing *Add* allows you to add a title block drawing to the available list.

Choosing *Delete* allows you to delete an entry from the available list.

Choosing *Redisplay* redisplays the list of title block options.

Choosing *Create* allows you to establish viewports. AutoCAD prompts:

> Enter option *(press* **Enter** *to create viewports, and AutoCAD lists the available viewport layout options)*

Choosing *Create Viewports* (default) allows you to create multiple viewports to one of the standard layouts. AutoCAD displays the following prompts:

> Available layout options:
>
> 0: None
>
> 1: Single
>
> 2: Std. Engineering
>
> 3: Array of Viewports
>
> Enter layout number to load or [Redisplay]: *(select one of the available layouts)*

Choosing *None* creates no viewports. Selecting 1 creates a single viewport whose size is determined during subsequent prompt responses. Selecting 2 creates four viewports with preset viewing angles by dividing a specified area into quadrants. The size is determined by responses to subsequent prompts. Selecting 3 creates a matrix of viewports along the X and Y axes.

Choosing *Delete Objects* deletes the existing viewports.

Choosing *Undo* reverses operations performed in the current MVSETUP session.

Choosing *Scale Viewports* adjusts the scale factor of the objects displayed in the viewports. The scale factor is specified as a ratio of paper space to model space. For example, 1:48 is 1 paper space unit for 48 model space units (scale for 1/4″ = 1′0″).

Choosing OPTIONS allows you to establish several different environment settings that are associated with your layout. AutoCAD displays the following prompt:

```
Enter an option (enter an option)
```

Choosing *Layer* allows you to specify a layer for placing the title block.

Choosing *Limits* allows you to specify whether or not to reset the limits to drawing extents after the title block has been inserted.

Choosing *Units* allows you to specify whether sizes and point locations will be translated to inch or millimeter paper units.

Choosing *Xref* allows you to specify whether the title block is to be inserted or externally referenced.

Choosing *Align* causes AutoCAD to pan the view in one viewport so that it aligns with a base point in another viewport. Whichever viewport the other point moves to becomes the active viewport. AutoCAD displays the following prompt:

```
Enter an option
```

Choosing *Angled* causes AutoCAD to pan the view in a viewport in a specified direction.

Choosing *Horizontal Alignment* causes AutoCAD to pan the view in one viewport, aligning it horizontally with a base point in another viewport.

Choosing *Vertical Alignment* causes AutoCAD to pan the view in one viewport, aligning it vertically with a base point in another viewport.

Choosing *Rotate View* causes AutoCAD to rotate the view in a viewport around a base point.

Choosing *Undo* causes AutoCAD to undo the results of the current MVSETUP command.

LAYER TRANSLATOR

The Layer Translator is used to make selected layers in the current drawing match layers in another drawing or in a CAD Standards file.

Invoke the LAYTRANS command from the Standards panel (Figure 14–33) located in the Manage tab by selecting **Layer Translator** and AutoCAD displays the Layer Translator dialog box (Figure 14–34).

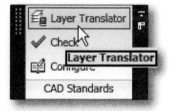

FIGURE 14–33 *Invoking the LAYTRANS command from the CAD Standards panel*

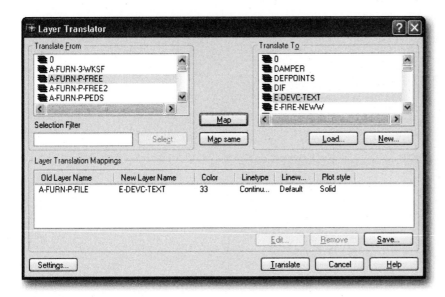

FIGURE 14–34 *Layer Translator dialog box*

The **Translate From** list box lists all the layer(s) in the current drawing that can be changed. A dark colored icon indicates that the layer is referenced in the drawing; a white icon indicates that the layer is unreferenced. You can delete unreferenced layers from the drawing by right-clicking in the **Translate From** list box and choosing *Purge Layers*.

Choose **Load** to load the layers to match from an existing drawing, template file, or CAD Standards file. AutoCAD lists the layers you can translate the current drawing's layers to in the **Translate To** list box. You can also create new layers and assign properties by choosing **New**, which will be added to the list in the **Translate To** list box. You cannot create a new layer with the same name as an existing layer.

Select one or more layers in the **Translate From** list box. You can also select layers by using the selection filter. Select layers in the **Translate To** list box to map to the selected layer(s). Choose **Map** to map the layers selected in **Translate From** to the layer selected in **Translate To.** Choose **Map same** to map all layers that have the same name in both lists.

AutoCAD lists each layer to be translated and the properties to which the layer will be converted in the Layer Translation Mappings section. You can select layers in this list and edit their properties using **Edit**.

Choose **Edit** to open the Edit Layer dialog box where you can edit the selected translation mapping (Figure 14–35). You can change the layer's linetype, color, and lineweight. If all drawings involved in translation use plot styles, you can also change the plot style for the mapping.

Choose **Remove** to remove the selected translation mapping from the Layer Translation Mappings list.

Choose **Save** to save the current layer translation mappings to a file for later use. Layer mappings are saved in the *.dwg* or *.dws* file format. You can replace an existing file or create a new file. The Layer Translator creates the referenced layers in the file and stores the layer mappings in each layer. All linetypes used by those layers are also copied into the file.

Choose **Settings** to open the Settings dialog box, where you can customize the process of layer translation (Figure 14–36).

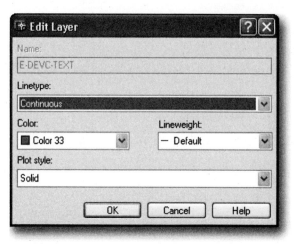

FIGURE 14–35 *Edit Layer dialog box*

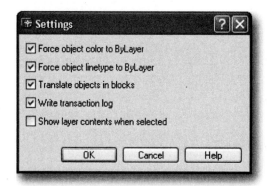

FIGURE 14–36 *Settings dialog box*

Selecting **Force object color to ByLayer** causes every object translated to take on the color assigned to its layer. If it is not checked, every object retains its original color.

Selecting **Force object linetype to ByLayer** causes every object translated to take on the linetype assigned to its layer. If it is not checked, every object retains its original linetype.

Selecting **Translate objects in blocks** causes objects nested within blocks, including nested blocks, to be translated. If it is not checked, nested objects in blocks are not translated.

Selecting **Write transaction log option** causes a log file to be created detailing the results of the translation. The log file is assigned the same name as the translated drawing, with a *.log* file name extension and is created in the same folder. If it is not checked, no log file is created.

Selecting **Show layer contents** causes only the layers selected in the Layer Translator dialog box to be displayed in the drawing area. If it is not checked, all layers in the drawing are displayed.

Choose **Translate** to start the translation of the layers you have mapped. If you have not saved the current layer translation mappings, you are prompted to save the mappings before translation begins.

TIME COMMAND

The TIME command displays the current time and date related to your current drawing session. In addition, you can find out how long you have been working in AutoCAD. This command uses the clock in your computer to keep track of the time functions and displays to the nearest millisecond using 24-hour military format.

Invoke the TIME command by selecting **Time** from the **Inquiry** panel of the Tools tab, and the following listing is displayed in the text screen, followed by a prompt:

```
Current time: Wednesday, February 16, 2009 11:05:04:406 PM
Times for this drawing:
Created: Sunday, February 06, 2009 6:55:59:140 AM
Last updated: Sunday, February 06, 2009 7:25:29:859 AM
Total editing time: 0 days 02:23:51:531
Elapsed timer (on): 0 days 02:23:51:047
Next automatic save in: <no modifications yet>
Enter option [Display/ON/OFF/Reset]: (select one of the
options)
```

The first line gives today's date and time.

The third line gives the date and time the current drawing was initially created. The drawing time starts when you initially begin a new drawing. If the drawing was created by means of the WBLOCK command, the date and time that the command was executed is displayed here.

The fourth line gives the date and time the drawing was last updated. Initially set to the drawing creation time, this is updated each time you use the END or SAVE command.

The fifth line gives the length of time you are in AutoCAD. This timer is continuously updated by AutoCAD while you are in the program, excluding plotting and printer plot time. This timer cannot be stopped or reset.

The sixth line provides information about the stopwatch timer. You can turn this timer ON or OFF and reset it to zero. This timer is independent of other functions.

The seventh line provides information about when the next automatic save will take place.

Choose *Display* to redisplay the time functions with updated times.

Choose ON to set the stopwatch timer to ON if it is OFF. By default it is set to ON.

Choose OFF to set the stopwatch timer to OFF and display the accumulated time.

Choose *Reset* to reset the stopwatch timer to zero.

To exit the TIME command at the prompt, press ESC or Enter.

AUDIT COMMAND

The AUDIT command serves as a diagnostic tool to correct any errors or defects in the database of the current drawing. AutoCAD generates an extensive report of the problems, and for every error detected, AutoCAD recommends an action to correct it.

Invoke the AUDIT command by selecting **Audit** from **Drawing Utilities** panel in the Tools **tab**, and AutoCAD displays the following prompt:

> Fix any errors detected? ⬇ *(choose* YES *or* NO *from the shortcut menu)*

If you respond with YES, AutoCAD will fix all the errors detected and display an audit report in the text screen with detailed information about the errors detected and fixing them. If you respond with NO, AutoCAD will just display a report and will not fix any errors.

In addition, AutoCAD creates an ASCII report file describing the problems and the actions taken. AUDITCLT system variable should be set to ON. It will save the file in the current directory, using the current drawing's name with the file extension *.adt*. You can use any ASCII editor to display the report file on the screen or print it on the printer, respectively.

NOTE

If a drawing contains errors that the AUDIT command cannot fix, open the drawing with the RECOVER command to retrieve the drawing and correct its errors.

OBJECT LINKING AND EMBEDDING (OLE)

Object linking and embedding (OLE) is a Microsoft® Windows® feature that combines various application data into one compound document. AutoCAD has client as well as server capabilities. As a client, AutoCAD permits you to have objects from other Windows applications either embedded in or linked to your drawing.

When an object is inserted into an AutoCAD drawing from an application that supports OLE, the object can maintain a connection with its source file. If you insert an object as an embedded object into AutoCAD (client), it is no longer associated with the source (server). If necessary, you can edit the embedded data from inside the AutoCAD drawing by using the original application. At the same time, this editing does not change the original file.

If, instead, you insert an object as a linked object into AutoCAD (client), the object remains associated with its source (server). When you edit a linked object in AutoCAD by using the original application, the original file changes as well as the object inserted into AutoCAD.

Linked or embedded objects appear on the screen in AutoCAD and can be printed or plotted using Windows system drivers.

Let's look at an example of object linking between an AutoCAD (server) drawing and Microsoft Word® (client). The following examples are shown using the AutoCAD 2D Drafting & Annotation Workspace. Figure 14–37 shows a drawing of a desk, a computer, and a chair that contains various attribute values. We are going to link this drawing to a Microsoft Word document.

From the Clipboard panel located in the Home tab, select **Copy**, and AutoCAD prompts you to select objects. Select the computer, the table, and the chair, and press

Enter . This action will copy the selection to the Windows clipboard. Minimize the
AutoCAD program. In order for objects and changes to objects to be linked, the
drawing must be saved first.

Instead of selecting specific objects, you can copy the current view into the Windows
Clipboard by invoking the COPYLINK command.

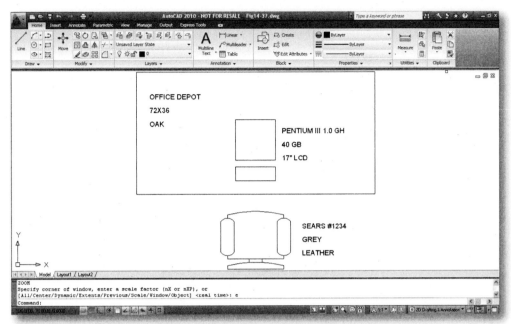

FIGURE 14–37 *Drawing of a desk, a computer, and a chair with attribute values*

Open the Microsoft Word program from the desktop by double-clicking the Word
program icon. The Microsoft Word program is displayed, as shown in Figure 14–38.

FIGURE 14–38 *Microsoft Word program*

From the Clipboard panel in Microsoft Word, select **Paste Special**, and Word displays the Paste Special dialog box, as shown in Figure 14–39. Choose **Paste Link** to insert the AutoCAD drawing object into Microsoft Word, as shown in Figure 14–40.

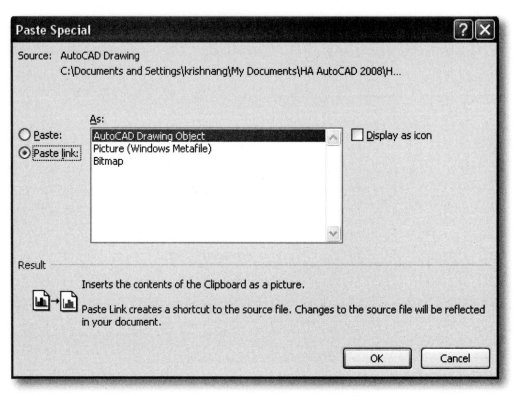

FIGURE 14–39 *Paste Special dialog box*

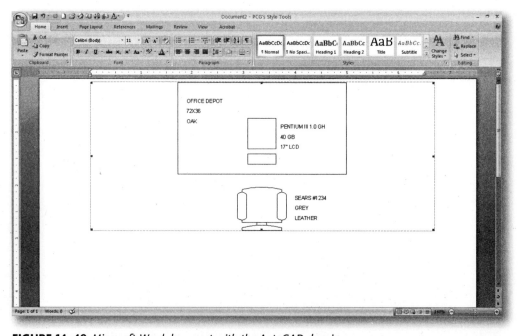

FIGURE 14–40 *Microsoft Word document with the AutoCAD drawing*

Minimize the Word program and maximize AutoCAD. If the AutoCAD program is not open, double-click the drawing image in the Word document; this will launch the AutoCAD program with the image drawing open. Edit the values of the attributes in the computer block to Pentium IV 2.5 GHz, 120.0 GB, 21″, which represent a Pentium IV computer with a 120.0-gigabyte hard drive and a 21″ monitor.

Switch back to Word, click the Microsoft Office Button, and select **Edit Links to Files** from the Prepare menu. Word will display the Links dialog box, as shown in Figure 14–41.

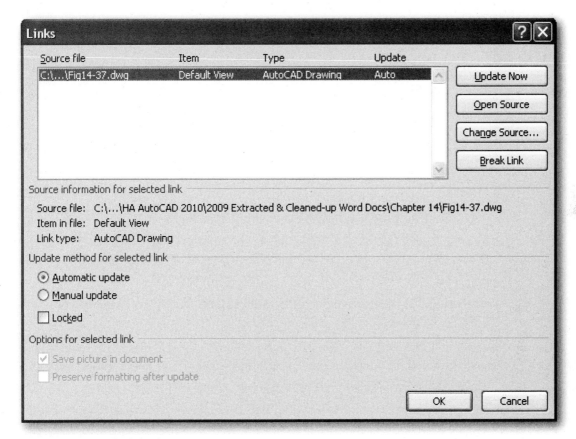

FIGURE 14–41 *Links dialog box*

Choose **Update Now,** and then choose **OK**. The image in the Word document is updated, as shown in Figure 14–42.

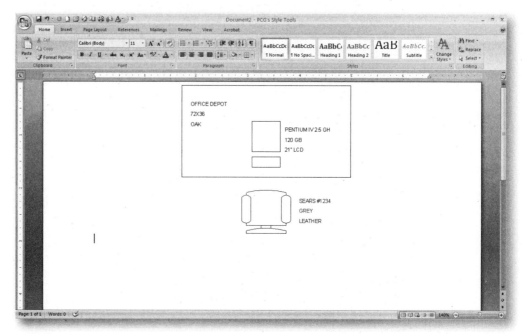

FIGURE 14–42 *AutoCAD drawing updated in Word (Word as the client)*

In our example, AutoCAD is the server and Microsoft Word is the client.

Conversely, you can place a linked object in AutoCAD, where AutoCAD is the client and another application is the server. Let's look at an example in which AutoCAD is the client and Microsoft Excel® is the server.

Start the Excel program and create a spreadsheet. Copy the contents into the Windows Clipboard.

From the Clipboard panel located in the Home tab in AutoCAD, select **Paste Special.** AutoCAD displays the Paste Special dialog box. Select **Paste Link** and **Microsoft Excel Worksheet** from the list box, and choose **OK.**

The Excel spreadsheet is inserted to the drawing, as shown in Figure 14–43. AutoCAD is now the client, and Excel is the server.

To edit the spreadsheet, double-click anywhere on the spreadsheet, which in turn will launch Excel with the spreadsheet document open. Any changes made to the spreadsheet will be reflected in the drawing. Figure 14–44 shows the changes (Gateway 2000 changed to Dell) that were made in the spreadsheet.

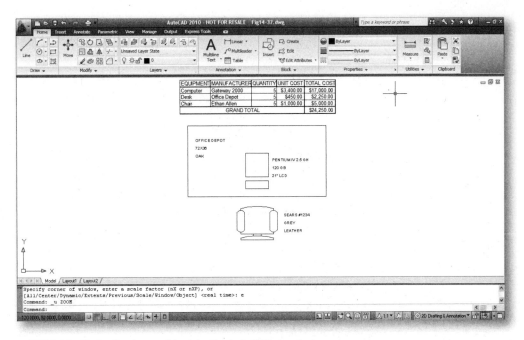

FIGURE 14–43 *An Excel spreadsheet in the AutoCAD drawing*

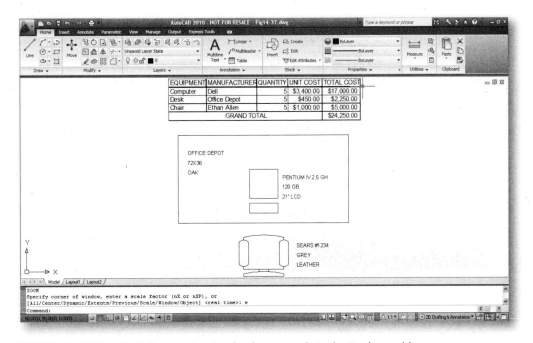

FIGURE 14–44 *AutoCAD drawing showing the changes made in the Excel spreadsheet*

Here is another example in which an AutoCAD drawing is the client, for both embedding from Word and linking from Excel.

Figure 14–45 shows both an AutoCAD screen and an Excel spreadsheet, in which the spreadsheet is being used for area calculations. The AREA cells are formulas that calculate the product of the corresponding WIDTH and LENGTH cells. In turn, the TOTAL cell is the sum of the AREA cells.

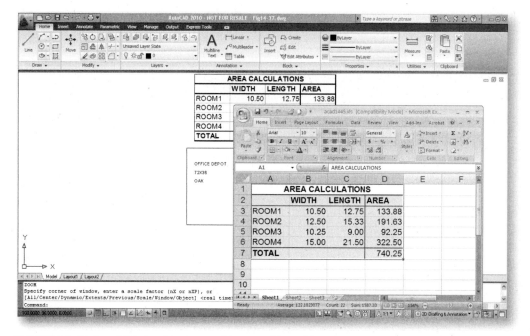

FIGURE 14–45 *AutoCAD drawing screen and an Excel spreadsheet*

Figure 14–46 shows the same AutoCAD screen and a Word document that was used for typing the GENERAL NOTES, which were in turn embedded into the AutoCAD drawing.

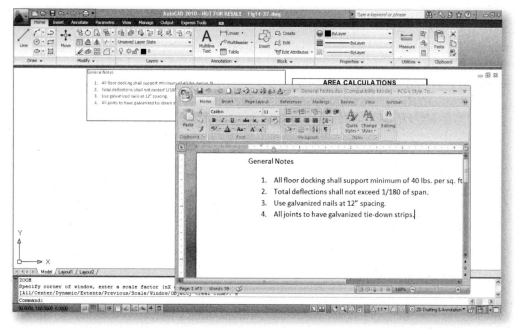

FIGURE 14–46 *AutoCAD drawing and a Word document*

Figure 14–47 shows how the cell value in the WIDTH for ROOM1 has been changed, resulting in changes in the AREA and TOTAL cells. Because this object, that is, the spreadsheet consisting of four columns and seven rows, including the title, was paste linked into the AutoCAD drawing, the linked object automatically reflects the changes. Similarly, if the Word document is also changed, the changes are automatically reflected in the AutoCAD drawing.

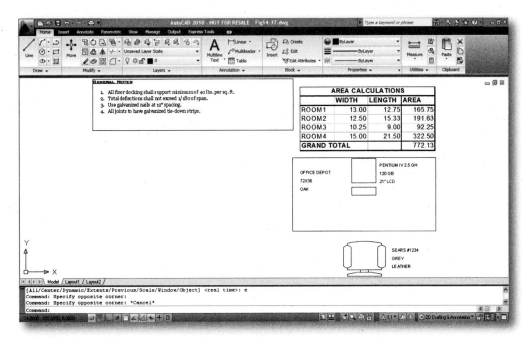

FIGURE 14–47 *Changes shown in the spreadsheet and the AutoCAD screen*

If necessary, you can change the size of the object by changing the height and width in drawing units, or you can enter a percentage of the current height or width in the Scale section. You can also change the text size and OLE plot quality. At the on-screen prompt invoke the OLESCALE command. AutoCAD displays the OLE Text Size dialog box, as shown in Figure 14–48, whenever you insert an OLE object into a drawing.

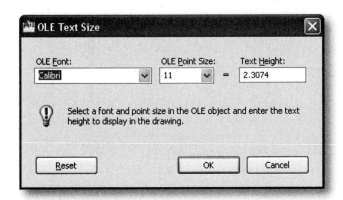

FIGURE 14–48 *OLE Text Size dialog box*

Choosing **OLE Font** displays a list of the fonts used in the OLE object.

Choosing **OLE Point Size** displays a list of the point sizes available for the selected font.

Selecting **Text Height** lets you specify a text height for the font at the selected point size.

Choosing **Reset** restores the OLE object to its size when it was inserted in the drawing.

If you need to change the OLE properties after pasting the objects into the current drawing, first select the OLE object, and from the shortcut menu select *Properties*.

SECURITY, PASSWORDS, AND ENCRYPTION

Electronic drawing files, like their paper counterparts, often need to have the information they contain protected from unauthorized viewing. AutoCAD provides password and encryption capabilities to achieve this. It might be necessary to determine that the person who last edited and saved the drawing is the person who was supposed to edit and save it. To make this determination, AutoCAD allows the use of digital signatures.

Passwords

AutoCAD's password protection makes it possible to prevent a drawing file from being opened without first entering the pre-assigned password.

To assign a password to a drawing, invoke the SECURITYOPTIONS command at the on-screen prompt. AutoCAD displays the Security Options dialog box (Figure 14–49).

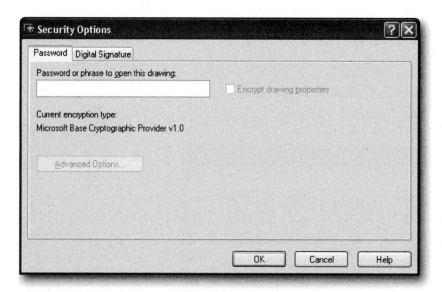

FIGURE 14–49 *Security Options dialog box with the Password tab displayed*

On the Password tab of the Security Options dialog box, enter a password in the **Password or phrase to open this drawing** text box. This action prevents the drawing from being opened without first entering the password specified. Passwords can be a single word or a phrase and are not case sensitive.

To view data in a password-protected drawing, open the drawing in a standard way, and enter the password in the **Enter password to open drawing** text box in the Password dialog box, as shown in Figure 14–50. Unless the title, author, subject, keywords, or other drawing properties were encrypted when the password was attached, you can view the properties in the Properties dialog box in Windows Explorer.

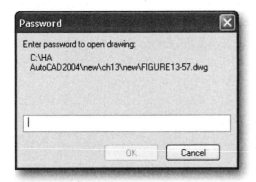

FIGURE 14–50 *Password dialog box*

Encryption

You can encrypt drawing properties, such as the title, author, subject, and keywords, and thus require a password to view the properties and thumbnail preview of the drawing. If you decide to specify an encryption type and key length, you can select them from the ones available on your computer. On the Password tab of the Security Options dialog box, after you have entered a password in the **Password or phrase to open this drawing** text box, set the **Encrypt drawing properties** to ON. Under the **Password or phrase to open this drawing** text box, AutoCAD displays the current encryption type. To change the encryption type, choose **Advanced Options.** AutoCAD displays the Advanced Options dialog box. Select one of the encryption providers listed, and specify the key length in the **Choose a key length** text box. The higher the key length, the higher the protection.

Digital Signature

AutoCAD provides a means to sign the drawing file electronically. This means that it is possible to verify that a drawing has had a distinct and unique digital signature attached to it when it was last saved. Along with positive electronic identification, you can also apply a time stamp and comments.

To attach a digital signature to a drawing, you must first obtain a digital ID. This can be done by contacting a certificate authority through a search engine in your Internet browser, using the term "digital certificate." Once a digital ID has been established on your computer, invoke the SECURITYOPTIONS command at the on-screen prompt. AutoCAD displays the Security Options dialog box (Figure 14–49). Select the Digital Signature tab, and AutoCAD displays various options available for a digital signature. If no valid digital IDs are installed, AutoCAD displays the Valid Digital ID Not Available dialog box, from which you can obtain a digital ID by choosing **Get a digital ID**. AutoCAD logs you onto a Web site where a digital ID can be obtained.

AutoCAD displays a list of digital IDs that you can use to sign files, which includes information about the organization or individual to whom the digital ID was issued, the digital ID vendor who issued the digital ID, and when the digital ID expires. Select one of the available digital IDs, and choose **Attach digital signature after**

saving drawing. From the Signature information section of the Digital Signature tab, you can select a time stamp to be attached with the digital ID from the **Get time stamp from** text box. You can also add comments to the digital ID in the **Comment** text box.

A digital ID has a name, expiration date, serial number, and certain certifying information. The certificate authority that you obtain the digital ID from can provide low, medium, and high levels of security.

From the digital signature feature, you can determine whether or not the file was changed since it was signed, whether or not the signers are who they claim to be, and if they can be traced. The digital signature is considered invalid if the file was corrupted when the digital signature was attached, it was corrupted in transit, or if the digital signature is no longer valid. In order to maintain validity of the digital signature you must not add a password to the drawing or modify or save it after the digital signature has been attached.

NOTE | The digital signature status is displayed when you open a drawing if the sigwarn system variable is set to ON. If it is set to OFF, the signature status is displayed only if the signature is invalid.

In the Open and Save tab of the Options dialog box, if **Display Digital Signature Information** is checked, when you open a drawing that has a digital signature attached, the Digital Signature Contents dialog box is displayed, providing information on the status of the drawing and the signer. In the **Other Fields** list, you can obtain information about the issuer, beginning and expiration dates, and the serial number of the digital signature.

CUSTOM SETTINGS WITH THE OPTIONS DIALOG BOX

The Options dialog box allows you to customize the AutoCAD settings. AutoCAD allows you to save and restore a set of custom AutoCAD settings called a profile. A profile can include custom settings that are not saved in the drawing, with the exception of pointer and printer driver settings. By default, AutoCAD stores your current settings in a profile named <Unnamed Profile>.

To open the Options dialog box, invoke the OPTIONS command by choosing *Options* from the shortcut menu that appears when you right-click in the drawing area with no commands active and no objects selected. AutoCAD displays the Options dialog box (Figure 14–51). From the Options dialog box, the user can control various aspects of the AutoCAD environment. The Options dialog box has nine tabs. To make changes to any of the sections, select the corresponding tab from the top of the Options dialog box.

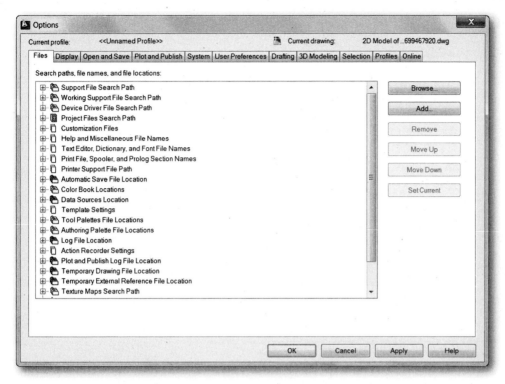

FIGURE 14–51 *Options dialog box with the Files tab selected*

Files

The Files tab of the Options dialog box specifies the directory in which AutoCAD searches for support files, driver files, project files, template drawing file location, temporary drawing file location, temporary external reference file location, and texture maps (Figure 14–51). It also specifies the location of menu, help, log, text editor, and dictionary files.

Choosing **Browse** causes AutoCAD to display the Browse for Folder or Select a File dialog box, depending on what was selected from the list.

Choosing **Add** causes AutoCAD to add a search path for the selected folder.

Choosing **Remove** causes AutoCAD to remove the selected search path or file.

Choosing **Move Up** causes AutoCAD to move the selected search path above the preceding search path.

Choosing **Move Down** causes AutoCAD to move the selected search path below the following search path.

Choosing **Set Current** causes AutoCAD to make the selected project or spelling dictionary current.

Display

The Display tab of the Options dialog box controls preferences that relate to AutoCAD performance (Figure 14–52).

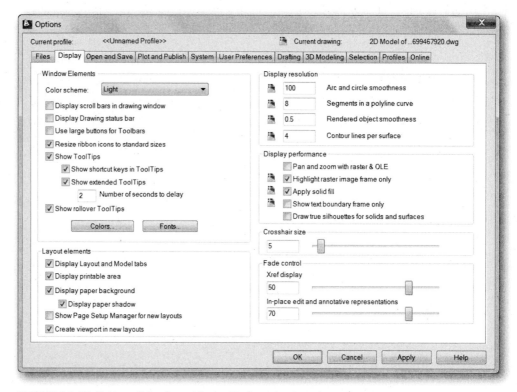

FIGURE 14–52 *Options dialog box with the Display tab selected*

The Window Elements section controls the parameters of the AutoCAD drawing window.

The **Color scheme** text box lets you control color settings in a dark or light color for elements such as the status bar, title bar, ribbon bar, and the Application Menu frame.

Selecting **Display scroll bars in drawing window** causes scroll bars at the bottom and right sides of the drawing window to be displayed.

Selecting **Display Drawing status bar** displays the drawing status bar, which displays several tools for scaling annotations. When the drawing status bar is turned off, the tools found on the drawing status bar are moved to the application status bar.

Selecting **Display screen menu** causes the screen menu on the right side of the drawing window to be displayed.

Selecting **Use Large Buttons for Toolbars** displays buttons in a larger format at 32 by 30 pixels. The default display size is 16 by 15 pixels. Selecting **Show Tooltip** causes Tooltips to be displayed when the cursor is near a command or option.

Selecting **Show shortcut keys in ToolTips**, available only when **Show ToolTips** is checked, causes the shortcut keys to be included when a tooltip is displayed.

Selecting **Show extended Tooltips** lets you control the display of extended tooltips.

The **Number of seconds to delay** text box lets you set the delay time between the display of basic tooltips and extended tooltips.

Selecting **Show rollover Tooltips** lets you control the display of rollover of tooltips for highlighted objects.

Choosing **Colors** causes AutoCAD to display the AutoCAD Window Colors dialog box, which can be used to set the colors for drawing area, screen menu, text window, and command line.

Choosing **Fonts** causes AutoCAD to display the Graphics Window Font dialog box, which can be used to specify the font AutoCAD uses for the screen menu and command line and in the text window.

The Display resolution section allows you to set the resolution in the following text boxes:

- Arc and circle smoothness
- Segments in a polyline curve
- Rendered object smoothness
- Contour lines per surface

The Layout elements section has check boxes to toggle ON and OFF the following:

- Display Layout and Model tabs
- Display printable area
- Display paper background
- Display paper shadow
- Show page setup dialog for new layouts
- Create viewport in new layouts

The Display performance section has check boxes to toggle ON and OFF the following:

- Pan and zoom with raster image
- Highlight raster image frame only
- Apply solid fill
- Show text boundary frame only
- Show silhouettes in wireframe

The **Crosshair size** text box lets you set the size of the crosshairs with a range of 1 to 100 percent of the screen.

The Fade Control section lets you control the fading intensity value for DWG xrefs and reference editing.

The **Xref Display** text box lets you specify the fading intensity value of externally referenced drawings.

The **In-place Edit and Annotative Representations** text box lets you specify the fading intensity value for objects during in-place reference editing. Objects that are not being edited are displayed at a lesser intensity.

Open and Save

The Open and Save tab of the Options dialog box allows you to determine formats and parameters for drawings, external references (xrefs), and ObjectARX applications as they are opened or saved (Figure 14–53).

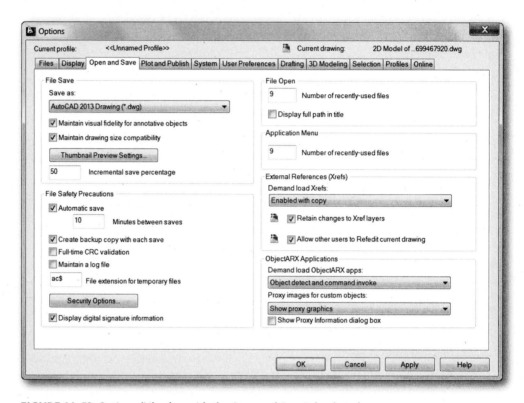

FIGURE 14–53 *Options dialog box with the Open and Save tab selected*

The File Save section controls settings related to saving a file.

The **Save as** text box allows you to select a default save format when you invoke the SAVEAS command. Choose from the following formats:

- AutoCAD 2013 Drawing (*.dwg)
- AutoCAD 2010 Drawing (*.dwg)
- AutoCAD 2007/LT2007 Drawing (*.dwg)
- AutoCAD 2004/LT2004 Drawing (*.dwg)
- AutoCAD 2000/LT2000 Drawing (*.dwg)
- AutoCAD R14/LT98/LT97 Drawing (*.dwg)
- AutoCAD Drawing Template File (*.dwt)
- AutoCAD 2010 DXF (*.dxf)
- AutoCAD 2007/LT2007 DXF (*.dxf)
- AutoCAD 2004/LT2004 DXF (*.dxf)
- AutoCAD 2000/LT2000 DXF (*.dxf)
- AutoCAD R12/LT2 DXF (*.dxf)

Selecting **Maintain visual fidelity for annotative objects** lets you specify whether or not drawings are saved with visual fidelity for annotative objects.

Selecting **Maintain drawing size compatibility** lets you specify whether or not the AutoCAD 2010 and earlier object size limits are used instead of those for AutoCAD 2011. Click the information icon to learn more about object size limits and how they affect opening and saving a drawing.

Choosing **Thumbnail Preview Settings** causes the Thumbnail Preview Settings dialog box to be displayed, which controls whether thumbnail previews are updated when the drawing is saved.

The **Incremental save percentage** sets the percentage of potentially wasted space in a drawing file. Full saves eliminate wasted space. Incremental saves are faster but they increase the size of your drawing. If you set Incremental Save Percentage to zero, every save is a full save. For optimum performance, set the value to 50. If hard disk space becomes an issue, set the value to 25. If you set the value to 20 or less, performance of the SAVE and SAVEAS commands slows significantly.

The File Open section controls settings that related to recently used files and open files. The **Number of recently-used files to List** text box controls the number of recently used files that are listed in the File menu for quick access. Valid values are 0 to 9.

Selecting **Display full path in title** causes AutoCAD to display the full path in the title.

The Application Menu section lets you control the number of recently used files listed in the menu browser's Recent Documents quick menu. Valid values are 0 to 50.

The External References (Xrefs) section controls the settings that relate to editing and loading external references.

The **Demand load Xrefs** text box controls demand loading of xrefs. Demand loading improves performance by loading only the parts of the referenced drawing needed to regenerate the current drawing. You can select **Disabled, Enabled,** or **Enabled with copy. Disabled** means that demand loading is not on in the current drawing. Someone else can open and edit an xref file except as it is being read into the current drawing. **Enabled** means that demand loading is ON in the current drawing. No one else can edit an xref file while the current drawing is open. However, someone else can reference the xref file. **Enabled with copy** means that demand loading is ON in the current drawing. An xref file can still be opened and edited by someone else. AutoCAD only uses a copy of the xref file, treating it as a completely separate file from the original xref.

Selecting **Retain changes to Xref layers** causes the properties of layers in xrefs to be saved as they are changed in the current drawing for reloading later.

Selecting **Allow other users to Refedit current drawing** allows the current drawing to be edited while being referenced by other drawing(s).

The File Safety Precautions section helps to detect errors and avoid losing data.

Selecting **Automatic save** with the **Minutes between saves** text box allows you to determine if and for what intervals periodic automatic saves will be performed.

Selecting **Create backup copy with each save** allows you to determine whether or not a backup copy is created when you save the drawing.

Selecting **Full-time CRC validation** allows you to determine whether or not a cyclic redundancy check (CRC) is performed when an object is read into the drawing. Cyclic redundancy check is a mechanism for error checking. If you suspect a hardware problem or that AutoCAD error is causing your drawings to be corrupted, check this box.

Selecting **Maintain a log file** allows you to determine whether or not the contents of the text window are written to a log file. Use the Files tab in the Options dialog box to specify the name and location of the log file.

The **File extension for temporary files** text box allows you to specify an extension for temporary files on a network. The default extension is *.ac$*. **Security Options** and **Display digital signature information** are described in the section on "Security, Passwords, and Encryption" earlier in this chapter.

The ObjectARX Applications section controls parameters for AutoCAD Runtime Extension applications and proxy graphics.

The **Demand load ObjectARX Apps** text box allows you to select when and if a third-party application is demand-loaded when a drawing has custom objects that were created in that application. Choosing **Disable load on demand** turns off demand-loading. Choosing **Custom object detect** demand-loads the source application when you open a drawing that contains custom objects. It does not demand-load the application when you invoke one of the application's commands. Choosing **Command invoke** demand-loads the source application when you invoke the application's command. This setting does not demand-load the application when you open a drawing that contains custom objects. Choosing **Object detect and command invoke** demand-loads the source application when you open a drawing that contains custom objects or when you invoke one of the applications' commands.

The **Proxy images for custom objects** text box controls how custom objects in the drawings are displayed. Choosing **Do not show proxy graphics** causes custom objects in drawings not to be displayed. Choosing **Show proxy** causes custom objects in drawings to be displayed. Choosing **Show proxy bounding box** causes a box to be displayed in place of custom objects in drawings. Choosing **Show Proxy Information dialog box** causes a warning to be displayed when you open a drawing that contains custom objects.

Plot and Publish

The Plot and Publish tab of the Options dialog box allows you to set the parameters for plotting your drawing (Figure 14–54).

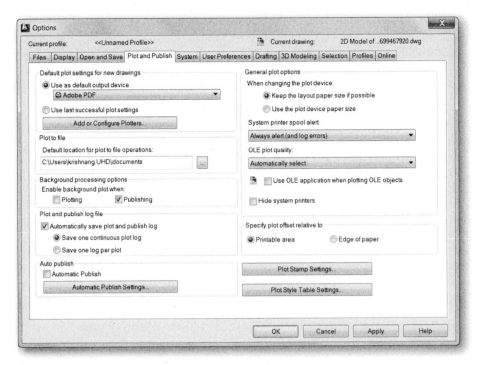

FIGURE 14–54 *Options dialog box with the Plot and Publish tab selected*

The Default plot settings for new drawings section determines plotting parameters for new drawings. It will also determine settings for drawings created in releases prior to AutoCAD 2000 that have never been saved in AutoCAD 2000 format.

Selecting **Use as default output device** causes the device selected in the list box to become the default output device for new drawings and for drawings created in an earlier release of AutoCAD that have never been saved in AutoCAD 2000 format. The list displays all plotter configuration files (*.eps3*) that are found in the plotter configuration search path. It also displays all system printers configured in the system.

Selecting **Use last successful plot settings** causes the settings of the last successful plot to be used for the current settings.

Choosing **Add or Configure Plotters** allows you to add or configure a plotter from the Plotters program window. See Chapter 8 for adding and configuring plotters.

The General plot options section allows you to set general parameters such as paper size settings, system printer alert parameters, and OLE objects.

Selecting **Keep the layout paper size if possible** applies the paper size in the **Layout Settings** tab in the Page Setup dialog box provided the selected output device is able to plot to this paper size. If it cannot, AutoCAD displays a warning message and uses the paper size specified either in the plotter configuration file (*.pc3*) or in the default system settings if the output device is a system printer.

Selecting **Use the plot device paper size** applies the paper size in either the plotter configuration file (*.pc3*) or in the default system settings if the output device is a system printer.

The **System printer spool alert** list box determines if a warning will be displayed if the plotted drawing is spooled through a system printer because of an input or output port conflict. Selecting **Always alert (and log errors)** causes a warning to be displayed and always logs an error when the plotted drawing spools through a system printer. Selecting **Alert first time only (and log errors)** causes a warning to be displayed once and always logs an error when the plotted drawing spools through a system printer. Selecting **Never alert (and log first error)** causes a warning not to be displayed and logs only the first error when the plotted drawing spools through a system printer. Selecting **Never alert (do not log errors)** causes a warning not to be displayed and does not log an error when the plotted drawing spools through a system printer.

The **OLE plot quality** list box determines the quality of plotted OLE objects. Options include Automatically select, Monochrome, such as spreadsheets, Low graphics, such as color text and pie charts, and High graphics, such as photographs.

Selecting **Use OLE application when plotting OLE objects** starts the application that creates the OLE object when you plot a drawing with OLE objects. This step will help optimize quality of OLE objects.

Selecting **Hide system printers** causes AutoCAD to hide system printers.

The Plot to File section specifies the default location for plot to file operations. You can enter a location or choose the **Browse** icon to specify a new location.

The Background processing options section specifies options for background plotting and publishing. You can use background plotting to start a job you are plotting or publishing and immediately return to work on your drawing while your job is plotted or published as you work.

The Plot and publish log file section controls options for saving a plot and publish log file as a comma-separated value (*.csv*) file that can be viewed in a spreadsheet program.

The Specify plot offset relative to section specifies whether the offset of the plot area is from the lower-left corner of the printable area or from the edge of the paper.

The Auto Publish section specifies whether or not drawings are published automatically and controls Auto Publish settings.

Choosing Plot Stamp Settings opens the Plot Stamp Settings dialog box, which allows you to specify information for the plot stamp.

Choosing **Plot Style Table Settings** opens the Plot Style Table Settings dialog box, which allows you to specify settings for plot style tables.

System

The System tab of the Options dialog box has sections for managing the 3D graphics display, pointing devices, dbConnect options, and general options (Figure 14–55).

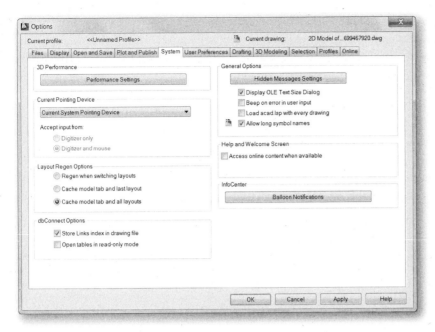

FIGURE 14–55 *Options dialog box with the System tab selected*

The 3D Performance section controls settings that relate to configuration of the 3D graphics display system.

Choosing **Properties Settings** displays a 3D Graphics System Configuration dialog box for the current 3D graphics display system. In the 3D Graphics System Configuration dialog box, you set options that affect the way objects are displayed and system resources are used in the 3D Orbit view. The options you set also affect the way objects are shaded with SHADEMODE.

The Current Pointing Device section determines the parameter for the pointing device(s) being used. The text box lists available pointing device drivers from which to choose. Selecting **Current system pointing device** causes the system pointing device to be the current pointing device. Selecting **Wintab Compatible Digitizer ADI 4.2 by Autodesk** causes the Wintab compatible digitizer to be the current pointing device.

The **Accept input from** option determines if the digitizer only is active or both the mouse and digitizer are active. These are determined by choosing either **Digitizer only** or **Digitizer and mouse.**

The Layout Regen Options section of the System tab allows you to specify how AutoCAD updates the display list in the Model and layout tabs. The display list for each tab is updated either by regenerating the drawing when you switch to that tab or by saving the display list to memory and regenerating only the modified objects when you switch to that tab.

Selecting **Regen when switching layouts** causes AutoCAD to regenerate the drawing each time you switch tabs.

Selecting **Cache model tab and last layout** saves the display list to memory for the Model tab and the last layout made current. When checked, it suppresses regenerations when you switch between the two tabs. Regenerations for all other layouts still occur when you switch to those tabs.

Selecting **Cache model tab and all layouts** causes AutoCAD to regenerate the drawing the first time you switch to each tab. For the remainder of the drawing session, when you switch to those tabs, the display list is saved to memory and regenerations are suppressed.

The dbConnect Options section of the System tab allows you to manage the options associated with database connectivity.

Selecting **Store links index in drawing file** causes AutoCAD to store the database index within the drawing file. When this option is checked, performance during link selection operations is enhanced. When this option is not checked, the drawing file size is decreased and the opening process is enhanced for drawings with database information.

Selecting **Open tables in read-only mode** causes AutoCAD to open database tables in read-only mode within the drawing file.

The General Options section of the System tab has check boxes relating to a variety of options for system parameter settings.

Choosing **Hidden Message Settings** displays the Hidden Message Settings dialog box where you can control the display of hidden messages.

Selecting **Display OLE Text Size Dialog** causes the OLE Properties dialog box to be displayed when inserting OLE objects into AutoCAD drawings.

Selecting **Beep on error in user input** causes an alarm beep when AutoCAD detects an invalid entry.

Selecting **Load acad.lsp with every drawing** causes AutoCAD to load the *acad.lsp* file into every drawing.

Selecting **Allow long symbol names** allows up to 255 characters for named objects.

The Live Enabler Options section of the System tab allows you to specify how AutoCAD checks for Object Enablers. Using Object Enablers, you can display and use custom objects in AutoCAD drawings even when the ObjectARX application that created them is unavailable.

The **Check Web for Live Enablers** check box, when checked, causes AutoCAD to check for Object Enablers on the Autodesk Web site.

The **Maximum number of unsuccessful checks** text box allows you to specify the number of times AutoCAD will continue to check for Object Enablers after unsuccessful attempts.

The InfoCenter section controls options related to Balloon Notifications (located in the upper-right corner of the application window) such as the content, frequency, and duration.

User Preferences

The User Preferences tab of the Options dialog box controls options that optimize the way you work in AutoCAD (Figure 14–56).

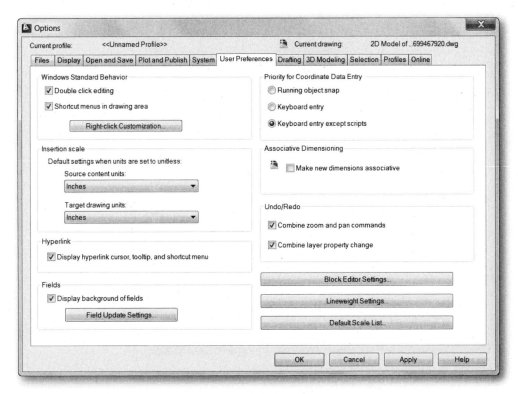

FIGURE 14–56 *Options dialog box with the User Preferences tab selected*

The Windows Standard Behavior section of the User Preferences tab allows you to apply Windows techniques and methods in AutoCAD.

Selecting **Double click editing** enables double-click editing functions in the drawing area.

Selecting **Shortcut menus in drawing area** causes shortcut menus to be displayed when the pointing device is right-clicked. Otherwise, right-clicking is the same as pressing [Enter].

Choosing **Right-click customization** causes the Right-Click Customization dialog box to be displayed (Figure 14–57). This dialog box provides various options related to Shortcut Menus.

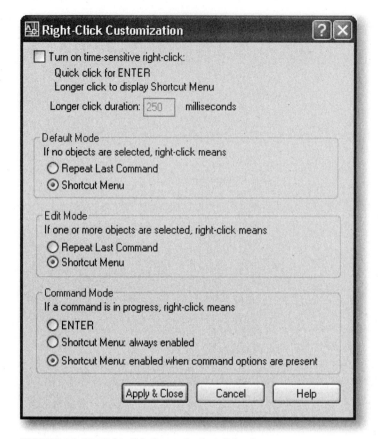

FIGURE 14–57 *Right-click Customization dialog box*

Choosing **Turn on time-sensitive right-click** controls right-click functions. A quick click is the same as pressing Enter. A longer click displays a shortcut menu. You can set the duration of the longer click in milliseconds.

The Default Mode section of the Right-Click Customization dialog box determines the effect of right-clicking when no objects are selected. Selecting **Repeat Last Command** causes right-clicking to be the same as pressing Enter. Selecting **Shortcut Menu** causes right-clicking to display a shortcut menu when applicable.

The Edit Mode section of the Right-Click Customization dialog box determines the effect of right-clicking when one or more objects are selected. Selecting **Repeat Last Command** causes right-clicking to be the same as pressing Enter. Selecting **Shortcut Menu** causes right-clicking to display the Edit shortcut menu.

The Command Mode section of the Right-Click Customization dialog box determines the effect of right-clicking when a command is in progress. Selecting **ENTER** causes right-clicking to be the same as pressing Enter when a command is in progress. Selecting **Shortcut Menu: always enabled** causes right-clicking to display the Command shortcut menu. Selecting **Shortcut Menu: enabled when command options are present** causes right-clicking to display the Command shortcut menu to be displayed only when options are currently available from the command line. Otherwise, right-clicking is the same as pressing Enter.

The Insertion scale section of the User Preferences tab allows you to control the default scale for dragging objects into a drawing using i-drop or DesignCenter.

The **Source content units** list box allows you to set the units AutoCAD uses for an object being inserted into the current drawing when no insert units are specified with the INSUNITS system variable.

The **Target drawing units** list box allows you to set the units AutoCAD uses in the current drawing when no insert units are specified with the INSUNITS system variable. The value is stored in the INSUNITSDEFTARGET system variable. The list box allow you to choose from the following units: **Inches, Feet, Miles, Millimeters, Centimeters, Meters, Kilometers, Microinches, Mills, Yards, Angstroms, Nanometers, Microns, Decimeters, Decameters, Hectometers, Gigameters, Astronomical Units, Light Years**, and **Parsecs**. If **Unspecified-Unitless** is selected, the object is not scaled when inserted.

The Fields section of the User Preferences tab sets preferences related to fields.

Selecting **Display background of fields** displays fields with a light gray background that is not plotted. When this option is cleared, fields are displayed with the same background as any text.

Choosing **Field Update Settings** displays the Field Update Settings dialog box, which allows you to set the fields that will be updated automatically.

The Priority for Coordinate Data Entry section of the User Preferences tab determines how input of coordinate data affects AutoCAD's actions.

Selecting **Running object snap** causes running object snaps to be used at all times instead of specific coordinates.

Selecting **Keyboard entry** causes the coordinates that you enter to be used at all times and overrides running object snaps.

Selecting **Keyboard entry except scripts** causes the specific coordinates that you enter to be used rather than running object snaps, except in scripts.

The Associative Dimensioning section of the User Preferences tab controls whether or not new dimensions are associative.

Selecting **Make new dimensions associative** causes new dimensions to be drawn as associative dimensions and will be associated with the objects being dimensioned.

The Hyperlink section of the User Preferences tab determines display property settings of hyperlinks.

Selecting **Display hyperlink cursor, tooltip, and shortcut menu** causes the hyperlink cursor, Tooltip, and shortcut menu to be displayed when the cursor is over an object that contains a hyperlink.

The Undo/Redo section of the User Preferences tab controls Undo and Redo for ZOOM and PAN.

Selecting **Combine zoom and pan commands** causes these two commands to act together.

Selecting **Combing Layer Property Changes** groups layer property changes made from the Layer Properties Manager.

Choosing **Block Editor Settings** causes the Block Editor Settings dialog box to be displayed. This action allows you to control the environment settings of the Block Editor.

Choosing **Lineweight Settings** causes the Lineweight Settings dialog box to be displayed. This action allows you to set lineweight options.

Choosing **Initial Setup** causes the Initial Setup dialog box to be displayed. This action allows you to customize how AutoCAD starts up.

Choosing **Edit Scale List** causes the Edit Scale List dialog box to be displayed, which controls the list of scales available for layout viewports, page layouts, and plotting.

Drafting

The Drafting tab of the Options dialog box allows you to customize drafting options in AutoCAD (Figure 14–58). In this tab are sections for **AutoSnap Settings, AutoSnap Marker Size, AutoSnap Marker Color, AutoTrack Settings, Alignment Point Acquisition, Aperture size, Object Snap Options**, and **Drafting Tooltip Appearance.** The options available in these sections are explained in Chapter 3 in the section on "Drafting Settings."

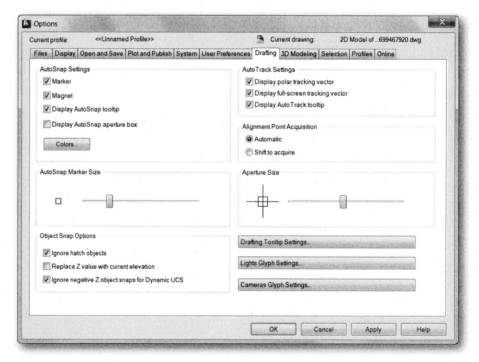

FIGURE 14–58 *Options dialog box with the Drafting tab selected*

3D Modeling

The 3D Modeling tab of the Options dialog box allows you to customize options related to working with solids and surfaces in AutoCAD (Figure 14–59). This tab includes sections for **3D Crosshairs, Display ViewCube or UCS Icon, 3D Objects, 3D Navigation, and Dynamic Input.** The options available in these sections are explained in Chapter 16.

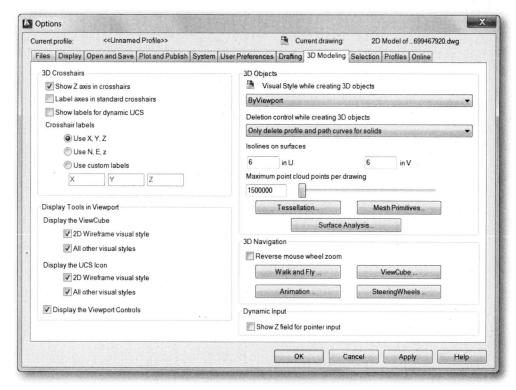

FIGURE 14–59 *Options dialog box with the 3D Modeling tab selected*

Selection

The Selection tab of the **Options** dialog box allows you to customize selection options in AutoCAD (Figure 14–60). This tab includes sections for **Pickbox Size, Selection Preview, Selection Modes, Grip Size,** and **Grips.** You can set object selection settings for the ribbon contextual tabs through the Ribbon Contextual Tab State Options dialog box by selecting the **Contextual Tab States button**. The options available for the **Selection Modes** are explained in Chapter 5 in the section on "Object Selection Modes." The options available for **Grips** are explained in Chapter 6 in the section on "Editing with Grips."

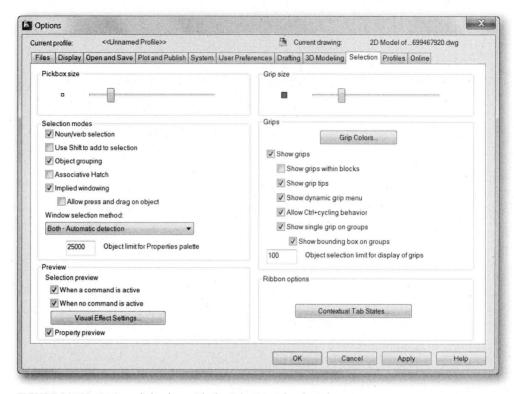

FIGURE 14–60 *Options dialog box with the Selection tab selected*

Profiles

The Profiles tab of the Options dialog box allows you to manage profiles (Figure 14–61). A profile is a named and saved group of environment settings. This profile can be restored as a group when desired. AutoCAD stores your current options in a profile named **Unnamed Profile.** AutoCAD displays the current profile name, as well as the current drawing name, in the Options dialog box. The profile data is saved in the system registry and can be written to a text file with an *.arg* extension file. AutoCAD organizes essential data and maintains changes in the registry as necessary.

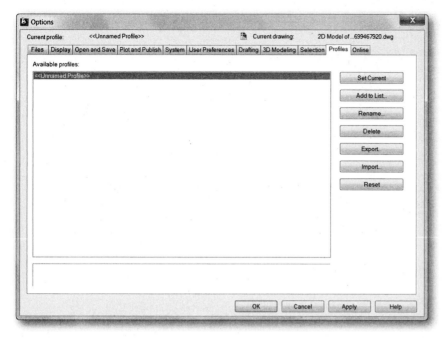

FIGURE 14–61 *Options dialog box with the Profiles tab selected*

A profile can be exported to or imported from different computers. If changes have been made to your current profile during an AutoCAD session and you want to save them in the *.arg* file, the profile must be exported. After the profile with the current profile name has been exported, AutoCAD updates the *.arg* file with the new settings. The profile can be imported again into AutoCAD, thus updating your profile settings.

Choosing **Set Current** makes the profile that is highlighted in the **Available profiles** list box the current profile.

Choosing **Add To List** allows you to name and save the current environment settings as a profile.

Choosing **Rename** allows you to rename the highlighted profile.

Choosing **Delete** allows you to delete the highlighted profile.

Choosing **Export** causes the Export Profiles dialog box to be displayed. This is a file manager dialog box in which the highlighted profile can be saved to the path you specify.

Choosing **Import** causes the Import Profiles dialog box to be displayed. This is a file manager dialog box in which you can select a profile from a saved path to be imported.

Choosing **Reset** causes the highlighted profile to be reset. The default profile name is listed in the description pane at the bottom of the dialog box.

After making changes, choose **Apply** to make the changes effective, or choose **OK** to save the settings and close the Options dialog box.

Online

The Online tab (Figure 14–62) of the Options dialog box sets options for working online with Autodesk 360, and provides access to the design documents that you have stored in your cloud account.

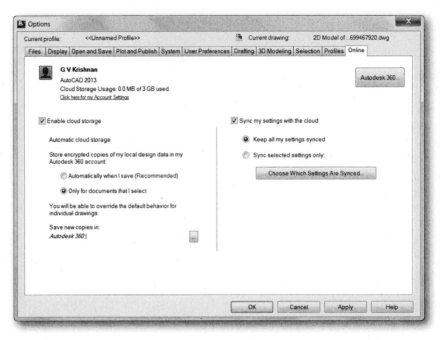

FIGURE 14–62 *Operations dialog with the Online tab selected*

SAVING OBJECTS IN OTHER FILE FORMATS (EXPORTING)

The EXPORT command allows you to save a selected object in other file formats such as *.eps*, *.sat*, *.3ds*, and *.wmf.* Invoke the EXPORT command by entering **export** at the on-screen prompt and AutoCAD displays the Export Data dialog box (Figure 14–63). In the **Files of type** list box, select the format type in which you wish to export objects. Enter the file name in the **File name** edit box. Select **Save,** and AutoCAD displays the following prompt:

Select objects: *(select the objects to export, and press* [Enter] *to complete object selection)*

AutoCAD exports the selected objects in the specified file format using the specified file name.

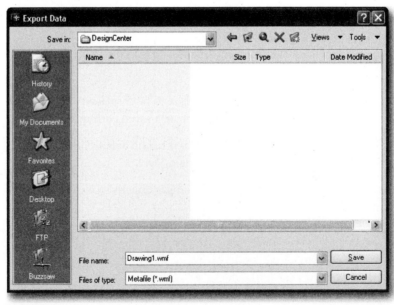

FIGURE 14–63 *Export Data dialog box*

Table 14–1 lists the format types available in AutoCAD for exporting the current drawing.

TABLE 14–1 *AutoCAD exportable format types*

Format type	Description
Dwf & Dwfx	Design Web format
bmp	Device-independent bitmap file
dwg	AutoCAD 2000 drawing file; same as invoking the `WBLOCK` command
dxf	AutoCAD attribute extract DXF file; same as invoking the `ATTEXT` command
eps	Encapsulated PostScript file
sat	ACIS solid-object file
stl	Solid object stereo-lithography file
wmf	Windows metafile
dgn	MicroStation design file
pdf	Portable Document Format created by Adobe
fbx	Import and export format for 3D objects & 2D objects with thickness, lights, cameras, and materials from one Autodesk program to another

IMPORTING VARIOUS FILE FORMATS

The `IMPORT` command allows you to import various file formats, such as *.3ds*, *.sat*, *.wmf*, and *.dgn*, into AutoCAD. Invoke the `IMPORT` command from the Import panel (Figure 14–64) located in the Insert panel, and AutoCAD displays the Import File dialog box (Figure 14–65).

FIGURE 14–64 *Invoking the* `IMPORT` *command from the Insert toolbar*

In the **Files of Type** list box, select the format type you wish to import into AutoCAD. Select the file from the appropriate directory from the list box, and choose **Open.** AutoCAD imports the file into the AutoCAD drawing.

Table 14–2 lists the format types available to import into AutoCAD.

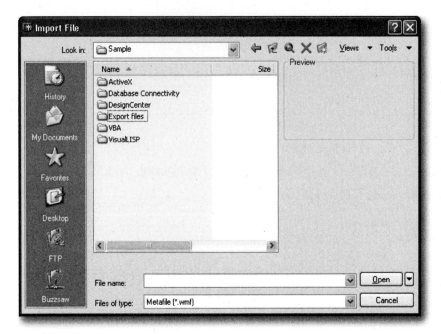

FIGURE 14–65 *Import File dialog box*

TABLE 14–2 *Format types available to import into AutoCAD*

Format type	Description
dgn	MicroStation design file
3ds	3D Studio file
sat	ACIS solid-object file
wmf	Windows Metafile

STANDARDS

AutoCAD has a feature that allows you to verify that the layers, dimension styles, linetypes, and text styles of the drawing you are working in conform to an accepted standard such as a company, trade, or client standard. To use this feature, your drawing must have some standard drawing(s) with which it is associated.

You can create a standards file from an existing drawing or you can create a new drawing and save it as a standards file with an extension of *.dws*. Open an existing drawing from which you want to create a standards file, invoke the SAVEAS command, and enter a name for the standards file in the Save Drawing As dialog box. Select AutoCAD Drawing Standards (*.dws) from the **Files of type** list, and then choose **Save**. You can also create a new drawing and set appropriate standards for layers, text styles, dimension styles, and linetypes. Invoke the SAVEAS command, enter a name for the standards file in the Save Drawing As dialog box, select AutoCAD Drawing Standards (*.dws) from the **Files of type** list, and then choose **Save**.

The STANDARDS command manages the association of standards files with drawings.

Invoke the STANDARDS command by choosing **Configure** from the CAD Standards panel of the Manage tab and AutoCAD displays the Configure Standards dialog box (Figure 14–66).

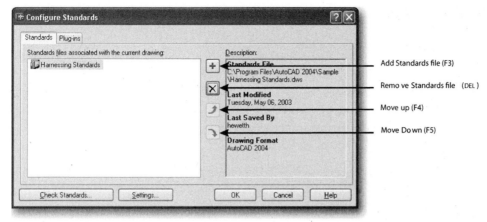

FIGURE 14–66 *Configure Standards dialog box with the Standards tab selected*

If a drawing has an associated standards file, an Associated Standards File(s) icon will display on the status bar tray at the bottom-right corner of the drawing area.

The Standards files associated with the current drawing section lists all standards (*.dws*) files that are associated with the current drawing. To add a standards file, choose the **Add Standards File** icon. AutoCAD displays the Select Standards File dialog box. Select a standards file from an appropriate folder. To remove a standards file from the current drawing, choose the **Remove Standards File** icon. If conflicts arise between multiple standards in the list—for example, if two standards specify layers of the same name but with different properties—the standard that appears first in the list takes precedence. To change the position of a standards file in the list, select it and choose the **Move Up** or **Move Down** icon.

In the Plug-ins tab of the Configure Standards dialog box, the Plug-ins used when checking standards section lists the standards plug-ins that are installed on the current system (Figure 14–67). For the CAD Standards Extension, a standards plug-in is installed for each of the named objects for which standards can be defined: layers, dimension styles, linetypes, and text styles. The Description section has descriptions of the Purpose, Version, and Publisher of the plug-in that is highlighted in the Plug-ins used when checking standards section.

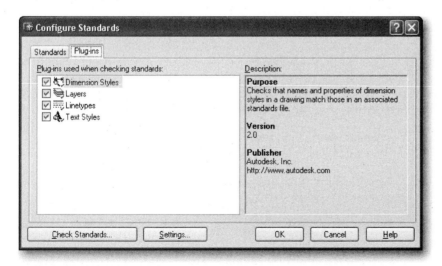

FIGURE 14–67 *Configure Standards dialog box with the Plug-ins tab selected*

The CHECKSTANDARDS command analyzes the current drawing for standards violations.

Invoke the CHECKSTANDARDS command by choosing **Check** from the **CAD Standards** panel of the Manage tab, and AutoCAD displays the Check Standards dialog box (Figure 14–68).

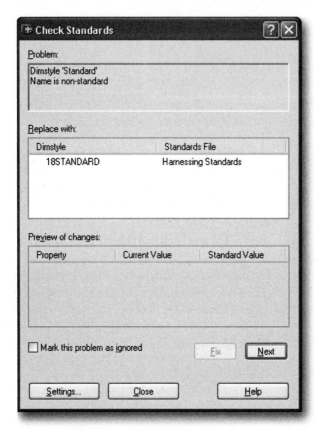

FIGURE 14–68 *Check Standards dialog box*

The Check Standards dialog box has sections titled **Problem, Replace with**, and **Preview of changes.**

The Problem section includes a description of a nonstandard object in the current drawing. To fix a problem, select a replacement from the **Replace with** list, and then choose **Fix**.

The Replace with section lists possible replacements for the current standards violation. If a recommended fix is available, it is preceded by a checkmark.

The Preview of changes section indicates the properties of the nonstandard AutoCAD object that will be changed if the fix currently selected in the **Replace with** list is applied.

Choose **Close** to close the Check Standards dialog box without applying a fix to the standards violation currently displayed in the Problem section.

You can use CAD Standards tools to check for violations as you work. You are immediately alerted whenever you create an object with a nonstandard name.

ACTION RECORDER

With action macros, you can automate repetitive tasks with no previous programming experience. The Action Recorder is used to record an action macro. After an action macro is recorded, you can save the recorded commands with the file extension ACTM and playback the repetitive tasks.

All the tools related to Action Recorder are available in the Action Recorder panel (see Figure 14–69), which is accessed from the Manage tab.

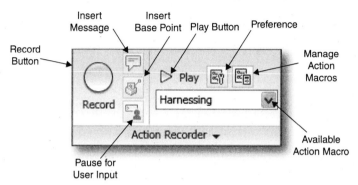

FIGURE 14–69 *Action Recorder Panel*

During the playback, editing, or recording of an action macro, you can expand the Action Recorder panel to access the individual actions of the current action macro from the Action tree.

In the Action tree, you can modify and delete the action nodes of an action macro. Action nodes represent the recorded commands or input values of an action macro. You can also insert user messages and request user input for a value node during playback. A value node in an action macro represents the input that was provided at a sub-prompt of a command during recording. Value nodes can contain acquired points, text strings, numbers, keywords, or other values that might be entered when recording a command.

You can also set the preferences for the Action Recorder from the Action Recorder Preferences dialog box. Select **Preference** from the Action Recorder panel and AutoCAD Displays the Action Recorder Preferences dialog box (Figure 14–70).

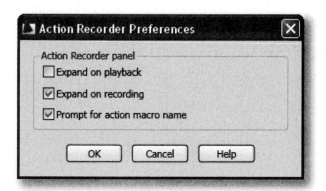

FIGURE 14–70 *Action Recorder Preferences dialog box*

In the Action Recorder Preferences dialog box you can control if the Action Recorder panel expands when recording or playing back an action macro, and you can also control if you are prompted to provide a command and file name for the action macro when recording is stopped.

In the Action Recorder Preferences dialog box, selecting **Expand on Playback** expands the Action Recorder panel during playback. Selecting **Expand on Recording** expands the Action Recorder panel during recording. Selecting **Prompt for Action Macro Name** displays the Action Macro dialog box when recording is stopped.

Choosing **Manage Action Macros** causes the Action Macro Manager dialog box to be displayed from which you can Copy, Rename, Modify, and Delete Action Macros. Choosing **Options** causes the Options dialog box to be displayed.

Recording

With the Action Recorder you can record most of the commands that can be used from the command line and from the user interface elements with which you are already familiar.

The Action Recorder is used to record commands and input values for an action macro. While recording an action macro, the Red Recording Circle icon is displayed near the crosshairs to indicate that the Action Recorder is active and when commands and input are being recorded. You can also insert messages for requesting user input. Choosing **Insert Base Point** inserts a base point in an action macro, which requests a user-defined point when the action macro is played back. A base point establishes a point that is used by the prompts that follow in the action macro.

Commands and input entered at the command line are recorded except for commands used to open or close drawing files. If a dialog box is displayed while recording an action macro, only the display of the dialog box is recorded and not the changes made to the dialog box. It is recommended that dialog boxes are not used when recording an action macro. Use the command line version of the command instead. For example, use the -ARRAY command instead of the ARRAY command, which displays the Array dialog box.

Once you have completed recording an action macro, you may save or cancel the recorded action macro (see Figure 14–71). If you save the action macro, you must specify a name. A description is optional. You can specify the playback settings for the action macro. The playback settings control the view prior to the playback of the action macro is restored when a request for user input is made or when playback is complete.

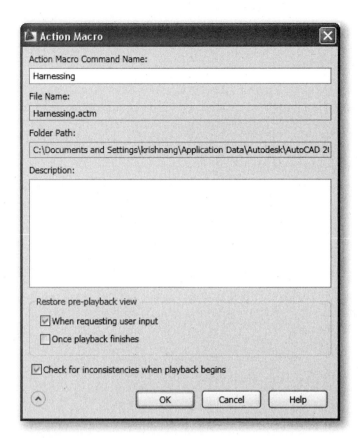

FIGURE 14–71 *Action Macro dialog box*

When the Action Recorder is set to check for inconsistencies, it compares the settings of the drawing environment when the action macro was recorded against the current settings of the drawing environment. For example, the Action Recorder checks the value of the INSUNITS system variable in the current drawing against the value that was used when the action macro was recorded. If an inconsistency is found, you are given the option to continue playing back the action macro or to stop playback. If you continue playing back the action macro, the action macro might produce unexpected results.

You can play back an action macro while you are recording an action macro; you can combine multiple action macros together to create a new action macro. To use an existing action macro while recording another action macro, the action macro that you want to play back needs to be present in one of the paths defined by the system variables ACTPATH or ACTRECPATH. If the action macro is in one of the defined paths, enter the name of the action macro at the Command prompt to play it back.

Playback

After the recording of an action macro is complete, it can be played back as a series of commands and input values.

The action macro can be played back from the Action Recorder panel. As an action macro is played back, you might be prompted for input or requested to respond to a message.

Based on the current action or request for user interaction in the action macro, an icon near the cursor is displayed to indicate when the action macro needs input in order to continue. A dialog box may be displayed where you enter a value or use the recorded value.

As an action macro is played back, the series of commands are performed is the sequence they were recorded until playback is complete or an error is encountered. Some of the reasons for the playback to stop or fail include an Invalid command, an empty selection, or a cancelled macro.

Using the Action Recorder

The drawing settings in effect when recording an action macro are not completely retained by an action macro. To make sure that certain drawing settings are used during playback, you can record them as part of the action macro using the system variable name or the SETVAR command.

Not all commands are recorded, such as those used for file operations, working with the Action Recorder, and grip editing. Commands not recorded include ACTSTOP, ACTUSERINPUT, ACTUSERMESSAGE, DXFIN, FILEOPEN, NEW, OPEN, PARTIAL OPEN, PRESSPULL, QNEW, RECOVER, TABLEDIT, VBALOAD, VBANEW, VBAMAN, VBAPREF, VBARUN, VBASTMT, XOPEN and VBAUNLOAD.

Actions recorded with the Action Recorder can only be used interactively with the Properties palette, Quick Properties panel, the Layer Modeless dialog box, and any user interface element that starts a command.

Commands defined with AutoLISP and ObjectARX can be recorded with an action macro, but the command must be loaded before the action macro is played back. VBA macros and .NET assemblies can also be recorded as part of an action macro, but they must be loaded in order to be played back.

Editing the Action Recorder

You can rename, copy, or delete an action macro. You can also modify an action macro by removing actions from the Action tree, inserting user messages and requests for user input, or editing and changing the behavior of an action. In the Action Recorder panel, you can modify the actions in an action macro and perform some basic file management on an action macro file. Modification and management of an action macro file are handled through the Action tree, which is displayed when the Action Recorder panel is expanded.

Action nodes that represent commands can be deleted and value nodes can be edited. Action nodes in an action macro are modified through the Action tree, which is displayed when the Action Recorder panel is expanded. The modify options are available when you right-click an action node in the Action tree.

SLIDES AND SCRIPTS

Slides are quickly viewable, noneditable views of a drawing or parts of a drawing. There are two primary uses for slides. One is to have a quick and ready picture to display symbols, objects, or written data for informational purposes only. The other useful application of slides is displaying a series of pictures, organized in a prearranged sequence for a timed slide show. This is a useful tool for demonstrations to clients or in a showroom. This feature supplements the time-consuming process of calling up views required when using the ZOOM, PAN, or other display commands. The "slide show" is implemented through the SCRIPT command, as described later in this chapter.

It should be noted that a slide merely masks the current display. Any cursor movement or editor functions employed while a slide is being displayed affect the current drawing under the slide and not the slide itself.

Making a Slide

The current display can be made into a slide with the MSLIDE command. The current viewport becomes the slide while you are working in model space. The entire display, including all viewports, becomes the slide when using MSLIDE while you are working in paper space. The MSLIDE command takes a picture of the current display and stores it in a file, so be sure it is the correct view.

Invoke the MSLIDE command by typing **mslide** at the on-screen prompt, and AutoCAD displays the Create Slide File dialog box.

The default is the drawing name, which you can use as the slide file name by pressing Enter. You can type any other name, as long as you are within the limitations of the OS file-naming conventions. AutoCAD automatically appends the extension *.sld*. Only objects that are visible in the screen drawing area, or in the current viewport when in model space, are made into the slide.

If you plan to show the slide on different systems, you should use a full-screen view with a high-resolution display for creating the slide.

Viewing a Slide

The VSLIDE command displays a slide in the current viewport.

Invoke the vslide command by typing VSLIDE from the on-screen prompt, and AutoCAD displays the Select Slide File dialog box. This is similar to most file-management dialog boxes. Select the slide to display in the current viewport.

Scripts

Of the many means available to enhance AutoCAD through customization, scripts are among the easiest to create. Scripts are similar to the macros that can be created to enhance word-processing programs. They permit you to combine a sequence of commands and data into one or two entries. Creating a script, like most enhancements to AutoCAD, requires that you use a text editor to write the script file, with the extension *.scr*, which contains the instructions and data for the SCRIPT command to follow.

Because script files are written for use at a later time, you must anticipate the conditions under which they will be used. Therefore, familiarity with sequences of prompts that will occur and the types of responses required is necessary for the script to function properly. Writing a script is a simple form of programming.

A script text file must be written in ASCII format. That is, it must have no embedded print codes or control characters that are automatically written in files when created with a word processor in the document mode. If you are not using one of the available line editors, be sure you are in the nondocument, programmer, or ASCII mode of your word processor when creating or saving the file. Save the script with the file extension *.scr*.

Each command can occupy a separate line, or you can combine several command/data responses on one line. Each space between commands and data is read as an [Enter], just as pressing [SPACEBAR] is read while in AutoCAD. The end of a line of text is also the same as an [Enter].

The following script file contains several commands and data. The commands are GRID, LINE, CIRCLE, LINE, and CIRCLE again. The data are the response ON, coordinates such as 0,0 and 5,5, and distances, such as radius 3.

```
GRID ON LINE 0,0 5,5 CIRCLE 3,3 3
LINE 0,5 5,0
CIRCLE 5,2.5 1
```

The first line includes the GRID, LINE, and CIRCLE commands and their responses. Note the two spaces after 5,5; these are required to simulate the double [Enter]. Not obvious is the extra space following the 5,0 response in the second LINE command and after the 1 in the third line. This extra space and the invisible CR-LF (carriage-return linefeed) code that ends every line in a text file combine to simulate pressing the SPACEBAR twice. Again, this is necessary to exit the LINE command.

Some text editors automatically remove blank spaces at the end of text lines. To guard against that action, an alternative is to have a blank line indicate the second [Enter], as follows:

```
GRID ON LINE 0,0 5,5 CIRCLE 3,3 3
LINE 0,5 5,0
CIRCLE 5,2.5 1
```

Invoke the SCRIPT command by choosing **Run Script** on the **Tools** menu, and AutoCAD displays the Select Script File dialog box. Select the appropriate script file from the list box, and choose the Open button. AutoCAD executes the command sequence from the script file.

Using a script to perform a repetitive task is illustrated in the following example. This application also offers some insight on changing the objects in an inserted block with attributes without affecting the attribute values.

Figure 14–72 shows a group of drawings all of which use a common block with attribute values in one insertion that are different from the attribute values of those in other drawings. In this case, the border/title block is a block named BRDR. It was originally drawn with the short lines around and outside of the main border line. It was discovered that these lines interfered with the rollers on the plotter and needed to be removed. The BRDR block definition is shown in Figure 14–73. Note that the inserted block has different attribute values in each drawing such as drawing number, date, and title.

You want to change objects in the block but maintain the attribute values as they are. There are two approaches to redefinition. One is to find a clear place in the drawing and insert the block with an asterisk (*). This is the same as inserting and exploding the block. You make the necessary changes in the objects, and make the revised group into a block with the same block name. This action redefines all insertions of blocks with that same name in the drawing. In this case, there is only one insertion. You must be attentive to how any changes to attributes might affect the already-inserted block of that name.

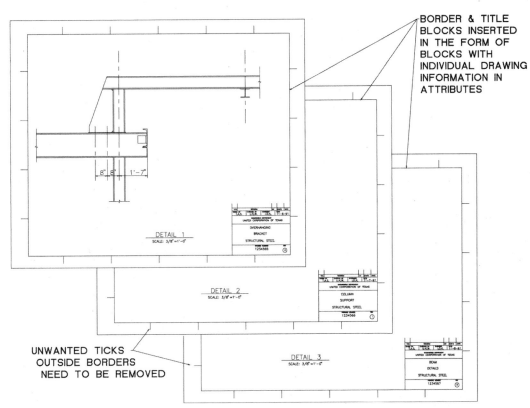

FIGURE 14–72 *Drawings using common block and attribute values that are different from those of other drawings*

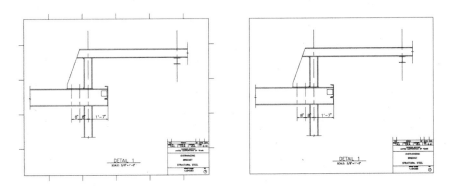

FIGURE 14–73 *Common block attributes modified: Border removed (left) and Ticks removed (right)*

The second method is to use the WBLOCK command to place a copy of the block in a file with the same name. This makes a new and separate drawing of the block. You exit the current drawing, call up the newly created drawing, make the required changes in the objects, and end the drawing that was created by the WBLOCK command. Reenter the drawing in which the block objects need to be changed. You now use the INSERT command respond with "blockname=," and the block is redefined without losing the attribute values. For example, if the block name is BRDR, the sequence would be as follows:

```
Command: insert
Enter block name or [?]: brdr=
Block "b" already exists. Redefine it? [Yes/No] <N>: y
Block BRDR redefined
Regenerating drawing
Specify insertion point or [Scale/X/Y/Z/Rotate/PScale/PX/
PY/PZ/PRotate]: (press Enter to cancel the command sequence)
```

The key to this sequence is the equals sign (=) following the block name. This causes AutoCAD to change the definition of the block named BRDR to be that of the drawing named BRDR, although it maintains the attribute values as long as attribute definitions remain unchanged.

If the preceding procedure must be repeated many times, a script file can be employed to automate the process. In the following example, we show how to apply the script to a drawing named PLAN_1. The sequence included an Enter as it was described to be used while in AutoCAD. This expedited the operation by not actually having the block inserted but only having its definition brought into the drawing. Because an Enter during the running of a script file causes the script to terminate, those keystrokes cannot be made in the middle of a script; besides, invoking in a SCRIPT command requires using the AutoLISP function "(command)." The script can be written in a file—called *BRDRCHNG.SCR* for this example—in ASCII format as follows:

```
INSERT
BRDR=
0,0 (the 0,0 is followed by six spaces)
ERASE L (the L is followed by one space)
REDRAW
```

From the on-screen prompt, you can apply the script to drawing PLAN_1 by responding as follows:

```
script
```

Several aspects of this sequence are important:

- *Lines 1–2—* This is where you might press Enter if you were not in the SCRIPT command and have the definition of the block named BRDR take on that of the drawing BRDR without actually having to continue with the insertion in the drawing.

- *Line 3—* In this case, 0,0 as the insertion point is arbitrary because the inserted block is going to be erased anyway. Special attention is given to the six spaces following the insertion point. These are the same as pressing SPACEBAR or Enter six times in response to the "X-scale," "Y-scale," and "Rotation angle" prompts, and the number of spaces that follow must correspond to the number of attributes that require responses for values. In this example there were three attributes. Again, the fact that the responses are null is immaterial, because this insertion will not be kept.

- *Line 4*— The ERASE L is self-explanatory, but note that after the "L," you must insert another space by pressing the SPACEBAR to terminate the object selection process and complete the ERASE command.
- *Line 5*— The REDRAW command is not really required except to show the user for a second time that the changes have been made before the script ends.

Following are the utility commands that may be used within a script file.

The DELAY command causes the script to pause for the number of milliseconds that have been specified by the command. To set the delay in the script for 5 seconds, you would write:

```
DELAY 5000
```

The RESUME command causes the script to resume running after the user has interrupted the script.

The GRAPHSCR and TEXTSCR commands are used to flip or toggle the screen to the graphics or text mode, respectively, while the script is running. These two screen toggle commands can be used transparently by preceding them with an apostrophe.

The RSCRIPT command, when placed at the end of a script, causes the script to repeat itself. With this feature, you can have a slide show run continuously until terminated by ESC key.

A repeating demonstration can be set up to show some sequences of commands and responses as follows:

```
GRID ON
LIMITS 0,0 24,24
ZOOM A
CIRCLE 12,12 4
DELAY 2000
COPY L M 12,12 18,12 12,18 6,12 12,6 (an extra space at the
end)
DELAY 5000
ERASE W 0,0 24,24 (an extra space at the end)
DELAY 2000
LIMITS 0,0 12,9
ZOOM A
GRID OFF
TEXT 1,1 .5 0 THAT'S ALL FOLKS!
ERASE L (an extra space at the end)
RSCRIPT
```

This script file utilizes the DELAY and RSCRIPT subcommands. Note the extra spaces where continuation of some actions must be terminated.

The SCRIPT command can be used to show a series of slides, as in the following sequence:

```
VSLIDE SLD_A
VSLIDE *SLD_B
DELAY 5000
```

```
VSLIDE
VSLIDE *SLD_C
DELAY 5000
VSLIDE
DELAY 10000
RSCRIPT
```

This script uses the asterisk (*) before the slide name prior to the delay. This causes AutoCAD to load the slide that is ready for viewing. Otherwise, there would be a blank screen between slides while the next one is being loaded. The RSCRIPT command repeats the slide show.

WORKSPACES

Any time you are in a drawing session, the combination of settings, content, and arrangements of the menus, toolbars, and dockable windows such as the Tool Palettes window or DesignCenter can be saved to a named workspace. Later, after some of these items have been changed, you can recall the workspace by its name and return to the same combination of settings, content, and arrangements that was in effect when you created the workspace. This is especially helpful when your work requires two or more types of workspaces. It is time consuming to make changes to the drawing environment each time you switch back and forth between different types of work. By naming and saving each combination that makes up the best suitable environment, you can save time.

AutoCAD has four built-in workspaces: 2D Drafting & Annotation (default), 3D Basics, 3D Modeling, and AutoCAD Classic.

Invoke the WORKSPACE command by typing it from the keyboard at the on-screen prompt, and AutoCAD displays the following prompt:

Enter workspace option *(choose the desired option)*

Choosing SETCURRENT sets an existing workspace current. AutoCAD displays the following prompt:

Enter name of workspace to make current [?] <current>:
(specify a name or? to list available workspaces)

You can set a workspace as current from the Workspaces toolbar (Figure 14–74) located in the top left of the AutoCAD application window or from the Workspace icon on the status bar.

FIGURE 14–74 *Selection of a workspace from the status bar*

Choosing Save Current As...saves a current environment configuration as a named workspace. AutoCAD prompts:

> Save workspace as <current>: (specify a name for the new workspace)

Choosing EDIT opens the Customize User Interface dialog box with the Customize tab displayed where you can modify a workspace.

Choosing RENAME allows you to rename a workspace. AutoCAD displays the following prompts:

> Enter workspace to rename ⏎ (specify the name of the workspace to be renamed)
>
> Enter new workspace name <current>: (specify a new name for the workspace)

Choosing **Delete** allows you to delete a workspace. AutoCAD displays the following prompts:

> Enter new workspace to delete <current>: (specify a name or? to list available workspaces)
>
> Workspace <name> deleted.
>
> Do you really want to delete the workspace "<specified>" ⏎ (choose Y)

Choosing **Settings** opens the Workspace Settings dialog box, which controls the display, menu order, and Save settings of a workspace.

Choosing **?** (list workspaces) displays a list of all workspaces defined in the main and enterprise CUI files in the AutoCAD Text Window.

NOTE

Workspaces differ from Profiles. Workspaces let you change how the menus, toolbars, and dockable windows in the drawing area are displayed. Profiles are a combination of user options, drafting settings, paths, and values. With workspaces, you can easily switch between them during a drawing session and manage them using the CUI dialog box. Profiles are updated each time an option, setting, or other value is changed. Profiles are managed from the Options dialog box.

Open the Exercise Manual PDF for Chapter 14 for discipline-specific exercises. Related files are downloaded from the student companion site mentioned in the Introduction (refer to page number xii for instructions).

REVIEW QUESTIONS

1. All of the following can be renamed using the RENAME command, except:
   a. A current drawing name
   b. Named views within the current drawing
   c. Block names within the current drawing
   d. Text style names within the current drawing

2. The PURGE command can be used:
   a. After an editing session
   b. At the beginning of the editing session
   c. At any time during the editing session
   d. a and b only

3. The most common setting for the current drawing color is:
   a. Red
   b. ByLayer
   c. ByBlock
   d. White
   e. None of the above

4. Which of the following are not valid color names in AutoCAD?
   a. Brown
   b. Red
   c. Yellow
   d. Magenta
   e. All are valid

5. To change the background color for the graphics drawing area, you should use:
   a. COLOR
   b. BGCOLOR
   c. OPTIONS
   d. SETTINGS
   e. CONFIG

6. AutoCAD 2006 can be used to edit a drawing saved as an ACAD2011 drawing.
   a. True
   b. False

7. AutoCAD 2009 can be used to edit a drawing that is saved in an AutoCAD 2011 drawing format.
   a. True
   b. False

8. All the following items can be purged from a drawing file, except:
   a. Text styles
   b. Blocks
   c. System variables
   d. Views
   e. Linetypes

9. The following can be deleted with the PURGE command except:

   a. Blocks not referenced in the current drawing

   b. Linetypes that are not being used in the current drawing

   c. Layer 0, if it is not being used

   d. All can be purged

10. With object linking and embedding (OLE), you can copy or move information from one application to another while retaining the ability to edit the information in the original application.

   a. True

   b. False

11. If the TIME command is not turned off during lunch break, the TIME command will:

   a. Include the lunch break time

   b. Exclude the lunch break time

   c. Turn itself off after 10 minutes of inactivity

   d. Automatically subtract 1 hour for lunch

   e. None of the above

12. The VIEW command:

   a. Serves a purpose similar to the PAN command

   b. Will restore previously saved views of your drawing

   c. Is normally used on very small drawings

   d. None of the above

13. When using a .X point filter, AutoCAD will request that you complete the point selection process by entering:

   a. An X coordinate

   b. Y and Z coordinates

   c. An OSNAP mode

   d. Nothing; .X is not a valid filter

14. The geometric calculator function MEE will:

   a. Calculate the midpoint of a selected line

   b. Calculate the midpoint between the endpoints of any two objects

   c. Return the node name for the computer you are working on

   d. Average a string of numbers

   e. None of the above

15. To load a script file called *sample.scr*, use:

   a. script, then sample

   b. load, then sample

   c. load, script, then sample

   d. None of the above

16. The AutoCAD command used for viewing a slide is:
    a. VSLIDE
    b. VIEWSLIDE
    c. SSLD
    d. SLIDE
    e. MSLIDE

17. If a REDRAW is performed while viewing a slide:
    a. The command will be ignored
    b. The current slide will be deleted
    c. The current drawing will be displayed
    d. AutoCAD will load the drawing the slide was created from
    e. None of the above

18. A script file is identified by the following extension:
    a. *scr*
    b. *bak*
    c. *dwk*
    d. *spt*
    e. None of the above

19. A slide file is identified by the following extension:
    a. *sld*
    b. *scr*
    c. *sle*
    d. *slu*
    e. None of the above

20. Slides can be removed from the display with the command:
    a. ZOOM ALL
    b. REGEN
    c. OOPS
    d. Both a and b
    e. None of the above

21. The SCRIPT command cannot be used to:
    a. Insert blocks
    b. Create layers
    c. Place text
    d. Create another script file
    e. All are possible

22. To cause a script file to execute in an infinite loop, what command should you place at the end of the file?

    a. REPEAT

    b. RSCRIPT

    c. GOTO:START

    d. BEGIN

    e. None of the above

23. If AutoCAD is executing in an infinite script file loop, how can you terminate the loop?

    a. Press BACKSPACE

    b. Press CTRL + C

    c. Press ALT + C

    d. Press F1

    e. None of the above

24. What command permits a user to work with just a portion of a drawing by loading geometry from specific views or layer?

    a. PLOAD

    b. PARTOPN

    c. PARTIALOPEN

    d. PLTOPN

25. What command permits additional geometry to be loaded into the current partially loaded drawing?

    a. PLOAD

    b. PARTIAL LOAD

    c. PARTADD

    d. ADDGEO

26. Can the PURGE command be used in partially open drawings?

    a. Yes

    b. No

27. What term can be added to draw, modify, or inquiry commands that will cause them to automatically repeat?

    a. REDO

    b. REPEAT

    c. RETURN

    d. MULTIPLE

    e. No such option

28. When one or more objects overlay each other, which command is used to control their order of display?

    a. DWGORDER

    b. DRAWORDER

    c. VIEWORDER

    d. ARRANGE

29. Which command permits you to execute operating system programs without exiting AutoCAD?

    a. RUN

    b. EXOPRG

    c. OPSYS

    d. SHELL

15

Internet Utilities and Drawing Sets

INTRODUCTION

The Internet is the way to convey digital information around the world. You are probably already familiar with the best-known uses of the Internet: electronic mail (e-mail) and surfing the "World Wide Web," also called the "the Web" or the Internet. E-mail lets users exchange messages and data at very low cost. The Web brings together text, graphics, audio, and video in an easy-to-use format. Other uses of the Internet include file transfer protocol (FTP), for effortless binary-file transfer; Gopher, which presents data in a structured, subdirectory-like format; and Usenet, a collection of news groups.

AutoCAD allows you to interact with the Internet in several ways. You can launch a Web browser from within AutoCAD. AutoCAD can create design Web format (DWF) files for viewing drawings in 2D and 3D format on Web pages. AutoCAD can open and insert drawings from and save drawings to the Internet.

This chapter also covers AutoCAD's drawing set feature that makes it possible to manage all the drawings that make up a design project.

OBJECTIVES

After completing this chapter, you will be able to do the following:

- Launch the default Web browser
- Use the Communication Center
- Open drawings from the Internet
- Save drawings to the Internet
- Create and use hyperlinks
- Create and view DWF files
- Use the eTransmit utility
- Publish to the Web
- Collect, sort, create, and manage drawing sheets
- Create layout views automatically

- Automate the numbering of sheets
- Archive a set of drawings
- Publish sets and subsets of drawing sheets

LAUNCHING THE DEFAULT WEB BROWSER

The BROWSER command allows you to start a Web browser from within AutoCAD. By default, the BROWSER command uses the Web browser program that is registered in your computer's Windows® operating system. The BROWSER command can be used in scripts, toolbar or menu macros, and AutoLISP routines to access the Internet automatically.

Invoke the BROWSER command from the on-screen prompt. AutoCAD opens the Web browser (Figure 15–1), which connects to the default website.

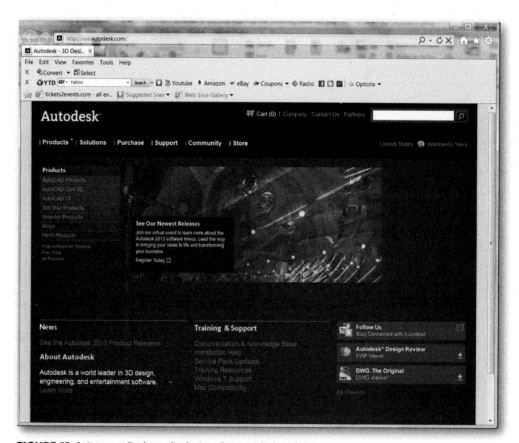

FIGURE 15–1 *Internet Explorer displaying the Autodesk website*

You can specify a default URL in AutoCAD's Options dialog box. Select the Files tab, choose **Menu, Help and Miscellaneous File Name,** select **Default Internet Location** (Figure 15–2), and specify the default URL Web location.

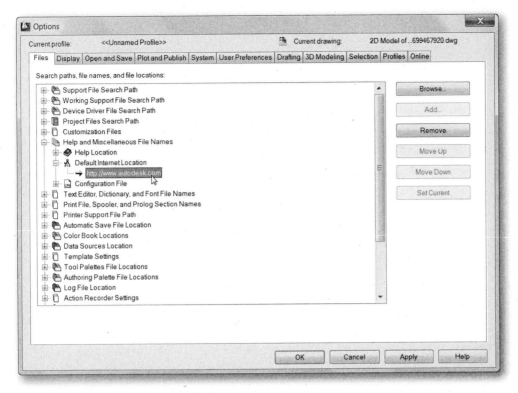

FIGURE 15–2 *Options dialog box with Files tab selected*

INFOCENTER AND AUTODESK EXCHANGE

InfoCenter (see Figure 15–3) provides a convenient way to search for topics in the Help system, sign in to Autodesk ID, open Autodesk Exchange, and display the options in the Help menu. It can also display product announcements, updates, and notifications.

FIGURE 15–3 *InfoCenter*

AutoCAD opens the Help window (Figure 15–4) when you type a keyword or phrase. AutoCAD Help window lists on the left side all the available topics related to the search word, and on the right side provides access to a wide a variety of content, including announcements, expert tips, videos, and links to blogs.

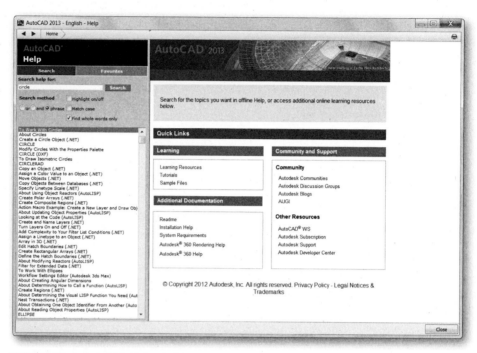

FIGURE 15–4 *AutoCAD Help window*

When you select one of the listed topics on the left side, AutoCAD provides detailed information on the right side of the window (Figure 15–5).

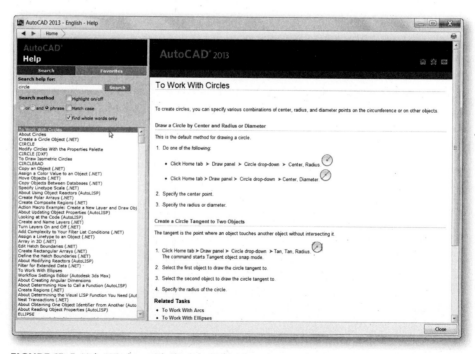

FIGURE 15–5 *Help Window with the selected topic*

OPENING AND SAVING DRAWINGS FROM THE INTERNET

AutoCAD allows you to open and save drawing files from the Internet or an intranet. You can also attach externally referenced drawings stored on the Internet/intranet to drawings stored locally on your system. Whenever you open a drawing file from an Internet or intranet location, it is first downloaded into your computer and then opened in the AutoCAD drawing area. You can edit the drawing and save it either locally or back to the Internet or intranet location for which you have appropriate access privileges.

To open an AutoCAD drawing from an Internet/intranet location, invoke the OPEN command from the Quick toolbar. AutoCAD displays the Select File dialog box.

Specify the URL of the file you wish to open in the **File name** text field, and choose **Open** to open the drawing from the specified Internet/intranet location. Be sure to specify the transfer protocol, such as ftp:// or http://, and the file extension, such as *.dwg* or *.dwt*. You can also choose **Search the Web,** located in the toolbar at the top of the dialog box, to open the Browse the Web dialog box. From there, you can navigate to the Internet location where the file is stored. You can also access the Buzzsaw.com and FTP locations by selecting the appropriate tabs provided on the left side of the dialog box.

To save an AutoCAD drawing to an Internet/intranet location, invoke the SAVEAS command from the File menu. AutoCAD displays the Save Drawing As dialog box.

Specify the URL of the file you wish to save in the **File name** text field and choose **Save** to save the drawing to the specified Internet/intranet location. Be sure to specify the transfer protocol and file extension, such as *.dwg* or *.dwt*. You can also choose **Search the Web** to open the Browse the Web dialog box. From there, you can navigate to the Internet location where the file is to be saved.

To attach an external reference (xref) to a drawing stored on the Internet/intranet location, invoke the XATTACH command. AutoCAD displays the Select Reference File dialog box.

Specify the URL of the file you wish to attach in the **File name** text field, and choose **Open** to attach the drawing as a reference file from the specified Internet/intranet location. Be sure to specify the transfer protocol, such as ftp:// or http://, in the URL. You can also choose **Search the Web** to open the Browse the Web dialog box. From there, you can navigate to the Internet location where the file is stored. You can also access the Buzzsaw.com and FTP locations by selecting the appropriate tabs provided on the left side of the dialog box.

WORKING WITH HYPERLINKS

AutoCAD allows you to create hyperlinks that jump to associated files. Hyperlinks provide a simple and powerful way to quickly associate a variety of documents with an AutoCAD drawing. For example, you can create a hyperlink that opens another drawing file from the local drive or network drive or from a website. You can also specify a named location to jump to within a file, such as a view name in an AutoCAD drawing, or a bookmark in a word-processing program. You can also attach a URL to jump to a specific website. You can attach hyperlinks to any graphical object in an AutoCAD drawing.

AutoCAD allows you to create both *absolute* and *relative* hyperlinks in your Auto-CAD drawings. Absolute hyperlinks store the full path to a file location, whereas relative hyperlinks store a partial path to a file location, relative to a default URL or name of the directory you specify using the HYPERLINKBASE system variable.

You can also specify the relative path for a drawing on the Summary tab of the Drawing Properties dialog box (Figure 15–6). The Drawing Properties dialog box is opened from the Applications menu.

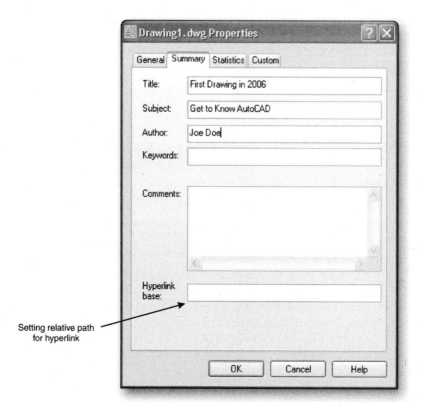

FIGURE 15–6 *Drawing Properties dialog box showing the Summary tab*

Whenever you attach a hyperlink to an object, AutoCAD provides feedback as you position the cursor over the object. To activate the hyperlink, first select the object; make sure the PICKFIRST system variable is set to 1. Right-click to display the shortcut menu and activate the link from the HYPERLINK submenu (Figure 15–7).

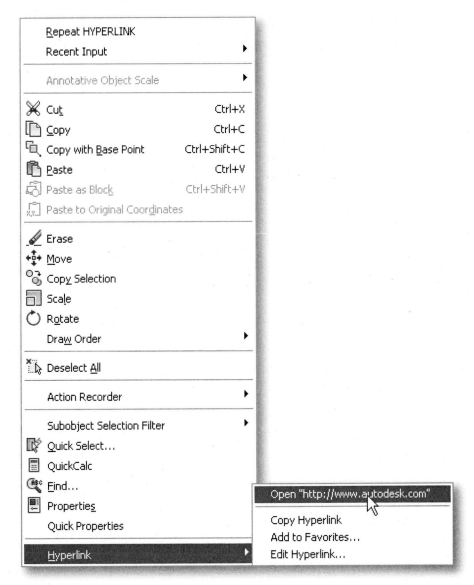

FIGURE 15–7 *HYPERLINK submenu on the shortcut menu*

When you create a hyperlink that points to an AutoCAD drawing template (*.dwt*) file, AutoCAD creates a new drawing file when you activate the hyperlink that is based on the selected template rather than opening the actual template. With this method, there is no risk of accidentally overwriting the original template.

When you create a hyperlink that points to an AutoCAD named view and activate the hyperlink, the named view that was created in the model space is restored in the Model tab, and the named view that was created in paper space is restored in the Layout tab.

To create a hyperlink, invoke the HYPERLINK command from the Data panel located in the Insert tab. AutoCAD displays the following prompt:

> Select objects: *(select one or more objects to attach the hyperlink and press* Enter*)*

AutoCAD displays the Insert Hyperlink dialog box (Figure 15–8).

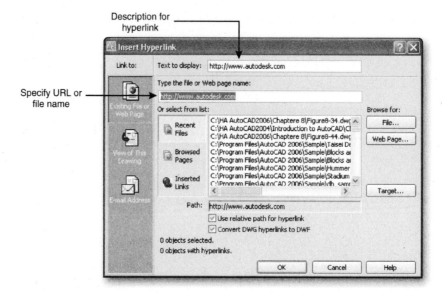

FIGURE 15–8 *Insert Hyperlink dialog box*

Specify a description for the hyperlink in the **Text to Display** text field. This is useful when the file name or URL is not helpful in identifying the contents of the linked file. Specify the URL or path with the name of the file that you wish to have associated with the selected objects in the **Type the File or Web Page Name** text field. Or, you can choose one of the **Browse for** buttons: **File, Web Page,** or **Target**. Choosing **File** opens the Browse the Web – Select Hyperlink dialog box, a standard file-selection dialog box. Use the dialog box to navigate to the file that you want associated with the hyperlink. Choosing **Web Page** opens the AutoCAD browser. Use the browser to navigate to a Web page to associate with the hyperlink. Choosing **Target** opens the Select Place in Document dialog box, in which you specify a link to a named location in a drawing. The named location that you select is the initial view that is restored when the hyperlink is executed. You can also select the path, with the name of the file or URL, from the list box categorized from **Recent Files, Browsed Pages,** and **Inserted Links.**

The **Path** text field displays the path to the file associated with the hyperlink.

The **Use relative path for hyperlink** check box toggles the use of a relative path for the current drawing. If this option is selected, the full path to the linked file is not stored with the hyperlink. AutoCAD sets the relative path to the value specified by the HYPERLINKBASE system variable or, if this variable is not set, the current draw-ing path. If this option is not selected, the full path to the associated file is stored with the hyperlink.

The **Convert DWG hyperlinks to DWF** check box specifies that the DWG hyperlink will convert to a DWF file hyperlink when you publish or plot the drawing to a DWF file.

Choosing **Existing File or Web Page,** located on the left side of the Insert Hyperlink dialog box, displays the options for creating a hyperlink to an existing file or Web page. Choosing **View of This Drawing** allows you to select a named view in the cur-rent drawing to link to. Choosing **E-mail Address** specifies an e-mail address to link to. When the hyperlink is executed, a new e-mail is created using the default system e-mail program.

Choose **OK** to create the hyperlink to the selected objects, and close the Insert Hyperlink dialog box.

To edit or remove a hyperlink, first select the object that has a hyperlink and, from the shortcut menu, choose *Edit Hyperlink*.

AutoCAD displays the Edit Hyperlink dialog box, similar to the Insert Hyperlink dialog box. Make necessary changes in the **Text to display** and/or **Type the file or Web page name** text fields. To remove the hyperlink, choose **Remove Link**. Choose **OK** to accept the changes and close the Edit Hyperlink dialog box.

AutoCAD allows you to attach hyperlinks to blocks, including nested objects contained within blocks. If the blocks contain any relative hyperlinks, the relative hyperlinks adopt the relative base path of the current drawing when you insert them.

Whenever you attach a hyperlink to a block reference, AutoCAD provides feedback when you position the cursor over the inserted block. To activate the hyperlink, first select the block reference; make sure the PICKFIRST system variable is set to 1. Right-click to display the shortcut menu and activate the hyperlink associated with the currently selected block element from the *Hyperlink* submenu.

Whenever you include objects that have hyperlinks in a block definition, you can activate the hyperlink from any of the block references. If you attach a hyperlink to a block reference, you will have the choice of activating the block hyperlink or selected object hyperlink. To remove or edit the hyperlinks of the objects within the block, you must explode the block reference and then proceed with removing or editing hyperlinks. You can remove or edit the hyperlink that was attached to a block reference without exploding the block.

> **NOTE**
> When a hyperlink is attached to a block reference that already contains an object with a hyperlink, the cursor feedback for that block will only show the block hyperlink. The object hyperlink can still be accessed through the *Hyperlink* submenu, as previously described.

DESIGN WEB FORMAT

Design Web Format (DWF/DWFx) is an open, secure file format developed by Autodesk for the transfer of drawings over networks including the Internet. DWF/DWFx files are highly compressed, so they are smaller—less than half the size of a *.dwg* file—and faster to transmit, enabling the communication of rich design data without the overhead associated with typical heavy CAD drawings. DWF/DWFx files are not a replacement for native CAD formats such as DWGs and don't allow editing of the data within the file. The sole purpose of DWF/DWFx is to allow designers, engineers, developers, and their colleagues to communicate design information to anyone needing to view, review, or print it.

Autodesk® Design Review 2012, the latest release of DWF/DWFx viewer is a free application that you can download from **www.autodesk.com**. It helps save time and money with easy-to-use tools for team members to review, mark up, and revise designs and 3D models. This tool is tightly integrated with all Autodesk design software, and it enables project teams to move to a two-way design review process and gain time-saving functionality in their markup and approval processes. In addition, Autodesk provides a free downloadable application called DWG TrueView that allows you to accurately view, plot, and publish to DWF files. You can also translate any AutoCAD or AutoCAD-based drawing file for compatibility with AutoCAD release 14 through AutoCAD 2012.

Using ePlot to Create DWF/DWFx Files

With AutoCAD's ePlot feature, you can generate electronic drawing files that are optimized for either plotting or viewing. The files you create are stored in either DWF or DWFx format. DWF/DWFx files can be opened, viewed, and plotted by anyone using Autodesk Design Review. You can view DWFx files on Microsoft Vista and Windows 7 without being required to install additional viewing software. It also allows users who have installed XPS viewing components to view the files. DWF/DWFx files support real-time panning and zooming and the display of layers and named views.

With ePlot, you can specify a variety of settings, such as pen assignments, rotation, and paper size, all of which control the appearance of plotted DWF/DWFx files. With ePlot, you can also create DWF/DWFx files that have rendered images and multiple viewports displayed in the **Layout** tab.

NOTE By default, AutoCAD plots all objects with a lineweight of 0.06, even if you haven't specified lineweight values in the Layer Properties Manager. If you want to plot without any lineweight, clear the **Plot Object Lineweight**s option from the *Plot Options* section in the Plot dialog box.

AutoCAD provides a preconfigured plotter driver file named DWF6 ePlot to create DWF files and DWFx ePlot (XPS Compatible) to create DWFx files. It generates electronic drawing files that are optimized for either printing or viewing.

DWF/DWFx files are created in a vector-based format, except for inserted raster image content, and are typically compressed. Compressed DWF/DWFx files can be opened and transmitted much faster than AutoCAD drawing files. Their vector-based format ensures that precision is maintained.

DWF/DWFx files are an ideal way to share AutoCAD drawing files with others who don't have AutoCAD. To create a DWF/DWFx file, invoke the PLOT command from the Quick Access toolbar, and AutoCAD displays the Plot dialog box (Figure 15–9).

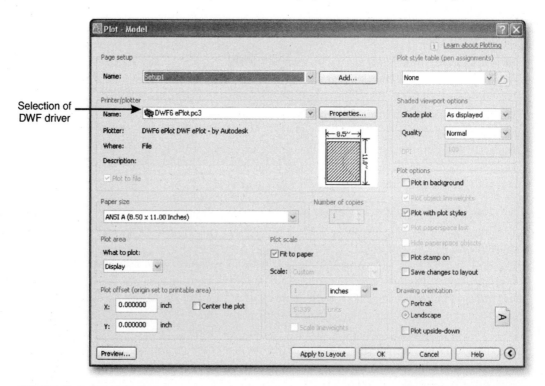

Selection of DWF driver

FIGURE 15–9 *Plot dialog box*

To create a DWF file, select *DWF6 ePlot.pc3* from the **Name** list box, in the **Printer/Plotter** configuration section. To create a DWFx file, select *DWFx ePlot (XPS compatible).pc3* from the **Name** list box in the **Printer/Plotter** configuration section.

Make necessary changes to **Paper size** and paper units, **Drawing orientation, Plot area, Plot scale, Plot offset, Plot options,** and choose **OK** to create the DWF file. AutoCAD will prompt for the name and location of the DWF/DWFx file.

When creating DWF/DWFx files, AutoCAD allows you to fine-tune custom plotting properties such as resolution, file compression, background color, inclusion of paper boundary, and related settings.

To modify custom plotting properties, open the Plot dialog box, select a DWF6 ePlot plotting device, and then choose **Properties.** AutoCAD displays the Plotter Configuration Editor dialog box (Figure 15–10).

Choose the **Device and Document Settings** tab, and then select **Custom Properties** from the tree window (Figure 15–10). In the Access Custom Dialog section, choose **Custom Properties.** AutoCAD displays the DWF6 ePlot Properties dialog box (Figure 15–11).

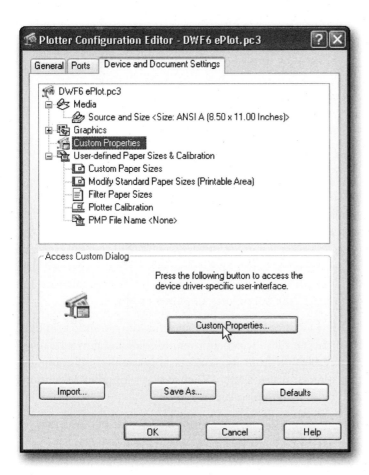

FIGURE 15–10 *Plotter Configuration Editor dialog box*

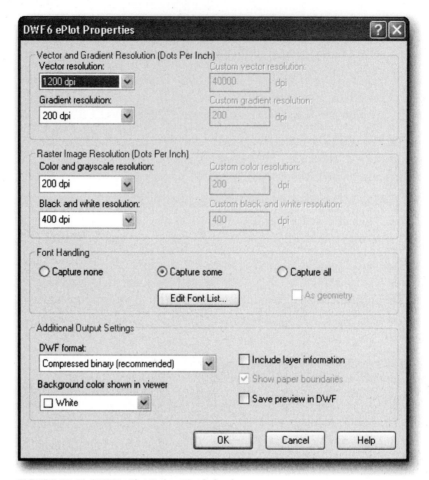

FIGURE 15–11 *DWF6 ePlot Properties dialog box*

The Vector and Gradient Resolution (Dots Per Inch) section specifies the resolution, in dots per inch, for vector graphics and gradients for DWF files. When you set a higher resolution, the file is more precise, but the file size is also larger. Select appropriate vector and gradient resolution from the Vector resolution menu and Gradient resolution menu, respectively. If there are large numbers of objects in the drawing, it is recommended that you create a DWF file with high resolution. When you create DWF files intended for plotting, select a resolution to match the output of your plotter or printer.

The Raster Image Resolution (Dots Per Inch) section specifies the resolution, in dots per inch, for raster images for DWF files. Select appropriate color and grayscale resolution and black-and-white resolution from the Color and grayscale resolution and Black and white resolution menus, respectively. When you set a higher resolution, the file is more precise, but the file size is also larger.

The Font Handling section specifies the inclusion and handling of fonts in DWF files.

Choosing **Capture none (all viewer supplied)** specifies that no fonts will be included in the DWF file. In order for the fonts used in the source drawing for the DWF file to be visible in the DWF file, the fonts must be present on the DWF viewer's system. If the fonts used to create the DWF file are not present on the viewer's system, other fonts will be substituted.

Choosing **Capture some (recommended)** specifies that fonts used in the source drawing for the DWF file that are selected in the Available True Type Fonts dialog box will be included in the DWF file. The selected fonts do not need to be available on the DWF viewer's system in order for them to appear in the DWF file.

Choose **Edit Font List** to open the Available True Type Fonts dialog box, where you can edit the list of fonts eligible for capture in the DWF file.

Choosing **Capture all** specifies that all fonts used in the drawing will be included in the DWF file. The **As geometry** check box specifies that all fonts used in the drawing will be included as geometry in the DWF file. If you select this option, you should plot your drawing at a scale factor of 1:1 or better to ensure the quality in the output file. This option is only available for DWF files created with the DWF6 ePlot model.

> The size of a DWF file can be affected by the font-handling settings, the amount of text, and the number and type of fonts used in the file. If the size of your DWF file seems too large, try changing the font-handling settings.

NOTE

Choose the compression format for DWF files from the **DWF Format** menu. Selection of **Compressed binary (recommended)** format plots the DWF file in a compressed, binary format; compression does not cause data loss. This is the recommended file format for most DWF files. Selection of **Zipped ASCII encoded 2D stream (advanced)** plots the DWF file in zipped ASCII Encoded 2D Stream (plain text) format. You can use WinZip to unzip the files.

Specify the background color from the **Background color shown in viewer** option menu. In addition, specify toggle settings for **Include layer information, Show paper boundaries,** and **Save preview in DWF.**

After making necessary changes, choose **OK** to close the DWF Properties dialog box. Choose **OK** to save the settings and close the Plotter Configuration Editor dialog box. AutoCAD displays the Changes to a Printer Configuration File dialog box (Figure 15–12).

FIGURE 15–12 *Changes to a Printer Configuration File dialog box*

If you want to apply the changes in the settings to the current plot, select **Apply changes to the current plot only.** If you want to save the settings and apply them to all plots, select **Save changes to the following file,** and specify the name of the file in the edit field. Choose **OK** to close the Changes to a Printer Configuration File dialog box.

Similarly, you can apply customized properties when creating a file in the DWFx format.

Viewing DWF/DWFx Files

AutoCAD cannot display DWF/DWFx files, nor can DWF/DWFx files be converted back to DWG format without using file translation software from a third-party vendor. In order to view a DWF/DWFx file, you need Autodesk Design Review.

Autodesk Design Review enables users to view and print complex 2D and 3D drawings, maps, and models published from Autodesk design applications. Figure 15–13 shows an example of DWF displayed in Autodesk Design Review.

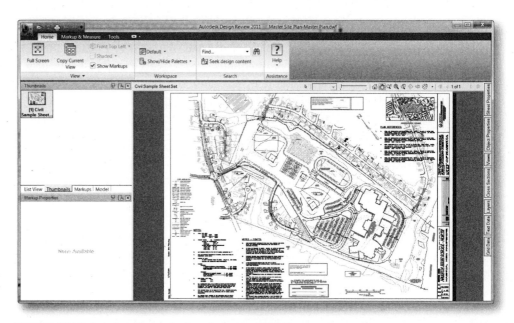

FIGURE 15–13 *Autodesk Design Review displaying DWF file*

PUBLISHING AUTOCAD DRAWINGS TO THE WEB

The Publish to Web Wizard allows you to easily and seamlessly publish AutoCAD drawings to the Web. Drawings are published in HTML format using three predefined image types: DWF, DWFx, JPEG, and PNG. The DWF/DWFx image type translates and publishes specified layouts into DWF/DWFx, which are easily viewed with Autodesk Design Review. With the JPEG and PNG image types, you specify a drawing perspective, and AutoCAD translates the specified layout into a JPEG or PNG raster image. Anyone with a standard browser can view JPEG or PNG content. To move your content to the Web or a company intranet, specify the server location and configuration, and the content uploads automatically. Once it's posted, updating it is simple and fast.

To publish AutoCAD drawings, invoke the PUBLISHTOWEB command at the on-screen prompt, AutoCAD displays the introductory text of the Publish to Web – Begin dialog box (Figure 15–14). Choose one of the following two options:

- **Create New Web Page**—Allows you to create a new Web page.
- **Edit Existing Web Page**—Allows you to edit an existing page by adding or removing Web pages.

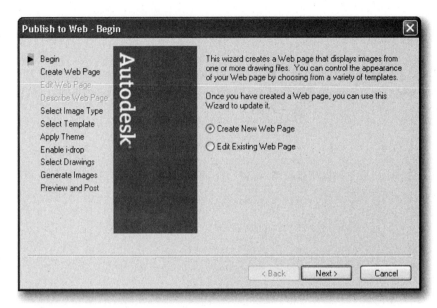

FIGURE 15–14 *Publish to Web – Begin dialog box*

Choose **Next,** and AutoCAD displays the Publish to Web – Create Web Page dialog box (Figure 15–15). Specify the name of your Web page in the first text field, and specify the parent directory in your file system where the Web page folder will be created. If desired, provide a description to appear on your Web page in the last text field in the wizard.

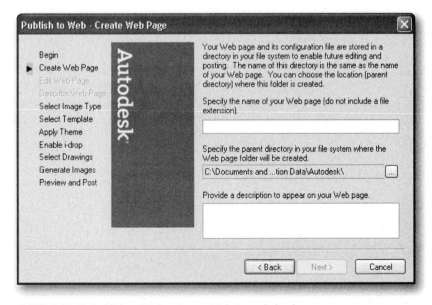

FIGURE 15–15 *Publish to Web – Create Web Page dialog box*

Choose **Next,** and AutoCAD displays the Publish to Web – Select Image Type dialog box (Figure 15–16). Choose one of the following three image types:

- DWF image type—The DWF template translates and publishes specified DWG files in DWF. DWF files are inserted into your completed Web page in a size that is optimized to display well with most browser settings.

- DWFx image type—The DWFx template translates and publishes specified DWG files in DWFx. DWFx files are inserted into your completed Web page in a size that is optimized to display well with most browser settings.

- JPEG image type—JPEG files are raster-based representations of AutoCAD drawing files. JPEGs are one of the most common formats used on the Web. Due to the compression mechanism used, this format is not suitable for large drawings or drawings that have a lot of text.

- PNG image type—Portable Network Graphics (PNG) are raster-based representations of drawing files. Most browsers now support the PNG image type, making it more suitable than JPEG for creating images of AutoCAD drawings.

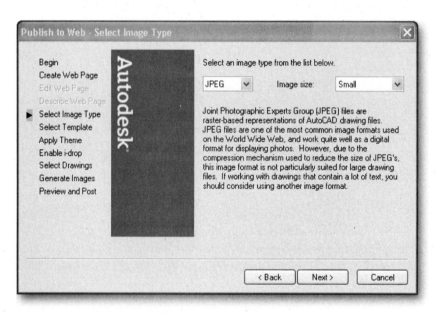

FIGURE 15–16 *Publish to Web – Select Image Type dialog box*

Choose **Next,** and AutoCAD displays the Publish to Web – Select Template dialog box (Figure 15–17). Select one of the available templates. The **Preview** pane demonstrates how the selected template will affect the layout of drawing images in your Web page.

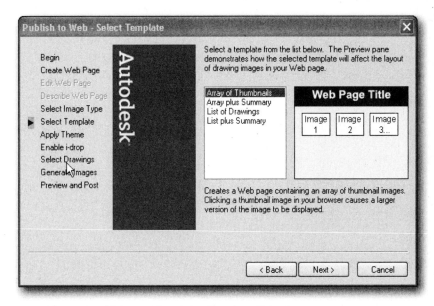

FIGURE 15–17 *Publish to Web – Select Template dialog box*

Choose **Next,** and AutoCAD displays the Publish to Web – Apply Theme dialog box (Figure 15–18). Choose one of the available themes. Themes are preset elements such as fonts and colors that control the appearance of elements of your completed Web page. The **Preview** pane demonstrates how the selected theme will display the layout of your Web page. The available themes are Autumn Fields, Classic, Cloudy Sky, Dusky Maize, Ocean Waves, Rainy Day, and Supper Club.

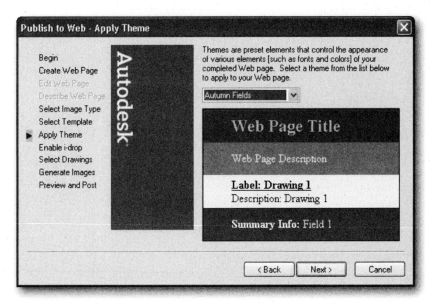

FIGURE 15–18 *Publish to Web – Apply Theme dialog box*

Choose **Next,** and AutoCAD displays the Publish to Web – Enable i-drop dialog box (Figure 15–19).

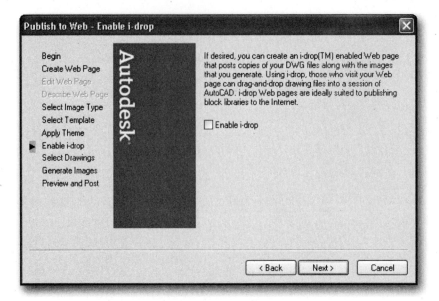

FIGURE 15–19 *Publish to Web – Enable i-drop dialog box*

If desired, you can create an i-drop-enabled Web page that posts copies of your DWG files along with the images. Using i-drop, visitors to your website can drag and drop drawing files into a session of AutoCAD.

Choose **Next,** and AutoCAD displays the Publish to Web – Select Drawings dialog box (Figure 15–20).

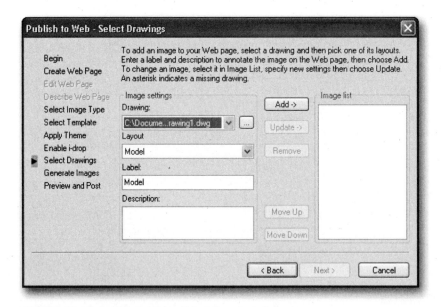

FIGURE 15–20 *Publish to Web – Select Drawings dialog box*

Select a drawing and then choose one of its layouts from the **Layout** menu. Specify a label in the **Label** text box and description in the **Description** text box to annotate the selected image on the Web page. Choose **Add** to add the selected image to the **Image list** box. If necessary, you can change the properties of the selected image and choose **Update** to apply the changes, or you can remove it from the selection by choosing **Remove.** Similarly, you can add additional drawings/layouts to the selection. Choosing **Move Up** and **Move Down** allows you to rearrange the selected images.

Choose **Next,** and AutoCAD displays the Publish to Web – Generate Images dialog box (Figure 15–21).

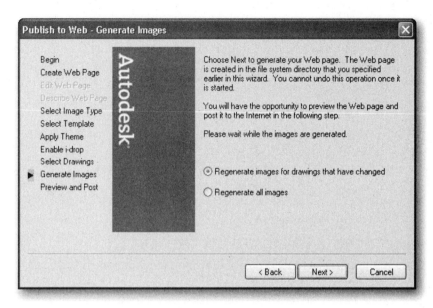

FIGURE 15–21 *Publish to Web – Generate Images dialog box*

Choose one of the following two options:

- **Regenerate images for drawings that have changed**—Generates the images for all the selected drawings that have changed.
- **Regenerate all images**—Generates the images for all the selected drawings.

Choose **Next,** and AutoCAD creates the Web pages and stores them in the file directory that you specified earlier in the wizard. You cannot undo this operation once it is started.

When finished, AutoCAD displays the Publish to Web – Preview and Post dialog box (Figure 15–22). To preview the Web pages, choose **Preview.** AutoCAD opens the default browser and displays the Web pages with appropriate links. To post the Web pages, close the browser, and choose **Post Now.** AutoCAD displays the Posting Web File Handling dialog box. Select the URL where you want to post it, and choose the **Save** button. If desired, choose **Send Email** to create and send an e-mail message that includes a hyperlinked URL to its location.

Choose **Finish** to close the Publish to Web Wizard.

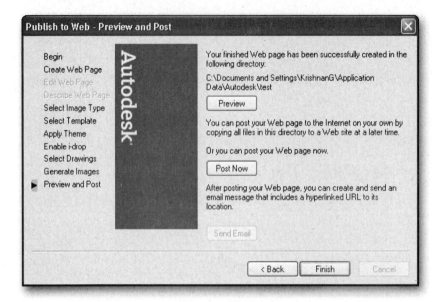

FIGURE 15–22 *Publish to Web – Preview and Post dialog box*

PUBLISHING AUTOCAD DRAWINGS TO DWF FORMAT

Publishing allows you to assemble a collection of drawings and plot directly to paper or publish to a DWF file. AutoCAD allows you to publish your drawing sets as either a single multisheet DWF format file or as multiple single-sheet DWF format files or to plot to the designated plotter in the page setup. You can publish to plotters or files specified in the page setups for each layout. When using Publish, you have the flexibility to create electronic or paper drawing sets for distribution. The recipients can then view or plot your drawing sets.

You can customize your drawing set for a specific user, and you can add and remove sheets in a drawing set as a project evolves. The PUBLISH command allows you to publish directly to paper or to an intermediate electronic format that can be distributed by e-mail, FTP sites, project websites, or CD. You can open DWF files with Autodesk Design Review.

Invoke the PUBLISH command from the Plot panel (Figure 15–23) located in the output tab, and AutoCAD displays the Publish dialog box (Figure 15–24).

FIGURE 15–23 *Invoking the PUBLISH command from the Plot panel*

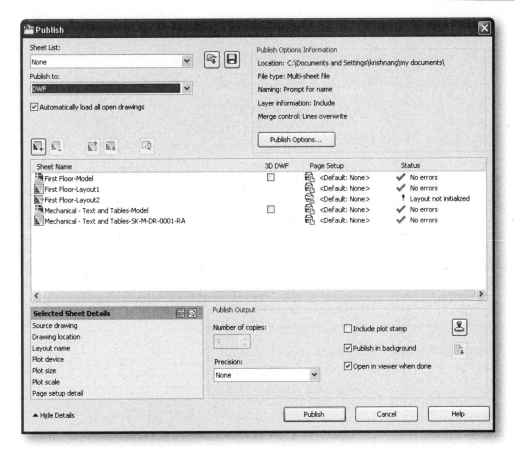

FIGURE 15–24 *Publish dialog box*

The **Sheet List:** field displays the current drawing set (DSD) or batch plot (BP3) file. Choose the **Load Sheet List** button to load an existing drawing set or batch plot file. You can either replace the existing list of drawing sheets with the new sheets or append the new sheets to the current list. Choose the Save Sheet List button to save the current list of drawings as a DSD file. DSD files are used to describe lists of drawing files and selected lists of layouts within those drawing files.

Selecting the **Automatically load all open drawings** toggle button includes all open documents (layouts and/or model space) in the publish list. When not selected, only the current document's contents are loaded in the publish list.

The current publish settings are listed in the Publish Options Information section. To customize publishing, choose **Publish Options,** and AutoCAD opens the Publish Options dialog box (Figure 15–25) in which you can specify options for publishing.

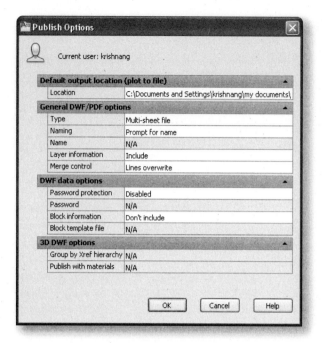

FIGURE 15–25 *Publish Options dialog box*

The **Current user:** field displays the name of the current user or current sheet set. When the name of the current user is shown, changes made in the dialog box are saved in the current user's profile. When the name of the current sheet set is shown, changes made in the dialog box are saved with the sheet set.

In the Default output location section, specify where DWF and plot files are saved when you publish drawing sheets.

The General DWF/PDF options section specifies options for creating a single-sheet DWF or DWFx file.

In the Type section, specify whether a single-sheet DWF or a single multisheet DWF is created.

In the Specify name section, specify the name and location in the name field of the multisheet DWF file. If you select **Prompt for name,** the name and location will be provided when prompted for published DWF files.

In the Layer information section, specify whether layer information is included in the published DWF, DWFx, or PDF file.

In the Merge Control section, specify whether overlapping lines merge (the colors of the lines blend together into a third color) or overwrite (the last plotted line obscures the line beneath it).

The DWF data section lists and allows you to specify the data that can optionally be included in the DWF file.

The Password protection section specifies options for protecting DWF files with passwords. A recipient of a DWF file that has a password applied to it must have the password to open the DWF file. Select **Specify password** and specify the password in the **Password** field that will be applied to the DWF file. DWF passwords are case sensitive. The password or phrase can be made of letters, numbers, punctuation, or non-ASCII characters. If you select **Prompt for password,** AutoCAD will display the DWF Password dialog box when you click Publish. If you select **Disabled,** the DWF will not require the password in order to open.

If you lose or forget the password, it cannot be recovered. Keep a list of passwords and their corresponding DWF file names in a safe place.

The Block information section specifies whether block property and attribute information is included in the published DWF or DWFx files.

The Block template file section provides options for creating a new block template (DXE) file, editing an existing block template file, or using the settings of a previously created block template file.

The 3D DWF Options section lists and allows you to specify the data that you can optionally include in 3D DWF publishing.

Choose **OK** to save the changes and close the Publish Options dialog box.

The **Sheets to publish** section lists the drawing sheets to be included for publishing.

- The **Sheet Name** column displays a combination of the drawing name and the layout name, separated by a dash (–). If necessary, you can rename using the *Rename Sheet* option available from the shortcut menu. Drawing sheet names must be unique within a single DWF file.
- The **Page Setup** column displays the named page setup for the sheet. You can change the page setup by clicking the page setup name and selecting another page setup from the list. Select *Import* to import page setups from another *.dwg* file through the Import Page Setups for Publishing dialog box. Only Model tab page setups can be applied to Model tab sheets, and only paper space page setups can be applied to paper space layouts.
- The **Status** column displays the status of the sheet when it is loaded to the list of sheets.

To add sheets to the existing selection, choose the **Add Sheets** icon. AutoCAD displays a standard file selection dialog box, where you can add sheets to the list of drawing sheets. The layout names from those files are extracted, and one sheet is added to the list of drawing sheets for each layout. New drawing sheets are always appended to the end of the current list.

To remove sheets, choose the **Remove Sheets** icon, and AutoCAD deletes the currently selected drawing sheet from the list of sheets.

To move the selected drawing sheet up one position in the list, choose the **Move Up** icon. To move the selected drawing sheet down one position in the list, choose the **Move Down** icon.

Specify the number of copies to publish in the **Number of copies** box. If the **DWF file** option is selected in the Publish to section, the **Number of copies** setting defaults to 1 and cannot be changed.

The Precision section optimizes the dpi of DWF, DWFx, and PDF files for your field: manufacturing, architecture, or civil engineering.

To include a plot stamp on a specified corner of each drawing and log it to a file, check the **Include plot stamp** check box. To customize the plot stamp settings, open the Plot Stamp dialog box by choosing the **Plot Stamp Settings** icon. AutoCAD displays the Plot Stamp dialog box, in which you can specify the information, such as drawing name and plot scale, that you want applied to the plot stamp.

The **Publish in Background** button toggles background publishing for the selected sheet(s).

Clicking **Open in viewer when done** toggles whether the DWF, DWFx, or PDF file opens in a viewer application when publishing completes.

Choose **Show Details** or **Hide Details** to display or hide the Selected sheet information and Publish Output sections in the Publish dialog box.

Choose **Publish** to publish the selected layouts. AutoCAD begins the publishing operation, creating one or more single-sheet DWF files or a single multisheet DWF file, or plotting to a device or file, depending on the option selected in the **Publish to** section and the options selected in the Publish Options dialog box.

If a drawing sheet fails to plot, PUBLISH continues plotting the remaining sheets in the drawing set. A log file is created that contains detailed information, including any errors or warnings encountered during the publishing process. You can stop publishing after a sheet has finished plotting. If you stop publishing a multisheet DWF file before it is complete, no output file is generated. After publishing is complete, the **Status** field is updated to show the results.

ETRANSMIT UTILITY

The eTransmit utility allows you to select and bundle together the drawing file and its related files. You can create a transmittal set of files as a compressed self-extracting executable file, as a compressed zip file, or as a set of uncompressed files in a new or existing folder. You can include all the reference files attached to the drawing file, word files, spreadsheet, and so on, to be part of the bundle. It is easier to transmit by e-mail one single compressed file consisting of a drawing file and several related files.

To create a transmittal set of a drawing and related files, invoke the ETRANSMIT command from the Application menu. AutoCAD displays the Create Transmittal dialog box, shown in Figure 15–26.

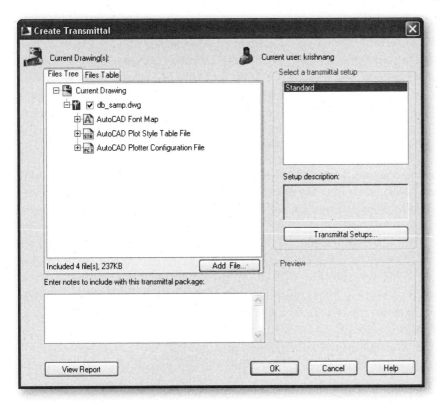

FIGURE 15-26 *Create Transmittal dialog box*

The Files Tree tab lists the files to be included in the transmittal package in a hierarchical tree format. By default, all files associated with the current drawing such as related external references, plot styles, and fonts are listed. You can add files to the transmittal package or remove existing files. To add files to the transmittal package, choose **Add File**. AutoCAD opens a standard file selection dialog box, in which you can select additional files to include in the transmittal package.

Enter notes related to a transmittal package in the text area under **Enter notes to be included with this transmittal package.** The notes are included in the transmittal report. You can specify a template of default notes to be included with all your transmittal packages by creating an ASCII text file called *etransmit.txt*. This file must be saved to a location specified by the **Support File Search Path** option on the Files tab in the Options dialog box.

Select the transmittal setup from the **Select a transmittal setup** list. AutoCAD lists previously saved transmittal setups. The default transmittal setup is named Standard. To create, modify, and delete transmittal setups, choose **Transmittal Setups.** AutoCAD displays the Transmittal Setup dialog box. You can create a new transmittal setup or modify an existing transmittal setup that specifies the type of transmittal package created.

To view the report that is included with the transmittal package, choose **View Report.** AutoCAD displays report information that includes any transmittal notes that you entered and distribution notes automatically generated by AutoCAD that detail what steps must be taken for the transmittal package to work properly.

Choose **OK** to create the transmittal set and close the Create Transmittal dialog box.

DRAWING SETS

AutoCAD's powerful Drawing Set Management feature allows you to handle sets of drawings for the purpose of plotting, publishing, and otherwise managing and tracking projects. Chapter 8 explains the use of AutoCAD's layout and paper space features to easily produce sheets for plotting. This chapter explains how to apply the Sheet Set Manager feature to collect, organize, and otherwise manage the assortment of layout sheets and views from different drawings so they can be plotted as sets and subsets of deliverables traditionally referred to as "blueprints."

A drawing set is just that, a set of drawings. A small design project might require only two or three sheets, while for a major construction project, a drawing set might consist of hundreds of sheets. There are often subsets for various disciplines: civil/survey, structural, mechanical, architectural, and electrical. When all of these sheets are scattered throughout one or more offices in one or more locations, and some sheets are just one layout or view in one drawing and other sheets are layouts from many different drawings, it becomes a difficult or almost impossible task to organize, manage, and continually update the sheets individually and as sets and subsets. With the Sheet Set Manager, you can manage drawings as sheet sets. A sheet set is an organized and named collection of sheets from several drawing files. A sheet is a selected layout from a drawing file. You can import a layout from any drawing into a sheet set as a numbered sheet, and you can manage, transmit, publish, and archive sheet sets as a unit.

Creating a New Sheet Set

The Create Sheet Set wizard contains a series of pages that step you through the process of creating a new sheet set. You can choose to create a new sheet set from existing drawings, or you can use an existing sheet set as a template on which to base your new sheet set. The following steps should be performed before creating a sheet set:

- Move the drawing files to be used in the sheet set into a minimum number of folders. This will make sheet set-management easier.
- Have only one layout tab in each drawing in the sheet set. This affects access to sheets by multiple users. Only one sheet in each drawing can be open at a time.
- Specify or create a drawing template (*.dwt*) file for use by the sheet set for creating new sheets. This template file is called the sheet creation template and can be specified in the Sheet Set Properties dialog box or the Subset Properties dialog box.
- Create a page setup overrides file. Specify or create another.*dwt* file for storing page setups for plotting and publishing. This file is called the page setup overrides file and can be used to apply a single page setup to each sheet in a sheet set. This action will override the individual page setups stored in each drawing.

Invoke the NEWSHEETSET command from the New menu (accessed from the Application Browser) to open the Create Sheet Set wizard (Figure 15–27).

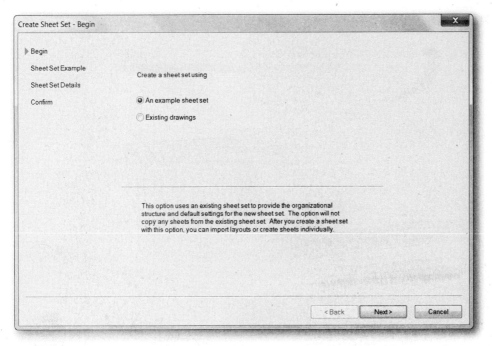

FIGURE 15–27 *The Create Sheet Set – Begin dialog box of the Create Sheet Set wizard*

On the Begin page of the Create Sheet Set wizard, select a method for creating a sheet set. Two methods are provided: **An example sheet set** and **Existing drawings** (Figure 15–27).

Choosing **An example sheet set** allows you to use a sample sheet set format, structure, and default settings without copying any of the sheets in the sample sheet set to the new set. After you create a sheet set using this option, you can import layouts or create sheets individually.

Choosing **Existing drawings** allows you to import layouts from existing drawings in the specified folder(s). The layouts from these drawings are imported into the newly created sheet set automatically.

Creating Sheet Sets from Examples

To create a sheet set from a sample sheet set, select **An example sheet set** in the Create Sheet Set – Begin wizard dialog box, and choose **Next.** AutoCAD displays the Create Sheet Set – Sheet Set Example dialog box (Figure 15–28).

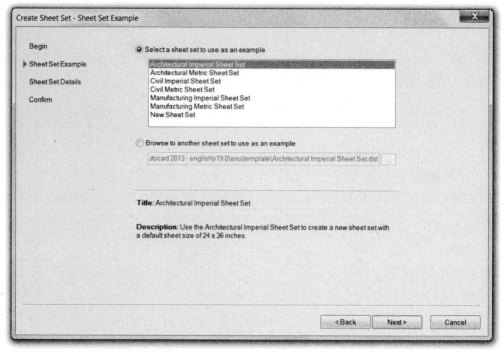

FIGURE 15–28 *Create Sheet Set – Sheet Set Example dialog box*

AutoCAD displays a list of sample sheet sets in the list box. From here you can choose one to use as the basis of the new sheet set. If you select **Browse to another sheet set to use as an example,** you can use the standard Windows-type browsing mechanism to search folders for a sheet set to use as the basis of a new sheet set. After selecting an existing sheet set from one of the optional methods, choose **Next.** AutoCAD displays the Create Sheet Set – Sheet Set Details dialog box (Figure 15–29).

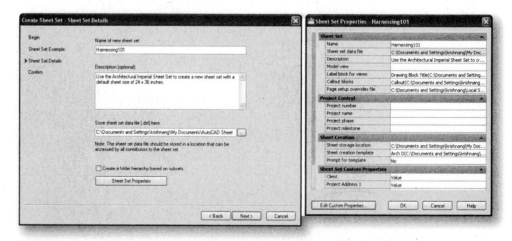

FIGURE 15–29 *Create Sheet Set – Sheet Set Details dialog box and Sheet Set properties*

Specify the name and description of the newly created sheet set in the **Name of new sheet set** box and **Description (optional)** box, respectively. Specify the folder in which to store the newly created sheet set in the **Store sheet set data file (.*dst*) here** box. The Sheet Set data file should be stored in a location that can be accessed by all contributors to the sheet set.

Choose **Sheet Set Properties** to view or edit the sheet set properties. AutoCAD displays the Sheet Set Properties dialog box (Figure 15–29). The Sheet Set Properties dialog box displays information specific to the sheet set selected. The information can be modified in the boxes to the right of individual data descriptions. This includes information such as the path and file name of the sheet set data (*.dst*) file, folder paths that contain drawing files included in the sheet set, and custom properties associated with the sheet set.

The Sheet Set section includes the name of the sheet set, location of the sheet set data file, description (if any) for the newly created sheet set, path for the resource drawings, name of the drawing that contains label blocks for views, callout block names associated with the sheet set, and name of the page setup override file.

The Project Control section contains information related to the project provided by the user.

The Sheet Creation section includes the paths of the folders that contain the drawing files associated with the sheet set, the name of the template for creating a new sheet, and a setting for "prompt for new template."

The Sheet Set Custom Properties section contains user-defined properties.

Choose **Edit Custom Properties** to add or remove custom properties associated with a sheet set. Custom properties can be used to store information such as a contract number, the name of the designer, or the release date.

Choose **OK** to return to the Create Sheet Set – Sheet Set Details wizard page, and then choose **Next**. AutoCAD displays the Create Sheet Set – Confirm dialog box, as shown in Figure 15–30.

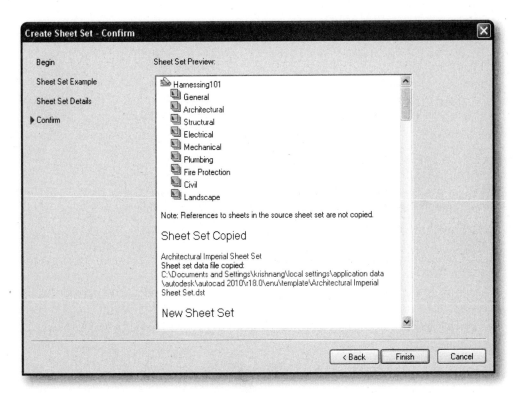

FIGURE 15–30 *Create Sheet Set – Confirm dialog box*

In the Create Sheet Set – Confirm wizard dialog box, the **Sheet Set Preview** box lists the information about the sheet set for review before you accept it. If the information is acceptable, choose **Finish** to complete the creation of the new sheet set. If any of the information needs to be modified, choose **Back** and make the necessary changes. AutoCAD creates a new sheet set in the form of a sheet set data file with the extension of *.dst* in the specified location with the specified properties.

Creating Sheet Sets from Existing Drawings

To create a sheet set from existing drawings, select **Existing drawings** on the Create Sheet Set – Begin wizard dialog box, and choose **Next.** AutoCAD displays the Create Sheet Set – Sheet Set Details dialog box similar to creating a sheet from an example sheet set. Specify the name and description of the newly created sheet set in the **Name of new sheet set** box and **Description (optional)** box, respectively. Specify the folder in which to store the newly created sheet set in the **Store sheet set data file (*.dst*) here** box. Choose **Sheet Set Properties** to view or edit the sheet set properties.

Choose **Next,** and AutoCAD displays the Create Sheet Set – Choose Layouts dialog box (Figure 15–31).

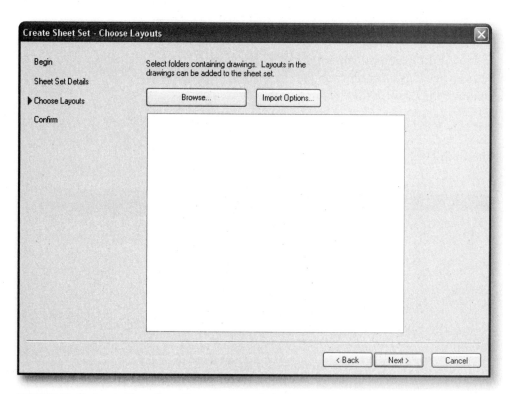

FIGURE 15–31 *Create Sheet Set – Choose Layouts dialog box*

Choose **Browse** to select a folder or folders to list drawing files from which Auto-CAD will import layouts in drawing files to create the sheet set. You can add additional folders containing drawings by choosing **Browse** for each additional folder. Choosing **Import Options** causes the Import Options dialog box to be displayed (Figure 15–32).

FIGURE 15–32 *The Import Options dialog box*

Selecting **Prefix sheet titles with file name** causes AutoCAD to automatically add the drawing file name to the beginning of the sheet title.

Selecting **Create subsets based on folder structure** causes AutoCAD to automatically create subsets in the newly created sheet sets based on folder structure.

Choose **OK** to close the dialog box and accept the settings.

On the Create Sheet Set – Choose Layouts dialog box, choose **Next,** and AutoCAD displays the Create Sheet Set – Confirm wizard dialog box, similar to creating a sheet from an example sheet set. The **Sheet Set Preview** box lists the information about the sheet set for review before you accept it. If the information is acceptable, choose **Finish** to complete the creation of the new sheet set. If any of the information needs to be modified, choose **Back** and make the necessary changes. AutoCAD creates a new sheet set in the form of a sheet set data file with the extension of *.dst* in the specified location with the specified properties.

Sheet Set Manager

The Sheet Set Manager provides the tools to organize, manage, and update a set of drawings. The Sheet Set Manager not only accesses the drawings associated with a project, but also allows you to access the layouts and views that become the plotted sheets making up the final set of plotted drawings. As mentioned earlier, sheet sets are stored in the form of a sheet set data file with the extension of *.dst*. Each sheet in a sheet set is a layout in a drawing (*.dwg*) file. Open the Sheet Set Manager from the Palettes panel located in the View tab. AutoCAD displays the Sheet Set Manager palette with the **Sheet List** tab displayed (Figure 15–33, left).

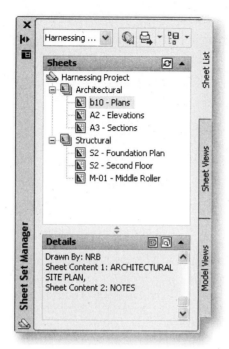 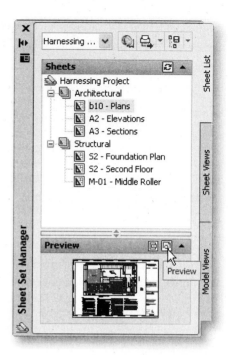

FIGURE 15–33 *Sheet Set Manager palette with the Sheet List tab displayed (left) and Sheet Set Manager palette with the preview image of the selected sheet (right)*

Choosing **Sheet Set Control** lists menu options to create a new sheet set, open an existing sheet set, or switch between open sheet sets.

The Sheet List tab displays an organized list of all sheets in the sheet set. Each sheet in a sheet set is a specified layout in a drawing file.

The Sheet Views tab displays an organized list of all sheet views in the sheet set. Only sheet views created with AutoCAD 2005 and later are listed.

The Model Views tab displays a list of folders, drawing files, and model space views available for the current sheet set. You can add and remove folder locations to control which drawing files are associated with the current sheet set.

Choosing **Details** displays descriptive information about the currently selected item in the tree view.

Choosing **Preview** displays a thumbnail preview of the currently selected item in the tree view.

Viewing and Modifying a Sheet Set

The **Sheets** box displays the name of the sheet set that is open (Figure 15–33, left). If no sheet set is open, choose **Open** to open an existing sheet set. To create a new sheet set, choose **New,** which in turn will start the wizard.

The Sheet Set Manager (Figure 15–33, left) contains two subsets called Architectural and Structural in the sheet set called Harnessing Project. The Architectural subset contains three numbered sheets: Plans, Elevations, and Sections. The Structural subset contains three numbered sheets: Foundation Plan, Second Floor, and Middle Roller. Each of these sheets is a layout in a *.dwg* file. When you select a sheet, information about it is displayed under the **Details** section of the palette. Figure 15–33 at the left shows details of the Plans sheet. To view a thumbnail preview of the selected sheet, choose **Preview.** Figure 15–33 at the right shows a preview image of the Plans sheet.

You can also open the Sheet Set Properties in the Sheet Set Properties dialog box from the shortcut menu that is displayed when the name of the sheet set is selected.

Instead of using the OPEN command to open a drawing, you can open it using the Sheet Set Manager. Double-click the name of the sheet, and it will open in a new window. When the sheet is open, it will be locked automatically, and no other user can open it at the same time. The lock status is indicated in the details section of the selected item (Figure 15–34). When you close the drawing, the status will be changed to Accessible.

FIGURE 15–34 *Sheet Set Manager palette with the Details section indicating the status of the selected item*

In a similar manner, you can create a new sheet. Right-click while the cursor is over a sheet set or sheet name in the Sheets list and choose *New Sheet*. AutoCAD displays the New Sheet dialog box, in which you can specify the name and number of the new sheet based on the default template set in the sheet set properties. When you create a new sheet, you create a new layout in a new drawing file, which is stored in the location specified in the sheet set properties. Instead of manually creating new drawing files, you can use the Sheet Set Manager.

To import a layout from an existing drawing, right-click while the cursor is over a sheet set or sheet name in the Sheets list, and choose *Import Layout as Sheet*. AutoCAD displays the Import Layouts as Sheets dialog box. In the **Select drawing file containing layouts** box, choose **Browse for Drawings** to find a drawing from which to import layouts. In the **Select Drawing** dialog box, select a drawing to import and choose **Open.** You can choose to have AutoCAD prefix sheet titles with the file name. Choose **Import Checked** to import the layout.

With a large sheet set, you will find it necessary to organize sheets and views in the tree view. On the Sheet List tab, sheets can be arranged into collections called subsets. To create a new subset, first select the name of the sheet set and choose *New Subset* from the shortcut menu. AutoCAD displays the Subset Properties dialog box, where you can create a new sheet subset for organizing the sheets. Figure 15–34 shows two subsets, Architectural and Structural, in the Harnessing Project sheet set.

If necessary, you can rename and renumber the sheet. To do so, choose *Rename & Renumber* from the shortcut menu. AutoCAD displays the Rename & Renumber dialog box (Figure 15–35), where you can specify the sheet number and title for the selected sheet.

The **Number** box specifies the sheet number of the selected sheet.

The **Sheet title** box specifies the sheet title of the selected sheet.

The **Layout Name** box specifies the name of the layout associated with the selected sheet.

The **Folder path** box displays the path to the selected sheet.

In the Rename options section, checking **Rename layout to match Sheet Title** changes the layout name to match the sheet title. Checking **Prefix with Sheet Number** changes the layout name to a new name formed by adding the sheet number to the beginning of the sheet title. Checking **Rename Drawing File to match Sheet Title** changes the drawing file name to match the sheet title. Checking **Prefix with Sheet Number** changes the drawing file name to a new name formed by adding the sheet number to the beginning of the sheet title.

The **Next** option loads the next sheet into this dialog box.

Choose **OK** to close the Rename & Renumber Sheet dialog box.

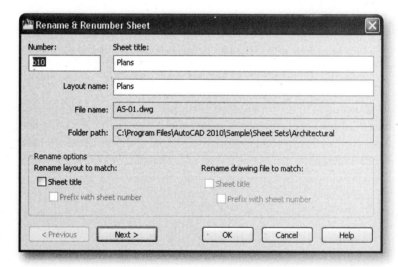

FIGURE 15–35 *The Rename and Renumber Sheet dialog box*

Choosing the **Sheet Selections** icon, in the upper-right corner of the Sheet List tab of the Sheet Set Manager, causes a menu to be displayed where you can save, manage, and restore sheet selections by name. This action makes it easy to specify a group of sheets for publish, transmit, or archive operations. To create a sheet selection, first select several sheets from the sheet list, and choose Create from the Sheet Selections menu (Figure 15–36). AutoCAD displays the New Sheet Selection dialog box, where you can specify a sheet selection name and choose **OK** to create the new sheet selection. To restore the selection, choose the name of the sheet selection from the Sheet Selections menu. Choosing **Manage** causes the Sheet Selections dialog box to be displayed, where you can rename or delete the selected Sheet Selection.

To remove the selected sheet from the sheet set, choose *Remove Sheet* from the shortcut menu. AutoCAD removes the currently selected sheet from the sheet set.

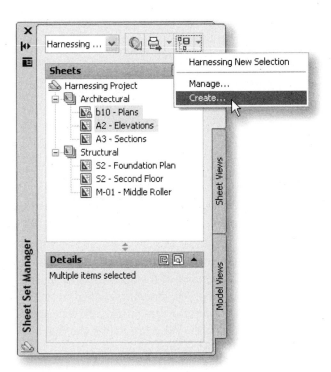

FIGURE 15–36 *Sheet Set Manager with Create selected from the Sheet Selections menu*

Placing a View on a Sheet

The Sheet Set Manager automates and enhances the process for placing a view of a drawing on a sheet. First open the sheet where you want to place the view. To find the view to add to this sheet, select the Model Views tab. On this tab, you can browse for drawings that contain the views you want to add to your sheet. Any drawings you want to use must be listed at this location. To add a folder that contains drawings to the list, choose the **Add New Location** icon, located in the upper-right corner of the Model Views tab, and select the folder from the Browse for Folder dialog box. When you select a view, information about it is displayed in the Details section of the palette. To view a thumbnail preview of the selected view, choose **Preview.** To place a view, first select the view and, from the shortcut menu, select *Place on sheet* (Figure 15–37). Before you place the view, right-click to view or change the scale of the view. Click anywhere on the sheet to insert it. A block label is also inserted. When you place a view on a sheet, AutoCAD attaches the drawing with the named view as an xref. The view is listed as a paper space view on the View List tab. From the View List tab, you can add a view number to the paper space view. Numbers added to the paper space view are updated when the drawing is regenerated.

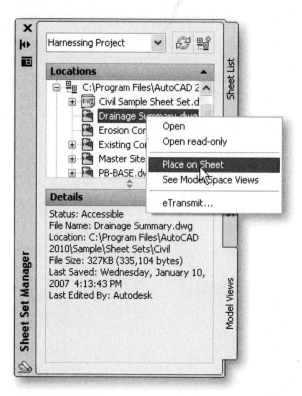

FIGURE 15–37 *Sheet Set Manager – Model Views tab with the shortcut menu to place a view*

Instead of using the OPEN command to open a drawing, you can also open the drawing using the Sheet Set Manager. Double-click the name of the drawing in the Model Views tab, and it will open in a new window. When the drawing is open, it will be locked automatically, and no other user can open it at the same time. The lock status is indicated in the details section of the selected item. When you close the drawing, the status will be changed to Accessible.

The Sheet Views tab (Figure 15–38) displays all named views, also called sheet views, on the layouts in your sheet set. You can use the view list to keep track of all sheet views in the sheet set. You can navigate to any sheet view in the sheet set. You can also link sheet views together for coordination across the sheet set.

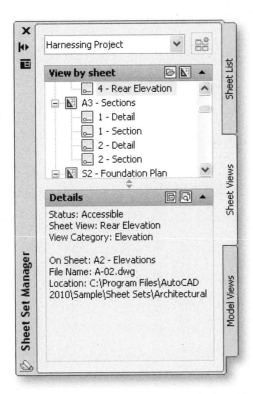

FIGURE 15–38 *Sheet Set Manager with Sheet Views tab selected*

You can create a sheet view by creating a named view in paper space on any layout by invoking the VIEW command. In addition, AutoCAD also automatically creates a sheet view whenever you place a view of a drawing on a sheet. For details on placing a view, see the earlier section on "Placing a View on a Sheet." With the Sheet Set Manager, you can apply label and callout blocks to views on sheets.

A callout block refers to other views in the sheet set. It is a symbol that shows, for example, a cross-reference to an elevation, a detail, or a section (Figure 15–39). With the Sheet Set Manager, you can automatically update the information in your label and callout blocks when the reference information changes. For example, when you renumber or rename a view, information on the label and callout blocks is updated when the drawing regenerates. To place a callout block for the selected view, select *Place Callout Block* from the shortcut menu, and choose the type of callout block you want to use (Figure 15–40). Click anywhere on the sheet to place the callout block. This callout block contains information about the sheet number of the drawing in which the view is saved and the view number (Figure 15–39).

FIGURE 15–39 *Example callout symbols*

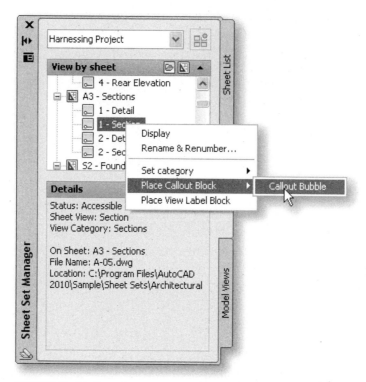

FIGURE 15–40 *Sheet Set Manager – Sheet Views tab with the shortcut menu to place a callout block*

If necessary, you can rename and renumber a view. To rename or renumber the selected view, choose *Rename & Renumber* from the shortcut menu. AutoCAD displays the Rename & Renumber View dialog box. Specify a view number and view title for the selected view. The **Number** box specifies the view number of the selected view. The **View title** option specifies the view title of the selected view. Choosing **Next** loads the next view into this dialog box. Choose **OK** to close the dialog box.

Creating a Sheet List Table

With the Sheet Set Manager, you can create a sheet list table and then update it to match changes in your sheet list. Start by opening the sheet in which you want to create the sheet list table. Select the Sheet List tab, select the sheet set, and, from the shortcut menu, select INSERT SHEET LIST TABLE (Figure 15–41, top). AutoCAD displays the Insert Sheet List Table dialog box (Figure 15–41, bottom). There are several ways to change the appearance of the table before you insert it. For example, when you select **Show Subheader**, the table includes rows displaying subheaders, or the subsets in the sheet set. After making necessary changes in the table, choose **OK.** Click anywhere on the sheet to place the sheet list table. This table lists all the sheets and subsheets in the sheet set (Figure 15–42). If you remove, add, or make any changes to the sheet number or sheet name, you can update the sheet list table by selecting the table and choosing *Update Sheet List table* from the shortcut menu. AutoCAD updates the sheet list table with the changes.

NOTE

If you modify the sheet list table manually, the changes are temporary and are lost when you update the table.

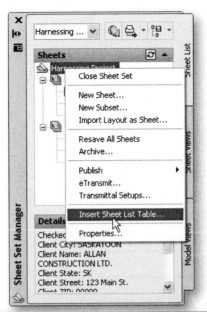

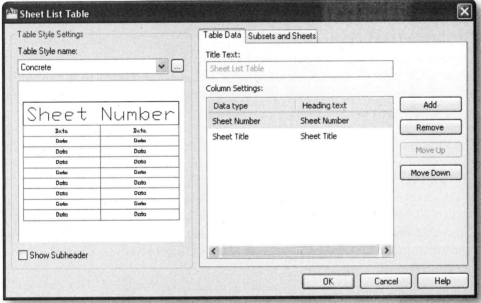

FIGURE 15–41 *Sheet Set Manager – Sheet List tab with the shortcut menu to insert a sheet list table (top); Insert Sheet List Table dialog box (bottom)*

Structural List of Drawings		
Sheet Number	Sheet Title	Author
Architectural		
A1	Plans	GVK
A2	Elevations	TOMS
A3	Sections	GVK
Structural		
S1	Foundation Plan	GVK
S2	Second Floor	TOMS
MS1	Middle Roller	GVK

FIGURE 15–42 *Example table created from the Sheet Set Manager*

Creating a Transmittal Package

The Sheet Set Manager allows you to eTransmit a sheet set, selected sheets, or a subset. eTransmit packages a set of files for Internet transmittal. In a transmittal package, sheet set data files, xrefs, plot configuration files, font files, and so on are automatically included. To create a transmittal package, first select the sheet set, one or more sheets, or subset you want to include in the transmittal package and select *Etransmit* from the shortcut menu. Figure 15–43 at the top shows the Architectural subset selection and ETRANSMIT selected from the shortcut menu. AutoCAD displays the Create Transmittal dialog box (Figure 15–43, bottom). The Sheets tab lists all the sheets in the transmittal package for the Architectural subset. The Files Tree tab lists all the xref, sheet set data, and template files for the transmittal package. Choose **Transmittal Setups** to customize the transmittal setup that will define how your transmittal is packaged. Choose **OK** to create the eTransmit package.

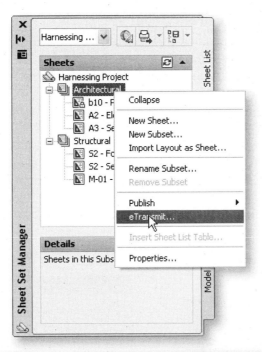

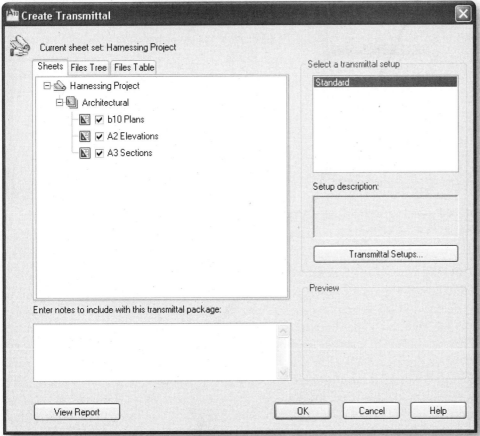

FIGURE 15–43 *Sheet Set Manager – Sheet List tab with the shortcut menu for eTransmit selection (top); Create Transmittal dialog box (bottom)*

Creating an Archive of the Sheet Set

The Sheet Set Manager allows you to archive the selected sheet set. The ARCHIVE option brings together for archiving purposes the files associated with the current sheet set. To create an archive package, first select the sheet set to include in the archive package, and select *Archive* from the shortcut menu (Figure 15–44, left). AutoCAD displays the Archive a Sheet Set dialog box (Figure 15–44, right). The Sheets tab lists all the sheets in the archive package for the selected sheet set. The Files Tree tab lists all the xref, sheet set data, and template files for the archive package. Choose **Modify Archive Setup** to customize the archive setup that will define how your archive is packaged. If necessary, enter information relative to the archive package in the **Enter notes to include with this archive** box. The information is included in the archive report. Choose **OK** to create the archive package. Be sure that the files to be archived are not open.

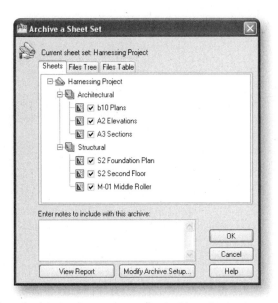

FIGURE 15–44 *Sheet Set Manager – Sheet List tab with the shortcut menu for Archive selection; Archive a Sheet Set dialog box*

Plotting the Sheet Set and Publishing to DWF

The Sheet Set Manager allows you to plot to the default plotter or printer, or publish to specified DWF format a sheet set, selected sheets, or subset. To plot, in the Sheet List tab, first select the sheet set, one or more sheets, or subset you want to include. Then select *Publish* from the shortcut menu and *Publish to Plotter* from the submenu, as shown in Figure 15–45. AutoCAD will plot the selected sheet(s) to the default plotter. For detailed information on plotting and its related settings, refer to Chapter 8. Similarly, to publish to a specified DWF format, first select the sheet set, one or more sheets, or subset you want to include. Select *Publish* from the shortcut menu and *Publish to dwf* from the submenu. AutoCAD will create the DWF file of the selected sheet(s).

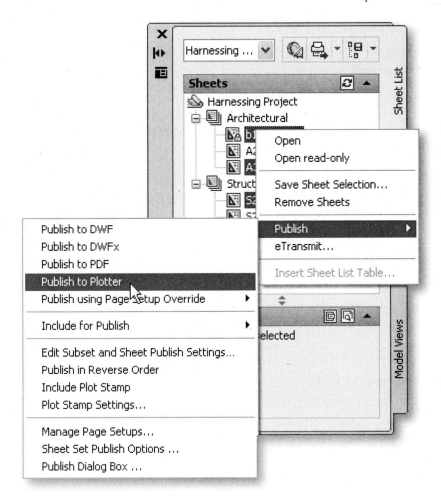

FIGURE 15–45 *Sheet Set Manager – Sheet List tab with the shortcut menu to Publish to Plotter selection*

REVIEW QUESTIONS

1. The command to invoke the Internet browser in AutoCAD is:

   a. Internet

   b. BROWSER

   c. HTTP

   d. WWW

   e. None of the above

2. DWF is short for:

   a. DraWing Format

   b. Design Web Format

   c. DXF Web Format

3. The purpose of DWF files is to view:

a. 2D drawings on the Internet

b. 3D drawings on the Internet

c. 3D drawings in another CAD system

d. a & b

4. URL is short for:

a. Union Region Lengthen

b. Earl

c. Useful Resource Line

d. Uniform Resource Locator

5. Which of the following URLs are valid?

a. www.autodesk.com

b. http://www.autodesk.com

c. All of the above

d. None of the above

6. What is the purpose of a URL?

a. To access files on computers, networks, and the Internet

b. A universal file-naming system for the Internet

c. To create a link to another file

d. All of the above

7. FTP is short for:

a. Forwarding Transfer Protocol

b. File Transfer Protocol

c. File Transference Protocol

d. File Transfer Partition

8. URLs are used in an AutoCAD drawing to browse the Internet.

a. True

b. False

9. The purpose of URLs is to let you create _____ between files.

a. Backups

b. Links

c. Copies

d. Partitions

10. You can attach a URL to rays and xlines.

a. True

b. False

11. Compression in the DWF file causes it to take _____ time to transmit over the Internet.

    a. More

    b. Less

    c. All of the above

    d. None of the above

12. A "plug-in" lets a Web browser:

    a. Plug in to the Internet

    b. Display a file format

    c. Log in to the Internet

    d. Display a URL

13. A Web browser can view DWG drawing files over the Internet.

    a. True

    b. False

14. Which AutoCAD command is used to start a Web browser?

    a. WEBSTART

    b. WEBDWG

    c. BROWSER

    d. LAUNCH

15. Before a URL can be accessed, you must always type the prefix "http://".

    a. True

    b. False

16. Which AutoCAD command is used to open an AutoCAD drawing file from an Internet site?

    a. LAUNCH

    b. OPEN

    c. START

    d. LAUNCHFILE

17. Once an Internet AutoCAD drawing file has been modified, it can only be saved back to the Internet site.

    a. True

    b. False

18. AutoCAD creates hyperlinks as being _____.

    a. Absolute

    b. Relative

    c. Polar

    d. Only a or b

19. Hyperlinks can only be attached to text objects within a drawing.
    a. True
    b. False

20. The system variable PICKFIRST must be set to a value of **1** before a hyperlink can be attached.
    a. True
    b. False

21. DWF files are compressed up to _____ of the original DWG file size.
    a. 1/8
    b. 1/4
    c. 1/2
    d. 1/3

AutoCAD 3D

INTRODUCTION

AutoCAD offers a comprehensive set of commands and features for creating and viewing objects in Three Dimensions. These include Solid, Surface, and Mesh modeling. Powerful modeling and editing tools allow you to generate complex 3D shapes from primitive solids and 2D objects. With the tools provided you can create 3D objects that can be analyzed mathematically and, with the addition of material, texture, lighting, and color (covered in the chapter on Rendering), produce realistic drawings.

OBJECTIVES

After completing this chapter, you will be able to do the following:

- View objects using the ViewCube tool

- Define a user coordinate system (UCS)

- Create 3D objects using Solid, Surface, and Mesh modeling

- Use the REGION command

- Use the 3DPOLY, 3DFACE, CONVTOSURF, and PLANESURF commands

- Edit in 3D using the 3DALIGN, 3DROTATE, MIRROR3D, 3DARRAY, EXTEND, and TRIM commands

- Create solids from existing 2D objects and regions

- Create composite solids

- Edit 3D solids using the CHAMFEREDGE, FILLETEDGE, SECTION, SLICE, and INTERFERE commands, in addition to face editing, body editing, and edge editing

- Obtain the mass properties of a solid

- Place a multiview in paper space

- Generate views in viewports

- Generate profiles

- Use Free-form design for a more flowing style

- Apply 3D printing

WHAT IS 3D AND WHY USE IT?

In two-dimensional drawings, you work in a single plane with two axes, X and Y. In three-dimensional drawings, you work with the Z axis also (see Figure 16–1). Plan views, sections, and elevations represent only two dimensions. Isometric, perspective, and axonometric drawings, on the other hand, represent all three axes.

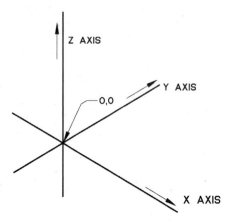

FIGURE 16–1 *The X, Y, and Z axes for a 3D drawing*

Drawing objects in 3D provides three major advantages:

An object drawn in 3D can be viewed and plotted from any viewpoint.

A 3D object provides mathematical information that can be used in engineering analysis such as mass properties, finite element analysis, and computer numerical control (CNC) machinery.

Lighting, shading, and application of color, texture, and material while rendering enhances the visualization of an object.

Whether you realize it or not, all the drawings done in previous chapters were created by AutoCAD in true 3D. This means that every line, circle, and arc that was drawn, even if you think it was drawn in 2D, was stored using three coordinates. By default, when a point was specified in a 2D work plane, AutoCAD used the current elevation as the Z value. What you think of as a 2D view is actually only one of an infinite number of possible drawing views in 3D space.

Whenever you wish to input a set of 3D coordinates whose Z coordinate is different from the current work plane's elevation, there are several methods you can use:

You can specify the three coordinate values using the numeric keyboard. The mouse normally supplies only two of the three coordinates at a time (see the exceptions below).

One exception to being limited to the current work plane is to use Object Snap (OSNAP) to specify a point on an object not in the current work plane. In addition to 2D Object Snap modes, the 3DOSNAP command has the following options: **ZVERtex** snaps to a vertex, **ZMIDpoint** snaps to the midpoint on a face edge, **ZCENter** snaps to the center of a face, **ZKNOt** snaps to a spline knot, **ZPERpendicular** snaps to a perpendicular face (planar faces only), **ZNEAr** snaps to an object nearest to face, and **ZNONe** turns off all 3D object snaps.

Another exception is to use the ± Z option available with the Ortho mode turned on. This allows specifying a point orthogonally in a direction along the Z axis. Also, when you are specifying a point relative to a point outside the current work plane, the Ortho mode allows you to specify the next point in the same plane as the previous point. For example, if you draw a line from 0,0,0 to 0,0,3 (using either keyboard input or the ± Z option), you can continue drawing the next line orthogonally from that last point to another point in the X- or Y-axis direction that is 3 units above the XY plane.

GOING FROM 2D TO 3D

In Chapter 2 we described how to specify points using the three major coordinate systems: Cartesian, Polar, and Cylindrical. In this chapter, we will be using the Cartesian coordinate system, even though the other two are viable systems. The projection will be parallel and not perspective. Perspective projection will be covered in Chapter 17 on Rendering.

AutoCAD provides a variety of workspaces and coordinate systems, but for these first examples the Drafting & Annotation workspace with the default World Coordinate System (WCS) will be used. You can cause the display to change to the SE Isometric option from the drop-down menu in the upper-left corner of the screen as shown in Figure 16–2a. Or, you can select the same view at the vertex of the Top, Front, and Right faces as shown in Figure 16–2b.

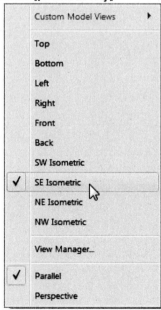

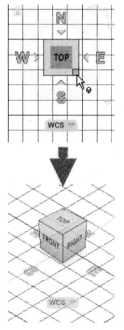

FIGURE 16–2a *Selecting the SE Isometric option from the drop-down menu*

FIGURE 16–2b *Selecting the SE Isometric option from the Vertex on the ViewCube*

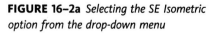

Drawing a line in the work (XY) plane while viewing from a view other than the plan view (viewpoint at coordinates 0,0,1) is similar to drawing from the plan view. However, if you change the viewpoint to be somewhere in the work plane (Z=0), you lose the ability to visualize changes in the direction of your viewpoint. You would be viewing your work plane on edge and using the cursor to specify points would be impractical.

When an object is drawn using 2D methods only, all views are usually drawn in the same work plane even though they represent the object from different points of view in space. The lever below (Figure 16–3a) is taken from the chapter on dimensioning. The top and side views are drawn in the same 2D work plane. These views represent projections from something that may only be a design in someone's imagination. You cannot view the object from any viewpoint other than the ones drawn separately in the 2D work plane. Trying to change the point of view (Figure 16–3b) only distorts the views because they are only projections of the solid object on a work plane, not the solid object.

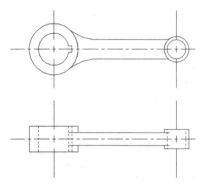

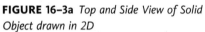

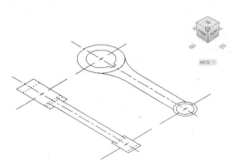

FIGURE 16–3a *Top and Side View of Solid Object drawn in 2D*

FIGURE 16–3b *2D Drawing viewed from Southeast Isometric corner*

To get multiple realistic views of the object without drawing multiple 2D views, AutoCAD provides a variety of 3D modeling tools that will be explained in this chapter. Figure 16–4 is one of an infinite number of views possible after the same object is drawn using these 3D modeling tools.

FIGURE 16–4 *Solid Object drawn using 3D Modeling*

SPECIFYING 3D POINTS

Before considering the 3D modeling commands and features, you need to become familiar with how to specify 3D points using AutoCAD's various methods and modes such as Snap, Object Snap, Polar Tracking, Polar Snap, and Ortho.

2D Primitive Objects

Drawing a line from 0,0 to 5,5 can be accomplished by cursor input with Snap set to the appropriate value or keyboard input or a combination of the two (Figure 16–5).

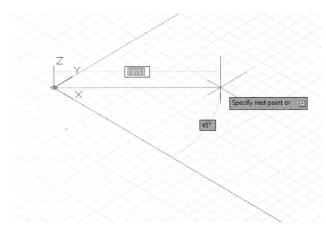

FIGURE 16–5 *Drawing a Line from 0,0 to 5,5 in the work (XY) plane*

If there is no part of an object outside the work plane for specifying a point using OSNAP, you can always enter the coordinates from the keyboard. For example, you can draw a line from 0,0,3.5 to 5,5,3.5 as shown in Figure 16–6. Once a 3D point is specified it is not necessary to include the Z coordinate in the second point. In this example the first point would be input as 0,0,3.5 and the second point could be input as simply 5,5. AutoCAD places the second point using the same Z coordinate value as the first.

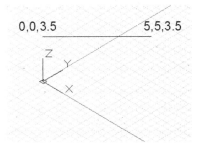

FIGURE 16–6 *Drawing a line with Keyboard input from 0,0,3.5 to 5,5,3.5 in 3D space*

AutoCAD allows you to use the cursor to specify a point that is not in the work plane. To facilitate this function Polar Tracking (F10) and Ortho (F8) should be turned ON. For example, you can begin a line at coordinates 2,4 as shown in Figure 16–7a (either with the cursor using the appropriate Snap values or with keyboard entry) and then move the cursor in the apparent vertical direction to a point above or below the work plane (positive or negative Z). AutoCAD assumes you wish to create a vertical line and the tooltip displays (in this example) "Polar 3.5000< +Z". Selecting this point specifies the coordinates 2,4,3.5 for the endpoint of the line as shown in Figure 16–7b.

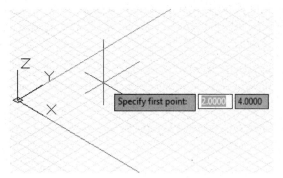

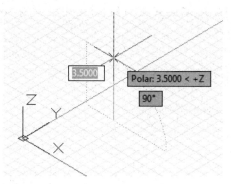

FIGURE 16–7a *Starting a Vertical Line at coordinates 2,4*   **FIGURE 16–7b** *Completing a Vertical Line using the ± Z Option*

If you do not wish to create a vertical line, turn the Polar Tracking and Ortho off. Under some conditions (if the cursor is exactly on the vertical line) the ± Z option might appear, forcing the line to be vertical. In that case you may have to toggle the Polar Tracking again to allow you to specify a point in the XY plane where Z=0.

Once a start point has been established with a non-zero Z coordinate, you can continue with the cursor on one of the orthogonal axes and AutoCAD will allow the next point to remain in the new plane. For example, if you drew a series of lines from 0,0 to 5,5 and then used the vertical (± Z) option to draw end of the second line at 5,5,3.5, then you could move the cursor to create a third line whose end would be 6 units along the X (or Y) axis as shown in Figure 16–8.

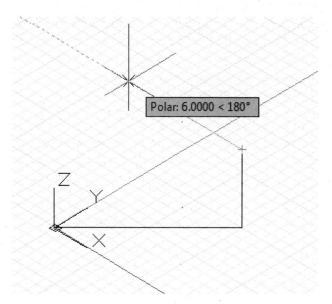

FIGURE 16–8 *Drawing a Horizontal Line using the Ortho option*

SPECIAL OBJECTS IN 3D

Regions and 3D Polylines are objects with properties that make them especially useful for working in 3D space.

Regions

Regions are 2D areas that you create from closed shapes or loops. The REGION command creates a region object from a selection set of objects. Closed polylines, lines, curves, circular arcs, circles, elliptical arcs, ellipses, and splines are valid selections. Once you create a region, you can extrude it with the EXTRUDE command to make a 3D solid. You can also create a composite region with the UNION, SUBTRACTION, and INTERSECTION commands. If necessary, you can hatch a region with the BHATCH command.

AutoCAD converts closed 2D and planar 3D polylines in the selection set to separate regions and then converts the polylines, lines, and curves that form closed planar loops. If more than two curves share an endpoint, the resultant region might be arbitrary. Each object retains its layer, linetype, and color. AutoCAD deletes the original objects after converting them to regions and, by default, does not hatch the regions. Invoke the REGION command from the Draw panel on the Home tab.

3D Polylines

The 3DPOLY command draws polylines with independent X-, Y-, and Z-axis coordinates using the continuous linetype. The 3DPOLY command works similarly to the PLINE command with a few exceptions. Unlike the PLINE command, 3DPOLY draws only straight-line segments without variable width. Editing a 3D polyline with the PEDIT command is similar to editing a 2D polyline except for some options. 3D polylines cannot be joined, curve-fit with arc segments, or given a width or tangent. The options available for 3DPOLY command are similar to those available for the PLINE command, as described in Chapter 4.

Helixes

The HELIX command creates an open 2D or 3D spiral. Invoke the HELIX command from the Draw panel on the Home tab. AutoCAD prompts for the center point of the base. The second prompt is for the base radius with a right-click option for using the diameter. The third prompt is for the top radius with a right-click option for using the diameter. The fourth prompt is for the helix height with the following options:

The *Axis Endpoint* option allows you to specify the endpoint location for the helix axis. The axis endpoint can be located anywhere in 3D space. The axis endpoint defines the length and orientation of the helix.

The *Turns* option allows you to specify the number of turns (revolutions) for the helix. The number of turns for a helix cannot exceed 500. Initially, the default value for the number of turns is three. During a drawing session, the default value for the number of turns is always the previously entered number of TURNS value.

The *Turn Height* option allows you to specify the height of one turn. The number of turns will adjust automatically to allow for the specified height. The turn height cannot be applied if the number of turns for the helix has been specified.

The *Twist* option allows you to specify whether the helix is drawn in the clockwise (CW) or the counterclockwise (CCW) direction. The default value for the helix twist is CCW.

The spiral created by the HELIX command allows you to draw a coil spring-like or thread-like solid by applying the SWEEP command described later in this chapter.

3D WORKSPACES AND TEMPLATES

AutoCAD provides two workspaces for using 3D commands, features, and viewing. The 3D Modeling Workspace is designed for more advanced 3D modeling. The 3D Basics Workspace combines the most common 3D modeling commands for users who are new to 3D modeling or for more basic 3D modeling.

> **NOTE**
>
> Instead of opening a new drawing for 3D modeling using the default *acad.dwt* template you can use the *acad3D.dwt* which has the 3D Modeling Workspace enabled with a custom viewpoint in the perspective mode and the realistic style. Most of the other environmental settings such as units, display colors, snap, and grid are the same as the default acad.dwt. However, using the 3D template with its perspective mode and realistic visual style is more appropriate for rendering than for 3D solid modeling.

The 3D Modeling Workspace

AutoCAD provides the 3D Modeling Workspace for 3D-related commands (see Figure 16–9). Panels on the Home tab include Modeling, Mesh, Solid Editing, and the 2D/3D panels of Draw, Modify, Section, Coordinates, View, Selection, Layers, and Groups. In addition to the tabs contained on the Drafting & Annotation workspace and the custom panels in the Home tab, the 3D Modeling Workspace has the special tabs for Solid, Surface, and Mesh modeling and Render commands.

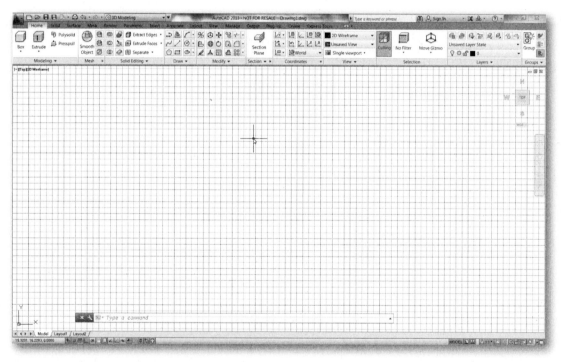

FIGURE 16–9 *The 3D Modeling Workspace*

The 3D Basics Workspace

AutoCAD provides the 3D Basics Workspace for a less advanced grouping of the 3D-related commands (see Figure 16–10). Panels included are Create, Edit, and the 2D/3D panels of Draw, Modify, Selection, Coordinates, and Layers.

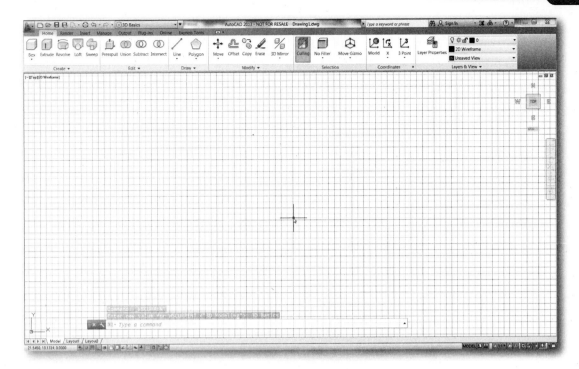

FIGURE 16–10 *The 3D Basics Workspace*

DISPLAYING 3D OBJECTS

AutoCAD provides three different modes for displaying objects in 3D: Wireframe, Surfaces, and Solids. Figure 16–11 shows how a box can be represented by Wireframe (left) with lines in 3D space, by Surfaces (middle) with four surfaces, or by Solids (right) as a solid box. The top and bottom surfaces are left out of the middle view for illustration purposes. The three display methods are used in the various Visual Styles covered in Chapter 17 on Rendering.

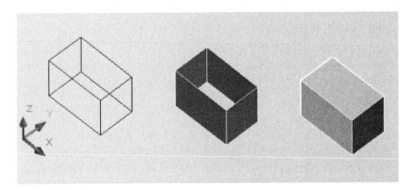

FIGURE 16–11 *A box displayed in Wireframe, Surface, and Solids modes*

Figure 16–12 shows a solid modeling drawing of a house. Walls and furniture have been drawn inside of it.

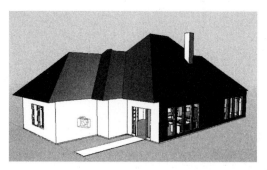

FIGURE 16–12 *House displayed in realistic visual style and perspective mode*

After changing to the 3D Wireframe visual style, the solid objects such as the roof, exterior and interior walls, and furniture become transparent. They are displayed only by showing the edges of their surfaces (see Figure 16–13).

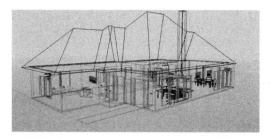

FIGURE 16–13 *House displayed in 3D Wireframe mode*

Curved 3D surfaces are difficult to display in a manner that clearly defines the shape, especially in Wireframe visual styles. For example, a cylinder only has circles on each end. These edges of the circular end planes don't change as your view rotates around the object. If you draw the lines that depict where the round vertical plane disappears around the curve of the cylinder, they are valid from that point of view only. As your point of view rotates, new lines are required. It is even more difficult to draw spheres or compound curves using the wireframe method if you wish the viewer to be able to perceive the "roundness" of the object from various points of view. Therefore, it is advisable to not use the wireframe modeling mode for objects with complex curved surfaces, especially when you can create them using Solid modeling commands but still use the 3D Wireframe visual style.

NOTE

When a curved object created with a Solid modeling method is displayed in the 3D Wireframe visual style, you can use the `ISOLINES` system variable to control the number of tessellation lines used to visualize curved portions of the wireframe. Figure 16–14 shows a cylinder with `ISOLINES` set to the default value of 4 (left) and a cylinder with isolines set to 16 (right). The cylinders were created with the `CYLINDER` command.

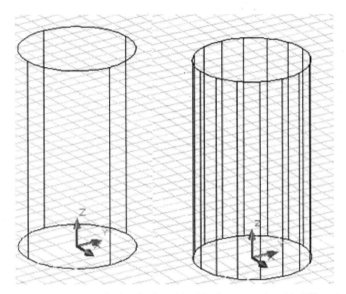

FIGURE 16–14 *Cylinder displayed in Wireframe mode with different values for the ISOLINES system variable*

Unlike the "real" surfaces created with surface-generating commands, you can display pseudo-surfaces that are visible but do not have substance. For example, when a line or curve is drawn with thickness, it will be displayed as a wide band, appearing to be a surface (Figure 16–15). It will even hide other lines behind it when the HIDE command is invoked or the Conceptual or Realistic visual styles are enabled.

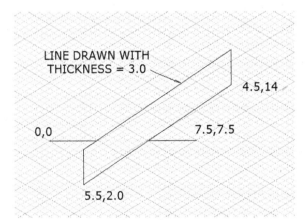

FIGURE 16–15 *A Line with thickness hides other objects*

You can create Wireframe objects with lines, arcs, circles, rectangles, and 3D polylines by drawing them using 3D coordinates and by creating UCSs to place them in the desired planes. The resulting planes and apparent volumes will have no characteristics of solids and regions such as centers of gravity and mass. AutoCAD will convert 3D objects such as Solids, Meshes, and Surfaces to Wireframe objects by using the XEDGES command. Wireframe objects created using the XEDGES command are lines, circles, arcs and 3D polylines that were the edges, intersections, and mesh and surface intermediate lines of the 3D objects from which they were created. Unlike the line with thickness shown in Figure 16–15, Wireframe objects will not obscure objects behind them when the HIDE command is used.

COORDINATE SYSTEMS

AutoCAD provides a single fixed coordinate system called the world coordinate system (WCS) and the capability of establishing an infinite number of user-defined coordinate systems available through the UCS command. Named UCSs can be deleted, but the WCS cannot.

The WCS is fixed and cannot be changed, as indicated in Chapter 2. In this system, when viewing the origin from 0,0,1, (plan view) the positive X axis starts at the origin (coordinates 0,0,0) and values increase as the point moves to the user's right. The positive Y axis starts at the origin and values increase as the point moves to the top of the screen. Finally, the positive Z axis starts at the origin and values increase as the point moves toward the viewer. All drawings from previous chapters were created using the WCS. The WCS is still the basic system used in virtually all 2D AutoCAD drawings. However, because of the complexity of specifying most 3D points, the WCS is not sufficient for most 3D applications.

The UCS allows you to change the location of the origin and orientation of the X, Y, and Z axes to reduce the number of calculations needed to create 3D objects. For example, if you need to draw details in the plane of the sloped roof of a house using the WCS, points on each object on the inclined roof plane must be meticulously calculated. If, however, a UCS is created so the new XY plane is coplanar to the plane of the roof, each object can be drawn in plan view in the newly created work plane. You can define any number of UCSs relative to the fixed WCS and, if desired, save them with user-determined names.

NOTE When you specify a point, be careful to note the coordinate system that it is relative to. In most cases, point coordinates are relative to the current coordinate system. Also note the status of the DYNPICOORDS system variable. To enter absolute coordinates when relative coordinates are displayed in the tooltip, enter # to temporarily override the DYNPICOORDS system variable. To enter relative coordinates when absolute coordinates are displayed, enter @ to temporarily override the DYNPICOORDS system variable. To enter absolute coordinates relative to the WCS, enter * as the prefix to the coordinates.

Only one coordinate system can be current at any given time, and all coordinate input and display is relative to it. This is unlike a rotated snap grid in which direction and coordinates are based on the WCS. If multiple viewports are active, AutoCAD allows you to assign a different UCS to each viewport. Each UCS can have different origins and orientations for various design/drafting requirements.

Right-Hand Rule

Unless you are just relocating the origin, the directions of at least two of the X, Y, and Z axes change with a change in orientation when a new coordinate system is created. Hence, the positive rotation direction of the axes may become difficult to determine. The right-hand rule helps in determining the rotation direction when changing the UCS or using commands that require object rotation. To remember the orientation of the axes, do the following:

- Hold your right hand with the thumb, forefinger, and middle finger pointing at right angles to each other (see Figure 16–16).
- Consider the thumb to be pointing in the positive direction of the X axis.
- The forefinger points in the positive direction of the Y axis.
- The middle finger points in the positive direction of the Z axis.

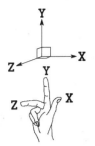

FIGURE 16–16 *The correct hand position when using the right-hand rule*

The UCS Icon

The UCS icon provides a visual indication of how the UCS axes are oriented, where the current UCS origin is, and the viewing direction relative to the UCS XY plane. AutoCAD provides three methods of displaying icons: 2D UCS Style, 3D UCS Style, and 3D Shaded Style. It displays different coordinate system icons in model space (see Figure 16–17a) and in paper space (see Figure 16–17b). When selected with the cursor, the icon displays grips for manipulation using the cursor (see Figure 16–17c).

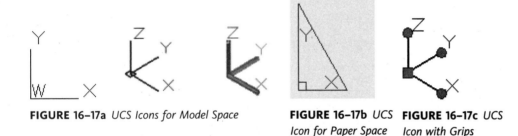

FIGURE 16–17a *UCS Icons for Model Space*   **FIGURE 16–17b** *UCS Icon for Paper Space*   **FIGURE 16–17c** *UCS Icon with Grips*

If the viewing angle is from a point in the XY plane, the 2D UCS Style icon will change to a "broken pencil" (see Figure 16–18a). When this icon is showing in a view, it is recommended that you avoid trying to use the cursor to specify points in that view, because the results are unpredictable. The 3D UCS icon does not use a broken pencil icon. When the viewpoint is from below the XY plane (from the negative Z direction) the Z axis is depicted as a dotted line (see Figure 16–18b).

2D UCS BROKEN PENCIL

3D UCS VIEWED FROM BELOW

FIGURE 16–18a *The UCS Icon when viewing from a point in the XY plane*   **FIGURE 16–18b** *The UCS Icon when viewing from below the XY plane*

The display and placement of the origin of the UCS icon is controlled by the UCSICON command which can be invoked by typing from the keyboard, and AutoCAD displays a drop-down menu (see Figure 16–19).

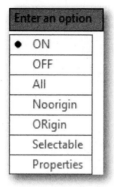

FIGURE 16–19 *Menu displayes at UCSICON command*

Choose *ON* to display the icon if it is OFF in the current viewport.

Choose *OFF* to hide the icon if it is ON in the current viewport.

Choose *All* to set whether or not the options that follow affect all of the viewports or just the current active viewport.

Choose *Noorigin* (default setting) to display the icon at the lower-left corner of the viewport, regardless of the location of the UCS origin. This is like "parking" the icon in the lower-left corner.

Choose *ORigin* to display the icon at the origin of the current coordinate system.

NOTE If the origin is off screen, the icon is displayed at the lower-left corner of the viewport.

You can also choose certain options (*Noorigin/ORigin, ON or OFF*) to control the display and placement of the UCS icon from Coordinates panel (see Figure 16–20).

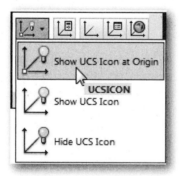

FIGURE 16–20 *Invoking the UCS icon display option from the Coordinates panel*

Choose *Selectable* to enable the grips that will be displayed and can be selected for manipulating the UCS Icon, thus creating a new UCS.

Choose *Properties* to display the UCS Icon dialog box (see Figure 16–21) from which you can control the style, visibility, and location of the UCS icon.

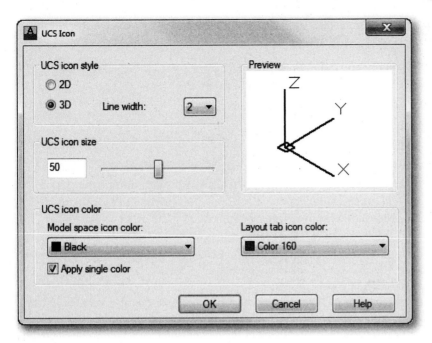

FIGURE 16–21 *UCS Icon dialog box*

The UCS icon style section controls display of either the 2D or the 3D UCS icon and its appearance. The 2D selection displays a 2D icon without a representation of the Z axis, and the 3D selection displays a 3D icon. **Line width** controls the line width of the UCS icon, and you can set the value to one of the three available selections: 1, 2, or 3 pixels, available only when 3D UCS icon is selected. The Preview section displays a preview of the selected UCS icon style in model space. The UCS icon size section controls the size of the UCS icon as a fraction of viewport size. The default value is 12, and the valid range is from 5 to 95. The UCS Icon color section allows you to select the color of the UCS icon in model space viewports and in Layout. Choosing **Apply single color** causes all axes of the 2D UCS icon to be the selected color. Choose **OK** to save the settings and close the UCS Icon dialog box. You can also display the UCS Icon Properties dialog box from the Coordinates panel.

Defining a New UCS

Working with user-defined coordinate systems is the key to easy and accurate 3D modeling in AutoCAD. Similar to specifying points and directions when drawing objects, you can use various methods in defining a UCS. For example, to specify the origin you can input the coordinates from the keyboard, snap to a location on the grid or use an OSNAP mode to specify a point on an existing object. To specify a direction for an axis, you can input a distance/angle vector or use the cursor to specify it. In this case, Ortho is helpful as is Polar Tracking.

Many AutoCAD commands are typically thought of as 2D commands. They are effective in 3D because they are always relative to the current UCS. For example, if you want to use the ROTATE command to rotate an object about a line that is not parallel to the Z axis, then you must change the UCS so that the XY plane is perpendicular to the desired axis of rotation. Then, you can use the ROTATE command. Otherwise you will need to use the 3DROTATE command.

AutoCAD provides numerous ways to define a UCS. They all involve the following steps:

- Establish an origin point or accept the existing one.
- Establish the direction of one of the axes (in most cases X).
- Establish the direction of a second axis.
- Perform all three above steps at once with one of the special options of the UCS command.

Each method handles the above steps in a unique way, offering an advantage when applied under the appropriate situations. Establishing the location of the origin and the direction of two axes is sufficient to define a UCS. The direction of the third axis is, by definition, perpendicular to the plane generated by the other two axes.

> AutoCAD provides a number of options to the UCS command which prompt you to specify just one or two of the above steps and then automatically calculates the remaining steps to complete the definition of a new coordinate system. These options are available through various menus and through the Coordinates panel on the Home tab of the Ribbon of the 3D workspaces. The Coordinates panel is hidden in the Drafting & Annotations workspace Ribbon but can be displayed by right-clicking in the Ribbon and checking it on the shortcut menu.

UCS Command

To establish a new coordinate system you must use the UCS command which lets you relocate the origin and reorient the axes of the coordinate system in your drawing by the following methods:

- Specify one point and press **Enter** or right-click to accept. AutoCAD establishes that point as the origin of the new coordinate system. The orientations of the X, Y, and Z axes remain the same.
- Specify two points and press **Enter** or right-click to accept. AutoCAD uses the first point as the origin of the new coordinate system and the second point establishes the orientation of the X axis rotated about the Z axis. The orientation of the Y axis, of course, rotates to be perpendicular to the new XZ plane.
- Specify three points. AutoCAD uses the first point as the origin of the new coordinate system and the second point establishes the orientation of the X axis rotated about the Z axis. The third point establishes the orientation of the Y axis rotated about the X axis and the Z axis, of course, rotates to be perpendicular to the new XY plane.
- Specify the origin and orientation relative to an existing object.
- Specify the origin and orientation by selecting a face.
- Specify the orientation by aligning with the current viewing direction.
- Specify the orientation by rotating the current UCS around one of its axes.

Shortcut Menus. Right-click after entering UCS at the command prompt and AutoCAD displays a shortcut menu (see Figure 16–22a). Right-click while the cursor is hovering over the UCS Icon and AutoCAD displays a shortcut menu (see Figure 16–22b).

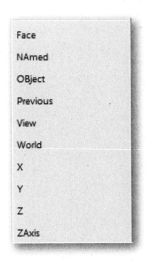

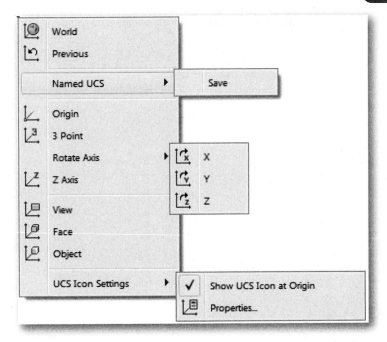

FIGURE 16–22a *UCS Command Shortcut Menu at the UCS Icon*

FIGURE 16–22b *UCS Command Shortcut Menu at the UCS Command*

The shortcut menu contains the major options to the UCS command for defining a new coordinate system. These include Face, NAmed, OBject, Previous, View, World, X, Y, and Z, and ZAxis. These are explained as they appear on the Coordinates panel below.

> Even though several different options might be used to create most UCSs, it is important to know how each of the various UCS command options works so that when a specific user-defined coordinate system is needed to make point input possible (not just easier), the needed UCS can be created.

NOTE

Coordinates Panel

In lieu of using the shortcut menu, the various options can be invoked from the Coordinates panel on the Home tab of the Ribbon in the 3D Modeling Workspace. The Coordinates panel offers a variety of options to the UCS command to create new UCSs or recall named, previous, or predetermined UCSs.

UCS Option. You can invoke the UCS command from the Coordinates panel (see Figure 16–23).

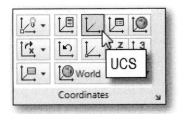

FIGURE 16–23 *Invoking the UCS Command from the Coordinates panel*

When selected, the UCS option first prompts for the new location of the origin. In the example below, the UCS icon is displayed at the WCS origin before invoking the option as shown in Figure 16–24a. After invoking the option, Figure 16–24b shows the selecting the coordinates -3,3, indicating where the new location of the UCS origin will be if accepted. The coordinates reported in the status are relative to the WCS which is the current coordinate system when the command was invoked.

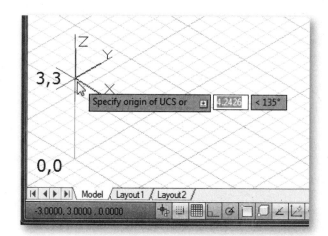

FIGURE 16–24a *WCS Origin*   **FIGURE 16–24b** *Selecting a Point for the Origin of the new UCS*

After accepting the new location for the origin, AutoCAD prompts you to specify a point on the new X axis. Figure 16–25 shows selecting a point that will be on the new X axis if accepted. If you press [Enter] without specifying a point on the new X axis, AutoCAD will just relocate the new origin and the X, Y, and Z axes will maintain their original directions relative to the current coordinate system.

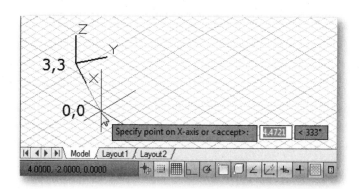

FIGURE 16–25 *Selecting a point to establish the direction of the X axis for the new UCS*

NOTE

Once you have specified the new origin, as you move the cursor the coordinates reported in the status bar are relative to that origin. AutoCAD assumes that the orientation of the axes has not changed until the point is accepted. In the above example, the coordinates reported as 4,−2,0 indicate that the new origin (which was −3,3 in the WCS) is now 0,0,0. The second point being prompted for will be on the new X axis and its Y coordinate value will be 0 in the final coordinate system. However, even though the anticipated orientation of the X axis will change the coordinates, that change will not be reported until the second point is accepted.

After accepting the second point (which establishes the orientation of the X axis), AutoCAD prompts you to "Specify point on the XY plane ..." This point will establish the orientation for the new Y axis. In the example below, the third point is being determined by the cursor location, unless forced to another plane, is in the XY plane. Therefore, the point you select will determine the direction of positive Y in relation to the XZ plane. If you force the positive Y direction to the right (see Figure 16–26a), you be viewing from above the XY plane from the positive Z side. If you move the cursor to the left in this example, you be viewing from below the XY plane from the negative Z side (see Figure 16–26b). Thus, when accepted, the Z axis of the UCS icon is displayed as a dashed line (see Figure 16–26c).

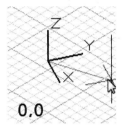

FIGURE 16–26a *Specifying the orientation of Y axis to the right*

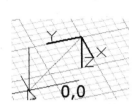

FIGURE 16–26b *Specifying the orientation of Y axis to the left*

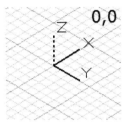

FIGURE 16–26c *Viewing from below the XY plane*

If some other result had been required, any of the points could have been entered above or below the original work plane by using keyboard entry or another method of specifying a non-zero Z coordinate value.

Named Option. Choose *Named* from the Coordinates panel to invoke the UCSMAN command. AutoCAD displays the UCS dialog box (see Figure 16–27).

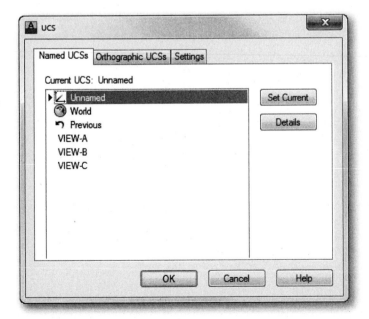

FIGURE 16–27 *UCS Dialog Box with the Named UCSs Tab selected*

When a new UCS has been created you can save it by clicking in the "Unnamed" box, entering a name and then selecting **OK**. If you are just using a newly created coordinate system temporarily and do not wish to save, its settings will be deleted after you have changed to other coordinate systems twice. If you wish to recall an unnamed UCS after changing to another, use the Previous option before you end the session. To set one of the saved UCSs to be current, select it from the listed UCS saved names, choose **Set Current**, or double-click on the listed name, or right-click and select *Set Current* from the shortcut menu.

Choose *Details* and AutoCAD displays the UCS Details dialog box (see Figure 16–28).

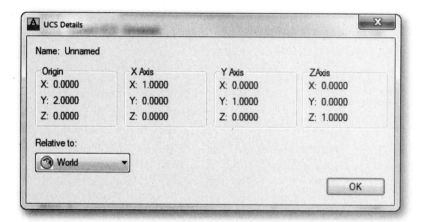

FIGURE 16–28 *UCS Details dialog box with the VIEW-C UCS details showing*

The UCS Details dialog box lists the coordinates of the Origin and of points on each of the axes of the UCS named. The listed coordinates are relative to the coordinate system selected in the Relative to: box.

World Option. Choose *World* to return the drawing to the WCS. Invoke *World* from the Coordinates panel.

Choosing the *World* option causes the WCS to be the current coordinate system.

> **NOTE**
>
> When you define a new coordinate system or set another one current (as in the case above recalling the WCS), the view of objects in 3D space does not change. However, the View-Cube does reorient itself to reflect the new view point with respect to the new coordinate system.

X, Y, or Z Options. Choosing one of the *X*, *Y*, or *Z* rotation options lets you define a new coordinate system by rotating the coordinate system about the axis selected. You can specify the desired angle by selecting a point on the screen, or you can enter the rotation angle from the keyboard. Or, as illustrated in the examples below, you can accept the 90° default angle by pressing Enter or right-clicking. In any case, the new angle is calculated in the plane of the other two axes in the current UCS. See Figure 16–29, Figure 16–30, and Figure 16–31 for examples of rotating the UCS around the X, Y, and Z axes, respectively.

FIGURE 16–29 *Example of rotating the UCS around the X axis*

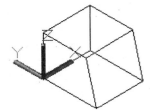

FIGURE 16–30 *Example of rotating the UCS around the Y axis*

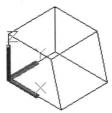

FIGURE 16–31 *Example of rotating the UCS around the Z axis*

Previous Option. Choose *Previous* to restore the previous origin. AutoCAD saves the last 10 coordinate systems in both model space and paper space. You can step back through them by using the *Previous* option repeatedly.

Origin Option. Choose *Origin* from the Coordinates panel. Specify a point to define a new UCS by relocating the origin of the current UCS, leaving the directions of the X, Y, and Z axes unchanged (see Figure 16–32).

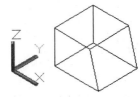

FIGURE 16–32 *Specifying a new origin point relative to the origin of the current UCS*

Z-Axis Vector Option. Choose the *Z-Axis Vector* option from the Coordinates panel. This option is useful when you wish to establish the UCS on the face of an existing object. AutoCAD first prompts for the new location of the origin. After specifying the point for the origin, AutoCAD prompts for a point on the positive portion of the Z axis, establishing the direction. Figure 16–33 shows selecting the face of a solid in response to the first point.

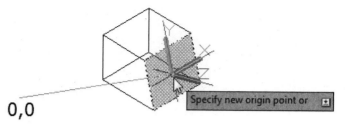

0,0

FIGURE 16–33 *Specifying a point for the Origin*

If, in response to the prompt to specify a new origin point, you have selected a 3D face and press **Enter** in response to the second prompt, the Z axis of the new coordinate system will be perpendicular to the face. The orientations of the X and Y axes are pre-determined by AutoCAD. If you have specified a point that is not on a 3D face and press **Enter** in response to the second prompt, the orientation of all of the axes are unchanged. This is similar to using the ORIGIN option.

If you press **Enter** in response to the first prompt, AutoCAD displays a shortcut menu from which you can select **Object**. Valid objects are 2D objects with endpoints such as lines and arcs. AutoCAD places the new origin at the endpoint nearest the selection point and aligns the Z axis tangent to the object at that endpoint.

3 Point Option. Choose *3point* to define three points that include the origin and the directions of the positive X and Y axes. The origin point acts as a base for the UCS rotation, and when a point is selected to define the direction of the positive X axis, the direction of the Y axis is limited because it is always perpendicular to the X axis. When the X and Y axes are defined, the Z axis is automatically placed perpendicular to the XY plane. You can invoke the UCS command from the Coordinates panel.

When selected, the 3 POINT option first prompts for the new location of the origin. In the example below, the UCS icon is displayed at the WCS origin before invoking the option, as shown in Figure 16–34a. After invoking the option, Figure 16–34b shows the selecting the near corner of the roof, using OSNAP, indicating where the new location of the UCS origin will be if accepted.

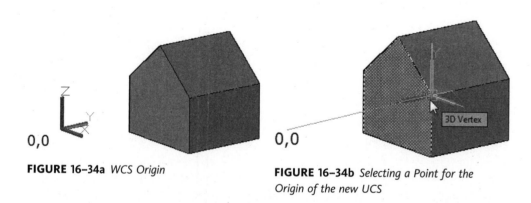

0,0

FIGURE 16–34a *WCS Origin*

0,0

FIGURE 16–34b *Selecting a Point for the Origin of the new UCS*

After accepting the new location for the origin, AutoCAD prompts you to specify a point on the new X axis. Figure 16–35 shows selecting the far corner of the roof indicating a point that will be on the new X axis if accepted. If you press **Enter** without specifying a point on the new X axis, AutoCAD will just relocate the new origin and the X and Y axes will maintain their original directions relative to the WCS.

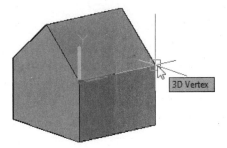

FIGURE 16–35 *Selecting a point to establish the direction of the X axis in the new UCS*

After accepting the new location for a point on the X axis, AutoCAD prompts you to specify a point on the new Y axis. Figure 16–36a shows selecting the apex of the roof indicating a point that will be on the new Y axis if accepted. If you press Enter without specifying a point on the new Y axis, AutoCAD will use the existing orientation of the Z axis and rotate the Y axis to be perpendicular to the new XZ plane. Figure 16–36b shows the resulting location and orientation of the new UCS after selecting the near corner, the rear corner, and the apex of the roof as the first, second, and third points, respectively.

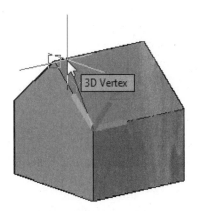

FIGURE 16–36a *WCS Origin*

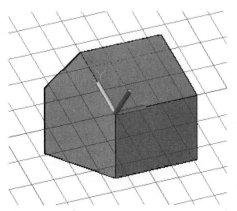

FIGURE 16–36b *Selecting a Point for the Origin of the new UCS*

View Option. Choose *View* to place the XY plane parallel to the screen and make the Z axis perpendicular. The UCS origin remains unchanged. This method is used mainly for labeling text, which should be aligned with the screen rather than with objects. Invoke *View* from the Coordinates panel.

Object Option. Choose *Object* to define a new coordinate system based on a selected object. The actual orientation of the UCS depends on how the object was created. When the object is selected, the UCS origin and positive direction of the X axis are determined by the point used to select the object. In the case of a line, the origin will be the closest endpoint; for a circle, it will be the center point of the circle. The X axis is determined by the direction from the origin to the second point used to define the object. The Z axis direction is placed perpendicular to the XY plane in which the object sits. Table 16–1 lists the location of the origin and its X axis for different types of objects.

TABLE 16–1 *Location of the origin and its X axis for different types of objects*

Object	Method of UCS determination
Line	The endpoint nearest the specified point becomes the new UCS origin. The new X axis is chosen so that the line lies in the XZ plane of the new UCS.
Point	The point becomes the new UCS origin.
Circle	The circle's center becomes the new UCS origin, and the X axis passes through the point specified.
Arc	The arc's center becomes the new UCS origin, and the X axis passes through the endpoint of the arc closest to the pick point.
2D polyline	The polyline's start point becomes the new UCS origin, with the X axis extending from the start point to the next vertex.
Solid	The first point of the solid determines the new UCS origin, and the X axis lies along the line between the first two points.
3D face	The new UCS origin is located at the first point, the X axis from the first two points, and the Y positive side from the first and fourth points. The Z axis is determined by the right-hand rule.
Shape, text, block reference, attribute definition	The new UCS origin is located at the insertion point of the object, and the new X axis is defined by the rotation of the object about its extrusion direction. The object you select to establish a new UCS has a rotation angle of zero in the new UCS.
Dimension	The new UCS origin is the middle point of the dimension text, and the direction of the X axis is parallel to the X axis of the UCS in effect when the dimension was drawn.
Trace	The "from" point of the trace becomes the UCS origin, with the X axis lying along its centerline.

You can invoke the *Object* option from the Coordinates panel.

Face Option. Choose *Face* to align the UCS to the selected face of a solid object. To select a face, click within the boundary of the face or on the edge of the face. The face is highlighted, and the UCS X axis is aligned with the closest edge of the first face found. Invoke *Face* from the Coordinates panel.

When you choose the Face option, AutoCAD prompts you to select the face of a solid, surface, or mesh (see Figure 16–37a). After selecting a face, AutoCAD displays a shortcut menu with the list of options requiring a response (see Figure 16–37b).

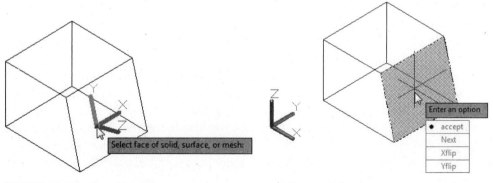

FIGURE 16–37a *Choosing a Face*   **FIGURE 16–37b** *Shortcut menu*

Choosing **Accept** accepts the location. Even though the display of the UCS icon reverts back to the previous location during this step, when you choose **Accept**, the location on the face will be where the origin, and icon, will be located. Choosing **Next** locates the UCS on either the adjacent face or the back face of the selected edge. Choosing **Xflip** rotates the UCS 180° around the X axis. Choosing **Yflip** rotates the UCS 180° around the Y axis.

Predefined Option. In addition to various methods explained earlier for defining a new UCS, AutoCAD allows you to change the UCS to one of the standard orthographic settings or user saved coordinate systems from the Named UCS Combo Control available in the Coordinate panel.

UCS Settings Option. You can also select one of the available standard orthographic configurations from the UCS dialog box. In addition, the UCS dialog box lists saved UCSs and allows you to modify UCS icon settings and UCS settings saved with a viewport. Choose the down arrow at the bottom of the Coordinates panel to open the UCS dialog box and choose the Orthographic UCSs tab (see Figure 16–38).

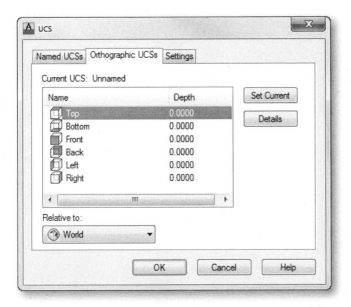

FIGURE 16–38 *UCS dialog box with the Orthographic UCSs tab selected*

AutoCAD displays the name of the current UCS view at the top of the dialog box. If the UCS setting has not been saved and named, the current UCS reads Unnamed. AutoCAD lists the standard orthographic coordinate systems in the current drawing. The orthographic coordinate systems are defined relative to the UCS specified in the Relative to list box. By default, it is set to the WCS. To set the UCS to one of the orthographic coordinate systems, select one of the six listed names, choose **Set Current,** double-click on the listed name, or right-click and select *Set Current* from the shortcut menu. The **Depth** column lists the distance between the orthographic coordinate system and the parallel plane passing through the origin of the UCS base setting, stored in the UCSBASE system variable. To change the depth, double-click on the Depth field, and AutoCAD displays the Orthographic UCS depth dialog box. Make the necessary changes to the depth, and choose **OK** to close the dialog box.

To set the UCS to one of the saved UCSs, select the **Named UCSs** tab, and Auto-CAD lists the saved UCS in the current drawing (see Figure 16–39). Select one of the listed UCS saved names, choose **Set Current**, and double-click on the listed name, or right-click and select *Set Current* from the shortcut menu.

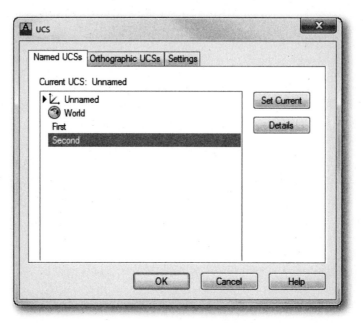

FIGURE 16–39 *UCS dialog box with the Named UCSs tab selected*

The Settings tab of the UCS dialog box displays and allows you to modify UCS icon settings and UCS settings saved with a viewport (see Figure 16–40).

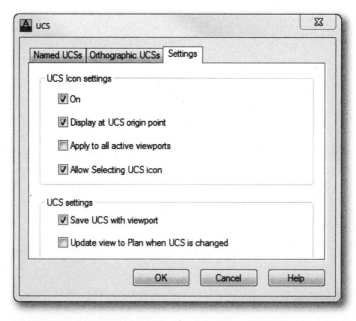

FIGURE 16–40 *UCS dialog box with the Settings tab selected*

The UCS icon Settings section allows you to control the UCS settings for the current viewport. Choose **ON** to display the UCS icon in the current viewport. **Display at UCS origin point** controls the display of the UCS icon at the origin of the current coordinate system for the current viewport. If this option is set to OFF, or if the origin of the coordinate system is not visible in the viewport, the UCS icon is displayed at the lower-left corner of the viewport. Choose **Apply all to active viewports** to apply the UCS icon settings to all active viewports in the current drawing. **Allow Selecting UCS icon** controls whether you can click the icon to select it and access grips and whether the UCS icon is highlighted when the cursor hovers over it.

The UCS settings section specifies the UCS settings for the current viewport. Choose **Save UCS with viewport** to save the coordinate system setting with the viewport. If this option is set to OFF, the viewport reflects the UCS of the viewport that is current. Choose **Update view to Plan when UCS is changed** to restore the plan view when the coordinate system in the viewport changes. Plan view is restored when the dialog box is closed and the selected UCS setting is restored.

Choose **OK** to close the UCS dialog box and accept the changes.

UCS Icon Grips

Select the UCS icon with the cursor and AutoCAD displays 4 grips; the origin and each of the axes. When you hover the cursor over the grip at the origin, a shortcut menu is displayed as shown in Figure 16–41a. Selecting the origin grip lets you move the icon, thus redefining the UCS with the new origin location or pressing CTRL and cycling through the options of **Move and Align**, **Move Origin Only** and **World** as shown in Figure 16–41b.

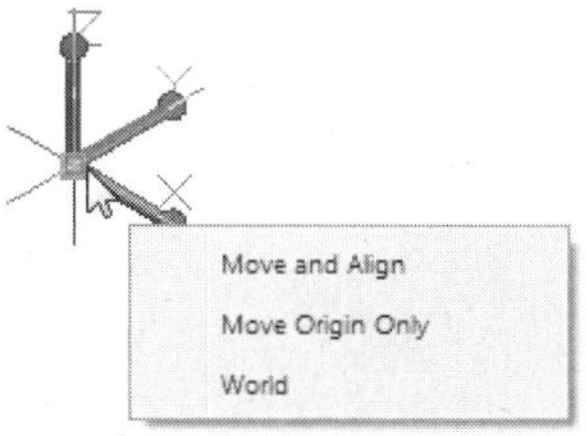

FIGURE 16–41a *Hovering the cursor at the Origin Grip*

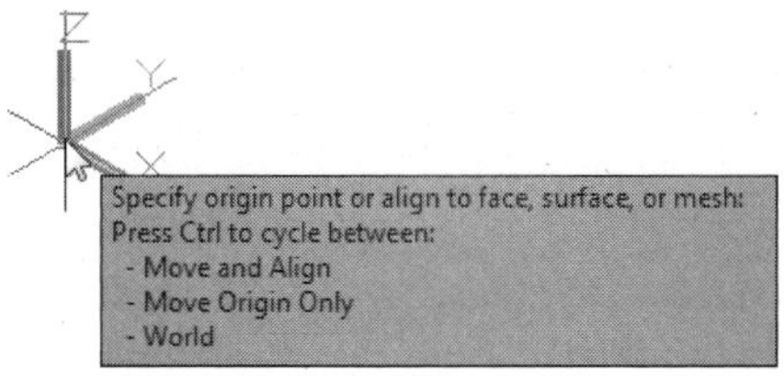

FIGURE 16–41b *Selecting the Origin Grip*

When you hover the cursor over the grip at one of the axes, a shortcut menu is displayed as shown in Figure 16–42a. Selecting the grip lets you rotate the icon around the axis selected, thus redefining the UCS with the new orientation of the axes not selected or pressing CTRL and cycling through the options of **X Axis Direction**, **Rotate Around Z Axis**, and **Rotate Around Y Axis** as shown in Figure 16–42b.

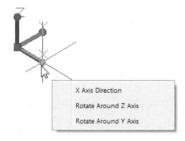

FIGURE 16–42a *Hovering the cursor at the X Axis Grip*

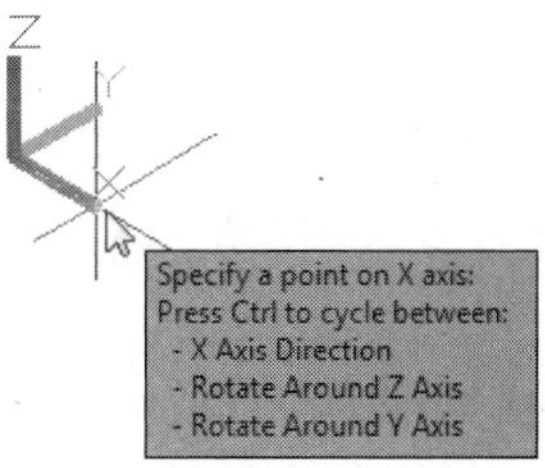

FIGURE 16–42b *Selecting the X Axis Grip*

Figure 16–43a shows the selected axis being rotated about the Z axis. Figure 16–43b shows the UCS icon after the rotation angle has been specified.

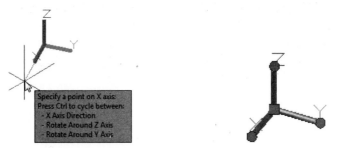

FIGURE 16–43a *Rotating X axis about Z axis*   **FIGURE 16–43b** *UCS Icon after rotation*

Dynamic UCS

The Dynamic UCS feature, located in the Status bar, when set to ON causes the current coordinate system to be temporarily realigned when the cursor passes over a skew plane of a solid object during a drawing command. The new XY plane will be coplanar with the skew plane being passed over. For example, to place a cylindrical solid on the roof of the house, select the CYLINDER command, place the cursor over the plane representing the roof slope, and the UCS is realigned as shown in Figure 16–44 (left). Notice how the cursor is realigned. After selecting the center for the base of the cylinder, the UCS icon is relocated and realigned as shown in Figure 16–44 (right top). The radius and height of the cylinder are created based on the realigned UCS. When the command is completed, the UCS reverts to the one in effect before the command was started as shown in Figure 16–44 (right bottom).

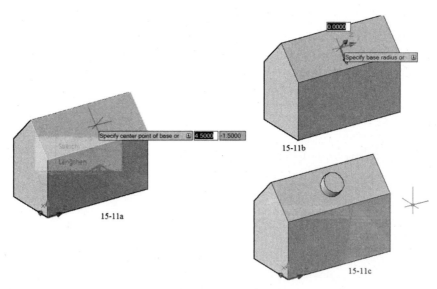

FIGURE 16–44 *Applying Dynamic UCS*

VIEWING IN 3D WITH THE VIEWCUBE TOOL

Until now, you have been working on the plan view or the XY plane. You have been looking down at the plan view from a positive distance along the Z axis. The direction from which you view your drawing or model is called the viewpoint. You can view a drawing from any point in model space. From your selected viewpoint, you can add objects, modify existing objects, or suppress the hidden lines from the drawing. The ViewCube tool (Figure 16–45) is a 3D navigation tool that is displayed when you are working in a 3D visual style. With the ViewCube tool, you can switch between standard and isometric views. The VeiwCube Tool is used to control viewing of a model from various points in model space.

FIGURE 16–45 *The ViewCube Tool*

ViewCube Settings

The location, size, opacity, and other settings of the ViewCube can be controlled by right-clicking on the ViewCube and selecting **ViewCube Settings** from the shortcut menu. AutoCAD displays the ViewCube Settings dialog box (Figure 16–46).

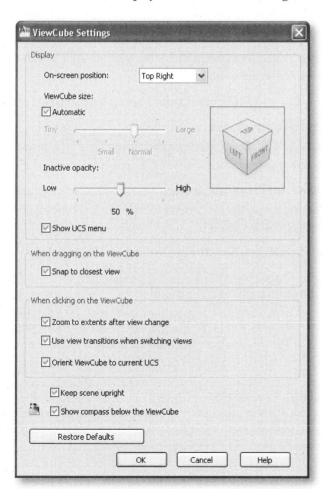

FIGURE 16–46 *The ViewCube Settings Dialog Box*

In the Display section you can control position, size, opacity, and display of UCS menu, and see a preview of the ViewCube.

In the **On-screen position** text box you can choose a position from the following: Top Right, Bottom Right, Top Left, and Bottom Left.

Under ViewCube size, selecting **Automatic** causes the ViewCube to adjust in size based on the current size of the active viewport, the zoom factor, or the active layout or drawing window. Moving the slider bar adjusts the size of the ViewCube.

Under **Inactive opacity**, moving the slider bar adjusts the opacity of the ViewCube.

Selecting **Show UCS Menu** controls the display of the UCS drop-down menu below the ViewCube.

The Preview box displays a real-time preview of the ViewCube based on the current settings.

In the When Dragging on the ViewCube section, choosing **Snap to Closest View** lets you specify whether the current view is adjusted to the closest preset view when changing the view by dragging the ViewCube tool.

In the When Clicking on the ViewCube section, you can control how the ViewCube is displayed when clicked on.

Choosing **Zoom to Extents After View Change** lets you specify if the model is forced to fit the current viewport after a view change.

Choosing **Use View Transitions When Switching Views** lets you control the use of smooth view transitions when switching between views.

Selecting **Orient ViewCube to Current UCS** orients the ViewCube tool based on the current coordinate system.

Selecting **Keep Scene Upright** lets you specify whether the viewpoint of the model can be turned upside-down or not.

Choosing **Show Compass Below the ViewCube** lets you control whether the compass is displayed below the ViewCube tool. The North direction indicated on the compass is the value defined by the NORTHDIRECTION system variable.

Selecting **Restore Defaults** applies the default settings for the ViewCube tool.

Changing Viewpoints

Viewing a model in 3D requires changing the viewpoint if you wish to view the objects from any other direction than the plan view. The default plan view viewpoint is 0,0,1; that is, you are looking at the model from 0,0,1 on the positive Z axis above the model to 0,0,0 (the origin).

NOTE   The value of 1 as the Z coordinate in the point 0,0,1 is not critical in a parallel view. The Z coordinate can be any positive value (0.5 or even 1,000). It establishes the distance from which you are viewing the objects. However, the Z coordinate can affect the appearance of objects drawn in 3D if you are using the PERSPECTIVE feature, which is discussed later.

Figure 16–47 shows an object being viewed in 3D from a viewpoint showing the front, right side, and bottom of the object. The ViewCube Tool is displayed when you are in a 3D visual style. The default position of the ViewCube is in the upper-right hand of the display area.

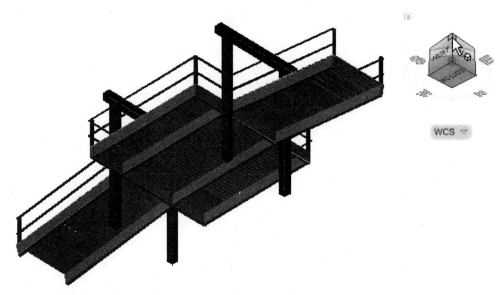

FIGURE 16–47 *Selecting the Upper Southeast corner of the ViewCube Tool*

Selecting the upper southeast corner of the ViewCube tool (as shown in Figure 16–47) changes the point of view to the true SE isometric view. Selecting this view is equivalent to setting the point of view coordinates X, Y, and Z to 1, −1, and 1, respectively. Figure 16–48 shows the result of the selection made in Figure 16–47.

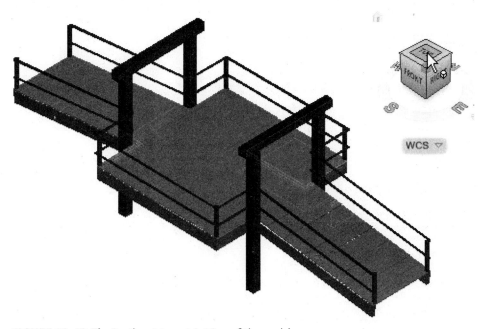

FIGURE 16–48 *The Southeast Isometric View of the model*

You can change to the plan view of the current UCS by selecting Top on the ViewCube. Selecting Top changes the point of view to the current UCS plan view (Figure 16–49).

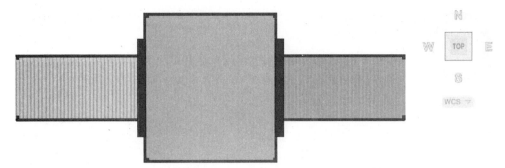

FIGURE 16–49 *The Top View of the model*

You can change to the front view of the current UCS by selecting Front on the View-Cube which changes the point of view to the current UCS front view (Figure 16–50).

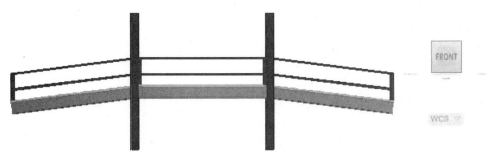

FIGURE 16–50 *The Front View of the model*

Parallel and Perspective Projection

You can switch between parallel and perspective projections by right-clicking on the ViewCube and selecting the appropriate item from the shortcut menu (Figure 16–51).

FIGURE 16–51 *Selecting Perspective from the Shortcut Menu*

Selecting **Perspective** changes the display to perspective projection as shown in Figure 16–52.

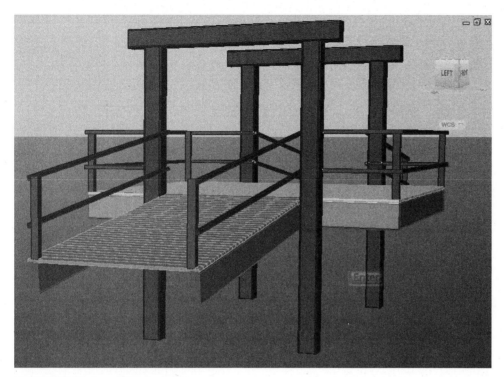

FIGURE 16–52 *Perspective Projection*

Dynamic View Manipulation

You can dynamically manipulate the viewpoint by placing the cursor on the View-Cube, holding the pick button down, and moving the cursor. The scene changes with the cursor movements as objects in the model are viewed from the changing viewpoints.

Placing the cursor on the ring with the N, E, S, and W, holding the pick button down, and moving the cursor causes the viewpoint to rotate around the origin of the current UCS. Clicking one of the compass direction letters causes the view to change, the viewpoint being from the direction chosen.

When the viewpoint is set to one of the orthographic projections (Left, Right, Front, Back, Top, or Bottom) there are two curved arrows displayed above and to the right of the ViewCube.

Selecting the Home icon above and to the left of the ViewCube returns the display to the previous view. Selecting one of the arrows near and pointing toward the View-Cube causes the display to change the view to the one on that side of the ViewCube. For example, if the Left View is in effect and you select the arrow to the right of the ViewCube, the view will change to the Front View.

Selecting the **WCS** icon below the ViewCube displays a shortcut menu from which coordinate systems can be managed (Figure 16–53).

FIGURE 16–53 *Shortcut menu from WCS selection*

Working with Multiple Viewports in 3D

As mentioned in Chapter 3, AutoCAD allows you to set multiple viewports to provide different views of your model. For example, you might set up viewports that display top, front, right side, and isometric views. You can accomplish this with the help of the ViewCube tool explained earlier. To facilitate editing objects in different views, you can define a different UCS for each view. Each time you make a viewport current, you can begin drawing using the same UCS you used the last time that viewport was current. The UCSVP system variable controls the setting for saving the UCS in the current viewport. When UCSVP is set to 1 (default) in a viewport, the UCS last used in that viewport is saved with the viewport and is restored when the viewport is made current again. When UCSVP is set to 0 in a viewport, its UCS is always the same as the UCS in the current viewport.

For example, you might set up four viewports: top view, front view, right-side view, and isometric view. If you set the UCSVP system variable to 0 in the isometric viewport and to 1 in the top view, front view, and right side view, when you make the front viewport current, the isometric viewport's UCS reflects the UCS front viewport. Likewise, making the top viewport current switches the isometric viewport's UCS to match that of the top viewport.

3D MODELING

AutoCAD supports three modes of 3D modeling: Solids, Meshes, and Surfaces. Using a combination of these three methods offers the designer a powerful tool for creating models of any 3D object imaginable. Objects can be converted from a Solid to a Mesh for creasing and smoothing and then converted to a Surface for working with associativity and NURBS (Non-Uniform Rational Basic Spline). Each method has its own distinct advantages in shaping the final object.

Solid modeling is the easiest type of 3D modeling and differs from Mesh or Surface modeling in two fundamental ways:

- The information is more complete.
- The method of construction of the model itself is inherently straightforward.

Mesh or Surface modeling objects are created by positioning lines or surfaces in 3D space. In Solid modeling, you build the model as you would with building blocks;

from beginning to end, you think, draw, and communicate in 3D. One of the main benefits of Solid modeling is its ability to be analyzed. You can calculate the mass properties of a Solid object such as its mass, center of gravity, surface area, and moments of inertia.

Modeling Workflow

Creating 3D objects requires planning. All of the points are no longer neatly confined to the XY work plane as they are in an orthogonal projection of the object. Once the object has been visualized, the shape defining tools and modeling method or combination of methods need to be decided. One or more coordinate systems will be involved. Like many tasks in AutoCAD, there is more than one sequence and set of commands that will achieve the desired end results in 3D modeling. The more the designer understands how to use the tools available in AutoCAD the better he or she will be able to plan and execute modeling in 3D.

The following example shows one way to begin the task of going from 2D orthogonal views of an object, a 6" 90° schedule 40 welding pipe elbow in this case, to a 3D model. Here we only generate the outer shell of the elbow and not the thickness or the bevel on the ends. This is to illustrate how to set up a UCS and apply one of the surface modeling commands.

Step 1 takes the 2D drawing (see Figure 16–54a) and turns off the layers with dimensions and center line on them (see Figure 16–54b).

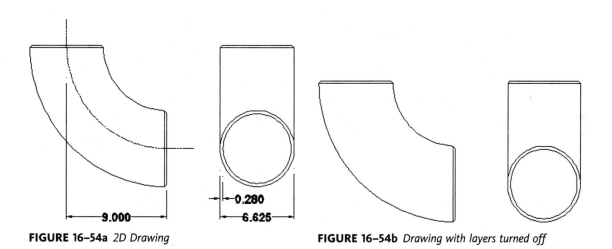

FIGURE 16–54a *2D Drawing*   **FIGURE 16–54b** *Drawing with layers turned off*

Step 2 involves changing the viewpoint to the SE Isometric (Figure 16–55a) and selecting the UCS Icon to display the grips (Figure 16–55b).

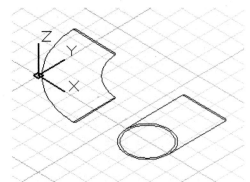

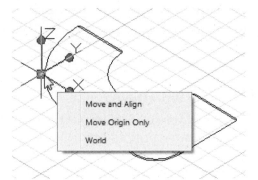

FIGURE 16–55a *Changing the Viewpoint to SE Isometric*

FIGURE 16–55b *Selecting the UCS Icon*

Step 3 involves relocating the origin of the UCS to the center of the end of the elbow (Figure 16–56a) and selecting the grip on the UCS Icon Z Axis (Figure 16–56b).

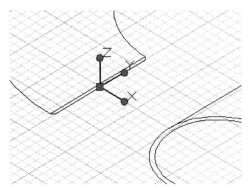

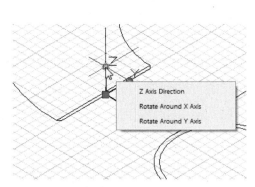

FIGURE 16–56a *Moving the UCS origin*

FIGURE 16–56b *Selecting the UCS Icon Z Axis Grip*

Step 4 involves rotating the UCS Icon Z Axis 90° clockwise about the Y axis (Figure 16–57a) and drawing a circle in the XY work plane of the newly defined UCS (Figure 16–57b). In this case, the diameter is 6 5/8", the outside diameter of 6" pipe.

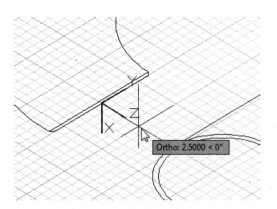

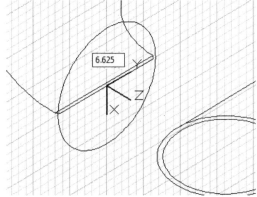

FIGURE 16–57a *Moving the UCS origin*

FIGURE 16–57b *Selecting the UCS Icon Z Axis Grip*

Step 5 involves resetting the coordinate system back to the WCS and drawing an arc along the center line of the elbow (Figure 16–58a) and then using arc as the path for applying the SWEEP command to the circle (see Figure 16–58b). The SWEEP command is found in the Create panel of the Surface tab on the Ribbon.

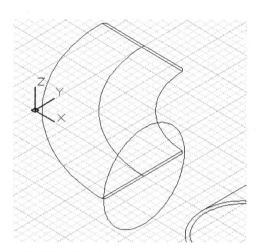

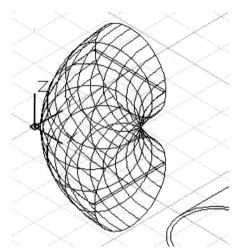

FIGURE 16–58a *Drawing a center line for a path*   **FIGURE 16–58b** *Using the Sweep Command*

To better view the object just created the View Style was changed to Shades of Gray (see Figure 16–59). If desired, all layers that do not contain the newly created surface can be turned off to view only the surface that represents the outer shell of the 90° elbow.

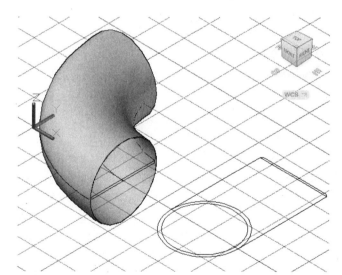

FIGURE 16–59 *Surface that represents the outer shell of the 90° elbow*

The steps in the example above are only the first toward creating a complete 3D model of the object. The surface needs to be converted into a solid object with the specified wall thickness and then the ends need to be beveled. After this the object can be analyzed for mass and center of gravity. It can have editing commands applied to it put holes in it if desired or have objects joined to it so they act as one solid object. These steps illustrate the planning and process that can be used when you start modeling in 3D.

Solid Modeling

Solids can be created as one of the basic solid shapes called primitives: box, cone, cylinder, pyramid, sphere, torus, or wedge. Solids can also be generated from 2D objects like lines, arcs, polylines, and regions by commands such as EXTRUDE, SWEEP, LOFT, or REVOLVE. New solids can be created from existing solids using the SLICE and SECTION commands and from Meshes and Surfaces using converting commands. In addition, you can create more complex solid shapes by combining solids together by performing a Boolean operation—UNION, SUBTRACTION, or INTERSECTION.

Solid Primitives

AutoCAD lets you create seven basic shapes called primitives: box, cone, cylinder, pyramid, sphere, torus, or wedge. Starting with one or more of these can be the basis for generating more complex solid shapes with the advanced editing commands.

Box. The BOX command creates a solid box or cube. The base of the box is defined parallel to the current UCS by default. The solid box can be drawn by providing a center point or a starting corner of the box.

Invoke the BOX command from the Solid Primitives drop-down menu of the Modeling panel on the Home tab (see Figure 16–60).

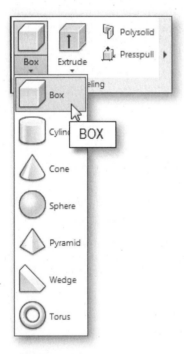

FIGURE 16–60 *Invoking the BOX command from the Modeling panel*

AutoCAD displays the following prompt:

> Specify corner of box or ⊡ *(specify a point, or choose* CENTER *from the shortcut menu)*

By default, you are prompted for the starting corner of the box. Once you provide the starting corner, the box's dimensions can be entered in one of three ways.

The default option lets you create a box by dragging the cursor to the opposite corner of its base rectangle and then dragging the cursor in the Z direction to specify its height. An alternative method is to enter the relative or absolute coordinates for the opposite corner of the base and for the height. The following command sequence defines a box using the second method (see Figure 16–61):

```
box
Specify corner of box or ⏎ 3,3
Specify corner or ⏎ 7,7
Specify height: 4
```

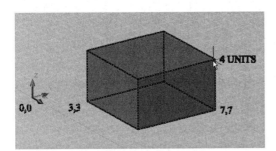

FIGURE 16–61 *Creating a solid box using the default option of the* BOX *command*

The *Cube* option allows you to create a box in which all edges are of equal length. The following command sequence defines a box using the CUBE option:

```
box
Specify corner of box or ⏎ 3,3
Specify corner or ⏎ (choose CUBE from the shortcut menu)
Specify length: 3
```

The *Length* option lets you create a box by defining its length, width, and height. The following command sequence defines a box using the LENGTH option:

```
box
Specify corner of box or ⏎ 3,3
Specify corner or (choose LENGTH from the shortcut menu)
Specify length: 3
Specify width: 4
Specify height: 3
```

The *Center* option allows you to create a box by locating its center point. Once you locate the center point, a line rubber-bands from this point to help you visualize the size of the rectangle. AutoCAD prompts you to define the size of the box:

```
Specify corner or ⏎ (select one of the options from the
shortcut menu)
```

Once you create a box you cannot stretch it or change its size. However, you can extrude the faces of a box with the SOLIDEDIT command.

NOTE

Cylinder. The `CYLINDER` command creates a cylinder of equal diameter on each end and similar to an extruded circle or an ellipse. The solid cylinder can be created by one of two methods: you can provide a center point for a circular base, or you can select the elliptical option to draw the base of the cylinder as an elliptical shape.

Invoke the `CYLINDER` command from the Solid Primitives drop-down menu of the Modeling panel. AutoCAD displays the following prompt:

> `Specify center point for base of cylinder or` `Enter` *(specify a point, or choose* `ELLIPTICAL` *from the shortcut menu)*

The prompts are identical to those for a cone. For example, the following command sequence lists the steps for drawing a cylinder using the default option (see Figure 16–62):

> `cylinder`
>
> `Specify center point for base of cylinder or` ⬇ `5,5`
>
> `Specify radius for base of cylinder or` ⬇ `3`
>
> `Specify height of cylinder or` ⬇ `4`

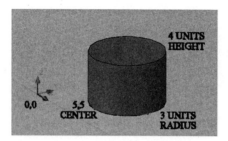

FIGURE 16–62 *Creating a solid cylinder using the default option of the* `CYLINDER` *command*

Cone. The `CONE` command creates a round or elliptical cone. By default, the base of the cone is parallel to the current UCS. Solid cones are symmetrical and come to a point along the Z axis. The solid cone can be drawn by providing a center point for a circular base or by selecting the elliptical option to draw the base of the cone as an elliptical shape. Invoke the `CONE` command from the Solid Primitives drop-down menu of the Modeling panel. AutoCAD displays the following prompt:

> `Specify center point for base of cone or` ⬇ *(specify a point, or choose* `ELLIPTICAL` *from the shortcut menu)*

By default, AutoCAD prompts you for the center point of the base of the cone and assumes the base to be a circle. You are prompted for the radius, or you can enter D for diameter. Enter the appropriate value or drag the cursor to a point on the base circumference. You are then prompted for the apex/height of the cone. The height of the cone is the default option, and it allows you to set the height of the cone—not the orientation—by dragging the cursor in the Z direction to specify its height or by entering a value. The base of the cone is parallel to the current base plane. The Apex option, in contrast, prompts you for a point. It then sets the height and orientation of the cone. The following command sequence lists the steps for drawing a cone using the default option (see Figure 16–63):

```
cone
Specify center point for base of cone or ⊡ 5,5
Specify radius base of cone or ⊡  3
Specify height of cone or ⊡ 4
```

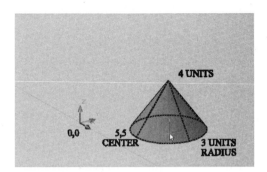

FIGURE 16–63 *Creating a solid cone using the default option of the CONE command*

Selecting the *Elliptical* option indicates that the base of the cone is an ellipse. The prompts are identical to the regular AutoCAD ELLIPSE command. For example, the following command sequence shows the steps for drawing a cone using the *Elliptical* option:

```
cone
Specify center point for base of cone or ⊡(choose ELLIPTICAL
from the shortcut menu)
Specify axis endpoint of ellipse for base of cone or ⊡ 3,3
Specify second axis endpoint of ellipse for base of cone: 6,6
Specify length of other axis for base of cone: 5,7
Specify height of cone or ⊡ 4
```

Sphere. The SPHERE command creates a 3D body in which all surface points are equidistant from the center. The sphere is drawn in such a way that its central axis is coincident with the Z axis of the current UCS.

Invoke the SPHERE command from the Solid Primitives drop-down menu of the Modeling panel. AutoCAD displays the following prompt:

```
Specify center of sphere <0,0,0>: (specify a point)
```

AutoCAD prompts for the center point of the sphere. You then provide the radius or diameter to define a sphere by dragging the cursor or by entering a value.

For example, the following command sequence shows the steps for drawing a sphere (see Figure 16–64):

```
sphere
Specify center of sphere <0,0,0>: 5,5
Specify radius of sphere or⊡ 3
```

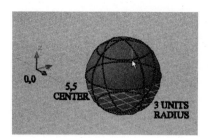

FIGURE 16–64 *Creating a solid sphere using the default option of the SPHERE command*

Pyramid. The PYRAMID command creates a pyramid with a polygonal base of the number of sides you specify. By default, the base of the pyramid is parallel to the current UCS. Solid pyramids are symmetrical and come to a point along the Z axis.

Invoke the PYRAMID command from the Solid Primitives drop-down menu of the Modeling panel. By default, AutoCAD prompts for the center point of the base of the pyramid and assumes the base to be a square circumscribed about the radius you specify. Enter the appropriate value or drag the cursor to a point on the base circumference. AutoCAD then prompts for the apex/height of the cone. The height of the pyramid is specified by dragging the cursor in the Z direction to specify its height or by entering a value. For example, the following command sequence lists the steps for drawing a pyramid (see Figure 16–65):

AutoCAD displays the following prompts:

```
pyramid
Specify center point of base or ⏎ 5,5
Specify base radius or ⏎ @2,2
Specify height or ⏎ 4
```

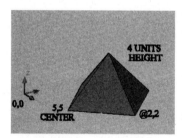

FIGURE 16–65 *Creating a pyramid*

Selecting the *Edge* option at the prompt for the center allows you select two points to specify one edge of the pyramid.

Selecting the *Sides* option at the prompt for the center allows you specify the number of sides for the base.

Selecting the *Inscribed* option at the prompt for the radius causes the polygonal base to be inscribed by a circle of the specified radius.

Selecting the *2point* option at the prompt for the height allows you specify the height by picking two points on the screen.

Selecting the *Axis Endpoint* option at the prompt for the height allows you specify the location of the apex by specifying a point on the screen or typing the coordinates.

Selecting the *Top Radius* option at the prompt for the height causes the subsequent specifications for the size of the plane to become the top of the pyramid, followed by prompts for the size of the base. Using this option permits you to create a truncated pyramid.

Wedge. The WEDGE command creates a solid like a box that has been a cut in half diagonally along one face. The face of the wedge is always drawn parallel to the current UCS, with the sloped face tapering along the Z axis. The solid wedge can be drawn by providing a center point of the base or by providing starting corner of the box.

Invoke the WEDGE command from the Solid Primitives drop-down menu of the Modeling panel. AutoCAD displays the following prompt:

> Specify first corner of wedge or ⊡ *(specify a point, or choose* CENTER OF THE WEDGE *from the shortcut menu)*

By default you are prompted for the starting corner of the box. Once you provide the starting corner, AutoCAD displays the following prompt:

> Specify corner or ⊡ *(specify the corner of the wedge or select one of the available options)*

The wedge dimensions can be specified by using one of the three options. The default option lets you create a wedge by dragging the cursor to the opposite corner of its base rectangle and then dragging the cursor in the Z direction to specify its height. An alternative to dragging the cursor is to type in the relative or absolute coordinates for the opposite corner of the base and for the height. The CUBE option allows you to create a wedge in which all edges are of equal length. The LENGTH option lets you create a box by defining its length, width, and height.

The *Center* option allows you to create a wedge by first locating its center point. Once you locate the center point, a line rubber-bands from this point to help you visualize the size of the rectangle. AutoCAD prompts you to define the size of the box by entering one of the following options:

> Specify corner or ⊡ *(specify corner of the wedge or select one of the options)*

Torus. The TORUS command creates a solid with a donut-like shape. If a torus were a wheel, the center point would be the hub. The torus is created lying parallel to and bisected by the XY plane of the current UCS.

Invoke TORUS from the Solid Primitives drop-down menu of the Modeling panel. AutoCAD displays the following prompt:

> Specify center of torus <0,0,0>: *(specify a point)*

AutoCAD prompts for the center point of the torus and subsequently for the diameter or radius of the torus and the diameter or radius of the tube (see Figure 16–66). You can also draw a torus without a center hole if the radius of the tube is defined as greater than the radius of the torus. A negative torus radius will create a football-shaped solid if the tube diameter is greater than the absolute value of the specified radius.

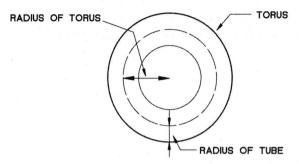

FIGURE 16–66 *Creating a solid torus with a center hole using the TORUS command*

For example, the following command sequence lists the steps for drawing a torus (see Figure 16–67):

```
torus
Specify center of torus <0,0,0>: 5,5
Specify radius of torus or ⏎ 3
Specify radius of tube or ⏎ 0.5
```

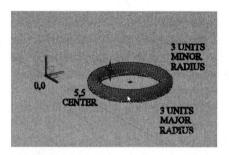

FIGURE 16–67 *Creating a torus by specifying the base plane and central axis direction using the TORUS command*

Creating a Polysolid

The POLYSOLID command creates a solid shape with preset width and height along a path that is created in a manner similar to creating a polyline. The polysolid is useful for creating walls in civil and architectural applications. The POLYSOLID command can be used to convert an existing line, 2D polyline, arc, or circle into a polysolid. Even though the polysolid might have curved segments, the profile will always be rectangular by default.

Invoke the POLYSOLID command from the Modeling panel. AutoCAD displays the following prompt:

```
Start point or ⏎ (specify a point, or choose an option from
the shortcut menu)
```

By default, you are prompted for the starting corner of the polysolid.

Selecting the *Object* option allows you to select a line, polyline, arc, or circle and convert it into a polysolid with the current width and height.

Selecting the *Height* option allows you to specify a new height.

Selecting the *Width* option allows you to specify a new width.

Selecting the *Justify* option allows you to choose between left, center, or right for the location of the polysolid in relation to the object selected or the new polysolid line drawn.

The following command sequence defines a polysolid using the second option (see Figure 16–68):

Polysolid
Specify start point or ⬇ 3,3
Specify next point or ⬇ @3,0
Specify next point or ⬇ @0,3
Specify next point or ⬇ (Enter)

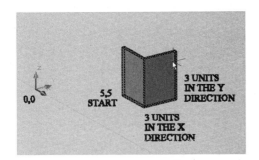

FIGURE 16–68 *Creating a polysolid*

Solids from Existing 2D Objects

AutoCAD has commands for creating 3D solid objects from existing 3D objects. These include the EXTRUDE, LOFT, REVOLVE, and SWEEP commands.

Solid Extrude. The EXTRUDE command creates a unique solid by extruding circles, closed polylines, polygons, ellipses, closed splines, donuts, and regions. Because a polyline can have virtually any shape, the EXTRUDE command allows you to create irregular shapes. In addition, AutoCAD allows you to taper the sides of the extrusion.

> **NOTE**
>
> A polyline must contain at least 3 but not more than 500 vertices, and none of the segments can cross each other. See Figure 16–69 for examples of shapes that cannot be extruded. If the polyline has width, AutoCAD ignores the width and extrudes from the center of the polyline path. If a selected object has thickness, AutoCAD ignores the thickness.

EXAMPLES OF SHAPES THAT CANNOT BE EXTRUDED (SHOWN IN PLAN VIEW)

FIGURE 16–69 *Shapes, shown in plan view, that cannot be extruded using the EXTRUDE command*

Invoke the EXTRUDE command from the Solid Creation drop-down menu of the Modeling panel (see Figure 16–70) on the Home tab.

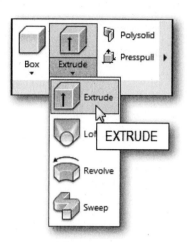

FIGURE 16–70 *Invoking the EXTRUDE command from the Modeling panel*

AutoCAD displays the following prompts:

> Select objects: *(select the objects to extrude and press* Enter *to complete the selection)*
>
> Specify height of extrusion or ⊕ *(specify height of extrusion or choose* PATH *from the shortcut menu)*

The height of extrusion is the distance for extrusion along the positive side of the Z axis of the current UCS. A negative value causes the object to be extruded along the negative axis.

The *Path* option allows you to select the extrusion path based on a specified curve object. All the profiles of the selected object are extruded along the chosen path to create solids. Lines, circles, arcs, ellipses, elliptical arcs, polylines, or splines can be paths. The path should not lie on the same plane as the profile, nor should it have areas of high curvature. The extruded solid starts from the plane of the profile and ends on a plane perpendicular to the path's endpoint. One of the endpoints of the path should be on the plane of the profile. Otherwise, AutoCAD moves the path to the center of the profile.

Once you specify the height of extrusion and path appropriately, AutoCAD displays the following prompt:

> Specify angle of taper for extrusion <0>: *(specify the angle)*

Specify an angle between −90° and +90°, or press Enter or SPACEBAR to accept the default value of 0°. If you specify 0° as the taper angle, AutoCAD extrudes a 2D object perpendicular to its 2D plane (see Figure 16–71). Positive angles taper in from the base object; negative angles taper out.

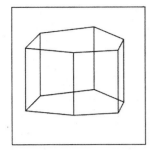 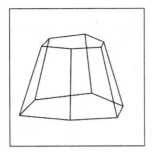

TAPER ANGLE 0 DEGREES   TAPER ANGLE 15 DEGREES

FIGURE 16–71 *Creating a solid with the EXTRUDE command with 0° and with 15° of taper angle*

> It is possible for a large taper angle or a long extrusion height to cause the object, or portions of the object, to taper to a point before reaching the extrusion height.

NOTE

Solid Loft. The LOFT command creates a new solid or surface by lofting, or drawing a solid or surface, through a set of two or more cross-section curves. The cross sections define the profile (shape) of the resulting solid or surface. The curves or lines that define the cross sections can be closed or open. The LOFT command allows you to draw a solid or surface in the space between the cross sections. At least two cross sections must be specified. Using closed objects for cross sections produces a solid. Using open objects produces a surface. Invoke the LOFT command from the Solid Creation drop-down menu of the Modeling panel.

Figure 16–72 shows a series of polylines to be used in creating a lofted solid. Figure 16–73 shows the result of applying the LOFT command by selecting, when prompted, the polylines from the bottom up.

AutoCAD displays the following prompts:

```
loft
Select cross sections in lofting order: (select bottom
polyline)
Select cross sections in lofting order: (select second
polyline)
Select cross sections in lofting order: (select third
polyline)
Select cross sections in lofting order: (select fourth
polyline)
Select cross sections in lofting order: ( Enter )
```

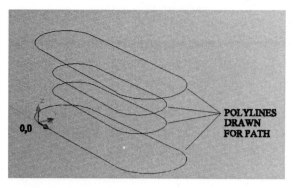

FIGURE 16–72 *Polyline paths to be used by the LOFT command*

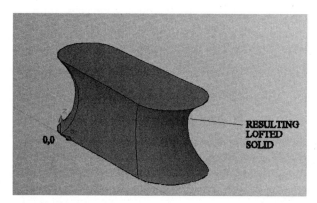

FIGURE 16–73 *Resulting solid created by lofting through the path defined by the polylines*

After pressing Enter to accept the resulting lofted solid, AutoCAD displays an option menu (see Figure 16–74).

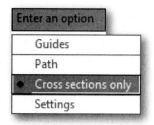

FIGURE 16–74 *Options menu*

The *Guides* option allows you to specify guide curves that control the shape of the lofted solid or surface. Guide curves are lines or curves that further define the form of the solid or surface by adding additional wireframe information to the object.

The *Path* option allows you to specify a single path for the lofted solid or surface.

The *Cross-Section Only* option creates lofted objects without using guides or paths.

The *Settings* option causes the Loft Settings dialog box to be displayed (see Figure 16–75).

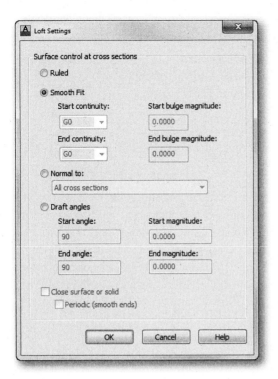

FIGURE 16–75 *Loft Settings dialog box*

The **Surface control at cross sections** section controls the contour of a lofted surface at its cross sections. It also allows you to close the surface or solid.

Choose **Ruled** to specify that the solid or surface is ruled (straight) between the cross sections and has sharp edges at the cross sections.

Choose **Smooth Fit** to specify that a smooth solid or surface is drawn between the cross sections and has sharp edges at the start and end cross sections. The **Start Continuity** text box lets you set the tangency and curvature of the first cross section. The **End Continuity** text box lets you set the tangency and curvature of the last cross section. The **Start Bulge Magnitude** lets you set the size of the curve of the first cross section. The **End Bulge Magnitude** lets you set the size of the curve of the last cross section.

Choose **Normal to** to control the surface normal of the solid or surface where it passes through the cross sections. Choosing START CROSS SECTION specifies that the surface normal is normal to the start cross section. Choosing END CROSS SECTION specifies that the surface normal is normal to the end cross section. Choosing START AND END CROSS SECTIONS specifies that the surface normal is normal to both the start and end cross sections. Choosing ALL CROSS SECTIONS specifies that the surface normal is normal to all cross sections.

Choose **Draft angles** allows you to control the draft angle of the first and last cross sections of the lofted solid or surface. The draft angle is the beginning direction of the surface. Zero is outward from the curve (cross section) on the plane of the curve. A value between 1 and 180 is inward toward the center of the curve of the surface. The Start textbox allows you to specify the draft angle for the start cross section. The Magnitude (Start) textbox allows you to control the relative distance of the surface from the start cross section in the direction of the draft angle before the surface starts to bend toward the next cross section. The End angle textbox allows you to specify the draft angle for the end cross section. The MAGNITUDE (END) controls the relative distance of the surface from the end cross section in the direction of the draft angle before the surface starts to bend toward the previous cross section.

Choosing **Close surface or solid** closes and opens a surface or solid.

Choosing **Preview changes** applies the current settings to the lofted solid or surface and displays a preview in the drawing area.

Solid Revolve. The REVOLVE command creates a unique solid by revolving or sweeping a closed polyline, polygon, circle, ellipse, closed spline, donut, and region. Polylines that have crossing or self-intersecting segments cannot be revolved. The REVOLVE command is similar to the REVSURF command. The REVSURF command creates a surface of revolution, whereas REVOLVE creates a solid of revolution. The REVOLVE command provides several options for defining the axis of revolution. Invoke the REVOLVE command from the Solid Creation drop-down menu of the Modeling panel.

AutoCAD displays the following prompts:

> Select objects: *(select the objects to revolve and press* ⊡ *to complete the selection)*
>
> Specify start point for axis of revolution or define axis by ⊡ *(specify start point for axis of revolution, or select one of the options from the shortcut menu)*

The **Start point of axis option** (default) allows you to specify two points for the start point and the endpoint of the axis, and the positive direction of rotation is based on the right-hand rule.

The *Object* option allows you select an existing line or single polyline segment that defines the axis about which to revolve the object. The positive axis direction is from the closest to the farthest endpoint of this line.

The *X Axis* option uses the positive X axis of the current UCS as the axis of the revolution.

The *Y Axis* option uses the positive Y axis of the current UCS as the axis of the revolution.

Once you specify the axis of revolution, AutoCAD displays the following prompt:

> Specify angle of revolution <360>: *(Specify the angle for revolution)*

The default angle of revolution is a full circle. You can specify any angle between 0 and 360°.

Solid Sweep. The SWEEP command creates a new solid or surface by sweeping an open or closed planar curve (profile) along an open or closed 2D or 3D path. Polylines that have crossing or self-intersecting segments cannot be revolved. The solid or surface created by the SWEEP command is in the shape of the specified profile, or the swept object, along the specified path. More than one object can be swept, but all must lie on the same plane. Invoke the SWEEP command from the Solid Creation drop-down menu of the Modeling panel.

AutoCAD displays the following prompts:

```
sweep
Select objects to sweep: (select the circle)
Select objects to sweep: (Enter)
Select sweep path or ⬇ (select polyline path)
```

The *Alignment* option is the default and causes the profile to be aligned normal to the tangent direction of the sweep path. AutoCAD displays the following prompt:

```
Align sweep object perpendicular to path before sweep [Yes/
No] <Yes>:
```

Press Enter to specify that the profile is aligned, or enter **No** to specify that the profile is not be aligned.

The *Base Point* option allows you to specify a base point for the objects to be swept.

The *Scale* option allows you to specify a scale factor for the sweep operation. The scale factor is uniformly applied to the objects that are swept. AutoCAD displays the following prompt:

```
Enter scale factor or [Reference] <1.0000>:
```

Press Enter to specify the default value or specify a scale factor, enter R for the reference option. The reference option allows you to determine the scale with reference to two given values.

The *Twist* option allows you to set a twist angle for the objects being swept. The twist angle determines the amount of rotation along the entire length of the sweep path.

Figure 16–76 shows a circle that is to be swept along the polyline that is the path. The circle is drawn in a plane perpendicular to the start of the polyline, but it may be drawn in a different plane and still operate properly. Figure 16–77 shows the result of applying the SWEEP command; the circle is selected as the object to be swept and the polyline is selected as the path.

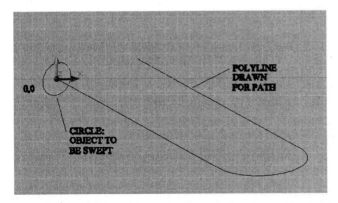

FIGURE 16–76 *Circle object and polyline path to be used by the* SWEEP *command*

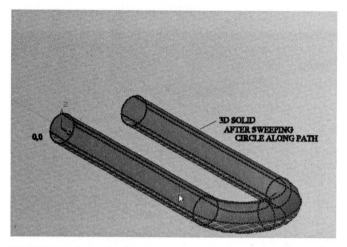

FIGURE 16–77 *Result of sweeping the circle along the polyline path*

Figure 16–78 shows a spiral with 3 turns, top and bottom radii of 2 units, and a height of 4 units created with the HELIX command. Figure 16–79 shows the result of applying the SWEEP command, selecting a 0.05 unit diameter circle for the object to be swept and selecting, the previously created spiral for the sweep path.

AutoCAD displays the following prompts:

```
helix
Specify center point of base: 0,0,0
Specify base radius or [Diameter] <2.0000>: 2
Specify top radius or [Diameter] <2.0000>: 2
Specify helix height or [Axis endpoint/Turns/turn Height/
tWist] <4.0000>: 4
```

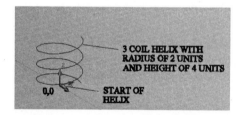

FIGURE 16–78 *Spiral created by the HELIX command*

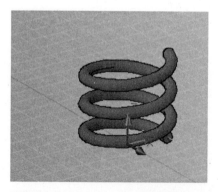

FIGURE 16–79 *Coil spring-like solid created by the SWEEP command applied to the spiral*

Solids from Surfaces with Surfsculpt

The SURFSCULPT command creates a 3D solid from an enclosed volume with no gaps at the edges. The watertight enclosure can be a set of surface objects or mesh objects. Invoke the SURFSCULPT command from the Edit panel of the Surface tab. Two surfaces are selected in Figure 16–80a. Figure 16–80b shows the sculpted solid created as a result.

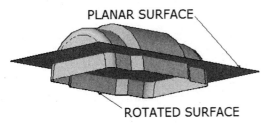

FIGURE 16–80a *Selecting Surfaces*   **FIGURE 16–80b** *Resulting Sculpted Solid*

Creating Composite Solids

As mentioned earlier in this chapter, you can create a new composite solid or region by combining two or more solids or regions via Boolean operations. Although the term Boolean implies that only two objects can be operated upon at once, AutoCAD lets you select many solid objects in a single Boolean command. Three basic Boolean operations can be performed in AutoCAD: union, subtraction, and intersection.

The UNION, SUBTRACT, and INTERSECT commands let you select both the solids and regions in a single use of the commands, but solids are combined only with solids, and regions are combined only with regions. You can make composite regions only with those that lie in the same plane. This means that a single command creates a maximum of one composite solid but might create many composite regions.

Union

Union is the process of creating a new composite object from one or more original objects. The union operation joins the original solids or regions in such a way that there is no duplication of volume. Therefore, the total resulting volume can be equal to or less than the sum of the volumes in the original solids or regions. Invoke the UNION command from the Solid Editing panel of the Home tab or the Boolean panel of the Solid tab and AutoCAD prompts you to select two or more solids.

You can select more than two objects at once. The objects (solids or regions) can be overlapping, adjacent, or nonadjacent.

The following example shows creating a composite solid by selecting and joining cylinders A and B (see Figure 16–81).

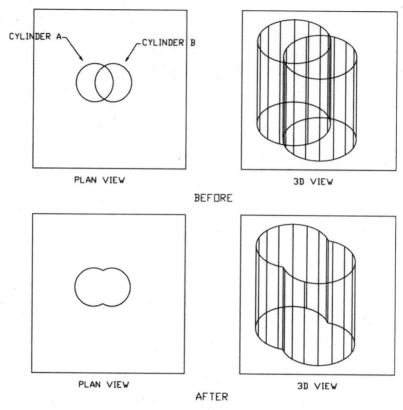

FIGURE 16–81 *Creating a composite solid by joining two cylinders using the* UNION *command*

Subtraction

Subtraction is the process of forming a new composite object by starting with one object and removing from it any volume that it has in common with a second object. In the case of solids, you subtract the volume of one set of solids from another set. If the entire volume of the second solid is contained in the first solid, what is left is the first solid minus the volume of the second solid. However, if only part of the volume of the second solid is contained within the first solid, only the part that is duplicated in the two solids is subtracted. Similarly, in the case of regions, you subtract the common area of one set of existing regions from another set. Invoke the SUBTRACT command from the Solid Editing panel of the Home tab or the Boolean panel of the Solid tab and AutoCAD prompts you to select solids to subtract from. After selection AutoCAD prompts you to select solids to subtract.

If necessary, you can select one or more objects to subtract from the source object. If you select several, they are automatically joined before they are subtracted from the source object.

Objects that are neither solids nor regions are ignored.

The following example shows creating a composite solid by selecting cylinder B to be subtracted from cylinder A (see Figure 16–82).

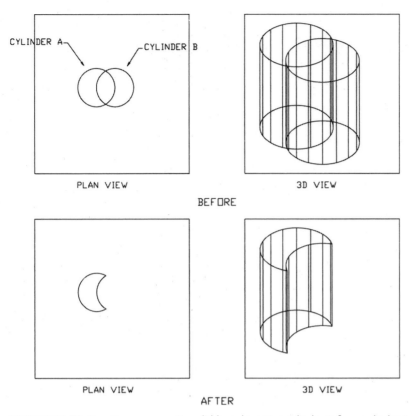

FIGURE 16–82 *Creating a composite solid by subtracting cylinder B from cylinder A using the* SUBTRACT *command*

Intersection

Intersection is the process of forming a composite object from only the volume that is common to two or more original objects. In the case of solids, you can create a new composite solid by calculating the common volume of two or more existing solids. In the case of regions, this step is done by calculating the overlapping area of two or more existing regions. Invoke the INTERSECT command from the Solid Editing panel of the Home tab or the Boolean panel of the Solid tab and AutoCAD prompts you to select solid objects.

The following example shows creating a composite solid by selecting cylinders A and B, resulting in the intersection of the two (see Figure 16–83).

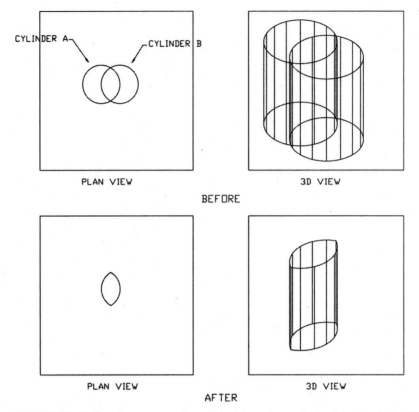

FIGURE 16–83 *Creating a composite solid by intersecting cylinder A from cylinder B using the* `INTERSECT` *command*

Editing Solids

AutoCAD makes the work of creating solids a little easier by providing editing tools, including chamfering or filleting the edges, creating a cross section through a solid, creating a new solid by cutting the existing solid and removing a specified side, and creating a composite solid from the interference of two or more solids. In addition, AutoCAD provides tools such as extrude faces, move faces, offset faces, delete faces, rotate faces, taper faces, color faces, copy faces, color and copy edges, imprint, clean, separate solids, shell, and check. You can always use the AutoCAD modify and construct commands such as `MOVE`, `COPY`, `ROTATE`, `SCALE`, and `ARRAY` to edit solids.

Chamfering Solids

The `CHAMFER` command, explained in Chapter 4, can be used to bevel the edges of an existing solid object. Invoke the `CHAMFER` command from the Solid Editing panel on the Solid tab and AutoCAD prompt you select an edge on a solid. If you pick an edge that is common to two surfaces, AutoCAD highlights one of the surfaces. If this is the surface you want, press [Enter] to accept it. If it is not, enter **N** (for next) to highlight the adjoining surface and then press [Enter]. AutoCAD displays the following prompts for the chamfer distances and then prompts for the base surface and then the other surface.

Once you provide the chamfer distances, AutoCAD prompts for the edge to champher.

The *Loop* option allows you to select one of the edges on the base surface, and AutoCAD automatically selects all edges on the base surface for chamfering.

The following example shows creating chamfered edges by selecting four edges (see Figure 16–84a). The distances were specified as 0.25 and 0.50 for the base and other surfaces, respectively. Figure 16–84b shows the resulting chamfered edges.

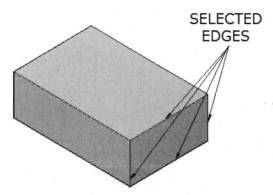

SELECTED EDGES

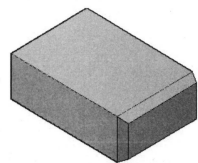

FIGURE 16–84a *Selecting edges for chamfer*

FIGURE 16–84b *Resulting chamfered edges*

The CHAMFEREDGE command can also be used to bevel the edges of an existing solid object and is accessible from the Solid Editing shortcut menu of the Modify menu in the AutoCAD Classic Workspace.

Filleting Solids

The FILLET command, explained in Chapter 4, can be used to round the edge of an existing solid object. Invoke the FILLET command from the Solid Editing panel on the Solid tab and AutoCAD prompts you select an edge on a solid. You can right-click and select *radius* from the shortcut menu to change the radius.

The following example shows creating filleted edges by selecting two edges (see Figure 16–85a). The radius was specified as 0.50. Figure 16–85b shows the resulting filleted edges (see Figure 16–92).

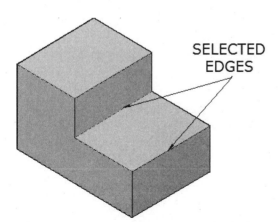

SELECTED EDGES

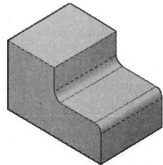

FIGURE 16–85a *Selecting edges for fillet*

FIGURE 16–85b *Resulting filleted edges*

The FILLETEDGE command can also be used to fillet the edges of an existing solid object and is accessible from the Solid Editing shortcut menu of the Modify menu in the AutoCAD Classic Workspace.

Sectioning Solids

The SECTION command creates a cross section of one or more solids. The cross section is created as one or more regions. The region is created on the current layer and is inserted at the location of the cross section. If necessary, you can use the MOVE command to move the cross section. Invoke the SECTION command by typing in **section** at the on-screen prompt and AutoCAD displays the following prompts:

> Select objects: *(select the objects from which you want the cross section to be generated)*
>
> Specify first point on Section plane by ⊡ *(specify one of the three points to define a plane or select one of the options from the shortcut menu)*

The *3points* option (default) allows you to define a section plane by locating three points. The first point is the origin, the second point determines the positive direction of the X axis for the section plane, and the third point determines the positive Y axis of the section plane. This option is similar to the *3point* option of the AutoCAD UCS command.

The *Object* option aligns the sectioning plane with a circle, ellipse, circular or elliptical arc, 2D spline, or 2D polyline segment.

The *Z-axis* option defines the section plane by locating its origin point and a point on the Z axis (normal) to the plane.

The *View* option aligns the section plane with the viewing plane of the current viewport. Specifying a point defines the location of the sectioning plane.

The *XY* option aligns the sectioning plane with the XY plane of the current UCS. Specifying a point defines the location of the sectioning plane.

The *YZ* option aligns the sectioning plane with the XY plane of the current UCS. Specifying a point defines the location of the sectioning plane.

The *ZX* option aligns the sectioning plane with the XY plane of the current UCS. Specifying a point defines the location of the sectioning plane.

Figure 16–86 shows a hatched cross section produced with the SECTION command.

NOTE | The section may be hatched using the hatching techniques described in Chapter 9.

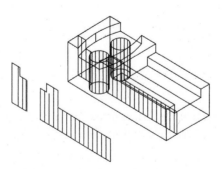

FIGURE 16–86 *Creating a 2D hatched cross section using the* SECTION *command*

The SECTIONPLANE command creates a section object that acts as a cutting plane through solids, surfaces, or regions. Turning ON **live sectioning** allows you to see a section in real-time as the section object moves throughout the 3D model of one or more solids. Invoke the SECTIONPLANE command from the Section panel of the Home tab in the 3D Modeling Workspace.

The SECTIONPLANEJOG command lets you add a jogged segment to a section object. Invoke the SECTIONPLANEJOG command from the Section panel of the Home tab in the 3D Modeling Workspace.

The SECTIONPLANETOBLOCK command lets you save selected section planes as 2D or 3D blocks. Invoke the SECTIONPLANETOBLOCK command from the Section panel of the Home tab in the 3D Modeling Workspace.

The FLATSHOT command lets you create a 2D representation of all 3D objects based on the current view. Invoke the FLATSHOT command from the Section panel of the Home tab in the 3D Modeling Workspace.

Choosing Section on the Section panel of the Home tab causes the Section Settings dialog box to be displayed, from which you can set display options for section planes.

In the Section Plane section choosing **Select Section Plane** temporarily closes the Section Settings dialog box so that you can select a section object in the drawing area.

Choosing **2D Section/Elevation Block Creation Settings** determines how a 2D section from a 3D object is displayed when generated.

Choosing **3D Section Block Creation Settings** determines how a 3D object is displayed when generated.

Choosing **Live Section Settings** determines how sectioned objects are displayed when live sectioning is turned on.

Choosing **Active Live Section** turns on live sectioning for the selected section object (only available when the Live Section Settings option is selected).

In the Properties section you can set the properties to be applied to the new section block.

Under the Intersection boundary window, you can set the Color, Layer, Linetype, Linetype Scale, Plot Style, and Lineweight, and whether division lines are displayed or if the intersection boundary is displayed.

Under the Intersection Fill window you can set the display options for the fill inside the boundary area of the cut surface where the section object intersects the 3D object. These include Face Hatch, Angle, Hatch Scale, Hatch Spacing, Color, Layer, Linetype, Linetype Scale, Plot Style, Lineweight, and Surface Transparency.

Under the Background Lines window you can set the display options for the background lines. These include Hidden Line, Color, Layer, Linetype, Linetype Scale, Plot Style, and Lineweight.

Under the Curve Tangency Lines window you can set the display options for the curved lines that are tangent to the section plane. These include Color, Layer, Linetype, Linetype Scale, and Lineweight.

Under the Cut-away Geometry window you can set the display options for the cut-away objects. These include Color, Layer, Linetype, Linetype Scale, Lineweight, Face Transparency, and Edge Transparency.

Choosing **Apply Settings to All Section Objects** lets you apply all the settings to all section objects in the drawing. When cleared, it applies settings to the current section object only.

Choosing **Reset** resets settings in the dialog box to their default values.

Slicing Solids

The SLICE command allows you to create a new solid by cutting the existing solid and removing a specified portion. If necessary, you can retain both portions of the sliced solid(s) or just the portion you specify. The sliced solids retain the layer and color of the original solids. Invoke the SLICE command from the Solid Editing panel on the Home tab and AutoCAD displays the following prompt:

Select objects: *(select the objects to create a new solid by slicing, and press* Enter *)*

After selecting the objects, press Enter, and AutoCAD prompts you to define the slice plane:

Specify first point on slicing plane by ⬇ *(specify one of the three points to define a plane or select one of the options)*

The options are the same as those for the SECTION command explained earlier in this chapter.

After defining the slicing plane, AutoCAD prompts you to indicate which part of the cut solid is to be retained:

Specify a point on desired side of the plane or ⬇

By default, AutoCAD allows you to select, with your pointing device, the side of the slice that is to be retained in your drawing.

The *Keep Both Sides* option allows you to retain both portions of the sliced solids.

Figure 16–87 shows two parts of a solid model that have been cut using the SLICE command and moved apart using the MOVE command.

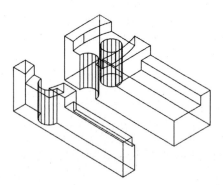

FIGURE 16–87 *Cutting a solid model into two parts using the* SLICE *command*

Solid Interference

The INTERFERE command checks the interference between two or more solids and creates a composite solid from their common volume.

There are two ways to determine the interference between solids:

- Select two sets of solids. AutoCAD determines the interference between the first and second sets of solids.

- Select one set of solids instead of two. AutoCAD determines the interference between all of the solids in the set. They are checked against each other.

Invoke the INTERFERE command from the Solid Editing panel on the Home tab and AutoCAD displays the following prompts:

> Select objects: *(select the first set of solids and press*
> [Enter] *)*
>
> Select objects: *(select the second set of solids or press*
> [Enter] *)*

The second selection set is optional. Press [Enter] if you do not want to define the second selection set. If the same solid is included in both the selection sets, it is considered part of the first selection set and is ignored in the second selection set. AutoCAD highlights all interfering solids and prompts:

> Create interference solids? *(choose an option)*

Entering **y** creates and highlights a new solid on the current layer that is the intersection of the interfering solids. If there are more than two interfering solids, AutoCAD displays the following prompt:

> Highlight pairs of interfering solids? *(choose an option)*

If you specify **y** for yes, and if there is more than one interfering pair, AutoCAD displays the following prompt:

> Enter an option *(specify* x *or* n *)*

Pressing [Enter] cycles through the interfering pairs of solids, and AutoCAD highlights each interfering pair of solids. Enter **x** to complete the command sequence.

Editing Faces of 3D Solids

AutoCAD allows you to edit faces of solid objects with the following options to the SOLIDEDIT command: *Extrude Faces, Copy Faces, Offset Faces, Move Faces, Rotate Faces, Taper Faces, Color Faces,* and *Delete Faces.* These options to the SOLIDEDIT command are accessible in the Solid Editing panel of the Home tab in the 3D Modeling Workspace.

Extruding Faces. AutoCAD allows you to extrude selected faces of a 3D solid object to a specified height or along a path. Specifying a positive value extrudes the selected face in its positive direction, usually outward. A negative value extrudes in the negative direction, usually inward. Tapering the selected face with a positive angle tapers the face inward, and a negative angle tapers the face outward. Tapering the selected face to 0° extrudes the face perpendicular to its plane.

Face extrusion along a path is based on a path curve such as lines, circles, arcs, ellipses, elliptical arcs, polylines, or splines. Invoke *Extrude Faces* from the Faces Editing dropdown menu of the Solid Editing panel (see Figure 16–88). Note that all of the face editing choices are options of the SOLIDEDIT command.

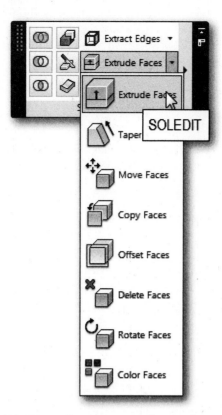

FIGURE 16–88 *Invoking Extrude Faces from the Faces Editing drop-down menu of the Solid Editing panel*

AutoCAD displays the following prompts:

> Select faces or ⊡ *(select faces and press* **Enter** *to complete the selection)*
>
> Specify height of extrusion or ⊡ *(specify height of extrusion or select path option to extrude along a path)*
>
> Specify angle of taper for extrusion <0>: *(press* **Enter** *to accept the default angle or specify the angle for taper for extrusion)*
>
> Enter a face editing option *(select the* EXIT *option to exit the face editing)*
>
> Enter a solids editing option *(select the* EXIT *option to exit solids editing)*

Figure 16–89 shows an example of a solid model in which one of the faces is extruded by a positive value with a 15° tapered angle with the *Extrude Faces* option of the SOLIDEDIT command.

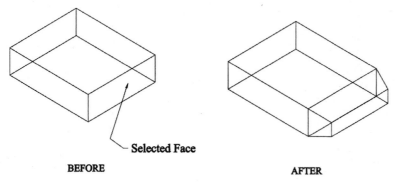

Selected Face

BEFORE   AFTER

FIGURE 16–89 *An example in extruding one of the faces in a solid model*

Copying Faces. The *Copy Faces* option of the SOLIDEDIT command allows you to copy selected faces of a 3D solid object. AutoCAD copies selected faces as regions or bodies. Prompts are similar to the regular COPY command. Invoke *Copy Faces* from the Faces Editing drop-down menu of the Solid Editing panel and AutoCAD displays the following prompts:

> Select faces or ⬇ *(select faces and press* **Enter** *to complete the selection)*
>
> Specify a base point or displacement: *(specify a base point)*
>
> Specify a second point of displacement: *(specify a second point of displacement, or press* **Enter** *to consider the original selection point as a base point)*
>
> Enter a face editing option *(select the* EXIT *option to exit the face editing)*
>
> Enter a solids editing option *(select the* EXIT *option to exit solids editing)*

Figure 16–90 shows an example of a solid model in which one of the faces is copied by the *Copy Faces* option of the SOLIDEDIT command.

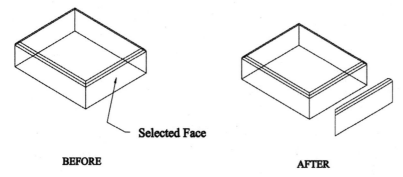

Selected Face

BEFORE   AFTER

FIGURE 16–90 *An example of copying one of the faces in a solid model*

Offsetting Faces. AutoCAD allows you to uniformly offset selected faces of a 3D solid object by a specified distance. New faces are created by offsetting existing ones inside or outside at a specified distance from their original positions. Specifying a positive value increases the size or volume of the solid; a negative value decreases the size or volume of the solid. Invoke *Offset Faces* from the Faces Editing drop-down menu of the Solid Editing panel and AutoCAD displays the following prompts:

```
Select faces or ⬇ (select faces and press Enter to complete
the selection)
Specify the offset distance: (specify the offset distance)
Enter a face editing option (select the EXIT option to exit
the face editing)
Enter a solids editing option (select the EXIT option to exit
solids editing)
```

Figure 16–91 shows an example of a solid model in which one of the faces is offset (positive value) by the *Offset Faces* option of the SOLIDEDIT command.

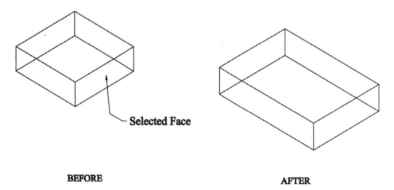

BEFORE   AFTER

FIGURE 16–91 *An example of offsetting one of the faces (positive value) in a solid model*

Moving Faces. The *Move Faces* option of the SOLIDEDIT command allows you to move selected faces of a 3D solid object. You can move holes from one location to another location in a 3D solid. Prompts are similar to the regular MOVE command. Invoke *Move Faces* from the Faces Editing drop-down menu of the Solid Editing panel and AutoCAD displays the following prompts:

```
Select faces or Enter (select faces and press Enter to
complete the selection)
Specify a base point or displacement: (specify a base point)
Specify a second point of displacement: (specify a second
point of displacement, or press Enter to consider the
original selection point as a base point)
Enter a face editing option (select the EXIT option to exit
the face editing)
Enter a solids editing option (select the EXIT option to exit
solids editing)
```

Figure 16–92 shows an example of a solid model in which an elliptical cylinder is moved by the *Move Faces* option of the SOLIDEDIT command.

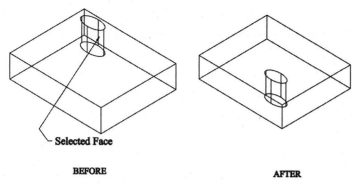

BEFORE   AFTER

FIGURE 16–92 *An example of moving the elliptical cylinder in a solid model*

Rotating Faces. The *Rotate Faces* option of the SOLIDEDIT command allows you to rotate selected faces of a 3D solid object by choosing a base point to relative or absolute angle. All 3D faces rotate about a specified axis. Invoke *Rotate Faces* from Faces Editing drop-down menu of the Solid Editing panel and AutoCAD displays the following prompts:

> Select faces or ⬇ (*select faces and press* [Enter] *to complete the selection*)
>
> Specify an axis point or ⬇ (*specify an axis point or select one of the available options*)
>
> Enter a face editing option (*select the* EXIT *option to exit the face editing*)
>
> Enter a solids editing option (*select the* EXIT *option to exit solids editing*)

Figure 16–93 shows an example of a solid model in which elliptical cylinder is rotated by the *Rotate Faces* option of the SOLIDEDIT command.

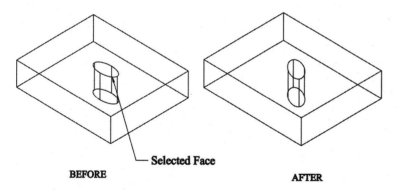

BEFORE   AFTER

FIGURE 16–93 *An example of rotating elliptical cylinder of a solid model*

Tapering Faces. The SOLIDEDIT command allows you to taper selected faces of a 3D solid object with a draft angle along a vector direction. Tapering the selected face with a positive angle tapers the face inward, and a negative angle tapers the face outward. Invoke *Taper Faces* from Faces Editing drop-down menu of the Solid Editing panel and AutoCAD displays the following prompts:

> Select faces or ⬇ (*select faces and press* [Enter] *to complete the selection*)

```
Specify the base point: (specify the base point)
Specify another point along the axis of tapering: (specify a
point to define the axis of tapering)
Specify the taper angle: (specify the taper angle and press
[Enter] to continue)
Enter a face editing option (select the EXIT option to exit
the face editing)
Enter a solids editing option (select the EXIT option to exit
solids editing)
```

Figure 16–94 shows an example of a solid model in which the cylinder tapered angle is changed by the *Taper Faces* option of the SOLIDEDIT command.

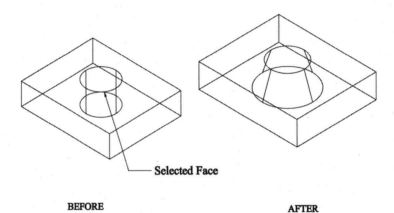

Selected Face

BEFORE AFTER

FIGURE 16–94 *An example of change in a tapered angle of the cylinder in a solid model*

Coloring Faces. The *Color Faces* option of the SOLIDEDIT command allows you to change the color of selected faces of a 3D solid object. You can choose a color from the Select Color dialog box. Setting a color on a face overrides the color setting for the layer on which the solid object resides. Invoke *Color Faces* from the Faces Editing drop-down menu of the Solid Editing panel and AutoCAD displays the following prompt:

```
Select faces or ⬇ (select faces and press [Enter] to complete
the selection)
```

AutoCAD displays the Select Color dialog box. Select the color to change for selected faces and choose the **OK** button. AutoCAD displays the following prompts:

```
Enter a face editing option (select the EXIT option to exit
the face editing)
Enter a solids editing option (select the EXIT option to exit
solids editing)
```

AutoCAD changes the color of the selected faces of the 3D solid model.

Deleting Faces. The *Delete Faces* option of the SOLIDEDIT command allows you to delete selected faces, holes, and fillets of a 3D solid object. Invoke *Delete Faces* from Faces Editing drop-down menu of the Solid Editing panel and AutoCAD displays the following prompts:

```
Select faces or ⬇ (select faces and press [Enter] to complete
the selection)
```

> If the object is a simple box, AutoCAD responds with the message, "Modeling Operation Error—Gap cannot be filled." This informs you that there is no way to join the opposite sides to fill the gap.

```
Enter a face editing option (select the EXIT option to exit
the face editing)
Enter a solids editing option (select the EXIT option to exit
solids editing)
```

AutoCAD deletes the selected faces of the 3D solid model.

Editing Edges of 3D Solids

AutoCAD allows you to copy individual edges and change color of edges on a 3D solid object. The edges are copied as lines, arcs, circles, ellipses, or spline objects. Functions available as options to the SOLIDEDIT command are *Copy Edges*, *Color Edges*, and *Imprint*. Also available is the XEDGES command to Extract Edges. These options are accessible in the Solid Editing panel of the Home tab in the 3D Modeling Workspace.

Copying Edges. The *Copy Edges* option of the SOLIDEDIT command allows you to copy selected edges of a 3D solid object. AutoCAD copies selected edges as lines, arcs, circles, ellipses, or splines. Prompts are similar to the regular COPY command.

Invoke *Copy Edges* from the Edges Editing drop-down menu of the Solid Editing panel (see Figure 16–95) on the Home tab. Note that the *Color Edges* and *Copy Edges* choices are options of the SOLIDEDIT command.

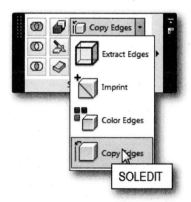

FIGURE 16–95 *Invoking Copy Edges from the Edges Editing drop-down menu of the Solid Editing panel*

AutoCAD displays the following prompts:

```
Select edges or ⬇ (select edges and press Enter to complete
the selection)
Specify a base point or displacement: (specify a base point)
Specify a second point of displacement: (specify a second
point of displacement or press Enter to consider the original
selection point as a base point)
```

> Enter an edge editing option *(select the* EXIT *option to exit the edge editing)*
>
> Enter a solids editing option *(select the* EXIT *option to exit solids editing)*

Figure 16–96 shows an example of a solid model in which the edges, at the right side of the model, are copied by the *Copy Edges* option of the SOLIDEDIT command.

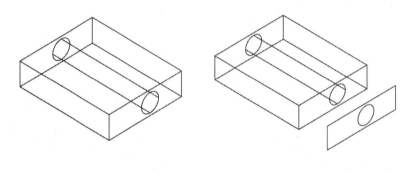

BEFORE   AFTER

FIGURE 16–96 *An example of copying edges of a solid model*

Coloring Edges. The *Color Edges* option of the SOLIDEDIT command allows you to change color of selected edges of a 3D solid object. You can choose a color from the Select Color dialog box. Setting a color on an edge overrides the color setting for the layer on which the solid object resides.

Invoke *Color Edges* from the Edges Editing drop-down menu of the Solid Editing panel and AutoCAD displays the following prompt:

> Select edges or ⬇ *(select faces, and press* Enter *to complete the selection)*

AutoCAD displays the Select Color dialog box. Select the color to change for selected faces and choose **OK**. AutoCAD displays the following prompts:

> Enter an edge editing option *(select the* EXIT *option to exit the edge editing)*
>
> Enter a solids editing option *(select the* EXIT *option to exit solids editing)*

AutoCAD changes the color of the selected edges of the 3D solid model.

Imprinting Edges. AutoCAD allows you to have an imprint of an object on the selected solid. The object to be imprinted must intersect one or more faces on the selected solid in order for imprinting to be successful. Imprinting is limited to the following objects: arcs, circles, lines, 2D and 3D polylines, ellipses, splines, regions, bodies, and 3D solids.

Invoke *Imprint* from the Edges Editing drop-down menu of the Solid Editing panel and AutoCAD displays the following prompts:

> Select a 3D solid: *(select a 3D solid object)*
>
> Select an object to imprint: *(select an object to imprint)*

Delete the source object *(press* Enter *to delete the source objects, or enter y to keep the source object)*

Select an object to imprint: *(select another object to imprint, or press* Enter *to complete the selection)*

Enter a body editing option *(select the* EXIT *option to exit body editing)*

Enter a solids editing option *(select the* EXIT *option to exit solids editing)*

AutoCAD creates an imprint of the selected object.

Extracting Edges. AutoCAD allows you to create wireframe geometry from the edges of a 3D solid, surface, mesh, region, or subobject.

Invoke XEDGES from the Edges Editing drop-down menu of the Solid Editing panel and AutoCAD displays the following prompts:

Select edges: *(select objects from which to extract wireframe geometry and press* Enter *)*

AutoCAD creates wireframe geometry from the selected edges.

Editing Bodies of 3D Solids

AutoCAD allows you to edit bodies of solid objects with the following options to the SOLIDEDIT command: *Separate, Clean, Shell,* and *Check.* These options to the SOLIDEDIT command are accessible in the Solid Editing panel of the Home tab in the 3D Modeling Workspace.

Separating Solids. AutoCAD separates solids from a composite solid, but it cannot separate solids if the composite 3D solid object shares a common area or volume. After separation of the 3D solid, the individual solids retain the layers and colors of the original. Invoke *Separate* from the Body Editing drop-down menu of the Solid Editing panel (see Figure 16–97) on the Home tab. Note that all of the body editing choices are options of the SOLIDEDIT command.

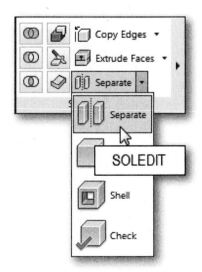

FIGURE 16–97 *Invoking* SEPARATE *from the Faces Editing drop-down menu of the Solid Editing panel*

AutoCAD displays the following prompts:

    Select a 3D solid: *(select a 3D solid object, and press* Enter
    to complete the selection)

    Enter a body-editing option *(select the* EXIT *option to exit
    body editing)*

    Enter a solids-editing option *(select the* EXIT *option to exit
    solids editing)*

AutoCAD separates the selected composite solid.

Shelling Solids. AutoCAD creates a shell or a hollow, thin wall with a specified thickness from the selected 3D solid object. AutoCAD creates new faces by offsetting existing ones inside or outside their original positions. AutoCAD treats continuously tangent faces as single faces when offsetting. A positive offset value creates a shell in the positive face direction; a negative value creates a shell in the negative face direction. Invoke *Shell* from the Faces Editing drop-down menu of the Solid Editing panel and AutoCAD displays the following prompts:

    Select a 3D solid: *(select a 3D solid object)*

    Remove faces or ⤓ *(select faces to be excluded from shelling,
    and press* Enter *to complete the selection)*

    Specify the shell offset value: *(specify the shell offset
    value)*

    Enter a body-editing option *(select the* EXIT *option to exit
    body editing)*

    Enter a solids-editing option *(select the* EXIT *option to exit
    solids editing)*

AutoCAD creates a shell with the specified thickness from the 3D solid object.

Cleaning Solids. AutoCAD allows you to remove edges or vertices if they share the same surface or vertex definition on either side of the edge or vertex. All redundant edges, imprinted as well as used, on the selected 3D solid object are deleted. Invoke *Clean* from the Faces Editing drop-down menu of the Solid Editing panel and AutoCAD displays the following prompts:

    Select a 3D solid: *(select a 3D solid object and press* Enter *to
    complete the selection)*

    Enter a body-editing option *(select the* EXIT *option to exit
    body editing)*

    Enter a solids-editing option *(select the* EXIT *option to exit
    solids editing)*

AutoCAD removes the selected edges or vertices of the selected 3D model.

Checking Solids. AutoCAD checks to see if the selected solid object is a valid 3D solid object. With a 3D solid model, you can modify the object without incurring ACIS failure error messages. If the selected solid 3D model is not valid, you cannot edit the object. Invoke *Check* from the Body Editing drop-down menu of the Solid Editing panel and AutoCAD displays the following prompts:

> Select a 3D solid: *(select a 3D solid object, and press* Enter *to complete the selection)*
>
> Enter a body-editing option *(select the* EXIT *option to exit body editing)*
>
> Enter a solids-editing option *(select the* EXIT *option to exit solids editing)*

AutoCAD checks the solid 3D model and displays with appropriate information about the selected solid.

Mesh Modeling/Free Form Design

Meshes provide the 3D modeler a more free-form shaping capability than is available with solids and surfaces. Meshes are made up of subobjects such as vertices, edges, and faces which can be manipulated to create unique shapes from basic geometric objects.

Meshes can be created as one of the basic shapes called primitives in the same way as solid primitives: box, cone, cylinder, pyramid, sphere, torus, or wedge. Therefore, the commands to create mesh primitives will not be covered here. Mesh primitives, however, have surfaces that are divided into facets. These facets with their vertices and edges are what gives meshes their special shaping capability.

Meshes can be created by using commands to create geometry-generated surfaces: RULESURF, REVSURF, TABSURF, and EDGESURF. The differences between these types of meshes depend on the types of objects connecting the surfaces. In addition, AutoCAD also provides commands for creating a polygon mesh: 3DMESH and PFACE. The key to using meshes effectively is to understand the purpose and requirement of each type of mesh and to select the appropriate one for the given condition.

A 3D mesh is a single object. It defines a flat surface or approximates a curved one by placing multiple 3D faces on the surface of an object. It is a series of lines consisting of columns and rows. AutoCAD lets you determine the spacing between rows (M) and columns (N).

AutoCAD offers a set of functions and commands for Mesh modeling and can be accessed on the Mesh tab (see Figure 16–98).

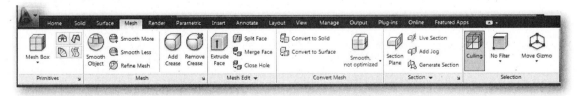

FIGURE 16–98 *Mesh tab*

Mesh models can be modified in ways that are not available for 3D solids or surfaces. For example you can apply creases, splits, and increasing levels of smoothness. You can drag mesh subobjects (faces, edges, and vertices) to reshape the object. To achieve more granular results, you can refine the mesh in specific areas before modifying it. Mesh models provide the hiding, shading, and rendering capabilities of a solid model without the physical properties such as mass and moments of inertia.

Mesh Primitives

AutoCAD allows you to create meshes similar to the primitives created for solid objects with the following options to the MESH command: Box, Cone, Cylinder, Pyramid, Sphere, Wedge, and Torus. Figure 16–99 shows primitives created by using the MESH command. These are options to the MESH command accessible in the Primitives panel (see Figure 16–100) of the Mesh tab in the 3D Modeling Workspace.

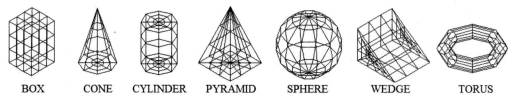

| BOX | CONE | CYLINDER | PYRAMID | SPHERE | WEDGE | TORUS |

FIGURE 16–99 *Primitives created by using the* MESH *command*

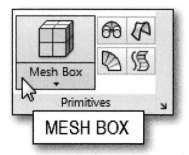

FIGURE 16–100 *Primitives panel*

Mesh objects consist of multiple subdivisions. These are called tessellations and define the editable faces. AutoCAD lets you control the appearance of the mesh primitives by setting the number of tessellations for the various faces of the primitives. Invoke the MESHPRIMITIVEOPTIONS command from the Primitive panel by selecting the down arrow (Figure 16–101).

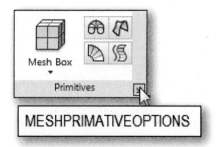

FIGURE 16–101 *Invoking the* MESHPRIMITIVEOPTIONS *command from the Primitives panel*

AutoCAD displays the Mesh Primitive Options dialog box (Figure 16–102).

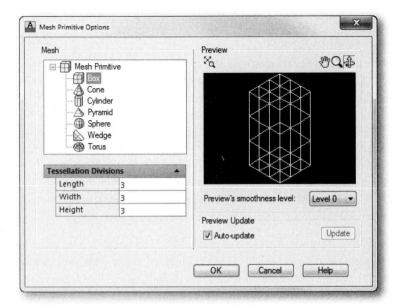

FIGURE 16–102 *Mesh Primitive Options dialog box*

The Mesh section lets you specify the initial mesh density for each type of mesh primitive by specifying the number of divisions per face. From the **Mesh Primitive** list choose the primitive whose tessellations you wish to set. Table 16–2 lists the primitives and the initial settings of the divisions for the sides for each.

TABLE 16–2 *Side Divisions*

Primitive	Length	Width	Height	Axis	Base	Slope	Radius	Sweep Path
Box	3	3	3					
Cone			3	8	3			
Cylinder			3	8	3			
Pyramid	3		3		3			
Sphere			6	12				
Wedge	4	3	3		3	3		
Torus							8	8

The Preview section lets you view the effects of the changes to the settings. Choosing the **Zoom Extents** button causes the image to fill the window. Selecting the **Pan** button lets you pan the image around in the window. Choosing **Zoom** lets you use the scroll wheel to zoom in and out in the window. Choosing the **Orbit** button lets you change the point of view with the 3D Orbit feature. Right-click and AutoCAD displays a shortcut menu from which you can choose viewing tools and Visual Styles.

In the **Preview Smoothness Level** text box you can change the preview image to reflect the level of smoothness you specify without affecting the default level of smoothness for new primitive meshes.

The Preview Update section lets you set how the preview image is updated. Choosing **Auto-update** causes the image to automatically update as you modify the options. If **Auto-update** is cleared you can update the preview with the **Update** button.

Meshes from Existing 2D Objects

Additional mesh objects can be created by using the REVSURF, EDGESURF, RULESURF, and TABSURF commands explained earlier in this chapter. These commands are also accessible in the Primitives panel of the Mesh tab.

Creating a Ruled Surface between Two Objects. The RULESURF command creates a polygon mesh between two objects. The two objects can be lines, points, arcs, circles, 2D polylines, or 3D polylines. If one object, such as a line or arc, is open, the other must be open as well. If one object, such as a circle, is closed, the other must also be closed. A point can be used as one object, regardless of whether the other object is open or closed, but only one of the objects can be a point.

RULESURF creates an M × N mesh: the value of mesh M is a constant 2. The value of mesh N can be changed depending on the required number of faces. This step can be done with the help of the SURFTAB1 system variable. By default, SURFTAB1 is set to 6.

The following command sequence shows how to change the value of SURFTAB1 from 6 to 20:

```
surftab1
Enter new value for SURFTAB1 <6>: 20
```

Invoke the RULESURF command from the Primitives panel of the Mesh tab and AutoCAD displays the following prompts:

```
rulesurf
Select first defining curve: (select the first defining
curve)
Select second defining curve: (select the second defining
curve)
```

Identify the two objects to which a mesh has to be created. See Figure 16–103, in which an arc (A1–A2) and a line (A3–A4) were selected, and a mesh was created with SURFTAB1 set to 16. Two lines (B1–B2 and B3–B4) were selected, and a mesh was created with SURFTAB1 set to 20. A cone was created by drawing a circle at an elevation of 0 and a point (C1) at an elevation of 5, followed by the application of the RULESURF command with SURFTAB1 set to 20.

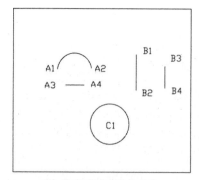 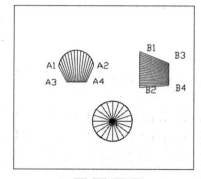

FIGURE 16–103 *Creating ruled surfaces with the RULESURF command*

When you identify the two objects, be sure to select on the same side of the objects, left or right. If you pick the left side of one of the sides and the right side of the other, you will get a "bow tie" effect.

Creating a Tabulated Surface. The TABSURF command creates a surface extrusion from an object with a length and direction determined by the direction vector. The object is called the defining curve, and it can be a line, arc, circle, 2D polyline, or 3D polyline. The direction vector can be a line or open polyline. The endpoint of the direction vector nearest the specified point will be swept along the path curve, describing the surface. Once the mesh is created, the direction vector can be deleted. The number of intervals along the path curve is controlled by the SURFTAB1 system variable, similar to the RULESURF command. By default, SURFTAB1 is set to 6.

Invoke the TABSURF command from the Primitives panel on the Mesh tab, and AutoCAD displays the following prompts:

```
tabsurf
Select object for path curve: (select the path curve)
Select object for direction vector: (select the direction
vector)
```

The location at which the direction vector is selected determines the direction of the constructed mesh. The mesh is created in the direction from the selection point to the nearest endpoint of the direction vector. In Figure 16–104, a mesh was created with SURFTAB1 set to 16 by identifying a polyline as the path curve and the line as the direction vector.

The length of the 3D mesh is the same as that of the direction vector.

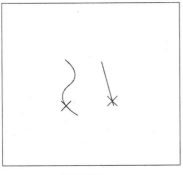
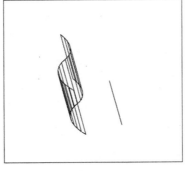

BEFORE   AFTER

FIGURE 16–104 *Creating a tabulated surface with the* TABSURF *command*

Creating a Revolved Surface. The REVSURF command creates a 3D mesh that follows the path defined by a path curve and is rotated around a centerline. The object used to define the path curve may be an arc, circle, line, 2D polyline, or 3D polyline. Complex shapes consisting of lines, arcs, or polylines can be joined into one object using the PEDIT command, and then you can create a single rotated mesh instead of several individual meshes.

The centerline can be a line or polyline that defines the axis around which the faces are constructed. The centerline can be of any length and at any orientation. If necessary, you can erase the centerline after the construction of the mesh. Thus it is recommended that you make the axis longer than the path curve so that it is easy to erase after the rotation.

In the case of REVSURF, both the mesh M size and the mesh N are controlled by the SURFTAB1 and SURFTAB2 system variables, respectively. The SURFTAB1 value determines how many faces are placed around the rotation axis, and it can be an integer value between 3 and 1,024. The SURFTAB2 determines how many faces are used to simulate the curves created by arcs or circles in the path curve. By default, SURFTAB1 and SURFTAB2 are set to 6.

The following command sequence shows how to change the value of SURFTAB1 from 6 to 20 and that of SURFTAB2 from 6 to 15:

```
surftab1
Enter new value for SURFTAB1 <6>: 20
surftab2
Enter new value for SURFTAB1 <6>: 15
```

Invoke the REVSURF command from the Primitives panel on the Mesh tab, and AutoCAD displays the following prompts:

```
revsurf
Select object to revolve: (select the path curve)
Select object that defines the axis of revolution: (select
the axis of revolution)
Specify Start angle<0>: (specify the start angle, or press
Enter to accept the default angle)
Specify Included angle (+=ccw,-=cw)<360>: (specify the
included angle, or press Enter to accept the default)
```

At the "Specify Start angle:" prompt, it does not matter if you are going to rotate the curve a full circle (360°). If you want to rotate the curve only a certain angle, you must provide the start angle in reference to 3:00 (the default) and then indicate the angle of rotation in the CCW (positive) or CW (negative) direction. See Figure 16–105, in which a mesh was created with SURFTAB1 set to 16 and SURFTAB2 set to 12. A polyline was identified as the path curve, the vertical line was identified as the axis of revolution, and the shape was rotated 360°.

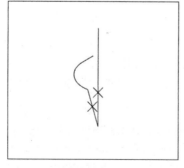 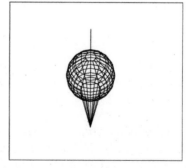

BEFORE   AFTER

FIGURE 16–105 *Creating a meshed surface with the* REVSURF *command*

Creating an Edge Surface with Four Adjoining Sides. The EDGESURF command allows a mesh to be created with four adjoining sides defining its boundaries. The EDGESURF command requires that the mesh have exactly four sides. The sides can be lines, arcs, or any combination of polylines and polyarcs. Each side must join the adjacent one to create a closed boundary.

In EDGESURF, both the mesh M size and mesh N can be controlled by the SURF-TAB1 and SURFTAB2 system variables, respectively, just as in REVSURF.

Invoke the EDGESURF command from the Primitives panel on the Mesh tab, and AutoCAD displays the following prompts:

```
edgesurf
Select edge 1 for surface edge: (select the first edge)
Select edge 2 for surface edge: (select the second edge)
Select edge 3 for surface edge: (select the third edge)
Select edge 4 for surface edge: (select the fourth edge)
```

When picking four sides, you must be consistent in picking the beginning of each polyline group. If you pick the beginning of one side and the end of another, the final mesh will cross and look strange. See Figure 16–106, in which a mesh was created with SURFTAB1 set to 25 and SURFTAB2 set to 20 by identifying four sides.

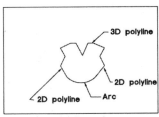

Basic Boundary

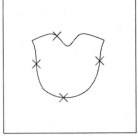

Boundary After Polylines are splined

Model at VPOINT 1,-3,1

FIGURE 16–106 *Creating a meshed surface with the* EDGESURF *command*

Smooth, Refine, and Crease Meshes

AutoCAD provides powerful free form design features that allow you to edit meshes by smoothing and refining them and by using creases. These features are accessible in the Mesh panel of the Mesh tab.

Smoothing and Refining Meshes. With the MESHSMOOTH, MESHSMOOTHMORE, MESHSMOOTHLESS, and MESHREFINE commands you can increase or decrease the roundness of mesh objects by adjusting the smoothness levels. As described above, tessellations define the editable faces of mesh primitives. Each face consists of underlying facets. When you increase smoothness, you increase the number of facets to provide a smoother, more rounded look.

The MESHSMOOTH command converts certain objects to mesh. You can convert 3D solids, surfaces, and legacy mesh objects to enhanced mesh object in order to take advantage of capabilities such as smoothing, refinement, creasing, and splitting. Invoke the MESHSMOOTH command from the Mesh panel (Figure 16–107).

FIGURE 16–107 *Invoking the MESHSMOOTH command from the Primitives panel*

Figure 16–108 shows a 3D Solid Box converted to a smoothed mesh. AutoCAD displays the following prompts:

```
Select objects to convert: (select a 3D solid object, and
press Enter to complete the selection)
```

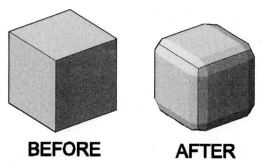

FIGURE 16–108 *Applying the MESHSMOOTH command to a 3D Solid Box*

The MESHSMOOTHMORE command increases the level of smoothness for mesh objects by one level. Invoke the MESHSMOOTHMORE command from the Mesh panel. Figure 16–109 shows the converted 3D Mesh Box after the MESHSMOOTH-MORE is applied. AutoCAD displays the following prompts:

```
Select mesh objects to increase the smoothness level:
(select an object, and press Enter to complete the selection)
```

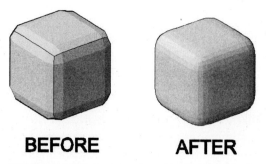

FIGURE 16–109 *Applying the MESHSMOOTHMORE command to a converted Mesh Box*

The MESHSMOOTHLESS command decreases the level of smoothness for mesh objects by one level. Invoke the MESHSMOOTHLESS command from the Mesh panel.

The MESHREFINE command multiplies the number of faces in selected mesh objects or faces. Invoke the MESHREFINE command from the Mesh panel. Figure 16–110 shows the MESHREFINE after being applied to the smoothed 3D Mesh Box. Auto-CAD displays the following prompts:

```
Select mesh object or face subobjects to refine: (select an
object, and press Enter to complete the selection)
```

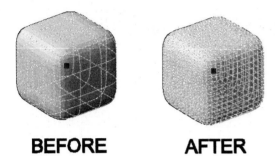

BEFORE   **AFTER**

FIGURE 16–110 *Applying the* MESHREFINE *command to a converted Mesh Box*

Refining a mesh object increases the number of editable faces, providing additional control over fine modeling details. It also consumes more computer memory. To preserve program memory, you can refine specific faces instead of the entire object.

Creasing Mesh Subobjects. With the MESHCREASE command you can sharpen or crease the edges of selected mesh subobjects of 3D Meshes. Invoke the MESHCREASE command from the Mesh panel (Figure 16–111).

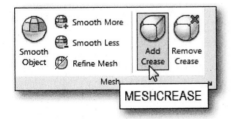

FIGURE 16–111 *Invoking the* MESHCREASE *command from the Mesh panel*

Figure 16–112 shows a 3D Mesh Box where 12 facets have been selected for creasing. AutoCAD displays the following prompts:

```
Select mesh subobjects to crease: (select facets on a 3D mesh
object, and press Enter to complete the selection)
```

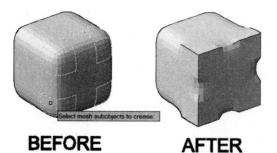

BEFORE   AFTER

FIGURE 16–112 *Applying the MESHCREASE command to a 3D Mesh Box*

Creasing deforms mesh faces and edges that are adjacent to the selected subobject. Creases added to a mesh that has no smoothness are not apparent until the mesh is smoothed. You can also apply creases to mesh subobjects by changing the crease type and crease level in the Properties palette.

With the MESHUNCREASE command you can remove the crease from selected mesh faces, edges, or vertices. Invoke the MESHUNCREASE command from the Mesh panel.

Editing Mesh Faces

AutoCAD provides commands for editing faces of meshes. These features are accessible in the Mesh Edit panel of the Mesh tab.

NOTE Selecting an individual facet from a 3D Solid or 3D Mesh requires that the CTRL key be depressed during the selection.

Splitting and Extruding Mesh Faces

With the MESHSPLIT and MESHEXTRUDE commands you can edit the faces of meshes.

Meshsplit. The MESHSPLIT command allows you to split a mesh face into two faces. You can split a face to add more definition to an area without having to refine it. Because you specify the start and endpoint of the split, this method provides greater control over the location of the split.

Invoke the MESHSPLIT command from the Mesh Edit panel (Figure 16–113).

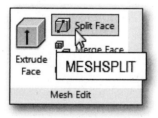

FIGURE 16–113 *Invoking the MESHSPLIT command from the Mesh panel*

Figure 16–114 shows applying the MESHSPLIT command to a 3D Mesh Box by first selecting a face and then specifying the two split points. AutoCAD displays the following prompts:

> Select a mesh face to split: *(hold down the CTRL key and select a face of a 3D mesh object)*

After the face is selected, AutoCAD prompts:

> Specify first split point: *(select the first point on the boundary of the face)*

> Specify second split point: *(select the second point on the boundary of the face)*

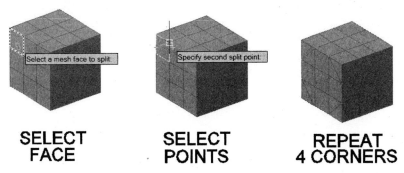

SELECT FACE **SELECT POINTS** **REPEAT 4 CORNERS**

FIGURE 16–114 *Applying the MESHSPLIT command to facets on a 3D Mesh Box*

Meshextrude. The MESHEXTRUDE command extends the dimensions of a 3D face. Invoke the MESHEXTRUDE command from the Mesh Edit panel (Figure 16–115).

FIGURE 16–115 *Invoking the MESHEXTRUDE command from the Mesh panel*

Figure 16–116 shows applying the MESHEXTRUDE command to 9 facets of a 3D Mesh Box by first selecting the faces and then specifying the extrusion distance. AutoCAD displays the following prompts:

> Select objects to extrude: *(select a face on the object)*

> Select objects to extrude: *(press* Enter *to complete the selection)*

> Specify height of extrusion ⊡ *(specify a distance)*

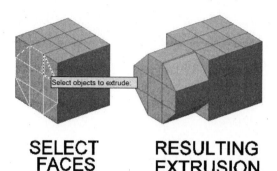

SELECT FACES   RESULTING EXTRUSION

FIGURE 16-116 *Applying the* MESHEXTRUDE *command to a 3D Mesh Box*

The selected faces are extruded for the distance and in the direction specified.

With the MERGE FACE command you can combine adjacent faces to become one face. With the CLOSE command you can close gaps in mesh objects. These features are best used for faces that are on the same plane.

Converting Meshes to Solids and Surfaces

With the CONVTOSOLID and CONVTOSURFACE commands you can change meshes into 3D Solids or into Surfaces.

The CONVTOSOLID command allows you to convert 3D meshes and polylines and circles with thickness to 3D solids. You can then use the solid modeling capabilities available for 3D solids. When you convert a mesh object into a solid, you can specify whether the converted objects are smoothed or faceted and whether the faces are merged.

Sectioning Meshes

The commands AutoCAD provides for sectioning 3D Solid objects (explained elsewhere in this chapter) are also available for 3D Mesh objects. Commands include SECTIONPLANE, LIVESECTION, SECTIONPLANEJOG, SECTIONPLANETO-BLOCK, FLATSHOT, and XEDGES. These commands are accessible in the Section panel.

Filters and Gizmos for Mesh Subobjects

With the SUBOBJSELECTIONMODE system variable AutoCAD lets you filter whether a face, edge, or vertex is selected when using CTRL+click to select a subobject. The SUBOBJSELECTIONMODE system variable is accessible in the Subobject panel.

AutoCAD provides the same Move, Rotate, and Scale Gizmos for 3D Mesh objects that are available for 3D Solid objects explained earlier in this chapter. The Gizmos are accessible in the Subobject panel.

Creating a Free-Form Polygon Mesh

You can define a free-form 3D polygon mesh by means of the 3DMESH command. Initially, the feature prompts you for the number of rows and columns, in terms of mesh M and mesh N, respectively. It then prompts you for the location of each vertex in the mesh. The product of M × N gives the number of vertices for the mesh.

Invoke the 3DMESH command at the on-screen prompt, and AutoCAD displays the following prompts:

> Enter size of mesh in M direction: *(specify an integer value between 2 and 256)*
>
> Enter size of mesh in N direction: *(specify an integer value between 2 and 256)*

The points for each vertex must be entered separately. The M value can be considered the number of lines that will be connected by faces, while the N value is the number of points each line consists of. Vertices may be specified as 2D or 3D points and may be any distance from each other.

The following command sequence creates a simple 5×4 polygon mesh. The mesh is created between the first point of the first line, the first point of the second line, and so on (see Figure 16–117).

```
3dmesh
Enter size of mesh in M direction: 5
Enter size of mesh in N direction: 4
Specify location for vertex (0,0): (select point A1)
Specify location for vertex (0,1): (select point A2)
Specify location for vertex (0,2): (select point A3)
Specify location for vertex (0,3): (select point A4)
Specify location for vertex (1,0): (select point B1)
Specify location for vertex (1,1): (select point B2)
Specify location for vertex (1,2): (select point B3)
Specify location for vertex (1,3): (select point B4)
Specify location for vertex (2,0): (select point C1)
Specify location for vertex (2,1): (select point C2)
Specify location for vertex (2,2): (select point C3)
Specify location for vertex (2,3): (select point C4)
Specify location for vertex (3,0): (select point D1)
Specify location for vertex (3,1): (select point D2)
Specify location for vertex (3,2): (select point D3)
Specify location for vertex (3,3): (select point D4)
Specify location for vertex (4,0): (select point E1)
Specify location for vertex (4,1): (select point E2)
Specify location for vertex (4,2): (select point E3)
Specify location for vertex (4,3): (select point E4)
```

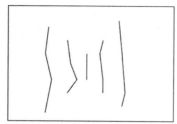

LINE DIAGRAM

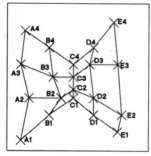

LINE DIAGRAM WITH 3D FACES

MODEL AT VPOINT 1,–1,1

FIGURE 16–117 *Creating a 3D mesh*

NOTE

Specifying a 3D mesh of any size can be time consuming and tedious. For geometry-generated surfaces, these commands are preferred: RULESURF, REVSURF, TABSURF, or EDGESURF. The 3DMESH command is designed primarily for AutoLISP and ADS applications.

Creating a 3D Polyface Mesh

The PFACE command allows you to construct a mesh of any topology you desire. This command is similar to the 3DFACE command, but it creates surfaces with invisible interior divisions. You can specify any number of vertices and 3D faces, unlike the other meshes. Producing this kind of mesh lets you conveniently avoid creating many unrelated 3D faces with the same vertices.

AutoCAD first prompts you to pick all the vertex points, and then you can create the faces by entering the vertex numbers that define their edges.

Invoke the PFACE command from the on-screen prompt, and AutoCAD displays the following prompts:

```
pface
Specify location for vertex 1: (specify a point)
```

Specify all the vertices used in the mesh one by one, keeping track of the vertex numbers shown in the prompts. You can specify the vertices as 2D or 3D points and place them at any distance from one another. Press [Enter] after specifying all the vertices, and AutoCAD prompts for a vertex number that must be assigned to each face. You can define any number of vertices for each face, and then press [Enter]. AutoCAD prompts for the next face. After all the vertex numbers for all the faces are defined, press [Enter], and AutoCAD draws the mesh.

The following command sequence creates a simple polyface for a given six-sided polygon, with a circle of 1 radius drawn at the center of the polygon at a depth of −2 (see Figure 16–118).

pface

Specify location for vertex 1: *(select point A1)*

Specify location for vertex 2 or <define faces>: *(select point A2)*

Specify location for vertex 3 or <define faces>: *(select point A3)*

Specify location for vertex 4 or <define faces>: *(select point A4)*

Specify location for vertex 5 or <define faces>: *(select point A5)*

Specify location for vertex 6 or <define faces>: *(select point A6)*

Specify location for vertex 7 or <define faces>: (Enter)

Face 1, Vertex 1:

Enter a vertex number or ⊥ 1

Face 1, Vertex 2:

Enter a vertex number or ⊥ 2

Face 1, Vertex 3:

Enter a vertex number or ⊥ 1

Face 1, Vertex 4:

Enter a vertex number or ⊥ 2

Face 1, Vertex 5:

Enter a vertex number or ⊥ 1

Face 1, Vertex 6:

Enter a vertex number or ⊥ 2

Face 1, Vertex 7:

Enter a vertex number or ⊥ (Enter)

Face 2, Vertex 1:

Enter a vertex number or ⊥ (Enter)

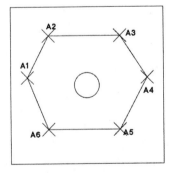

POLYGON WITH PFACE

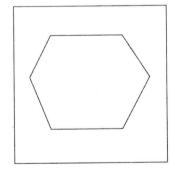

POLYGON WITH CIRCLE HIDDEN

FIGURE 16–118 *Creating a polyface for a given six-sided polygon with a circle at the center*

If necessary, you can make an edge of the polyface mesh invisible by entering a negative number for the beginning vertex of the edge. By default, the faces are drawn on the current layer and with the current color. However, you can create the faces in layers and colors different from the original object. You can assign a layer or color by responding to the "Enter a vertex number or ⬇" prompt with **L** for layer or **C** for color. AutoCAD then prompts for the name of the layer or color, as appropriate. It will continue to prompt for vertex numbers. The layer or color you enter is used for the face that you are currently defining and for any subsequent faces created.

Editing Polymesh Surfaces

As with blocks, polylines, hatch, and dimensioning, you can explode a mesh. When you explode a mesh, it separates into individual 3D faces. Meshes can also be altered by invoking the PEDIT command, similar to editing polylines using the PEDIT command. Most of the options under the PEDIT command can be applied to meshes except when giving width to the edges of the polymesh. For a detailed explanation of the PEDIT command, refer to Chapter 5.

Surface Modeling

Surface modeling creates a thin (zero thickness) shell that can be flat (in one plane) or be a simple or compound curve. Surfaces do not have mass or volume. There are two types of surfaces; procedural and NURBS and they can be planar or network.

NURBS surfaces are not associative, but are 3D equivalents to splines with control vertices that let you define a more natural curvature to surfaces. Control vertices are only available with NURBS surfaces.

AutoCAD provides a variety of methods for creating procedural and NURBS surfaces. Profiles can be used to create surfaces with the EXTRUDE, LOFT, PLANESURF, REVOLVE, SURFNETWORK, and SWEEP commands. Surfaces can be created from other surfaces with the SURFBLEND, SURFEXTEND, SURFFILLET, SURFOFFSET, and SURFPATCH commands.

Objects can be converted to procedural surfaces with the CONVTOSURFACE command. Procedural surfaces can be converted to NURBS surfaces with the CONVTO-NURBS command.

Surface continuity affects the way that curves meet. Continuity can be set to one of three degrees from G0 which means the curves meet in a collinear edge, to G1 in which the curves are tangent where they adjoin to G2 where the curvature of the adjoining curves is the smoothest.

Surfaces can be created as networks.

Procedural surfaces offer associative modeling and NURBS surfaces can be manipulated and shaped using control vertices. Neither of these capabilities is possible with solid or mesh modeling. Figure 16–119 shows examples of surfaces.

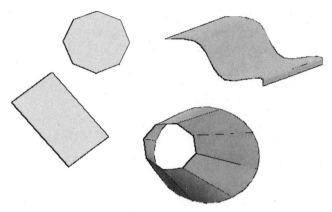

FIGURE 16–119 *Surfaces created using PLANESURF, SURFNETWORK, and LOFT commands*

When the SURFACEMODELINGMODE system variable is set to 0, procedural surfaces are created. When it is set to 1, NURBS surfaces are created as shown in Figure 16–120. The CVSHOW command was invoked in order to display the control vertices.

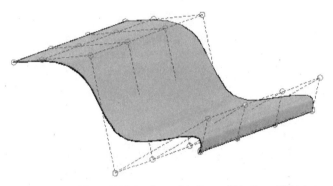

FIGURE 16–120 *NURBS Surface with Control Vertices (CVs) being displayed*

Planar Surfaces

You can use the PLANESURF command to create a planar surface by selecting one or more objects that form one or more enclosed areas or by specifying the opposite corners of a rectangle. When you specify the corners of a rectangle to create the surface, the surface is created parallel to the work plane. If you select a closed object, it does not have to be in the current plane. If you select multiple closed objects, they do not have to be in the same plane. Each will be a separate surface.

Planesurf Rectangle. Invoke the PLANESURF command from the Create panel of the Surfaces tab (see Figure 16–121a). By default, AutoCAD prompts for the first corner of a rectangle. AutoCAD then prompts for the second (diagonally opposite) corner. AutoCAD draws a planar surface in the defined rectangle (see Figure 16–121b).

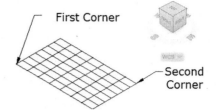

FIGURE 16–121a *Selecting Planar from the Create panel*

FIGURE 16–121b *Selecting Corners for Planar Surface*

The Isolines shown on the surfaces are in two directions, U and V, and are for visual purposes. They cannot be used to manipulate the surface like the facets of a mesh. They can be changed in the 3D Objects section on the 3D Modeling tab of the Options dialog box. The default for both directions is 6.

Planesurf Object. At the first prompt of the PLANESURF command, right-click and select *Object* from the shortcut menu and AutoCAD prompts you to select an object. The selection must be closed and made up of the following valid objects: lines, circles, arcs, ellipses, elliptical arcs, 2D polylines, planar 3D polylines, and 2D splines or a combination of these. The selected objects do not have to be joined but their endpoints must be coincidental. They cannot overlap. Figure 16–122a shows a combination of objects selected in response to the prompt to select an object. Figure 16–122b shows the resultant planar surface.

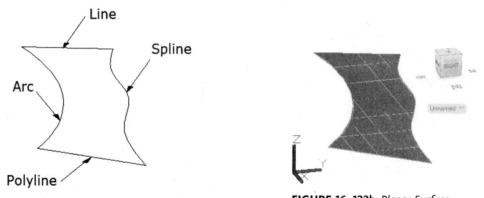

FIGURE 16–122a *Objects selected for creating a Surface*

FIGURE 16–122b *Planar Surface*

The setting of the DELOBJ system variable determines whether the source objects (selected to create the surface) are retained or deleted when the surface is created.

Associativity

Procedural surfaces can be associative, which means that they are created as a member of an associated group of objects. When a source object of an associated group is modified then all members are subject to modification. When applied properly, associativity provides advantages that can only be achieved with procedural surfaces. If needed, you can switch off associativity. Enable associativity from the Create panel of the Surface tab as shown in Figure 16–123.

FIGURE 16–123 *Surface Associativity selected*

Figure 16–124a shows a rectangle created with the RECTANG command with the corners filleted. Using the PEDIT command, the two long sides have each had four control points added for future manipulation (see Figure 16–124b).

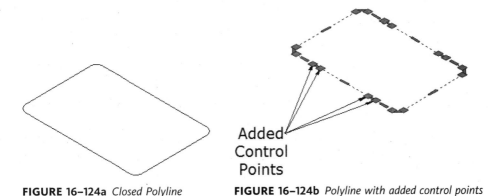

Added
Control
Points

FIGURE 16–124a *Closed Polyline*   **FIGURE 16–124b** *Polyline with added control points*

Figure 16–125a shows a planar surface created by invoking the PLANESURF command, right-clicking to select *Object* from the shortcut menu, and selecting the rectangle. Two of the control points on each side of the source rectangle have been moved out and the planar surface is reshaped to conform to the new shape of the source object (see Figure 16–125b).

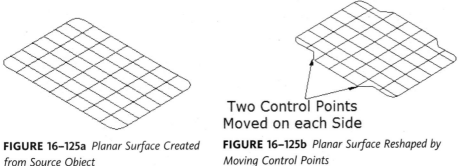

Two Control Points
Moved on each Side

FIGURE 16–125a *Planar Surface Created from Source Object*   **FIGURE 16–125b** *Planar Surface Reshaped by Moving Control Points*

In the Properties palette you can remove existing associativity of a selected object by choosing **remove** from the Maintain Associativity list. You can control the highlighted display of the source object in the Show Associativity list.

If you attempt to modify the surface that is associated with a source object AutoCAD displays a warning message (see Figure 16–126).

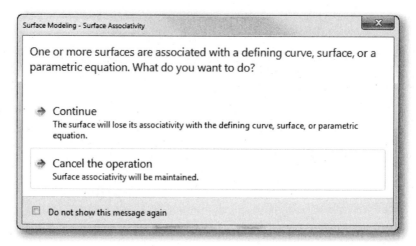

FIGURE 16–126 *Warning Message regarding Associativity*

It is the source object that can be modified and the associated surface that conforms to the modification, not the reverse.

Surfaces from 2D Objects

Profiles consisting of 2D objects can be used to create procedural and NURBS surfaces with the EXTRUDE, LOFT, REVOLVE, and SWEEP commands. Source profiles can be lines and/or curves.

NOTE | If the NURBS Creation option is selected, the surfaces created will not be associative.

Surface Extrude. The EXTRUDE command creates a 3D surface from an existing open or closed 2D object. Figure 16–124b (from the previous example of creating a planar surface) shows the rectangle used now to create a procedural surface by invoking the EXTRUDE command from the Create panel of the Surface tab. In response to the prompt to specify the height of the extrusion the value of 12 was entered and the resulting surface was created (see Figure 16–127a). Figure 16–127b shows the associative procedural surface after the source object has been edited in the same manner as for the planar surface from Figure 16–125b.

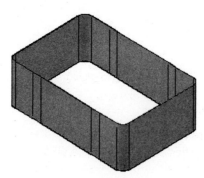

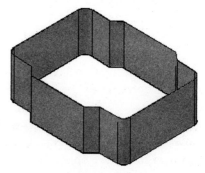

FIGURE 16–127a *Procedural Surface created from rectangle*

FIGURE 16–127b *Surface after Source has been edited*

Surface Extrude with Taper, Path, and Direction. Figure 16–128a shows a closed rectangle to be used as the source object for creating a 3D procedural surface and a vertical line to be used as a path to determine the direction and height of the surface. After invoking the EXTRUDE command, selecting the rectangle, and in response to the prompt to specify the height, right-click and select *Taper angle* from the shortcut menu (see Figure 16–128b).

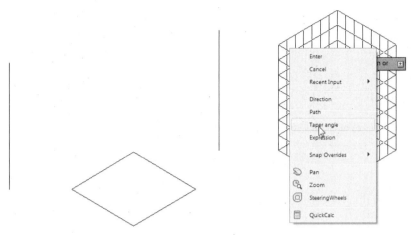

FIGURE 16–128a *Source rectangle and vertical line*

FIGURE 16–128b *Source rectangle selected for object to Extrude*

Figure 16–129a shows the taper angle being specified as 5°. Right-click again and select *Path* from the shortcut menu (see Figure 16–129b).

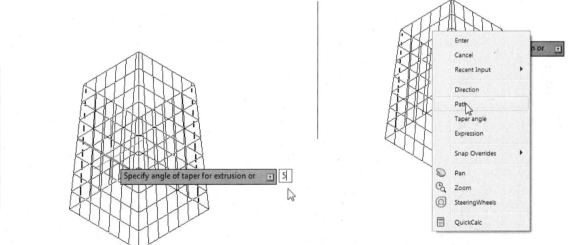

FIGURE 16–129a *Taper Angle specified*

FIGURE 16–129b *Path selected from Shortcut menu*

Figure 16–130a shows the vertical line being selected as the path. AutoCAD draws a procedural surface at height and direction of the vertical line with a taper angle of 5° (see Figure 16–130b).

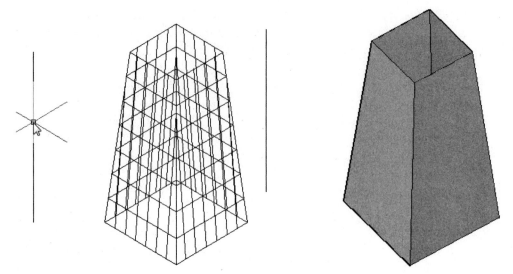

FIGURE 16–130a *Vertical line selected as Path*　　　　**FIGURE 16–130b** *Resulting Procedural Surface*

The examples above illustrate using the *Taper angle* option by input from the keyboard and the *Path* option by selecting a profile (the vertical line). The *Direction* option can be applied by using the cursor (or keyboard entry) to specify two points. AutoCAD will not allow you to enter two points that are in a plane that is parallel to the same plane that the source object is in.

Associativity at Work. The previous examples explaining associativity showed how a procedural surface automatically changes to conform to changes in the source closed object from which it was created. Figure 16–131a shows the vertical line that was the path for the procedural surface (see Figure 16–130b) height and direction being selected for modification. When the height of the vertical line is shortened, the height of the procedural surface height is automatically shortened (see Figure 16–131b).

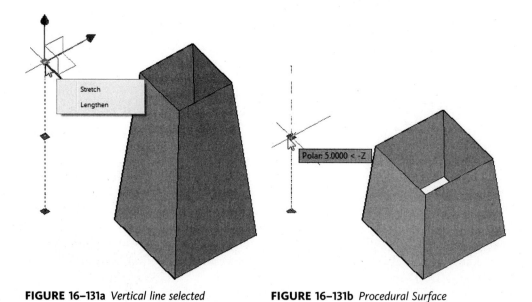

FIGURE 16–131a *Vertical line selected*
for modification　　　　**FIGURE 16–131b** *Procedural Surface*
automatically modified

Figure 16–132a shows the vertical line used as a path for a surface that was extruded without taper. When the vertical line is changed to a slanted line, the procedural surface is automatically slanted (see Figure 16–132b).

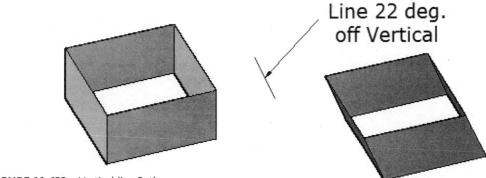

FIGURE 16–132a *Vertical line Path and Procedural Surface*

FIGURE 16–132b *Resulting Procedural Surface*

Surface Loft. The LOFT command creates a 3D surface by specifying two or more open or closed profiles. Invoke the LOFT command from the Create panel of the Surface tab. Two closed polylines are selected in Figure 16–133a. Figure 16–133b shows the lofted surface created as a result.

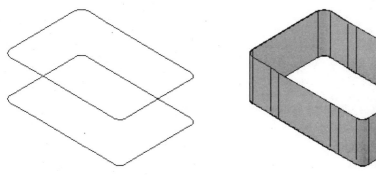

FIGURE 16–133a *Two polyline profiles selected*   **FIGURE 16–133b** *Resulting lofted surface*

Four closed polylines are selected in sequence from bottom to top (or top to bottom) in Figure 16–134a. Figure 16–134b shows the lofted surface created as a result.

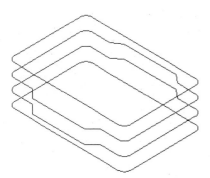
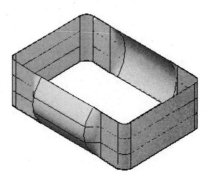

FIGURE 16–134a *Four polyline profiles selected*   **FIGURE 16–134b** *Resulting lofted surface*

Two closed polylines are selected as cross sections in Figure 16–135a. A preview of the anticipated surface is shown in Figure 16–135b as two polylines are selected as guides. Figure 16–136 shows the lofted surface created as a result.

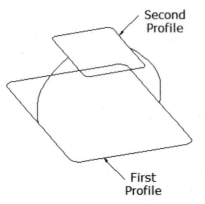

FIGURE 16–135a *Two polyline profiles selected*

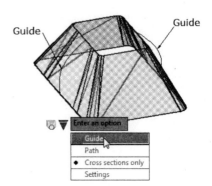

FIGURE 16–135b *Two splines selected as guides*

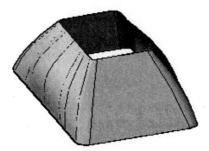

FIGURE 16–136 *Result of using cross sections and guides*

Two open polylines are selected as cross sections in Figure 16–137a. In Figure 16–137b the same two polylines are selected and the two straight and one curved polylines are selected as guides.

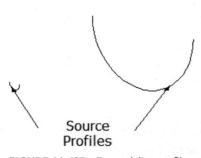

FIGURE 16–137a *Two polyline profiles selected*

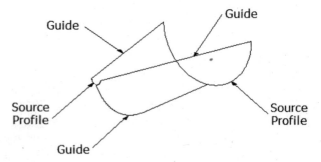

FIGURE 16–137b *Adding three polylines selected as guides*

Figure 16–138a shows the lofted surface created as a result of selecting cross-section polylines only. Figure 16–138b shows the lofted surface created as a result of selecting cross-section polylines and the two straight and one curved polylines as guides also.

FIGURE 16–138a *Surface without guides*   **FIGURE 16–138b** *Surface with guides*

> In the previous examples the series of sources selected for a particular surface were either all closed or all open. You cannot mix open cross sections with closed ones. Also, if one of the ends needs to be a point the other cross sections must be closed objects. A point can only be used at the ends of the surface.

Objects that can be used as cross sections include 2D polylines, 2D solids, 2D splines, Arcs, Circles, Edge subobjects, Ellipses, Elliptical arcs, Helices, Lines, Planar or non-planar faces of solids, Planar or non-planar surfaces, Points (first and last cross sections only), Regions and Traces.

Objects that can be used as a loft path include Splines, Helices, Arcs, Circles, Edge subobjects, Ellipses, Elliptical arcs, 2D polylines, Lines and 3D polylines.

Objects that can be used as guides include 2D and 3D splines, Arcs, 2D polylines (only if they contain 1 segment), Edge subobjects, 3D polylines, and Elliptical arcs.

Loft Settings. After selecting the cross-section profiles, right-click and AutoCAD displays a shortcut menu similar to the one for the Solids (see Figure 16–74).Choose *Settings* to display the Loft Settings Dialog Box (see Figure 16–75).

Surface Revolve. The REVOLVE command creates a 3D surface by sweeping an object around an axis. Invoke the REVOLVE command from the Create panel of the Surface tab. AutoCAD first prompts you to select an object to revolve and then prompts for the axis start point and axis endpoint. You are then prompted to specify the start angle and angle of revolution. After selecting the profile object and specifying the endpoints of the axis and angle of rotation AutoCAD creates the surface.

Revolve around Object. In response to the prompt to select the object to revolve the polyline is selected (see Figure 16–139a). When prompted to specify axis start point, right-click and from the shortcut menu select *Object* and then select the line as the axis for revolution. When prompted for the angle, specify 180°. Figure 16–139b shows the resulting revolved surface.

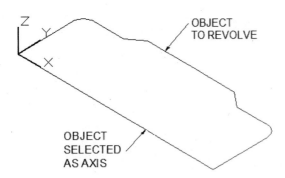

FIGURE 16–139a *Selecting the object for revolving and for the axis*

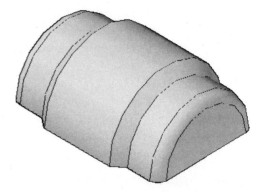

FIGURE 16–139b *Resulting revolved surface*

Open profiles generate surfaces only, but closed profiles can generate either surfaces or solids. Objects that cannot be revolved are those contained within a block or those that will be self-intersecting.

Revolve around Axis. In response to the prompt to select the object to revolve the polyline is selected (see Figure 16–140a). When prompted to specify axis start point, right-click and from the shortcut menu select *Y Axis*. When prompted for the angle, press [Enter] and the angle defaults to 360°. Figure 16–140b shows the resulting revolved surface.

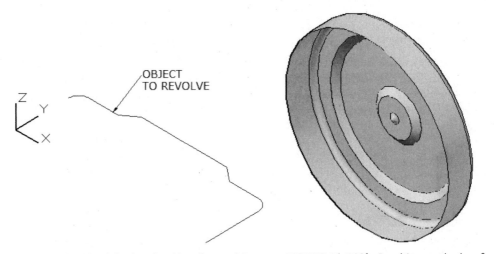

FIGURE 16–140a *Selecting the object for revolving*

FIGURE 16–140b *Resulting revolved surface*

Surface Sweep. The SWEEP command creates a 3D surface by sweeping an object along a path. Invoke the SWEEP command from the Create panel of the Surface tab. AutoCAD first prompts you to select an object to sweep and then prompts for the path (see Figure 16–141a). After selecting the profile object and specifying the path object AutoCAD creates the surface (see Figure 16–141b).

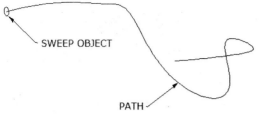

FIGURE 16–141a *Selecting the object and path for sweeping*   **FIGURE 16–141b** *Resulting sweeped surface*

Sweep Alignment, Base Point, Scale, and Twist. After AutoCAD prompts for the sweep path, if you right-click click, a menu appears with the options of *Alignment, Base Point, Scale,* and *Twist.*

Alignment control whether the profile is aligned to be perpendicular to the tangent direction of the path. **Base Point** lets you specify a base point for the profile objects. **Scale** lets you specify a scale factor that will be uniformly applied to the profile from the beginning to the end of the path. **Twist** lets you specify an angle for the profile to be rotated around the path from the beginning to the end.

Sweep Objects and Limits

NOTE

There are limits to the size of the profile object relative to the curvature of the path. If the profile is too large, AutoCAD displays a warning:

"Modeling Operation Error:

The sweep results in a self-intersecting body.

Unable to sweep selected objects."

Objects that can be swept include 2D and 3D splines, 2D polylines, 2D solids, 3D solid face subobjects, Arcs, Circles, Ellipses, Elliptical arcs, Lines, Regions, Traces, and Solid, surface, and mesh edge subobjects.

Objects that can be used as a path include 2D and 3D splines, 2D and 3D polylines, Helices, Arcs, Circles, Ellipses, Elliptical arcs, Lines, and Solid, surface, and mesh edge subobjects.

Network Surfaces

The SURFNETWORK command creates a non-planar surface by selecting two sets of objects. One set of objects determines the curvature in one direction and the second set determines the curvature in the other direction. The directions are labeled U and V. The number of lines displayed in each direction are saved in the SURFU and SURFV system variables.

Surfnetwork. The SURNETWORK command creates a surface conforming to the curves determined by two sets of selected profiles. Invoke the SURNETWORK command from the Create panel of the Surface tab and AutoCAD prompts for the first set of profile objects (U). After specifying the first set of profile objects press **Enter** and AutoCAD prompts for the second set of profile objects (V). Figure 16–142a shows selecting the 5 splines in the U direction and then the three splines in the V direction as profiles. Figure 16–142b shows the resulting network surface.

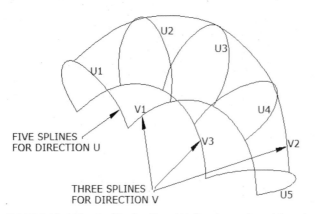

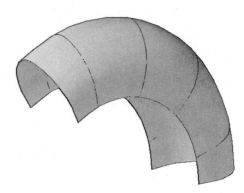

FIGURE 16–142a *Profiles for U and V directions selected in order*　　**FIGURE 16–142b** *Resulting Network Surface*

NURBS Surfaces

NURBS surfaces let you reshape the contours of the surface to create unique smooth flowing forms. NURBS surfaces are based on splines (or Bezier curves). Settings like degree, fit points, weights, control vertices, and knot parameterization are involved in defining NURBS surfaces or curvatures. These characteristics are best controlled when you use splines to create NURBS surfaces.

NURBS Loft. The previous example (Figures 16–142a & b) has been redone with similar splines with the NURBS option enabled, using the LOFT command. With the LOFT command only the 5 splines (U1 through U5) need to be selected. In this case U and V directions do not apply like they do in the SURFNETWORK command. Figure 16–143 shows the control vertices being displayed after invoking the CVSHOW command from the Control Vertices panel and selecting the surface.

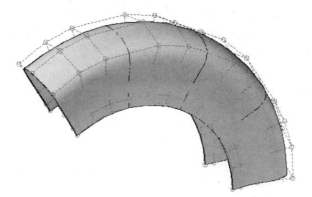

FIGURE 16–143 *NURBS surface with control vertices showing*

CVREBUILD. The density of the control vertices can be changed by using the CVRE-BUILD command. Invoke the CVREBUILD command from the Control Vertices panel and AutoCAD displays the Rebuild Surface dialog box (see Figure 16–144).

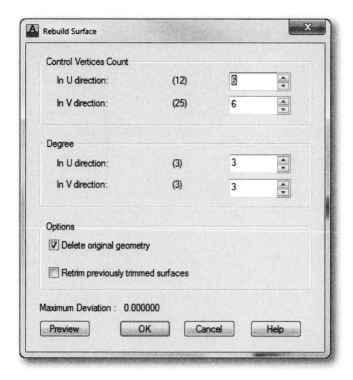

FIGURE 16–144 *Rebuild Surface dialog box*

After changing the number of Control Vertices in the V direction to 5 the rebuilt surface becomes more rounded as shown in Figure 16–145. Also, with fewer CVs, a larger radius curve is easier to obtain if that is what is needed.

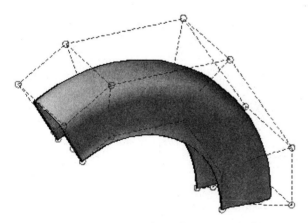

FIGURE 16–145 *NURBS surface after using Rebuild*

Select a control vertex (see Figure 16–146a) and drag to a new location to achieve the desired curvature as shown in Figure 16–146b.

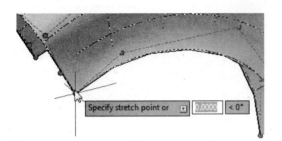

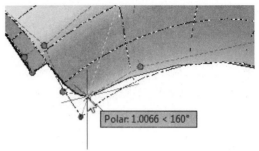

FIGURE 16–146a *Selecting Control Vertex*

FIGURE 16–146b *Move CV to change curvature*

All four sharp corners can be rounded by relocating the appropriate CVs as shown in Figure 16–147.

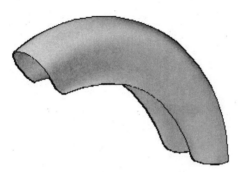

FIGURE 16–147 *NURBS surface after manipulating CVs*

You can use the CVADD and CVREMOVE commands to add and remove individual CVs as needed.

Surfaces from other Surfaces

AutoCAD provides commands that let you create surfaces from other surfaces. These include the SURFBLEND, SURFEXTEND, SURFFILLET, SURFOFFSET, SURFPATCH, and SLICE commands.

Surfblend. The SURFBLEND command creates a surface that continuously blends between the edges of two existing surfaces. Invoke the SURFBLEND command from the Create panel of the Surface tab and AutoCAD prompts for the first surface edge(s) to blend. After specifying one or more edges press Enter and AutoCAD prompts for the second surface edge(s) to blend. Figure 16–148a shows selecting the edges of two surfaces to blend. Figure 16–148b shows the resulting blended surface.

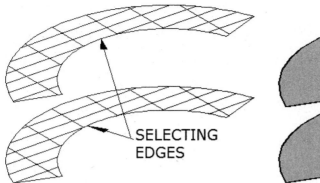

FIGURE 16–148a *Selecting Edges*

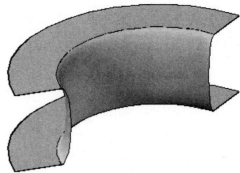

FIGURE 16–148b *Resulting Blended Surface*

After pressing [Enter] for the second edge AutoCAD displays a shortcut menu for controlling the Continuity or Bulge magnitude (see Figure 16–149).

> Press Enter to accept the blend surface or
> CONtinuity
> Bulge magnitude

FIGURE 16–149 *Shortcut menu for Continuity and Bulge Magnitude*

Continuity lets you choose between G0, G1, and G2 which determines the smoothness of the connection of the blend surface to the source surface. *Bulge magnitude* lets you control the roundness of the blend surface edge at the source surface. Values are from 0 to 1 with 0.5 being the default.

Surfextend. The SURFEXTEND command creates an extension to a surface for a specified length. Invoke the SURFEXTEND command from the Edit panel of the Surface tab and AutoCAD prompts for the surface edge(s) to extend. After specifying the edge press [Enter]. Figure 16–150a shows selecting the edges of two surfaces to blend. Figure 16–150b shows the resulting extended surface.

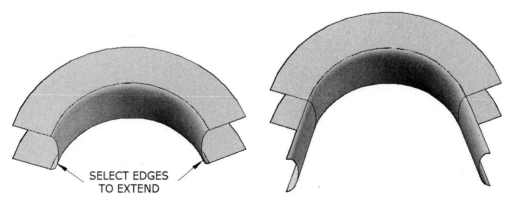

FIGURE 16–150a *Selecting Edges*

FIGURE 16–150b *Resulting Extended Surfaces*

After selecting the edge to extend, right-click and AutoCAD displays a shortcut menu with the Mode options of *Extend* or *Stretch*. Choose **Extend** to simulate the source edge shape. Choose **Stretch** to extend without simulating the source edge shape. After selecting the Mode, AutoCAD displays another shortcut menu with the Creation Type options of Merge or *Append*. Choosing **Merge** extends the source surface. Choosing **Append** attaches a separate surface. If the source surface is NURBS then the extension will also be NURBS.

Surffillet. The SURFFILLET command connects two surfaces with a filleted surface. Invoke the SURFFILLET command from the Edit panel of the Surface tab and AutoCAD prompts for the surface edge(s) to fillet. After specifying the edges press Enter Figure 16–151a shows selecting the edges of three pairs of surfaces to fillet. Figure 16–151b shows the resulting filleted surfaces.

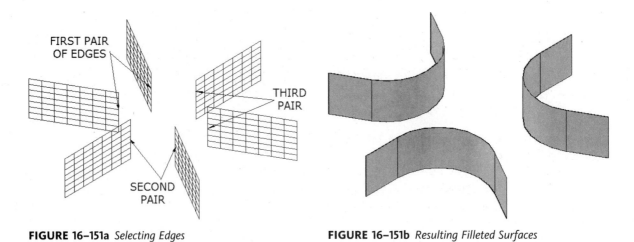

FIGURE 16–151a *Selecting Edges*   **FIGURE 16–151b** *Resulting Filleted Surfaces*

NOTE   The value of the radius cannot be smaller than the distance between the selected surfaces.

Surfoffset. The SURFOFFSET command creates a new surface that is offset from a selected source surface by a specified distance. Invoke the SURFOFFSET command from the Create panel of the Surface tab and AutoCAD prompts for the surface to offset. After specifying the surface press Enter. Figure 16–152a shows selecting the three surfaces to offset. Figure 16–152b shows the resulting offset surfaces.

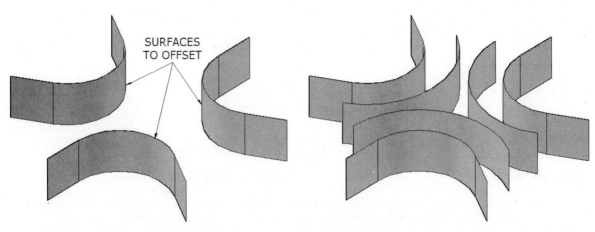

FIGURE 16–152a *Selecting Surfaces*   **FIGURE 16–152b** *Resulting Offset Surfaces*

After selecting the surfaces and pressing `Enter` AutoCAD displays arrows pointing from the surface(s) indicating the side of the source surface that the new surface will be created. Right-click and from the shortcut menu you can select from the options *Flip Direction*, *Both Sides*, *Solid*, and *Connect*. You are also prompted for the offset distance. If you press `Enter` AutoCAD will offset the new surface in the direction of the arrows for the default distance. You can change the distance before pressing `Enter`. Selecting **FlipDirection** changes the arrows so the surface will be the offset on the opposite side. Selecting **Both Sides** causes new surfaces to be created on both sides of the source surface. Selecting **Solid** causes a solid to be created between the source and new surfaces. Selecting **Connect** causes multiple new sources to be connected if the source surfaces are connected.

Surfpatch. The `SURFPATCH` command creates a new surface that fills an opening in a selected source surface. Invoke the `SURFPATCH` command from the Create panel of the Surface tab and AutoCAD prompts for the surface edge to patch. After specifying the edge press `Enter` Figure 16–153a shows selecting the surface edge to patch. Figure 16–153b shows the resulting surface patch.

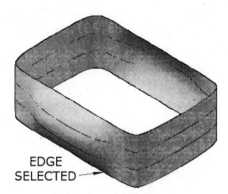

FIGURE 16–153a *Selecting Surface Edge*   **FIGURE 16–153b** *Resulting Patch Surface*

After selecting the surface edge and pressing `Enter` AutoCAD displays the shortcut menu you can select from the options *Continuity*, *Bulge Magnitude*, and *Guides*. If you press `Enter` AutoCAD will create the new surface patch. Selecting **Continuity** lets you choose between G0, G1, and G2 which determines the smoothness of the connection of the blend surface to the source surface. Selecting **Bulge magnitude** lets you control the roundness of the blend surface edge at the source surface. Values are from 0 to 1 with 0.5. Selecting **Guides** lets you select object profiles to use as guides for the new patch surface.

Surfaces from other Objects

AutoCAD provides commands to convert non-surface and non-NURBS objects to procedural and NURBS surfaces.

Converting to Procedural Surfaces. The `CONVTOSURF` command creates a new surface from 2D solids, meshes, regions, planar 3D faces, and polylines, lines, and arcs with thickness. Invoke the `CONVTOSURF` command from the Solid Editing panel of the Home tab and AutoCAD prompts you to select the object(s) to convert to a procedural surface.

As you convert objects to surfaces, you can determine in the resulting surface is faceted or smooth. The SMOOTHMESHCONVERT system variable is used to control the smoothness and number of faces when converting meshes.

Converting to NURBS Surfaces. The CONVTONURBS command creates a new NURBS surface from 2D solids, meshes, regions, planar 3D faces, and polylines, lines, and arcs with thickness. Invoke the CONVTONURBS command from the Control Vertices panel of the Surface tab and AutoCAD prompts you to select the object(s) to convert to a NURBS surface. To convert meshes to NURBS surfaces, first convert them to a solid or surface and then convert them to NURBS surfaces.

Editing Surfaces

AutoCAD provides commands to edit procedural and NURBS surfaces. Some of the commands work also on solids and meshes.

Trim and Untrim Surfaces. The SURFTRIM command trims surfaces at the intersection of other surfaces. You can also project geometry onto a surface as a trimming edge. Invoke the SURFTRIM command from the Edit panel of the Surface tab. In Figure 16–154a the surface to trim, cutting surface, and area to trim are selected. Figure 16–154b shows the trimmed surface created as a result after the cutting surface has been removed (either by erasing or turning off its layer).

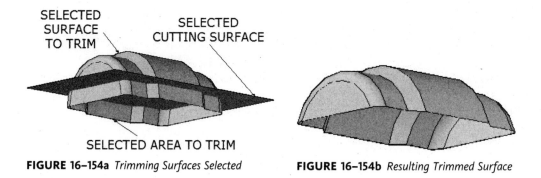

FIGURE 16–154a *Trimming Surfaces Selected*   **FIGURE 16–154b** *Resulting Trimmed Surface*

Figure 16–155a shows a surface to trim with a cutting surface (circle) above it. After selecting the surface to trim, right-click and choose **PROJection direction** from the shortcut menu and then choose **Automatic** from the next shortcut menu. Then select the circle as the cutting surface or region. Next select inside the projected circle as the area to trim. Figure 16–155b shows the trimmed surface created as a result.

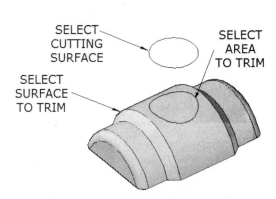

SELECT CUTTING SURFACE

SELECT AREA TO TRIM

SELECT SURFACE TO TRIM

FIGURE 16–155a *Trimming Surfaces Selected*

FIGURE 16–155b *Resulting Trimmed Surface*

The **Automatic** option makes a projection on the surface to be trimmed depending on the viewing environment. In an orthogonal view, the cutting geometry is projected in the view direction. With a planar surface in an angular view, the cutting geometry is projected perpendicular to the curve plane of the surface to be trimmed. With a curved surface the projection acts like the UCS option. Other options to the Projection Direction include: **UCS** projects the geometry along the Z axis of the current UCS. **View** projects the geometry according to the current view. **None** will only trim the surface if the cutting curve lies on the surface.

Edit NURBS Surfaces. NURBS surfaces can be edited with the 3D Edit Bar or by manipulating individual control vertices. Invoke the 3D Edit Bar from the Control Vertices panel of the Surface tab and AutoCAD prompts you to select a NURBS surface to edit. Figure 16–156a shows the surface to be selected being displayed in an isometric view. The view was changed to the bottom in order to place the point along the center of the bottom of the surface. After selecting the surface, you are then prompted to select a point on the surface. Figure 16–156b shows the point being selected. AutoCAD displays crossed red lines to indicate the axes along which the editing will take effect.

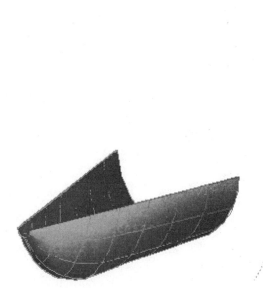

FIGURE 16–156a *Surface to be Selected*

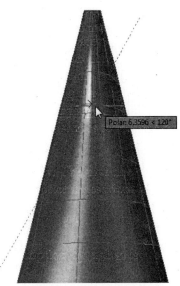

FIGURE 16–156b *Selecting a Point on the Surface*

Figure 16–157a shows the surface reoriented in order to visualize the changes to the surface as the point is manipulated using the dynamic directional arrows. The arrow in the vertical direction was dragged slightly downward first and then the front to back arrow was dragged toward the front creating a protuberance in the bottom of the surface (see Figure 16–157b).

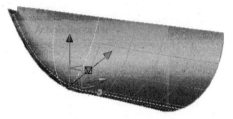

FIGURE 16–157a *Trimming Surfaces Selected*   **FIGURE 16–157b** *Resulting Trimmed Surface*

Analyzing Surfaces

AutoCAD provides commands to analyze surfaces. These include Zebra, Curvature, and Draft analyses.

Zebra Surface Analysis. Zebra Analysis projects stripes onto a 3D model to analyze surface continuity. The way the stripes line up where two surfaces meet, helps analyze the tangency and curvature of the intersection. Invoke the Zebra Analysis from the Analysis panel of the Surface tab.

Curvature Surface Analysis. Curvature Analysis displays a color gradient onto NURBS surfaces to evaluate areas of high, low, and Gaussian curvature. Invoke the Curvature Analysis from the Analysis panel of the Surface tab.

Draft Surface Analysis. Draft Analysis evaluates whether a part has adequate taper to allow it to be removed from its mold. Invoke the Draft Analysis from the Analysis panel of the Surface tab.

EDITING IN 3D

This section describes how to perform various 3D editing operations such as selecting, aligning, rotating, mirroring, arraying, extending, and trimming.

Culling Selected Objects and Subobjects

AutoCAD provides a feature to allow you to control whether 3D objects or subobjects that are hidden from view can be highlighted or selected. The CULLINGOBJ system variable applies to 3D subobjects. Invoke the CULLINGOBJ system variable from the Selection panel of the Home tab. AutoCAD prompts you to enter the new value for the CULLINGOBJ system variable.

> Entering **0** specifies that there is no subobject culling. Rolling over 3D objects causes all 3D subobjects, including hidden, to be highlighted. Selecting 3D objects by dragging causes all 3D subobjects, including hidden, to be selected.

> Entering **1** applies subobject culling during selection. Rolling over 3D objects causes only subobjects that are normal in the current view to be highlighted. Selecting 3D objects by dragging causes only subobjects that are normal in the current view to be selected.

Invoke the `CULLINGOBJSELECTION` command by typing at the command prompt. AutoCAD prompts you to enter the new value for the CULLINGOBJ-SELECTION system variable.

Entering **0** specifies that there is no object culling. Rolling over 3D objects causes all 3D objects, including hidden, to be highlighted. Selecting 3D objects by dragging causes all 3D objects, including hidden, to be selected.

Entering **1** applies object culling during selection. Rolling over 3D objects causes only objects that are normal in the current view to be highlighted. Selecting 3D objects by dragging causes only objects that are normal in the current view to be selected.

Aligning Objects

The `ALIGN` command allows you to translate and rotate objects in 3D space regardless of the position of the current UCS. Three source points and three destination points define the move. The `ALIGN` command lets you select the objects to move and subsequently prompts for three source points and three destination points.

Invoke the `ALIGN` command from the Modify panel on the Home tab of the 3D Modeling Workspace. AutoCAD displays the following prompt:

```
align
Select objects: (select the objects)
```

Select the objects to move and press Enter. AutoCAD prompts for three source points and three destination points. Temporary lines are displayed between the matching pairs of source and destination points. If you enter all six points, the move consists of a translation and two rotations based on the six points. The translation moves the first source point to the first destination point. The first rotation aligns the line defined by the first and second source points with the line defined by the first and second destination points. The second rotation aligns the plane defined by the three source points with the plane defined by the three destination points.

If instead of entering three pairs of points, you enter two pairs of points, the transformation reduces to a translation from the first source point to the first destination point and a rotation such that the line passing through the two source points aligns with the line passing through the two destination points. The transformation occurs in either 2D or 3D, depending on your response to the following prompt:

```
<2D> or 3D transformation:
```

If you enter 2D or press Enter, the rotation is performed in the XY plane of the current UCS. If you enter 3D, the rotation is in the plane defined by the two destination points and the second source point.

If you enter only one pair of points, the transformation reduces to a simple translation from the source to the destination point. This is similar to using the AutoCAD regular `MOVE` command without the dynamic dragging.

Rotating Objects about a 3D Object

The `3DROTATE` command lets you rotate an object about an arbitrary 3D axis.

Invoke the `3DROTATE` command by selecting 3D Rotate from the Modify panel on the Home tab. AutoCAD displays the following prompt:

```
Select objects: (select object and press Enter when done)
```

AutoCAD lists the options for selecting the axis of rotation:

> Specify first point on axis or define axis by ⬇ (specify a point or select one of the options from the shortcut menu)

Choose *2points* from the shortcut menu to define the axis of rotation. The axis of rotation is the line that passes through the two points, and the positive direction is from the first to the second point.

Choose *Object* from the shortcut menu to select an object and then derive the axis of rotation based on the type of object selected. Valid objects include line, circle, arc, and spline.

Choose *Last* from the shortcut menu to select the last-used axis. If there is no last axis, a message to that effect is displayed, and the axis selection prompt is redisplayed.

Choose *View* from the shortcut menu to base the rotation on a selected view. The axis of rotation is perpendicular to the view direction and passes through the selected point. The positive axis direction is toward the viewer.

Choose *X, Y,* or *Z-Axis* option from the shortcut menu to select a point for the axis of rotation parallel to the standard axis of the current UCS and passing through the selected point.

Once you have selected the axis of rotation, AutoCAD displays the following prompt:

> Specify by rotation angle or ⬇ (specify the rotation angle, or Enter , for reference)

AutoCAD rotates the selected object(s) to the specified rotation angle. The REFERENCE option allows you to specify the current orientation as the reference angle. Or you can specify the angle by pointing to the two endpoints of a line to be rotated and then specifying the desired new rotation. AutoCAD automatically calculates the rotation angle and rotates the selected object appropriately. See Figure 16–158 for an example of rotating a cylinder around the Z axis.

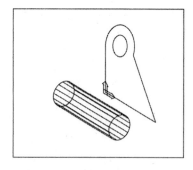
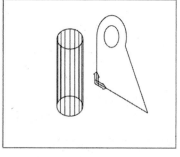

BEFORE   AFTER

FIGURE 16–158 *Rotating a cylinder about the Z axis with the 3DROTATE command*

Mirroring about a Plane

The MIRROR3D command lets you mirror a selected object about a plane.

Invoke the MIRROR3D command from the Modify panel on the Home tab. Auto-CAD displays the following prompt:

        Select objects: *(select the objects)*

Select the objects to mirror and press [Enter]. AutoCAD then lists the options for selecting the mirroring plane:

        Specify first point of mirror plane (3 points) or ⊡ *(select one of the options from the shortcut menu)*

Choose *3points* from the shortcut menu to define three points. The mirroring plane is the plane that passes through the three selected points.

Choose *Object* from the shortcut menu to align the mirroring plane with the plane of the object selected. Valid objects include circle, arc, and spline.

Choose *Last* from the shortcut menu to mirror the object based on the last used plane. If there is no previous mirror plane, the on-screen prompt repeats the default prompt.

Choose *Z axis* from the shortcut menu to define the mirroring plane by two points. The first point is the point at which the object will rotate about, and the second defines the angle of rotation on the Z plane's normal, or perpendicular to the plane.

Choose *View* from the shortcut menu to create a mirroring plane perpendicular to the view direction and passing through the selected point.

Choose *XY*, *YZ*, or *XZ* from the shortcut menu to select a point to create the mirroring plane parallel to the standard plane of the current UCS and passing through the selected point.

See Figure 16–159 for an example of mirroring a cylinder aligned with the plane of the object selected (spline).

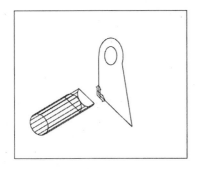 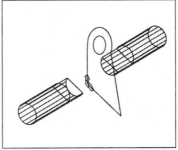

BEFORE   AFTER

FIGURE 16–159 *Mirroring a cylinder about a polyline object*

Creating a 3D Array

The 3DARRAY command is used to make multiple copies of selected objects in either rectangular or polar array in 3D. In the rectangular array, specify the number of columns (X direction), the number of rows (Y direction), the number of levels (Z direction), and the spacing between columns, rows, and levels. In the polar array, specify the number of items to array, the angle that the arrayed objects are to fill, the start point and endpoint of the axis about which the objects are to be rotated, and whether or not the objects are rotated about the center of the group.

Invoke the 3DARRAY command by selecting 3D Array from the Modify panel on the Home tab. AutoCAD displays the following prompts:

> Select objects: *(select the objects to array and press* Enter *)*
>
> Enter type of array: *(select one of the two available options)*

To generate a rectangular array, choose *Rectangular*, and AutoCAD displays the following prompts:

> Enter the number of rows(—)<1>: *(specify the number of rows or press* Enter *)*
>
> Enter the number of columns <1>: *(specify the number of columns or press* Enter *)*
>
> Number of levels(…)<1>: *(specify the number of levels or press* Enter *)*
>
> Specify the distance between rows(—)<1>: *(specify a distance)*
>
> Specify the distance between columns <1>: *(specify a distance)*
>
> Specify the distance between levels(…)<1>: *(specify a distance)*

Any combination of whole numbers of columns, rows, and levels may be entered. AutoCAD includes the original object in the number you enter. An array must have at least two columns, two rows, or two levels. Specifying one row requires that more than one column be specified or vice versa. Specifying one level creates a 2D array. Column, row, and level spacing can be different from one another. They can be entered separately when prompted, or you can select two points and let AutoCAD measure the spacing. Positive values for spacing generate the array along the positive X, Y, and Z axes. Negative values generate the array along the negative X, Y, and Z axes.

To generate a polar array, choose POLAR, and AutoCAD displays the following prompts:

> Enter the number of items in the array: *(specify the number of items in the array; include the original object)*
>
> Specify the angle to fill (+=ccw, —=cw) <360>: *(specify an angle, or press* Enter *for 360°)*
>
> Rotate arrayed objects? *(enter* y *to rotate the objects as they are copied, or enter* n *not to rotate the objects as they are copied)*
>
> Specify center point of array: *(specify a point)*
>
> Specify second point on axis of rotation: *(specify a point for the axis of rotation)*

Extend, Presspull, and Trim in 3D

AutoCAD allows you to extend any object in 3D space by means of the EXTEND command, explained in Chapter 4. Trim an object to any other 3D space by means of the TRIM command, explained in Chapter 4, regardless of whether or not the objects are on the same plane or parallel to the cutting or boundary edges. Before you select an object to extend or trim in 3D space, specify one of the three available projection modes: *None*, *UCS*, or *View*. The *None* option specifies no projection. AutoCAD extends/trims only objects that intersect with the boundary/cutting edge in 3D space. The *UCS* option specifies projection onto the XY plane of the current UCS. AutoCAD extends/trims objects that do not intersect with the boundary/cutting objects in 3D space. The *View* option specifies projection along the current view direction. The PROJMODE system variable allows you to set one of the available projection modes. You can also set the projection mode by selecting the *Project* option available in the EXTEND command.

In addition to specifying the projection mode, you must specify one of the two available options for the edge. The edge determines whether the object is extended or trimmed to another object's implied edge or only to an object that actually intersects it in 3D space. The available options are *Extend* and *No Extend*. The *Extend* option extends the boundary/cutting object/edge along its natural path to intersect another object or its implied edge in 3D space. The *No Extend* option specifies that the object is extended or trimmed only to a boundary/cutting object/edge that actually intersects it in 3D space. The *Extedge* system variable allows you to set one of the available modes. You can also set the edge by selecting the EDGE option available in the EXTEND command.

The PRESSPULL command operates similar to the EXTEND command, but allows you to extend the selected face or bounded area in both directions while giving visual feedback. Invoke the PRESSPULL command from the Modeling panel of the Home tab.

Converting 3D Objects to Solids

You can use the CONVTOSOLID command to convert the following objects into extruded 3D solids: uniform-width wide polylines with thickness, closed, zero-width polylines with thickness, and circles with thickness. However, you cannot use CONVTOSOLID with polylines that contain vertices with zero width or that contain segments of variable width.

Using the 3D Move, Rotate, and Scale Gizmos

Gizmos are icons that you use in a 3D view to easily constrain the movement, rotation, or scale of a selection set of objects to an axis or a plane. The three types of Gizmos are 3D Move (3DMOVE command), 3D Rotate (3DROTATE command), and 3D Scale (3DSCALE command).

To use the Move Gizmo, first select the objects you want to manipulate. Click the 3D Move button on the Modify panel on the Home tab, and place the Gizmo at the location of the desired base point. In Figure 16–160, the base point is selected as the origin. From there, the second point can be selected on the screen or specified as coordinates to determine the displacement. The second point can be anywhere in 3D space.

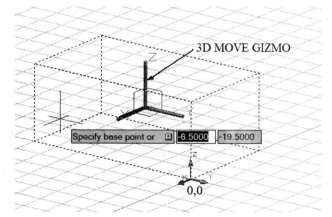

FIGURE 16–160 *Selecting the Move grip tool*

When prompted to specify the base point, you can also click on a plane to constrain the movement to be in the selected plane (see Figure 16–161). Or you can click on an axis to constrain the movement to be in the direction of the selected axis (see Figure 16–162). After selecting the restraint to be planar or linear, the second point can be selected on the screen or specified as coordinates to determine the displacement. The second point must be in the selected plane, in the case of planar restraint, or on the selected axis, in the case of linear restraint.

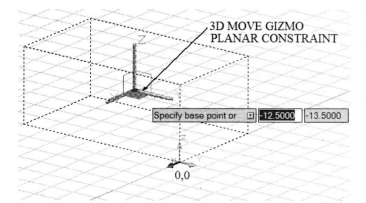

FIGURE 16–161 *Applying the Planar Constraint to the Move grip tool*

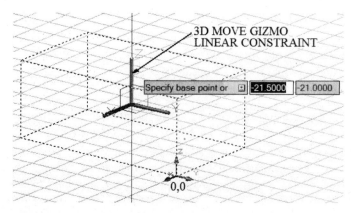

FIGURE 16–162 *Applying the Linear Constraint to the Move grip tool*

To use the Rotate Gizmo, select the objects you want to manipulate. Click the 3D Rotate button on the Modify panel on the Home tab. AutoCAD displays the 3D Rotate Gizmo (see Figure 16–163).

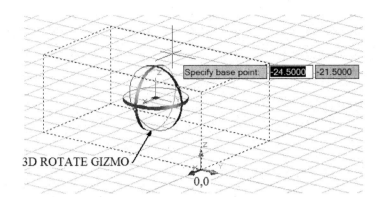

FIGURE 16–163 *Selecting the Rotate Gizmo*

You can place the Gizmo at another location if needed (see Figure 16–164).

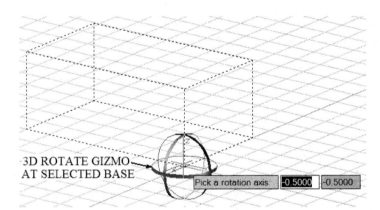

FIGURE 16–164 *Resetting the Base Point for the Rotate Gizmo*

You can click on an axis handle to constrain the rotation. The circles that represent the axes are red, green, and blue. When you select one to be the axis about which the selected objects are to be rotated, the selected axis turns yellow. Specify the start and endpoints of the rotation angle. The object rotates.

To use the Scale Gizmo, select the objects you want to manipulate. Click the 3D Scale button on the Modify panel on the Home tab. AutoCAD displays the 3D Scale Gizmo (see Figure 16–165).

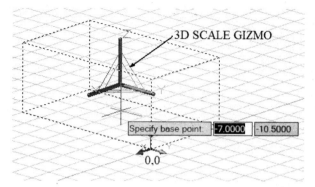

FIGURE 16–165 *Selecting the Scale Gizmo*

You can place the Gizmo at another location if needed (see Figure 16–166).

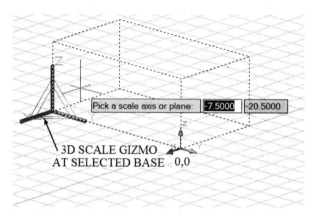

FIGURE 16–166 *Resetting the Base Point for the Scale Gizmo*

The 3DSCALE command lets you change the size of 3D objects. Objects created by the Solid Modeling methods (Box, Cylinder, Cone, etc.) can be scaled up and down in size uniformly in all three axes. Other objects, such as those created by the Mesh Modeling methods can be scaled in a single plane or single axis additionally. You can click on a plane or an axis handle to constrain the scaling. If the objects you select are restricted to uniform scaling only and you attempt to use a plane icon or axis icon, you will be prompted: "Only uniform scale is allowed for selected objects".

MASS PROPERTIES OF A SOLID

The MASSPROP command calculates and displays the mass properties of selected solids and regions. The mass properties displayed for solids are mass, volume, bounding box, centroid, moments of inertia, products of inertia, radii of gyration, and principal moments with corresponding principal directions. The mass properties are calculated based on the current UCS.

Invoke the MASSPROP command by typing **massprop** at the on-screen prompt.

AutoCAD displays the following prompt:

> Select objects: *(select the objects, and press* Enter *to complete the selection)*

The MASSPROP command displays the object mass properties of the selected objects in the text window, as shown in Figure 16–167. AutoCAD displays the following prompt:

Write analysis to a file? *(type y to create a file or n to not create a file)*

```
AutoCAD Text Window                                        _ □
Edit

Mass:                    105.66
Volume:                  105.66
Bounding box:      X:   0.00  --   8.00
                   Y:   0.00  --   7.00
                   Z:  -3.00  --   2.00
Centroid:          X:   3.55
                   Y:   3.50
                   Z:  -1.25
Moments of inertia:  X:  1989.68
                     Y:  2177.49
                     Z:  3640.82
Products of inertia: XY: 1312.68
                     YZ: -461.91
                     ZX: -600.08
Radii of gyration:   X:  4.34
                     Y:  4.54
                     Z:  5.87
Press ENTER to continue:
Principal moments and X-Y-Z directions about centroid:
                  I:  497.04 along [0.97 0.00 -0.25]
                  J:  681.40 along [0.00 1.00  0.00]
                  K: 1048.65 along [0.25 0.00  0.97]

Write to a file ? <N>: |                            ◄ □   ►
```

FIGURE 16–167 *Mass properties listing*

If you enter **y,** AutoCAD prompts for a file name and saves the file in ASCII format.

You can also use the AutoCAD list and AREA commands to obtain information about coordinates and areas of solid(s).

HIDING OBJECTS

The HIDE command hides, or displays in different colors, objects that are behind other objects in the current viewport. Complex models are difficult to read in wireframe form, and they become clearer when the model is displayed with hidden lines removed. HIDE considers circles, solids, traces, wide polyline segments, 3D faces, polygon meshes, and the extruded edges of objects with a thickness to be opaque surfaces hiding objects that lie behind them. The HIDE command remains active only until the next time the display is regenerated. Depending on the complexity of the model, hiding may take from a few seconds to several minutes.

Invoke the HIDE command from the View menu on the Menu Browser.

There are no prompts to be answered. The current viewport goes blank for a period of time, depending on the complexity of the model, and is then redrawn with hidden lines removed temporarily. Hidden-line removal is lost during plotting unless you specify that AutoCAD remove hidden lines in the plotting configuration.

PLACING MULTIVIEWS IN PAPER SPACE

The SOLVIEW command creates untiled viewports using orthographic projection to lay out orthographic views and sectional views. View-specific information is saved with each viewport as you create it. The information that is saved is used by the SOLDRAW command, which does the final generation of the drawing view. The SOLVIEW command automatically creates a set of layers that the SOLDRAW command uses to place the visible lines, hidden lines, and section hatching for each view. In addition, the SOLVIEW command creates a specific layer for dimensions that are visible in individual viewports. The SOLVIEW command applies the following conventions in naming the layers:

Layer	Name
For visible lines	VIEW NAME-VIS
For hidden lines	VIEW NAME-HID
For dimensions	VIEW NAME-DIM
For hatch patterns (sections)	VIEW NAME-HAT

The viewport objects are drawn on the VPORTS layer. The SOLVIEW command creates the VPORTS layer if it does not already exist. All the layers created by SOLVIEW are assigned the color white and the linetype continuous. The stored information is deleted and updated when you run the SOLDRAW command, so do not draw any objects on these layers. Invoke the SOLVIEW command from the Modeling panel on the Home tab and AutoCAD prompts:

> Enter an option *(select one of the available options)*

Choose *UCS* to create a profile view relative to a UCS. AutoCAD displays the following prompt:

> Enter an option *(select one of the available options)*

Choose *Named* to use the XY plane of a named UCS to create a profile view. After prompting for the name of the view, AutoCAD displays the following prompts:

> Enter view scale <1.0>: *(specify the scale factor for the view to be displayed)*
>
> Specify View center: *(specify the center location for the viewport to be drawn in the paper space, and press* Enter *)*
>
> Specify first corner of viewport: *(specify a point for one corner of the viewport)*
>
> Specify opposite corner of viewport: *(specify a point for the opposite corner of the viewport)*
>
> Enter view name: *(specify a name for the newly created view)*

Choose *World* to use the XY plane of the WCS to create a profile view. The prompts are the same as just described for the Named option.

Choose *?* to list the names of existing UCSs. After the list is displayed, press any key to return to the first prompt.

Choose *Current* (default option) to use the XY plane of the current UCS to create a profile view. The prompts are the same as explained earlier for the Named option.

If no untiled viewports exist in your drawing, the UCS option allows you to create an initial viewport from which other views can be created.

Choose *Ortho* to create a folded orthographic view from an existing view. AutoCAD displays the following prompts:

> Specify side of viewport to project: *(select the one of the edges of a viewport)*

Specify view center: *(specify the center location for the viewport to be drawn in the paper space, and press* Enter *)*

Specify first corner of viewport: *(specify a point for one corner of the viewport)*

Specify opposite corner of viewport: *(specify a point for the opposite corner of the* viewport*)*

Enter view name: *(specify a name for the newly created view)*

Choose *Auxiliary* to create an auxiliary view from an existing view. An auxiliary view is one that is projected onto a plane perpendicular to one of the orthographic views and inclined in the adjacent view. AutoCAD displays the following prompts:

Specify first point of inclined plane: *(specify a point)*

Specify second point of incline plane: *(specify a point)*

Specify side to view from: *(specify a point that determines the side from which you will view the plane)*

Specify view center: *(specify the center location for the viewport to be drawn in the paper space, and press* Enter *)*

Specify first corner of viewport: *(specify a point for one corner of the viewport)*

Specify opposite corner of viewport: *(specify a point for the opposite corner of the viewport)*

Enter view name: *(specify a name for the newly created view)*

Choose *Section* to create a sectional view of solids with crosshatching. AutoCAD displays the following prompts:

Specify first point of cutting plane: *(specify a point to define the first point of the cutting plane)*

Specify second point of cutting plane: *(specify a point to define the second point of the cutting plane)*

Specify side to view from: *(specify a point to define the side of the cutting plane from which to view)*

Enter view scale: *(specify the scale factor for the view to be displayed)*

Specify view center: *(specify the center location for the viewport to be drawn in the paper space, and press* Enter *)*

Specify first corner of viewport: *(specify a point for one corner of the viewport)*

Specify opposite corner of viewport: *(specify a point for the opposite corner of the viewport)*

Enter view name: *(specify a name for the newly created view)*

To exit the command sequence, press Enter.

GENERATING VIEWS IN VIEWPORTS

The SOLDRAW command generates sections and profiles in viewports that have been created with the SOLVIEW command. Visible and hidden lines representing the silhouette and edges of solids in the viewports are created and then projected to a plane perpendicular to the viewing direction. AutoCAD deletes any existing profiles and sections in the selected viewports, and new ones are generated. In addition,

AutoCAD freezes all the layers in each viewport except those required to display the profile or section. Invoke the SOLDRAW command from the Modeling panel on the Home tab and AutoCAD prompts:

Select objects: *(select the viewports to be drawn, and press* Enter *)*

AutoCAD generates views in the selected viewports.

GENERATING PROFILES

The SOLPROF command creates a profile image of a solid, including all of its edges, according to the view in the current viewport. The profile image is created from lines, circles, arcs, and/or polylines. SOLPROF will not give correct results in perspective view; it is designed for parallel projections only.

The SOLPROF command will work only when you are working in layout tab and are in model space. Invoke the SOLPROF command from the Modeling panel on the Home tab and AutoCAD prompts:

Select objects: *(select one or more objects)*

Select one or more solids, and press Enter. The next prompt lets you decide the placement of hidden lines of the profile on a separate layer:

Display hidden profile lines on separate layer?

Enter **Y** or **N**. If you answer **Y** (default), two block inserts are created: one for the visible lines in the same linetype as the original and the other for hidden lines in the hidden linetype. The visible lines are placed on a layer whose name is PV, which is the viewport handle of the current viewport. The hidden lines are placed on a layer whose name is PH, which is the viewport handle or the current viewport. If these layers do not exist, AutoCAD will create them. For example, if you create a profile in a viewport whose handle is 6, visible lines will be placed on a layer PV-6 and hidden lines on layer PH-6. To control the visibility of the layers, you can turn the appropriate layers ON and OFF.

The next prompt determines whether 2D or 3D entities are used to represent the visible and hidden lines of the profile:

Project profile lines onto a profile?

Enter **Y** or **N**. If you answer **Y** (default), AutoCAD creates the visible and hidden lines of the profile with 2D AutoCAD entities. A response of **N** creates the visible and hidden lines of the profile with 3D AutoCAD entities.

Finally, AutoCAD prompts to delete tangential edges. A tangential edge is an imaginary edge at which two facets meet and are tangent. In most of the drafting applications, the tangential edges are not shown. The prompt sequence is as follows for deleting the tangential edges:

Delete tangential edges?

Enter **Y** to delete the tangential edges or **N** to retain them.

Open the Exercise Manual PDF file for Chapter 16 on the accompanying CD for project and discipline specific exercises.

REVIEW QUESTIONS

1. What type of coordinate system does AutoCAD use?

   a. Lagrangian system

   b. Right-hand system

   c. Left-hand system

   d. Maxwellian system

   e. None of the above

2. If the origin of the current coordinate system is off the screen, the UCSICON, if it is set to Origin, will:

   a. Appear in the lower-left corner of the screen

   b. Attempt to be at the origin and will therefore not be on the screen

   c. Appear at the center of the screen

   d. Force AutoCAD to perform a ZOOM so that it will be on the screen

3. Which of the following UCS options will perform a translation only on the coordinate system?

   a. Rotate

   b. Z axis

   c. Origin

   d. View

   e. None of the above

4. Which of the following UCS options will perform only a rotation of the coordinate system?

   a. Rotate

   b. Z axis

   c. Origin

   d. View

   e. None of the above

5. Which of the following UCS options will perform both a rotation and a translation of the coordinate system?

   a. Rotate

   b. Z axis

   c. View

   d. None of the above

6. The command used to generate a perspective view of a 3D model is:

   a. VPOINT

   b. PERSPECT

   c. VPORT

   d. DVIEW

   e. None of the above

7. The VPOINT command will allow you to specify the viewing direction in all of the following methods except:

   a. 2 points

   b. 2 angles

   c. 1 point

   d. Dynamically dragging the X, Y, and Z axes

   e. None of the above; all are valid

8. Objects can be assigned a negative thickness.

   a. True

   b. False

9. Regions:

   a. Are 3D objects

   b. Can be created from any group of objects

   c. Can be used in conjunction with Boolean operations

   d. Cannot be crosshatched

10. The PFACE command:

    a. Generates a mesh

    b. Requests the vertexes of the object

    c. Can create an opaque flat polygon

    d. All of the above

11. To align two objects in 3D space, invoke the command called:

    a. ALIGN

    b. ROTATE

    c. 3DROTATE

    d. TRANSFORM

    e. 3DALIGN

12. 3D arrays are limited to rectangular patterns—that is, rows, columns, and layers.

    a. True

    b. False

13. The UNION command will allow you to select more than two solid objects concurrently.

    a. True

    b. False

14. Which of the following is not a standard solid primitive shape used by AutoCAD?

    a. SPHERE

    b. BOX

    c. DOME

    d. WEDGE

    e. CONE

15. Fillets on solid objects are limited to planar edges.

    a. True

    b. False

16. To split a solid object into two solids, use:

    a. TRIM

    b. SLICE

    c. CUT

    d. SPLIT

    e. DIVIDE

17. To display a more realistic view of a **3**D object, use:

    a. HIDE

    b. VIEW

    c. DISPLAY3D

    d. MAKE3D

    e. VIEWEDIT

18. To force invisible edges of **3**D faces to display, what system variable should be set to **1**?

    a. 3DEGE

    b. 3DFRAME

    c. SPLFRAME

    d. EDGEFACE

    e. EDGEVIEW

19. Advantages of solid modeling include:

    a. Creation of objects that can be manufactured

    b. Interfaces with computer-aided manufacturing (CAM)

    c. Analysis of physical properties of objects

    d. All of the above

20. The SECTION command:

    a. Creates a region

    b. Creates a polyline

    c. Crosshatches the area where a plane intersects a solid

    d. Is an alias for the HATCH command

    e. Will give you a choice of either A or B

21. Which AutoCAD command permits a **3D** model to be viewed interactively in the current viewport?

    a. 3DVIEW

    b. 3DPAN

    c. 3DROTATE

    d. 3DORBIT

22. When the cursor is positioned inside the arcball of the 3DORBIT command, the **3D** object can be viewed:

    a. Horizontally

    b. Vertically

    c. Diagonally

    d. All of the above

23. When the cursor is positioned outside the arcball of the 3DORBIT command, the **3D** object can be viewed:

    a. Horizontally

    b. Vertically

    c. Diagonally

    d. All of the above

    e. Both A and B

24. Which of the following is not an option in the 3DORBIT command?

    a. Pan

    b. Zoom

    c. Tilt

    d. None of the above

25. AutoCAD allows you to display a perspective or a parallel projection of the view while 3DORBIT is active.

    a. True

    b. False

26. AutoCAD allows you to view shaded view while 3DORBIT is active.

    a. True

    b. False

27. Which of the following AutoCAD commands allows you to set clipping planes for the objects in **3D** orbit view?

    a. CLIPLANE

    b. 3DCLIP

    c. 3DFACED

    d. CLIPVIEW

28. The Continuous Orbit option of 3DORBIT permits the **3D** model to be set into motion, even in shaded mode.

    a. True

    b. False

29. AutoCAD allows you to edit solid objects by:

    a. Extruding faces

    b. Copying faces

    c. Offsetting faces

    d. Moving faces

    e. All the above

30. When offsetting a face of a **3D** solid, a positive value will increase the volume of the solid.

    a. True

    b. False

31. A hole in a **3D** solid model can be relocated using the Move Faces option within the SOLIDEDIT command.

    a. True

    b. False

32. Colors of individual edges of a **3D** solid can be changed using SOLIDEDIT.

    a. True

    b. False

33. When an edge is copied using the SOLIDEDIT command, it is copied as an:

    a. Arc

    b. Spline

    c. Circle

    d. Line

    e. All of the above

34. When a composite **3D** solid is separated into individual solids, each solid element will retain the layer and color settings of the original solid.

    a. True

    b. False

35. AutoCAD refers to the removal of specified edges or vertices as:

    a. Extraction

    b. Cleaning

    c. Removing

    d. Erasing

CHAPTER
17

Rendering

INTRODUCTION

Shading or rendering turns your three-dimensional model into a realistic, eye-catching image. The AutoCAD SHADEMODE command allows you to produce quick, shaded models, and the AutoCAD RENDER command gives you control over the appearance of the final image. You can add lights and control lighting in your drawing and define the reflective qualities of surfaces in the drawing, making objects dull or shiny. Rendering creates a 2D image based on your 3D model. It shades the scene's geometry using the lighting you have set up, the materials you have applied, and environmental settings such as background and fog.

OBJECTIVES

After completing this chapter, you will be able to do the following:

- Manage visual styles
- Render a 3D model
- Create and modify ambient light, distant light, point lights, and spotlights
- Add and modify the settings for the Sun feature
- Create and modify a material
- Save an image
- Create a walkthrough or flythrough

VISUAL STYLES

AutoCAD provides a Visual Styles Manager to control the display of edges and shading of objects on the screen, primarily in 3D models. The collective settings can be saved in a Visual Style and applied when needed. Effects are visible as soon as a visual style is applied or one of its settings is changed. The selected visual style in the Visual Style Manager is indicated by a yellow border, and its settings are displayed in the panel below the sample images. You can make the necessary changes to the settings for the selected image sample and then apply them to the model.

You can also change some frequently used settings.

Open the Visual Styles Manager by choosing the down arrow from the Visual Styles panel (Figure 17–1) in the View tab.

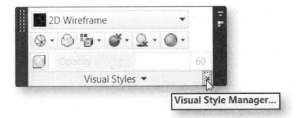

FIGURE 17–1 *Choosing the Visual Style Manager from the Visual Style panel*

AutoCAD opens the Visual Style Manager, as shown in Figure 17–2. The information displayed below the top pane varies with the selection of the visual style.

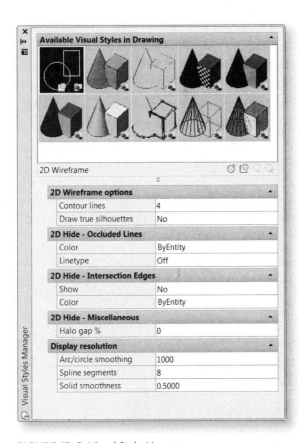

FIGURE 17–2 *Visual Style Manager*

By default, AutoCAD provides 10 default visual styles, and you can apply any one of them to the selected model:

- **2D Wireframe** uses lines and curves to represent the edges and intersections of surfaces. Linetypes, lineweights, raster, and OLE objects are visible.
- **Conceptual** applies shading to the objects and smoothes the edges between polygon faces. Shading uses the Gooch face style—a transition between cool

and warm colors rather than dark to light. The effect is less realistic, but it can make the details of the model easier to see.

- **Hidden** uses a 3D wireframe representation and hides lines and curves that are behind surfaces.
- **Realistic** applies shading to the objects and smoothes the edges between polygon faces. Materials attached to objects are displayed.
- **Shaded** displays a shaded image of the drawing and hidden lines are removed.
- **Shaded with edges** displays a smooth shaded image of objects with their edges.
- **Shades of Gray** displays a monochromatic gray shaded image.
- **Sketchy** displays a hand-sketched appearance of images using Jitter edge and Line Extensions modifiers.
- **Wireframe** also uses lines and curves to represent the edges and intersections of surfaces and adds gridlines.

Visual styles that are shaded have the faces lighted by two default distant light sources that stay with the viewpoint as you orbit the model. This lighting illuminates all faces in the model so that they can be seen. This lighting is only available when other lights, including the sun, are OFF.

RENDERING A MODEL

The AutoCAD RENDER command allows you to create realistic pictures from your AutoCAD drawings. With the RENDER command, you can adjust lighting factors, material finishes, and camera placement. All these options give you a great deal of flexibility.

Rendering tools allow you to adjust the type and quality of the rendering, set up lights and scenes, save, and replay images, but you can always use the RENDER command without any other AutoCAD render setup. By default, RENDER uses the current view if no scene or selection is specified. If no lights are specified, the RENDER command assumes a default over-the-shoulder distant light source with an intensity of 1.

Rendering creates a photographic type of image from your 3D model, shading the objects using environmental settings, such as background and fog, the lighting you've set up, and the materials you've applied. The rendering mechanism includes ray-traced reflections, refractions, and global illumination.

Invoke the RENDER command from the Render panel (Figure 17–3) in the Visualize tab:

FIGURE 17–3 *Invoking the command from the Render panel*

AutoCAD displays the Render Window (Figure 17–4) with the view rendered.

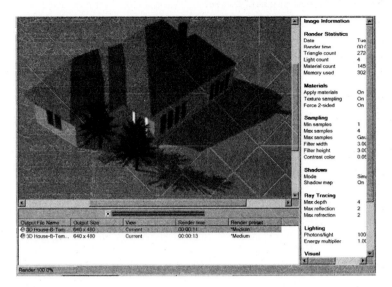

FIGURE 17–4 *Render Window*

The Render Window shows the result of the RENDER command on the current model.

The Render Window has three panes. The Image pane displays the rendered image. The Statistics pane, on the right, lists the current settings, and the History pane, at the bottom left, shows the image history from the current model.

The Render Window allows you to save or copy the image to a file, check the render settings, see the rendering history, purge and/or delete an image, and zoom or pan in, out, or around the image.

The upper bar contains the following menus:

- Under the File menu, choose **Save** to save rendered images to a bitmap file. Choose **Save Copy** to save a copy of an image to a new location without affecting the location stored with the current entry. Choose **Exit** to close the Render Window.
- Under the View menu, choose **Status Bar** to display the status bar below the History pane. Choose **Statistics Pane** to display the entire Statistics pane.
- Under the Tools menu, choose **Zoom +** to enlarge the rendering in the Image pane. Choose **Zoom –** to decrease the rendering in the Image pane.

The upper-left portion of the Render Window is the Image pane where the rendered image is displayed.

The lower-left portion of the Render Window contains the History pane, listing information about each rendering.

- Under Output File name, you will find the name of the rendering with an indicator that identifies the type of rendering.
- Under Output Size, you will find the resolution of the image.
- Under View, you will find the name of the view that is being rendered.
- Under Render time, you will find the hours, minutes, and seconds.
- Under Render preset, you will find the name of the preset used for the rendering.

Right-clicking on a history entry displays a shortcut menu that contains the following options: *Render Again, Save, Save Copy, Make Render Settings Current, Remove From The List*, and *Delete Output File*.

The right-hand Image Information portion of the Render Window contains the Statistics Pane, where the render settings are displayed.

This information is derived from settings made in the Presets Manager dialog box along with information that is generated at the time of the rendering.

In addition to the Statistics, the right-hand Image Information pane includes the following data groups:

The Materials group reports how materials are handled by the renderer.

- The **Apply Materials** setting, when set to ON, allows you to apply the surface materials that you define for an object. If **Apply Materials** is not ON, all objects in the drawing assume the color, ambient, diffuse, reflection, roughness, transparency, refraction, and bump map attribute values defined for the GLOBAL material (default).
- The **Texture Sampling** setting, when set to ON, causes texture maps to be filtered.
- The **Force 2-Sided** option, when set to ON, causes both sides of faces to be rendered.

The Sampling group reports how the renderer performs sampling.

- The **Min Samples** setting value determines the minimum sample rate, which is the number of samples per pixel.
- The **Max Samples** setting value determines the maximum sample rate. The value of Min Samples cannot exceed the value of Max Samples.
- The **Filter Type** setting determines how multiple samples are combined into a single pixel value. The filter types include Box, Gauss, Triangle, Mitchell, and Lanczos.
- The **Filter Width** and **Filter Height** settings determine the size of the filtered area.
- The **Contrast Color** settings determine the red, green, and blue threshold values.

The Shadows group reports how shadows appear in the rendered image.

- The **Mode** settings include Simple, Sort, and Segments. **Simple** means that shadow shaders are generated in random order. **Sort** means that shadow shaders are generated in order from the object to the light. **Segments** means that shadow shaders are generated in order along the light ray from the volume shaders to the segments of the light ray between the object and the light.
- The **Shadow Map** settings report if shadow mapping is used to render shadows. When set to ON, the renderer will render to the shadows. When OFF, all shadows are ray traced.

The Ray Tracing group reports settings that affect the shading of a rendered image.

- The **Max Depth** settings show the limits of the combination of reflection and refraction. Ray tracing stops when the total number of reflections and refractions reaches the maximum depth.

- The **Max Reflection** settings determine the number of times a ray can be reflected. At 0, there is no reflection. At 1, the ray can be reflected only once. At *N*, the ray can be reflected *n* times.

- The **Max Refraction** settings determine the number of times a ray can be refracted. At 0, there is no refraction. At 1, the ray can be refracted only once. At *N*, the ray can be refracted *N* times.

The Lighting group reports how lights behave when calculating indirect illumination. By default, the energy and photon settings apply to all lights in a scene.

- The **Photons/Light** setting determines the number of photons emitted by each light for use in global illumination. Increasing this value increases the accuracy of global illumination, but it also increases the amount of memory used and the length of render time. Decreasing this value improves memory usage and render time, and it can be useful for previewing global-illumination effects.

- The **Energy Multiplier** setting multiplies the global illumination, indirect light, and intensity of the rendered image.

The Visual group reports renderer behavior settings.

- The **Grid** setting shows the coordinate space of objects, the world, or the camera.

- The **Grid Size** setting determines the size of the grid.

- The **Photon** setting, when set to ON, renders the effect of a photon map. This action requires that a photon map be present. If no photon map is present, the photon rendering looks like the nondiagnostic rendering of the scene: the renderer first renders the shaded scene and then replaces it with the pseudocolor image.

- The **BSP** setting, when set to ON, renders a visualization of the parameters used by the tree in the BSP ray-trace acceleration method.

The Processing group reports how the image is processed.

- The **Tile Size** setting determines the tile size for rendering.

- The **Tile Order** setting determines the method used, or render order, for tiles as an image is rendered.

- The **Memory Limit** setting determines the memory limit for rendering. If the memory limit is reached, the geometry for some objects is discarded in order to allocate memory for other objects.

SETTING UP LIGHTING

AutoCAD's render facility gives you control over four types of lights in your renderings:

- **Ambient light** can be thought of as background light that is constant and distributed equally among all objects.

- **Distant light** gives off a fairly straight beam of light that radiates in one direction. Another property of distant light is that its brilliance remains constant, so an object close to the light receives as much light as a distant object.

- **Point light** can be thought of as a ball of light. A point light radiates beams of light in all directions. Point lights also have more natural characteristics. Their brilliance may be diminished as the light moves away from its source. An object that is near a point light appears brighter; an object that is farther away will appear darker.

- **Spotlight** is very much like the kind of spotlight you might be accustomed to seeing at a theater or auditorium. Spotlights produce a cone of light toward a target that you specify.

The Lights panel (Figure 17–5) in the Render tab allows you to manage existing and default lighting and to create new light sources.

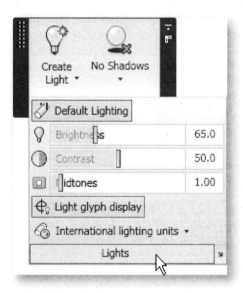

FIGURE 17–5 *Lights panel*

The Create Light drop-down menu lets you create **Point**, **Spot**, **Distant**, and **Weblight** light sources (see Figure 17–6).

The Point light creates a point light that radiates light in all directions from its location. Choose **Point** and AutoCAD prompts:

```
Specify source location <0,0,0>: (specify a point to locate a
new point light)
```

The Spot light creates a spotlight that emits a directional cone of light. Choose **Spot** and AutoCAD prompts:

```
Specify source location <0,0,0>: (specify a point to locate a
new spotlight)

Specify target location <0,0,-10>: (specify a point to
locate a new target)
```

The Distant light creates a distant light. Choose **Distant** and AutoCAD prompts:

```
Specify light direction FROM <0,0,0> or [Vector]: (specify a
point to locate a new distant light)

Specify light direction TO <1,1,1>: (specify a point
establish the light direction)

Enter an option to change [Name/Intensity/Status/shadoW/
Color/eXit] <eXit>: (press Enter or choose an option from the
shortcut menu)
```

The Weblight option creates a web light. Choose **Weblight** and AutoCAD prompts:

Specify source location: *(specify source location)*

Specify target location: *(specify a point to locate the target)*

Enter an option to change [Name/Intensity factor/Status/Photometry/WeB/shadoW/filterColor/eXit] <eXit>: *(press* Enter *or choose an option from the shortcut menu)*

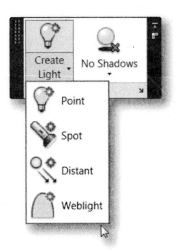

FIGURE 17–6 *Selection of Light sources from Lights panel*

The Shadows drop-down menu lets you create **No Shadows, Ground Shadows,** and **Full Shadows** generated from the specified light sources.

Choosing **Default Lighting** lets you switch between default lighting and User light/Sunlight sources.

Slider bars let you adjust the **Brightness, Contrast,** and **Midtones** in the rendering.

Choose **Light Glyphs display** to switch between ON and OFF. Light glyphs are the icons that show the location(s) of point lights and spotlights.

Clicking the arrow at the right of the Lights panel title bar causes the Lights in Model window to be displayed, which lists the lights in the model (Figure 17–7). You can select one or more lights, right-click, select Delete to remove the light(s), or select Properties to display the Properties palette.

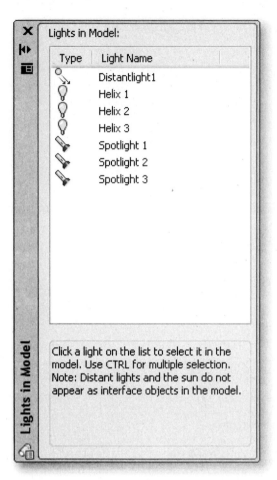

FIGURE 17–7 *Lights in Model window*

SUN & LOCATION

The Sun & Location panel of the Render tab allows you adjust the effects of the sun, sunlight, and light from the sky on your model. You can also set the location of your model on the globe to compensate for the affect of the sun at the specific location.

In the Sun & Location panel (Figure 17–8) in the Render tab, select one of the icons for managing the sunlight in the model.

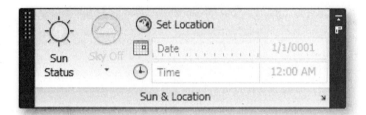

FIGURE 17–8 *Materials panel*

Choosing **Sun Status** switches between the sun source being ON or OFF.

The **Date** slider bar allows you to set the date, which determines in part the location of the Sun source.

The **Time** slider bar allows you to set the time of day, which determines in part the location of the Sun source.

Choose **Set Location** to determine the latitude and longitude of the model.

Clicking the arrow at the right of the Sun & Location title bar causes the Sun Properties window to be displayed, from which you can change the settings for the Sun source (Figure 17–9).

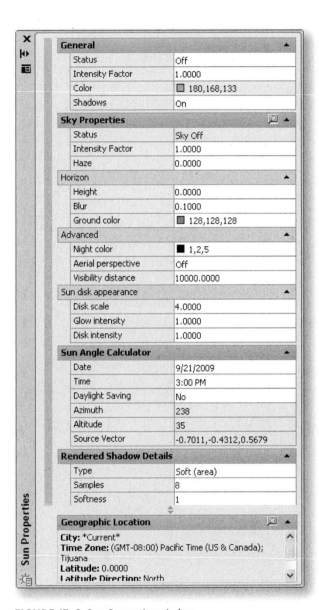

FIGURE 17–9 *Sun Properties window*

The General section lets you set the general properties of the sun.

- The **Status** setting turns the sun ON and OFF.
- The **Intensity Factor** setting lets you set the intensity or brightness of the sun. The range is from 0 (no light) to maximum. Higher settings brighten the light.

- The **Color** setting lets you control the color of the light. Enter a color name or number or click Select Color to open the Select Color dialog box.
- The **Shadows** setting turns the shadows for the sun ON and OFF.

The Sky Properties section lets you set the general properties of the sky.

- **Status** lets you determine if the sky illumination is computed at render time. It makes the sky available as a gathered light source for rendering.
- The **Intensity Factor** setting lets adjust the effect of the skylight.
- The **Haze** setting lets you set the intensity of the scattering effects of the atmosphere.

The Horizon section lets you control the appearance and location of the ground plane.

- The **Height** setting lets you determine the absolute position of the ground plane relative to the world zero.
- The **Blur** setting lets you determine the amount of blurring between the ground plane and sky.
- The **Ground Color** setting lets you determine the color of the ground plane.

The Advanced section lets you apply various artistic effects.

- The **Night Color** setting lets you specify the color of the night sky.
- The **Aerial Perspective** setting lets you specify if aerial perspective is applied.
- The **Visibility Distance** setting lets you specify the distance at which 10% haze occlusion results.

The Sun Disk Appearance section lets you control the appearance of the sun disk.

- The **Disk Scale** setting lets you specify the scale of the sun disk.
- The **Glow Intensity** setting lets you specify the intensity of the sun.
- The **Disk Intensity** setting lets you specify the intensity of the sun disk.

The Sun Angle Calculator lets you set the angle of the sun.

- **Date** displays the current date setting.
- **Time** displays the current time setting.
- **Daylight Savings** displays the current daylight saving time setting.
- **Azimuth** displays the angle of the sun along the horizon clockwise from due north.
- **Altitude** displays the angle of the sun vertically from the horizon.
- **Source Vector** displays the direction of the sun.

The Rendered Shadow Details section lets you specify the properties of shadows.

The Geographic Location section displays the current geographic location settings.

ATTACHING MATERIALS

The Materials control panel of the Render tab allows you attach materials to surfaces and to modify the light reflection characteristics of the objects that you render. By modifying these characteristics, you can make objects appear rough or shiny. These finish characteristics are stored in the drawing via surface property blocks.

In the Materials panel (Figure 17–10) in the Render tab, select one of the icons for managing materials in the model.

FIGURE 17–10 *Materials panel*

Choose **Materials Browser** to navigate and manage materials.

Choose **Material/Textures On** to turn ON the display of materials.

Choose **Material Mapping** to cause the materials to be mapped to planar surfaces.

Choose **Attach by Layer** to cause the materials to be attached to all objects on the layer whose Material property is set to BYLAYER.

- Choose **Copy Mapping Coordinates** to apply the mapping from the original object or face to the selected objects. This duplicates a texture map along with any adjustments to other objects

- Choose **Copy Mapping Coordinates** to reset the UV coordinates to the default values for the map. This reverses the effects of all previous adjustments made with the mapping gizmo to the position and orientation of the map.

Materials Browser Window

The logical place to begin attaching materials to objects in the drawing is by going to a library of materials. AutoCAD has a large library of materials that is accessible by way of the Materials Browser. When the Material Browser is selected, AutoCAD displays the Materials Browser window as shown in Figure 17–11.

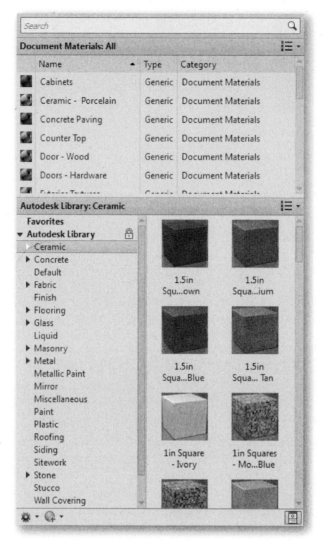

FIGURE 17–11 *Materials Browser Window*

The Materials Browser window lets you find and manage materials that have been saved in libraries. The top section is a search window to find materials that appear in various libraries. The **Document Materials** panel displays the materials saved in the current drawing. The **Library** panel lists the categories of materials in the selected library. A category selected in list on the left will cause the materials included to be displayed in the panel on the right. Click on the right end of the **Document Materials** panel title bar (see Figure 17–12) and AutoCAD displays a dropdown menu (see Figure 17–13) to access options for filtering displayed materials.

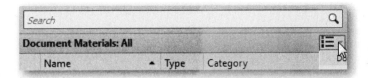

FIGURE 17–12 *Invoking Filter Menu*

FIGURE 17–13 *Filter Menu*

Materials Editor Window

To edit materials loaded in the current drawing, double-click the name of the material in the **Document Materials** panel and AutoCAD displays the Materials Editor Window (see Figure 17–14) showing the Cabinets materials properties.

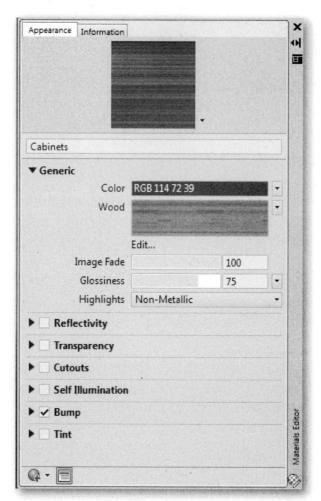

FIGURE 17–14 *Materials Editor Window*

The Materials Editor Window allows you to make adjustments in **Reflectivity, Transparency, Cutouts, Self Illumination, Bump** and **Tint.** Also, settings to **Image Fade, Glossiness** and **Highlights** can be made.

In the bottom left corner of both the Materials Browser and Materials Editor windows are icons for creating new materials.

CREATING A WALKTHROUGH

The 3DWALK command allows you to simulate a walkthrough of your 3D model. The path that the walkthrough, or flythrough, will follow is controlled using the Walk and Fly Settings dialog box.

Navigation

With the 3DWalk and 3DFly modes, you can walk or fly through 3D space in and around your model by pressing a set of shortcut keys such as the arrow keys. The camera can be moved forward, backward, right, and left using the arrow keys, and the camera target can be changed with the mouse. In 3DFly mode, motion is not confined to the XY plane as in 3DWalk mode.

Animation

A walkthrough animation of your scene can be recorded, saved, and played back by placing a camera on a path, setting the height for the camera, turning the camera, and viewing a preview.

Other parameters and settings are described in the AutoCAD Help facility for the 3DWALK command.

REVIEW QUESTIONS

1. Text is ignored by the SHADEMODE command.

   a. True

   b. False

2. Approximately where is the light source for the SHADE command?

   a. It is pure ambient light, or all around

   b. Just behind your right shoulder

   c. Directly behind the object

   d. In the upper-right corner of the screen

3. It is possible to make a slide of a shaded image.

   a. True

   b. False

4. If you save your drawing while a shaded view is displayed, the preview of the drawing will appear shaded.

   a. True

   b. False

5. Which of the following is not a valid type of light to add to a drawing with the RENDER command?

   a. Point

   b. Spot

   c. Distant

   d. Ambient

   e. All are valid

6. It is possible to make a slide of a rendered image.

   a. True

   b. False

7. If a single-point light source exists in your drawing, objects will be darker if you set the attenuation for that light to:

a. None

b. Inverse linear

c. Inverse square

d. Both b and c

8. Distant light sources do not attenuate, or get dimmer with distance.

a. True

b. False

9. Material finishes can be assigned based on the color of the object in AutoCAD.

a. True

b. False

10. Which of the following file types is not available to the SAVEIMG command?

a. .eps

b. .gif

c. .tga

d. .eps

11. By default, all light types emit white light.

a. True

b. False

12. The smoothing angle is the minimum angle:

a. Between faces that AutoCAD will ignore

b. Between the viewing angle and the face that AutoCAD will display

c. Between the incident light and the face that AutoCAD will still reflect

d. None of the above

13. Both 2D and 3D models can be rendered using the SHADEMODE command.

a. True

b. False

14. The image option allows which type of file to be used as a background image?

a. .gif

b. .eps

c. .jpg

d. .eps

15. Which AutoCAD command is used to control the position from which a **3**D model is viewed in model space?

    a. VPORTS

    b. VIEWRES

    c. RENDER

    d. VPOINT

16. Which AutoCAD command is used to adjust the light-reflection characteristics of an object?

    a. REFL

    b. LGTRFL

    c. LIGHTREF

    d. RMAT

17. Which of the following functions depicts the point at which a light's value fades to zero?

    a. Fade

    b. Highlight

    c. Falloff

    d. Diminish

Customizing AutoCAD

INTRODUCTION

Off the shelf, AutoCAD is extremely powerful, but, like many popular engineering and business software programs, it does not automatically do all things for all users. It does—probably better than any software available—permit users to make changes and additions to the core program to suit individual needs and applications. Word processors offer a feature by which you can save a combination of many keystrokes and invoke them at any time with just one or a combination of two keystrokes or by choosing a button on a user-created toolbar. This automation is known as a *macro*. Spreadsheet and database management programs, as well as other types of programs, have their own library of user functions that can be combined and saved as user-named, custom-designed commands. Using these features to make your copy of a generic program unique and more powerful for your particular application is known as *customizing*.

Customizing AutoCAD can include several facets requiring various skill levels. The topics in this chapter include creating custom linetypes, shapes, and hatch patterns. With the customize user interface (CUI) dialog box, you can customize the interface features used in a drawing session such as ribbon tabs and panels, workspaces, toolbars, menus, shortcut menus, Quick Properties, and keyboard shortcuts.

OBJECTIVES

After completing this chapter, you will be able to customize the following:

- Ribbon Tabs and Panels
- Workspaces
- Toolbars
- Menus
- Shortcut menus
- Keyboard shortcuts
- Quick Properties
- Mouse buttons

Other customizations include the following:

- Loading partial CUIx files

- Creating custom linetypes, hatch patterns, shapes, and fonts

CUSTOMIZING THE USER INTERFACE

To customize workspaces, toolbars, menus, shortcut menus, keyboard shortcuts, or mouse buttons, select User Interface from the Customization panel on the Manage tab while in the 2D Drafting & Annotation or 3D Modeling workspace. This invokes the `CUI` command. AutoCAD displays the Customize User Interface (CUI) dialog box (Figure 18–1). The Customize User Interface (CUI) Editor is used to modify customization that is in the XML-based CUIx file. The editor allows you to create and manage commands that are used in the CUIx file in a centralized location. Along with commands, you are able to customize many of the different user interface elements.

NOTE

If you are in the AutoCAD Classic workspace, you can right-click on any panel or toolbar and choose *Customize* from the shortcut menu and AutoCAD displays the Customize User Interface Editor in a collapsed state.

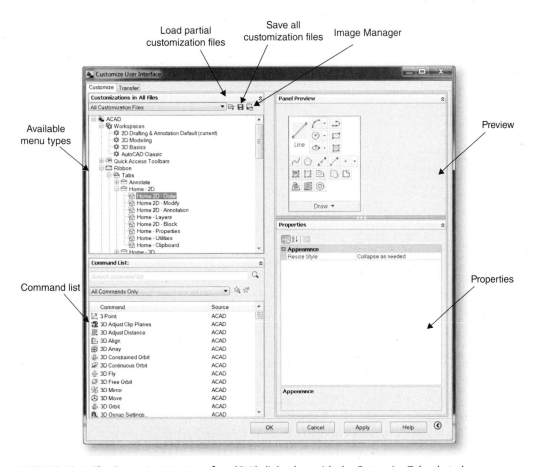

FIGURE 18–1 *The Customize User Interface (CUI) dialog box with the Customize Tab selected*

Panes shown on this tab include the **Customizations in All CUI Files, Preview, Command List,** and **Properties** panes. Various panes are displayed, depending on the selection made in the list box, and they will be explained along with the node selection with which they are associated.

The buttons to the right of the textbox in the **Customizations in All CUI Files** pane include options to **Load partial customization file, Save all current customization files,** and **Image Manager.**

Workspaces

The CUI dialog box can be used to edit existing workspaces and create new ones. Application of the workspace concept and using the WORKSPACE command are discussed in Chapter 14.

To customize workspaces, right-click on any panel or toolbar and choose CUSTOM-IZE from the shortcut menu. AutoCAD displays the CUI dialog box with the **Customize** tab chosen and the node named **Workspaces** selected in the list of nodes (Figure 18–2). The **Information** pane, when a node is selected, gives a description of the node—Workspaces, in this case—and how it can be used and customized.

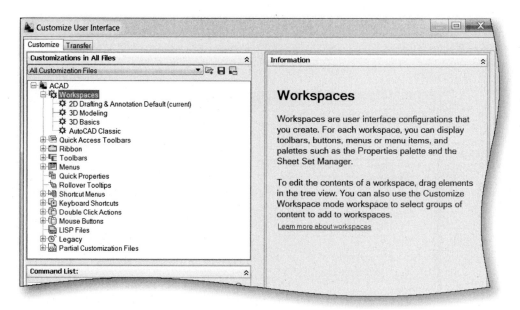

FIGURE 18–2 *Upper half of the Customize Tab of the CUI dialog box with Workspaces selected and the Information Pane displayed*

In the CUI dialog box, open the **Workspaces** node: double-click or select the + in the square next to the Workspace node. The named workspaces are displayed in tree mode. If you select the 2D Drafting & Annotation workspace (Figure 18–3), its contents are displayed in the **Workspace Contents** pane. Other panes displayed include the **Command List** and **Properties** panes.

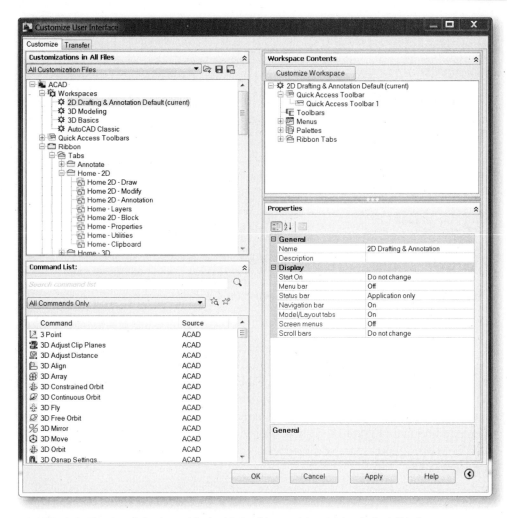

FIGURE 18–3 *The CUI dialog box with the 2D Drafting and Annotation Default Workspace selected and the Workspace Contents, Command List, and Properties panes displayed*

To create a new workspace, select the Workspace node or one of the individual workspaces in the **Customizations in All Files** pane, and right-click to display the shortcut menu. Choose *New Workspace* from the shortcut menu.

AutoCAD adds the new workspace to the list and assigns it the name Workspace1. It is in an editable textbox where you can change its name. The **Workspace Contents** pane displays the new workspace with a list of empty nodes—Quick Access Toolbar, Toolbars, Menus, Palettes, and Ribbon Tabs—in a tree view (Figure 18–4).

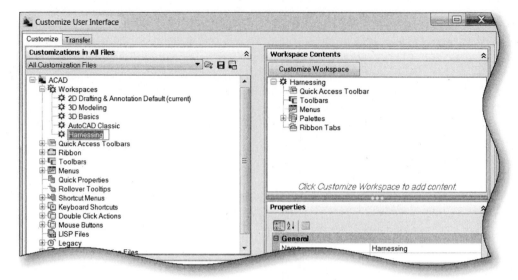

FIGURE 18–4 *New Workspace Harnessing displayed in the CUI dialog box panes*

To populate the workspace with nodes, choose **Customize Workspace** button at the top of the **Workspace Contents** pane and then from the list in the **Customizations in All Files** pane, expand a node such as Ribbon Tabs by selecting the adjacent box or double-clicking. AutoCAD lists the available tabs with empty check boxes beside them. To add a ribbon tab to the newly named workspace, check the box next to the desired ribbon tab. Figure 18–5 shows the addition of the Insert tab to the newly created Harnessing workspace and it appears under the Ribbon Tabs in the **Workspace Contents** pane.

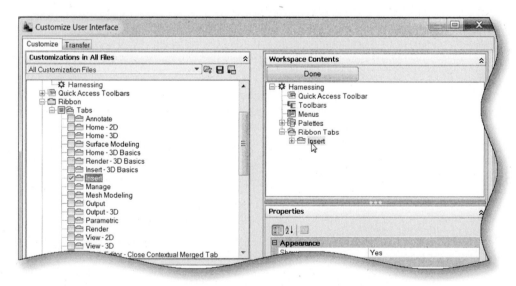

FIGURE 18–5 *The CUI dialog box with the Blocks and References ribbon tab added to a new Workspace*

The **Properties** pane allows you to customize the appearance properties, such as orientation and size of the selected panel. These properties can be edited in this pane by selecting the value of the property and then changing its value in place. After the workspace has been named and the nodes arranged, choose **Done** and then **Apply** at the bottom of the CUI dialog box to save the new workspace. Changing the orientation, location, and shape of a toolbar in this manner may not be as exact as it is when you place the toolbar by dragging it and releasing it with the cursor. See the explanation in Chapter 14 of how to save a workspace by using the WORKSPACE command for more exact control of the toolbar and panel location.

NOTE

At the top of all *Properties* panes are three buttons to sort properties according to categories, sort properties alphabetically, and toggle the display of the tips box below the properties grid. When selected, these buttons act in accordance with their respective titles.

To make the new workspace current, right-click on its name in the list in the **Customizations in All Files** pane, and choose *Set Current*. Other options in the shortcut menu include *Customize Workspace, Duplicate, Rename, Delete, Find,* and *Replace.* The *Duplicate* option lets you create a new workspace by using the selected one as a basis and then you can make necessary changes.

Ribbon Panels

You can customize the ribbon by creating and modifying ribbon panels and organizing ribbon panels into groups of task-based tools with ribbon tabs.

As explained in Chapter 1, panels contain the buttons that invoke a command, mode, or custom macro, or display a dialog box when chosen.

Customizing panels can make your drawing tasks easier and more efficient. Frequently used buttons can be consolidated on one panel, and rarely used buttons can be removed or hidden. You can also specify a text string to be displayed when the cursor is near a button. You can create custom panels and create or change the button image associated with a command.

Ribbon panels are organized by rows, sub-panels, and panel separators. A ribbon can have one or more rows. A row, similar to a toolbar, determines the order and position that commands and controls appear on the ribbon panel. Rows run horizontally on a ribbon panel. If all the commands and controls cannot be displayed on the ribbon panel, a gray down arrow is displayed for expanding the ribbon panel. Rows can be divided using a sub-panel, which holds rows to order and position commands and controls.

A panel separator is automatically added to each ribbon panel and controls which rows are shown by default. Rows located below a panel separator are displayed only when a ribbon panel is expanded. Figure 18–6 shows the various parts of the Draw panel.

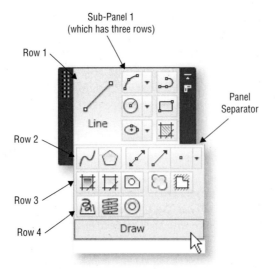

FIGURE 18–6 *Parts of the Draw panel*

To create a new panel, select the Panels node under the Ribbon category or one of the individual panel in the **Customizations in All Files** pane of the CUI dialog box. Right-click and choose *New Panel* from the shortcut menu (Figure 18–7).

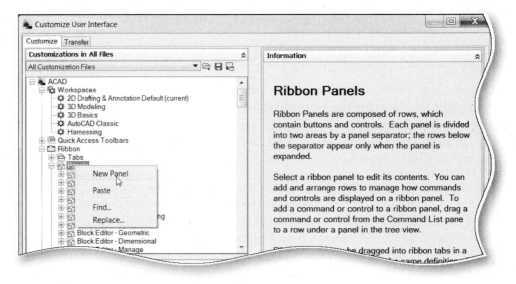

FIGURE 18–7 *Creating a new panel from the flyout menu*

AutoCAD creates a new panel with the name Panel1 in an editable textbox where you can enter a name for the new panel. Initially the newly created panel has no commands or buttons assigned to it. By default, with the new panel, AutoCAD creates Row 1 and Panel Separator. To create additional rows, select the newly created panel, right-click and choose NEW ROW from the shortcut menu. Similarly, you can also add sub-panel to the selected row.

If necessary, you can make changes to the properties of the newly created panel, row, and/or sub-panel.

Select the new panel in the tree view, and make necessary changes in the Properties pane:

The General section displays and lets you change the **Name, Display Text,** and **Description** of the newly created panel.

The Appearance section displays the number of rows that are on a ribbon panel.

The Advanced section displays the aliases and element IDs that you can define for each user interface element.

Adding Commands to a Panel Using the CUI Dialog Box

AutoCAD displays a list of commands that are loaded in the program in the **Command List** pane for the selected category. From the **Categories** list box, you can select a filter so that only commands in the selected category will be displayed in the main list box below. Categories include **All Commands Only, All Commands and Controls, ACAD Commands, EXPRESS Commands, AUTODESKSEEK Commands, Custom Commands, Toolbar Control Elements, Ribbon Control Elements, File, Edit, View, Insert, Format, Tools, Draw, Dimension, Modify, Window, Help,** and **Legacy** (Figure 18–8). The **Search command list** text box allows you to search for commands by typing in one or more characters and the list will display commands containing those characters.

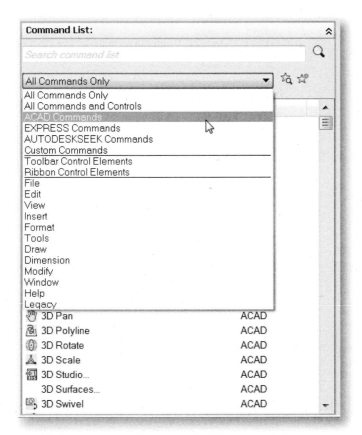

FIGURE 18–8 *Command categories in the Command List pane*

To add a command to the selected row, drag the command from the **Command List** pane to a location just below the selected name in the **Customizations in All Files** pane.

When you choose an individual command that is added to the newly created row in the **Customizations in All Files** pane, the modified panel is displayed in the **Panel Preview** pane. Under the **Properties** pane, AutoCAD displays the lists of properties for the selected command under the headings of Display, Appearance, **Command, Advanced,** and **Images** (Figure 18–9).

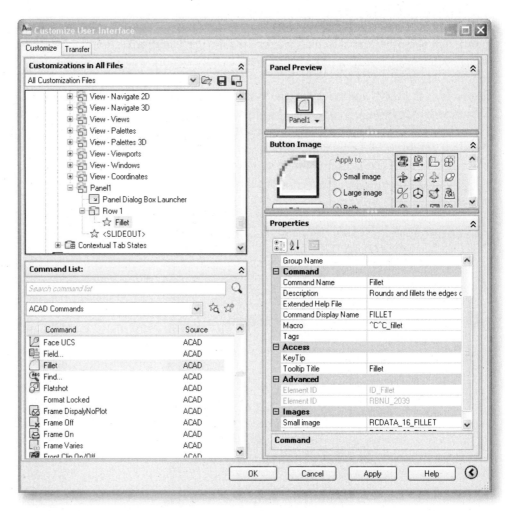

FIGURE 18–9 *The CUI dialog box with the selected command properties*

The Display section determines which user interface elements are displayed after start up or when a workspace is set current.

The Appearance section controls the appearance of commands on a ribbon panel. **Orientation** specifies the orientation of the command (Horizontal or Vertical). The **Size** specifies the button size for a command (Standard, Medium, or Large). **Show Label** specifies if a label is displayed near the command (Yes or No).

The Command section displays the properties assigned to a command. The **Name** you enter is the label or tooltip name displayed in the program. The **Description** you enter is displayed in the status bar or in a tooltip. **Extended Help File** displays the file name and ID from the extended help file that will be displayed when the cursor continues to hover over a panel button for a specified period of time. **Command Display Name** displays the command line text string that is shown in the command tooltip. **Macro** displays the macro assigned to the selected command. The **Tags** displays the user-defined keywords that are associated with a command. Tags can be used to search for commands in the Application Menu. The Advanced section displays the element IDs that you can define for each user interface element.

The Images section determines which images are assigned to a command when displayed on a user interface element. **Small Image** specifies the small image file to use when a command is added to a panel. **Large Image** specifies the large image file to use when a command is added to a panel.

In addition to adding commands to the selected row, you can also delete the commands from the selected row. After making necessary changes, choose **Apply** to apply the changes to user interface.

Tabs

Ribbon tabs control the display and order of ribbon panels on the ribbon. You can add ribbon tabs to a workspace to control which ribbon tabs are displayed on the ribbon.

Ribbon tabs do not contain any commands or controls like a ribbon panel does; instead, they manage the display of ribbon panels on the ribbon. Once a ribbon tab is created, you can then add references to any of the ribbon panels from the Ribbon Panels node. References to a ribbon panel can be created by using copy and paste, or by dragging a ribbon panel from the Ribbon Panels node onto the ribbon tab.

After references to ribbon panels are added to a ribbon tab, you can change the order in which the ribbon panels appear by dragging them up or down under the ribbon tab's node in the Workspace Contents pane. The initial display order is set by the order in which panels appear under the ribbon tab's node. New ribbon tabs are added automatically to all workspaces, just like new toolbars and menus are. If you do not want the ribbon tab on the ribbon for a particular workspace, remove it from the workspace using the Workspace Contents pane.

To create a new tab, select the Tabs node under the Ribbon category or one of the individual tabs in the **Customizations in All Files** pane of the CUI dialog box. Right-click and choose NEW TAB from the shortcut menu (Figure 18–10).

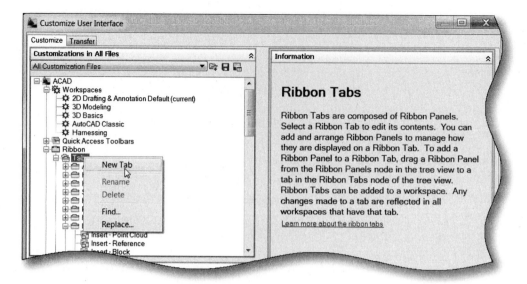

FIGURE 18–10 *Creating a new tab from the flyout menu*

AutoCAD creates a new tab with the name New Tab in an editable textbox where you can enter a name for the new tab. Initially the newly created tab has no panels assigned to it. If necessary, you can make changes to the properties of the newly created tab.

Select the new tab in the tree view, and make necessary changes in the Properties pane:

The General section displays and lets you change the **Name, Display Text,** and **Description** of the newly created tab.

The Advanced section displays the aliases and element IDs that you can define for each user interface element.

To add a panel to the selected tab, drag the ribbon panel from the **Ribbon Panels** pane to a location just below the selected tab in the **Customizations in All Files** pane.

Quick Access Toolbar

The Quick Access Toolbar is located to the right of the Application Menu and provides direct access to a defined set of commands. You can locate the Quick Access toolbar above or below the Ribbon. You can add, remove, and reposition commands on the Quick Access Toolbar. You can add as many commands as you want. If there is no room available, commands roll into a flyout button.

You can define multiple Quick Access toolbars, but only one can be assigned to a workspace at a time. To designate which toolbar is displayed as the active Quick Access toolbar, select a workspace and assign the desired toolbar to the Quick Access toolbar node and set the workspace current.

To create a new Quick Access toolbar, select the Quick Access Toolbars node or one of the individual Quick Access toolbars in the **Customizations in All Files** pane of the CUI dialog box. Right-click and choose new *New Access Quick Toolbar* from the shortcut menu (Figure 18–11).

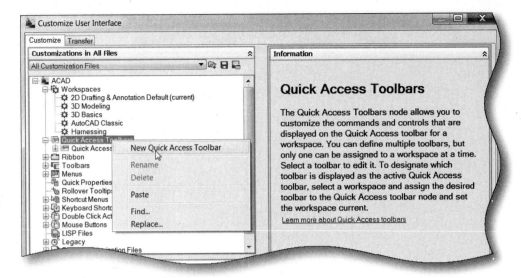

FIGURE 18–11 *Available commands in the Quick Access Toolbar*

Under the **Command List** pane, drag the command you want to add to the Quick Access Toolbar in the **Customizations in All Files** pane. A splitter bar will be displayed under the Quick Access Toolbar node to indicate the location where the command will be added when the pointing device button is released. Once the splitter bar is in the location you want to insert the command, release the pointing device button. If you want to remove one of the commands from the Quick Access Toolbar, right-click the command you want to remove from the Quick Access Toolbar and click *Remove*. AutoCAD removes the selected command from the Quick Access Toolbar.

You can also add commands directly to the Quick Access toolbar from the flyout menu displayed by choosing the down arrow (see Figure 18–12). If you do not see the command that you want to add in the flyout menu, then choose *More Commands ...* and select one of the available commands to add to the Quick Access toolbar from the command list. Drag and drop the selected command onto the Quick Access toolbar. To remove a command, right-click on the icon to be removed and choose *Remove from Quick Access Toolbar* from the shortcut menu.

FIGURE 18–12 *Quick Access toolbar with the flyout menu*

Toolbars

Toolbars can be customized similar to ribbon panels. Commands can be selected and dragged onto a toolbar in the same manner as dragging them onto a panel.

Custom Commands (Macros)

In addition to placing standard AutoCAD commands on toolbars and panels, you can create custom commands from macro syntax and AutoLISP or combinations of the two. These custom commands can then be added to existing or custom-created panels and toolbars.

Creating New Commands Using the CUI Dialog Box

AutoCAD allows you to create a new command from scratch or edit the properties of an existing command. When you create or edit a command, the properties you can define are the command name, description, macro, large or small image, and element ID for new commands. When you change any properties of a command in the **Command List** pane, the command is updated for all interface items that reference that command.

To create a new command, choose **New Command** from the **Command List** pane (see Figure 18–13). AutoCAD creates a new command in the list with the name Command1. In Figure 18–14, three new commands were added to the Furniture toolbar. They are named **Chair, Computer,** and **Desk.**

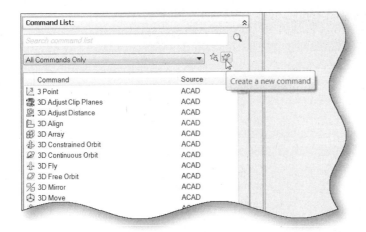

FIGURE 18–13 *Choosing Create a new command from the Command List pane*

Choose the newly created command and AutoCAD displays under the **Properties** pane the lists of properties for the selected command under the headings of **Command, Advanced,** and **Images.**

The Command section displays the name and description of the selected command. In the **Name** field, enter a name for the command. The name will be displayed as a Tooltip or menu name when you select this command. In the **Description** field, enter a description for the command. The description will be displayed on the status bar when the cursor hovers over the menu item or toolbar button.

In the **Macro** field, you can create a macro or edit an existing macro. See the explanation of Menu Macro Syntax later in this chapter. The macro line "^C^C-insert;chair; \;;\\ \ \" (Figure 18–14) uses the INSERT command to insert a block named Chair and pauses for you to specify an insertion point and rotation angle on the screen. This command macro requires that a block named Chair be available for insertion or an error message will result. It does not pause for you to specify the X and Y scale factors, but it automatically invokes the Enter key for these prompts.

The Advanced section lets you enter an element ID for the command for new commands. You cannot modify the element ID of an existing command.

Provide an image resource file as the button's icon in the Images section.

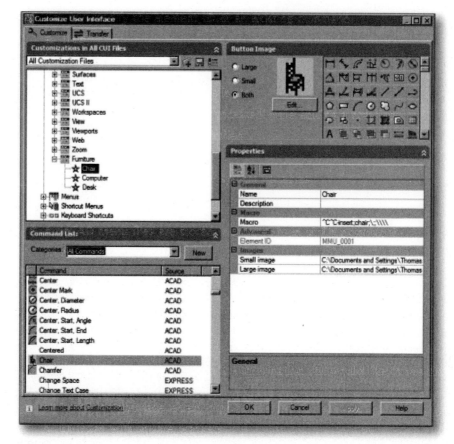

FIGURE 18–14 *The CUI dialog box with the custom command named "Chair" highlighted*

In the **Button Image** pane, you can select one of the available icon images from the list and, if it is not suitable, you can edit it or start from scratch. Choose **Edit,** and AutoCAD displays the Button Editor in which you can modify or create button images.

The **Button Image** window displays a thumbnail of the image as it is edited.

Choosing **Pencil** lets you edit one pixel at a time using the selected color.

Choosing **Line** lets you draw lines using the selected color. Click to set the start point, draw the line, drag, and then release to complete the line.

Choosing **Circle** lets you draw circles using the selected color. Click to set the center and drag to set the radius, and then release to complete the circle.

Choosing **Erase** lets you erase pixels. Click and drag over colored pixels to erase them.

The **Color Palette** lets you select the color palette used by the editing tools. Choosing **More** opens the **True Color** tab in the Select Color dialog box. When you select a color, it is displayed in the color swatch.

In the **Editing Area,** you can create the button image using the selected editing tool with the specified color.

Choosing **Grid** causes a grid to be displayed in the editing area. Each grid square represents a single pixel.

Choosing **Clear** erases the editing area.

Choosing **Open** opens an existing button image file for editing. Button images are stored as bitmap (*.eps*) files.

Choosing **Undo** undoes the last action.

Choosing **Save** saves the customized button image.

Choosing **Save As** saves the customized button image using a different name or location.

Choosing **Close** closes the Button Editor dialog box.

> **NOTE**
>
> Creating a custom command macro, such as Chair in the above example, does not actually create an AutoCAD command named chair that can be entered at the command prompt. The name "Chair" is merely the word that appears on the button that, when chosen, invokes the INSERT command in a macro that calls for the block named Chair. It is not even significant that the name of the button and the name of the block are the same. In order to create and name a command that can be entered as that name, you must use one of the higher-level programming languages available in AutoCAD such as AutoLISP.

Menus

Drop-down menus are displayed as a list under a menu bar. In the CUI dialog box, the **Menus** node displays menus defined in all workspaces. Drop-down menus should have one alias in the range of POP1 through POP499. Menus with an alias of POP1 through POP16 are loaded by default when a menu loads. All other menus must be added to a workspace to be displayed.

To edit properties of an existing menu, select the menu name in the **Customizations in AllFiles** pane and AutoCAD lists the corresponding properties for the selected menu under the headings of **General** and **Advanced**. The General section displays the **Name** and **Description** of the selected menu. The Advanced section displays the **Alias** name and **Element ID** (read-only). Here you may make changes to the properties of an existing menu.

To edit properties of an individual command listed under an existing menu, first choose the menu name, and then choose the name of the command in the **Customizations in All Files** pane (Figure 18–15). AutoCAD lists the corresponding properties for the selected command in the Properties section under the headings of **Display, Command, Advanced,** and **Images.** Here you may make changes to the properties of an individual command. If desired, you can edit the image of the selected command in the Button Image section.

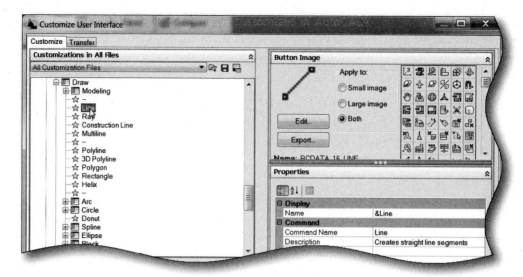

FIGURE 18–15 *The CUI dialog box with Line selected*

To create a new menu, right-click **Menus** in the **Customizations in All Files** pane and from the shortcut menu, choose *Menu* from the NEW flyout menu. AutoCAD creates a new menu, named Menu1, and places it at the bottom of the **Menus** tree. Rename it with an appropriate name. Select the new menu in the tree view, and update the **Properties** pane as follows:

In the **Description** box, enter a description for the menu.

In the **Aliases** box, an alias is automatically assigned to the new menu, based on the number of menus already loaded. For example, if the alias assignment is POP14, 13 menus are already loaded. Here you can edit the alias.

To add the commands to the newly created menu, from the **Command List** pane, drag the command to a location just below the newly created menu in the **Customizations in All Files** pane. If necessary, make the necessary changes to the properties of the newly added command to the menu.

You can also create a submenu in the same way that you create a menu.

Shortcut Menus

Shortcut menus are displayed at your cursor location when you right-click. The shortcut menu and the options it provides depend on the pointer location and whether an object is selected or a command is in progress. Context-sensitive shortcut menus display options that are relative to the current command or the selected object. Shortcut menus are referenced by their aliases and are used in specific situations.

To create a new shortcut menu, right-click **Shortcut Menus** in the **Customizations in All Files** pane and from the shortcut menu, choose *New Shortcut Menu*. AutoCAD creates a new menu, named ShortcutMenu1, and places it at the bottom of the **Shortcut Menu** tree. Rename it with an appropriate name. Select the new shortcut menu in the tree view, and update the **Properties** pane as follows:

In the **Description** box, enter a description for the shortcut menu.

In the **Aliases** box, enter additional aliases for this menu. An alias is automatically assigned, and defaults to the next available POP number, based on the number of shortcut menus already loaded in the program.

To add the commands to the newly created shortcut menu, from the **Command List** pane, drag the command to a location just below the newly created menu in the **Customizations in All Files** pane. Here you may make changes to the properties of the newly added command.

You can also create a submenu in the same way that you create a shortcut menu.

Quick Properties Panel

The Quick Properties panel lists the most commonly used properties for each object type or a set of objects. You can easily customize the quick properties for any object in the Customize User Interface (CUI) editor.

The properties (displayed when an object is selected) are common to all object types and specific to the object that was selected. The available properties are the same as those on the Properties palette and for rollover tooltips.

The Quick Properties panel is displayed when objects are selected, if the object type of the selected objects is supported for Quick Properties, and if the QPMODE system variable is set to 1 or 2.

When customizing the Quick Properties panel, you control which object types display properties on the Quick Properties panel and which properties are displayed. You use the Objects pane to add and remove the object types that are set to display properties on the Quick Properties panel. Once an object type is selected from the Objects pane, you can then decide which properties from the Properties pane you would want displayed when that object type is selected in the drawing window.

When an object is added to the list on the Objects pane, the state of the general properties used for all object types is assigned to the general properties of the newly added object. For example, if the properties Color and Layer are checked under the General properties, the same checked properties are assigned to the object that is being added to the Object list.

To override the general properties for all object types, select Quick Properties in the **Customizations in All Files** pane of the CUI dialog box. A list of the general properties that can be used to override the properties of all individual object types is displayed along with the **Reset Overrides** button. Select the general properties that you want displayed for all object types and click **Reset Overrides** to share the on/off state of selected general properties with all defined object types in the current CUIx file (Figure 18–16).

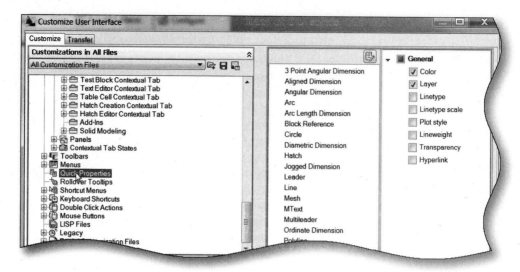

FIGURE 18–16 *Quick Properties – General Properties settings*

NOTE

When an object is added to the list on the Objects pane, the state of the general properties used for all object types is assigned to the general properties of the newly added object.

To make changes to the list of properties for an object type, select object type in the Objects pane. In the Properties pane, select the properties you want displayed for the selected object type on the Quick Properties panel. If a property is checked, the property is displayed on the Quick Properties panel when an object of the same type is selected in the drawing window. Clear the check mark next to a property to remove the property for the selected object type from the Quick Properties panel (see Figure 18–17).

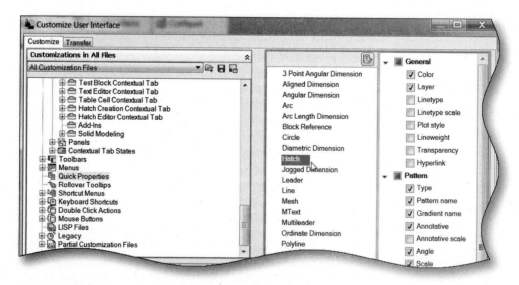

FIGURE 18–17 *List of properties available for selected object type*

To add object types to the Quick Properties panel, choose **Edit Object Type List** (see Figure 18–18). AutoCAD opens **Edit Object Type List** dialog box listing all the available object types. If an object type is checked, the properties for the object type are displayed on the Quick Properties panel when an object of the same type is selected in the drawing window. Select the object type you want to display properties in the Quick Properties panel. Clear the check mark next to an object type to remove the object type from the Quick Properties panel. After making necessary changes, choose **OK** to save the changes. After adding the object types, if necessary, make the changes to the list of properties to the newly added object types.

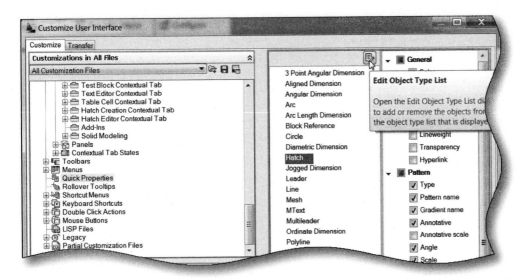

FIGURE 18–18 *Choosing Edit Object Type List to add object types*

Keyboard Shortcuts

Keyboard shortcuts include shortcut keys and temporary override keys. Keyboard shortcuts are used to assign commands to custom keystroke combinations. You can assign shortcut keys, sometimes called accelerator keys, to commands that you use frequently. Shortcut keys are keys or key combinations that start commands. For example, you can press CTRL + N to create a new drawing file and CTRL + S to save a file, which is the same result as choosing **New** and **Save** from the **File** menu. Shortcut keys can be associated with any command in the command list, and you can create new shortcut keys or modify existing shortcut keys. You can create temporary override keys to execute a command or change a setting when a key is pressed. Temporary override keys are keys that temporarily turn ON or OFF one of the drawing aids that are set in the Drafting Settings dialog box such as Ortho mode, Object Snap (OSNAP), or Polar mode.

Temporary override keys execute a command or change a setting only while the key or combination of keys is being pressed. For example, if you are prompted for a second point of a line and the ORTHO mode is ON, you can press and hold the SHIFT key, and the ORTHO mode will be temporarily turned OFF. For some of the common overrides, you use a temporary override key for the left hand, for right-handed mouse operators, and another one for the right hand. For example, to temporarily invoke the Endpoint Object Snap mode with the left hand, the combination of keys to hold down at the same time is SHIFT + E. For the right hand, it is SHIFT + P.

Creating a New Shortcut Key Using the CUI Dialog Box

You can create and edit keyboard shortcuts for a selected command in the **Properties** pane. In the CUI dialog box, expand the **Shortcut Keys** node in the **Customizations**

in All Files pane. Then drag the command to be invoked by the new shortcut combination of strokes from those listed on the **Command List** pane in the CUI dialog box to the list of **Shortcut Keys** in the **Customizations in All Files** pane. In the **Properties** pane, the properties for the newly created shortcut key are displayed.

In the **Key(s)** field, choose the [...] button to open the Shortcut Keys dialog box. In the Shortcut Keys dialog box, in the **Press new shortcut key** field, hold a modifier key, such as CTRL, ALT, or SHIFT and press a letter, number, or function key. Valid modifier keys include the following:

- Function (Fn) keys containing no modifiers
- CTRL + letter, CTRL + number, CTRL + function key
- CTRL + ALT + letter, CTRL + ALT + number, CTRL + ALT + function key
- SHIFT + CTRL + letter, SHIFT + CTRL + number, SHIFT + CTRL + function key
- SHIFT + CTRL + ALT + letter, SHIFT + CTRL + ALT + number, SHIFT + CTRL + ALT + function key

The Currently Assigned To section displays any current assignments for the shortcut key. If you do not want to replace the current assignment, use a different shortcut key. Otherwise, choose **Assign**.

NOTE More than one command can share the same shortcut, but only the last command assigned will be active.

Choose **OK** to assign the shortcut key and close the Shortcut Keys dialog box.

Creating a New Temporary Override Key Using the CUI Dialog Box

In the CUI dialog box (Figure 18–19), select the **Temporary Override** node in the **Customizations in All Files** pane, right-click, and choose *New Temporary Override* from the shortcut menu. AutoCAD creates a new temporary override in the list with the name Temporary Override1. Rename it with an appropriate name.

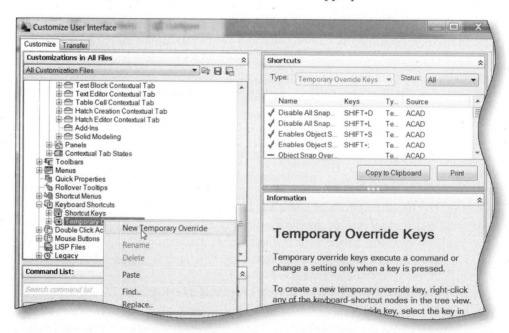

FIGURE 18–19 *The CUI dialog box creating a New Temporary Override*

Image(s) © Cengage Learning 2013

The **Properties** pane displays the lists of properties for the new toolbar under the headings of **General** and **Shortcut.**

The General section displays and lets you change the **Name** and **Description** of the new toolbar.

In the Shortcut section, click in the **Key(s)** textbox, and then choose the button on the right. AutoCAD displays the Shortcut Keys dialog box where you can enter the new shortcut key combination after clicking in the **Press new shortcut key** textbox. Any current keys assigned to an existing temporary override selected to be edited will be shown in the **Current Keys** list box. Choose **Assign** to assign the selected item to the override or **Remove** to remove the assignment.

In the **Macro 1 (Key Down)** textbox, enter the macro code for the new key down mode of the new override or edit the existing override. See the explanation of Menu Macro Syntax later in this chapter.

In the **Macro 2 (Key Up)** textbox, enter the macro code for when the temporary override key is released. If no code is defined, releasing the key restores the application to the state that existed before the temporary override was executed.

Mouse Buttons

Mouse buttons control the functions of a Windows® system pointing device. You can customize these functions in the CUI dialog box. For a pointing device with more than two buttons, you can change the functions of the second and third button. The first button on any pointing device cannot be changed in the CUI dialog box.

Different functions of the pointing device can be created by using various combinations of SHIFT and CTRL keys with clicking. The number of possible functions depends on the number of assignable buttons. The combinations include click, SHIFT + click, CTRL + click, and CTRL & SHIFT + click. The tablet buttons are numbered sequentially.

Adding Commands to Mouse Button Combinations Using the CUI Dialog Box

To add a command to a mouse button combination, drag the command from the **Command List** pane to the name of the mouse button combination in the **Customizations in All CUI Files** pane.

The **Properties** pane displays the lists of properties for the button combination under the headings of **General, Macro,** and **Advanced.**

The General section displays and lets you change the **Name** and **Description** of a button combination.

In the Macro section textbox, enter the macro code for a button combination. See the explanation of Menu Macro Syntax later in this chapter.

The Advanced section displays the button combination **Element ID** as read-only.

Legacy

The **Legacy** node of the **Customizations in All Files** pane lets you make changes in features that are considered obsolete or of minimal usage in the latest version of Auto-CAD. These include **Tablet Menus, Tablet Buttons, Screen Menus,** and **Image Tile Menus.**

Partial CUIx Files

The **Partial CUIx Files** node of the **Customizations in All Files** pane lets you load partial CUIx files. Partial CUIx files are loaded on top of the main CUIx file. They allow you to create and modify most interface elements, such as toolbars, menus, and so on, in an external CUIx file without having to import the customizations to your main CUIx file.

> **NOTE**
>
> If you define a workspace in a partial CUIx file, you need to transfer the contents of the workspace to the main CUIx file so that AutoCAD can display the customized workspace. Use the *Transfer* tab in this dialog box to move the workspace.

Loading a Partial CUIx File Using the Customize Tab

In the Customize User Interface dialog box, to the right of the drop-down list, choose **Load partial customization file** icon. AutoCAD displays the Open dialog box; choose the partial CUIx file you want to open, and choose **Open**.

> **NOTE**
>
> You need to change the customization group name if the partial CUIx file you are attempting to load has the same customization group name as the main CUIx file. To create a new customization group name, open the CUIx file in the Customize dialog box, select the file name, and right-click to rename it. Choose the main CUIx file from the drop-down list in the *Customizations in All Files* pane to check that the file has been loaded into the main CUIx file.

In the tree view of the main customization file, click the plus sign (+) next to the **Partial CUIx Files** node to expand it. Any partial menus loaded in the main CUIx file are displayed.

Choose **OK** to save the changes and close the CUI dialog box.

If you open the CUIx file with a text editor, you will see the following message:

> **WARNING**
>
> Do not edit the contents of this file.

If you attempt to edit this file using an XML editor, you could lose customization and migration functionality. If you need to change information in the customization file, use the Customize User Interface dialog box in the product. To access the Customize User Interface dialog box, click the Tools menu > Customize > Interface, or enter **CUI** on the command line.

> **NOTE**
>
> If the loaded file configuration is not satisfactory, you can restore the menus and other environmental components by using the menu command. When prompted to select a CUIx file, choose acad.cui.

EXTERNAL COMMANDS AND ALIASES

A command alias is an abbreviation that you enter at the on-screen prompt instead of entering the entire command name. For example, you can enter **l** instead of **line** to start the LINE command. An alias is not the same as a keyboard shortcut, which is a combination of keystrokes, such as [CTRL] + N for NEW.

An alias can be defined for any AutoCAD command, device driver command, or external command. A command alias is defined in the *acad.pgp* file. You can change existing aliases or add new ones by editing the *acad.pgp* file in an ASCII text editor such as Notepad. To open the *.pgp* file, choose **Edit Program Parameters** (*acad.pgp*) from the **Customize** flyout of the **Tools** menu. The file can also contain comment lines preceded by a semicolon (;).

Prior to modifying the *acad.pgp* file, it is recommended that you make a backup copy of the file, such as *xacad.pgp*, so that if you make a mistake, you can restore the original version.

To open the *acad.pgp* file, choose **Edit Aliases** from the Customization panel (Figure 18–20) located in the Manage tab, and AutoCAD opens the *acad.pgp* file in the Notepad program.

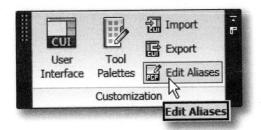

FIGURE 18–20 *Choosing Edit Aliases from the Customization panel*

The External Commands are defined at the top of the *acad.pgp* file.

The format for a command line is as follows:

```
<Command name>, <executable>,flags,[*]<Prompt>, <Return code>
```

The format to define an alias is as follows:

```
<Alias>,*<Full command name>
```

The abbreviation preceding the comma is the character or characters to be entered at the on-screen prompt. The asterisk (*) must precede the command you wish to be invoked. It may be a standard AutoCAD command name, a custom command name that has been defined in and loaded with AutoLISP or ADS, or a display or machine driver command name. Aliases cannot be used in scripts. You can prefix a command in an alias with the hyphen that causes a command-line version to be used instead of a dialog box, as shown here:

```
BH, *-BHATCH
```

See Appendix H for list of available aliases from the default *acad.pgp* file.

CUSTOMIZING MENUS WITH MACROS

Selecting an item from a menu or toolbar might execute a command, an AutoLISP routine, or a macro, or cause another menu to be displayed. If you perform an application-specific task on a regular basis that requires multiple steps to accomplish this task, you can place this in a menu macro as shown in the examples of Macro lines in the Macro textbox in the **Properties** pane of the CUI dialog box. Refer to the discussion earlier in this chapter regarding how to have AutoCAD complete all the required procedures in a single step while pausing for input if necessary. Menu macros are similar to script files, or files ending with *.scr. Script files are also capable of executing many commands in sequence, but they have no decision-making capability and cannot pause for interactive user input.

AutoCAD menu macros can be used in the following kinds of menus:

- Drop-down menus
- Shortcut menus
- Toolbars
- Keyboard shortcuts

Menu Macro Syntax

Following is a partial list of the special characters you will encounter in menu macros:

Syntax	Description
;	Semicolon; equivalent to pressing Enter on the keyboard
^M	Caret M; equivalent to pressing Enter on the keyboard
^I	Caret I; equivalent to pressing TAB on the keyboard
space	A space character; equivalent to pressing Enter on the keyboard
\	Pause for user input
_	Translates AutoCAD commands and options that follow
=*	Displays the current top-level drop-down, shortcut, or image menu
*^C^C	Repeats a command until another command is chosen
$	Introduces a conditional DIESEL macro expression ($M=)
^B	Repeats a command until another command is chosen
^C	Cancels a command; equivalent to ESC
^D	Turns Coords on or off; equivalent to CTRL + D
^E	Sets the next isometric plane; equivalent to CTRL + E
^G	Turns Grid on or off; equivalent to CTRL + G
^H	Equivalent to pressing BACKSPACE
^O	Turns Ortho ON or OFF
^P	Turns MENUECHO ON or OFF
^Q	Echoes all prompts, status listings, and input to the printer
^T	Turns tablet on or off; equivalent to CTRL + T
^V	Changes the current viewport
^Z	Null character that suppresses the automatic addition of SPACEBAR at the end of a command

The following example macro will create a layer called "EL_OFFEQ" for "Electrical Office Equipment," assign the color "RED," and make it the current layer:

```
^C^C-LAYER;M;EL-OFFEQ;C;RED;;;
```

The equivalent action, if typed, would proceed as follows: type the -LAYER command, select the MAKE option, type **EL-OFFEQ** as the desired layer name, select the Color option to assign the color red, and finally, press three Enter times to exit the command. Selecting this macro will execute all of these steps in one operation.

Use the MAKE option of the -LAYER command in case the layer does not yet exist. If the layer does exist, it will become the current layer. Note that there is no space after the ˆCˆC.

CUSTOMIZING HATCH PATTERNS

Certain concepts about hatch patterns should be understood before learning to create one.

1. Hatch patterns are made of lines or line segment/space combinations. There are no circles or arcs available in hatch patterns as there are in shapes and fonts, which will be covered next.

2. A hatch pattern may be one or more series of repeated parallel lines, repeating dots, or line segment/space combinations. That is, each line in one so-called family is like every other line in that same family. Each line has the same offset and stagger, if it is a segment/space combination, relative to its adjacent sibling, as does every other line.

3. One hatch pattern can contain multiple families of lines. One family of lines may or may not be parallel to other families. With properly specified basepoints, offsets, staggers, segment/space combinations, lengths, and relative angles, you can create a hatch pattern from multiple families of segment/space combinations that will display repeated closed polygons.

4. Each family of lines is drawn with offsets and staggers based on its own specified basepoint and angle.

5. All families of lines in a particular hatch pattern will be located, with a base point, rotated, and scaled as a group. The factors of location, angle of rotation, and scale are determined when the hatch pattern is loaded by the HATCH command and used to fill a closed polygon in a drawing.

6. The pattern can be achieved by specifying parameters in different ways.

Hatch patterns are created by including their definition in a file whose extension is *.pat*. This action can be performed by using a text editor such as Notepad or a word processor in the nondocument, or programmer, mode, which will save the text in ASCII format. Your hatch pattern definition can also be added to the *acad.pat* file. You can also create a new file specifically for a pattern.

Each pattern definition has one header line, which gives the pattern name and description, and a separate specification line describing each family of lines in the pattern.

The header line has the following format:

*pattern-name[,description]

The pattern name will be the name for which you will be prompted when using the HATCH command. The description is optional, and it is there so that someone reading the *.pat* file can identify the pattern. The description has no effect, nor will it be displayed while using the HATCH command. The leading asterisk denotes the beginning of a hatch pattern.

The format for a line family is as follows:

```
angle, x-origin, y-origin, delta-x, delta-y [,dash-1, dash-2…]
```

The brackets "[ ]" denote optional segment/space specifications used for noncontinuous-line families. Note also that any text following a semicolon (;) is for comment only and will be ignored. In all definitions, the angle, origins, and deltas are mandatory, even if their values are zero.

An example of continuous lines that are rotated at 30° and separated by 0.25 units could be written as follows (Figure 18–21):

```
*P30, 30° continuous
30, 0,0, 0,.25
```

The 30° specifies the angle.

The first and second zero specify the coordinates of the origin.

The third zero, though required, is meaningless for continuous lines.

The 0.25 specifies the distance between lines.

A pattern of continuous lines crossing at 60° to each other could be written as follows (Figure 18–22):

```
*PX60,x-ing @ 60
30, 0,0, 0,.25
330, 0,0, 0,.25
```

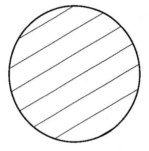

FIGURE 18–21 *A hatch pattern with continuous lines rotated at 30° and separated by 0.25 units*

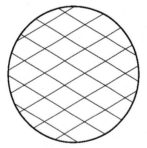

FIGURE 18–22 *A hatch pattern with continuous lines crossing at 60° to each other and separated by 0.25 units*

A pattern of lines crossing at 90° but having different offsets is as follows (Figure 18–23):

```
*PX90, x-ing @ 90 w/ 2:1 rectangles
0, 0,0, 0,.25
90, 0,0, 0,.5
```

Note the effect of the delta-Y. It is the amount of offset between lines in one family. Hatch patterns with continuous lines do not require a value, other than zero, for delta-X. Orthogonal continuous lines also do not require values for the X origin unless they are used in a pattern that includes broken lines.

To illustrate the use of a value for the Y origin, two parallel families of lines can be written to define a hatch pattern for steel as follows (Figure 18–24):

```
*steel
45, 0,0, 0,1
45, 0,.25, 0,1
```

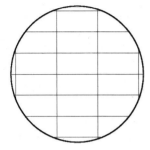

FIGURE 18–23 *A hatch pattern with lines crossing at 90° and having different offsets*

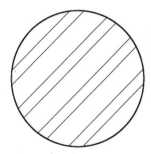

FIGURE 18–24 *Defining a hatch pattern for steel*

The following three concepts are worthy of notation in this example:

- If the families are not parallel, then specifying origins other than zero will serve no purpose.
- Parallel families of lines should have the same delta-Y offsets. Different offsets will serve little purpose.
- Most important, the delta-Y is at a right angle to the angle of rotation, but the Y origin is in the Y direction of the coordinate system. The steel pattern as written in the example would fill a polygon, as shown in Figure 18–25. Note the dimensions when used with no changes at a scale factor of 1.0 or a rotation angle of zero.

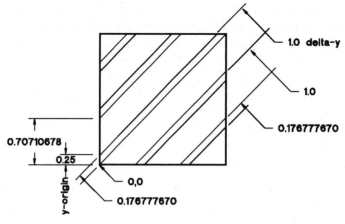

FIGURE 18–25 *The steel hatch pattern with the delta-Y at a right angle to the angle of rotation and the Y origin in the Y direction of the coordinate system*

Custom Hatch Patterns and Trigonometry

The dimensions in the hatch pattern resulting from a 0.25 value for the delta-Y of the second line-family definition may not be what you expected (Figure 18–26). If you wished to have a 0.25 separation between the two line families (Figure 18–27), then you must either know enough trigonometry/geometry to predict accurate results or must burden the user to reply to prompts with the correct responses to achieve those results. For example, you could write the definition as follows:

```
*steel
0, 0,0, 0,1
0, 0,.25, 0,1
```

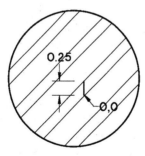

FIGURE 18–26 *Hatch pattern dimensions resulting from a 0.25 value for the delta Y of the second line-family definition*

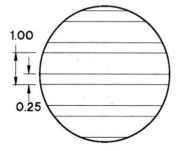

FIGURE 18–27 *Steel hatch pattern defined with a 0.25 separation between the two line families*

In order to use this pattern as shown, the user will have to specify a 45° rotation when using it. This will maintain the ratio of 1 to 25 between the offset (delta-Y) and the spacing between families (Y origin). If you wish to avoid this inconvenience to the user but still wish to have the families separated by .25, you can write the definition as follows:

```
*steel
45, 0,0, 0,1
45, 0,.353553391, 0,1
```

The value for the Y origin of .353553391 was obtained by dividing 25 by the sine, or cosine, of 45°, which is .70710678. The X origin and Y origin specify the coordinates of a point. Therefore, setting the origins of any family of continuous lines merely tells AutoCAD that the line must pass through that point. See Figure 18–28 for the trigonometry used.

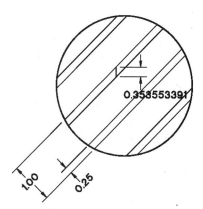

FIGURE 18–28 *Steel hatch pattern defined with a 45° rotation to maintain a 1:25 offset ratio*

For families of lines that have segment/space distances, the point determined by the origins can tell AutoCAD not only that the line passes through that point but that one of the segments will begin at that point. A dashed pattern can be written as follows (Figure 18–29):

```
*dashed
0, 0,0, 0,.25, .25,-.25
```

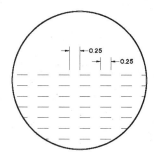

FIGURE 18–29 *Writing a dashed pattern*

Note that the value of the X origin is zero, thus causing the dashes of one line to line up with the dashes of other lines. Staggers can be produced by giving a value to the X origin as follows (Figure 18–30):

```
*dashstagger
0, 0,0, .25,.25, .25,-.25
```

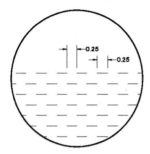

FIGURE 18–30 *Writing a staggered dashed pattern*

In a manner similar to defining linetypes, you can cause lines in a family to have several lengths of segments and spaces (Figure 18–31).

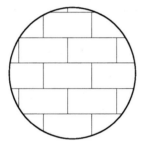

FIGURE 18–31 *A pattern with several lengths of segments and spaces*

```
*simple
0, 0,0, 0,.5
90, 0,0, 0,1, .5,-.5
```

A similar, but more complex, hatch pattern could be written as follows (Figure 18–32):

```
*complex
45, 0,0, 0,.5
-45, 0,0, 0,1.414213562, 0,1.41421356
```

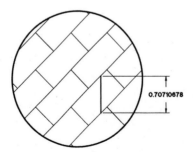

FIGURE 18–32 *A pattern with more complex hatch patterns*

Repeating Closed Polygons

Creating hatch patterns with closed polygons requires planning. For example, a pattern of 45°, 90°, and 45° triangles, as shown in Figure 18–33, should be started by first creating construction lines, as shown at (a) of Figure 18–33. Extend the construction lines through points of the object parallel to other lines of the object. A grid pattern will emerge when you place the lines based on distances obtained from the object to form a pattern.

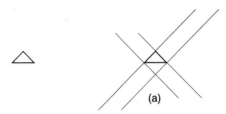

FIGURE 18–33 *Closed polygons*

It is also helpful to sketch construction lines that are perpendicular to the object lines. This action will assist you in specifying segment/space values. In the example, two of the lines are perpendicular to one another, thus making this easier. Figure 18–34 through Figure 18–37 illustrate potential patterns of triangles. Once the pattern is selected, the grid, and some knowledge of trigonometry, will assist you in specifying all of the values in the definition for each line family.

For the following pattern (PA), the horizontal line families can be written as follows (Figure 18–38):

```
0, 0,0, 0,1, 1,-1
```

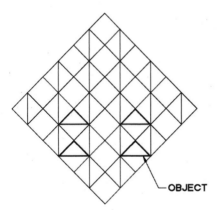

FIGURE 18–34 *Creating triangular hatch patterns—method #1*

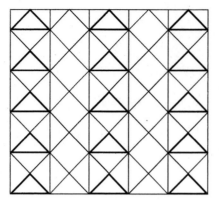

FIGURE 18–35 *Creating triangular hatch patterns—method #2*

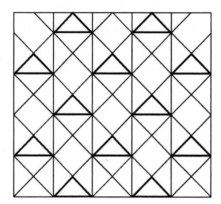

FIGURE 18–36 *Creating triangular hatch patterns—method #3*

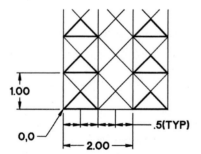

FIGURE 18–37 *Creating triangular hatch patterns—method #4*

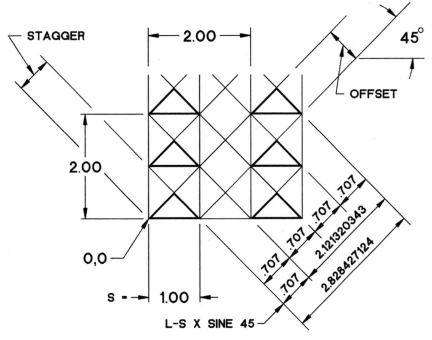

FIGURE 18–38 *Triangular patterns—pattern PA*

The specifications for the 45° family of lines can be determined by using the following trigonometry:

```
sin 45 degrees = 0.70710678
S = 1
L = S times sin 45 degrees
L = 1 times sin 45 degrees = 0.70710678
```

Note that the trigonometry function is applied to the hypotenuse of the right triangle. In the example, the hypotenuse is 1 unit. A different value would simply produce a proportional result; that is, a hypotenuse of 5 would produce $L = S \times 0.70710678 = 0.353553391$. The specifications for the 45° family of lines could be written as follows:

45,	0,0	0.70710678,	0.70710678,	0.70710678,	−2.121320343
angle,	origin,	offset,	stagger,	segment,	Space

For the 135° family of lines, the offset, stagger, segment, and space have the same values (absolute) as for the 45° family. Only the angle, the X origin, and the sign (+ or −) of the offset or stagger may need to be changed.

The 135° family of lines could be written as follows:

```
135,1,0,-0.70710678,-0.70710678,0.70710678,-2.121320343
```

Putting the three families of lines together under a header could be written as follows and as shown in Figure 18–39.

```
*PA,45/90/45 triangles stacked
0, 0,0, 0,1, 1,-1
45, 0,0 0, 0.70710678,0.70710678, 0.70710678,-2.121320343
135, 1,0, -0.70710678,0.70710678, 0.70710678,-2.121320343
```

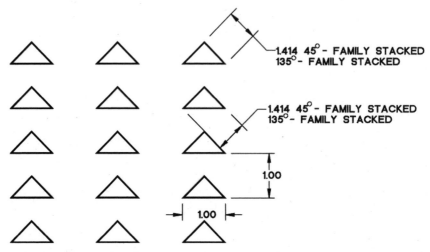

FIGURE 18–39 *Families of triangular patterns—pattern PA*

In the preceding statement, "could be written" tells you that there may be other ways to write the definitions. As an exercise, write the descriptions using 225° instead of 45° and 315 instead of 135 for the second and third families, respectively. As a hint, you determine the origin values of each family of lines from the standard coordinate system. However, to visualize the offset and stagger, orient the layout grid so that the rotation angle coincides with the zero angle of the coordinate system. The signs and the values of delta-X and delta-Y will be easier to establish along the standard plus for right/up and negatives for left/down directions. Examples of two hatch patterns, PB and HONEYCOMB, follow.

The PB pattern can be written as follows and as shown in Figure 18–40.

```
*PB, 45/90/45 triangle staggered
0, 0,0, 1,1, 1,-1
45, 0,0, 0,1.414213562, 0.70710678,-0.70710678
135, 1,0, 0,1.414213562, 0.70710678,-0.70710678
```

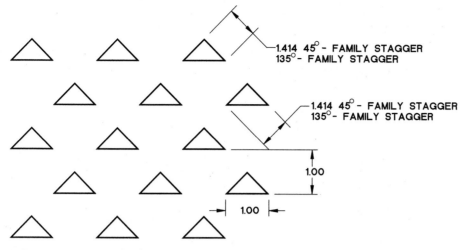

FIGURE 18–40 *Example of pattern PB*

Note that this alignment simplifies the definitions of the second and third families of lines over the PA pattern.

The HONEYCOMB pattern can be written as follows and as shown in Figure 18–41.

```
*HONEYCOMB
90, 0,0, 0.866025399,0.5, 0.577350264,-1.154700538
330, 0,0, 0.866025399,0.5, 0.577350264,-1.154700538
30, 0.5, -0.288675135, 0.866025399, 0.5,
0.577350264,-1.154700538
```

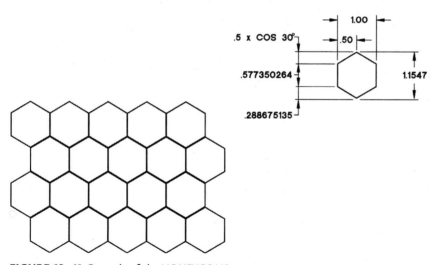

FIGURE 18–41 *Example of the HONEYCOMB pattern*

CUSTOMIZING SHAPES AND TEXT FONTS

Shapes and fonts are written in the same manner, and both are stored in files with the *.shp* file extension. The *.shp* files must be compiled into *.shx* files. This section covers how to create and save *.shp* files and how to compile *.shp* files into *.shx* files. To compile an *.shp* file into an *.shx* file for shapes or fonts, enter **compile** at the prompt.

From the Select Shape or Font File dialog box, select the file to be compiled. If the file has errors they will be reported. Otherwise, the Command window will show the following:

```
Compilation successful
Output file name .shx contains nnn bytes
```

The main difference between shapes and fonts is in the commands used to place them in a drawing. Shapes are drawn by using the SHAPE command, and fonts are drawn using commands that insert text such as TEXT or DIM. Whether or not an object in an *.shp* or *.shx* file can be used with the SHAPE command or as a font character is partly determined by whether its shape name is written in uppercase or lowercase, as explained below.

Each shape or character in a font in an *.shp* or *.shx* file is made of simplified objects. These objects are simplified lines, arcs, and circles. They are referred to as simplified because in specifying their directions and distances, you cannot use decimals or architectural units. You must use only integers or integer fractions. For example, if the line distance needs to be equal to 1 divided by the square root of 2 (or .7071068), the fraction 70 divided by 99, which equals .707070707, is as close as you can get. Rather than call the simplified lines and arcs "objects," we will refer to them as "primitives."

Individual shapes and font characters are written and stored in ASCII format. Both *.shp* and *.shx* files may contain up to 255 Shape-Characters. Each Shape-Character definition contains a header line as follows:

```
*shape number, defbytes, shapename
```

The codes that describe the Shape-Character may take one or more lines following the header. Most of the simple shapes can be written on one or two lines. The meaning of each item in the header is as follows:

1. The shape number may be from 1 to 255 with no duplications within one file.

2. *Defbytes* is the number of bytes used to define the individual Shape-Character, including the required zero that signals the end of a definition. The maximum allowable bytes in a Shape-Character definition is 2,000. Defbytes (the bitcodes) in the definition are separated by commas. You may enclose pairs of bitcodes within parentheses for clarity of intent, but this does not affect the definition.

3. The *shapename* should be in uppercase if it is to be used by the SHAPE command. Like a block name is used in the BLOCK command, you enter the shapename when prompted to do so during the SHAPE command. If the shape is a character in a font file, you may make any or all of the shapename characters lowercase, thereby causing the name to be ignored when compiled and stored in memory. It will serve for reference only in the *.shp* file for someone reading that file.

Pen Movement Distances and Directions

The specifications for pen movement distances and directions, that is, whether the pen is up or down, for drawing the primitives that will make up a Shape-Character are written in bitcodes. Each bitcode is considered one defbyte. Codes 0 through 16 are not Distance-Direction codes but special instructions-to-AutoCAD codes. This will be explained after the discussion of Distance-Direction codes.

Distance-Direction codes have three characters. They begin with a zero. The second character specifies distance. More specifically, it specifies vector length, which may be affected by a scale factor. Vector length and scale factor combine to determine actual distances. The third character specifies direction. Sixteen standard directions are available through use of the Distance-Direction bitcodes, or defbytes. Vectors 1 unit in length are shown in the 16 standard directions in Figure 18–42.

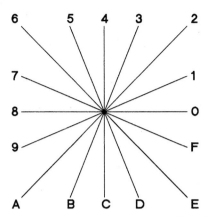

FIGURE 18–42 *Distance-Direction bitcodes*

Directions 0, 4, 8, and C are equivalent to the default 0°, 90°, 180°, and 270°, respectively. Directions 2, 6, A, and E are 45°, 135°, 225°, and 315°, respectively. The odd-numbered direction codes are not increments of 22.5°, as you might think. They are directions that coincide with a line whose delta-X and delta-Y ratio are 1 unit to 2 units. For example, the direction specified by code 1 is equivalent to drawing a line from 0,0 to 1,.5. This equates to approximately 26.56505118°, or the arctangent of 0.5. The direction specified by code 3 equates to 63.434494882°, or the arctangent of 2, and is the same as drawing a line from 0,0 to .5,1.

Distances specified will be measured on the nearest horizontal or vertical axis. For example, 1 unit in the 1 direction specifies a vector that will project 1 unit on the horizontal axis. Three units in the D direction will project 3 units on the vertical axis (downward). The vector specified as 1 unit in the 1 direction will actually be 1.118033989 units long at an angle of 26.65606118°, and the vector specified as 3 units in the D direction will be 3.354101967 units long at an angle of 296.5650512°. See Figure 18–43 for examples of specifying direction.

To illustrate the codes specifying the Distance-Direction vector, the following example is a definition called "oddity" that will draw the shape shown in Figure 18–44.

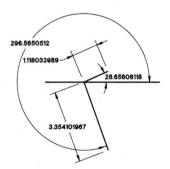

FIGURE 18–43 *Specified directions*

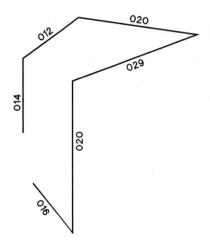

FIGURE 18–44 *Distance-Direction vector-specifying code*

The "oddity" shape could be written as follows:

```
*200,7,ODDITY
014,012,020,029,02C,016,0
```

To draw the shape named "oddity," you would first load the shape file that contains the definition and then use the SHAPE command as follows:

```
shape
Enter shape name or ⬇ oddity
Specify insertion point: (specify a point)
Specify height <default>: (specify a scale factor)
Specify rotation angle <default>: (specify a rotation angle)
```

 NOTE An alternative to the standard Distance-Direction codes is to use codes 8 and 9 to move the pen by paired (delta-X, delta-Y) ordinate displacements. This process is explained in the following section on "Special Codes."

Special Codes

Special codes can be written in decimal or hexadecimal. You can specify a special code as 0 through 16 or as 000 through 00E. A three-character defbyte with two leading zeros will be interpreted as a hexadecimal special code. Code 10 is a special code in decimal. However, 010 is equivalent to decimal 16. More importantly, it will be interpreted by AutoCAD as a Distance-Direction code with a vector length of 1 and a direction of 0. The hexadecimal equivalent to 10 is 00A. The code functions are as follows:

Code 0: End of Shape—The end of each separate shape definition must be marked with the code 0.

Code 1 and Code 2: Pen Up and Down—The pen down (or Draw) mode is ON at the beginning of each shape. Code 2 turns the Draw mode OFF or lifts the pen. This permits moving the pen without drawing. Code 1 turns the Draw mode ON.

Note the relationship between the insertion point specified during the SHAPE command and where you wish the object and its primitives to be located. If you wish for

AutoCAD to begin drawing a primitive in the shape at a point remote from the insertion point, then you must lift the pen with Code 2 and move the pen, with the proper codes, and then lower the pen with Code 1. Movement of the pen, directed by other codes, after a Pen Down Code 1 causes AutoCAD to draw primitives in a shape.

Codes 3 and Code 4: Scale Factors—Individual and groups of primitives within a shape can be increased or decreased in size by integer factors as follows. Code 3 tells AutoCAD to divide the subsequent vectors by the number that immediately follows the Code 3. Code 4 tells AutoCAD to multiply the subsequent vectors by the number that immediately follows the code 4.

Scale factors are cumulative. The advantage of this is that you can specify a scale factor that is the quotient of two integers. A two-thirds scale factor can be achieved by a Code 4, followed by a factor of 2, followed by a Code 3, followed by a factor of 3. The effects of scale factor codes must be reversed when they are no longer needed. They do not go away by themselves. Therefore, at the end of the definition, or when you wish to return to normal or other scaling within the definition, the scale factor must be countered. For example, when you wish to return to the normal scale from a two-thirds scale, you must use Code 3, followed by a factor of 3, followed by a Code 4, followed by a factor of 2. There is no law that states you must always return to normal from a scaled mode. You can, with Code 3 and 4 and the correct factors, change from a two-thirds scale to a one-third scale for drawing additional primitives within the shape. You should *always*, however, return to the normal scale at the end of the definition. A scale factor in effect at the end of one shape will carry over to the next shape.

Code 5 and Code 6: Saving and Recalling Locations—Each location in a shape definition is specified relative to a previous location. However, once the pen is at a particular location, you can store that location for later use within that shape definition before moving on. This is handy when an object has several primitives starting or ending at the same location. For example, a wheel with spokes would be easier to define by using Code 5 to store the center location, drawing a spoke, and then using Code 6 to return to the center.

Storing and recalling locations are known as *pushing* and *popping* them, respectively, in a stack. The stack storage is limited to four locations at any one time. The order in which they are popped is the reverse of the order in which they were pushed. Every location pushed must be popped.

More pushes than pops will result in the following Command line error message:

    Position stack overflow in shape nnn

More pops than pushes will result in the following Command line error message:

    Position stack underflow in shape nnn

Code 7: Subshape—One shape in an *.shp* or *.shx* file can be included in the definition of another shape in the same file by using Code 7, followed by the inserted shape's number.

Code 8 and Code 9: X-Y Displacements—Normal vector lengths range from 1 to 15 and can be drawn in one of the 16 standard directions unless you use a Code 8 or Code 9 to specify X-Y displacements. A Code 8 tells AutoCAD to use the next two bytes as the X and Y displacements, respectively. For example, 8, (7,–8) tells

AutoCAD to move the pen a distance that is 7 in the X direction and 8 in the Y direction. The parentheses are optional. After the displacement bytes, specifications revert to normal.

Code 9—Code 9 tells AutoCAD to use all following pairs of bytes as X-Y displacements until terminated by a pair of zeros. For example, 9,(7,–8),(14,9), (–17,3),(0,0) tells AutoCAD to use the three pairs of values for displacements for the current mode and then revert to normal after the (0,0) pair.

Code 00A: Octant Arc—Code 00A (or 10) tells AutoCAD to use the next two bytes to define an arc. It is referred to as an octant arc, or an increment of 45°. Octant arcs start and end on octant boundaries. Figure 18–45 shows the code numbers for the octants. The specification is written in the following format:

```
10, radius, (-)OSC
```

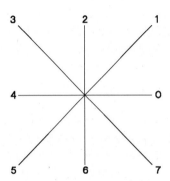

FIGURE 18–45 *Code numbers for octants*

The radius may range from 1 to 255. The second byte begins with 0 and specifies the direction by its sign—clockwise if negative, counterclockwise otherwise—the starting octant (S), and the number of octants it spans (C), which may be written as 0 to 7, with 0 being 8, or a full circle. Figure 18–46 shows an arc drawn with the following codes:

```
10,(2,-043)
```

The arc has a radius of 2, begins at octant arc 4, and turns 135° (3 octants) clockwise.

Code 00B: Fractional Arc—Code 00B (11) can be used to specify an arc that begins and ends at points other than the octants. The definition is written as follows:

```
11, start-offset, end-offset, high-radius, low-radius, (-)OSC
```

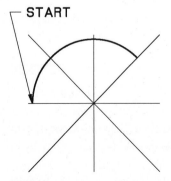

FIGURE 18–46 *An arc drawn with code 10,(2,–043)*

Start and end offsets specify how far from an octant the arc starts and ends. The high-radius, if not 0, specifies a radius greater than 255. The low-radius is specified in the same manner as the radius in a Code 10 arc, as are the starting octant and octants covering specifications in the last byte. The presence of the negative also signifies a clockwise direction.

The units of offset from an octant are a fraction of 1° multiplied by 45 divided by 256, or approximately .17578125°. For example, if you wish to specify the starting value near 60°, the equation would be as follows:

```
offset = (60-45)*(256/45) = 85.333333
```

The specification value would be 85.

To end the arc at 102 degrees, the equation would be as follows:

```
offset = (102-90)*(256/45) = 68.2666667
```

The specification value would be 68.

To draw an arc with a radius of 2 that starts near 60° and ends near 102°, the specifications would be as follows:

```
11,(85,68,0,2,012)
```

The last byte (012) specifies the starting octant to be 1 (45°), and the ending octant to be 2 (90°).

Code 00C and Code 00D: Bulge-Specified Arc—Codes 00C and 00D (12 and 13) are used to specify arcs in a different manner from octant codes. Codes 00C and 00D call out bulge factors to be applied to a vector displacement. The effect of using Code 00C or Code 00D involves specifying the endpoints of a flexible line by the X-Y displacement method and then specifying the bulge. The bulge determines the distance from the straight line between the endpoints and the extreme point on the arc. The bulge can range from −127 to 127. The maximum or minimum values, 127 or −127, define a 180° arc, or a half circle. Smaller values define proportionately smaller-degree arcs. That is, an arc of a given value, say X, will be X times 180 divided by 127 degrees. A bulge value of zero will define a straight line.

Code 00C precedes single-bulge-defined arcs, and Code 00D precedes multiple arcs. This is similar to the way Code 008 and Code 009 work on X-Y displacement lines. Code 00D, like Code 009, must be terminated by a 0,0 byte pair. You can specify a series of bulge arcs and lines without exiting Code 00D by using the zero bulge value for the lines.

Code 00E: Flag Vertical Text Command—Code 00E (14) is only for dual-orientation text font descriptions, where a font might be used in either horizontal or vertical orientations. When Code 00E is encountered in the shape definition, the next code will be ignored if the text is horizontal.

Text Fonts

Text fonts are special shape files written for use with AutoCAD text drawing commands. The shape numbers should correspond to ASCII codes for characters. Table 18–1 shows the ASCII codes. Codes 1 through 31 are reserved for special control characters. Only Code 10 (line feed) is used in AutoCAD. In order to be used as a font, the file must include a special shape number, 0, to describe the font. Its format is as follows:

```
*0,4,fontname
above, below, modes, 0
```

NOTE	Codes 1 to 31 are for control characters, only one of which is used in AutoCAD text fonts.

"Above" specifies the number of vector lengths that uppercase letters extend above the baseline, and "below" specifies the number of vector lengths that lowercase letters extend below the baseline. A modes byte value of 0 defines a horizontal (normal) mode, and a value of 2 defines a dual orientation, either horizontal or vertical. A value of 2 must be present in order for the special Code 00E (14) to operate.

Standard AutoCAD fonts include special shape numbers 127, 128, and 129 for the degrees symbol, plus/minus symbol, and diameter dimensioning symbol, respectively.

The definition of a character from the *txt.shp* file is as follows:

```
*65,21,uca
2,14,8,(-2,-6),1,024,043,04D,02C,2,047,1,040,2,02E,14,8,
(-4,-3),0
```

Note that the number 65 corresponds to the ASCII character that is an uppercase "A." The name "uca" (for uppercase a) is in lowercase to avoid taking up memory. As an exercise, you can follow the defbytes to see how the character is drawn. The given character definition starts by lifting the pen. A font containing the alphanumeric characters must take into consideration the spaces between characters. This is done by having similar starting and stopping points based on each character's particular width.

TABLE 18–1 *ASCII codes for text font*

Code	Character	Code	Character	Code	Character
32	space	64	@	96	' left apostrophe
33	!	65	A	97	a
34	" double quote	66	B	98	b
35	#	67	C	99	c
36	$	68	D	100	d
37	%	69	E	101	e
38	&	70	F	102	f
39	' apostrophe	71	G	103	g
40	(	72	H	104	h
41	)	73	I	105	i
42	*	74	J	106	j
43	+	75	K	107	k
44	, comma	76	L	108	l

(continued)

TABLE 18–1 *ASCII codes for text font (continued)*

Code	Character	Code	Character	Code	Character
45	- hyphen	77	M	109	m
46	. period	78	N	110	n
47	/	79	O	111	o
48	0	80	P	112	p
49	1	81	Q	113	q
50	2	82	R	114	r
51	3	83	S	115	s
52	4	84	T	116	t
53	5	85	U	117	u
54	6	86	V	118	v
55	7	87	W	119	w
56	8	88	X	120	x
57	9	89	Y	121	y
58	: colon	90	Z	122	z
59	; semicolon	91	[	123	{
60	<	92	\ backslash	124	&bar; vertical bar
61	=	93	]	125	}
62	>	94	ˆ caret	126	~ tilde
63	?	95	_ underscore		

CUSTOM LINETYPES

Linetype definitions are stored in files with a *.lin* extension. Approximately 40 standard linetype definitions are stored for use in the *acad.lin* file. The definitions are in ASCII format and can be edited. You can add new definitions of your own by using either a text editor or the *Create* option of the LINETYPE command, or you can save new or existing linetype definitions in another *filename .lin* file.

Simple linetypes consist of series of dashes, dots, and spaces. Their definitions are considered the in-line Pen Up/Pen Down type. Complex linetypes have repeating "out-of-line" objects, such as text and shapes, along with the optional in-line dashes, dots, and spaces. These are used in mapping or surveying drawings for such things as topography lines, fences, utilities, and many other descriptive lines. Instrumentation or control drawings also use many lines with repeating shapes to indicate the purpose of each line.

Each linetype definition in a file comprises two lines. The first line must begin with an asterisk, followed by the linetype name and an optional description, in the following format:

```
*ltname,description
```

The second line gives the alignment and description by using proper codes and symbols, in the following format:

```
alignment,patdesc-1,patdesc-2,…
```

A simple linetype definition for two dashes and a dot, called DDD, could be written as follows:

```
*DDD,_ _ . _ _ . _ _ .
A,.75,-.5,.75,-.5,0,-.5
```

The linetype name is DDD. A graphic description of underscores, spaces, and periods follow. The dashes are given as .75 in length—positive for Pen Down—separated by spaces −.5 in length—negative for Pen Up—with the 0 specifying a dot. No character other than the A should be entered for the alignment; it is the only one applicable at this time. This type of alignment causes the lines to begin and end with dashes, except for linetypes with dots only.

The complex linetype definitions include a descriptor, enclosed in square brackets, in addition to the alignment and dash/dot/space specification. A shape descriptor will include the shape name, shape file, and optional transform specification, as follows:

```
[shapename,filename,transform]
```

A text descriptor will include the actual text string in quotes, the text style, and optional transform specification, as follows:

```
["string",textstyle,transform]
```

Transform specifications, if included, can be one or more of the following:

```
A=⊡# absolute rotation X=⊡# x offset
R=⊡# relative rotation Y=⊡# Y offset
S=⊡# scale
```

The ⊡# for rotation is in decimal degrees, plus or minus. For scale and offset, it is in decimal units.

The following example of an embedded shape in a line for an instrument air line with repeating circles could be written as follows:

```
*INSTRAIR, _ [CIRC] [CIRC]
A,2.0,-.5,[CIRC,ctrls.shx],-.5
```

If the *ctrls.shx* file contains a proper shape description of the desired circle, it will be repeated in the broken line, with spaces on each side, when applied as the INSTRAIR linetype. If the scale of the circle needs to be doubled in order to have the proper appearance, it could be written as follows:

```
*INSTRAIR, _ [CIRC] [CIRC]
A,2.0,-.5,[CIRC,ctrls.shx,S=2],-.5
```

The following example of an embedded text string in a line for a storm sewer, with repeating SSs, could be written as follows:

```
*STRMSWR, SS SS
A,3.0,-1.0,["SS",simplex,S=1,R=0,X=0,Y=-0.125],-1.0
```

EXPRESS TOOLS

In addition to the customization tools, AutoCAD provides a set of tools under the category of Express tools that will make your tasks easier. You can install AutoCAD Express tools from the install program provided in the AutoCAD 2007 CD. Before installing the Express tools, AutoCAD must already be installed on your system.

See Appendix B for a list of commands available with a brief description.

CUSTOMIZING AND PROGRAMMING LANGUAGE

There are various other topics related to customization of AutoCAD, some of which are beyond the scope of this book. AutoCAD provides a programming language called AutoLISP. AutoLISP is a structured programming language similar in a number of ways to other programming languages. Refer to Chapter 18. A compiled language is first converted, or compiled, into object code or machine language and then linked with various other compiled object code modules, which is called linking, to form an executable file. AutoCAD supports a number of compiled languages. C/C++ and ARx, or AutoCAD Runtime Extension, are examples of compiled programming languages.

REVIEW QUESTIONS

1. The customize user interface (CUI) dialog box can be used to customize:
   a. Workspaces
   b. Toolbars
   c. Menus
   d. Shortcuts
   e. All of the above

2. AutoCAD contains no workspaces unless you create one.
   a. True
   b. False

3. A workspace contains:
   a. Rows
   b. Columns
   c. Orientation
   d. Docking location
   e. None of the above

4. A toolbar can contain:
   a. Rows
   b. Columns
   c. Orientation
   d. Flyouts
   e. All of the above

5. Customization can include:
   a. Macro language
   b. AutoLISP functions
   c. DIESEL expressions
   d. Script files
   e. All of the above

6. You can modify the element ID of an existing command.
   a. True
   b. False

7. The purpose of the *acad.pgp* file is to:
   a. Allow other programs to be accessed while editing a drawing
   b. Enable shape files to be compiled
   c. Store system configurations
   d. Serve as a "file manager" for system variables
   e. None of the above

8. AutoCAD menu files are stored with what type of file extension?
   a. *dwg*
   b. CUIx
   c. *mnu*
   d. *men*
   e. None of the above

9. When developing screen menus, the information you would like to see displayed in the SCREEN menus should be:
   a. Typed in uppercase letters only
   b. Enclosed with square brackets "[ ]"
   c. Longer than four characters but shorter than ten characters
   d. All of the above

10. What file defines external commands and their parameters?
    a. *acad.dwk*
    b. *acad.ext*
    c. *acad.pgp*
    d. *acad.lsp*
    e. *acad.cmd*

11. The system variable used in conjunction with DIESEL to modify the contents of the status line is:
    a. statusline
    b. macro
    c. modemacro
    d. modestatus

12. AutoCAD allows you to create your own icons for use in toolbar menus.
    a. True
    b. False

13. Command aliases are stored in what file?

    a. *acad.ini*

    b. *acad.als*

    c. *acad.pgp*

    d. *acad.mnu*

    e. None of the above

14. The first line of a custom crosshatching definition always begins with:

    a. *

    b. **

    c. * * *

    d. !

    e. !!

15. To define a custom linetype with three elements—a **1**–unit dash, a **0.5**–unit dash, and a dot, all separated by **0.25** unit spaces—what would the definition look like?

    a. A,1,0.25,0.5,0.25,0

    b. A,1,.25,0.5,0.25,0,0.25

    c. A,1,–0.25,0.5,–0.25,0

    d. A,1,–0.25,0.5,–0.25,0,–0.25

    e. A,1,0.5,0,–0.25

16. The bitcode that specifies the direction of **12:00** in a shape file is:

    a. 0

    b. 4

    c. 8

    d. 12

    e. C

17. When specifying the name of a shape character, what should you do to conserve memory?

    a. Use uppercase

    b. Use lowercase

    c. Preface the name with an*

    d. Use a short name

    e. It doesn't matter; all shapes require the same amount of memory regardless of their names

18. A macro permits the user to save and invoke a series of keystrokes with a combination of only one or two keystrokes.

    a. True

    b. False

19. Which of the following can be customized in AutoCAD?
    a. Aliases
    b. Linetypes
    c. Shapes
    d. Hatch patterns
    e. All of the above

20. Command aliases can include command options.
    a. True
    b. False

21. Which of the following is the correct format to define an alias using L for the LINE command?
    a. Line = L
    b. L = <line>
    c. L, *line
    d. L = *line

22. Which of the following is not a valid AutoCAD menu type?
    a. Drop-down
    b. Screen
    c. Tablet
    d. Tooltip
    e. All of the above

23. When you're writing a script file, command aliases cannot be used.
    a. True
    b. False

24. How many drop-down menus does AutoCAD have available to use?
    a. 12
    b. 256
    c. 498
    d. 1,024
    e. Unlimited

INDEX

B

Q